# Prealgebra
## Second Edition

**Margaret L. Lial**
*American River College*

**Diana L. Hestwood**
*Minneapolis Community and Technical College*

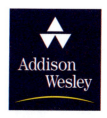

Boston   San Francisco   New York
London   Toronto   Sydney   Tokyo   Singapore   Madrid
Mexico City   Munich   Paris   Cape Town   Hong Kong   Montreal

| | |
|---:|:---|
| Publisher | Greg Tobin |
| Editor-in-Chief | Maureen O'Connor |
| Editorial Project Management | Ruth Berry and Suzanne Alley |
| Editorial Assistant | Melissa Wright and Jolene Lehr |
| Managing Editor/Production Supervisor | Ron Hampton |
| Text and Cover Design | Dennis Schaefer |
| Supplements Production | Sheila C. Spinney |
| Production Services | Elm Street Publishing Services, Inc. |
| Media Producer | Lorie Reilly |
| Software Development | Jozef Kubit, InterAct Math; David Malone, TestGen-EQ |
| Associate Producer | Rebecca Martin |
| Marketing Manager | Dona Kenly |
| Marketing Coordinator | Heather Rosefsky |
| Prepress Services Buyer | Caroline Fell |
| Technical Art Supervisor | Joseph K. Vetere |
| First Print Buyer | Hugh Crawford |
| Art Creation & Composition Services | Pre-Press Company, Inc. |
| Cover Photo Credits | Mick Roessler/Index Stock Imagery; Walter Bibikow/Index Stock Imagery; Tina Buckman/Index Stock Imagery |
| Photo Credits | All photos from PhotoDisk except the following: |

Bill Aron/PhotoEdit, pp. 197, 280; Bettmann/CORBIS, p. 322 bottom; Bongarts/A. Hassenstein/Sportschrome, pp. 573, 611; Cleve Brayant/PhotoEdit, p. 8 top left; Robert Brenner/PhotoEdit, p. 539 left; Cindy Charles/PhotoEdit, p. 758; CORBIS, p. 605 right; Bob Daemmrich/The Image Works, pp. 369, 408; © 2000 Daimler Chrysler. All rights reserved., p. 57; T. Dickenson/The Image Works, pp. 91 right, 117 left; Paul Dols/Press Publications, p. 395 left; Duomo/CORBIS, p. 351; Najlah Feanny/Stock Boston, pp. 184 left, 551; Myrleen Ferguson Cate/PhotoEdit, p. 234 bottom; Owen Franken/Stock Boston, p. 269 left; Tony Freeman/PhotoEdit, p. 167 right; Stephen Frisch/Stock Boston, p. 143; Michael P. Gadomski/Photo Researchers, Inc., p. 582 right; Judy Gelles/Stock Boston, p. 552 left; Robert Ginn/PhotoEdit, pp. 305, 313; Earl Glass/Stock Boston, p. 92 left; Spencer Grant/PhotoEdit, pp. 8 top right, 499; Spencer Grant/Stock Boston, p. 314; John Griffin/The Image Works, p. 233; Carol Kaelson/CORBIS, p. 599 left; Layne Kennedy/CORBIS, p. 614; Earl & Nazima Kowall/CORBIS, p. 13; Tom & Pat Leeson/Photo Researchers, Inc., p. 147; Craig Lovell/CORBIS, p. 514; Chris Marona/Photo Researchers, Inc., pp. 2, 352; Phil Martin/PhotoEdit, p. 269 right; Felicia Martinez/PhotoEdit, pp. 8 bottom left, 92 right, 144; Sven Martson/The Image Works, p. 384; Tom McCarthy/PhotoEdit, p. 539 right; Will & Deni McIntyre/Photo Researchers, Inc., pp. 1, 14; Wendell D. Metzen/Photo Researchers, Inc., p. 234 top; NASA, pp. 359, 542 right, 719 top, 719 bottom; Bill Nation/CORBIS-Sygma, p. 322 top; Michael Newman/PhotoEdit, pp. 8 bottom right, 394, 540 left, 584 bottom, 622 left; Boyd Norton/The Image Works, p. 178 right; Richard Pasley/Stock Boston, p. 631; PhotoEdit, p. 601; Pictor/Uniphoto, p. 395 right; Neal Preston/CORBIS, p. 205; Pascal Quittemelle/Stock Boston, p. 552 right; A. Ramey/PhotoEdit, pp. 91 left, 117 right, 149,158; Roger Ressmeyer/CORBIS, p. 290; Gary Retherford/Photo Researchers, Inc., p. 584 top; Mark Richards/PhotoEdit, p. 599 right; N. Richmond/The Image Works, p. 540 right; Elena Rooraid/PhotoEdit, p. 23; Steven Rubin/The Image Works, p. 542 left; Manny Rubio/SportsChrome USA, p. 435; David A. Sams/Stock Boston, p. 178 left; Phil Schermeister/CORBIS, p. 582 left; SportsChrome USA, p. 778 left; Barbara Stitzer/PhotoEdit, p. 184 right; Norm Thomas/Photo Researchers, Inc., p. 564; Kent Wood/Photo Researchers, Inc., pp. 83, 94; David Young-Wolff/PhotoEdit, pp. 592, 622 right; Michael Zito/SportsChrome USA, p. 24.

**Library of Congress Cataloging-in-Publication Data**

Lial, Margaret L.
    Prealgebra.—2nd ed./Margaret L. Lial, Diana L. Hestwood.
       p.   cm.
    Includes index.
    ISBN 0-321-06460-7 (Student Edition)
    ISBN 0-321-08933-2 (Annotated Instructor's Edition)
    ISBN 0-321-09731-9 (Hardback)
  1. Mathematics.   I. Hestwood, Diana L.   II. Prealgebra.
QA39.3 .L53 2001
513'.1—dc21                                                             2001046382

Copyright © 2002 Pearson Education, Inc.

All rights reserved. No part of this publication may be reproduced, stored in a retrieval system, or transmitted, in any form or by any means, electronic, mechanical, photocopying, recording, or otherwise, without the prior written permission of the publisher. Printed in the United States of America. For information on obtaining permission for the use of material from this work, please submit a written request to Pearson Education, Inc., Rights and Contracts Department, 75 Arlington Street, Suite 300, Boston, MA 02116.

1 2 3 4 5 6 7 8 9 10   WC   04 03 02

# Contents

| | | |
|---|---|---|
| List of Applications | | vi |
| List of Focus on Real-Data Applications | | x |
| Preface | | xi |
| Feature Walk-Through | | xvii |
| An Introduction to Calculators | | xxi |
| To the Student | | xxvii |
| Diagnostic Pretest | | xxix |

### Chapter 1   Introduction to Algebra: Integers   1

| 1.1 | Place Value | 2 |
|---|---|---|
| 1.2 | Introduction to Signed Numbers | 9 |
| 1.3 | Adding Integers | 15 |
| 1.4 | Subtracting Integers | 25 |
| 1.5 | Problem Solving: Rounding and Estimating | 29 |
| 1.6 | Multiplying Integers | 39 |
| 1.7 | Dividing Integers | 49 |
| 1.8 | Exponents and Order of Operations | 59 |
| | *Summary Exercises on Operations with Integers* | 69 |
| | Chapter 1 Summary | 71 |
| | Chapter 1 Review Exercises | 77 |
| | Chapter 1 Test | 81 |

### Chapter 2   Understanding Variables and Solving Equations   83

| 2.1 | Introduction to Variables | 84 |
|---|---|---|
| 2.2 | Simplifying Expressions | 95 |
| 2.3 | Solving Equations Using Addition | 107 |
| 2.4 | Solving Equations Using Division | 119 |
| 2.5 | Solving Equations with Several Steps | 127 |
| | Chapter 2 Summary | 137 |
| | Chapter 2 Review Exercises | 143 |
| | Chapter 2 Test | 145 |
| | Cumulative Review Exercises: Chapters 1–2 | 147 |

### Chapter 3   Solving Application Problems   149

| 3.1 | Problem Solving: Perimeter | 150 |
|---|---|---|
| 3.2 | Problem Solving: Area | 159 |
| 3.3 | Solving Application Problems with One Unknown Quantity | 169 |
| 3.4 | Solving Application Problems with Two Unknown Quantities | 179 |
| | Chapter 3 Summary | 185 |
| | Chapter 3 Review Exercises | 191 |
| | Chapter 3 Test | 193 |
| | Cumulative Review Exercises: Chapters 1–3 | 195 |

### Chapter 4   Rational Numbers: Positive and Negative Fractions   197

| 4.1 | Introduction to Signed Fractions | 198 |
|---|---|---|
| 4.2 | Writing Fractions in Lowest Terms | 211 |
| 4.3 | Multiplying and Dividing Signed Fractions | 223 |
| 4.4 | Adding and Subtracting Signed Fractions | 235 |
| 4.5 | Problem Solving: Mixed Numbers and Estimating | 247 |
| | *Summary Exercises on Fractions* | 261 |
| 4.6 | Exponents, Order of Operations, and Complex Fractions | 263 |
| 4.7 | Problem Solving: Equations Containing Fractions | 271 |
| 4.8 | Geometry Applications: Area and Volume | 281 |
| | Chapter 4 Summary | 291 |
| | Chapter 4 Review Exercises | 299 |
| | Chapter 4 Test | 301 |
| | Cumulative Review Exercises: Chapters 1–4 | 303 |

## Chapter 5 — Rational Numbers: Positive and Negative Decimals — 305

| | | |
|---|---|---|
| 5.1 | Reading and Writing Decimal Numbers | 306 |
| 5.2 | Rounding Decimal Numbers | 315 |
| 5.3 | Adding and Subtracting Signed Decimal Numbers | 323 |
| 5.4 | Multiplying Signed Decimal Numbers | 333 |
| 5.5 | Dividing Signed Decimal Numbers | 341 |
| 5.6 | Fractions and Decimals | 353 |
| 5.7 | Problem Solving with Statistics: Mean, Median, Mode, and Variability | 361 |
| 5.8 | Geometry Applications: Pythagorean Theorem and Square Roots | 371 |
| 5.9 | Problem Solving: Equations Containing Decimals | 379 |
| 5.10 | Geometry Applications: Circles, Cylinders, and Surface Area | 385 |
| | **Chapter 5 Summary** | 399 |
| | **Chapter 5 Review Exercises** | 407 |
| | **Chapter 5 Test** | 413 |
| | **Cumulative Review Exercises: Chapters 1–5** | 415 |

## Chapter 6 — Ratio, Proportion, and Line/Angle/Triangle Relationships — 417

| | | |
|---|---|---|
| 6.1 | Ratios | 418 |
| 6.2 | Rates | 429 |
| 6.3 | Proportions | 437 |
| 6.4 | Problem Solving with Proportions | 449 |
| 6.5 | Geometry: Lines and Angles | 457 |
| 6.6 | Geometry Applications: Congruent and Similar Triangles | 471 |
| | **Chapter 6 Summary** | 481 |
| | **Chapter 6 Review Exercises** | 489 |
| | **Chapter 6 Test** | 495 |
| | **Cumulative Review Exercises: Chapters 1–6** | 497 |

## Chapter 7 — Percent — 499

| | | |
|---|---|---|
| 7.1 | The Basics of Percent | 500 |
| 7.2 | The Percent Proportion | 515 |
| 7.3 | The Percent Equation | 523 |
| 7.4 | Problem Solving with Percent | 533 |
| 7.5 | Consumer Applications: Sales Tax, Tips, Discounts, and Simple Interest | 543 |
| | *Summary Exercises on Percent* | 555 |
| | **Chapter 7 Summary** | 557 |
| | **Chapter 7 Review Exercises** | 563 |
| | **Chapter 7 Test** | 567 |
| | **Cumulative Review Exercises: Chapters 1–7** | 569 |

## Chapter 8 — Measurement — 573

| | | |
|---|---|---|
| 8.1 | Problem Solving with English Measurement | 574 |
| 8.2 | The Metric System—Length | 585 |
| 8.3 | The Metric System—Capacity and Weight (Mass) | 593 |
| 8.4 | Problem Solving with Metric Measurement | 603 |
| 8.5 | Metric–English Conversions and Temperature | 607 |
| | **Chapter 8 Summary** | 615 |
| | **Chapter 8 Review Exercises** | 621 |
| | **Chapter 8 Test** | 625 |
| | **Cumulative Review Exercises: Chapters 1–8** | 627 |

## Chapter 9 — Graphs — 631

| | | |
|---|---|---|
| 9.1 | Problem Solving with Tables and Pictographs | 632 |
| 9.2 | Reading and Constructing Circle Graphs | 641 |
| 9.3 | Bar Graphs and Line Graphs | 651 |
| 9.4 | The Rectangular Coordinate System | 659 |
| 9.5 | Introduction to Graphing Linear Equations | 665 |
| | **Chapter 9 Summary** | 679 |
| | **Chapter 9 Review Exercises** | 687 |
| | **Chapter 9 Test** | 693 |
| | **Cumulative Review Exercises: Chapters 1–9** | 697 |

## Chapter 10   Exponents and Polynomials   701

| | | |
|---|---|---|
| 10.1 | The Product Rule and Power Rules for Exponents | 702 |
| 10.2 | Integer Exponents and the Quotient Rule | 709 |
| 10.3 | An Application of Exponents: Scientific Notation | 715 |
| 10.4 | Adding and Subtracting Polynomials | 721 |
| 10.5 | Multiplying Polynomials: An Introduction | 729 |
| | **Chapter 10 Summary** | 733 |
| | **Chapter 10 Review Exercises** | 735 |
| | **Chapter 10 Test** | 739 |
| | **Cumulative Review Exercises: Chapters 1–10** | 741 |

**Whole Numbers Computation: Pretest**   743

## Chapter R   Whole Numbers Review   745

| | | |
|---|---|---|
| R.1 | Adding Whole Numbers | 745 |
| R.2 | Subtracting Whole Numbers | 753 |
| R.3 | Multiplying Whole Numbers | 761 |
| R.4 | Dividing Whole Numbers | 769 |
| R.5 | Long Division | 779 |
| | **Chapter R Summary** | 785 |
| | **Chapter R Review Exercises** | 787 |
| | **Chapter R Test** | 789 |

**Appendix: Inductive and Deductive Reasoning**   A–1

**Answers to Selected Exercises**   A–7

**Index**   I–1

# List of Applications

## Astronomy

| | |
|---|---|
| Distance between Venus and the sun | 736 |
| Distance light travels in one year | 742 |
| Houston Space Center | 788 |
| Hubble telescope repairs | 359 |
| Mass of Earth | 720 |
| Mass of the moon | 720 |
| Mass of the sun | 740 |
| Speed of light | 720 |

## Automotive

| | |
|---|---|
| Cars in a parking lot | 368 |
| Drivers wearing seat belts | 539 |
| Driving speed | 158 |
| Gas mileage | 351, 367, 434, 449–450, 489, 498, 540, 768 |
| Gas tank capacity | 259 |
| Handicapped parking spaces | 537 |
| Imported Japanese cars | 82 |
| Producing license plates | 790 |
| Skidding distance | 728 |

## Biology

| | |
|---|---|
| Animal's speed | 178, 584, 760 |
| Alligator laying eggs | 234 |
| Average length of sharks and whales | 489 |
| Endangered species | 564, 693 |
| Growing sunflowers | 580 |
| Height of a starflower | 582 |
| Horse's height | 428 |
| Hummingbirds | 452, 613, 768 |
| Kelp plant growth | 768 |
| Largest meat-eating dinosaur | 581 |
| Mice in a medical experiment | 359 |
| Mouse vs. elephant breaths | 742 |
| Prehistoric rhino | 331 |
| Samples taken | 369 |
| Tracking a Siberian tiger | 147 |
| Vitamin supplement for cats | 493 |
| Water in a human body | 455 |
| Weight of a spider | 718 |
| Weights of animals | 87–88, 234, 331, 581, 602, 605, 768 |
| World's longest insect | 605 |

## Business

| | |
|---|---|
| Advertising | 221, 694 |
| Assembling products | 778, 784, 790 |
| Business expenses | 682, 695 |
| Carpeting apartments | 260 |
| Chicken noodle soup sold during the cold-and-flu season | 541 |
| Child care | 47, 300, 367 |
| Companies receiving patents | 540 |
| Copy machine paper | 339 |
| Custom hat bands | 260 |
| Customers | 47, 363 |
| Day care center | 584 |
| Defective products | 44, 454 |
| Deliveries | 206, 368, 758, 788 |
| Dinner check amounts | 367 |
| Flats needed for packing eggs | 784 |
| Gasoline sales | 656 |
| Hair salon expenditures | 690 |
| Hair stylist | 541 |
| Income losses | 47, 80 |
| Insurance claims processed | 409 |
| Inventory | 80 |
| Loss due to shoplifting | 57 |
| Mail order catalog | 340 |
| Manufacturing | 260, 542 |
| Monthly commissions of sales people | 364 |
| Monthly expenses | 80 |
| Moving company | 579 |
| Newspaper carrier's customers | 788 |
| Ordering products | 91, 92, 178, 192, 245 |
| Packaging products | 422, 784 |
| Phone orders for precision parts | 314 |
| Producing candy canes | 584 |
| Profit | 80, 658 |
| Purchasing land | 245 |
| Realtor's fee | 339 |
| Sale price | 546, 551, 552, 556, 560, 565, 568, 572, 630 |
| Sales staff | 752 |
| Sales | 80, 177, 352, 361, 366, 367, 368, 369, 409, 489, 540, 642, 649, 657, 658, 681, 683, 689 |
| Shipping | 82 |
| White Water Rafting Company | 649 |

## Chemistry

| | |
|---|---|
| Acid for an experiment | 75, 606 |
| Chlorine in a swimming pool | 605 |
| Diameter of a hydrogen atom | 720 |
| Lab experiments | 172–173 |
| Molecules in a gram of hydrogen | 717 |
| Ordering supplies for a chemistry lab | 143, 584 |
| Plastic tubing for a science lab | 605 |
| Temperature in a laboratory experiment | 38, 52 |

## Construction/Engineering

| | |
|---|---|
| Anasazi multistory apartment towns | 418–419 |
| Brass trim | 260 |
| Building a structure | 79, 493, 606 |
| Clear gloss wood finish | 453 |
| Cost of a studio addition | 645 |
| Fencing | 85–86, 119, 154, 164, 168, 184, 192, 193, 245, 269, 732, 788 |
| Floor plan | 454 |
| Hauling sand | 255 |
| Height difference between structures | 760 |
| Height of a farm silo | 377 |
| Home builder buying land | 262 |
| Hours to build a home | 788 |
| Installing cabinets | 245, 262 |
| Ladder leaning against a building | 374, 377 |
| Landscaping | 259 |
| Latex enamel paint coverage | 453 |
| Length of a board or pipe | 259, 422, 629 |
| Length of a loading ramp | 377 |
| Length of a support wire | 374 |
| Making shelves | 259 |
| Measuring a board | 604, 618, 624 |
| Oak trim | 258 |
| Roofing | 377 |
| Scale model of a building | 495 |

List of Applications **vii**

Stained glass window installation 260
"The ultimate ruler" 590
Weight of floor tiles 605

### Consumer

Baby carrier imported from Sweden 613
Best buy 396, 430–431, 432, 434, 435, 484, 490, 493, 495, 569, 606, 630
Buying 36, 145, 171–172, 340, 416, 700
Cost 183, 262, 320, 339, 350, 351, 409, 410, 412, 413, 416, 435, 436, 453, 491, 552, 553, 556, 572, 584, 603, 605, 606, 622, 630, 689, 732, 752, 766, 768, 788, 790
Directions for a self-stick clothes hook 613
Gender buys 654
Insurance policies 367, 559
Long-distance phone service and cost 435, 656
Lottery 631
Parking fees 118, 147, 362
Price 400, 428, 541, 578–579, 583
Rental costs 147, 234, 340, 384, 537, 790
Tips 500, 544–545, 550, 552, 53, 556, 560, 565, 568
Total cost 340, 553, 556, 567, 752, 766, 768, 788

### Economics

Cost of printing money 352
Sales tax 543–544, 552, 560, 565, 742
Simple interest 547, 548, 551, 560, 565, 568
Stocks 13, 36, 52, 147, 409, 455, 541
Taxes 234, 500, 533–534, 539

### Education

Absent students 542
American Sign Language class 205
Ceramics class A-4
College cheerleading squad 205
College classes taken A-4, A-6
College enrollment 221, 498, 535, 694
College students drinking coffee 456
College tuition 47, 541
Cost of a math textbook 414
Cost of college education 233
Credits 368, 539, 630
Elementary school's average daily attendance 538
Enrollment increase or decrease 47, 57, 559, 567
Final exam scores 367
Financial aid 221, 451, 533
Grade point average 363, 368
Graduates taking advanced courses 651
Graduation ceremony 91, 117
Grant money 416, 700
Ordering math books 84, 85, 107
Points earned in a math class 48, 500, 534
Quiz scores 367, 368, 534
Ratio of math students to English students 489
Refresher math course 454
Scholarships 58, 172, 416, 540
School expenses 648, 694
Student groups 196
Student letter grades 683
Student-to-teacher ratio 455
Students enrolled in Algebra 368
Students per class 368
Students' income 694
Students' time 246, 641
Students' voting 491
Studying time 665
Test scores 92, 361, 365, 370, 403, 412, 536, 540, 572, 700
Total points on a screening exam 364
U.S. colleges who accept the lowest percent of student applications 572
Working students 302, 572

### Environment

Acres of seedlings 245
Average depth of the world's oceans 621
Bottom of Lake Baykal 13
Distance from a thunderstorm 94
Earthquakes 395, 738
Flood level 23, 28, 36
Fog 57
Hazardous waste dump 246
Insect spray 259
Land on the shore of the Dead Sea 78
Length of a lake 450, 480
Rainfall 359, 535, 626, 742
Recycling 258, 621
Salt in sea water 602
Shortest river in the world 592
Snowfall 584, 687
Tallest mountain in the world 13
Temperature 9, 10, 13, 17, 23, 36, 48, 91, 117, 147, 184, 366, 369, 370, 497, 613
Trash pick-up 259
Tree planting project 564
Trout stocked in a river 652
Water in Lake Natoma 689

### Finance

Budgeting 38, 52, 53, 427, 540, 564
Car payment 339
Change received 331, 340, 408
Charge account balance 350
Checking account balance 9, 10, 13, 17, 23, 33, 38, 57, 58, 80, 82, 177, 192, 760, 788, 790
Check total 408
Church donations 230
Deposits 10, 36, 331
Down payment 567
Estate money 183, 778
Household expenses 57, 171
Income 80, 234
Investing money 348, 554, 778
Money in a purse 188, 194
Monthly bills 361, 367, 369
Monthly expenses 332, 365
Monthly payments 57, 75, 568, 766, 784
Paying off a loan 351
Payment preferences 647
Raising money 788
Retirement income 230
Savings 234
Shopping for a loan 556
Spending 178, 234, 339, 416, 419, 493, 626, 768
Total amount due on a loan 548, 551, 552, 553, 565, 568, 630

### Geometry

Area of a circle 390, 394, 395, 412, 699, 742
Area of a parallelogram 163, 187, 191
Area of a piece of land 233
Area of a rectangle 160, 161, 165, 168, 233, 269, 300, 572
Area of a skating rink 396
Area of a square 161–162, 165, 166, 184, 191, 192, 193, 196, 262, 269
Area of a triangle 282, 288
Base or height of a parallelogram 163–164, 166, 191, 193
Circumference of a circle 389, 394, 395, 412, 699, 742

Decorative strip 168
Diameters 246, 262, 395, 411
Height of an object 455, 476, 479, 480
Length of a side of a pentagon 126
Length of one side of a square 151, 155, 162, 187, 191, 192, 193, 194
Length of one side of a triangle 126
Length or width of a rectangle 152–153, 156, 161, 165, 166, 168, 180–181, 184, 186, 191, 192, 194, 196, 304, 416, 428, 454, 497, 629, 699
Lengths of a cut object 183, 184, 194, 629
Perimeter of a pentagon 91
Perimeter of a rectangle 152, 156, 167, 168, 184, 193, 196, 269, 290, 572, 732
Perimeter of a square 186, 151, 155, 186, 191, 193, 196, 262
Perimeter of a triangle 91
Perimeter of an irregular shape 154, 742
Radius of a can 411
Surface area of a box 398, 405, 411
Surface area of a cylinder 398, 406
Volume of a box 289, 411, 699
Volume of a cylinder 392, 398, 412, 630, 699
Volume of a pyramid 290

## Government

College students ranking issues they'd like presidential candidates to address 512
City council candidates 183
United States Congress 184

## Health/Life Sciences

Americans eating more fish 541
Antibiotics 428, 493
Artificial heart valve 359
Blood alcohol level 540
Blood vessels 592
Bone lengths 330
Breathing 601
Burning calories 455, 491, 638, 760
Calcium 359, 693
Calories and dietary fiber 456
Calories consumed 47, 58, 82, 768
Child's growth 254, 255
Copper in the body 720
Dentists recommending sugarless gum 451
Diet 352
Donated blood 535–536
Drinking water 778
Exercise 456
Eye drop dispensers 233
Fat calories 536, 538
Hair and nail growth 590
Hearing aids 435
Heart pumping blood 601, 605
Height 251, 280
Late patients 564
Maximum safe heart rate 384
Medicine 358, 384, 419, 425, 453, 483, 491, 495, 604, 768
Nerve signals in the human body 611
Nonsmokers 700
Normal body temperature 13
Patients admitted to the hospital 369
Patient's T scores 14
Premature baby 582
Recommended daily iron intake 178
Recommended maximum daily amount of dietary fat 542
Sodium in a Subway sandwich 626
Sweat glands 601
Systolic blood pressure 87, 274–275, 279
Types of teeth in an adult's mouth 512
Vaccine 302
Weight 178, 624
Weight gain and loss 10, 13, 23, 78, 177, 435, 537–538, 622
Weight of the human brain 542, 601

## Labor

Earnings 181, 183, 188–189, 338, 351, 409, 453, 490, 540, 778
Gross pay 33, 338
Hours worked 179–180, 255, 259, 300, 331, 368, 400, 784
Money earned working part time 36
Monthly income 221
Number of employees 144, 752
Part time work schedule 541
Pay for painting a house 571
Pay rate per hour 435
Salaries 367, 370, 760
Unemployment 655
Wages 331, 536–537
Work perks 688

## Miscellaneous

Age 178, 183, 194, 364, 365, 367, 369, 370, 415
Animal shelter 491
Animals received in a Humane Society 566
Bags of rice 778
Beverage consumption 680
Bird seed 177, 304
Bottle of contact lens solution 262
Box of imported Belgium chocolates 613
Cards in a deck 788
Collecting General Mills box tops 352
Cookie sales at the Minnesota State Fair 624
Cookies consumed 177
Dishwashers 611, 623
Disposable diapers 571
Fabric 604, 606, 626, 630
Fertilizer 450, 788
Film width 592
Fundraising 53, 75, 339, 605
Gallons of beverage 778
Gold wedding bands 592
Grams of turkey per person 622
Greeting cards 426
Gum wrapper chain 586
Halloween candy 178
Harvesting wild rice 582
Hot tub temperatures 414
How Americans fall asleep 650
Ingredients 456
Length of an arrow shaft 259
Loaning books 414
Ordering doughnuts 92
Pets 234, 455, 612
Plant food 498
Planting grass seed 453, 485–486
Preparing hamburgers 582
Prize money 13, 192
Reasons for eating out 646
Recipes 198, 255, 259, 262, 300, 435, 572, 579, 624
Replacing lightbulbs 177
Serving punch at a wedding reception 622
Servings 453
Shadow length 455
Station's broadcast area 395
Tea bag from Scotland 612
Tent rental 53
Theft 23
Thickness of a sheet of paper 717–718
Time spent on household chores 194, 221
Toddler shoe sizes 742
Tomato plants 768
Typing rate 490, 495
Waiting in line 196

List of Applications    ix

Water concentration in shampoo    531
Weights of objects    38, 58, 602, 606, 630, 784
Window cleaner spray bottle    230
Words in the Webster's dictionary    539
Wrapper from a ball of twine    612
Yellow Pages directories    78

## Social Sciences

Americans with Disabilities Act    572, 647, 784
Catering    58
Charity    498
Contributions to a food pantry    362
Coordinator of Toys for Tots    778
Donations    173, 177
Favorite ice cream flavor    454
Flood victims    58
Homeless shelter    38, 630
Left-handed people    495
Low income housing project    533
Number of people liking poached eggs    454
Number of people wanting to lose weight    450–451
Parenting survey    556

## Sports/Hobbies/Entertainment

400-meter event    435
Americans favorite recreational activity    408
Americans going to a video store each week    36
Americans who play various instruments    426
Baseball    44, 183, 233, 367, 378, 742, 778
Basketball    435, 454, 637
Batting average    339, 446, 491
Bowling scores    57, 88
Card game    10, 23, 28
Cost of sod for a playing field    396
Depth of a scuba diver    9, 38
Fans buying drinks at the ballpark    493

Field goal attempts    534
Filming time for a movie    428
Fishing line    313, 360
Free throw attempts    540
Garlic Festival Fun Run    233
Golf    24, 370
Knitting a scarf    630
Knot tying    616
Lengths of swimming events    614
Making holiday wreaths    259
Making necklaces    491
Michael Johnson's Olympic track shoes    611
Model railroad    491
Money spent on attending motion pictures    720
Music club dues    177
Olympic long jumps    351
Olympic medals    584
Participation in selected NCAA sports    687
Points scored in a game    365–366, 454, 493
Recording artists    205
Running    413, 498
Seats in an auditorium    495
Sewing    233, 259, 603–604, 626
Sketching cartoon strips    453
Soccer    196, 369, 542, 642–644, 760
Softball diamond    378
Television    36, 451, 455, A-6
Time to record an album    453
Wheelchair race    302, 408
Yardage gained or lost    9, 13, 15, 17, 23, 39, 40, 57, 82

## Statistics/Demographics

Average length of a newborn baby    358
Number of people 100 years old or older    556
People per square mile in China    720
Population of States    36, 752
Population of U.S. metropolitan areas    634–635
Survey about UFOs    262

U.S. population    331, 540, 736
World population    78, 542

## Technology

Air-to-ground cell phone calls    382
Cell phones    57, 58, 436, 624
Computer memory    602
Computer service calls    368
Cost of laptop computers    766
Digital subscriber line installations    651–652
E-mail messages received    368
Electronic/computer games    518
Learning about computers    245
Sales of analog and digital cell phones    653
Why people shop on-line    655
Weight of a Toshiba Protege laptop computer    612

## Transportation

Airplanes    9, 57, 369, 377, 455, 760
Buses    53, 92, 171
Concorde jet's speed    404
Distance between two locations    412, 572, 749, 752, 768
Distance    158, 260, 369, 377
Freight train    205
Japanese bullet train    788
Luggage on airline flights    612
Mach speed of a car    352
Maximum taxicab fares    633
Miles driven    36, 58, 80
Motorcycles    205
Passenger arrival and departures at selected U.S. airports    639
Passengers on a plane    760
Performance data for the largest U.S. airlines    632–633, 640
Submarine depth    10, 36, 48, 57
Tanker for the British Petroleum Company    245
Time driving    158
Tour of the west    444
Trips needed to deliver wood    233

# List of Focus on Real-Data Applications

| | |
|---|---|
| Auto Aide | 20 |
| 'Til Debt Do You Part! | 54 |
| Expressions | 102 |
| Algebraic Expressions and Tuition Costs | 142 |
| Formulas | 174 |
| Connections: Arithmetic to Algebra | 182 |
| Music | 242 |
| Recipes | 256 |
| Heart-Rate Training Zone | 266 |
| Hotel Expenses | 276 |
| Quilt Patterns | 286 |
| Lawn Fertilizer | 320 |
| Life Insurance Benefits | 336 |
| Dollar-Cost Averaging | 348 |
| Historical Ratios | 424 |
| Tour of the West | 444 |
| Feeding Hummingbirds | 452 |
| Decimals, Percents, and Quilt Patterns | 508 |
| Make Your Investments Grow—Compound Interest | 554 |
| Educational Tax Incentives | 562 |
| Growing Sunflowers | 580 |
| Measuring Up | 590 |
| Currency Exchange | 636 |
| Grocery Shopping | 654 |
| Surfing the Net | 686 |
| Earthquake Intensities Measured by the Richter Scale | 738 |

# Preface

The second edition of *Prealgebra* continues our ongoing commitment to provide the best possible text and supplements package that will help instructors teach and students succeed. To that end, we have addressed the diverse needs of today's students through an attractive design, updated applications and graphs, helpful features, careful explanation of concepts, and an expanded package of supplements and study aids. We have also responded to the suggestions of users and reviewers and have added many new examples and exercises based on their feedback.

The text is designed for mathematics students who are new to algebra, relearning the algebra they studied in the past, or anxious about their ability to learn algebra. The text interweaves arithmetic review and geometry topics, as appropriate, into the algebraic themes of integers, variables and expressions, equations, solving application problems, positive and negative fractions and decimals, proportions, percents, measurements, graphing, and polynomials. The emphasis is on building a solid understanding of the foundations of algebra. This is accomplished by tying the content to students' experiences and previous knowledge, explaining important terminology in everyday English, showing *why* things work the way they do, and providing carefully sequenced exercises.

This text is part of a series that also includes the following books:

- *Essential Mathematics* by Lial and Salzman
- *Basic College Mathematics,* Sixth Edition, by Lial, Salzman, and Hestwood
- *Introductory Algebra,* Seventh Edition, by Lial, Hornsby, and McGinnis
- *Intermediate Algebra with Early Functions and Graphing,* Seventh Edition, by Lial, Hornsby, and McGinnis
- *Introductory and Intermediate Algebra,* Second Edition, by Lial, Hornsby, and McGinnis.

## WHAT'S NEW IN THIS EDITION?

We believe students and instructors will welcome the following new features.

▶ *New Real-Life Applications* We are always on the lookout for interesting data to use in real-life applications. As a result, we have included many new or updated examples and exercises throughout the text that focus on real-life applications of mathematics. Students are often asked to find data in a table, chart, graph, or advertisement. (See pp. 9, 167, and 274.) These applied problems provide an up-to-date flavor that will appeal to and motivate students. A comprehensive List of Applications appears on page vi.

▶ *New Figures and Photos* Today's students are more visually oriented than ever. Thus, we have made a concerted effort to add mathematical figures, diagrams, tables, and graphs whenever possible. (See pp. 8, 234, and 262.) Many of the graphs use a style similar to that seen by students in today's print and electronic media. Photos have been incorporated to enhance applications in examples and exercises. (See pp. 197 and 269.)

▶ *Increased Emphasis on Problem Solving* Chapter 3 introduces students to our six-step process for solving application problems algebraically: *Read, Assign a Variable, Write an Equation, Solve, State the Answer,* and *Check.* By devoting an entire chapter to this process, students build a strong foundation for problem solving, which is then reinforced through specific problem-solving examples in Chapters 4, 5, 6, 7, and 8. (See pp. 170, 180–181, and 274.) The same six steps are also used throughout the other algebra titles in this textbook series.

▶ *Study Skills Component* Poor study skills are a major reason why students do not succeed in math. A few generic tips sprinkled here and there are not enough to help students change their behavior. So, in this text, a desk-light icon at key points in the text directs students to a separate *Study Skills Workbook* containing carefully designed activities that correlate directly to the text. (See pp. 84, 165, and 243.) This unique workbook explains *how* the brain actually learns and remembers so students understand *why* the study skills activities will help them succeed in the course. Students are introduced to the workbook in a new To the Student section at the beginning of the text.

▶ *Focus on Real-Data Applications* Each one-page activity presents a relevant and in-depth look at how mathematics is used in the real world. Designed to help instructors answer the often-asked question, "When will I ever use this stuff?," these activities ask students to read and interpret data from newspaper articles, the Internet, and other familiar, real-world sources. (See pp. 142, 256, and 336.) The activities are well-suited to collaborative work or they can be completed by individuals or used for open-ended class discussions. Instructor teaching notes and activity extensions are provided in the *Printed Test Bank and Instructor's Resource Guide.*

▶ *Diagnostic Pretest* A diagnostic pretest is now included on p. xxix of the text and covers all the material in the book, much like a sample final exam. This pretest can be used to facilitate student placement in the correct chapter according to skill level. The pretest also exposes students to the scope of the course content.

▶ *Chapter Openers* New chapter openers feature real-world applications of mathematics that are relevant to students and tied to specific material within the chapters. Examples of topics include finding the best buy on cell phone service, home improvements, recipes, medical tests, fishing, and work/career applications. (See pp. 1, 83, and 149.)

▶ *Test Your Word Power* This new feature, incorporated into each chapter summary, helps students understand and master mathematical vocabulary. Key terms from the chapter are presented along with four possible definitions in a multiple-choice format. Answers and examples illustrating each term are provided. (See pp. 72, 137, and 292.)

## What Familiar Features Have Been Retained?

We have retained the popular features of previous editions of the text, including the following:

▶ *Learning Objectives* Each section begins with clearly stated, numbered objectives, and the material within sections is keyed to these objectives so that students know exactly what concepts are covered. (See pp. 25, 169, and 247.)

▶ *Cautions and Notes* These color-coded and boxed comments, one of the most popular features of previous editions, warn students about common errors and emphasize important ideas throughout the exposition. (See pp. 86, 107, 199, and 236.) There are more of these in the second edition, and the new text design makes them easier to spot; Cautions are highlighted in bright yellow and Notes are highlighted in green.

▶ ***Margin Problems*** Margin problems, with answers immediately available on the bottom of the page, are found in every section of the text. (See pp. 17, 163, and 284.) This key feature allows students to immediately practice the material covered in the examples in preparation for the exercise sets.

▶ ***Calculator Tips*** These optional tips, marked with a calculator icon, offer basic information and instruction for students using calculators in the course. (See pp. 87, 216, and 235.) In addition, an Introduction to Calculators is included in the student material that precedes Chapter 1.

▶ ***Ample and Varied Exercise Sets*** The text contains a wealth of exercises to provide students with opportunities to practice, apply, connect, and extend the skills they are learning. Numerous illustrations, tables, graphs, and photos have been added to the exercise sets to help students visualize the problems they are solving. Problem types include skill building, writing, estimation, and calculator exercises as well as applications and correct-the-error problems. In the *Annotated Instructor's Edition* of the text, the writing and estimation exercises are marked with icons for writing and for estimation ≈ so that instructors may assign these problems at their discretion. Exercises suitable for calculator work are marked in both the student and instructor editions with a calculator icon. (See pp. 21, 165, and 257.)

▶ ***Relating Concepts Exercises*** These sets of exercises help students tie concepts together and develop higher level problem-solving skills as they compare and contrast ideas, identify and describe patterns, and extend concepts to new situations. (See pp. 14, 94, and 158.) These exercises make great collaborative activities for pairs or small groups of students.

▶ ***Summary Exercises*** There are three sets of in-chapter summary exercises: operations with integers, fractions, and percents. These exercises provide students with the all-important *mixed* practice they need at these critical points in their skill development. (See pp. 69 and 261.)

▶ ***Ample Opportunity for Review*** Each chapter ends with a Chapter Summary featuring: Key Terms with definitions and helpful graphics, New Formulas, Test Your Word Power, and a Quick Review of each section's content with additional examples. Also included is a comprehensive set of Chapter Review Exercises keyed to individual sections, a set of Mixed Review Exercises, and a Chapter Test. Beginning with Chapter 2, each chapter concludes with a set of Cumulative Review Exercises. (See pp. 71–82, 185–196, and 291–304.)

## What Content Changes Have Been Made?

We have worked hard to fine-tune and polish presentations of topics throughout the text based on user and reviewer feedback. Some of the content changes include the following:

- In Chapter 5, Positive and Negative Decimals, surface area has been added to Section 5.10 using rectangular solids and cylinders.

- In Chapter 6, Ratio, Proportion, and Line/Angle/Triangle Relationships, the material on lines and angles is now presented in a new Section 6.5 rather than being placed in an appendix. Section 6.6 has been expanded to include congruent triangles as an important precursor to studying similar triangles.

- In Chapter 7, Percent, the material on simple interest is now incorporated into Section 7.5, Consumer Applications, rather than being treated in a separate section.

- In Chapter 8, Measurement, Section 8.1 on English measurement now focuses on solving application problems as well as converting among units.

- In Chapter 9, Graphs, a new Section 9.1 focuses on higher-order problem solving with tables and pictographs. While students read information from simple tables in earlier chapters, in this section they must make comparisons, interpolate, and analyze.

- Chapter 10, Exponents and Polynomials, has been added in response to reviewer and user requests. The material on exponents covers the product, power, and quotient rules; negative and zero exponents; and scientific notation applications. The final two sections include adding and subtracting polynomials and an introduction to multiplying polynomials.

## WHAT SUPPLEMENTS ARE AVAILABLE?

Our extensive supplements package includes an *Annotated Instructor's Edition*, testing materials, solutions manuals, tutorial software, videotapes, and a state-of-the-art Web site. For more information about any of the following supplement descriptions, please contact your Addison-Wesley sales consultant.

### FOR THE STUDENT

*Student's Solutions Manual* **(ISBN 0-321-09145-0)** The *Student's Solutions Manual* provides detailed solutions to the odd-numbered section exercises and to all margin, Relating Concepts, Summary, Chapter Review, Chapter Test, and Cumulative Review exercises.

*Study Skills Workbook* **(ISBN 0-321-09256-2)** A desk-light icon at key points in the text directs students to correlated activities in this unique workbook by Diana Hestwood and Linda Russell. The activities in the workbook teach students how to use the textbook effectively, plan their homework, take notes, make mind maps and study cards, manage study time, review a chapter, prepare for and take tests, evaluate test results, and prepare for a final exam. Students find out *how* their brains actually learn and remember, and what research tells us about ways to study effectively. A new To the Student section at the beginning of the text introduces students to the *Study Skills Workbook*.

*Addison-Wesley Math Tutor Center* The Addison-Wesley Math Tutor Center is staffed by qualified college mathematics instructors who tutor students on examples and exercises from their textbook. Tutoring is provided via toll-free telephone, toll-free fax, e-mail, and the Internet. White Board technology allows tutors and students to actually see problems being worked while they "talk" in real time over the Internet during tutoring sessions. The Math Tutor Center is accessed through a registration number that may be bundled free with a new textbook or purchased separately.

*Web Site:* **www.MyMathLab.com** Ideal for lecture-based, lab-based, and on-line courses, MyMathLab.com provides students with a centralized point of access to the wide variety of on-line resources available with this text. The pages of the actual book are loaded into MyMathLab.com, and as students work through a section of the on-line text, they can link directly from the pages to supplementary resources (such as tutorial software, interactive animations, and audio and video clips) that provide instruction, exploration, and practice beyond what is offered in the printed book. MyMathLab.com generates personalized study plans for students and allows instructors to track all student work on tutorials, quizzes, and tests.

*InterAct Math® Tutorial Software* **(ISBN 0-321-09150-7)** This interactive tutorial software provides algorithmically generated practice exercises that are correlated at the objective level to the content of the text. Every exercise in the program is accompanied by an example and a guided solution designed to involve students in the solution process. For Windows users, selected problems also include a video clip to help students visualize concepts. The software tracks student activity and scores and can generate printed summaries of students' progress. Instructors can use the InterAct Math® Plus course-management software to create, administer, and track tests and monitor student performance during practice sessions. (See For the Instructor.)

◯ **InterAct MathXL: www.mathxl.com** InterAct MathXL is a Web-based tutorial system that enables students to take practice tests and receive personalized study plans based on their results. Practice tests are correlated directly to the section objectives in the text, and once a student has taken a practice test, the software scores the test and generates a study plan that identifies strengths, pinpoints topics where more review is needed, and links directly to InterAct Math® tutorial software for additional practice and review. A course-management feature allows instructors to create and administer tests and view students' test results, study plans, and practice work. Students gain access to the InterAct MathXL Web site through a password-protected subscription, which can either be bundled free with a new copy of the text or purchased separately with a used book.

◯ **Real-to-Reel Videotape Series (ISBN 0-321-08934-0)** This series of videotapes, created specifically for *Prealgebra*, Second Edition, features an engaging team of math instructors who provide comprehensive lectures on every objective in the text. The videos include a stop-the-tape feature that encourages students to pause the video, work the presented example on their own, and then resume play to watch the video instructor go over the solution.

◯ **Digital Video Tutor (ISBN 0-321-08935-9)** This supplement provides the entire set of Real-to-Reel videotapes for the text in digital format on CD-ROM, making it easy and convenient for students to watch video segments from a computer, either at home or on campus. Available for purchase with the text at minimal cost, the Digital Video Tutor is ideal for distance learning and supplemental instruction.

### FOR THE INSTRUCTOR

◯ **Annotated Instructor's Edition (ISBN 0-321-08933-2)** The *Annotated Instructor's Edition* provides immediate access to the answers for all text exercises by printing them in color next to the corresponding problems. To assist instructors in assigning homework problems, icons identify writing , estimation ≈ , and calculator  exercises.

◯ **Instructor's Solutions Manual (ISBN 0-321-09144-2)** The *Instructor's Solutions Manual* provides complete solutions to all even-numbered section exercises.

◯ **Answer Book (ISBN 0-321-09146-9)** The *Answer Book* provides answers to all the exercises in the text.

◯ **Printed Test Bank and Instructor's Resource Guide (ISBN 0-321-09147-7)** The *Printed Test Bank* portion of this manual contains two diagnostic pretests, six free-response and two multiple-choice test forms per chapter, and two final exams. The *Instructor's Resource Guide* portion of the manual contains teaching suggestions for each chapter, additional practice exercises for every objective of every section, a correlation guide from the fifth to the sixth edition, phonetic spellings for all key terms in the text, and teaching notes and extensions for the Focus on Real-Data Applications pages in the text.

◯ **TestGen-EQ with QuizMaster-EQ (ISBN 0-321-09148-5)** This fully networkable software enables instructors to create, edit, and administer tests using a computerized test bank of questions organized according to the chapter content of the text. Six question formats are available, and a built-in question editor allows the user to create graphs, import graphics, and insert mathematical symbols and templates, variable numbers, or text. An "Export to HTML" feature allows practice tests to be posted to the Internet, and instructors can use QuizMaster-EQ to post quizzes to a local computer network so that students can take them on-line and have their results tracked automatically.

**Web Site: www.MyMathLab.com** In addition to providing a wealth of resources for lecture-based courses, MyMathLab.com gives instructors a quick and easy way to create a complete on-line course based on *Prealgebra*, Second Edition. MyMathLab.com is hosted nationally at no cost to instructors, students, or schools, and it provides access to an interactive learning environment where all content is keyed directly to the text. Using a customized version of Blackboard™ as the course-management platform, MyMathLab.com lets instructors administer preexisting tests and quizzes or create their own, and it provides detailed tracking of all student work as well as a wide array of communication tools for course participants. Within MyMathLab.com, students link directly from on-line pages of their text to supplementary resources such as tutorial software, interactive animations, and audio and video clips.

## ACKNOWLEDGMENTS

The comments, criticisms, and suggestions of users, nonusers, instructors, and students have positively shaped this textbook. We especially wish to thank these individuals who provided invaluable suggestions for this edition:

Carla Ainsworth, *Salt Lake Community College*
Kim Brown, *Tarrant County College—Northeast Campus*
John Close, *Salt Lake Community College*
Ky Davis, *Muskingum Area Technical College*
Randy Gallaher, *Lewis and Clark Community College*
Rosemary Karr, *Collin County Community College*
Lou Ann Mahaney, *Tarrant County College—Northeast Campus*
Elizabeth Morrison, *Valencia Community College—West Campus*
Faith Peters, *Broward Community College*
Janalyn Richards, *Idaho State University*
Julia Simms, *Southern Illinois University—Edwardsville*
Sounny Slitine, *Palo Alto College*
Sharon Testone, *Onondaga Community College*
Shae Thompson, *Montana State University*
Cheryl Wilcox, *Diablo Valley College*
Kevin Yokoyama, *College of the Redwoods*
Karl Zilm, *Lewis and Clark Community College*

Our sincere thanks go to these dedicated individuals at Addison-Wesley who worked long and hard to make this revision a success: Maureen O'Connor, Ruth Berry, Ron Hampton, Dennis Schaefer, Dona Kenly, Suzanne Alley, Melissa Wright, and Jolene Lehr. Steven Pusztai of Elm Street Publishing Services provided his customary excellent production work. We are most grateful to Peg Crider for researching and writing the Focus on Real-Data Applications feature; Paul Van Erden for his accurate and useful index; Becky Troutman for preparing the comprehensive List of Applications; Abby Tanenbaum for writing the new Diagnostic Pretest; and Ellen Sawyer and Shannon d'Hemecourt for accuracy checking the manuscript.

Much of the material in Chapter 10, Exponents and Polynomials, was adapted from *Introductory Algebra*, Seventh Edition, by Lial, Hornsby, and McGinnis. Many thanks to Marge, John, and Terry for their contribution.

Special thanks also go to Barbara Brown, *Anoka-Ramsey Community College*, who reviewed all the new material for this edition in great detail. Her expertise in teaching developmental algebra students resulted in greater clarity, consistency, and creativity. Finally, it is the wonderful collaboration with Linda Russell, reading and study skills specialist at *Minneapolis Community and Technical College*, which resulted in the Study Skills Workbooks to accompany this text and the others in the series.

The ultimate measure of this textbook's success is whether it helps students master algebra skills, develop problem-solving techniques, and increase their confidence in learning and using mathematics. Please let us know how we are doing by sending an e-mail to math@awl.com.

*This book is dedicated to my husband, Earl Orf, who is the wind beneath my wings, and to my students at Minneapolis Community and Technical College, from whom I have learned so much.*

*Diana L. Hestwood*

# Feature Walk-Through

**New! Chapter Openers** New chapter openers feature real-world applications of mathematics that are relevant to students and tied to specific material within the chapters (page 1).

### Introduction to Algebra: Integers

1.1 Place Value
1.2 Introduction to Signed Numbers
1.3 Adding Integers
1.4 Subtracting Integers
1.5 Problem Solving: Rounding and Estimating
1.6 Multiplying Integers
1.7 Dividing Integers
1.8 Exponents and Order of Operations
Summary Exercises on Operations with Integers

Osteoporosis is a bone disease in which bones become thin and brittle. One in four women over age 50, and one in eight men over 50, have the disease, but it can strike people in their 20s who have eating disorders. (*Source:* Osteoporosis online.) More than 1.5 million osteoporosis-related bone fractures occur each year in the United States. (*Source:* Mayo Clinic.) An ultrasound of the heel bone is a quick, inexpensive way to screen people for osteoporosis. The patients receive a "T score" that indicates their level of risk. Both the ultrasound technician and the patient must understand positive and negative numbers to interpret the score. (See Section 1.2, Exercises 45–48.)

---

234 Chapter 4 Rational Numbers: Positive and Negative Fractions

**59.** An adult male alligator may weigh 400 pounds. A newly hatched alligator weighs only $\frac{1}{8}$ pound. The adult weighs how many times the hatchling? (*Source:* St. Marks NWR.)

**60.** A female alligator lays about 35 eggs in a nest of marsh grass and mud. But $\frac{4}{5}$ of the eggs or hatchlings will fall prey to raccoons, wading birds, or larger alligators. How many of the 35 eggs will hatch and survive? (*Source:* St. Marks NWR.)

*The table below shows the earnings for the Gomez family last year and the circle graph shows how they spent their earnings. Use this information to work Exercises 61–64.*

| Month | Earnings | Month | Earnings |
|---|---|---|---|
| January | $3050 | July | $3160 |
| February | $2875 | August | $2355 |
| March | $3325 | September | $2780 |
| April | $3020 | October | $3675 |
| May | $2880 | November | $3310 |
| June | $3265 | December | $4305 |

**FAMILY EXPENSES**
Clothing $\frac{1}{8}$
Taxes $\frac{1}{4}$
Savings $\frac{1}{16}$
Other $\frac{1}{20}$
Rent $\frac{1}{5}$
Food $\frac{5}{16}$

**61. (a)** What was the family's total income for the year?
**(b)** Find the amount of the family's rent for the year.

**62. (a)** How much did the family pay in taxes during the year?
**(b)** How much more did the family spend on taxes than on rent?

**63.** How much did the family spend for food and clothing last year?

**64.** Find the amount the family saved during the year.

---

**Figures and Photos** Today's students are more visually oriented than ever. Thus, a concerted effort has been made to add mathematical figures, diagrams, tables, and graphs whenever possible. Many of the graphs use a style similar to that seen by students in today's print and electronic media. Photos have been incorporated to enhance applications in examples and exercises (page 234).

xvii

**New! Relating Concepts** These sets of exercises help students tie together topics and develop problem-solving skills as they compare and contrast ideas, identify and describe patterns, and extend concepts to new situations. These exercises make great collaborative activities for pairs or small groups of students (page 158).

**New! Focus on Real-Data Applications** These one-page activities, found throughout the text, present even more relevant and in-depth looks at how mathematics is used in the real world. Designed to help instructors answer the often-asked question, "When will I ever use this stuff?," these activities ask students to read and interpret data from newspaper articles, the Internet, and other familiar, real sources. The activities are well suited to collaborative work and can also be completed by individuals or used for open-ended class discussions (page 686).

**Calculator Tips** These optional tips, marked with calculator icons, offer basic information and instruction for students using calculators in the course (page 87).

# Feature Walk-Through

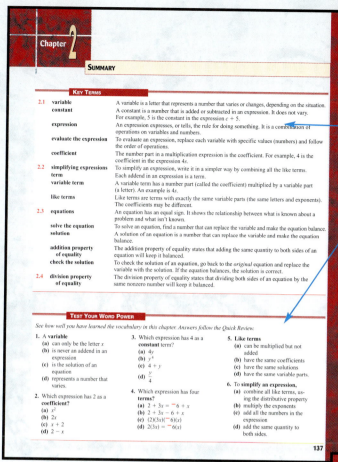

**End-of-Chapter Material** One of the most admired features of the Lial textbooks is the extensive and well-thought-out end of chapter material. At the end of each chapter, students will find:

**Key Terms** are listed, defined, and referenced back to the appropriate section number (page 137).

**New! Test Your Word Power** To help students understand and master mathematical vocabulary, Test Your Word Power has been incorporated in each Chapter Summary. Students are quizzed on Key Terms from the chapter in a multiple-choice format. Answers and examples illustrating each term are provided.

**Quick Review** sections give students not only the main concepts from the chapter (referenced back to the appropriate section), but also an adjacent example of each concept (page 138).

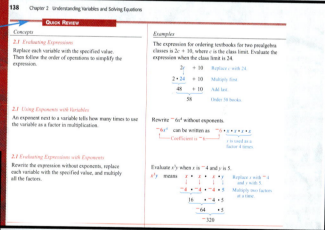

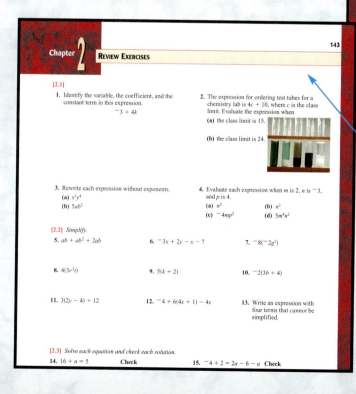

**Review Exercises** are keyed to the appropriate sections so that students can refer to examples of that type of problem if they need help (page 143).

# xx Feature Walk-Through

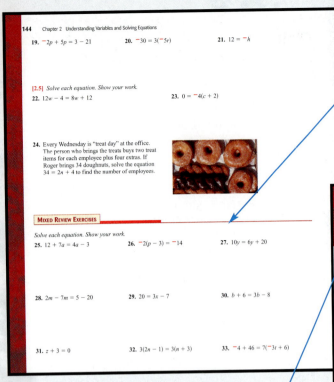

**Mixed Review Exercises** require students to solve problems without the help of section reference (page 144).

**A Chapter Test** helps students practice for the real thing (page 145).

**New! Study Skills Component** A desk-light icon at key points in the text directs students to a separate *Study Skills Workbook* containing activities correlated directly to the text. This unique workbook explains how the brain actually learns, so students understand *why* the study tips presented will help them succeed in the course.

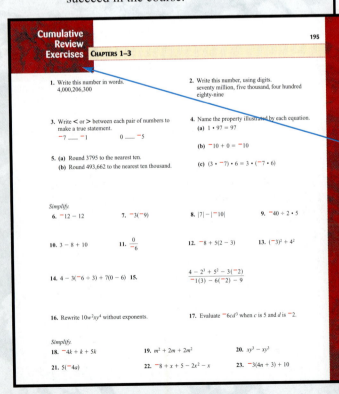

**Cumulative Review Exercises** gather various types of exercises from preceding chapters to help students remember and retain what they are learning throughout the course (page 195).

# An Introduction to Calculators

## SCIENTIFIC CALCULATORS

**Calculators** are among the more popular inventions of the last four decades. Each year better calculators are developed and costs drop. One of the first consumer models available was the Texas Instruments SR-10, which sold for about $150 in 1973. It could perform the four operations of arithmetic and take square roots, but could do very little more. Today, a $5 calculator the size of a credit card can do all these operations, plus provide percent and memory functions.

Calculators now come in a large array of different types, sizes, and prices. *For the course for which this textbook is intended, the most appropriate type is the scientific calculator,* which costs $10–$20.

In this introduction, we explain some of the more common features of scientific calculators. However, remember that calculators vary among manufacturers and models, and that while the methods explained here apply to many of them, they may not apply to your specific calculator. For this reason, it is important to remember that *this introduction is only a guide and is not intended to take the place of your owner's manual.*

**CAUTION**

*Be sure to keep the instruction booklet supplied with your calculator! Keys and functions vary from model to model, so always refer to the instruction booklet for details.*

### OBJECTIVES

1. Use the arithmetic computation keys.
2. Understand the "clear" and "clear entry" keys.
3. Understand the floating decimal point.
4. Use the % (percent) key.
5. Use the $x^2$ (squaring) and $\sqrt{x}$ (square root) keys.
6. Use the $y^x$ (exponent or power) key.
7. Use the $a^{b/c}$ (fraction) key.
8. Solve problems with negative numbers.
9. Use the calculator memory function.
10. Solve chain calculations using the order of operations.
11. Use the parentheses keys.

**1** **Use the arithmetic computation keys.** Most calculators use *algebraic logic* (such as models sold by Texas Instruments, Sharp, Casio, and Radio Shack). For example, enter 14 + 28 in the order it is written.

14 (+) 28 (=)    Answer is 42.

Enter 387 − 62 as follows.

387 (−) 62 (=)    Answer is 325.

If your calculator does *not* work problems in this way, check its instruction book to see how to proceed. (For example, Hewlett-Packard calculators use RPN, Reverse Polish Notation.)

**2** **Understand the "clear" and "clear entry" keys.** All calculators have a (C), (ON/C), or (ON/AC) key. Pressing this key erases everything in the calculator and prepares the calculator to begin a new problem. Some calculators also have a (CE) key. Pressing this key erases *only* the number displayed and allows you to correct a mistake without having to start the problem over.

Many calculators combine the Ⓒ key and the CE key into an ON/C key. This key turns the calculator on and is also used to erase the calculator display. If the ON/C key is pressed after the ⊜ or one of the operation keys (⊕, ⊖, ⊗, ⊘), everything in the calculator is erased. If you happen to press the wrong operation key, immediately press the correct key to cancel the error. For example, 7 ⊕ ⊖ 3 ⊜ 4. Pressing the ⊖ key cancels out the previous ⊕ key entry.

### CAUTION

Be sure to look at the directions that come with your calculator to see how to clear the memory.

**3** **Understand the floating decimal point.** Most calculators have a *floating decimal* that locates the decimal point in the final result. For example, to buy 55.75 square yards of carpet at $18.99 per square yard, proceed as follows.

55.75 ⊗ 18.99 ⊜    Answer is 1058.6925.

The decimal point is automatically placed in the answer. You should *round* money answers to the nearest cent. Start by drawing a cutoff line after the hundredths place.

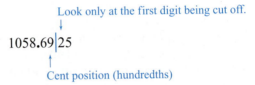

Because the first digit being cut off is 4 or less, the part you are keeping remains the same. The answer is rounded to $1058.69. If the first digit being cut off had been 5 or more, you would have rounded up by adding 1 to the cent position (see **Section 5.2**).

When using a calculator with a floating decimal, enter the decimal point as needed. For example, enter $47 by pressing 47 with no decimal point, but enter 95¢ as ⊙ 95 with a decimal point.

When you add $21.38 and $1.22, the answer is $22.60, but the calculator does *not* show the final 0.

21.38 ⊕ 1.22 ⊜    Answer is 22.6.

You must remember that the problem dealt with money and write the final 0, making the answer $22.60.

**4** **Use the % (percent) key.** The % key moves the decimal point two places to the left when pressed following multiplication or division. Calculate 8% of $4205 as follows.

4205 ⊗ 8 % ⊜    Answer is 336.4.

Because the problem involved money, write the answer as $336.40.

**5** **Use the $x^2$ (squaring) and the $\sqrt{x}$ (square root) keys.** The squaring key, $x^2$, squares the number in the display (multiplies the number by itself). For example, find $7^2$ (which means 7 × 7) as shown below. You do *not* need to press the ⊜ key.

7 $x^2$    Answer is 49.

Because $7^2 = 49$, the number 7 is called the *square root* of 49. (See **Section 5.8.**) Square roots are written with the symbol $\sqrt{\phantom{x}}$. Use the $\boxed{\sqrt{x}}$ key to find $\sqrt{49}$ and $\sqrt{20}$ as shown below.

        49 $\boxed{\sqrt{x}}$    Answer is 7.

        20 $\boxed{\sqrt{x}}$    Answer is 4.472135955.
                      Round to desired position.

Notice that you do *not* need to press the $\boxed{=}$ key.

**6**   **Use the $\boxed{y^x}$ (exponent or power) key.** The $\boxed{y^x}$ key raises a base to any desired power. (See **Section 1.8.**) In $3^5$, the 3 is the base and the exponent, 5, tells how many times the base is multiplied by itself ($3 \times 3 \times 3 \times 3 \times 3$). To find $3^5$, use these keystrokes.

        3 $\boxed{y^x}$ 5 $\boxed{=}$    Answer is 243.

**7**   **Use the $\boxed{a^{b/c}}$ (fraction) key.** Use the $\boxed{a^{b/c}}$ key to solve problems involving fractions or mixed numbers. Enter $\frac{3}{4} + \frac{6}{11}$ as follows.

      3 $\boxed{a^{b/c}}$ 4 $\boxed{+}$ 6 $\boxed{a^{b/c}}$ 11 $\boxed{=}$   $\boxed{\text{1\_13\_44}}$  Answer is $1\frac{13}{44}$.

         $\underbrace{\phantom{XXXX}}_{\frac{3}{4}}$      $\underbrace{\phantom{XXXX}}_{\frac{6}{11}}$

Enter the mixed number problem $4\frac{7}{8} \div 3\frac{4}{7}$ as shown below.

    4 $\boxed{a^{b/c}}$ 7 $\boxed{a^{b/c}}$ 8 $\boxed{\div}$ 3 $\boxed{a^{b/c}}$ 4 $\boxed{a^{b/c}}$ 7 $\boxed{=}$  $\boxed{\text{1\_73\_200}}$  Answer is $1\frac{73}{200}$.

     $\underbrace{\phantom{XXXXXXXX}}_{4\frac{7}{8}}$     $\underbrace{\phantom{XXXXXXXX}}_{3\frac{4}{7}}$

**NOTE**

The calculator automatically shows fractions in lowest terms and as mixed numbers when possible.

**8**   **Solve problems with negative numbers.** To enter a negative number, first enter the number and then press the $\boxed{+/-}$ or $\boxed{+\subset-}$ key. This changes the number to a negative number. For example, enter $-10 + 6 - 8$ as follows.

        10 $\boxed{+/-}$ $\boxed{+}$ 6 $\boxed{-}$ 8 $\boxed{=}$   Answer is $-12$.
                        ↑
         $-10$      Subtract.

**9**   **Use the calculator memory function.** Many calculators have memory keys, which are a sort of electronic scratch paper. These memory keys store intermediate steps in a calculation. On some basic calculators, the $\boxed{M}$ key is used to store the numbers in the display, with the $\boxed{MR}$ key used to recall the numbers from memory.

    Other basic calculators have $\boxed{M+}$ and $\boxed{M-}$ keys. The $\boxed{M+}$ key adds the number in the calculator display to the number already in memory. At the beginning of a problem, the memory contains the number 0. If the calculator display contains the number 29.4 for example, pressing $\boxed{M+}$ will cause 29.4

to be stored in the memory (the result of adding 0 + 29.4). If 57.8 is then entered into the display, pressing (M+) will cause 87.2 to be stored (the result of adding 29.4 + 57.8). If 11.9 is then entered into the display, and (M−) is pressed, the memory will contain 75.3 (the result of subtracting 87.2 − 11.9). The (MR) key is used to recall the number in memory, and (MC) is used to clear the memory.

*Scientific* calculators typically have one or more *registers* in which to store numbers. The memory keys are usually labeled as (STO) for store and (RCL) for recall. For example, you can store 25.6 in register 1 by pressing 25.6 (STO) 1, or you can store it in register 2 by pressing 25.6 (STO) 2, and so on for other registers. To recall numbers from a particular memory register, use the (RCL) key followed by the number of the register. For example, pressing (RCL) 2 recalls the number from register 2.

With a scientific calculator, a number stays in memory until it is replaced by another number or until the memory is cleared. With some calculators, the contents of the memory is saved even when the calculator is turned off (check your calculator's instruction booklet).

Here is an example of a problem that uses the memory keys. An elevator technician wants to find the average weight of a person using an elevator. She counts the number of people entering an elevator and also measures the weight of each group of people.

| Number of People | Total Weight |
|---|---|
| 6 | 839 pounds |
| 8 | 1184 pounds |
| 4 | 640 pounds |

First, find the total weight of all three groups and store the result in memory register 1.

839 (+) 1184 (+) 640 (=) (STO) 1    Stores 2663 in register 1

Then, find the total number of people and store the result in memory register 2.

6 (+) 8 (+) 4 (=) (STO) 2    Stores 18 in register 2

Finally, divide the contents of memory register 1 (total weight) by the contents of memory register 2 (18 people).

(RCL) 1 (÷) (RCL) 2 (=) 147.9444444 pounds    Round as needed.

The average weight of a person using the elevator is about 148 pounds (rounded to the nearest whole number).

**10** Solve chain calculations using the order of operations. Chain calculations, which involve several different operations, must be done in a specific sequence called the *order of operations* (see **Section 1.8**).

### Order of Operations

*Step 1*  Work inside **parentheses** or **other grouping symbols.**

*Step 2*  Simplify expressions with **exponents** and find any **square roots.**

*Step 3*  Do the remaining **multiplications and divisions** as they occur from left to right.

*Step 4*  Do the remaining **additions and subtractions** as they occur from left to right.

The logic of the order of operations is built into most scientific calculators. To check your calculator, try entering 3 + 5 × 2 in the order it is written.

   Answer should be 13.

**Using Order of Operations**

3 + 5 × 2    Multiply first.

3 + 10       Then add.

13 ←——— Correct

**Working from Left to Right**

3 + 5 × 2

8 × 2

16 ←——— Incorrect

If your calculator uses the order of operations, it will automatically multiply 5 × 2 *before* adding 3. If your calculator gives the *incorrect* answer of 16, it does *not* follow the order of operations. You would have to know that multiplication is done ahead of addition and enter the problem as 5 ⓧ 2 ⊕ 3 ⊜ to get the correct result.

**CAUTION**

*Scientific* calculators keep track of the order of operations for us. However, the basic four-function calculator is *not* programmed to observe the order of operations and will calculate correctly *only* if you enter numbers in the proper order.

**11** **Use the parentheses keys.** The parentheses keys allow you to group numbers in a chain calculation. For example, $\frac{24}{5+7}$ can be written as $\frac{24}{(5+7)}$ and entered as follows.

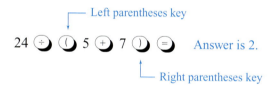

Answer is 2.

Without the parentheses the calculator would have automatically divided 24 by 5 *before* adding 7, giving an *incorrect* answer of 11.8.

Here is a more complicated example.

To solve $\frac{16 - 2.5}{55 - 29.4 \div 0.6}$, write it as $\frac{(16 - 2.5)}{(55 - 29.4 \div 0.6)}$

Use parentheses to set off the numerator and the denominator.

Answer is 2.25.

Numerator       Denominator

# To the Student: Success in Mathematics

There are two main reasons why students have difficulty with mathematics:

- Students start in a course for which they do not have the necessary background knowledge.
- Students don't know how to study mathematics effectively.

Your instructor can help you decide whether this is the right course for you. We can give you some study tips.

Studying mathematics *is* different from studying subjects like English and history. The key to success is regular practice. This should not be surprising. After all, can you learn to play the piano or ski well without a lot of regular practice? The same is true for learning mathematics. Working problems nearly every day is the key to becoming successful. Here is a list of things that will help you succeed in studying mathematics.

1. *Attend class regularly.* Pay attention to what your instructor says and does in class, and take careful notes. In particular, note the problems the instructor works on the board and copy the complete solutions. Keep these notes separate from your homework to avoid confusion when you review them later.

2. Don't hesitate to *ask questions in class.* It is not a sign of weakness but of strength. There are always other students with the same question who are too shy to ask.

3. *Read your text carefully.* Many students read only enough to get by, usually only the examples. Reading the complete section will help you solve the homework problems. Most exercises are keyed to specific examples or objectives that will explain the procedures for working them.

4. Before you start on your homework assignment, *rework the problems the teacher worked in class.* This will reinforce what you have learned. Many students say, "I understand it perfectly when you do it, but I get stuck when I try to work the problem myself."

5. Do your homework assignment only *after reading the text* and reviewing your notes from class. Check your work against the answers in the back of the book. If you get a problem wrong and are unable to understand why, mark that problem and ask your instructor about it. Then practice working additional problems of the same type to reinforce what you have learned.

6. *Work as neatly as you can.* Write your symbols clearly, and make sure the problems are clearly separated from each other. Working neatly will help you to think clearly and also make it easier to review the homework before a test.

7. After you complete a homework assignment, *look over the text again.* Try to identify the main ideas that are in the lesson. Often they are clearly highlighted or boxed in the text.

8. *Use the chapter test at the end of each chapter as a practice test.* Work through the problems under test conditions, without referring to the text or the answers until you are finished. You may want to time yourself to see how long it takes you. When you finish, check your answers against those in the back of the book, and study the problems you missed.

9. *Keep all quizzes and tests that are returned to you,* and use them when you study for future tests and the final exam. These quizzes and tests indicate what concepts your instructor considers to be most important. Be sure to correct any problems on these tests that you missed, so you will have the corrected work to study.

10. *Don't worry if you do not understand a new topic right away.* As you read more about it and work through the problems, you will gain understanding. Each time you review a topic you will understand it a little better. Few people understand each topic completely right from the start.

---

Reading a list of study tips is a good start, but you may need some help actually *applying* the tips to your work in this math course.

Watch for this icon as you work in this textbook, particularly in the first few chapters. It will direct you to one of 12 activities in the *Study Skills Workbook* that comes with this text. Each activity helps you to actually *use* a study skills technique. These techniques will greatly improve your chances for success in this course.

- Find out *how your brain learns new material.* Then use that information to set up effective ways to learn math.

- Find out *why short-term memory is so short* and what you can do to help your brain remember new material weeks and months later.

- Find out *what happens when you "blank out" on a test* and simple ways to prevent it from happening.

All the activities in the *Study Skills Workbook* are brain-friendly ways to enjoy and succeed at math. Whether you need help with note taking, managing homework, taking tests, or preparing for a final exam, you'll find specific, clearly explained ideas that really work because they're based on research about how the brain learns and remembers.

# Diagnostic Pretest

 *Study Skills Workbook*
Activity 1

**[Chapter 1]**

1. Write this number using digits: twenty-five million, three thousand, seven hundred one

2. Simplify $-3(-4)^2 \div 12 - (-10)$.

3. Round 4,356,028 to the nearest ten thousand.

4. Max had $185 in his checking account. He deposited his $372 paycheck, and then, wrote a $575 check for rent. What is the new balance in his account?

**[Chapter 2]**

5. Evaluate $-5a^3b^2$ when $a$ is 2 and $b$ is $-1$.

6. Simplify $-5 + 4(r - 7)$.

7. Solve $-18 = -2(t + 4)$.

8. Solve $7n - 6 = -2n + 3$.

**[Chapter 3]**

9. Find the perimeter.

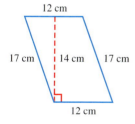

10. Find the area.

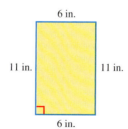

11. Translate this sentence into an equation and solve it. If 13 is added to four times a number, the result is 8 less than the number. What is the number?

12. Write an equation and solve it to answer this problem. A rope is 238 cm long. Marc cut it into two pieces, with one piece 12 cm longer than the other. Find the lengths of both pieces.

1. _____
2. _____
3. _____
4. _____
5. _____
6. _____
7. _____
8. _____
9. _____
10. _____
11. _____
12. _____

xxix

## [Chapter 4]

13. Write $\dfrac{39}{52}$ in lowest terms.

14. Add: $-\dfrac{5}{6} + \dfrac{7}{12}$

15. Solve $\dfrac{4}{5}x + 9 = -7$

16. Find the area.

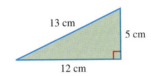

## [Chapter 5]

17. Anita bought 2.6 pounds of salmon at $6.99 per pound. How much did she pay for the salmon to the nearest cent?

18. Solve $-3r + 5.8 = -1.1$.

19. The hourly wages for six employees of a small company are $8.50, $9.40, $7.30, $13.75, $11, and $12. Find the median hourly wage for these employees.

20. Find the circumference of this circle. Use 3.14 for $\pi$ and round your answer to the nearest tenth.

## [Chapter 6]

21. Find the unknown number.

$$\dfrac{9}{8} = \dfrac{x}{96}$$

22. On a road map, 1 inch represents 32 miles. If two towns are 3.8 inches apart on the map, what is the actual distance between them?

23. Find the supplement of a 63° angle.

24. Find the unknown lengths in these similar triangles.

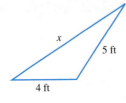
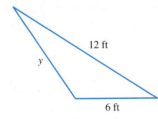

[Chapter 7]

25. (a) Write 250% as a decimal.

    (b) Write 0.3 as a percent.

25. (a) _____

    (b) _____

26. (a) Write 38% as a fraction in lowest terms.

    (b) Write $\dfrac{3}{40}$ as a percent.

26. (a) _____

    (b) _____

27. The price of a digital camera is $235 plus sales tax of 6%. Find the total cost of the camera including sales tax.

27. _____

28. Suppose that you go out to lunch with three friends and the bill comes to $48. You decide to add a 15% tip, and then, split the bill evenly. How much should each of you pay?

28. _____

[Chapter 8]

29. Convert each measurement.

    (a) 432 cm to meters

    (b) 0.08 kg to grams

29. (a) _____

    (b) _____

30. The average annual precipitation over the period 1961–1990 is 61.88 inches in New Orleans, Louisiana, and 16.18 inches in Salt Lake City, Utah. (*Source: World Almanac and Book of Facts, 2000.*) What was the difference in precipitation between the two cities in feet? Round your answer to the nearest tenth.

30. _____

31. Each serving of punch at a graduation party is to be 160 mL. How many liters of punch are needed for 65 guests?

31. _____

32. Write the most reasonable metric unit in each blank. Choose from km, m, cm, mm, L, mL, kg, g, and mg.

    (a) The length of Rosie's bedroom is about 4 _____.

    (b) Satish takes a pill containing 500 _____ of vitamin C every morning.

32. (a) _____

    (b) _____

## [Chapter 9]

33. This circle graph shows the budget for the Patel family. The total budget for one month is $3600. Find the monthly amount budgeted for savings.

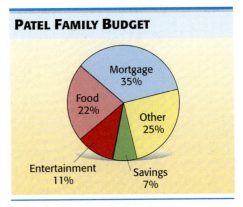

34. This graph shows the number of sections of Prealgebra and Intermediate Algebra at River Bend Community College over a four-year period.

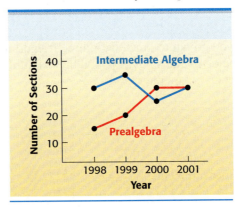

(a) In what year(s) were there more sections of Intermediate Algebra than of Prealgebra?
(b) In what year(s) were there the same number of sections of both courses? How many sections of each course were there in that year?

35. Name the quadrant (if any) in which each point is located.
    (a) $(-3, 5)$
    (b) $(6, 0)$
    (c) $(-2, -6)$
    (d) $(10, 10)$

36. Graph $y = 2x - 1$ on the grid at the left. Make your own table using 0, 1, and 2 as the values of $x$.

## [Chapter 10]

37. Evaluate $5^0 - 3^{-1}$.

38. Simplify $-4(a^4 b^3)^2$.

39. (a) Write 54,500,000 in scientific notation.
    (b) Write $4.07 \times 10^{-3}$ without exponents.

40. Multiply $(3x - 5)(2x + 1)$

# Introduction to Algebra: Integers

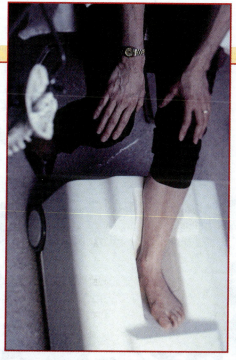

1.1 Place Value
1.2 Introduction to Signed Numbers
1.3 Adding Integers
1.4 Subtracting Integers
1.5 Problem Solving: Rounding and Estimating
1.6 Multiplying Integers
1.7 Dividing Integers
1.8 Exponents and Order of Operations
   Summary Exercises on Operations with Integers

Osteoporosis is a bone disease in which bones become thin and brittle. One in four women over age 50, and one in eight men over 50, have the disease, but it can strike people in their 20s who have eating disorders. (*Source: Osteoporosis online.*) More than 1.5 million osteoporosis-related bone fractures occur each year in the United States. (*Source:* Mayo Clinic.) An ultrasound of the heel bone is a quick, inexpensive way to screen people for osteoporosis. The patients receive a "T score" that indicates their level of risk. Both the ultrasound technician and the patient must understand positive and negative numbers to interpret the score. (See Section 1.2, Exercises 45–48.)

# 1.1 PLACE VALUE

**OBJECTIVES**

1. Identify whole numbers.
2. Identify the place value of a digit through hundred trillions.
3. Write a whole number in words or digits.

It would be nice to earn millions of dollars like some of our favorite entertainers or sports stars. But how much is a million? If you received $1 every second, 24 hours a day, day after day, how many days would it take for you to receive a million dollars? How long to receive a billion dollars? Or a trillion dollars? Make some guesses and write them here.

It would take _____ to receive a million dollars.

It would take _____ to receive a billion dollars.

It would take _____ to receive a trillion dollars.

The answers are at the bottom left of the page. Later, in the exercises for **Section 1.7,** you'll find out how to calculate the answers.

**1 Identify whole numbers.** First we have to be able to write the number that represents *one million*. We can write *one* as 1. How do we make it 1 *million*? Our number system is a **place value system.** That means that the location, or place, in which a number is written gives it a different value. Using money as an example, you can see that

$1 is one dollar.

$10 is ten dollars.

$100 is one hundred dollars.

$1000 is one thousand dollars.

Each time the 1 moved to the left one place, it was worth *ten times* as much. Can you keep moving it to the left? Yes, as many times as you like.

The chart below shows the *value* of each *place*. In other words, you write the 1 in the correct place to represent the number you want to express. It is important to memorize the place value names shown on the chart.

**Whole Number Place Value Chart**

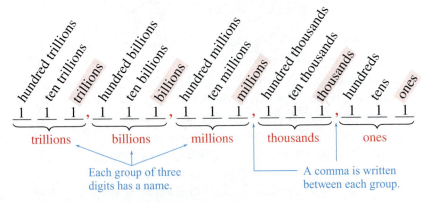

If there had been more room on this page, we could have continued to the left with quadrillions, quintillions, sextillions, septillions, octillions, and more.

Of course, we can use other *digits* besides 1. In our decimal system of writing numbers we can use these ten **digits:** 0, 1, 2, 3, 4, 5, 6, 7, 8, and 9. In this section we will use the digits to write **whole numbers.**

**ANSWERS**
About 11½ days to receive a million dollars; nearly 32 *years* to receive a billion dollars; about 31,710 *years* to receive a trillion dollars.

| These are whole numbers. | These are *not* whole numbers. |
|---|---|
| 0    8    37    100    24,014 | $-6 \quad \frac{3}{4} \quad 7.528 \quad 0.3 \quad 5\frac{2}{3}$ |

❶ Circle the whole numbers.

0.8    −14    502

$\frac{7}{9}$    3    $\frac{3}{2}$

14    0    $6\frac{4}{5}$

9.082    $-\frac{8}{3}$    60,005

### Example 1  Identifying Whole Numbers

Circle the whole numbers in this list.

75    −4    0    1.5    $\frac{5}{8}$    300    0.666    $7\frac{1}{2}$    2

Whole numbers *do* include zero. If we started a list of *all* the whole numbers, it would look like this: 0, 1, 2, 3, 4, 5, . . . with the three dots indicating that the list goes on and on. So the whole numbers in this example are: 75, 0, 300, and 2.

*Work Problem ❶ at the Side.*

**2** **Identify the place value of a digit through hundred trillions.** The number of Americans who voted in the presidential election in November 2000 was 105,380,935. (*Source:* Committee for the Study of the American Electorate.) There are two 5s in 105,380,935, but the value of each 5 is very different. The 5 on the right is in the ones place, so its value is simply 5. But the 5 on the left is worth a great deal more because of the *place* where it is written. Looking back at the place value chart, we see that this 5 is in the millions place, so its value is 5 *million*.

1 0 **5** , 3 8 0 , 9 3 **5**   ballots cast in the election
      ↑                       ↑
   Value of              Value of
  5 *million*             5 *ones*

❷ Give the place value of the digit 8 in each number.

(a) 45,628,665

(b) 800,503,622

### Example 2  Identifying Place Value

Give the place value of each 3 in the number of ballots cast.

1 0 5 , **3** 8 0 , 9 **3** 5
Hundred thousands place ⤴         ⤴ Tens place

*Work Problem ❷ at the Side.*

(c) 428,000,000,000

**3** **Write a whole number in words or digits.** To write a whole number in words, or to say it aloud, begin at the left. Write or say the number in each group of three, followed by the name for that group. When you get to the ones group, do *not* include the group name. Hyphens (dashes) are used whenever you write a number from 21 to 99, like twenty-one thousand, or thirty-seven million, or ninety-four billion.

(d) 2,385,071

### Example 3  Writing Numbers in Words

(a) Write 6,058,120 in words.
Start at the left.

6 , **058** , **120**

**six million**, **fifty-eight thousand**, **one hundred twenty**
   Group               Group              Do not use group
    name                name                name "ones."

*Continued on Next Page*

**ANSWERS**
1. 502; 3; 14; 0; 60,005
2. (a) thousands  (b) hundred millions
   (c) billions   (d) ten thousands

**3** Write these numbers in words.

(a) 23,605

(b) 400,033,007

(c) 193,080,102,000,000

**4** Write each number using digits.

(a) Eighteen million, two thousand, three hundred five

(b) Two hundred billion, fifty million, six hundred sixteen

(c) Five trillion, forty-two billion, nine million

(d) Three hundred six million, seven hundred thousand, nine hundred fifty-nine

**ANSWERS**
3. (a) twenty-three *thousand*, six hundred five
   (b) four hundred *million*, thirty-three *thousand*, seven
   (c) one hundred ninety-three *trillion*, eighty *billion*, one hundred two *million*
4. (a) 18,002,305  (b) 200,050,000,616
   (c) 5,042,009,000,000  (d) 306,700,959

---

(b) Write 50,588,000,040,000 in words.
Start at the left.

$$50,\underline{588},000,\underline{040},000$$

**fifty trillion, five hundred eighty-eight billion, forty thousand**  No millions and no ones in this number.

Group name · Group name · Group name

### CAUTION

You often hear people say "and" when reading a group of three digits. For example, you may hear 120 as "one hundred *and* twenty," but this is *not* correct. The word *and* is used only when reading a **decimal point**, which we do not have here. The correct wording for 120 is "one hundred twenty."

**Work Problem 3 at the Side.**

When you read or hear a number and want to write it in digits, look for the group names: **trillion, billion, million,** and **thousand**. Write the number in each group, followed by a comma. Do *not* put a comma at the end of the ones group.

### Example 4  Writing Numbers in Digits

Write each number using digits.

(a) Five hundred sixteen **thousand**, nine

The first group name is *thousand,* so you need to fill *two groups* of three digits: thousands and ones.

$$\underbrace{5\ 1\ 6}_{\text{thousands}},\underbrace{0\ 0\ 9}_{\text{ones}}$$

comma between groups

The number is 516,009.

(b) Seventy-seven **billion**, thirty **thousand**, five hundred

The first group name is *billion,* so you need to fill *four groups* of three digits: billions, millions, thousands, and ones.

$$\underbrace{0\ 7\ 7}_{\text{billions}},\underbrace{0\ 0\ 0}_{\text{millions}},\underbrace{0\ 3\ 0}_{\text{thousands}},\underbrace{5\ 0\ 0}_{\text{ones}}$$

There are no millions, so fill the millions group with zeros.
When writing the number, you can omit the leading **0** in the billions group. The number is 77,000,030,500.

**Work Problem 4 at the Side.**

## 1.1 Exercises

**Circle the whole numbers. See Example 1.**

1. 15   $8\frac{3}{4}$   0   3.781
   83,001   −8   $\frac{7}{16}$   $\frac{9}{5}$

2. 33.7   −5   457   $\frac{8}{5}$
   0   6   $1\frac{3}{4}$   −14.1

3. 5.8   −6   7   $\frac{5}{4}$
   $\frac{1}{10}$   362,049   0.1   $7\frac{7}{8}$

4. 75,039   $\frac{1}{3}$   −87   6.49
   −0.5   $2\frac{7}{10}$   $\frac{15}{8}$   4

**Give the place value of the digit 2 in each number. See Example 2.**

5. 61,284

6. 82,110

7. 284,100

8. 823,415

9. 725,837,166

10. 442,653,199

11. 253,045,701,000

12. 823,000,419,567

13. From left to right, name the place value for each 0 in this number: 302,016,450,098,570.

14. From left to right, name the place value for each 0 in this number: 810,704,069,809,035.

**Write each number in words. See Example 3.**

15. 8421

16. 1936

**17.** 46,205

**18.** 75,089

**19.** 3,064,801

**20.** 7,900,408

**21.** 840,111,003

**22.** 304,008,401

**23.** 51,006,888,321

**24.** 99,046,733,214

**25.** 3,000,712,000,000

**26.** 50,918,000,000,600

*Write each number using digits. See Example 4.*

**27.** Forty-six thousand, eight hundred five

**28.** Seventy-nine thousand, forty-six

**29.** Five million, six hundred thousand, eighty-two

**30.** One million, thirty thousand, five

**31.** Two hundred seventy-one million, nine hundred thousand

**32.** Three hundred eleven million, four hundred

**33.** Twelve billion, four hundred seventeen million, six hundred twenty-five thousand, three hundred ten

**34.** Seventy-five billion, eight hundred sixty-nine million, four hundred eighty-eight thousand, five hundred six

**35.** Six hundred trillion, seventy-one million, four hundred

**36.** Four hundred forty trillion, thirty-six thousand, one hundred two

### RELATING CONCEPTS (Exercises 37–40)  FOR INDIVIDUAL OR GROUP WORK

*Use your knowledge of place value to* **work Exercises 37–40 in order.**

**37.** Here is a group of digits: 6, 0, 9, 1, 5, 0, 7, 1. Using each digit exactly once, arrange them to make the largest possible whole number and the smallest possible whole number. Then write each number in words.

**38.** Write these numbers in digits and in words.
 (a) Your house number or apartment building number
 (b) Your phone number, including the area code
 (c) The approximate cost of your tuition and books for this quarter or semester
 (d) Your zip code

Now tell how you *usually* say each of the numbers in parts (a)–(d) above. Why do you think we ignore the rules when saying these numbers in everyday situations?

**39.** Look again at the Whole Number Place Value Chart at the beginning of this section. As you move to the left, each place is worth *ten* times as much as the previous place. Computers work on a *binary* system where each place is worth *two* times as much as the previous place. Complete this place value chart based on 2s.

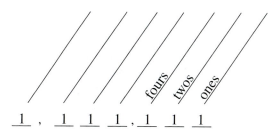

_1_ , _1_  _1_  _1_ , _1_  _1_  _1_

Try writing some numbers using the 2s (binary) place value chart.
 (a) 5 = _____
 (b) 10 = _____
 (c) 15 = _____

The *only* digits you may use in the binary system are 0 and 1. Here is an example.

$$6 = 110$$

one 4 + one 2 + zero 1s = 4 + 2 + 0 = 6

**40.** (a) Explain in your own words why our number system is called a *place value system*. Include an example as part of your explanation.

 (b) Find information on the Roman numeral system. Write these numbers using Roman numerals.
   8 = _____   38 = _____
   275 = _____   3322 = _____

 (c) Explain why the Roman system is *not* a place value system. What are the disadvantages of the Roman system?

*In Exercises 41–48, if the number is given in digits, write it in words. If the number is given in words, write it in digits. See Exercises 3 and 4. (Source: Minneapolis Star Tribune.)*

**41.** In the United States, 6567 couples get married every day.

**42.** There are 3582 pairs of bowling shoes purchased each day.

**43.** One hundred one million, two hundred eighty thousand adults are on diets at any one time in the United States.

**44.** Two million, five thousand Americans suffer from heartburn on any given day.

**45.** The amount spent each day on exercise equipment in the United States is $2,021,018.

**46.** Americans spend an average of $434,206,600 daily on toys.

**47.** People eat twenty-four million, five hundred hot dogs each day.

**48.** Five hundred twenty-four million servings of cola drinks are consumed every day.

## 1.2 INTRODUCTION TO SIGNED NUMBERS

**1** **Write positive and negative numbers used in everyday situations.** The whole numbers in **Section 1.1** were either 0 or greater than 0. Numbers *greater* than 0 are called *positive numbers*. But many everyday situations involve numbers that are *less* than 0, called *negative numbers*. Here are a few examples.

**OBJECTIVES**

**1** Write positive and negative numbers used in everyday situations.

**2** Graph signed numbers on a number line.

**3** Use the < and > symbols to compare integers.

**4** Find the absolute value of integers.

*Study Skills Workbook*
**Activity 2**

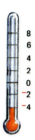
At midnight the temperature dropped to 4 degrees below zero.

$-4$ **degrees**

Jean had $30 in her checking account. She wrote a check for $40.75. We say she is now $10.75 "in the hole," or $10.75 "in the red," or overdrawn by $10.75.

$-\$10.75$

The Packers gained 6 yards on the first play. On the second play they lost 9 yards. We can write the results using a positive and a negative number.

$+6$ **yards**    and    $-9$ **yards**

A plane took off from the airport and climbed to 20,000 feet above sea level. We can write this using a positive number.

$+20,000$ **feet**

A scuba diver swam down to $25\frac{1}{2}$ feet below the surface. We can write this using a negative number.

$-25\frac{1}{2}$ **feet**

To write a negative number, put a negative sign (a dash) in front of it: $-10$. Notice that the negative sign looks exactly like the subtraction sign, as in $5 - 3 = 2$. The two signs do *not* mean the same thing (more on that in the next section). To avoid confusion for now, we will write negative signs in **red** and put them up higher than subtraction signs.

$^-10$ means **negative** 10     $14 - 10$ means $14$ **minus** $10$
└── Raised dash

Later, in Chapters 4 and 5, we will start writing negative signs in the traditional way. However, if you use a graphing calculator, it may show negative signs in the raised position.

*Positive numbers* can be written two ways:

1. Write a positive sign in front of the number: $^+2$ is positive 2. We will write the sign in the raised position to avoid confusion with the sign for addition, as in $6 + 3 = 9$.

2. Do not write any sign. For example, 16 is assumed to be *positive* 16.

**1** Write each negative number with a raised negative sign. Write each positive number in two ways.

(a) The temperature is $5\frac{1}{2}$ degrees below zero.

(b) Cameron lost 12 pounds on a diet.

(c) I deposited $210.35 in my checking account.

(d) I wrote too many checks, so my account is overdrawn by $65.

(e) The submarine dived to 100 feet below the surface of the sea.

(f) In this round of the card game, I won 50 points.

**2** Graph each set of numbers.

(a) $^-2$  (b) 2  (c) 0
(d) $^-4$  (e) 4

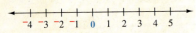

(f) $^-3\frac{1}{2}$  (g) $\frac{1}{2}$
(h) $^-1$  (i) 3

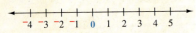

**ANSWERS**

1. (a) $^-5\frac{1}{2}$ degrees  (b) $^-12$ pounds
   (c) $210.35 or $^+$210.35  (d) $^-$$65
   (e) $^-100$ feet  (f) 50 points or $^+50$ points
2.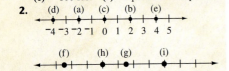

### Example 1 — Writing Positive and Negative Numbers

Write each negative number with a raised negative sign. Write each positive number in two ways.

(a) The river rose to 8 feet above flood stage.

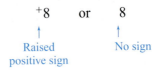

(b) Michael lost $500 in the stock market.

**Work Problem** ❶ **at the Side.**

**2** **Graph signed numbers on a number line.** Mathematicians often use a **number line** to show how numbers relate to each other. A number line is like a thermometer turned sideways. Zero is the dividing point between the positive and negative numbers.

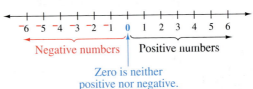

The number line could be shown with positive numbers on the left side of 0 instead of the right side. But it helps if everyone draws it the same way, as shown above. This method will also match what you do when graphing points and lines in Chapter 9.

### Example 2 — Graphing Numbers on a Number Line

Graph each number on the number line.

(a) $^-5$  (b) 3  (c) $1\frac{1}{2}$  (d) 0  (e) $^-1$

Draw a dot at the correct location for each number.

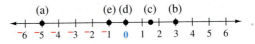

**Work Problem** ❷ **at the Side.**

**3** **Use the < and > symbols to compare integers.** In Chapters 4 and 5 you will work with fractions and decimals. For the rest of this chapter, you will work only with **integers**. A list of integers can be written like this:

$$\ldots, ^-6, ^-5, ^-4, ^-3, ^-2, ^-1, 0, 1, 2, 3, 4, 5, 6, \ldots$$

The dots show that the list goes on forever in both directions.

We can use the number line to compare two integers.

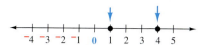

1 is to the *left* of 4.
1 is *less than* 4.
Use < to mean "is less than."

$$1 < 4$$

1 **is less than** 4

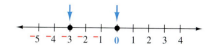

0 is to the *right* of ⁻3.
0 is *greater than* ⁻3.
Use > to mean "is greater than."

$$0 > {}^-3$$

0 **is greater than** ⁻3

**NOTE**

One way to remember which symbol to use is that the "small end of the symbol" points to the "smaller number" (the number that is less).

$$1 < 4 \qquad 0 > {}^-3$$

Smaller number — Small end of symbol    Small end of symbol — Smaller number

### Example 3  Comparing Integers, Using the < and > Symbols

Write < or > between each pair of numbers to make a true statement.

**(a)** 0 ___ 2

0 is to the *left* of 2 on the number line, so 0 is *less than* 2. Write 0 < 2.

**(b)** 1 ___ ⁻4

1 is to the *right* of ⁻4, so 1 is *greater than* ⁻4. Write 1 > ⁻4.

**(c)** ⁻4 ___ ⁻2

⁻4 is to the *left* of ⁻2, so ⁻4 is *less than* ⁻2. Write ⁻4 < ⁻2.

— Work Problem ❸ at the Side.

**4** Find the absolute value of integers.  In order to graph a number on the number line, you need to know two things:

1. Which *direction* it is from 0. It can be in a positive direction or a negative direction. You can tell the direction by looking for a positive sign or a negative sign (or no sign, which is positive).

2. How *far* it is from 0. The distance from 0 is the *absolute value* of a number.

### Absolute Value

The **absolute value** of a number is its distance from 0 on the number line. *Absolute value* is indicated by two vertical bars. For example,

$$|6| \text{ is read } \text{"the absolute value of 6."}$$

The absolute value of a number will *always* be positive (or 0), because it is the *distance* from 0. A distance is never negative. (You wouldn't say that your living room is ⁻16 feet long.) So absolute value concerns only *how far away* the number is from 0; we don't care which direction it is from 0.

❸ Write < or > between each pair of numbers to make a true statement.

**(a)** 5 ___ 4

**(b)** 0 ___ 2

**(c)** ⁻3 ___ ⁻2

**(d)** ⁻1 ___ ⁻4

**(e)** 2 ___ ⁻2

**(f)** ⁻5 ___ 1

ANSWERS
3. (a) >  (b) <  (c) <  (d) >
   (e) >  (f) <

**4** Find each absolute value.

(a) $|13|$

(b) $|{-7}|$

(c) $|0|$

(d) $|{-350}|$

(e) $|6000|$

---

**Example 4  Finding Absolute Values**

Find each absolute value.

(a) $|4|$   The distance from 0 to 4 on the number line is 4 spaces. So, $|4| = 4$.

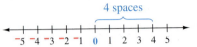

(b) $|{-4}|$   The distance from 0 to $-4$ on the number line is also 4 spaces. So, $|{-4}| = 4$.

(c) $|0|$   $|0| = 0$ because the distance from 0 to 0 on the number line is 0 spaces.

**Work Problem 4 at the Side.**

---

**ANSWERS**

**4.** (a) 13  (b) 7  (c) 0  (d) 350  (e) 6000

## 1.2 Exercises

*Write each negative number with a raised negative sign. Write each positive number in two ways. See Example 1.*

**1.** Mount Everest, the tallest mountain in the world, rises 29,035 feet above sea level. (*Source: World Almanac,* 2001.)

**2.** The bottom of Lake Baykal in central Asia is 5315 feet below the surface of the water. (*Source: World Almanac,* 2001.)

**3.** The coldest temperature ever recorded on Earth is 128.6 degrees below zero in Antarctica. (*Source: Fact Finder.*)

**4.** Normal body temperature is 98.6 degrees Fahrenheit, although it varies slightly for some people. (*Source:* Mayo Clinic *Health Letter.*)

**5.** During the first three plays of the football game, the Trojans lost a total of 18 yards.

**6.** The Jets gained 25 yards on a pass play.

**7.** Angelique won $100 in a prize drawing at the shopping mall.

**8.** Derice overdrew his checking account by $37.

**9.** Keith lost $6\frac{1}{2}$ pounds while he was sick with the flu.

**10.** The price of Mathtronic stock went up $2\frac{1}{2}$ dollars yesterday.

*Graph each set of numbers. See Example 2.*

**11.** $^-3, 3, 0, ^-5$

**12.** $^-2, 2, 0, 5$

**13.** $^-1, 4, ^-2, 5$

**14.** $3, ^-4, 1, ^-5$

**15.** $^-4\frac{1}{2}, \frac{1}{2}, 0, ^-8$

**16.** $^-7, 1\frac{1}{2}, -\frac{1}{2}, ^-9$

**14** Chapter 1 Introduction to Algebra: Integers

*Write < or > between each pair of numbers to make a true statement. See Example 3.*

17. 10 ___ 2

18. 6 ___ 0

19. ⁻1 ___ 0

20. ⁻3 ___ ⁻1

21. ⁻10 ___ 2

22. ⁻9 ___ 7

23. ⁻3 ___ ⁻6

24. 0 ___ ⁻1

25. ⁻10 ___ ⁻2

26. ⁻1 ___ ⁻5

27. 0 ___ ⁻8

28. 6 ___ ⁻4

29. 10 ___ ⁻2

30. ⁻2 ___ 1

31. ⁻4 ___ 4

32. 9 ___ ⁻9

*Find each absolute value. See Example 4.*

33. |15|

34. |10|

35. |⁻3|

36. |⁻8|

37. |0|

38. |100|

39. |200|

40. |⁻99|

41. |⁻75|

42. |⁻6320|

43. |⁻8042|

44. |0|

**RELATING CONCEPTS (Exercises 45–48)    FOR INDIVIDUAL OR GROUP WORK**

*At the beginning of this chapter, you saw a photo of an ultrasound machine used to screen patients for osteoporosis (brittle bone disease). Use the information on the Patient Report Form to **work Exercises 45–48 in order.***

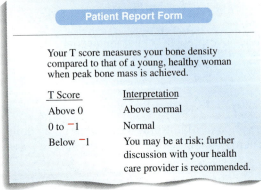

Source: Health Partners, Inc.

45. Here are the T scores for four patients. Draw a number line and graph the four scores.

   Patient A:    ⁻1.5
   Patient B:     0.5
   Patient C:    ⁻1
   Patient D:     0

46. List the patients' scores in order from lowest to highest.

47. What is the interpretation of each patient's score?

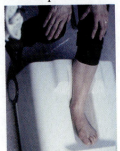

48. (a) What could happen if patient A did not understand the importance of a negative sign?

   (b) For which patient does the sign of the score make no difference? Explain your answer.

# 1.3 Adding Integers

**OBJECTIVES**
1. Add integers.
2. Identify properties of addition.

**1** **Add integers.** Numbers that you are adding are called **addends**. The result is called the **sum**. You can add integers while watching a football game. On each play, you can use a *positive* integer to stand for the yards *gained* by your team, and a *negative* integer for yards *lost*. Zero indicates no gain or loss. For example,

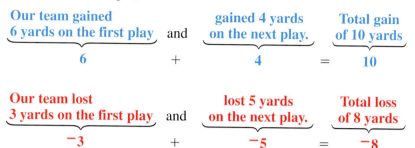

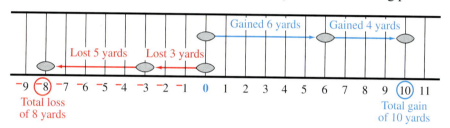

A drawing of a football field can help you add integers. Notice how similar the drawing is to a number line. Zero marks your team's starting point.

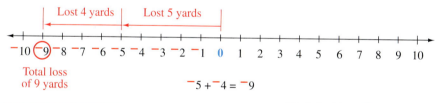

### Example 1  Using a Number Line to Add Integers

Use a number line to find $^-5 + ^-4$.

Think of the number line as a football field. Your team starts at zero. On the first play it lost 5 yards. On the next play it lost 4 yards. The total loss is 9 yards.

```
      Lost 4 yards   Lost 5 yards
  ←―――――――――――  ←―――――――――――
―+――+――+――+――+――+――+――+――+――+―0―+――+――+――+――+――+――+――+――+――+―→
 ⁻10 ⁻9 ⁻8 ⁻7 ⁻6 ⁻5 ⁻4 ⁻3 ⁻2 ⁻1  0  1  2  3  4  5  6  7  8  9  10
Total loss
of 9 yards              ⁻5 + ⁻4 = ⁻9
```

**Work Problem 1 at the Side.**

**1** Find each sum. Use the number line to help you.

(a) $^-2 + ^-2$

(b) $2 + 2$

(c) $^-10 + ^-1$

(d) $10 + 1$

(e) $^-3 + ^-7$

(f) $3 + 7$

Do you see a pattern in the margin problems you just did? The answers to the first two problems are the same, *except for the sign*. The same is true for the next two problems and for the last two problems. This pattern leads to a rule for adding two integers when the signs are the same.

### Adding Two Integers with the Same Sign

**Step 1**  *Add* the absolute values of the numbers.

**Step 2**  Use the *common sign* as the sign of the sum. If both numbers are positive, the sum is positive. If both numbers are negative, the sum is negative.

**ANSWERS**
1. (a) $^-4$  (b) 4  (c) $^-11$
   (d) 11  (e) $^-10$  (f) 10

## ❷ Find each sum.

(a) $^-6 + {}^-6$

(b) $9 + 7$

(c) $^-5 + {}^-10$

(d) $^-12 + {}^-4$

(e) $13 + 2$

### Example 2 — Adding Two Integers with the Same Sign

Add.

(a) $^-8 + {}^-7$

*Step 1* Add the absolute values.

$$|{}^-8| = 8 \quad \text{and} \quad |{}^-7| = 7$$

Add $8 + 7$ to get 15.

*Step 2* Use the *common sign* as the sign of the sum. Both numbers are negative, so the sum is negative.

$$^-8 + {}^-7 = {}^-15$$

Both negative — Sum is negative.

(b) $3 + 6 = 9$

Both positive — Sum is positive.

**Work Problem ❷ at the Side.**

You can also use a number line (or drawing of a football field) to add integers with *different* signs. For example, suppose that your team gained 2 yards on the first play and then lost 7 yards on the next play.

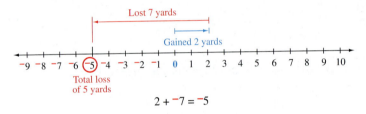

$2 + {}^-7 = {}^-5$

Or, try this one. On the first play your team gained 10 yards, but then it lost 4 yards on the next play.

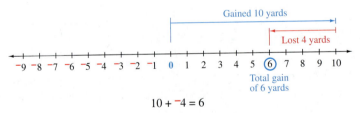

$10 + {}^-4 = 6$

These examples illustrate the rule for adding two integers with unlike, or different, signs.

### Adding Two Integers with Unlike Signs

*Step 1* **Subtract** the smaller absolute value from the larger absolute value.

*Step 2* Use the sign of the number with the *larger absolute value* as the sign of the sum.

**ANSWERS**

2. (a) $^-12$ (b) 16 (c) $^-15$
   (d) $^-16$ (e) 15

> **Example 3** Adding Two Integers with Unlike Signs
>
> Add.
>
> (a) $^-8 + 3$
>
> Step 1  $|^-8| = 8$   and   $|3| = 3$
>
> Subtract $8 - 3$ to get 5.
>
> Step 2  $^-8$ has the *larger absolute value* and is negative, so the sum is also negative.
>
> $$^-8 + 3 = ^-5$$
>
> (b) $^-5 + 11$
>
> Step 1  $|^-5| = 5$   and   $|11| = 11$
>
> Subtract $11 - 5$ to get 6.
>
> Step 2  11 has the *larger absolute value* and is positive, so the sum is also positive.
>
> $$^-5 + 11 = {^+}6 \text{ or } 6$$

*Work Problem ❸ at the Side.*

> **Example 4** Adding Several Integers
>
> A football team has to gain at least 10 yards during four plays in order to keep the ball. Suppose that your college team lost 6 yards on the first play, gained 8 yards on the second play, lost 2 yards on the third play, and gained 7 yards on the fourth play. Did the team gain enough to keep the ball?
>
> When you're adding several integers, work from left to right.
>
> 1st play   2nd play   3rd play   4th play
>
> $^-6$ yards $+$ 8 yards $+$ $^-2$ yards $+$ 7 yards   First add $^-6 + 8$.
>
> 2 yards $+$ $^-2$ yards $+$ 7 yards   Add $2 + {^-2}$.
>
> 0 yards $+$ 7 yards   Add $0 + 7$.
>
> 7 yards
>
> No, the team didn't gain enough yards to keep the ball.

*Work Problem ❹ at the Side.*

**2** **Identify properties of addition.** In our football model for adding integers, we said that 0 indicated no gain or loss, that is, no change in the position of the ball. This example illustrates one of the *properties of addition*. A property of addition is something that applies to all addition problems, regardless of the specific numbers you use.

### Addition Property of 0

Adding 0 to any number leaves the number unchanged.
Some examples are shown below.

$$0 + 6 = 6 \qquad ^-25 + 0 = {^-25} \qquad 72{,}399 + 0 = 72{,}399$$
$$0 + {^-100} = {^-100}$$

❸ Find each sum.

(a) $^-3 + 7$

(b) $6 + {^-12}$

(c) $12 + {^-7}$

(d) $^-10 + 2$

(e) $5 + {^-9}$

(f) $^-8 + 9$

❹ Write an addition problem and solve it for each situation.

(a) The temperature was $^-15$ degrees this morning. It rose 21 degrees during the day, then dropped 10 degrees. What is the new temperature?

(b) Andrew had $60 in his checking account. He wrote a $20 check for gas and a $75 check for groceries. Later in the day he deposited an $85 tax refund in his account. What is the balance in his account?

**Answers**
3. (a) 4   (b) $^-6$   (c) 5
   (d) $^-8$   (e) $^-4$   (f) 1
4. (a) $^-15 + 21 + {^-10} = {^-4}$ degrees
   (b) $60 + {^-20} + {^-75} + 85 = \$50$

Chapter 1 Introduction to Algebra: Integers

**5** Rewrite each sum using the commutative property of addition. Check that the sum is unchanged.

(a) $175 + 25 =$ __ $+$ __

Both sums are _____.

(b) $7 + {}^-37 =$ __ $+$ __

Both sums are _____.

(c) ${}^-16 + 16 =$ __ $+$ __

Both sums are _____.

(d) ${}^-9 + {}^-41 =$ __ $+$ __

Both sums are _____.

Another property of addition is that you can change the *order* of the addends and still get the same sum. For example,

Gaining 2 yards, then losing 7 yards, gives a result of ${}^-5$ yards.
Losing 7 yards, then gaining 2 yards, also gives a result of ${}^-5$ yards.

### Commutative Property of Addition

Changing the *order* of two addends does *not* change the sum. Here are some examples.

$84 + 2 = 2 + 84$     Both sums are 86.
${}^-10 + 6 = 6 + {}^-10$     Both sums are ${}^-4$.

**Example 5** Using the Commutative Property of Addition

Rewrite each sum, using the commutative property of addition. Check that the sum is unchanged.

(a) $65 + 35$

$$65 + 35 = 35 + 65$$
$$100 = 100$$

Both sums are 100, so the sum is unchanged.

(b) ${}^-20 + {}^-30$

$${}^-20 + {}^-30 = {}^-30 + {}^-20$$
$${}^-50 = {}^-50$$

Both sums are ${}^-50$, so the sum is unchanged.

**Work Problem 5 at the Side.**

When there are three addends, parentheses may be used to tell you which pair of numbers to add first, as shown here.

$(3 + 4) + 2$   First add $3 + 4$.
$\phantom{(}7\phantom{)} + 2$   Then add $7 + 2$.
$\phantom{(7)\ +\ }9$

$3 + (4 + 2)$   First add $4 + 2$.
$3 + \phantom{(}6\phantom{)}$   Then add $3 + 6$.
$\phantom{3\ +\ (}9$

Both sums are 9. This example illustrates another property of addition.

### Associative Property of Addition

Changing the *grouping* of addends does *not* change the sum. Some examples are shown below.

$({}^-5 + 5) + 8 = {}^-5 + (5 + 8)$
$\phantom{(}0\phantom{)} + 8 = {}^-5 + \phantom{(}13\phantom{)}$
$\phantom{(0)\ +\ }8 = \phantom{{}^-5\ +\ (}8$

Both sums are 8.

$3 + ({}^-4 + {}^-6) = (3 + {}^-4) + {}^-6$
$3 + \phantom{(}{}^-10\phantom{)} = \phantom{(}{}^-1\phantom{)} + {}^-6$
$\phantom{3\ +\ (}{}^-7 = \phantom{(3\ +}{}^-7$

Both sums are ${}^-7$.

**ANSWERS**
5. (a) $175 + 25 = 25 + 175$; 200
   (b) $7 + {}^-37 = {}^-37 + 7$; ${}^-30$
   (c) ${}^-16 + 16 = 16 + {}^-16$; 0
   (d) ${}^-9 + {}^-41 = {}^-41 + {}^-9$; ${}^-50$

We can use the associative property to make addition problems easier. Notice in the first example in the box on the previous page that it is easier to group $^-5 + 5$ (which is 0) and then add 8. In the second example in the box, it is helpful to group $^-4 + {^-6}$ because the sum is $^-10$, and it is easy to work with multiples of 10.

### Example 6  Using the Associative Property of Addition

In each addition problem, pick out the two addends that would be easiest to add. Write parentheses around those addends. Then find the sum.

**(a)** $6 + 9 + {^-9}$

Group $9 + {^-9}$ because the sum is 0.

$$6 + (9 + {^-9})$$
$$6 + 0$$
$$6$$

**(b)** $17 + 3 + {^-25}$

Group $17 + 3$ because the sum is 20, which is a multiple of 10.

$$(17 + 3) + {^-25}$$
$$20 + {^-25}$$
$$^-5$$

Work Problem ❻ at the Side.

❻ In each problem, write parentheses around the two addends that would be easiest to add. Then find the sum.

**(a)** $^-12 + 12 + {^-19}$

**(b)** $31 + 75 + {^-75}$

**(c)** $16 + {^-1} + {^-9}$

**(d)** $^-8 + 5 + {^-25}$

**Answers**

**6. (a)** $(^-12 + 12) + {^-19}$
$0 + {^-19}$
$^-19$

**(b)** $31 + (75 + {^-75})$
$31 + 0$
$31$

**(c)** $16 + (^-1 + {^-9})$
$16 + {^-10}$
$6$

**(d)** $^-8 + (5 + {^-25})$
$^-8 + {^-20}$
$^-28$

# Focus on Real-Data Applications

## Auto Aide

The *Auto Aide Service* is a hypothetical business located in downtown Houston, Texas. It specializes in assisting motorists who have minor mechanical problems, such as a flat tire or a dead battery. The I-10 corridor in the greater Houston area is a major highway with constant heavy traffic, so the company assigns five vans to concentrate on I-10 assistance needs. The owner is a retired mathematics teacher who believes in using integer arithmetic sentences to track the actions of the auto aides along their service routes.

A schematic map of the I-10 corridor is shown below. Each tick mark represents 1 mile. The home office is located at 0. Locations to the west are represented by negative integers, and locations to the east are represented by positive integers.

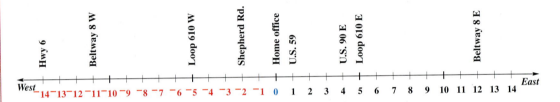

A radio dispatcher notifies the aide to assist a motorist at a given location. (Only a few locations are identified in this hypothetical problem, for simplicity.) Each aide must include four calculations on the daily report. First, the aide must translate the route instructions into an integer sentence that gives the direction and miles between locations. Second, the aide must record the displacement at the end of the route (location relative to home office). Third, the aide must record the distance from the home office at the end of the route. Fourth, the aide must record the total distance traveled on the route.

1. Aide Anne is shown as an example. Complete the table for the remaining aides.

| Aide | Route Instructions | Integer Sentence | Displacement (final location) | Distance from Home Office | Total Distance |
|---|---|---|---|---|---|
| Anne | Home office to U.S. 90 E; to Loop 610 W; to Hwy 6; to Shepherd Rd. | 4 + ⁻9 + ⁻9 + 12 | ⁻2 (Shepherd) (*Hint:* Find the integer sum.) | 2 miles (*Hint:* Find the absolute value of displacement.) | 34 miles (*Hint:* Find sum of absolute values of trip segments.) |
| Bill | Home office to Hwy 6; to U.S. 59; to Shepherd Rd.; to Loop 610 W | ⁻14 + 17 + ⁻5 + ⁻3 | ⁻5 (Loop 610 W) | 5 miles | 39 miles |
| Carlos | Home office to Shepherd Rd.; to Loop 610 W; to Beltway 8 W; to Hwy 6 | ⁻2 + ⁻3 + ⁻6 + ⁻3 | ⁻14 (Hwy 6) | 14 miles | 14 miles |
| Dylan | Home office to Beltway 8 E; to U.S. 90 E; to Home office | 11 + ⁻7 + ⁻4 | 0 (Home office) | 0 miles | 22 miles |
| Ellen | Home office to U.S. 59; to Shepherd Rd.; to U.S. 59; to Loop 610 W; to U.S. 59 | 3 + ⁻5 + 5 + ⁻8 + 8 | 3 (U.S. 59) | 3 miles | 29 miles |

## 1.3 Exercises

*Add by using the number line. See Example 1.*

1. ⁻2 + 5

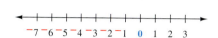

2. ⁻3 + 4

3. ⁻5 + ⁻2

4. ⁻2 + ⁻2

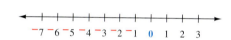

5. 3 + ⁻4

6. 5 + ⁻1

*Add. See Example 2.*

7. (a) ⁻5 + ⁻5
   (b) 5 + 5

8. (a) ⁻9 + ⁻9
   (b) 9 + 9

9. (a) 7 + 5
   (b) ⁻7 + ⁻5

10. (a) 3 + 6
    (b) ⁻3 + ⁻6

11. (a) ⁻25 + ⁻25
    (b) 25 + 25

12. (a) ⁻30 + ⁻30
    (b) 30 + 30

13. (a) 48 + 110
    (b) ⁻48 + ⁻110

14. (a) 235 + 21
    (b) ⁻235 + ⁻21

15. What pattern do you see in your answers to Exercises 7–14? Explain why this pattern occurs.

16. In your own words, explain how to add two integers that have the same sign.

## Chapter 1  Introduction to Algebra: Integers

*Add. See Example 3.*

17. (a) $^-6 + 8$

    (b) $6 + {^-8}$

18. (a) $^-3 + 7$

    (b) $3 + {^-7}$

19. (a) $^-9 + 2$

    (b) $9 + {^-2}$

20. (a) $^-8 + 7$

    (b) $8 + {^-7}$

21. (a) $20 + {^-25}$

    (b) $^-20 + 25$

22. (a) $30 + {^-40}$

    (b) $^-30 + 40$

23. (a) $200 + {^-50}$

    (b) $^-200 + 50$

24. (a) $150 + {^-100}$

    (b) $^-150 + 100$

25. What pattern do you see in your answers to Exercises 17–24? Explain why this pattern occurs.

26. In your own words, explain how to add two integers that have different signs.

*Add. See Examples 2–4.*

27. $^-8 + 5$

28. $^-3 + 2$

29. $^-1 + 8$

30. $^-4 + 10$

31. $^-2 + {^-5}$

32. $^-7 + {^-3}$

33. $6 + {^-5}$

34. $11 + {^-3}$

35. $4 + {^-12}$

36. $9 + {^-10}$

37. $^-10 + {^-10}$

38. $^-5 + {^-20}$

39. $^-17 + 0$

40. $0 + {^-11}$

41. $1 + {^-23}$

42. $13 + {^-1}$

43. $^-2 + {^-12} + {^-5}$

44. $^-16 + {^-1} + {^-3}$

45. $8 + 6 + {^-8}$

**46.** ⁻5 + 2 + 5

**47.** ⁻7 + 6 + ⁻4

**48.** ⁻9 + 8 + ⁻2

**49.** ⁻3 + ⁻11 + 14

**50.** 15 + ⁻7 + ⁻8

**51.** 10 + ⁻6 + ⁻3 + 4

**52.** 2 + ⁻1 + ⁻9 + 12

**53.** ⁻7 + 28 + ⁻56 + 3

**54.** 4 + ⁻37 + 29 + ⁻5

*Write an addition problem for each situation and find the sum.*

**55.** The football team gained 13 yards on the first play and lost 17 yards on the second play. How many yards did the team gain or lose in all?

**56.** At penguin breeding grounds on Antarctic islands, temperatures routinely drop to ⁻15 °C. Temperatures in the interior of the continent may drop another 60 °C below that. What is the temperature in the interior?

**57.** Nick's checking account was overdrawn by $62. He deposited $50 in his account. What is the balance in his account?

**58.** $88 was stolen from Jay's car. He got $35 of it back. What was his net loss?

**59.** Red River flood waters rose 8 feet on Monday, dropped 3 feet on Tuesday, and dropped 1 more foot on Wednesday. What was the new flood level?

**60.** Marion lost 4 pounds in April, gained 2 pounds in May, and gained 3 pounds in June. How many pounds did she gain or lose in all?

**61.** While playing a card game, Jeff first lost 20 points, won 75 points, and then lost 55 points. What was his point total?

**62.** Cynthia had $100 in her checking account. She wrote a check for $83 and was charged $17 for overdrawing her account last month. What is her account balance?

*At the 2001 Masters Golf Tournament in Augusta, Georgia, 72 strokes was the "par" score for each round of play. A negative number indicates that the player had fewer than 72 strokes for that round, and a positive number indicates more than 72 strokes. Find the total score for each of these players in the tournament.*

Tiger Woods

|    | Player | Round 1 | Round 2 | Round 3 | Round 4 | Total |
|----|--------|---------|---------|---------|---------|-------|
| 63. | Tiger Woods | −2 | −6 | −4 | −4 | |
| 64. | David Duval | −1 | −6 | −2 | −5 | |
| 65. | Toshi Izawa | −1 | −6 | 2 | −5 | |
| 66. | Chris DiMarco | −7 | −3 | 0 | 2 | |

*Source:* Associated Press.

*Rewrite each sum, using the commutative property of addition. Show that the sum is unchanged. See Example 5.*

**67.** $^-18 + {}^-5 =$ _____ + _____

Both sums are _____.

**68.** $^-12 + 20 =$ _____ + _____

Both sums are _____.

**69.** $^-4 + 15 =$ _____ + _____

Both sums are _____.

**70.** $17 + 1 =$ _____ + _____

Both sums are _____.

*In each addition problem, write parentheses around the two addends that would be easiest to add. Then find the sum. See Example 6.*

**71.** $6 + {}^-14 + 14$

**72.** $9 + {}^-9 + {}^-8$

**73.** $^-14 + {}^-6 + {}^-7$

**74.** $^-18 + 3 + 7$

**75.** Make up three of your own examples that illustrate the addition property of 0.

**76.** Make up three of your own examples that illustrate the associative property of addition. Show that the sum is unchanged.

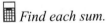 *Find each sum.*

**77.** $^-7081 + 2965$

**78.** $^-1398 + 3802$

**79.** $^-179 + {}^-61 + 8926$

**80.** $36 + {}^-6215 + 428$

**81.** $86 + {}^-99{,}000 + 0 + 2837$

**82.** $^-16{,}719 + 0 + 8878 + {}^-14$

## 1.4 SUBTRACTING INTEGERS

**OBJECTIVES**

1. Find the opposite of a signed number.
2. Subtract integers.
3. Combine adding and subtracting of integers.

**1** **Find the opposite of a signed number.** Look at how the integers match up on this number line.

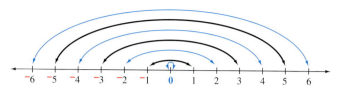

Each integer is matched with its *opposite*. **Opposites** are the same *distance* from 0 on the number line but are on *opposite sides* of 0.

$^+2$ is the opposite of $^-2$     and     $^-2$ is the opposite of $^+2$

When you add opposites, the sum is always 0. The opposite of a number is also called its *additive inverse*.

$2 + {}^-2 = \mathbf{0}$     and     $^-2 + 2 = \mathbf{0}$

**Example 1** Finding the Opposites of Signed Numbers

Find the opposite (additive inverse) of each number. Show that the sum of the number and its opposite is 0.

(a)  6      The opposite of 6 is $^-6$ and $6 + {}^-6 = \mathbf{0}$.
(b) $^-10$   The opposite of $^-10$ is 10 and $^-10 + 10 = \mathbf{0}$.
(c)  0      The opposite of 0 is 0 and $0 + 0 = \mathbf{0}$.

Work Problem **1** at the Side.

**2** **Subtract integers.** Now that you know how to add integers and how to find opposites, you can subtract integers. Every subtraction problem has the same answer as a related addition problem. The problems below illustrate how to change subtraction problems into addition problems.

$6 - 2 = 4$     Same answer     $8 - 3 = 5$     Same answer
$6 + {}^-2 = 4$                  $8 + {}^-3 = 5$

### Subtracting Two Integers

To subtract two numbers, *add* the first number to the *opposite* of the second number. Remember to change *two* things:

*Step 1*  Make one pencil stroke to change the subtraction symbol to an addition symbol.

*Step 2*  Make a second pencil stroke to change the *second* number to its *opposite*. If the second number is positive, change it to negative. If the second number is negative, change it to positive.

**CAUTION**

When changing a subtraction problem to an addition problem, do *not* make any change in the *first* number. The pattern is

1st number − 2nd number = 1st number + opposite of 2nd number.

---

**1** Find the additive inverse (opposite) of each number. Show that the sum of the number and its additive inverse is 0.

(a) 5

(b) 48

(c) 0

(d) $^-1$

(e) $^-24$

**ANSWERS**
1. (a) $^-5$; $5 + {}^-5 = 0$
   (b) $^-48$; $48 + {}^-48 = 0$
   (c) 0; $0 + 0 = 0$   (d) 1; $^-1 + 1 = 0$
   (e) 24; $^-24 + 24 = 0$

Chapter 1 Introduction to Algebra: Integers

**❷ Subtract by changing subtraction to adding the opposite. (Make *two* pencil strokes.)**

(a) $^-6 - 5$

(b) $3 - {^-10}$

(c) $^-8 - {^-2}$

(d) $0 - 10$

(e) $^-4 - {^-12}$

(f) $9 - 7$

**❸ Simplify.**

(a) $6 - 7 + {^-3}$

(b) $^-2 + {^-3} - {^-5}$

(c) $7 - 7 - 7$

(d) $^-3 - 9 + 4 - {^-20}$

**ANSWERS**
2. (a) $^-11$ (b) $13$ (c) $^-6$ (d) $^-10$
 (e) $8$ (f) $2$
3. (a) $^-4$ (b) $0$ (c) $^-7$ (d) $12$

---

**Example 2  Subtracting Two Integers**

Make *two* pencil strokes to change each subtraction problem into an addition problem. Then find the sum.

(a) $4 - 10 = 4 + {^-10} = {^-6}$   *Change 10 to $^-10$. Change subtraction to addition.*

(b) $^-9 - {^-6} = {^-9} + {^+6} = {^-3}$   *Change $^-6$ to $^+6$. Change subtraction to addition.*

(c) $3 - {^-5} = 3 + {^+5} = 8$   Make *two* pencil strokes. *Change $^-5$ to $^+5$. Change subtraction to addition.*

(d) $^-2 - 9 = {^-2} + {^-9} = {^-11}$   Make *two* pencil strokes. *Change 9 to $^-9$. Change subtraction to addition.*

**Work Problem ❷ at the Side.**

---

**3  Combine adding and subtracting of integers.** When adding and subtracting more than two signed numbers, first change all subtractions to additions. Then add from left to right.

**Example 3  Combining Addition and Subtraction**

Simplify by completing all the calculations.

$^-5 - 10 - 12 + 1$   Change all subtractions to addition. Change 10 to $^-10$. Change 12 to $^-12$.
$^-5 + {^-10} + {^-12} + 1$   Add from left to right. First add $^-5 + {^-10}$.
$^-15 + {^-12} + 1$   Then add $^-15 + {^-12}$.
$^-27 + 1$   Finally, add $^-27 + 1$.
$^-26$

**Work Problem ❸ at the Side.**

---

🖩 **Calculator Tip**   You can use the *change of sign* key +/− or +⌐ on your *scientific* calculator to enter negative numbers. To enter $^-5$, press ⑤ +/−. To enter $^+5$, just press ⑤. To enter Example 3 above, press the following keys.

5 +/− − 10 − 12 + 1 =   The answer is $^-26$.

$^-5$   Subtract

When using a calculator, you do *not* need to change subtraction to addition.

## 1.4 Exercises

*Find the opposite (additive inverse) of each number. Show that the sum of the number and its opposite is 0. See Example 1.*

**1.** 6         **2.** 10         **3.** ⁻13

**4.** ⁻3         **5.** 0         **6.** 1

*Subtract by changing subtraction to addition. See Example 2.*

**7.** 19 − 5         **8.** 24 − 11         **9.** 10 − 12         **10.** 1 − 8

**11.** 7 − 19         **12.** 2 − 17         **13.** ⁻15 − 10         **14.** ⁻10 − 4

**15.** ⁻9 − 14         **16.** ⁻3 − 11         **17.** ⁻3 − ⁻8         **18.** ⁻1 − ⁻4

**19.** 6 − ⁻14         **20.** 8 − ⁻1         **21.** 1 − ⁻10         **22.** 6 − ⁻1

**23.** ⁻30 − 30         **24.** ⁻25 − 25         **25.** ⁻16 − ⁻16         **26.** ⁻20 − ⁻20

**27.** 13 − 13         **28.** 19 − 19         **29.** 0 − 6         **30.** 0 − 12

**31.** (a) 3 − ⁻5         **32.** (a) 9 − 6         **33.** (a) 4 − 7         **34.** (a) 8 − ⁻2
     (b) 3 − 5              (b) ⁻9 − 6              (b) 4 − ⁻7              (b) ⁻8 − ⁻2
     (c) ⁻3 − ⁻5             (c) 9 − ⁻6              (c) ⁻4 − 7              (c) 8 − 2
     (d) ⁻3 − 5              (d) ⁻9 − ⁻6             (d) ⁻4 − ⁻7             (d) ⁻8 − 2

*Simplify. See Example 3.*

**35.** ⁻2 − 2 − 2         **36.** ⁻8 − 4 − 8         **37.** 9 − 6 − 3 − 5

**38.** 12 − 7 − 5 − 4         **39.** 3 − ⁻3 − 10 − ⁻7         **40.** 1 − 9 − ⁻2 − ⁻6

**41.** ⁻2 + ⁻11 − ⁻3         **42.** ⁻5 − ⁻2 + ⁻6         **43.** 4 − ⁻13 + ⁻5

**44.** 6 − ⁻1 + ⁻10         **45.** 6 + 0 − 12 + 1         **46.** ⁻10 − 4 + 0 + 18

**47.** Use the score sheet to find each player's point total after three rounds in a card game.

|         | Jeff        | Terry       |
|---------|-------------|-------------|
| Round 1 | Lost 20 pts | Won 42 pts  |
| Round 2 | Won 75 pts  | Lost 15 pts |
| Round 3 | Lost 55 pts | Won 20 pts  |

**48.** Use the information in the table on flood water depths to find the new flood level for each river.

|           | Red River | Mississippi |
|-----------|-----------|-------------|
| Monday    | Rose 8 ft | Rose 4 ft   |
| Tuesday   | Fell 3 ft | Rose 7 ft   |
| Wednesday | Fell 5 ft | Fell 13 ft  |

**49.** Find, correct, and explain the mistake made in this subtraction.

$$^{-}6 - 6$$
$$\downarrow \quad \downarrow$$
$$^{-}6 + 6 = 0$$

**50.** Find, correct, and explain the mistake made in this subtraction.

$$^{-}7 - 5$$
$$\downarrow \quad \downarrow$$
$$^{+}7 + {}^{-}5 = 2$$

*Simplify. Begin each exercise by working inside the absolute value bars or the parentheses.*

**51.** $^{-}2 + {}^{-}11 + |{}^{-}2|$

**52.** $5 - |{}^{-}3| + 3$

**53.** $0 - |{}^{-}7 + 2|$

**54.** $|1 - 8| - |0|$

**55.** $^{-}3 - ({}^{-}2 + 4) + {}^{-}5$

**56.** $5 - 8 - (6 - 7) + 1$

### RELATING CONCEPTS (Exercises 57–58)  FOR INDIVIDUAL OR GROUP WORK

*Use your knowledge of the properties of addition to* **work Exercises 57 and 58 in order.**

**57.** Look for a pattern in these pairs of subtractions.

$^{-}3 - 5 = $ \_\_\_\_     $^{-}4 - {}^{-}3 = $ \_\_\_\_

$5 - {}^{-}3 = $ \_\_\_\_     $^{-}3 - {}^{-}4 = $ \_\_\_\_

(a) Explain what happens when you try to apply the commutative property to subtraction.

(b) Write a rule that tells how to get the answer when the order of the numbers is switched in a subtraction problem.

**58.** Recall the addition property of 0. Can 0 be used in a subtraction problem without changing the other number? Explain what happens and give several examples. (*Hint:* Think about *order* in a subtraction problem.)

## 1.5 Problem Solving: Rounding and Estimating

One way to get a rough check on an answer is to *round* the numbers in the problem. **Rounding** a number means finding a number that is close to the original number, but easier to work with.

For example, a superintendent of schools in a large city might be discussing the need to build new schools. In making her point, it probably would not be necessary to say that the school district has 152,807 students—it probably would be sufficient to say that there are 153,000 students, or even 150,000 students.

### Objectives

1. Locate the place to which a number is to be rounded.
2. Round integers.
3. Use front end rounding to estimate answers in addition and subtraction.

**1** **Locate the place to which a number is to be rounded.** The first step in rounding a number is to locate the *place to which the number is to be rounded*.

### Example 1  Finding the Place to Which a Number Is to Be Rounded

Locate and draw a line under the place to which each number is to be rounded. Then answer the question.

**(a)** Round ⁻23 to the nearest ten. Is ⁻23 closer to ⁻2̲0 or ⁻3̲0?

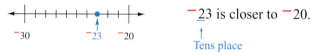

⁻23 is closer to ⁻20.
↑
Tens place

**(b)** Round $381 to the nearest hundred. Is it closer to $3̲00 or $4̲00?

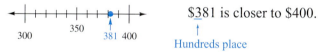

$381 is closer to $400.
↑
Hundreds place

**(c)** Round ⁻54,702 to the nearest thousand. Is it closer to ⁻5̲4,000 or ⁻5̲5,000?

⁻54,702 is closer to ⁻55,000.
↑
Thousands place

━━━━━━━━━━━━━━━ Work Problem **1** at the Side.

**2** **Round integers.** Use the following steps for rounding integers.

### Rounding an Integer

**Step 1** Locate the *place* to which the number is to be rounded. Draw a line under that place.

**Step 2** Look only at the next digit to the right of the one you underlined. If the next digit is *5 or more*, increase the underlined digit by 1. If the next digit is *4 or less*, do *not* change the digit in the underlined place.

**Step 3** Change all digits to the right of the underlined place to zeros.

### CAUTION

If you are rounding a negative number, be careful to write the negative sign in front of the rounded number. For example, ⁻79 rounds to ⁻80.

**1** Locate and draw a line under the place to which the number is to be rounded. Then answer the question.

**(a)** ⁻746 (nearest ten)

Is it closer to ⁻740 or ⁻750? _____

**(b)** 2412 (nearest thousand)

Is it closer to 2000 or 3000? _____

**(c)** ⁻89,512 (nearest hundred)

Is it closer to ⁻89,500 or ⁻89,600? _____

**(d)** 546,325 (nearest ten thousand)

Is it closer to 540,000 or 550,000? _____

**Answers**
1. (a) ⁻74̲6 is closer to ⁻750
   (b) 2̲412 is closer to 2000
   (c) ⁻89,5̲12 is closer to ⁻89,500
   (d) 54̲6,325 is closer to 550,000

**2** Round to the nearest ten.

(a) 34

(b) ⁻61

(c) ⁻683

(d) 1792

**3** Round to the nearest thousand.

(a) 1725

(b) ⁻6511

(c) 58,829

(d) ⁻83,904

---

**Example 2** Using the Rounding Rule for 4 or Less

Round 349 to the nearest hundred.

*Step 1* Locate the place to which the number is being rounded. Draw a line under that place.

$$3\underline{4}9$$
Hundreds place

*Step 2* Because the next digit to the right of the underlined place is 4, which is *4 or less*, do *not* change the digit in the underlined place.

Next digit is 4 or less.
$$3\underline{4}9$$
3 remains 3.

*Step 3* Change all digits to the right of the underlined place to zeros.

Change to 0.
$$3\underline{4}9 \text{ rounded to the nearest hundred is } 3\underline{0}0.$$
Leave 3 as 3.

In other words, 349 is closer to 300 than to 400.

**Work Problem 2 at the Side.**

**Example 3** Using the Rounding Rule for 5 or More

Round 36,833 to the nearest thousand.

*Step 1* Find the place to which the number is to be rounded. Draw a line under that place.

$$3\underline{6},833$$
Thousands place

*Step 2* Because the next digit to the right of the underlined place is 8, which is *5 or more*, add 1 to the underlined place.

Next digit is 5 or more.
$$3\underline{6},833$$
Change 6 to 7.

*Step 3* Change all digits to the right of the underlined place to zeros.

Change to 0.
$$3\underline{6},833 \text{ rounded to the nearest thousand is } 3\underline{7},000.$$
Change 6 to 7.

In other words, 36,833 is closer to 37,000 than to 36,000.

**Work Problem 3 at the Side.**

---

**ANSWERS**

2. (a) 30  (b) ⁻60  (c) ⁻680  (d) 1790
3. (a) 2000  (b) ⁻7000  (c) 59,000  (d) ⁻84,000

Section 1.5 Problem Solving: Rounding and Estimating  31

### Example 4  Using the Rules for Rounding

**(a)** Round ⁻2382 to the nearest ten.

Step 1  ⁻23<u>8</u>2
        └── Tens place

Step 2  The next digit to the right is 2, which is *4 or less*.

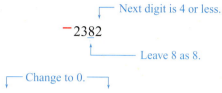

Step 3  ⁻23<u>8</u>2 rounds to ⁻2380.

⁻2382 rounded to the nearest ten is ⁻2380.

**(b)** Round 13,961 to the nearest hundred.

Step 1           13,<u>9</u>61
                    └── Hundreds place

Step 2  The next digit to the right is 6, which is *5 or more*.

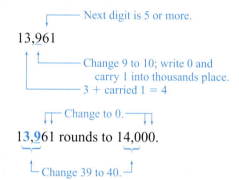

Step 3

13,961 rounded to the nearest hundred is 14,000.

> **NOTE**
>
> In Step 2, when you added 1 to the hundreds place, notice that the first three digits increased from 139 to 140.
>
> 13,**9**61   rounded to   14,**0**00

═══════════════ Work Problem ❹ at the Side.

**❹** Round as indicated.

(a) ⁻6036 to the nearest ten

(b) 31,968 to the nearest hundred

(c) ⁻73,077 to the nearest thousand

(d) 4952 to the nearest thousand

(e) 85,949 to the nearest hundred

(f) 40,387 to the nearest thousand

**ANSWERS**
4. (a) ⁻6040  (b) 32,000  (c) ⁻73,000
   (d) 5000  (e) 85,900  (f) 40,000

**5** Round as indicated.

(a) ⁻14,679 to the nearest ten thousand

(b) 724,518,715 to the nearest million

(c) ⁻49,900,700 to the nearest million

(d) 306,779,000 to the nearest hundred million

### Example 5  Rounding Large Numbers

(a) Round ⁻37,892 to the nearest ten thousand.

Step 1   ⁻3̲7,892
              ↑
              └── Ten thousands place

Step 2   The next digit to the right is 7.

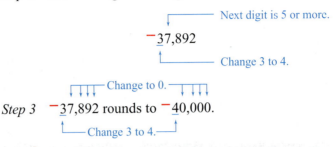

Step 3   ⁻3̲7,892 rounds to ⁻40,000.
              └── Change 3 to 4. ─┘

⁻37,892 rounded to the nearest ten thousand is ⁻40,000.

(b) Round 528,498,675 to the nearest million.

Step 1   528̲,498,675
            ↑
            └── Millions place

Step 2   528̲,498,675
                ↑
                │   ┌── Next digit is 4 or less.
                │   ↓
           528̲,498,675
            └── Leave 8 as 8.

Step 3   528̲,498,675 rounds to 528,000,000
                  └── Change to 0. ─┘

528,498,675 rounded to the nearest million is 528,000,000.

**Work Problem 5 at the Side.**

**3** Use front end rounding to estimate answers in addition and subtraction. In many everyday situations, we can round numbers and **estimate** the answer to a problem. For example, suppose that you're thinking about buying a sofa for $988 and a chair for $209. You can round the prices and estimate the total cost as $1000 + $200 ≈ $1200. The ≈ symbol means "approximately equal to." The estimated total of $1200 is close enough to help you decide whether you can afford both items. Of course, when it comes time to pay the bill, you'll want the *exact* total of $988 + $209 = $1197.

**Front end rounding** is often used to estimate answers. Each number is rounded to the highest possible place, so all the digits become 0 except the first digit. Once the numbers have lots of zeros, working with them is easy.

**ANSWERS**
5. (a) ⁻10,000  (b) 725,000,000
   (c) ⁻50,000,000  (d) 300,000,000

## Example 6 Using Front End Rounding

Use front end rounding to round each number.

**(a)** $^-216$

Round to the highest possible place, that is, the leftmost digit. In this case, the leftmost digit, 2, is in the hundreds place, so round to the nearest hundred.

┌─ Next digit is 4 or less.              ┌─ Change to 0. ─┐
$^-\underline{2}16$                                    $^-\underline{2}16$ rounds to $^-\underline{2}00$.
└─ Leave 2 as 2.                          └─ Leave 2 as 2. ─┘

The rounded number is $^-200$. Notice that all the digits in the rounded number are 0, except the first digit.

**(b)** 97,203

The leftmost digit, 9, is in the ten thousands place, so round to the nearest ten thousand.

┌─ Next digit is 5 or more.              ┌─ Change to 0. ─┐
$\underline{9}7,203$                                      $\underline{9}7,203$ rounds to $\underline{10}0,000$.
└─ Change 9 to 10.                        └─ Change 9 to 10. ─┘
                                          Carry 1 into the hundred thousands place.

The rounded number is 100,000. Notice that all the digits in the rounded number are 0, except the first digit.

*Work Problem* ❻ *at the Side.*

## Example 7 Using Front End Rounding to Estimate an Answer

Use front end rounding to estimate an answer. Then find the exact answer.

Meisha's paycheck showed gross pay of $823. It also listed deductions of $291. What is her net pay after deductions?

*Estimate:* Use front end rounding to round $823 and $291.

┌─ Next digit is 4 or less.              ┌─ Next digit is 5 or more.
$\$\underline{8}23$ rounds to $\$\underline{8}00$.         $\$\underline{2}91$ rounds to $\$\underline{3}00$.
└─ Leave 8 as 8.                          └─ Change 2 to 3.

Use the rounded numbers and subtract to estimate Miesha's net pay.

$\$800 - \$300 = \$500$ ←── Estimate

*Exact:*    $\$823 - \$291 = \$532$ ←── Exact

Meisha's paycheck will show the exact amount of $532. Because $532 is fairly close to the estimate of $500, Meisha can quickly see that the amount shown on her paycheck probably is correct. She might also use the estimate when talking to a friend, saying, "My net pay is about $500."

---

❻ Use front end rounding to round each number.

**(a)** $^-94$

**(b)** 508

**(c)** $^-2522$

**(d)** 9700

**(e)** 61,888

**(f)** $^-963,369$

ANSWERS
6. (a) $^-90$  (b) 500  (c) $^-3000$
   (d) 10,000  (e) 60,000  (f) $^-1,000,000$

**7** Use front end rounding to estimate an answer. Then find the exact answer.

Pao Xiong is a bookkeeper for a small business. The company checking account is overdrawn by $3881. He deposits a check for $2090. What is the balance in the account?

*Estimate:*

*Exact:*

**CAUTION**

Always estimate the answer first. Then, when you find the exact answer, check that it is close to the estimate. If your exact answer is very far off, rework the problem because you probably made an error.

**Calculator Tip** It's easy to press the wrong key when using a calculator. If you use front end rounding and estimate the answer *before* entering the numbers, you can catch many such mistakes. For example, a student thought that he entered this problem correctly.

$$7836 \; (+) \; 5060 \; (=) \quad\boxed{2776}$$

Front end rounding gives an estimated answer of $8000 + 5000 = 13{,}000$, which is very different from 2776. Can you figure out which key the student pressed incorrectly? (Answer: The student pressed $(-)$ instead of $(+)$.)

Work Problem **7** at the Side.

**ANSWERS**

7. *Estimate:* $^-\$4000 + \$2000 = {^-\$2000}$
   *Exact:* $^-\$3881 + \$2090 = {^-\$1791}$
   The account is overdrawn by $1791, which is fairly close to the estimate of $^-\$2000$.

## 1.5 Exercises

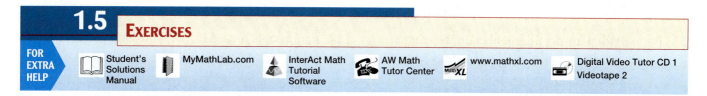

*Round each number to the indicated place. See Examples 1–5.*

1. 625 to the nearest ten
2. 206 to the nearest ten
3. −1083 to the nearest ten
4. −2439 to the nearest ten
5. 7862 to the nearest hundred
6. 6746 to the nearest hundred
7. −86,813 to the nearest hundred
8. −17,211 to the nearest hundred
9. 42,495 to the nearest hundred
10. 18,273 to the nearest hundred
11. −5996 to the nearest hundred
12. −8451 to the nearest hundred
13. 15,758 to the nearest hundred
14. 28,065 to the nearest hundred
15. −78,499 to the nearest thousand
16. −14,314 to the nearest thousand
17. 5847 to the nearest thousand
18. 49,706 to the nearest thousand
19. 53,182 to the nearest thousand
20. 13,124 to the nearest thousand
21. 595,008 to the nearest ten thousand
22. 725,182 to the nearest ten thousand
23. −8,906,422 to the nearest million
24. −13,713,409 to the nearest million
25. 139,610,000 to the nearest million
26. 609,845,500 to the nearest million

*Use front end rounding to round each number. See Example 6.*

**27.** Tyrone's truck shows this number on the odometer.

**28.** Ezra bought a used car with this odometer reading.

**29.** From summer to winter the average temperature drops 56 degrees.

**30.** The flood waters fell 42 inches yesterday.

**31.** Jan earned $9942 working part time.

**32.** Carol deposited $285 in her checking account.

**33.** 60,950,000 Americans go to a video store each week. (*Source:* Video Software Dealer's Assoc.)

**34.** 99,375,000 U.S. households have at least one TV. (*Source:* Nielsen Media Research.)

**35.** The submarine will dive to 255 feet below the surface of the ocean.

**36.** DeAnne lost $1352 in the stock market.

**37.** The population of Alaska is 626,932 people. (*Source:* U.S. Bureau of the Census.)

**38.** The population of California is 33,871,648 people. (*Source:* U.S. Bureau of the Census.)

**39.** Explain in your own words how to do front end rounding. Also show two examples of numbers and how you round them.

**40.** Describe two situations in your own life when you might use rounded numbers. Describe two situations in which exact numbers are important.

*Use your estimation skills to pick the most reasonable answer for each addition. Do **not** solve the problems. Circle your choices. See Example 7.*

**41.** ⁻42 + 89

  Estimate: _____ + _____ = _____

  Exact:  131   ⁻47   47

**42.** ⁻66 + 25

  Estimate: _____ + _____ = _____

  Exact:  ⁻91   ⁻41   ⁻21

**43.** 16 + ⁻97

  Estimate: _____ + _____ = _____

  Exact:  ⁻81   ⁻113   ⁻41

**44.** 58 + ⁻19

  Estimate: _____ + _____ = _____

  Exact:  39   ⁻39   ⁻77

**45.** ⁻273 + ⁻399

  Estimate:

  Exact:  ⁻126   ⁻672   ⁻992

**46.** ⁻311 + ⁻582

  Estimate:

  Exact:  893   ⁻271   ⁻893

**47.** 3081 + 6826

  Estimate:

  Exact:  3745   9907   15,907

**48.** 4904 + 1181

  Estimate:

  Exact:  3723   9025   6085

*Change subtractions to addition. Then use your estimation skills to pick the most reasonable answer for each problem. Circle your choices. See Example 7.*

**49.** 23 − 81

  Estimate:

  Exact:  58   104   ⁻58

**50.** 72 − 84

  Estimate:

  Exact:  12   ⁻12   ⁻156

**51.** ⁻39 − 39

  Estimate:

  Exact:  ⁻78   0   78

**52.** ⁻91 − 91

  Estimate:

  Exact:  0   182   ⁻182

**38** Chapter 1 Introduction to Algebra: Integers

**53.** ⁻106 + 34 − ⁻72

*Estimate:*

*Exact:*   ⁻143    68    0

**54.** 52 − ⁻87 − 139

*Estimate:*

*Exact:*   0    ⁻104    ⁻174

*First use front end rounding to estimate the answer to each application problem. Then find the exact answer. See Example 7.*

**55.** The community has raised $52,882 for the homeless shelter. If the amount needed for the shelter is $78,650, how much more needs to be collected?

*Estimate:*

*Exact:*

**56.** A truck weighs 9250 pounds when empty. After being loaded with firewood, it weighs 21,375 pounds. What is the weight of the firewood?

*Estimate:*

*Exact:*

**57.** Dorene Cox decided to establish a budget. She will spend $485 for rent, $325 for food, $320 for child care, $182 for transportation, and $150 for other expenses, and she will put the remainder in savings. If her monthly take-home pay is $1920, find her monthly savings.

*Estimate:*

*Exact:*

**58.** Jared Ueda had $2874 in his checking account. He wrote checks for $308 for auto repairs, $580 for child support, and $778 for tuition. Find the amount remaining in his account.

*Estimate:*

*Exact:*

**59.** In a laboratory experiment, a mixture started at a temperature of ⁻102 degrees. First the temperature was raised 37 degrees and then raised 52 degrees. What was the final temperature?

*Estimate:*

*Exact:*

**60.** A scuba diver was photographing fish at 65 feet below the surface of the lagoon. She swam up 24 feet and then swam down 49 feet. What was her final depth?

*Estimate:*

*Exact:*

**61.** A riding lawn mower costs $525 more than a self-propelled lawn mower. If a self-propelled mower costs $380, find the cost of a riding mower.

*Estimate:*

*Exact:*

**62.** The price of the least expensive rear-bagging lawn mower used in a recent test was $175. If this was $475 less than the most expensive model, find the price of the most expensive mower.

*Estimate:*

*Exact:*

# 1.6 MULTIPLYING INTEGERS

**OBJECTIVES**
1. Use a raised dot or parentheses to express multiplication.
2. Multiply integers.
3. Identify properties of multiplication.
4. Estimate answers to application problems involving multiplication.

**1** **Use a raised dot or parentheses to express multiplication.** In arithmetic we usually use "×" when writing multiplication problems. But in algebra, we use a raised dot or parentheses to show multiplication. The numbers being multiplied are called **factors** and the answer is called the **product.**

| Arithmetic | Algebra |
|---|---|
| 3 × 5 = 15 | 3 • 5 = 15   or   3(5) = 15   or   (3)(5) = 15 |
| Factors  Product | Factors  Product    Factors  Product    Factors  Product |

### Example 1  Expressing Multiplication in Algebra

Rewrite each multiplication in three different ways, using a dot or parentheses. Also identify the factors and the product.

(a) 10 × 7

   Rewrite it as 10 • 7 (Raised dot)   or   10(7)   or   (10)(7).

The factors are 10 and 7. The product is 70.

(b) 4 × 80

   Rewrite it as 4 • 80   or   4(80)   or   (4)(80).

The factors are 4 and 80. The product is 320.

⟹ Work Problem **1** at the Side.

**1** Rewrite each multiplication in three different ways using a dot or parentheses. Also identify the factors and the product.

(a) 100 × 6

(b) 7 × 12

**NOTE**

Parentheses are used to show several different things in algebra. When we discussed the associative property earlier in this chapter, we used parentheses in this way.

$$6 + \underbrace{(9 + {}^-9)}_{} \leftarrow \text{Parentheses show which numbers to add first.}$$
$$\underbrace{6 + \quad 0}_{6}$$

Now we are using parentheses to indicate multiplication, as in 3(5) or (3)(5).

**2** **Multiply integers.** Suppose that our football team gained 5 yards on the first play, gained 5 yards again on the second play, and gained 5 yards again on the third play. We can add to find the result.

   5 yards + 5 yards + 5 yards = 15 yards

A quick way to add the same number several times is to multiply.

| Our team made 3 plays | and | gained 5 yards each time. | | Our team gained a total of 15 yards. |
|---|---|---|---|---|
| 3 | • | 5 | = | 15 |

**ANSWERS**
1. (a) 100 • 6 or 100(6) or (100)(6)
       The factors are 100 and 6; the product is 600.
   (b) 7 • 12 or 7(12) or (7)(12)
       The factors are 7 and 12; the product is 84.

Here are the rules for multiplying two integers.

> **Multiplying Two Integers**
>
> If two factors have *different signs,* the product is *negative.*
> For example,
>
> $$^-2 \cdot 6 = {^-12} \qquad \text{and} \qquad 4 \cdot {^-5} = {^-20}.$$
>
> If two factors have the *same sign,* the product is *positive.*
> For example,
>
> $$7 \cdot 3 = 21 \qquad \text{and} \qquad {^-3} \cdot {^-10} = 30$$

There are several ways to illustrate these rules. First we'll continue with football. Remember, you are interested in the results for *our* team. We will designate **our team** with a **positive sign** and **their team** with a **negative sign**.

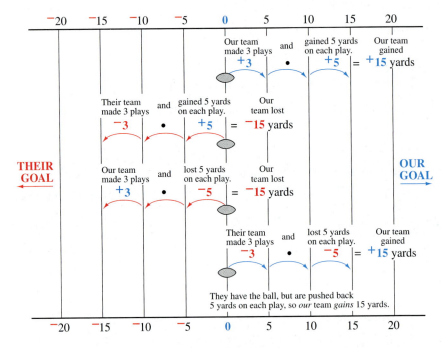

Here is a summary of the football examples.

When two factors have the *same* sign, the product is *positive.*

$$\text{Both positive}$$
$$3 \cdot 5 = 15$$
$$^-3 \cdot {^-5} = 15$$
$$\text{Both negative}$$

Product is positive.

When two factors have *different* signs, the product is *negative.*

$$^-3 \cdot 5 = {^-15}$$
$$3 \cdot {^-5} = {^-15}$$

Product is negative.

Section 1.6 Multiplying Integers   41

There is another way to look at these multiplication rules. In mathematics, the rules or patterns must always be consistent.

Look for a pattern in this list of products.

$$4 \cdot 2 = 8$$
$$3 \cdot 2 = 6$$
$$2 \cdot 2 = 4$$
$$1 \cdot 2 = 2$$
$$0 \cdot 2 = 0$$
$$^-1 \cdot 2 = \;?$$

Blue numbers decrease by 1.
Red numbers decrease by 2.

To keep the red pattern going, replace the ? with a number that is 2 *less than* 0, which is $^-2$.
 So, $^-1 \cdot 2 = \;^-2$. This pattern illustrates that the product of two numbers with *different* signs is *negative*.

Look for a pattern in this list of products.

$$4 \cdot \;^-2 = \;^-8$$
$$3 \cdot \;^-2 = \;^-6$$
$$2 \cdot \;^-2 = \;^-4$$
$$1 \cdot \;^-2 = \;^-2$$
$$0 \cdot \;^-2 = \;\;0$$
$$^-1 \cdot \;^-2 = \;?$$

Blue numbers decrease by 1.
Red numbers increase by 2.

To keep the red pattern going, replace the ? with a number that is 2 *more than* 0, which is $^+2$.
 So, $^-1 \cdot \;^-2 = \;^+2$. This pattern illustrates that the product of two numbers with the *same* sign is *positive*.

### Example 2  Multiplying Two Integers

(a) $^-2 \cdot 8 = \;^-16$   The factors have *different signs,* so the product is *negative.*
 Negative · Positive

(b) $^-10\,(^-6) = 60$   The factors have the *same sign,* so the product is *positive.*
 Both negative

(c) $(9)(^-11) = \;^-99$   The factors have *different signs,* so the product is *negative.*
 Positive · Negative

= Work Problem ❷ at the Side.

Sometimes there are more than two factors in a multiplication problem. If there are parentheses around two of the factors, multiply them first. If there aren't any parentheses, start at the left and work with two factors at a time.

### Example 3  Multiplying Several Factors

Multiply.

(a) $^-3 \cdot (4 \cdot \;^-5)$   Parentheses tell you to multiply $4 \cdot \;^-5$ first. The factors
    $^-3 \cdot \;^-20$         have *different* signs, so the product is *negative.*
       $60$                  Then multiply $^-3 \cdot \;^-20$. Both factors have the *same* sign, so the product is *positive.*

(b) $^-2 \cdot \;^-2 \cdot \;^-2$   There are no parentheses, so multiply $^-2 \cdot \;^-2$ first. The
     $\;\;4\;\; \cdot \;^-2$       factors have the *same* sign, so the product is *positive.*
        $^-8$                  Then multiply $4 \cdot \;^-2$. The factors have *different* signs, so the product is *negative.*

❷ Multiply.

(a) $7(^-2)$

(b) $^-5 \cdot \;^-5$

(c) $^-1(14)$

(d) $10 \cdot 6$

(e) $(^-4)(^-9)$

**ANSWERS**
2. (a) $^-14$  (b) 25  (c) $^-14$
   (d) 60  (e) 36

## 3 Multiply.

(a) $5 \cdot (^-10 \cdot 2)$

(b) $^-1 \cdot 8 \cdot ^-5$

(c) $^-3 \cdot ^-2 \cdot ^-4$

(d) $^-2 \cdot (7 \cdot ^-3)$

(e) $^-1 \cdot ^-1 \cdot ^-1$

## 4 Multiply. Then name the property illustrated by each example.

(a) $819 \cdot 0$

(b) $1(^-90)$

(c) $25 \cdot 1$

(d) $(0)(^-75)$

**ANSWERS**
3. (a) $^-100$ (b) 40 (c) $^-24$
   (d) 42 (e) $^-1$
4. (a) 0; multiplication property of 0
   (b) $^-90$; multiplication property of 1
   (c) 25; multiplication property of 1
   (d) 0; multiplication property of 0

---

**CAUTION**

In Example 3(b) you may be tempted to think that the final product will be *positive* because all the factors have the *same* sign. Be careful to work with just two factors at a time and keep track of the sign at each step.

**Calculator Tip** You can use the *change of sign* key for multiplication and division, just as you did for adding and subtracting. To enter Example 3(b) on your scientific calculator, press the following keys.

$$\underbrace{2 \; \boxed{+/-}}_{-2} \; \boxed{\times} \; \underbrace{2 \; \boxed{+/-}}_{-2} \; \boxed{\times} \; \underbrace{2 \; \boxed{+/-}}_{-2} \; \boxed{=} \qquad \text{The answer is } ^-8.$$

**Work Problem 3 at the Side.**

**3 Identify properties of multiplication.** Addition involving 0 is unusual because adding 0 does *not* change the number. For example, $7 + 0$ is still 7. (See **Section 1.3**.) But what happens in multiplication? Let's use our football team as an example.

$$\underbrace{\text{Our team made 3 plays}}_{3} \quad \text{and} \quad \underbrace{\text{didn't gain or lose yards on any play.}}_{0} \quad = \quad \underbrace{\text{Altogether our team didn't gain or lose any yards.}}_{0}$$

This example illustrates one of the properties of multiplication.

### Multiplication Property of 0

Multiplying any number by 0 gives a product of 0.
Some examples are shown below.

$$^-16 \cdot 0 = 0 \qquad (0)(5) = 0 \qquad 32{,}977 \cdot 0 = 0$$

So, can you multiply a number by something that will *not* change the number?

$$6 \cdot ? = 6 \qquad ^-12 \cdot ? = ^-12 \qquad ? \cdot 5876 = 5876$$

The number 1 can replace the ? in each example. This illustrates another property of multiplication.

### Multiplication Property of 1

Multiplying a number by 1 leaves the number unchanged.
Some examples are shown below.

$$6 \cdot 1 = 6 \qquad ^-12 \cdot 1 = ^-12 \qquad 1 \cdot 5876 = 5876$$

### Example 4 Using Properties of Multiplication

Multiply. Then name the property illustrated by each example.

(a) $0 \cdot ^-48 = 0$   Illustrates the multiplication property of 0.

(b) $615(1) = 615$   Illustrates the multiplication property of 1.

**Work Problem 4 at the Side.**

When adding, we said that changing the *order* of the addends did not change the sum (commutative property of addition). We also found that changing the *grouping* of addends did not change the sum (associative property of addition). These same ideas apply to multiplication.

## Commutative Property of Multiplication

Changing the *order* of two factors does not change the product.

## Associative Property of Multiplication

Changing the *grouping* of factors does not change the product.

**Example 5** Using the Commutative and Associative Properties

Show that the product is unchanged and name the property that is illustrated in each case.

(a) $^-7 \cdot {}^-4 = {}^-4 \cdot {}^-7$

   $28 \ \ \ = \ \ \ 28$  Both products are 28.

This example illustrates the commutative property of multiplication.

(b) $5 \cdot (10 \cdot 2) = (5 \cdot 10) \cdot 2$

   $5 \cdot \ 20 \ \ = \ \ 50 \ \ \cdot 2$

   $100 \ \ = \ \ 100$  Both products are 100.

This example illustrates the associative property of multiplication.

— Work Problem ❺ at the Side.

Now that you are familiar with multiplication and addition, we can look at a property that involves both operations.

## Distributive Property

Multiplication distributes over addition.
An example is shown below.

$$3(6 + 2) = 3 \cdot 6 + 3 \cdot 2$$

What is the *distributive property* really saying? Notice that there is an understood multiplication symbol between the 3 and the parentheses. To "distribute" the 3 means to multiply 3 times each number inside the parentheses.

   Understood to be     $3(6 + 2)$
     *multiplying* by 3      ↓
                         $3 \cdot (6 + 2)$

Using the distributive property,

   $3 \cdot (6 + 2)$   can be rewritten as   $3 \cdot 6 + 3 \cdot 2$.

Check that the product is unchanged. Either way the result is 24.

---

❺ Show that the product is unchanged and name the property that is illustrated in each case.

(a) $(^-3 \cdot {}^-3) \cdot {}^-2 =$
    $^-3 \cdot ({}^-3 \cdot {}^-2)$

(b) $11 \cdot 8 = 8 \cdot 11$

(c) $0 \cdot {}^-15 = {}^-15 \cdot 0$

(d) $4 \cdot ({}^-1 \cdot {}^-5) =$
    $(4 \cdot {}^-1) \cdot {}^-5$

**ANSWERS**
5. (a) $^-18 = {}^-18$; associative property of multiplication
   (b) $88 = 88$; commutative property of multiplication
   (c) $0 = 0$; commutative property of multiplication
   (d) $20 = 20$; associative property of multiplication

**44** Chapter 1 Introduction to Algebra: Integers

**6** Rewrite each product, using the distribution property. Show that the result is unchanged.

(a) 3(8 + 7)

(b) 10(6 + ⁻9)

(c) ⁻6(4 + 4)

**7** Use front end rounding to estimate an answer. Then find the exact answer.

An average of 27,095 baseball fans attended each of the 81 home games during the season. What was the total home game attendance for the season?

*Estimate:*

*Exact:*

**ANSWERS**
6. (a) 3 • 8 + 3 • 7; both results are 45.
   (b) 10 • 6 + 10 • ⁻9; both results are ⁻30.
   (c) ⁻6 • 4 + ⁻6 • 4; both results are ⁻48.
7. *Estimate:* 30,000 • 80 = 2,400,000 fans
   *Exact:* 27,095 • 81 = 2,194,695 fans

### Example 6  Using the Distributive Property

Rewrite each product, using the distributive property. Show that the result is unchanged.

(a) 4(3 + 7)

$$4(3 + 7) = 4 \cdot 3 + 4 \cdot 7$$
$$4 \cdot (10) = 12 + 28$$
$$40 = 40 \quad \text{Both results are 40.}$$

(b) ⁻2(⁻5 + 1)

$$^-2(^-5 + 1) = ^-2 \cdot {}^-5 + {}^-2 \cdot 1$$
$$^-2 \cdot (^-4) = 10 + {}^-2$$
$$8 = 8 \quad \text{Both results are 8.}$$

**Work Problem 6 at the Side.**

**4** Estimate answers to application problems involving multiplication. Front end rounding can be used to estimate answers in multiplication, just as we did when adding and subtracting (see **Section 1.5**). Once the numbers have been rounded so that there are lots of zeros, we can use the multiplication shortcut described in the Review chapter (see **Section R.3**). As a brief review, look at the pattern in these examples.

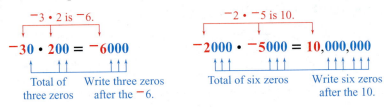

⁻3 • 2 is ⁻6.
⁻30 • 200 = ⁻6000
Total of three zeros   Write three zeros after the ⁻6.

⁻2 • ⁻5 is 10.
⁻2000 • ⁻5000 = 10,000,000
Total of six zeros   Write six zeros after the 10.

### Example 7  Using Front End Rounding to Estimate an Answer

Use front end rounding to estimate an answer. Then find the exact answer.

Last year the Video Land store had to replace 392 defective videos at a cost of $19 each. How much money did the store lose on defective videos? (*Hint:* Because it's a loss, use a negative number for the cost.)

*Estimate:* Use front end rounding: 392 rounds to 400 and ⁻$19 rounds to ⁻$20. Use the rounded numbers and multiply to estimate the total amount of money lost.

4 • ⁻2 is ⁻8.
400 • ⁻$20 = ⁻$8000   Estimate
Total of three zeros   Write three zeros after the ⁻8.

*Exact:* 392 • ⁻$19 = ⁻$7448

Because ⁻$7448 is fairly close to the estimate of ⁻$8000, you can see that ⁻$7448 probably is correct. The store manager could also use the estimate to say, "We lost about $8000 on defective videos last year."

**Work Problem 7 at the Side.**

## 1.6 Exercises

*Multiply. See Examples 1–4.*

1. (a) 9 · 7
   (b) ⁻9 · ⁻7
   (c) ⁻9 · 7
   (d) 9 · ⁻7

2. (a) ⁻6 · 9
   (b) 6 · ⁻9
   (c) ⁻6 · ⁻9
   (d) 6 · 9

3. (a) 7(⁻8)
   (b) ⁻7(8)
   (c) 7(8)
   (d) ⁻7(⁻8)

4. (a) 8(6)
   (b) ⁻8(⁻6)
   (c) ⁻8(6)
   (d) 8(⁻6)

5. ⁻5 · 7

6. ⁻10 · 2

7. (⁻5)(9)

8. (⁻9)(4)

9. 3(⁻6)

10. 8(⁻9)

11. 10(⁻5)

12. 5(⁻11)

13. (⁻1)(40)

14. (75)(⁻1)

15. ⁻56 · 1

16. 1 · ⁻87

17. ⁻8(⁻4)

18. ⁻3(⁻9)

19. 11 · 7

20. 4 · 25

21. 25 · 0

22. 0 · 30

23. ⁻19(⁻7)

24. ⁻21(⁻3)

25. ⁻13(⁻1)

26. ⁻1(⁻31)

27. (0)(⁻25)

28. (⁻50)(0)

29. ⁻4 · ⁻6 · 2

30. ⁻9 · 3 · ⁻3

31. ⁻4 · ⁻2 · ⁻7

32. ⁻6 · ⁻2 · ⁻3

33. 5 · ⁻8 · 4

34. 5 · 4 · ⁻6

*Write an integer in each blank to make a true statement.*

35. (⁻3)(____) = ⁻15

36. 6 · ____ = ⁻24

37. ____ · 10 = ⁻30

38. (____)(⁻4) = 16

39. ⁻17 = 17(____)

40. 29 = ⁻29(____)

41. (____)(⁻350) = 0

42. ____ · 99 = 99

43. 5 · ⁻4 · ____ = ⁻100

44. ____ · 2 · ⁻2 = ⁻24

45. (____)(⁻5)(⁻2) = ⁻40

46. (⁻3)(____)(⁻3) = ⁻27

**47.** In your own words, explain the difference between the commutative and associative properties of multiplication. Show an example of each.

**48.** A student did this multiplication.
$$^-3 \cdot {}^-3 \cdot {}^-3 = 27$$
He knew that $3 \cdot 3 \cdot 3$ is 27. Since all the factors have the same sign, he made the product positive. Do you agree with his reasoning? Explain.

---

**RELATING CONCEPTS (Exercises 49–50)** — **FOR INDIVIDUAL OR GROUP WORK**

*Look for patterns as you **work Exercises 49 and 50 in order.***

**49.** Write three numerical examples for each of these situations:
   (a) A positive number multiplied by $^-1$
   (b) A negative number multiplied by $^-1$

Now write a rule that explains what happens when you multiply a signed number by $^-1$.

**50.** Do these multiplications.
$$^-2 \cdot {}^-2 = \underline{\phantom{xx}}$$
$$^-2 \cdot {}^-2 \cdot {}^-2 = \underline{\phantom{xx}}$$
$$^-2 \cdot {}^-2 \cdot {}^-2 \cdot {}^-2 = \underline{\phantom{xx}}$$
$$^-2 \cdot {}^-2 \cdot {}^-2 \cdot {}^-2 \cdot {}^-2 = \underline{\phantom{xx}}$$

Describe the pattern in the products. Then find the next three products without multiplying all the $^-2$s.

---

*Rewrite each multiplication, using the stated property. Show that the result is unchanged. See Examples 5 and 6.*

**51.** Distributive property
$9(^-3 + 5)$

**52.** Distributive property
$^-6(4 + {}^-5)$

**53.** Commutative property
$25 \cdot 8$

**54.** Commutative property
$^-7 \cdot {}^-11$

**55.** Associative property
$^-3 \cdot ({}^-2 \cdot {}^-5)$

**56.** Associative property
$(5 \cdot 5) \cdot 10$

*First use front end rounding to estimate the answer to each application problem. Then find the exact answer. See Example 7.*

**57.** Alliette receives $324 per week for doing child care in her home. How much income will she have for an entire year? There are 52 weeks in a year.

*Estimate:*

*Exact:*

**58.** Enrollment at our community college has increased by 875 students each of the last four semesters. What is the total increase?

*Estimate:*

*Exact:*

**59.** A new computer software store had losses of $9950 during each month of its first year. What was the total loss for the year?

*Estimate:*

*Exact:*

**60.** A long-distance phone company estimates that it is losing 95 customers each week. How many customers will it lose in a year?

*Estimate:*

*Exact:*

**61.** Tuition at the state university is $182 per credit for undergraduates. How much tuition will Wei Chen pay for 13 credits?

*Estimate:*

*Exact:*

**62.** Pat ate a dozen crackers as a snack. Each cracker had 17 calories. How many calories did Pat eat?

*Estimate:*

*Exact:*

**63.** There are 24 hours in one day. How many hours are in one year (365 days)?

*Estimate:*

*Exact:*

**64.** There are 5280 feet in one mile. How many feet are in 17 miles?

*Estimate:*

*Exact:*

*Simplify.*

**65.** $^-8 \cdot |^-8 \cdot 8|$

**66.** $^-7 \cdot |7| \cdot |^-7|$

**67.** $^-37 \cdot {}^-1 \cdot 85 \cdot 0$

**68.** $^-1 \cdot 9732 \cdot {}^-1 \cdot {}^-1$

**69.** $|6 - 7| \cdot {}^-355{,}299$

**70.** $987 \cdot {}^-65{,}432 \cdot |9 - 9|$

*Each of these application problems requires several steps and may involve addition and subtraction as well as multiplication.*

**71.** Each of Maurice's four cats needs a $24 rabies shot and a $29 shot to prevent respiratory infections. There will also be one $35 office visit charge. What will be the total amount of Maurice's bill?

**72.** In Ms. Zubero's math class there are six tests of 100 points each, eight quizzes of six points each, and 20 homework assignments of five points each. There are also four "bonus points" on each test. What is the total number of possible points?

**73.** There is a 3-degree drop in temperature for every thousand feet that an airplane climbs into the sky. (*Source:* Lands' End.) If the temperature on the ground is 50 degrees, what will be the temperature when the plane reaches an altitude of 24,000 feet?

**74.** An unmanned research submarine descends to 150 feet below the surface of the ocean. Then it continues to go deeper, taking a water sample every 25 feet. What is its depth when it takes the 15th sample?

## 1.7 DIVIDING INTEGERS

**OBJECTIVES**
1. Divide integers.
2. Identify properties of division.
3. Combine multiplying and dividing of integers.
4. Estimate answers to application problems involving division.
5. Interpret remainders in division application problems.

**1** **Divide integers.** In arithmetic we usually use $\overline{)}$ to write division problems so that we can do the problem by hand. Calculator keys use the ÷ symbol for division. In algebra we usually show division by using a fraction bar, a slash mark, or the ÷ symbol. The answer to a division problem is called the **quotient**.

**Arithmetic**

Divisor → $2\overline{)16}$ ← Quotient
  ← Dividend

**Calculator and Algebra**

Dividend ↓   Divisor
$16 ÷ 2 = 8$
    ↑
   Quotient

Dividend → $\dfrac{16}{2} = 8$
Divisor →
    ↑
   Quotient

For every division problem, we can write a related multiplication problem. (See **Section R.4**.) Because of this relationship, the rules for dividing integers are the same as the rules for multiplying integers. For example,

$$\dfrac{16}{8} = 2 \quad \text{because} \quad 2 \cdot 8 = 16.$$

$$\dfrac{^-16}{^-8} = 2 \quad \text{because} \quad 2 \cdot {}^-8 = {}^-16.$$

$$\dfrac{^-16}{8} = {}^-2 \quad \text{because} \quad {}^-2 \cdot 8 = {}^-16.$$

$$\dfrac{16}{^-8} = {}^-2 \quad \text{because} \quad {}^-2 \cdot {}^-8 = 16.$$

**Dividing Two Integers**

If two numbers have *different signs*, the quotient is *negative*. Some examples are shown below.

$$\dfrac{^-18}{3} = {}^-6 \qquad \dfrac{40}{^-5} = {}^-8$$

If two numbers have the *same sign*, the quotient is *positive*. Some examples are shown below.

$$\dfrac{^-30}{^-6} = 5 \qquad \dfrac{48}{8} = 6$$

### Example 1  Dividing Two Integers

Divide.

(a) $\dfrac{^-20}{5}$ — Numbers have *different* signs, so the quotient is *negative*. $\quad \dfrac{^-20}{5} = {}^-4$

(b) $\dfrac{^-24}{^-4}$ — Numbers have the *same* sign, so the quotient is *positive*. $\quad \dfrac{^-24}{^-4} = 6$

(c) $60 ÷ {}^-2$ — Numbers have *different* signs, so the quotient is *negative*. $\quad 60 ÷ {}^-2 = {}^-30$

**Work Problem 1 at the Side.**

**1** Divide.

(a) $\dfrac{40}{^-8}$

(b) $\dfrac{49}{7}$

(c) $\dfrac{^-32}{4}$

(d) $\dfrac{^-10}{^-10}$

(e) ${}^-81 ÷ 9$

(f) ${}^-100 ÷ {}^-50$

**ANSWERS**
1. (a) ${}^-5$  (b) 7  (c) ${}^-8$  (d) 1
   (e) ${}^-9$  (f) 2

**❷ Divide. Then state the property illustrated by each division.**

(a) $\dfrac{-12}{0}$

(b) $\dfrac{0}{39}$

(c) $\dfrac{-9}{1}$

(d) $\dfrac{21}{21}$

**ANSWERS**
2. (a) undefined; division by 0 is undefined.
   (b) 0; 0 divided by any nonzero number is 0.
   (c) $-9$; any number divided by 1 is the number.
   (d) 1; any nonzero number divided by itself is 1.

## ❷ Identify properties of division.

You have seen that 0 and 1 are used in special ways in addition and multiplication. This is also true in division.

| Examples | | | Pattern (Division Property) |
|---|---|---|---|
| $\dfrac{5}{5}=1$ | $\dfrac{-18}{-18}=1$ | $\dfrac{-793}{-793}=1$ | When a nonzero number is divided by itself, the quotient is 1. |
| $\dfrac{5}{1}=5$ | $\dfrac{-18}{1}=-18$ | $\dfrac{-793}{1}=-793$ | When a number is divided by 1, the quotient is the number. |
| $\dfrac{0}{5}=0$ | $\dfrac{0}{-18}=0$ | $\dfrac{0}{-793}=0$ | When 0 is divided by any other number (except 0), the quotient is 0. |
| $\dfrac{5}{0}$ is undefined. | $\dfrac{-18}{0}$ is undefined. | | Division by 0 is *undefined*. There is no answer. |

The most surprising property is that division by 0 *cannot be done*. Let's review the reason for that by rewriting this division problem as a related multiplication problem.

$$\dfrac{-18}{0} = ? \quad \text{can be written as the multiplication} \quad ? \cdot 0 = -18$$

If you thought the answer to $\dfrac{-18}{0}$ should be 0, try replacing **?** with 0. It doesn't work in the related multiplication problem! Try replacing **?** with any number you like. The result in the related multiplication problem is always 0 instead of $-18$. That is how we know that dividing by 0 cannot be done. Mathematicians say that it is *undefined*.

### Example 2  Using the Properties of Division

Divide. Then state the property illustrated by each example.

(a) $\dfrac{-312}{-312} = 1$    Any nonzero number divided by itself is 1.

(b) $\dfrac{75}{1} = 75$    Any number divided by 1 is the number.

(c) $\dfrac{0}{-19} = 0$    Zero divided by any nonzero number is 0.

(d) $\dfrac{48}{0}$ is undefined.    Division by 0 is undefined.

**Calculator Tip**  Try Examples 2(c) and 2(d) on your calculator. Use the change of sign key to enter $-19$ on a *scientific* calculator.

  Answer is 0.

  Calculator shows "Error" or "ERR" or "E" for error because it cannot divide by 0.

**Work Problem ❷ at the Side.**

**3** **Combine multiplying and dividing of integers.** When a problem involves both multiplying and dividing, first check to see if there are any parentheses. Do what is inside parentheses first. Then start at the left and work toward the right, using two numbers at a time.

### Example 3  Combining Multiplication and Division of Integers

Simplify.

(a) $6 \cdot {}^-10 \div ({}^-3 \cdot 2)$
$\phantom{(a)\ }6 \cdot {}^-10 \div {}^-6$     Work inside parentheses first: $^-3 \cdot 2$ is 6.
$\phantom{(a)\ \ \ \ }{}^-60 \div {}^-6$     Start at the left: $6 \cdot {}^-10$ is $^-60$.
$\phantom{(a)\ \ \ \ \ \ \ }10$     Finally, $^-60 \div {}^-6$ is 10.

(b) $^-24 \div {}^-2 \cdot 4 \div {}^-6$     There are no parentheses, so start at the left:
$\phantom{(b)\ }12 \cdot 4 \div {}^-6$     $^-24 \div {}^-2$ is 12.
$\phantom{(b)\ \ \ \ }48 \div {}^-6$     Next, $12 \cdot 4$ is 48.
$\phantom{(b)\ \ \ \ \ \ \ }{}^-8$     Finally, $48 \div {}^-6$ is $^-8$.

(c) $^-50 \div {}^-5 \div {}^-2$     There are no parentheses, so start at the left:
$\phantom{(c)\ }10 \div {}^-2$     $^-50 \div {}^-5$ is 10. The signs are the *same*, so the quotient is *positive*.
$\phantom{(c)\ \ \ \ }{}^-5$     Now, $10 \div {}^-2$ is $^-5$. The signs are *different*, so the quotient is *negative*.

**Work Problem ❸ at the Side.**

**4** **Estimate answers to application problems involving division.** Front end rounding can be used to estimate answers in division just as you did when multiplying (see **Section 1.6**). Once the numbers have been rounded so that there are lots of zeros, you can use the division shortcut described in the Review chapter (see **Section R.5**). As a brief review, look at the pattern in these examples.

$400\cancel{0} \div {}^-5\cancel{0} = 400 \div {}^-5 = {}^-80$

Drop one 0 from both dividend and divisor.

$^-6\cancel{000} \div {}^-3\cancel{000} = {}^-6 \div {}^-3 = 2$

Drop three zeros from both dividend and divisor.

---

❸ Simplify.

(a) $60 \div {}^-3 \div 4 \cdot {}^-5$

(b) $^-6 \cdot ({}^-16 \div {}^-8) \cdot 2$

(c) $^-8 \cdot 10 \div 4 \cdot {}^-3 \div {}^-6$

(d) $56 \div {}^-8 \div {}^-1$

**ANSWERS**
**3.** (a) 25   (b) $^-24$   (c) $^-10$   (d) 7

**4** First use front end rounding to estimate an answer. Then find the exact answer.

Laurie and Chuck Struthers lost $2724 on their stock investments last year. What was their average loss each month?

*Estimate:*

*Exact:*

---

**Example 4** Using Front End Rounding to Estimate an Answer in Division

First use front end rounding to estimate an answer. Then find the exact answer.

During a 24-hour laboratory experiment, the temperature of a solution dropped 96 degrees. What was the average drop in temperature each hour?

*Estimate:* Use front end rounding: $^-96$ degrees rounds to $^-100$ degrees and 24 hours rounds to 20 hours. To estimate the average, divide the rounded number of degrees by the rounded number of hours.

$^-100$ degrees ÷ 20 hours = $^-5$ degrees each hour ← Estimate

*Exact:*  $^-96$ degrees ÷ 24 hours = $^-4$ degrees each hour ← Exact

Because $^-4$ degrees is close to the estimate of $^-5$ degrees, you can see that $^-4$ degrees probably is correct.

---

**Calculator Tip** The answer in Example 4 "came out even." In other words, the quotient was an integer. Suppose that the drop in temperature had been 97 degrees. Do the division on your calculator.

97 [+/−] [÷] 24 [=]   Calculator shows −4.041666667
 $^-97$

The quotient is *not* an integer. We will work with numbers like these in Chapter 5, Positive and Negative Decimals.

Work Problem **4** at the Side.

---

**5** Interpret remainders in division application problems. In arithmetic, division problems often have a remainder, as shown below.

$$\begin{array}{r} 14 \text{ R}10 \\ 25\overline{)360} \\ \underline{25}\phantom{0} \\ 110 \\ \underline{100} \\ 10 \leftarrow \text{Remainder} \end{array}$$

But what does **R10** really mean? Let's look at this same problem by using money amounts.

---

**Example 5** Interpreting Remainders in Division Applications

Divide; then interpret the remainder in each application.

**(a)** The math department at Lake Community College has $360 in its budget to buy scientific calculators for the math lab. If the calculators cost $25 each, how many can be purchased? How much money will be left over?

We can solve this problem by using the same division as shown above. But this time we can decide what the remainder really means.

$$\begin{array}{r} 14 \leftarrow \text{Number of calculators purchased} \\ \text{Cost of one calculator} \rightarrow \$25\overline{)\$360} \leftarrow \text{Budget} \\ \underline{25}\phantom{0} \\ 110 \\ \underline{100} \\ \$10 \leftarrow \text{Money left over} \end{array}$$

The department can buy 14 calculators. There will be $10 left over.

*Continued on Next Page*

---

**ANSWERS**

4. *Estimate:* $^-$$3000 ÷ 10 months = $^-$$300 each month
 *Exact:* $^-$$2724 ÷ 12 months = $^-$$227 each month

**Calculator Tip** You can use your calculator to solve Example 5(a). Recall that digits on the *right* side of the decimal point show *part* of one whole. You cannot order *part* of one calculator, so ignore those digits and use only the *whole number part* of the quotient.

$$360 \div 25 = 14.4$$

Write whole number → 14

$$25 \overline{)360}$$
$$\phantom{25)}350 \quad \text{Subtract } 360 - 350$$
$$\phantom{25)360}10 \quad \text{to get remainder.}$$

(ON/AC) 14 × 25 = 350

Clear 14.4 from calculator before multiplying by pressing (C), (ON/C), or (ON/AC).

**(b)** Luke's son is going on a Scout camping trip. There are 135 Scouts. Luke is renting tents that sleep 6 people each. How many tents should he rent?

We again use division to solve the problem. There is a remainder, but this time it must be interpreted differently than in the calculator example.

$$\begin{array}{r} 22 \\ 6\overline{)135} \\ 12 \\ \hline 15 \\ 12 \\ \hline 3 \end{array}$$

← Number of tents with 6 Scouts each
Each tent holds →    ← Total number of Scouts

← Scouts left over

If Luke rents 22 tents, 3 Scouts will have to sleep out in the rain. He must rent 23 tents to accommodate all the Scouts. (One tent will have only 3 Scouts in it.)

**Work Problem ❺ at the Side.**

❺ Divide; then interpret the remainder in each of these applications.

**(a)** Chad and Martha are baking cookies for a fund-raiser. They baked 116 cookies and are putting them into packages of a dozen each. How many packages will they have for the fund-raiser? How many cookies will be left over for them to eat?

**(b)** Coreen is a dispatcher for a bus company. A group of 249 senior citizens is going to a baseball game. If the buses will each hold 44 people, how many buses should she send to pick up the seniors?

**ANSWERS**
**5. (a)** 9 packages, with 8 cookies left over to eat
**(b)** 6 buses, because 5 buses would leave 29 seniors standing on the curb

# Focus on Real-Data Applications

## 'Til Debt Do You Part!

*Bride's* magazine released statistics comparing average wedding costs in the United States from 1990 and 1997. In 1990, the cost of the average wedding was $15,208. The average number of wedding guests had grown to 200 in 1997 compared to 170 in 1990.

| Category | Average Cost in 1997 |
|---|---|
| Miscellaneous expenses (stationery, clergy, gifts, limousine) | $1260 |
| Bouquets and other flowers | $ 756 |
| Photography and videography | $1311 |
| Music | $ 830 |
| Engagement and wedding rings (bride and groom) | $4060 |
| Rehearsal dinner | $ 698 |
| Bride's wedding dress and headpiece | $ 989 |
| Bridal attendants' dresses (average of 5 bridesmaids) | $ 790 |
| Mother of the bride's apparel | $ 231 |
| Men's formalwear (ushers, best man, groom) | $ 544 |
| Wedding reception | $7635 |
| Grand Total | |

Source: *Bride's* 1997 Millennium Report: Wedding Love & Money.

1. What is the grand total of expenses shown in the chart?
2. How much more expensive was a wedding in 1997 compared to 1990?
3. The groom pays for the bouquets and flowers, the rehearsal dinner, the bride's engagement and wedding rings ($3500), the clergy ($232), and the groom's formalwear ($95). What is the total amount spent by the groom?
4. If you budgeted $38 per person for the wedding reception and you invited 200 guests to a wedding in 2002, how much money would you have spent compared to the 1997 wedding reception costs?
5. If you budget $5000 for the reception and the caterer charges $38 per person, how many guests can you invite? How much of your budget is left over?
6. If you budget $8500 for the reception and the caterer charges $38 per person, how many guests can you invite? How much of your budget is left over?
7. What type of arithmetic problem did you work to get the answers to Problems 6 and 7? What is the mathematical term for the "left over" budget?
8. What is the average cost of each bridesmaid's dress? When you solved this problem, which number was the dividend? the divisor? the quotient? Rewrite the division as a related multiplication.

## 1.7 Exercises

*Divide. See Examples 1 and 2.*

1. (a) $14 \div 2$
   (b) $-14 \div -2$
   (c) $14 \div -2$
   (d) $-14 \div 2$

2. (a) $-18 \div -3$
   (b) $18 \div 3$
   (c) $-18 \div 3$
   (d) $18 \div -3$

3. (a) $-42 \div 6$
   (b) $-42 \div -6$
   (c) $42 \div -6$
   (d) $42 \div 6$

4. (a) $45 \div 5$
   (b) $45 \div -5$
   (c) $-45 \div -5$
   (d) $-45 \div 5$

5. (a) $\dfrac{35}{35}$
   (b) $\dfrac{35}{1}$
   (c) $\dfrac{-13}{1}$
   (d) $\dfrac{-13}{-13}$

6. (a) $\dfrac{-23}{1}$
   (b) $\dfrac{-23}{-23}$
   (c) $\dfrac{17}{1}$
   (d) $\dfrac{17}{17}$

7. (a) $\dfrac{0}{50}$
   (b) $\dfrac{50}{0}$
   (c) $\dfrac{-11}{0}$
   (d) $\dfrac{0}{-11}$

8. (a) $\dfrac{-85}{0}$
   (b) $\dfrac{0}{-85}$
   (c) $\dfrac{6}{0}$
   (d) $\dfrac{0}{6}$

9. $\dfrac{-8}{2}$

10. $\dfrac{-14}{7}$

11. $\dfrac{21}{-7}$

12. $\dfrac{30}{-6}$

13. $\dfrac{-54}{-9}$

14. $\dfrac{-48}{-6}$

15. $\dfrac{55}{-5}$

16. $\dfrac{70}{-7}$

17. $\dfrac{-28}{0}$

18. $\dfrac{-40}{0}$

19. $\dfrac{14}{-1}$

20. $\dfrac{25}{-1}$

21. $\dfrac{-20}{-2}$

22. $\dfrac{-80}{-4}$

23. $\dfrac{-48}{-12}$

24. $\dfrac{-30}{-15}$

25. $\dfrac{-18}{18}$

26. $\dfrac{50}{-50}$

27. $\dfrac{0}{-9}$

28. $\dfrac{0}{-4}$

29. $\dfrac{-573}{-3}$

30. $\dfrac{-580}{-5}$

31. $\dfrac{163{,}672}{-328}$

32. $\dfrac{-69{,}496}{1022}$

**Simplify.** See Example 3.

33. $^-60 \div 10 \div {}^-3$

34. $36 \div {}^-4 \div 3$

35. $^-64 \div {}^-8 \div {}^-2$

36. $^-72 \div {}^-9 \div {}^-4$

37. $100 \div {}^-5 \cdot {}^-2$

38. $^-80 \div 4 \cdot {}^-5$

39. $48 \div 3 \cdot (12 \div {}^-4)$

40. $^-2 \cdot ({}^-3 \cdot {}^-7) \div 7$

41. $^-5 \div {}^-5 \cdot {}^-10 \div {}^-2$

42. $^-9 \cdot 4 \div {}^-36 \cdot 50$

43. $64 \cdot 0 \div {}^-8 \cdot 10$

44. $^-88 \div {}^-8 \div {}^-11 \cdot 0$

### RELATING CONCEPTS (Exercises 45–50)   FOR INDIVIDUAL OR GROUP WORK

*Use your knowledge of the properties of multiplication as you* **work Exercises 45–50 in order.**

45. Explain whether or not division is commutative like multiplication. Start by doing these two divisions on your calculator: $2 \div 1$ and $1 \div 2$.

46. Explain whether or not division is associative like multiplication. Start by doing these two divisions: $(12 \div 6) \div 2$ and $12 \div (6 \div 2)$.

47. Explain what is different and what is similar about multiplying and dividing two signed numbers.

48. In your own words, describe at least three division properties. Include examples to illustrate each property.

49. Write three numerical examples for each situation.
    (a) A negative number divided by $^-1$
    
    (b) A positive number divided by $^-1$
    
    Now write a rule that explains what happens when you divide a signed number by $^-1$.

50. Explain why $\dfrac{0}{-3}$ and $\dfrac{-3}{0}$ do not give the same result.

*Solve these application problems by using addition, subtraction, multiplication, or division. First use front end rounding to estimate the answer. Then find the exact answer. See Example 4.*

**51.** The greatest ocean depth is 36,198 feet below sea level. (*Source: Factfinder.*) If an unmanned research sub dives to that depth in 18 equal steps, how far does it dive in each step?

*Estimate:*

*Exact:*

**52.** Our college enrollment dropped by 3245 students over the last 11 years. What was the average drop in enrollment each year?

*Estimate:*

*Exact:*

**53.** When Ashwini discovered that her checking account was overdrawn by $238, she quickly transferred $450 from her savings to her checking account. What is the new balance in her checking account?

*Estimate:*

*Exact:*

**54.** The Tigers offensive team lost a total of 48 yards during the first half of the football game. During the second half they gained 191 yards. How many yards did they gain or lose during the entire game?

*Estimate:*

*Exact:*

**55.** The foggiest place in the United States is Cape Disappointment, Washington. It is foggy there an average of 106 days each year. (*Source:* National Weather Service.) How many days is it not foggy each year?

*Estimate:*

*Exact:*

**56.** The number of cellular phone users worldwide in 1993 was 34 million. The number is expected to reach 298 million users by 2001. (*Source:* MNSCU *Performance.*) What increase in the number of users is expected during this 8-year period?

*Estimate:*

*Exact:*

**57.** A plane descended an average of 730 feet each minute during a 37-minute landing. How far did the plane descend during the landing?

*Estimate:*

*Exact:*

**58.** A discount store found that 174 items were lost to shoplifting last month. The average value of each item was $24. What was the total loss due to shoplifting?

*Estimate:*

*Exact:*

**59.** Mr. and Mrs. Martinez drove on the Interstate for five hours and traveled 315 miles. What was the average number of miles they drove each hour?

*Estimate:*

*Exact:*

**60.** Rochelle has a 48-month car loan for $9072. How much is her monthly payment?

*Estimate:*

*Exact:*

*Find the exact answer in Exercises 61–66. Solving these problems requires more than one step.*

**61.** Clarence bowled four games and had scores of 143, 190, 162, and 177. What was his average score? (*Hint:* To find the average, add all the scores and divide by the number of scores.)

**62.** Sheila kept track of her grocery expenses for six weeks. The amounts she spent were $84, $111, $82, $110, $98, and $79. What was the average weekly cost of her groceries?

**63.** On the back of an oatmeal box it says that one serving weighs 40 grams and that there are 13 servings in the box. On the front of the box it says that the weight of the contents is 510 grams. What is the difference in the total weight on the front and the back of the box? (*Source:* Quaker Oats.)

**64.** A 2000-calorie-per-day diet recommends that you eat no more than 65 grams of fat. If each gram of fat is 9 calories, how many calories can you consume in other types of food? (*Source:* U.S. Dept. of Agriculture.)

**65.** Stephanie had $302 in her checking account. She wrote a $116 check for day care and a $548 check for rent. She also deposited her $347 paycheck. What is the balance in her account?

**66.** Gary started a new checking account with a $500 deposit. The bank charged him $18 to print his checks. He also wrote a $193 check for car repairs and a $289 check to his credit card company. What is the balance in his account?

*Divide; then interpret the remainder in each application. See Example 5.*

**67.** A cellular phone company is offering 1000 free minutes of air time to new subscribers. How many hours of free time will a new subscriber receive?

**68.** Nikki is catering a large party. If one pie will serve eight guests, how many pies should she make for 100 guests?

**69.** Hurricane victims are being given temporary shelter in a hotel. Each room can hold five people. How many rooms are needed for 163 people?

**70.** A college has received a $250,000 donation to be used for scholarships. How many $3500 scholarships can be given to students?

*Simplify.*

**71.** $|{-8}| \div {-4} \cdot |{-5}| \cdot |1|$

**72.** ${-6} \cdot |{-3}| \div |9| \cdot {-2}$

**73.** ${-6} \cdot {-8} \div ({-5} - {-5})$

**74.** ${-9} \div {-9} \cdot ({-9} \div 9) \div (12 - 13)$

**75.** Look back at the first page of this chapter. You guessed how many days it would take to receive a million dollars if you got $1 each second. Here's how to use your calculator to get the answer. If you get $1 per second, it would take 1,000,000 seconds to receive $1,000,000. Press these keys.

1000000 ÷ 60 = 16666.66667 ÷ 60 = 277.7777778 ÷ 24 = 11.57407407

There are 60 seconds in one minute.
About 16,667 minutes
There are 60 minutes in one hour.
About 278 hours
There are 24 hours in one day.
About $11\frac{1}{2}$ days (11.5 is equivalent to $11\frac{1}{2}$.)

Notice that you do *not* have to re-enter the intermediate answers. When the answer 16666.66667 appears on your calculator display, just go ahead and enter ÷ 60.

Now use your calculator to find how long it takes to receive a *billion* dollars. Start by entering 1000000000. Then follow the pattern shown above. You will need to do one more division step to get the number of years. (Assume that there are 365 days in one year.)

## 1.8 Exponents and Order of Operations

**OBJECTIVES**
1. Use exponents to write repeated factors.
2. Simplify expressions containing exponents.
3. Use the order of operations.
4. Simplify expressions with fraction bars.

*Study Skills Workbook* Activity 4

**1** Use exponents to write repeated factors. An **exponent** is a quick way to write repeated multiplication. Here is an example.

$$2 \cdot 2 \cdot 2 \cdot 2 \cdot 2 \quad \text{can be written} \quad 2^5 \leftarrow \text{Exponent}$$
$$\uparrow$$
$$\text{Base}$$

The *base* is the number being multiplied over and over, and the exponent tells how many times to use the number as a factor. This is called *exponential notation* or *exponential form*.

To simplify $2^5$, actually do the multiplication.

$$2^5 = 2 \cdot 2 \cdot 2 \cdot 2 \cdot 2 = 32$$

Here are some more examples, using 2 as the base.

$2 = 2^1$ is read "2 to the **first power**."
$2 \cdot 2 = 2^2$ is read "2 to the **second power**" or, more commonly, "2 **squared**."
$2 \cdot 2 \cdot 2 = 2^3$ is read "2 to the **third power**" or, more commonly, "2 **cubed**."
$2 \cdot 2 \cdot 2 \cdot 2 = 2^4$ is read "2 to the **fourth power**."
$2 \cdot 2 \cdot 2 \cdot 2 \cdot 2 = 2^5$ is read "2 to the **fifth power**."

and so on.

We usually don't write an exponent of 1, so if no exponent is shown, you can assume that it is 1. For example, 6 is actually $6^1$, and 4 is actually $4^1$.

**NOTE**
Exponents can also be negative numbers or 0, for example, $2^{-3}$ and $2^0$. You will learn more about these exponents in **Chapter 10**.

**Example 1** Using Exponents

Rewrite each multiplication using exponents. Also indicate how to read the exponential form.

(a) $5 \cdot 5 \cdot 5$ can be written as $5^3$, which is read "5 cubed" or "5 to the third power."

(b) $4 \cdot 4$ can be written as $4^2$, which is read "4 squared" or "4 to the second power."

(c) 7 can be written as $7^1$, which is read "7 to the first power."

**Work Problem 1 at the Side.**

**2** Simplify expressions containing exponents. Exponents are also used with signed numbers, as shown below.

$(^-3)^2 = {^-3} \cdot {^-3} = 9$    The factors have the same sign, so the product is positive.

$(^-4)^3 = \underbrace{{^-4} \cdot {^-4}}_{16} \cdot {^-4}$    Multiply two numbers at a time.

$\underbrace{16 \cdot {^-4}}_{^-64}$    First, $^-4 \cdot {^-4}$ is positive 16.

Then, $16 \cdot {^-4}$ is $^-64$.

**1** Write each multiplication using exponents. Indicate how to read the exponential form.

(a) $3 \cdot 3 \cdot 3 \cdot 3$

(b) $6 \cdot 6$

(c) 9

(d) $2 \cdot 2 \cdot 2 \cdot 2 \cdot 2 \cdot 2$

**ANSWERS**
1. (a) $3^4$ is read "3 to the fourth power."
   (b) $6^2$ is read "6 squared" or "6 to the second power."
   (c) $9^1$ is read "9 to the first power."
   (d) $2^6$ is read "2 to the sixth power."

**2** Simplify.

(a) $(-2)^3$

(b) $(-6)^2$

(c) $2^4 \cdot (-3)^2$

(d) $(-4)^2 \cdot 3^3$

Simplify exponents before you do other multiplications, as shown below. Notice that the exponent applies only to the *first* thing to its *left*.

Exponent applies only to the 2.    $2^3 \cdot 5 \cdot 4^2$    Exponent applies only to the 4.

$2 \cdot 2 \cdot 2$ is 8. → $8 \cdot 5 \cdot 16$ ← $4 \cdot 4$ is 16.

$40 \cdot 16$

$640$

### Example 2  Using Exponents with Negative Numbers

Simplify.

(a) $(-5)^2 = -5 \cdot -5 = 25$

(b) $(-5)^3 = -5 \cdot -5 \cdot -5$
$\phantom{(-5)^3 =} \underbrace{25} \cdot -5$
$\phantom{(-5)^3 =} -125$

(c) $(-2)^4 = -2 \cdot -2 \cdot -2 \cdot -2 = 16$

(d) $(-3)^2 \cdot 2^3 = \underbrace{-3 \cdot -3} \cdot \underbrace{2 \cdot 2 \cdot 2}$
$\phantom{(-3)^2 \cdot 2^3 =} 9 \cdot 8$
$\phantom{(-3)^2 \cdot 2^3 =} 72$

**Calculator Tip**  On a *scientific* calculator, use the exponent key $y^x$ to enter exponents. To enter $5^8$, press the following keys.

5  $y^x$  8  =    Answer is 390,625.
↑        ↑
Base   Exponent

Be careful when using your calculator's exponent key with a negative number, such as $(-5)^3$. Different calculators use different keystrokes, so check the instruction manual, or experiment to see how your calculator works.

**Work Problem** ❷ **at the Side.**

**3**   **Use the order of operations.**   In **Sections 1.4** and **1.7** you worked examples that mixed either addition and subtraction or multiplication and division. In those situations you worked from left to right. Example 3 is a review.

### Example 3  Working from Left to Right

Simplify.

(a) $-8 - -6 + -11$   Do additions and subtractions from left to right.
$\phantom{(a)} \underbrace{-2} + -11$
$\phantom{(a)} -13$

*Continued on Next Page*

**A**NSWERS

2. (a) $-2 \cdot -2 \cdot -2 = -8$
   (b) $-6 \cdot -6 = 36$
   (c) $16 \cdot 9 = 144$
   (d) $16 \cdot 27 = 432$

**(b)** $\underbrace{\overline{\phantom{1}}15 \div \overline{\phantom{1}}3}_{5} \cdot 6$  Do multiplications and divisions from left to right.

$\underbrace{\phantom{xx}5\phantom{xx} \cdot 6}_{30}$

===== Work Problem ❸ at the Side. =====

Now we're ready to do problems that use a mix of the four operations, parentheses, and exponents. Let's start with a simple example: 4 + 2 • 3.

| If we work from left to right | If we multiply first |
|---|---|
| $\underbrace{4 + 2}_{6} \cdot 3$ | $4 + \underbrace{2 \cdot 3}_{6}$ |
| $\underbrace{6 \cdot 3}_{18}$ | $\underbrace{4 + 6}_{10}$ |

To be sure that everyone gets the same answer to a problem like this, mathematicians have agreed to do things in a certain order. The following order of operations shows that multiplying is done ahead of adding, so the correct answer is 10.

### Order of Operations

*Step 1*   Work inside **parentheses** or **other grouping symbols.**

*Step 2*   Simplify expressions with **exponents.**

*Step 3*   Do the remaining **multiplications and divisions** as they occur from left to right.

*Step 4*   Do the remaining **additions and subtractions** as they occur from left to right.

**Calculator Tip**   Enter the example above in your calculator.

$$4 \; \oplus \; 2 \; \otimes \; 3 \; \circledcirc$$

Which answer do you get? If you have a scientific calculator, it automatically uses the order of operations and multiplies first to get the correct answer of 10. Some standard, four-function calculators may *not* have the order of operations built into them and will give the *incorrect* answer of 18.

❸ Simplify.

(a) $\overline{\phantom{1}}9 + \overline{\phantom{1}}15 - 3$

(b) $\overline{\phantom{1}}4 - 2 + \overline{\phantom{1}}6$

(c) $3 \cdot \overline{\phantom{1}}4 \div \overline{\phantom{1}}6$

(d) $\overline{\phantom{1}}18 \div 9 \cdot \overline{\phantom{1}}4$

ANSWERS

3. (a) $\overline{\phantom{1}}27$   (b) $\overline{\phantom{1}}12$   (c) 2   (d) 8

④ Simplify.

(a) $8 + (14 \div 2) \cdot 6$

(b) $4(1) + 8(9 - 2)$

(c) $3(5 + 1) + 20 \div 4$

### Example 4  Using the Order of Operations with Whole Numbers

Simplify.

$9 + 3(20 - 4) \div 8$ — Work inside parentheses first: $20 - 4$ is 16. Bring down the other numbers and signs that you haven't used.

$9 + 3(16) \div 8$ — Look for exponents: none. Move from left to right, looking for multiplying and dividing.

$9 + 3(16) \div 8$ — Yes, here is multiplying: $3(16)$ is 48.

$9 + 48 \div 8$ — Here is dividing. $48 \div 8$ is 6. There is no other multiplying or dividing, so look for adding and subtracting.

$9 + 6$ — Add last: $9 + 6$ is 15.

$15$

**Work Problem ④ at the Side.**

### Example 5  Using the Order of Operations with Integers

Simplify.

(a) $\;^-8 \div (7 - 5) - 9$ — Work inside parentheses first: $7 - 5$ is 2. Bring down the other numbers and signs you haven't used.

$\;^-8 \div (2) - 9$ — Look for exponents: none. Move from left to right, looking for multiplying and dividing.

$\;^-8 \div (2) - 9$ — Here is dividing: $\;^-8 \div 2$ is $\;^-4$. No other multiplying or dividing, so look for adding and subtracting.

$\;^-4 - 9$ — Change subtracting to adding. Change 9 to its opposite.

$\;^-4 + \;^-9$ — Add $\;^-4 + \;^-9$.

$\;^-13$

**Continued on Next Page**

---

ANSWERS

4. (a) 50  (b) 60  (c) 23

(b) $3 + 2(6 - 8) \cdot (15 \div 3)$   Work inside first set of parentheses.
Change $6 - 8$ to $6 + {}^-8$ to get ${}^-2$.

$3 + 2\ ({}^-2)\ \cdot\ (15 \div 3)$   Work inside second set of parentheses:
$15 \div 3$ is $5$.

$3 + 2\ ({}^-2)\ \cdot\ 5$   Multiply and divide from left to right.
First multiply $2({}^-2)$ to get ${}^-4$.
$3 +\ {}^-4\ \cdot\ 5$   Then multiply ${}^-4 \cdot 5$ to get ${}^-20$.

$3 +\ {}^-20$   Add last: $3 + {}^-20$ is ${}^-17$.

${}^-17$

=== Work Problem ❺ at the Side.

**⑤** Simplify.

(a) $2 + 40 \div ({}^-5 + 3)$

(b) ${}^-5 \cdot 5 - (15 + 5)$

(c) $({}^-24 \div 2) + (15 - 3)$

(d) ${}^-3(2 - 8) - 5 \cdot (4 - 3)$

(e) $3 \cdot 3 - (10 \cdot 3) \div 5$

(f) $6 - (2 + 7) \div ({}^-4 + 1)$

### Example 6  Using the Order of Operations with Exponents

Simplify.

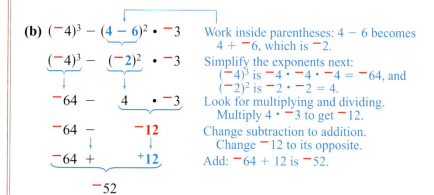

(a) $4^2 - ({}^-3)^2$   The only parentheses are around ${}^-3$, but there is no work to do inside these parentheses.

$4^2 - ({}^-3)^2$   Simplify the exponents: $4^2 = 4 \cdot 4 = 16$, and $({}^-3)^2 = {}^-3 \cdot {}^-3 = 9$.
$16 -\ 9$   There is no multiplying or dividing, so add and subtract: $16 - 9$ is $7$.
$7$

(b) $({}^-4)^3 - (4 - 6)^2 \cdot {}^-3$   Work inside parentheses: $4 - 6$ becomes $4 + {}^-6$, which is ${}^-2$.

$({}^-4)^3 - ({}^-2)^2 \cdot {}^-3$   Simplify the exponents next:
$({}^-4)^3$ is ${}^-4 \cdot {}^-4 \cdot {}^-4 = {}^-64$, and
$({}^-2)^2$ is ${}^-2 \cdot {}^-2 = 4$.

${}^-64 -\ 4\ \cdot\ {}^-3$   Look for multiplying and dividing.
Multiply $4 \cdot {}^-3$ to get ${}^-12$.

${}^-64 -\ {}^-12$   Change subtraction to addition.
Change ${}^-12$ to its opposite.

${}^-64 +\ {}^+12$   Add: ${}^-64 + 12$ is ${}^-52$.

${}^-52$

**⑥** Simplify.

(a) $2^3 - 3^2$

(b) $6^2 \div ({}^-4) \cdot ({}^-3)$

(c) $({}^-4)^2 - 3^2 \cdot (5 - 2)$

(d) $({}^-3)^3 + (3 - 9)^2$

### CAUTION

To help in remembering the order of operations, you may have memorized the letters **PEMDAS**, or the phrase "Please Excuse My Dear Aunt Sally."

**P**lease   **E**xcuse   **M**y   **D**ear   **A**unt   **S**ally

**P**arentheses; **E**xponents; **M**ultiply and **D**ivide; **A**dd and **S**ubtract

Be careful! Do *not* automatically do all multiplication before division. Multiplying and dividing are done *from left to right* (after parentheses and exponents).

Work Problem ❻ at the Side.

**ANSWERS**
5. (a) ${}^-18$  (b) ${}^-45$  (c) $0$  (d) $13$
   (e) $3$  (f) $9$
6. (a) ${}^-1$  (b) $27$  (c) ${}^-11$  (d) $9$

**7** Simplify.

(a) $\dfrac{-3(2^3)}{-10 - 6 + 8}$

(b) $\dfrac{(-10)(-5)}{-6 \div 3 \cdot 5}$

(c) $\dfrac{6 + 18 \div (-2)}{(1 - 10) \div 3}$

(d) $\dfrac{6^2 - 3^2 \cdot 4}{5 + (3 - 7)^2}$

**4** **Simplify expressions with fraction bars.** A fraction bar indicates division, as in $\dfrac{-6}{2}$, which means $-6 \div 2$. In an expression such as

$$\dfrac{-5 + 3^2}{16 - 7(2)}$$

the fraction bar also acts as a grouping symbol, like parentheses. It tells us to do the work in the numerator (above the bar) and then the work in the denominator (below the bar). The last step is to divide the results.

$$\dfrac{-5 + 3^2}{16 - 7(2)} \longrightarrow \dfrac{-5 + 9}{16 - 14} \longrightarrow \dfrac{4}{2} \longrightarrow 4 \div 2 = 2$$

The final result is 2.

**Example 7** Using the Order of Operations with Fraction Bars

Simplify $\dfrac{-8 + (4 - 6) \cdot 5}{4 - 4^2 \div 8}$.

First do the work in the numerator.

$-8 + \underline{(4 - 6)} \cdot 5$    Work inside the parentheses.

$-8 + \underline{-2 \cdot 5}$    Multiply.

$\underline{-8 + (-10)}$    Add.

Numerator $\longrightarrow -18$

Now do the work in the denominator.

$4 - \underline{4^2} \div 8$    There are no parentheses; simplify the exponent.

$4 - \underline{16 \div 8}$    Divide.

$\underline{4 - 2}$    Subtract.

Denominator $\longrightarrow 2$

The last step is the division.

Numerator $\longrightarrow \dfrac{-18}{2} = -9$
Denominator $\longrightarrow$

**Work Problem** ❼ **at the Side.**

---

**ANSWERS**

7. (a) $\dfrac{-24}{-8} = 3$   (b) $\dfrac{50}{-10} = -5$

(c) $\dfrac{-3}{-3} = 1$   (d) $\dfrac{0}{21} = 0$

## 1.8 Exercises

*Complete this table. See Example 1.*

| | Exponential Form | Factored Form | Simplified | Read as |
|---|---|---|---|---|
| 1. | $4^3$ | | 64 | |
| 2. | $10^2$ | | 100 | |
| 3. | | 2 · 2 · 2 · 2 · 2 · 2 · 2 | | |
| 4. | | 3 · 3 · 3 · 3 · 3 | | |
| 5. | | 5 · 5 · 5 · 5 | | |
| 6. | | 2 · 2 · 2 · 2 · 2 · 2 | | |
| 7. | | | | 7 squared |
| 8. | | | | 6 cubed |
| 9. | | | | 10 to the first power |
| 10. | | | | 4 to the fourth power |

*Simplify. See Examples 1 and 2.*

11. (a) $10^1$
    (b) $10^2$
    (c) $10^3$
    (d) $10^4$

12. (a) $5^1$
    (b) $5^2$
    (c) $5^3$
    (d) $5^4$

13. (a) $4^1$
    (b) $4^2$
    (c) $4^3$
    (d) $4^4$

14. (a) $3^1$
    (b) $3^2$
    (c) $3^3$
    (d) $3^4$

15. $5^{10}$

16. $4^9$

17. $2^{12}$

18. $3^{10}$

19. $(-2)^2$

20. $(-4)^2$

21. $(-5)^2$

22. $(-10)^2$

23. $(-4)^3$

24. $(-2)^3$

25. $(-3)^4$

26. $(-2)^4$

27. $(-10)^3$

28. $(-5)^3$

29. $1^4$

30. $1^5$

**31.** $3^3 \cdot 2^2$  **32.** $4^2 \cdot 5^2$  **33.** $(^-5)^2 \cdot 3^2$  **34.** $3^2 \cdot (^-2)^2$

**35.** $(^-5)^3 \cdot 6^1$  **36.** $7^1 \cdot (^-4)^3$  **37.** $(^-2)^4 \cdot {}^-2$  **38.** $^-6 \cdot (^-6)^2$

**39.** Simplify.

$(^-2)^2 = $ _____   $(^-2)^6 = $ _____
$(^-2)^3 = $ _____   $(^-2)^7 = $ _____
$(^-2)^4 = $ _____   $(^-2)^8 = $ _____
$(^-2)^5 = $ _____   $(^-2)^9 = $ _____

(a) Describe the pattern you see in the signs of the answers.

(b) What would be the sign of $(^-2)^{15}$ and the sign of $(^-2)^{24}$?

**40.** Explain why it is important to have rules for the order of operations. Why do you think our "natural instinct" is to just work from left to right?

*Simplify. See Examples 3–7.*

**41.** $6 + 3 \cdot {}^-4$  **42.** $10 - 30 \div 2$  **43.** $^-1 + 15 + {}^-7 \cdot 2$

**44.** $9 + {}^-5 + 2 \cdot {}^-2$  **45.** $10 - 7^2$  **46.** $5 - 5^2$

**47.** $2 - {}^-5 + 3^2$  **48.** $6 - {}^-9 + 2^3$  **49.** $3 + 5(6 - 2)$

**50.** $4 + 3(8 - 3)$  **51.** $^-7 + 6(8 - 14)$  **52.** $^-3 + 5(9 - 12)$

**53.** $2(^-3 + 5) - (9 - 12)$  **54.** $3(2 - 7) - (^-5 + 1)$  **55.** $^-5(7 - 13) \div {}^-10$

**56.** $^-4(9-17) \div {}^-8$

**57.** $9 \div (^-3)^2 + {}^-1$

**58.** $^-48 \div (^-4)^2 + 3$

**59.** $2 - {}^-5 \cdot (^-2)^3$

**60.** $1 - {}^-10 \cdot (^-3)^3$

**61.** $^-2(^-7) + 3(9)$

**62.** $4(^-2) + {}^-3(^-5)$

**63.** $30 \div {}^-5 - 36 \div {}^-9$

**64.** $8 \div {}^-4 - 42 \div {}^-7$

**65.** $2 \cdot 5 - 3 \cdot 4 + 5 \cdot 3$

**66.** $9 \cdot 3 - 6 \cdot 4 + 3 \cdot 7$

**67.** $4 \cdot 3^2 + 7(3+9) - {}^-6$

**68.** $5 \cdot 4^2 - 6(1+4) - {}^-3$

**69.** $(^-4)^2 \cdot (7-9)^2 \div 2^3$

**70.** $(^-5)^2 \cdot (9-17)^2 \div (^-10)^2$

**71.** $\dfrac{^-1 + 5^2 - {}^-3}{^-6 - 9 + 12}$

**72.** $\dfrac{^-6 + 3^2 - {}^-7}{7 - 9 - 3}$

73. $\dfrac{{}^-2 \cdot 4^2 - 4(6 - 2)}{{}^-4(8 - 13) \div {}^-5}$

74. $\dfrac{3 \cdot 3^2 - 5(9 - 2)}{8(6 - 9) \div {}^-3}$

75. $\dfrac{2^3 \cdot ({}^-2 - 5) + 4({}^-1)}{4 + 5({}^-6 \cdot 2) + (5 \cdot 11)}$

76. $\dfrac{3^3 + ({}^-1 - 2) \cdot 4 - 25}{{}^-4 + 4(3 \cdot 5) + ({}^-6 \cdot 9)}$

77. $5^2(9 - 11)({}^-3)({}^-3)^3$

78. $4^2(13 - 17)({}^-2)({}^-2)^3$

79. $|{}^-12| \div 4 + 2 \cdot |({}^-2)^3| \div 4$

80. $6 - |2 - 3 \cdot 4| + ({}^-5)^2 \div 5^2$

81. $\dfrac{{}^-9 + 18 \div {}^-3({}^-6)}{32 - 4(12) \div 3(2)}$

82. $\dfrac{{}^-20 - 15({}^-4) - {}^-40}{14 + 27 \div 3({}^-2) - {}^-4}$

## Summary Exercises on OPERATIONS WITH INTEGERS

*Simplify each expression.*

1. $2 - 8$

2. $(^-16)(0)$

3. $^-14 - {}^-7$

4. $\dfrac{^-42}{6}$

5. $^-9(^-7)$

6. $\dfrac{^-12}{12}$

7. $(1)(^-56)$

8. $1 + {}^-23$

9. $5 - {}^-7$

10. $^-88 \div {}^-11$

11. $^-18 + 5$

12. $\dfrac{0}{^-10}$

13. $^-40 - {}^-40$

14. $^-17 + 0$

15. $8(^-6)$

16. $^-1 - 9$

17. $^-5(10)$

18. $\dfrac{30}{0}$

19. $0 - 14$

20. $\dfrac{18}{^-3}$

21. $^-13 + 13$

22. $\dfrac{^-16}{^-1}$

23. $20 - 50$

24. $\dfrac{^-7}{0}$

25. $(^-4)(^-6)(2)$

26. $^-2 + {}^-12 + {}^-5$

27. $^-60 \div 10 \div {}^-3$

28. $^-8 - 4 - 8$

29. $64 \cdot 0 \div {}^-8$

30. $2 - {}^-5 + 3^2$

31. $^-9 + 8 + {}^-2$

32. $(^-6)(^-2)(^-3)$

33. $8 + 6 + {}^-8$

34. $3^2 - 2^4$

35. $(^-5)^2 \div {}^-5$

36. $(^-2)^5 + 1^3$

**37.** $^-72 \div {}^-9 \div {}^-4$

**38.** $^-7 + 28 + {}^-56 + 3$

**39.** $9 - 6 - 3 - 5$

**40.** $^-6(^-8) \div (^-5 - 7)$

**41.** $^-1(9732)(^-1)(^-1)$

**42.** $^-80 \div 4(^-5)$

**43.** $^-10 - 4 + 0 + 18$

**44.** $^-7 \cdot |7| \cdot |^-7|$

**45.** $5 - |^-3| + 3$

**46.** $^-2(^-3)(7) \div {}^-7$

**47.** $^-3 - (^-2 + 4) - 5$

**48.** $0 - |^-7 + 2|$

**49.** $(^-4)^2(7 - 9)^2 \div 2^3$

**50.** $12 \div 4 + 2(^-2)^2 \div {}^-4$

**51.** $\dfrac{^-1(5^2) - {}^-3}{(^-3)^3 + 4^2}$

**52.** $\dfrac{^-6 \div 3(^-2) + 4}{2 + 8 \div 4 - 6}$

**53.** $\dfrac{^-9 + 24 \div (^-4)(^-6)}{32 - 4(12) \div 3(2)}$

**54.** $\dfrac{5 - |2 - 4(4)| + (^-5)^2 \div 5^2}{^-9 \div 3(2 - 2) - {}^-8}$

# Chapter 1

## SUMMARY

 *Study Skills Workbook*
Activity 5

### KEY TERMS

| | | |
|---|---|---|
| **1.1** | place value system | A place value system is a number system in which the location, or place, where a digit is written gives it a different value. |
| | digits | The 10 digits in our number system are 0, 1, 2, 3, 4, 5, 6, 7, 8, and 9. |
| | whole numbers | The whole numbers are 0, 1, 2, 3, and so on. |
| **1.2** | number line | A number line is like a thermometer turned sideways. It is used to show how numbers relate to each other. |
| | integers | Integers are the whole numbers and their opposites. |
| | absolute value | The absolute value of a number is its distance from 0. Absolute value is indicated by two vertical bars and is always positive (or 0) but never negative. |
| **1.3** | addends | In an addition problem, the numbers being added are called addends. |
| | sum | The answer to an addition problem is called the sum. |
| | addition property of 0 | Adding 0 to any number leaves the number unchanged. |
| | commutative property of addition | Changing the *order* of two addends does not change the sum. |
| | associative property of addition | Changing the *grouping* of addends does not change the sum. |
| **1.4** | opposite | The opposite of a number is the same distance from 0 on the number line but on the opposite side of 0. It is also called the *additive inverse* because a number plus its opposite equals 0. |
| **1.5** | rounding | Rounding a number means finding a number that is close to the original number but easier to work with. |
| | estimate | Use rounded numbers to get an approximate answer, or estimate. |
| | front end rounding | Front end rounding is rounding numbers to the highest possible place, so all the digits become 0 except the first digit. |
| **1.6** | factors | In a multiplication problem, the numbers being multiplied are called factors. |
| | product | The answer to a multiplication problem is called the product. |
| | multiplication property of 0 | Multiplying any number by 0 gives a product of 0. |
| | multiplication property of 1 | Multiplying a number by 1 leaves the number unchanged. |
| | commutative property of multiplication | Changing the *order* of two factors does not change the product. |
| | associative property of multiplication | Changing the *grouping* of factors does not change the product. |
| | distributive property | Multiplication distributes over addition. For example, $3(6 + 2) = 3 \cdot 6 + 3 \cdot 2$. |
| **1.7** | quotient | The answer to a division problem is called the quotient. |
| **1.8** | exponent | An exponent tells how many times a number is used as a factor in repeated multiplication. |

## Test Your Word Power

See how well you have learned the vocabulary in this chapter. Answers follow the Quick Review.

1. In $(4)(^-6) = ^-24$, the 4 and $^-6$ are called
   (a) products
   (b) factors
   (c) addends
   (d) opposites.

2. A list of **whole numbers** is
   (a) 1, 2, 3, 4, …
   (b) …, $^-4$, $^-3$, $^-2$, $^-1$, 0, 1, 2, 3, …
   (c) 0, 1, 2, 3, 4, 5, 6, 7, 8, 9
   (d) 0, 1, 2, 3, 4, … .

3. The **absolute value** of a number is
   (a) its distance from 0
   (b) never positive
   (c) used when multiplying
   (d) less than 0.

4. An **exponent**
   (a) tells how many times a number is added
   (b) is the number being multiplied
   (c) tells how many times a number is multiplied
   (d) applies only to the first thing to its right.

5. The **opposite** of a number is
   (a) never negative
   (b) called the additive inverse
   (c) called the absolute value
   (d) at the same point on the number line.

6. **Front end rounding** is rounding numbers
   (a) to the nearest thousand
   (b) so there are no zeros
   (c) to the highest possible place
   (d) so they become integers.

7. A list of **integers** is
   (a) 1, 2, 3, 4, …
   (b) …, $^-4$, $^-3$, $^-2$, $^-1$, 0, 1, 2, 3, …
   (c) 0, 1, 2, 3, 4, 5, 6, 7, 8, 9
   (d) 0, 1, 2, 3, 4, … .

8. The **associative property of multiplication** says that
   (a) changing the order of two factors does not change the product
   (b) multiplication distributes over addition
   (c) multiplying a number by 0 is undefined
   (d) changing the grouping of factors does not change the product.

## Quick Review

| Concepts | Examples |
|---|---|
| **1.1 Reading and Writing Whole Numbers**<br>Do not use the word *and* when reading whole numbers. Commas separate groups of three digits. The first few group names are ones, thousands, millions, billions, trillions. | Write 3, 008, 160 in words.<br><br>three **million**, eight **thousand**, one hundred sixty<br><br>Write this number, using digits: twenty **billion**, sixty-five **thousand**, eighteen.<br><br>2 0 , 0 0 0 , 0 6 5 , 0 1 8<br>billions    millions    thousands    ones |
| **1.2 Graphing Signed Numbers**<br>Place a dot at the correct location on the number line. | Graph: (a) $^-4$   (b) 0   (c) $^-1$   (d) $\frac{1}{2}$   (e) 2<br> |
| **1.2 Comparing Integers**<br>When comparing two integers, the one that is farther to the left on the number line is less than the other.<br>Use the < symbol for "is less than" and the > symbol for "is greater than." | Write < or > between each pair of numbers to make a true statement.<br>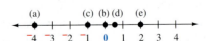<br>$^-3$ **is less than** $^-2$    0 **is greater than** $^-4$<br>because $^-3$ is to the    because 0 is to the<br>*left* of $^-2$ on the    *right* of $^-4$ on the<br>number line.    number line. |

| Concepts | Examples |
|---|---|
| **1.2 Finding the Absolute Value of a Number**<br>Find the distance on the number line from 0 to the number. The absolute value is always positive (or 0) but never negative. | Find each absolute value.<br>$\|{-5}\| = 5$  because $-5$ is 5 steps away from 0 on the number line.<br>$\|3\| = 3$  because 3 is 3 steps away from 0 on the number line. |
| **1.3 Adding Two Integers**<br>When both integers have the *same sign,* add the absolute values and use the common sign as the sign of the sum.<br>When the integers have *different signs,* subtract the smaller absolute value from the larger absolute value. Use the sign of the number with the larger absolute value as the sign of the sum. | Add.<br>(a) $-6 + {-7}$<br>Add the absolute values.<br>$\|{-6}\| = 6$   and   $\|{-7}\| = 7$<br>Add $6 + 7 = 13$ and use the common sign as the sign of the sum: $-6 + {-7} = -13$.<br>(b) $-10 + 4$<br>Subtract the smaller absolute value from the larger.<br>$\|{-10}\| = 10$   and   $\|4\| = 4$<br>Subtract $10 - 4 = 6$; the number with the larger absolute value is negative, so the sum is negative: $-10 + 4 = -6$. |
| **1.3 Using Properties of Addition**<br>*Addition Property of 0:* Adding 0 to any number leaves the number unchanged.<br>*Commutative Property of Addition:* Changing the *order* of two addends does not change the sum.<br>*Associative Property of Addition:* Changing the *grouping* of addends does not change the sum. | Name the property illustrated by each case.<br>(a) $-16 + 0 = -16$<br>(b) $4 + 10 = 10 + 4$    Both sums are 14.<br>(c) $2 + (-6 + 1) = (2 + {-6}) + 1$    Both sums are $-3$.<br>(a) Addition property of 0<br>(b) Commutative property of addition<br>(c) Associative property of addition |
| **1.4 Subtracting Two Integers**<br>To subtract two numbers, add the first number to the opposite of the second number.<br>*Step 1*  Make one pencil stroke to change the subtraction symbol to an addition symbol.<br>*Step 2*  Make a second pencil stroke to change the *second* number to its *opposite.* | Subtract.<br>(a) $7 - {-2}$     Change subtraction to addition.<br>$\phantom{7}\;\downarrow\;\downarrow$     Change $-2$ to its opposite, $+2$.<br>$7 + {+2}$<br>$\phantom{0}9$<br>(b) $-9 - 12$     Change subtraction to addition.<br>$\phantom{-9}\;\downarrow\;\downarrow$     Change 12 to its opposite, $-12$.<br>$-9 + {-12}$<br>$-21$ |

| Concepts | Examples |
|---|---|
| **1.5 Rounding Integers**<br>*Step 1* Draw a line under the place to which the number is to be rounded.<br>*Step 2* Look only at the next digit to the right of the underlined place. If the next digit is 5 or more, increase the underlined digit by 1. If the next digit is 4 or less, do not change the digit in the underlined place.<br>*Step 3* Change all digits to the right of the underlined place to zeros. | Round 36,833 to the nearest thousand.<br>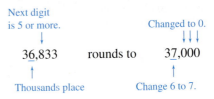<br>Round −3582 to the nearest ten.<br>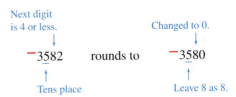 |
| **1.5 Front End Rounding**<br>Round to the highest possible place so that all the digits become 0 except the first digit. | Use front end rounding.<br>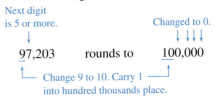 |
| **1.5 Estimating Answers in Addition and Subtraction**<br>Use front end rounding to round the numbers in a problem. Then add or subtract the rounded numbers to estimate the answer. | First use front end rounding to estimate the answer. Then find the exact answer.<br>The temperature was 48 degrees below zero. During the morning it rose 21 degrees. What was the new temperature?<br>*Estimate:* −50 + 20 = −30 degrees<br>*Exact:* −48 + 21 = −27 degrees |
| **1.6 Multiplying Two Integers**<br>If two factors have *different signs,* the product is *negative.*<br>If two factors have the *same sign,* the product is *positive.* | Multiply.<br>(a) −5(6) = −30   The factors have *different* signs, so the product is *negative.*<br>(b) (−10)(−2) = 20   The factors have the *same* sign, so the product is *positive.* |
| **1.6 Using Properties of Multiplication**<br>*Multiplication property of 0:* Multiplying any number by 0 gives a product of 0.<br>*Multiplication property of 1:* Multiplying any number by 1 leaves the number unchanged.<br>*Commutative property of multiplication:* Changing the *order* of two factors does not change the product.<br>*Associative property of multiplication:* Changing the *grouping* of factors does not change the product.<br>*Distributive property:* Multiplication distributes over addition. | Name the property illustrated by each case.<br>(a) −49 · 0 = 0<br>(b) 1(675) = 675<br>(c) −8 · 2 = 2 · −8   Both products are −16.<br>(d) (−3 · −2) · 4 = −3 · (−2 · 4)   Both products are 24.<br>(e) 5(2 + 4) = 5 · 2 + 5 · 4   Both results are 30.<br>(a) Multiplication property of 0<br>(b) Multiplication property of 1<br>(c) Commutative property of multiplication<br>(d) Associative property of multiplication<br>(e) Distributive property |

| Concepts | Examples |
|---|---|
| **1.6 Estimating Answers in Multiplication**<br>First use front end rounding. Then multiply the rounded numbers using a shortcut: Multiply the nonzero digits in each factor; count the total number of zeros in the two factors and write that number of zeros in the product. | First use front end rounding to estimate the answer. Then find the exact answer.<br><br>At a PTA fund-raiser, Lionel sold 96 photo albums at $22 each. How much money did he take in?<br><br>*Estimate:* 100 • $20 = $2000<br><br>*Exact:* 96 • $22 = $2112 |
| **1.7 Dividing Two Integers**<br>Use two same rules as for multiplying two integers. If two numbers have *different signs,* the quotient is *negative.* If two numbers have the *same sign,* the quotient is *positive.* | Divide.<br><br>$\dfrac{-24}{6} = -4$  Numbers have *different* signs, so the quotient is *negative.*<br><br>$-72 \div -8 = 9$  Numbers have the *same* sign, so the quotient is *positive.*<br><br>$\dfrac{50}{-5} = -10$  Numbers have *different* signs, so the quotient is *negative.* |
| **1.7 Using Properties of Division**<br>(a) When a nonzero number is divided by itself, the quotient is 1.<br>(b) When a number is divided by 1, the quotient is the number.<br>(c) When 0 is divided by any other number (except 0), the quotient is 0.<br>(d) Division by 0 is *undefined*. There is no answer. | State the property illustrated by each case.<br><br>(a) $\dfrac{-4}{-4} = 1$   (b) $\dfrac{65}{1} = 65$<br><br>(c) $\dfrac{0}{9} = 0$   (d) $\dfrac{-10}{0}$ is undefined.<br><br>The examples are in the same order as the properties listed at the left. Note that division is *not* commutative or associative. |
| **1.7 Estimating Answers in Division**<br>First use front end rounding. Then divide the rounded numbers, using a shortcut: Drop the same number of zeros in both the divisor and the dividend. | First use front end rounding to estimate the answer. Then find the exact answer.<br><br>Joan has one year to pay off a $1020 loan. What is her monthly payment?<br><br>*Estimate:* $1000 ÷ 10 = $100<br><br>*Exact:* $1020 ÷ 12 = $85 |
| **1.7 Interpreting Remainders in Division**<br>In some situations the remainder tells you how much is left over. In other situations, you must increase the quotient by 1 in order to accommodate the "left over." | Divide; then interpret the remainder.<br><br>Each chemistry student needs 35 milliliters of acid for an experiment. How many students can be served from a bottle holding 500 milliliters of acid?<br><br>$\phantom{35)}\,\,\underline{\mathbf{14}}$ → 14 students served<br>$35\overline{)500}$<br>$\phantom{35)}\,\,\underline{490}$<br>$\phantom{35)500}\,\,\mathbf{10}$ → 10 milliliters of acid left over |

| Concepts | Examples |
|---|---|
| **1.8 Using Exponents**<br>An exponent tells how many times a number is used as a factor in repeated multiplication. An exponent applies only to its base (the first thing to the left of the exponent). | Simplify.<br>Exponent ↓<br>(a) $2^5 = 2 \cdot 2 \cdot 2 \cdot 2 \cdot 2 = 32$<br>(b) $(^-3)^2 = {}^-3 \cdot {}^-3 = 9$ |
| **1.8 Order of Operations**<br>Mathematicians have agreed to follow this order.<br>*Step 1* Work inside parentheses or other grouping symbols.<br>*Step 2* Simplify expressions with exponents.<br>*Step 3* Do the remaining multiplications and divisions as they occur from left to right.<br>*Step 4* Do the remaining additions and subtractions as they occur from left to right. | Simplify.<br>$(^-2)^4 + 3(^-4 - {}^-2)$    Work inside parentheses.<br>$(^-2)^4 + 3(^-2)$    Simplify exponents.<br>$16 + 3(^-2)$    Multiply.<br>$16 + {}^-6$    Add.<br>$10$ |
| **1.8 Using the Order of Operations with Fraction Bars**<br>When there is a fraction bar, do all the work in the numerator. Then do all the work in the denominator. Finally, divide numerator by denominator. | Simplify.<br>$\dfrac{^-10 + 4^2 - 6}{2 + 3(1 - 4)} = \dfrac{^-10 + 16 - 6}{2 + 3(^-3)} = \dfrac{0}{^-7} = 0$ |

> **ANSWERS TO TEST YOUR WORD POWER**
>
> 1. **(b)** *Example:* In $7(10) = 70$, the factors are 7 and 10.
> 2. **(d)** *Example:* 12, 0, 710, and 89,475 are all whole numbers.
> 3. **(a)** *Example:* $|{}^-3| = 3$ and $|3| = 3$ because both $^-3$ and 3 are 3 steps away from 0.
> 4. **(c)** *Example:* In $2^5$, the exponent is 5, which indicates that 2 is multiplied 5 times, so $2^5 = 2 \cdot 2 \cdot 2 \cdot 2 \cdot 2 = 32$.
> 5. **(b)** *Example:* The opposite of 5 is $^-5$; it is the additive inverse because $5 + (^-5) = 0$.
> 6. **(c)** *Example:* Using front end rounding, round 48,299 to the highest possible place, which is ten thousands. So, 48,299 rounds to 50,000. All digits are 0 except the first digit.
> 7. **(b)** *Example:* $^-10$, 0, and 6 are all integers.
> 8. **(d)** *Example:* $4 \cdot (^-2 \cdot 3)$ can be rewritten as $(4 \cdot {}^-2) \cdot 3$; both products are $^-24$.

# Chapter 1

## REVIEW EXERCISES

If you need help with any of these Review Exercises, look in the section indicated in the red brackets.

[1.1] **1.** Circle the whole numbers:    86    2.831    ⁻4    0    $\frac{2}{3}$    35,600

*Write these numbers in words.*

**2.** 806

**3.** 319,012

**4.** 60,003,200

**5.** 15,749,000,000,006

*Write these numbers using digits.*

**6.** Five hundred four thousand, one hundred

**7.** Six hundred twenty million, eighty thousand

**8.** Ninety-nine billion, seven million, three hundred fifty-six

[1.2] **9.** Graph these numbers: $-3\frac{1}{2}$, 2, ⁻5, 0.

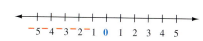

*Write < or > between each pair of numbers to make a true statement.*

**10.** 0 ___ ⁻4      **11.** ⁻3 ___ ⁻1      **12.** 2 ___ ⁻2      **13.** ⁻2 ___ 1

*Find each absolute value.*

**14.** |⁻5|      **15.** |9|      **16.** |0|      **17.** |⁻125|

[1.3] *Add.*

**18.** ⁻9 + 8      **19.** ⁻8 + ⁻5      **20.** 16 + ⁻19      **21.** ⁻4 + 4

**22.** 6 + ⁻5      **23.** ⁻12 + ⁻12      **24.** 0 + ⁻7      **25.** ⁻16 + 19

**26.** 9 + ⁻4 + ⁻8 + 3      **27.** ⁻11 + ⁻7 + 5 + ⁻4

**[1.4]** *Find the opposite (additive inverse) of each number. Show that the sum of the number and its opposite is 0.*

28. ⁻5

29. 18

*Subtract by changing subtraction to addition.*

30. 5 − 12

31. 24 − 7

32. ⁻12 − 4

33. 4 − ⁻9

34. ⁻12 − ⁻30

35. ⁻8 − 14

36. ⁻6 − ⁻6

37. ⁻10 − 10

38. ⁻8 − ⁻7

39. 0 − 3

40. 1 − ⁻13

41. 15 − 0

*Simplify.*

42. 3 − 12 − 7

43. ⁻7 − ⁻3 + 7

44. 4 + ⁻2 − 0 − 10

45. ⁻12 − 12 + 20 − ⁻4

**[1.5]** *Round each number as indicated.*

46. 205 to the nearest ten

47. 59,499 to the nearest thousand

48. 85,066,000 to the nearest million

49. ⁻2963 to the nearest hundred

50. ⁻7,063,885 to the nearest ten thousand

51. 399,712 to the nearest thousand

*Use front end rounding to round each number.*

52. The combined weight loss of 10 dieters was 197 pounds.

53. The land on the shore of the Dead Sea in the Middle East is 1312 feet below sea level. (*Source: Goode's World Atlas,* 2000.)

54. There are 362,000,000 Yellow Pages directories published in the United States each year. (*Source:* Yellow Pages Publishers Association.)

55. By 2050, the Census Bureau predicts a world population of 9,346,399,000 people. (*Source:* U.S. Bureau of the Census.)

**[1.6]** *Multiply.*

**56.** $^-6(9)$  **57.** $(^-7)(^-8)$  **58.** $10(^-10)$  **59.** $^-45 \cdot 0$

**60.** $^-1(^-24)$  **61.** $17 \cdot 1$  **62.** $4(^-12)$  **63.** $(^-5)(^-25)$

**64.** $^-3 \cdot {^-4} \cdot {^-3}$  **65.** $^-5(2) \cdot {^-5}$  **66.** $^-8 \cdot {^-1}(^-9)$

**[1.7]** *Simplify.*

**67.** $\dfrac{^-63}{^-7}$  **68.** $\dfrac{70}{^-10}$  **69.** $\dfrac{^-15}{0}$  **70.** $^-100 \div {^-20}$

**71.** $18 \div {^-1}$  **72.** $\dfrac{0}{12}$  **73.** $\dfrac{^-30}{^-2}$  **74.** $\dfrac{^-35}{35}$

**75.** $^-40 \div {^-4} \div {^-2}$  **76.** $^-18 \div 3 \cdot {^-3}$  **77.** $0 \div {^-10} \cdot 5 \div 5$

**78.** Divide; then interpret the remainder. It took 1250 hours to build the set for a new play. How many work days of eight hours each did it take to build the set?

**[1.8]** *Simplify.*

**79.** $10^4$  **80.** $2^5$  **81.** $3^3$  **82.** $(^-4)^2$

**83.** $(^-5)^3$  **84.** $8^1$  **85.** $6^2 \cdot 3^2$  **86.** $(^-2)^3 \cdot 5^2$

**87.** $^-30 \div 6 - 4 \cdot 5$  **88.** $6 + 8(2 - 3)$  **89.** $16 \div 4^2 + (^-6 + 9)^2$

**90.** $^-3(4) - 2(5) + 3(^-2)$  **91.** $\dfrac{^-10 + 3^2 - {^-9}}{3 - 10 - 1}$

**92.** $\dfrac{^-1(1 - 3)^3 + 12 \div 4}{^-5 + 24 \div 8 \cdot 2(6 - 6) + 5}$

## Mixed Review Exercises

*Name the property illustrated by each case.*

**93.** $^-3 + (5 + 1) = (^-3 + 5) + 1$  **94.** $^-7(2) = 2(^-7)$  **95.** $0 + 19 = 19$

**96.** $^-42 \cdot 0 = 0$  **97.** $2(^-6 + 4) = 2 \cdot {}^-6 + 2 \cdot 4$  **98.** $(^-6 \cdot 3) \cdot {}^-1 = {}^-6 \cdot (3 \cdot {}^-1)$

*First use front end rounding to estimate each answer. Then find the exact answer.*

**99.** Last year, 192 Elvis jukeboxes were sold at a price of $11,900 each. What was the total value of the jukeboxes?

Estimate:

Exact:

**100.** Chad had $185 in his checking account. He deposited his $428 paycheck and then wrote a $706 check for car repairs. What is the balance in his account?

Estimate:

Exact:

**101.** Georgia's car used 24 gallons of gas on her 840-mile vacation trip. What was the average number of miles she drove on each gallon of gas?

Estimate:

Exact:

**102.** When inventory was taken at Mathtronic Company, 19 calculators and 12 computer modems were missing. Each calculator is worth $39 and each modem is worth $85. What is the total value of the missing items?

Estimate:

Exact:

*Elena Sanchez opened a shop that does alterations and designs custom clothing. Use the table of her income and expenses to answer Exercises 103–106.*

| Month | Income | Expenses | Profit or Loss |
|-------|--------|----------|----------------|
| Jan.  | $2400  | $3100    |                |
| Feb.  | $1900  | $2000    |                |
| Mar.  | $2500  | $1800    |                |
| Apr.  | $2300  | $1400    |                |
| May   | $1600  | $1600    |                |
| June  | $1900  | $1200    |                |

**103.** Complete the table by finding Elena's profit or loss for each month.

**104.** Which month had the greatest loss? Which month had the greatest profit?

**105.** What was Elena's average monthly income?

**106.** What was the average monthly amount of expenses?

# Chapter 1 TEST

*Study Skills Workbook*
**Activity 6**

1. Write this number in words: 20,008,307

2. Write this number using digits:
   thirty billion, seven hundred thousand, five

3. Graph the numbers $3, {}^-2, 0, -\dfrac{1}{2}$ on the number line at the right.

4. Write < or > between each pair of numbers to make a true statement.
   
   $0 \underline{\phantom{xx}} {}^-3$ $\qquad\qquad {}^-2 \underline{\phantom{xx}} {}^-1$

5. Find $|10|$ and $|{}^-14|$.

*Add, subtract, multiply, or divide.*

6. $3 - 9$  

7. ${}^-12 + 7$  

8. $\dfrac{{}^-28}{{}^-4}$

9. ${}^-1(40)$  

10. ${}^-5 - {}^-15$  

11. $({}^-8)({}^-8)$

12. ${}^-25 + {}^-25$  

13. $\dfrac{17}{0}$  

14. ${}^-30 - 30$

15. $\dfrac{50}{{}^-10}$  

16. $5 \cdot {}^-9$  

17. $0 - {}^-6$

*Simplify.*

18. ${}^-35 \div 7 \cdot {}^-5$  

19. ${}^-15 - {}^-8 + 7$

20. $3 - 7({}^-2) - 8$  

21. $({}^-4)^2 \cdot 2^3$

22. $\dfrac{5^2 - 3^2}{(4)({}^-2)}$  

23. ${}^-2({}^-4 + 10) + 5 \cdot 4$

24. ${}^-3 + ({}^-7 - {}^-10) + 4(6 - 10)$

25. Explain how an exponent is used. Include two examples.

81

26. _____

**26.** Explain the commutative and associative properties of addition. Also give an example to illustrate each property.

*Round each number as indicated.*

27. _____

**27.** 851 to the nearest hundred

28. _____

**28.** 36,420,498,725 to the nearest million

29. _____

**29.** 349,812 to the nearest thousand

*First use front end rounding to estimate the answer to each application problem. Then find the exact answer.*

30. *Estimate:* _____

   *Exact:* _____

**30.** In 1988, the number of Japanese cars imported to the United States was 2,123,051. In 1998, the number was 1,454,581 cars. What was the decrease in the number of imported Japanese cars? (*Source:* U.S. Bureau of the Census.)

31. *Estimate:* _____

   *Exact:* _____

**31.** Lorene had $184 in her checking account. She deposited her $293 paycheck and then wrote a $506 check for tuition. What is the balance in her account?

32. *Estimate:* _____

   *Exact:* _____

**32.** The Cardinals football team had a bad year. It lost a total of 1140 yards in 12 games. What was the average loss in each game?

33. *Estimate:* _____

   *Exact:* _____

**33.** One kind of cereal has 220 calories in each serving. Another kind has 110 calories in each serving. (*Source:* General Mills and Post Cereals.) During a month with 31 days, how many calories would you save by eating the second kind of cereal each morning for breakfast?

*Divide; then interpret the remainder.*

34. _____

**34.** Anthony has a part-time job as a shipping clerk. He is sending 1276 pounds of books to a bookstore. Each shipping carton can safely hold 48 pounds. What is the minimum number of cartons he will need?

# Understanding Variables and Solving Equations

2.1 Introduction to Variables
2.2 Simplifying Expressions
2.3 Solving Equations Using Addition
2.4 Solving Equations Using Division
2.5 Solving Equations with Several Steps

At any given moment, there are approximately 2000 thunderstorms in progress around the world. (*Source: Guinness Book of Amazing Nature.*) When a thunderstorm is approaching, you can use an algebraic expression to estimate how far away it is. (See Section 2.1, Exercises 59 and 60.) Knowing the distance between you and the storm will help you decide what precautions to take to protect yourself from danger.

# 2.1 INTRODUCTION TO VARIABLES

**OBJECTIVES**

1. Identify variables, constants, and expressions.
2. Evaluate variable expressions for given replacement values.
3. Write properties of operations using variables.
4. Use exponents with variables.

*Study Skills Workbook* Activity 7

**1 Identify variables, constants, and expressions.** You probably know that algebra uses letters, especially the letter $x$. But why use letters when numbers are easier to understand? Here is an example.

Suppose that you run your college bookstore. When deciding how many books to order for a certain class, you first find out the class limit, that is, the maximum number of students allowed in the class. You will need at least that many books. But you decide to order 5 extra copies for emergencies.

Rule for ordering books: Order the class limit + 5 extra

How many books would you order for a prealgebra class with a limit of 25 students?

Class limit ⟶ ⟵ Extra
$25 + 5$    You would order 30 prealgebra books.

How many books would you order for a geometry class that allows 40 students to register?

Class limit ⟶ ⟵ Extra
$40 + 5$    You would order 45 geometry books.

You could set up a table to keep track of the number of books to order for various classes.

| Class | Rule for Ordering Books: Class Limit + 5 Extra | Number of Books to Order |
|---|---|---|
| Prealgebra | 25 + 5 | 30 |
| Geometry | 40 + 5 | 45 |
| College algebra | 35 + 5 | 40 |
| Calculus 1 | 50 + 5 | 55 |

A shorthand way to write your rule is shown below.

$c + 5$
↑
$c$ stands for class limit.

You can't write your rule by using just numbers because the class limit *varies*, or changes, depending on which class you're talking about. So you use a letter, called a **variable**, to represent the part of the rule that varies. Notice the similarity in the words *vari*es and *vari*able. When part of a rule does *not* change, it is called a **constant**.

The variable, or the part of the rule that varies or changes
↓
$c + 5$
↑
The constant, or the part of the rule that does *not* change

$c + 5$ is called an **expression**. It expresses (tells) the rule for ordering books. You could use any letter you like for the variable part of the expression, such as $x + 5$, or $n + 5$, and so on. But one suggestion is to use a letter that reminds you of what it stands for. In this situation, the letter $c$ reminds us of "**c**lass limit."

Section 2.1 Introduction to Variables  85

**Example 1** Writing an Expression and Identifying the Variable and Constant

Write an expression for this rule. Identify the variable and the constant.
  Order the class limit minus 10 books because some students will buy used books.

$$\underset{\uparrow}{c} - \underset{\uparrow}{10}$$
  Variable      Constant

Work Problem ❶ at the Side.

**2** **Evaluate variable expressions for given replacement values.** When you need to figure out how many books to order for a particular class, you use a specific value for the class limit, like 25 students in prealgebra. Then you **evaluate the expression,** that is, you follow the rule.

**Ordering Books for a Prealgebra Class**

$c + 5$    Expression (rule for ordering books) is $c + 5$.
          Replace $c$ with 25, the class limit for prealgebra.
$25 + 5$    Follow the rule. Add $25 + 5$.
$30$    Order 30 prealgebra books.

**Example 2** Evaluating an Expression

Use this rule for ordering books: Order the class limit minus 10. The expression is $c - 10$.

(a) Evaluate the expression when the class limit is 32.

$c - 10$    Replace $c$ with 32.
$32 - 10$    Follow the rule. Subtract to find $32 - 10$.
$22$    Order 22 books.

(b) Evaluate the expression when the class limit is 48.

$c - 10$    Replace $c$ with 48.
$48 - 10$    Follow the rule. Subtract to find $48 - 10$.
$38$    Order 38 books.

Work Problem ❷ at the Side.

In any career you choose, there will be many useful "rules" that need to be written using variables (letters) because part of the rule changes depending on the situation. This is one reason why algebra is such a powerful tool. Here is another example.

Suppose that you work in a landscaping business. You are putting a fence around a square-shaped garden. Each side of the garden is 6 feet long. How much fencing material should you bring to finish the job? You could add the lengths of the four sides.

6 feet + 6 feet + 6 feet + 6 feet = 24 feet of fencing

❶ Write an expression for this rule. Identify the variable and the constant.

Order the class limit plus 15 extra books because it is a very large class.

❷ Use this expression for ordering books: $c + 3$.

(a) Evaluate the expression when the class limit is 25.

(b) Evaluate the expression when the class limit is 60.

**ANSWERS**
1. The expression is $c + 15$. The variable is $c$ and the constant is 15.
2. (a) $25 + 3$ is 28; order 28 books
   (b) $60 + 3$ is 63; order 63 books

**86** Chapter 2 Understanding Variables and Solving Equations

❸ **(a)** Evaluate the expression 4s when the length of one side of a square table is 3 feet.

Or, recall that multiplication is a quick way to do repeated addition. The square garden has 4 sides, so multiply by 4.

$$4 \cdot 6 \text{ feet} = 24 \text{ feet of fencing}$$

So the rule for calculating the amount of fencing for a square garden is

$$4 \cdot \text{length of one side.}$$

Other jobs may require fencing for larger or smaller square shapes. The following table shows how much fencing you will need.

| Length of One Side of Square Shape | Expression (Rule) to Find Total Amount of Fencing: 4 • Length of One Side | Total Amount of Fencing Needed |
|---|---|---|
| 6 feet | 4 • 6 feet | 24 feet |
| 9 feet | 4 • 9 feet | 36 feet |
| 10 feet | 4 • 10 feet | 40 feet |
| 3 feet | 4 • 3 feet | 12 feet |

The expression (rule) can be written in shorthand form as shown below.

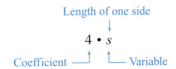

The number part in a *multiplication* expression is called the numerical coefficient, or just the **coefficient**. We usually don't write multiplication dots in expressions, so we do the following.

$$4 \cdot s \quad \text{is written as} \quad 4s$$

**(b)** Evaluate the expression 4s when the length of one side of a square park is 7 miles.

You can use the expression 4s any time you need to know the *perimeter* of a square shape, that is, the total distance around all four sides of the square.

**CAUTION**

If an expression involves adding, subtracting, or dividing, then you **do** have to write +, −, or ÷. It is *only* multiplication that is understood without writing a symbol.

| $4 + s$ | $4 - s$ | $4 \div s$ | $4s$ |
|---|---|---|---|
| ↑ | ↑ | ↑ | |
| Add s. | Subtract s. | Divide by s. | Multiply by s. |

**Example 3** Evaluating an Expression with Multiplication

The expression (rule) for finding the perimeter of a square shape is 4s. Evaluate the expression when the length of one side of a square parking lot is 30 yards.

4s     Replace s with 30 yards.

4 • **30 yards**     There is no operation symbol between the 4 and the s, so it is understood to be multiplication.

120 yards ←— Total distance (perimeter) around the lot

**Work Problem** ❸ **at the Side.**

ANSWERS
3. **(a)** 4 • 3 feet; 12 feet
    **(b)** 4 • 7 miles; 28 miles

Some expressions (rules) involve several different steps. An expression for finding the approximate systolic blood pressure of a person of a certain age is shown below.

$$100 + \frac{a}{2} \leftarrow \text{Age of person (the variable)}$$

Remember that a fraction bar means division, so $\frac{a}{2}$ is the person's age divided by 2. You also need to remember the order of operations, which means doing division before addition.

**Example 4** **Evaluating an Expression with Several Steps**

Evaluate the expression $100 + \frac{a}{2}$ when the age of the person is 24.

$100 + \dfrac{a}{2}$ — Replace $a$ with 24, the age of the person.

$100 + \dfrac{24}{2}$ — Follow the rule using the order of operations. First divide: $24 \div 2$ is 12.

$\underbrace{100 + 12}$ — Now add: $100 + 12$ is 112.

$112$ — The approximate systolic blood pressure is 112.

**Work Problem** ❹ **at the Side.**

❹ Evaluate the expression $100 + \frac{a}{2}$ when the age of the person is 40.

**Calculator Tip**  If you like to fish, you can use an expression (rule) like the one below to find the approximate weight (in pounds) of a fish you catch. Measure the length of the fish (in inches) and then use the correct expression for that type of fish. For a northern pike, the weight expression is shown below.

Variable (length of fish) — $\dfrac{l^3}{3600}$

where $l$ is the length of the fish in inches. (*Source: InFisherman.*)

To evaluate this expression for a fish that is 43 inches long, follow the rule by calculating as follows.

$\dfrac{43^3}{3600}$   Replace $l$ with 43, the length of the fish in inches.

In the numerator, you can multiply $43 \cdot 43 \cdot 43$ or use the $y^x$ key on your calculator. Then divide by 3600.

Enter 43 $y^x$ 3 $\div$ 3600 $=$     **Calculator shows 22.08527778**
     ↑    ↑
    Base Exponent

The fish weighs about 22 pounds.

Now use the expression to find the approximate weight of a northern pike that is 37 inches long. (Answer: about 14 pounds.)

(*continued*)

**ANSWERS**

4.  $100 + \dfrac{40}{2}$ ← Replace $a$ with 40.

   $\underbrace{100 + 20}$

   $120$    Approximate systolic blood pressure is 120.

**88** Chapter 2 Understanding Variables and Solving Equations

**5** (a) Use the expression for finding your average bowling score. Evaluate the expression if your total score for 4 games is 532.

Notice that variables are used on your calculator keys. On the $y^x$ key, $y$ represents the base and $x$ represents the exponent. You evaluated $y^x$ by entering 43 as the base and 3 as the exponent for the first fish. Then you evaluated $y^x$ by entering 37 as the base and 3 as the exponent for the second fish.

Some expressions (rules) involve several variables. For example, if you bowl three games and want to know your average score, you can use this expression.

$$\frac{t}{g}$$ 

$t$ ← Total score for all games (variable)
$g$ ← Number of games (variable)

### Example 5  Evaluating Expressions with Two Variables

(a) Find your average score if you bowl three games and your total score for all three games is 378.
Use the expression (rule) for finding your average score.

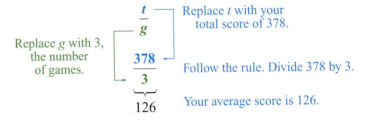

Replace $t$ with your total score of 378.
Replace $g$ with 3, the number of games.
Follow the rule. Divide 378 by 3.
Your average score is 126.

(b) Complete this table.

| Value of x | Value of y | Expression x − y |
|---|---|---|
| 16 | 10 | |
| 3 | 7 | |
| 8 | 0 | |

(b) Complete these tables to show how to evaluate each expression.

| Value of x | Value of y | Expression (Rule) x + y |
|---|---|---|
| 2 | 5 | 2 + 5 is 7 |
| ⁻6 | 4 | + is |
| 0 | 16 | is |

| Value of x | Value of y | Expression (Rule) xy |
|---|---|---|
| 2 | 5 | 2 • 5 is 10 |
| ⁻6 | 4 | • is |
| 0 | 16 | is |

The expression (rule) is to *add* the two variables. So the completed table is:

| Value of x | Value of y | Expression (Rule) x + y |
|---|---|---|
| 2 | 5 | 2 + 5 is 7 |
| ⁻6 | 4 | ⁻6 + 4 is ⁻2 |
| 0 | 16 | 0 + 16 is 16 |

The expression (rule) is to *multiply* the two variables. We know that it's multiplication because there is no operation symbol between the $x$ and $y$. So the completed table is:

| Value of x | Value of y | Expression (Rule) xy |
|---|---|---|
| 2 | 5 | 2 • 5 is 10 |
| ⁻6 | 4 | ⁻6 • 4 is ⁻24 |
| 0 | 16 | 0 • 16 is 0 |

Work Problem **5** at the Side.

**Answers**

5. (a) $\frac{532}{4}$; average score is 133.
   (b) 16 − 10 is 6.
   3 − 7 is ⁻4.
   8 − 0 is 8.

**3** **Write properties of operations using variables.** Now you can use variables as a shorthand way to express the properties you learned about in **Sections 1.3, 1.6,** and **1.7.** We'll use the letters *a* and *b* to represent any two numbers.

| Commutative Property of Addition | Commutative Property of Multiplication |
|:---:|:---:|
| $a + b = b + a$ | $a \cdot b = b \cdot a$ |

To get specific examples, you can pick values for *a* and *b*. For example, if *a* is $^-3$, replace every *a* with $^-3$. If *b* is 5, replace every *b* with 5.

$$a + b = b + a$$
$$^-3 + 5 = 5 + {}^-3$$
$$2 = 2$$
Both sums are 2.

$$a \cdot b = b \cdot a$$
$$^-3 \cdot 5 = 5 \cdot {}^-3$$
$$^-15 = {}^-15$$
Both products are $^-15$.

Of course, you could pick many different values for *a* and *b*, because the commutative "rule" will always work for adding *any* two numbers or multiplying *any* two numbers.

### Example 6  Writing Properties of Operations Using Variables

Use the variable *b* to state this property: When any number is divided by 1, the quotient is the number.

Use the letter *b* to represent any number.

$$\frac{b}{1} = b$$

=== **Work Problem 6 at the Side.**

**4** **Use exponents with variables.** In **Section 1.8** we used an exponent as a quick way to write repeated multiplication. For example,

$\underbrace{3 \cdot 3 \cdot 3 \cdot 3 \cdot 3}_{\text{3 is used as a factor 5 times.}}$ can be written $3^5$ ← Exponent, Base

The meaning of an exponent remains the same when a variable (a letter) is the base.

$\underbrace{c \cdot c \cdot c \cdot c \cdot c}_{\text{c is used as a factor 5 times.}}$ can be written $c^5$ ← Exponent, Base

$m^2$ means $m \cdot m$  Here *m* is used as a factor 2 times.

$x^4 y^3$ means $x^4 \cdot y^3$ or $\underbrace{x \cdot x \cdot x \cdot x}_{x^4} \cdot \underbrace{y \cdot y \cdot y}_{y^3}$

$7b^2$ means $7 \cdot b \cdot b$  The exponent applies *only* to *b*.

$^-4xy^2z$ means $^-4 \cdot x \cdot y \cdot y \cdot z$  The exponent applies *only* to *y*.

---

**6** **(a)** Use the variable *a* to state this property: Multiplying any number by 0 gives a product of 0.

**(b)** Use the variables *a*, *b*, and *c* to state the associative property of addition: Changing the grouping of addends does not change the sum.

**ANSWERS**
**6. (a)** $0 \cdot a = 0$  or  $a \cdot 0 = 0$
  **(b)** $(a + b) + c = a + (b + c)$

**7** Rewrite each expression without exponents.

(a) $x^5$

(b) $4a^2b^2$

(c) $-10xy^3$

(d) $s^4tu^2$

**8** Evaluate each expression.

(a) $y^3$ when $y$ is $-5$

(b) $r^2s^2$ when $r$ is 6 and $s$ is 3

(c) $10xy^2$ when $x$ is 4 and $y$ is $-3$

(d) $-3c^4$ when $c$ is 2

**ANSWERS**
7. (a) $x \cdot x \cdot x \cdot x \cdot x$
   (b) $4 \cdot a \cdot a \cdot b \cdot b$
   (c) $-10 \cdot x \cdot y \cdot y \cdot y$
   (d) $s \cdot s \cdot s \cdot s \cdot t \cdot u \cdot u$
8. (a) $(-5)^3$ is $-125$.  (b) $(6)^2 (3)^2$ is 324.
   (c) $10(4)(-3)^2$ is 360.  (d) $-3(2)^4$ is $-48$.

---

**Example 7** Understanding Exponents Used with Variables

Rewrite each expression without exponents.

(a) $y^6$ can be written as $y \cdot y \cdot y \cdot y \cdot y \cdot y$

$y$ is used as a factor 6 times.

(b) $12bc^3$ can be written as $12 \cdot b \cdot c \cdot c \cdot c$

Coefficient is 12. The exponent applies *only* to $c$. ($c^3$)

(c) $-2m^2n^4$ can be written as $-2 \cdot m \cdot m \cdot n \cdot n \cdot n \cdot n$

Coefficient is $-2$. ($m^2$, $n^4$)

**Work Problem 7 at the Side.**

To evaluate an expression with exponents, multiply all the factors.

**Example 8** Evaluating Expressions with Exponents

Evaluate each expression.

(a) $x^2$ when $x$ is $-3$

$x^2$ means $x \cdot x$   Replace each $x$ with $-3$.

$-3 \cdot -3$   Multiply $-3$ times $-3$.

9

So $x^2$ becomes $(-3)^2$, which is $(-3)(-3)$, or 9.

(b) $x^3y$ when $x$ is $-4$ and $y$ is $-10$

$x^3y$ means $x \cdot x \cdot x \cdot y$   Replace $x$ with $-4$, and replace $y$ with $-10$.

$-4 \cdot -4 \cdot -4 \cdot -10$   Multiply two factors at a time.

$16 \cdot -4 \cdot -10$

$-64 \cdot -10$

640

So $x^3y$ becomes $(-4)^3(-10)$, which is $(-4)(-4)(-4)(-10)$, or 640.

(c) $-5ab^2$ when $a$ is 5 and $b$ is 3

$-5ab^2$ means $-5 \cdot a \cdot b \cdot b$   Replace $a$ with 5, and replace $b$ with 3.

$-5 \cdot 5 \cdot 3 \cdot 3$   Multiply two factors at a time.

$-25 \cdot 3 \cdot 3$

$-75 \cdot 3$

$-225$

So $-5ab^2$ becomes $-5(5)(3)^2$, which is $-5(5)(3)(3)$, or $-225$.

Coefficient is $-5$.

**Work Problem 8 at the Side.**

## 2.1 EXERCISES

*Identify the parts of each expression. Choose from these labels:* variable, constant, *and* coefficient. *See Example 1.*

1. $c + 4$
2. $d + 6$
3. $5h$
4. $3s$

5. $^-3 + m$
6. $^-4 + n$
7. $2c + 10$
8. $6b + 1$

9. $x - y$
10. $xy$
11. $^-6g$
12. $^-10k$

*Evaluate each expression. See Examples 2–5.*

13. The expression (rule) for ordering robes for the graduation ceremony at West Community College is $g + 10$, where $g$ is the number of graduates. Evaluate the expression when

    (a) there are 654 graduates.

    (b) there are 208 graduates.

14. The expression (rule) for finding the approximate temperature (in degrees Fahrenheit) is $c + 40$, where $c$ is the number of chirps made in 15 seconds by a field cricket. (*Source:* Dial U Insect and Plant Information.) Evaluate the expression when the cricket

    (a) chirps 45 times.

    (b) chirps 33 times.

15. The expression for finding the perimeter of a triangle with sides of equal length is $3s$, where $s$ is the length of one side. Evaluate the expression when

    (a) the length of one side is 11 inches.

    (b) the length of one side is 3 feet.

16. The expression for finding the perimeter of a pentagon with sides of equal length is $5s$, where $s$ is the length of one side. Evaluate the expression when

    (a) the length of one side is 25 meters.

    (b) the length of one side is 8 inches.

**17.** The expression for ordering brushes for an art class is $3c - 5$, where $c$ is the class limit. Evaluate the expression when

(a) the class limit is 12.

(b) the class limit is 16.

**18.** The expression for ordering doughnuts for the office staff is $2n - 4$, where $n$ is the number of people at work. Evaluate the expression when

(a) there are 13 people at work.

(b) there are 18 people at work.

**19.** The expression for figuring a student's average test score is $\frac{p}{t}$, where $p$ is the total points earned on all the tests and $t$ is the number of tests. Evaluate the expression when

(a) 332 points were earned on 4 tests.

(b) there were 7 tests and 637 points earned.

**20.** The expression for deciding how many buses are needed for a group trip is $\frac{p}{b}$, where $p$ is the total number of people and $b$ is the number of people that one bus will hold. Evaluate the expression when

(a) 176 people are going on a trip and one bus holds 44 people.

(b) A bus holds 36 people and 72 people are going on a trip.

*Complete each table by evaluating the expressions. See Example 5.*

**21.**

| Value of x | Expression x + x + x + x | Expression 4x |
|---|---|---|
| ⁻2 | ⁻2 + ⁻2 + ⁻2 + ⁻2 is ⁻8 | 4 • ⁻2 is ⁻8 |
| 12 | | |
| 0 | | |
| ⁻5 | | |

**22.**

| Value of y | Expression 3y | Expression y + 2y |
|---|---|---|
| ⁻6 | 3 • ⁻6 is ⁻18 | ⁻6 + 2 • ⁻6 is ⁻6 + ⁻12, or ⁻18 |
| 10 | | |
| ⁻3 | | |
| 0 | | |

**23.**

| Value of x | Value of y | Expression ⁻2x + y |
|---|---|---|
| 3 | 7 | ⁻2 • 3 + 7 is ⁻6 + 7, or 1 |
| ⁻4 | 5 | |
| ⁻6 | ⁻2 | |
| 0 | ⁻8 | |

**24.**

| Value of x | Value of y | Expression ⁻2xy |
|---|---|---|
| 3 | 7 | ⁻2 • 3 • 7 is ⁻42 |
| ⁻4 | 5 | |
| ⁻6 | ⁻2 | |
| 0 | ⁻8 | |

**25.** Explain the words *variable* and *expression*.

**26.** Explain the words *coefficient* and *constant*.

*Use the variable b to express each of these properties. See Example 6.*

**27.** Multiplying a number by 1 leaves the number unchanged.

**28.** Adding 0 to any number leaves the number unchanged.

**29.** Any number divided by 0 is undefined.

**30.** Multiplication distributes over addition. (Use *a*, *b*, and *c* as the variables.)

*Rewrite each expression without exponents. See Example 7.*

**31.** $c^6$

**32.** $d^7$

**33.** $x^4 y^3$

**34.** $c^2 d^5$

**35.** $^{-}3a^3 b$

**36.** $^{-}8m^2 n$

**37.** $9xy^2$

**38.** $5ab^4$

**39.** $^{-}2c^5 d$

**40.** $^{-}4x^3 y$

**41.** $a^3 bc^2$

**42.** $x^2 yz^6$

*Evaluate each expression when r is $^{-}3$, s is 2, and t is $^{-}4$. See Example 8.*

**43.** $t^2$

**44.** $r^2$

**45.** $rs^3$

**46.** $s^4 t$

**47.** $3rs$

**48.** $6st$

**49.** $-2s^2t^2$

**50.** $-4rs^4$

**51.** $r^2s^5t^3$

**52.** $r^3s^4t^2$

**53.** $-10r^5s^7$

**54.** $-5s^6t^5$

*Evaluate each expression when x is 4, y is $-2$, and z is $-6$.*

**55.** $|xy| + |xyz|$

**56.** $x + |y^2| + |xz|$

**57.** $\dfrac{z^2}{-3y + z}$

**58.** $\dfrac{y^2}{x + 2y}$

### RELATING CONCEPTS (Exercises 59–60)   FOR INDIVIDUAL OR GROUP WORK

*At the beginning of this chapter, you read that there is an expression that tells your approximate distance from a thunderstorm. First, count the number of seconds from the time you see a lightning flash until you start to hear the thunder. Then, to estimate the distance (in miles), use the expression $\frac{s}{5}$, where s is the number of seconds. Use this expression as you **work Exercises 59 and 60 in order**. (This expression is based on the fact that light travels faster than sound.)*

**59.** Evaluate the thunderstorm expression for each number of seconds. How far away is the storm?

  **(a)** 15 seconds

  **(b)** 10 seconds

  **(c)** 5 seconds

**60.** Explain how you can use your answers from Exercise 59 to:

  **(a)** Estimate the distance when the time is $2\frac{1}{2}$ seconds.

  **(b)** Find the number of seconds when the distance is $1\frac{1}{2}$ miles.

  **(c)** Find the number of seconds when the distance is $2\frac{1}{2}$ miles.

## 2.2 SIMPLIFYING EXPRESSIONS

**1** **Combine like terms, using the distributive property.** In **Section 2.1**, the expression for ordering math textbooks was $c + 5$. This expression was simple and easy to use. Sometimes expressions are *not* written in the simplest possible way. For example:

**OBJECTIVES**

**1** Combine like terms, using the distributive property.
**2** Simplify expressions.
**3** Use the distributive property to multiply.

Evaluate this expression when $c$ is 20.

$$c + 5$$
$$\downarrow$$
$$20 + 5 \quad \text{Replace } c \text{ with 20.}$$
$$25 \quad \text{Add } 20 + 5.$$

Evaluate this expression when $c$ is 20.

$$2c - 10 - c + 15$$
$$\downarrow \quad\quad\quad \downarrow$$
$$2 \cdot 20 - 10 - 20 + 15 \quad \text{Replace } c \text{ with 20.}$$
$$40 - 10 - 20 + 15 \quad \text{Multiply } 2 \cdot 20.$$
$$40 + {}^-10 + {}^-20 + 15 \quad \text{Change subtraction to adding the opposite.}$$
$$\underbrace{30} + {}^-20 + 15 \quad \text{Add from left to right.}$$
$$\underbrace{10} + 15$$
$$25$$

These two expressions are actually equivalent. When you evaluate them, the final result is the same, but it takes a lot more work when you use the right-hand expression. To save a lot of work, you need to learn how to *simplify expressions*. Then you can rewrite $2c - 10 - c + 15$ in the simplest way possible, which is $c + 5$.

The basic idea in **simplifying expressions** is to *combine,* or *add,* like terms. Each addend in an expression is a **term.** Here are two examples.

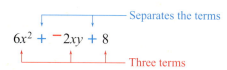

In $6x^2 + {}^-2xy + 8$, the 8 is the *constant term*. There are also two *variable terms* in the expression: $6x^2$ is a variable term, and ${}^-2xy$ is a variable term. A **variable term** has a number part (coefficient) and a letter part (variable).

$6x^2$ — Variable part is $x^2$. Coefficient is 6.

${}^-2xy$ — Variable part is $xy$. Coefficient is ${}^-2$.

If no coefficient is shown, it is assumed to be 1. Remember from **Section 1.6** that multiplying any number by 1 does *not* change the number.

Also, ${}^-c$ can be written ${}^-1 \cdot c$. The coefficient of ${}^-c$ is understood to be ${}^-1$.

### Like Terms

**Like terms** are terms with exactly the same variable parts (the same letters and exponents). The coefficients do *not* have to match.

**96** Chapter 2 Understanding Variables and Solving Equations

**❶** List the like terms in each expression. Then identify the coefficients of the like terms.

(a) $3b^2 + {}^-3b + 3 + b^3 + b$

| Like Terms | | Unlike Terms | |
|---|---|---|---|
| $5x$ and $3x$ | Variable parts match; both are $x$. | $3x$ and $3x^2$ | Variable parts do *not* match; exponents are different. |
| ${}^-6y^3$ and $y^3$ | Variable parts match; both are $y^3$. | ${}^-2x$ and ${}^-2y$ | Variable parts do *not* match; letters are different. |
| $4a^2b$ and $5a^2b$ | Variable parts match; both are $a^2b$. | $a^3b$ and $a^2b$ | Variable parts do *not* match; exponents are different. |
| ${}^-8$ and $4$ | There are no variable parts; numbers are like terms. | ${}^-8c$ and $4$ | Variable parts do *not* match; one term has a variable part, but the other term does not. |

(b) ${}^-4xy + 4x^2y + {}^-4xy^2 + {}^-4 + 4$

### Example 1   Identifying Like Terms and Their Coefficients

List the like terms in each expression. Then identify the coefficients of the like terms.

(a) ${}^-5x + {}^-5x^2 + 3xy + x + {}^-5$

The like terms are ${}^-5x$ and $x$.

The coefficient of ${}^-5x$ is ${}^-5$, and the coefficient of $x$ is understood to be 1.

(b) $2yz^2 + 2y^2z + {}^-3y^2z + 2 + {}^-6yz$

The like terms are $2y^2z$ and ${}^-3y^2z$.

The coefficients are 2 and ${}^-3$.

(c) $5r^2 + 2r + {}^-2r^2 + 5 + 5r^3$

(c) $10ab + 12 + {}^-10a + 12b + {}^-6$

The like terms are 12 and ${}^-6$.

The like terms are constants (there are no variable parts).

**Work Problem ❶ at the Side.**

The distributive property (see **Section 1.6**) can be used "in reverse" to combine like terms. Here is an example.

$$\underbrace{3 \cdot x}_{3x} + \underbrace{4 \cdot x}_{4x} \quad \text{can be written as} \quad \underbrace{(3 + 4) \cdot x}_{\underbrace{7 \cdot x}_{7x}}$$

(d) ${}^-10 + {}^-x + {}^-10x + {}^-x^2 + {}^-10y$

Thus, $3x + 4x$ can be written in *simplified form* as $7x$. To check, evaluate each expression when $x$ is 2.

$$3x + 4x \qquad\qquad 7x$$
$$\downarrow \quad \downarrow \qquad\qquad \downarrow$$
$$3 \cdot 2 + 4 \cdot 2 \qquad\qquad 7 \cdot 2$$
$$\underbrace{6 + 8}_{14} \qquad\qquad 14$$

**ANSWERS**

1. (a) ${}^-3b$ and $b$; the coefficients are ${}^-3$ and 1.
   (b) ${}^-4$ and 4; they are constants.
   (c) $5r^2$ and ${}^-2r^2$; the coefficients are 5 and ${}^-2$.
   (d) ${}^-x$ and ${}^-10x$; the coefficients are ${}^-1$ and ${}^-10$.

Both results are 14, so the expressions are equivalent. But you can see how much easier it is to work with $7x$, the simplified expression.

> **CAUTION**
> Notice that $3x + 4x$ is simplified to $7x$, **not** to $7x^2$.
> Variable part is unchanged. Do *not* change $x$ to $x^2$.

### Combining Like Terms

*Step 1* If there are any variable terms with no coefficient, write in the understood 1.

*Step 2* If there are any subtractions, change each one to adding the opposite.

*Step 3* Find *like* terms (the variable parts match).

*Step 4* Add the coefficients (number parts) of like terms. *The variable part stays the same.*

**Example 2   Combining Like Terms**

Combine like terms.

**(a)** $2x + 4x + x$

$2x + 4x + x$     No coefficient; write understood 1.
                   There are no subtractions to change.

$2x + 4x + 1x$    Find like terms: $2x$, $4x$, and $1x$ are like terms, so add the coefficients, $2 + 4 + 1$.

$(2 + 4 + 1)x$    The variable part, $x$, stays the same.

$7x$

Therefore, $2x + 4x + x$ can be written as $7x$.

**(b)** $^-3y^2 - 8y^2$

$^-3y^2 - 8y^2$           Both coefficients are shown. Change subtraction to adding the opposite.

$^-3y^2 + {^-8y^2}$       Find like terms: $^-3y^2$ and $^-8y^2$ are like terms, so add the coefficients, $^-3 + {^-8}$.

$(^-3 + {^-8})y^2$        The variable part, $y^2$, stays the same.

$^-11y^2$

Therefore, $^-3y^2 - 8y^2$ can be written as $^-11y^2$.

=========== Work Problem ❷ at the Side.

**2** ▭ **Simplify expressions.** When simplifying expressions, be careful to combine only *like* terms—those having variable parts that match. You *cannot* combine terms if the variable parts are different.

❷ Combine like terms.

**(a)** $10b + 4b + 10b$

**(b)** $y^3 + 8y^3$

**(c)** $^-7n - n$

**(d)** $3c - 5c - 4c$

**(e)** $^-9xy + xy$

**(f)** $^-4p^2 - 3p^2 + 8p^2$

**(g)** $ab - ab$

**ANSWERS**

**2. (a)** $24b$   **(b)** $9y^3$   **(c)** $^-8n$   **(d)** $^-6c$
   **(e)** $^-8xy$   **(f)** $1p^2$, or just $p^2$
   **(g)** 0 because $1ab + {^-1}ab$ is $(1 + {^-1})ab$, or $0ab$, and 0 times anything is 0.

**3** Simplify each expression by combining like terms.

(a) $3b^2 + 4d^2 + 7b^2$

(b) $4a + b - 6a + b$

(c) $^-6x + 5 + 6x + 2$

(d) $2y - 7 - y + 7$

(e) $^-3x - 5 + 12 + 10x$

---

> **Example 3** **Simplifying Expressions**
>
> Simplify each expression by combining like terms.
>
> (a) $6xy + 2y + 3xy$
>
> The *like* terms are $6xy$ and $3xy$. We can use the commutative property to rewrite the expression so that the like terms are next to each other. This helps to organize our work.
>
> $6xy + 3xy + 2y$ — Combine *like* terms only.
>
> $(6 + 3)xy + 2y$ — Add the coefficients, $6 + 3$. The variable part, $xy$, stays the same.
>
> $9xy + 2y$ — Keep writing $2y$, the term that was *not* combined; it is still part of the expression.
>
> The simplified expression is $9xy + 2y$.
>
> (b) Here is the expression from the first page in this section.
>
> $2c - 10 - c + 15$ — Write the understood 1 as the coefficient of $c$.
>
> $2c - 10 - 1c + 15$ — Change subtractions to adding the opposite.
>
> $2c + ^-10 + ^-1c + 15$ — Rewrite the expression so that like terms are next to each other.
>
> $2c + ^-1c + ^-10 + 15$ — Combine $2c + ^-1c$. Also combine $^-10 + 15$.
>
> $(2 + ^-1)c + 5$
>
> $1c + 5$
>
> The simplified expression is $1c + 5$ or just $c + 5$.
>
> $1c$ is the same as $c$.

**NOTE**

In this book, when combining like terms we will usually write the variable terms in alphabetical order. A constant term (number only) will be written last. So, in Examples 3(a) and 3(b) above, the preferred and alternative ways of writing the expressions are as follows.

Simplified expression is $9xy + 2y$ (alphabetical order). However, by the commutative property of addition, $2y + 9xy$ is also correct.

Simplified expression is $c + 5$ (constant written last). However, by the commutative property of addition, $5 + c$ is also correct.

**Work Problem 3 at the Side.**

We can use the associative property of multiplication to simplify an expression such as $4(3x)$.

$4(3x)$ can be written as $4 \cdot (3 \cdot x)$

Understood multiplications

---

**Answers**

**3.** (a) $10b^2 + 4d^2$ (b) $^-2a + 2b$
(c) 7 (d) $y$ (e) $7x + 7$

Using the associative property, we can regroup the factors.

$4 \cdot (3 \cdot x)$ can be written as $(4 \cdot 3) \cdot x$  To simplify, multiply $4 \cdot 3$.
$\phantom{4 \cdot (3 \cdot x) \text{ can be written as }} 12 \cdot x$  Write $12 \cdot x$ without the multiplication dot.
$\phantom{4 \cdot (3 \cdot x) \text{ can be written as }} 12x$

The simplified expression is $12x$.

### Example 4  Simplifying Multiplication Expressions

Simplify.

**(a)** $5(10y)$

Use the associative property.

$5 \cdot (10 \cdot y)$ can be written as $(5 \cdot 10) \cdot y$  Multiply $5 \cdot 10$.
$\phantom{5 \cdot (10 \cdot y) \text{ can be written as }} 50 \cdot y$  Write $50 \cdot y$ without the multiplication dot.
$\phantom{5 \cdot (10 \cdot y) \text{ can be written as }} 50y$

So, $5(10y)$ simplifies to $50y$.

**(b)** $^-6(3b)$

Use the associative property.

$^-6(3b)$ can be written as $(^-6 \cdot 3)b$
$\phantom{^-6(3b) \text{ can be written as }} ^-18b$

So, $^-6(3b)$ simplifies to $^-18b$.

**(c)** $^-4(^-2x^2)$

Use the associative property.

$^-4(^-2x^2)$ can be written as $(^-4 \cdot {}^-2)x^2$
$\phantom{^-4(^-2x^2) \text{ can be written as }} 8x^2$

So, $^-4(^-2x^2)$ simplifies to $8x^2$.

**Work Problem ❹ at the Side.**

**❸ Use the distributive property to multiply.** The distributive property can also be used to simplify expressions such as $3(x + 5)$. You *cannot* add the terms inside the parentheses because $x$ and $5$ are *not* like terms. But notice the understood multiplication dot between the 3 and the parentheses.

$$3(x + 5)$$
$$3 \cdot (x + 5)$$

Thus you can distribute multiplication over addition, as you did in **Section 1.6**. That is, multiply 3 times each term inside the parentheses.

$3 \cdot (x + 5)$ can be written as $3 \cdot x + 3 \cdot 5$
$\phantom{3 \cdot (x + 5) \text{ can be written as }} 3x + 15$

So, $3(x + 5)$ simplifies to $3x + 15$.
$\phantom{\text{So, }3(x + 5)\text{ simplifies to }}$ — Stays as addition —

❹ Simplify.

**(a)** $7(4c)$

**(b)** $^-3(5y^3)$

**(c)** $20(^-2a)$

**(d)** $^-10(^-x)$

**ANSWERS**
**4. (a)** $28c$  **(b)** $^-15y^3$  **(c)** $^-40a$  **(d)** $10x$

**5** Simplify.

(a) $7(a + 10)$

(b) $3(x - 3)$

(c) $4(2y + 6)$

(d) $^-5(3b + 2)$

(e) $^-8(c + 4)$

Multiplication also distributes over subtraction.

$$4 \cdot (y - 2) \text{ can be written as } 4 \cdot y - 4 \cdot 2$$
$$4y - 8$$

So, $4(y - 2)$ simplifies to $4y - 8$. Notice that we did *not* need to change subtraction to adding the opposite.
(Stays as subtraction)

### Example 5  Using the Distributive Property

Simplify.

(a) $6(y - 4)$  can be written as  $6 \cdot y - 6 \cdot 4$
$$6y - 24$$
Stays as subtraction

So, $6(y - 4)$ simplifies to $6y - 24$.

(b) $5(3x + 2)$  can be written as  $5 \cdot 3x + 5 \cdot 2$
$$5 \cdot 3 \cdot x + 10$$
$$15 \cdot x + 10$$
$$15x + 10$$

So, $5(3x + 2)$ simplifies to $15x + 10$.

(c) $^-2(4a + 3)$  can be written as  $^-2 \cdot 4a + {}^-2 \cdot 3$
$$^-2 \cdot 4 \cdot a + {}^-6$$
$$^-8 \cdot a + {}^-6$$
$$^-8a + {}^-6$$

Now we will use the definition of subtraction "in reverse" to rewrite $^-8a + {}^-6$.

Change $^-6$ to its opposite, $^+6$.

Write $^-8a + {}^-6$  as  $^-8a - 6$

Change addition to subtraction.

Think back to the way we changed subtraction to adding the opposite. Here we are "working backward." From now on, whenever addition is followed by a negative number, we will change it to subtracting a positive number.

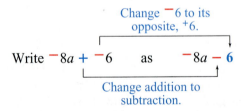

So, $^-2(4a + 3)$ simplifies to $^-8a - 6$.

**Work Problem 5 at the Side.**

ANSWERS
5. (a) $7a + 70$  (b) $3x - 9$  (c) $8y + 24$
   (d) $^-15b - 10$  (e) $^-8c - 32$

## Section 2.2 Simplifying Expressions

Sometimes you need to do several steps to simplify an expression.

**Example 6** Simplifying a More Complex Expression

Simplify: $8 + 3(x - 2)$.

| | |
|---|---|
| $8 + 3(x - 2)$ | Do *not* add $8 + 3$. Use the distributive property first because multiplying is done *before* adding. |
| $8 + 3 \cdot x - 3 \cdot 2$ | Do the multiplications. |
| $8 + 3x - 6$ | Rewrite so that like terms are next to each other. |
| $3x + 8 - 6$ | Subtract to find $8 - 6$ or change to adding $8 + {}^-6$. |
| $3x + 2$ | |

The simplified expression is $3x + 2$.

**CAUTION**

Do *not* add $8 + 3$ as the first step in Example 6 above. Remember that the order of operations tells you to add *last*.

Work Problem ❻ at the Side.

❻ Simplify.

(a) ${}^-4 + 5(y + 1)$

(b) $2(3w + 4) - 5$

(c) $5(6x - 2) + 3x$

(d) $21 + 7(a^2 - 3)$

(e) ${}^-y + 3(2y + 5) - 18$

**ANSWERS**
6. (a) $5y + 1$  (b) $6w + 3$  (c) $33x - 10$
   (d) $7a^2$  (e) $5y - 3$

# Focus on Real-Data Applications

## Expressions

A college with four campuses uses an auditorium for graduation that has 1500 seats. Each campus hosts its own graduation ceremony. Students are allocated a whole number of tickets. Round each result **down** to the nearest whole number.

1. How many tickets are allocated to each of 458 graduates at the **North** Campus graduation?

2. How many tickets are allocated to each of 297 graduates at the **East** Campus graduation?

3. How many tickets are allocated to each of 315 graduates at the **South** Campus graduation?

4. How many tickets are allocated to each of 186 graduates at the **West** Campus graduation?

5. Write an *expression* that represents the number of tickets that are allocated to each of $g$ graduates.

6. What recommendations do you have for unallocated tickets?

Traffic engineers have to decide how long to have the red, yellow, and green lights showing on a traffic signal. To decide the number of seconds that a yellow light should be on, the engineers use the expression

$$\frac{5v}{100} + 1$$

where $v$ is the speed limit in miles per hour (mph).

7. How many seconds should the yellow light be on if the speed limit is 20 mph? 40 mph? 60 mph?

8. Based on the answers you just calculated, how could you estimate the time for a yellow light if the speed limit is 30 mph? 50 mph?

9. Use the given expression to find the number of seconds that the yellow light should be on if the speed limit is 30 mph and 50 mph. Did you get the same result as in Problem 8?

To estimate the number of words in a child's vocabulary, a pediatrician uses the expression

$$60A - 900,$$

where $A$ is the child's age in months.

10. Estimate the number of words that a child aged 20 months knows.

11. Estimate the number of words that a child aged 2 years knows. (*Hint:* How many months are in two years?)

12. How many words does a child learn between the ages of 20 months and 2 years?

13. Estimate the number of words in the vocabulary of a 3-year-old child.

14. How many words does a child learn between the ages of 2 years and 3 years?

15. Evaluate the expression for a child who is 15 months old. Do you think that the answer is reasonable? Explain why or why not.

16. Evaluate the expression for a child who is 12 months old. Do you think that the answer makes sense? Explain why or why not.

## 2.2 EXERCISES

*Circle the like terms in each expression. Then identify the coefficients of the like terms. See Example 1.*

1. $2b^2 + 2b + 2b^3 + b^2 + 6$

2. $3x + x^3 + 3x^2 + 3 + 2x^3$

3. $^-x^2y + {}^-xy + 2xy + {}^-2xy^2$

4. $ab^2 + {}^-a^2b + 2ab + {}^-3a^2b$

5. $7 + 7c + 3 + 7c^3 + {}^-4$

6. $4d + {}^-5 + 1 + {}^-5d^2 + 4$

*Simplify each expression. See Example 2.*

7. $6r + 6r$

8. $4t + 10t$

9. $x^2 + 5x^2$

10. $9y^3 + y^3$

11. $p - 5p$

12. $n - 3n$

13. $^-2a^3 - a^3$

14. $^-10x^2 - x^2$

15. $c - c$

16. $b^2 - b^2$

17. $9xy + xy - 9xy$

18. $r^2s - 7r^2s + 7r^2s$

19. $5t^4 + 7t^4 - 6t^4$

20. $10mn - 9mn + 3mn$

21. $y^2 + y^2 + y^2 + y^2$

22. $a + a + a$

23. $^-x - 6x - x$

24. $^-y - y - 3y$

*Simplify by combining like terms. Write each answer with the variables in alphabetical order and any constant term last. See Example 3.*

**25.** $8a + 4b + 4a$

**26.** $6x + 5y + 4y$

**27.** $6 + 8 + 7rs$

**28.** $10 + 2c^2 + 15$

**29.** $a + ab^2 + ab^2$

**30.** $n + mn + n$

**31.** $6x + y - 8x + y$

**32.** $d + 3c - 7c + 3d$

**33.** $8b^2 - a^2 - b^2 + a^2$

**34.** $5ab - ab + 3a^2b - 4ab$

**35.** $^-x^3 + 3x - 3x^2 + 2$

**36.** $a^2b - 2ab - ab^3 + 3a^3b$

**37.** $^-9r + 6t - s - 5r + s + t - 6t + 5s - r$

**38.** $^-x - 3y + 4z + x - z + 5y - 8x - y$

*Simplify by using the associative property of multiplication. See Example 4.*

**39.** $3(10a)$

**40.** $8(4b)$

**41.** $^-4(2x^2)$

**42.** $^-7(3b^3)$

**43.** $5(^-4y^3)$

**44.** $2(^-6x)$

**45.** $^-9(^-2cd)$

**46.** $^-6(^-4rs)$

**47.** $7(3a^2bc)$

**48.** $4(2xy^2z^2)$

**49.** $^-12(^-w)$

**50.** $^-10(^-k)$

*Use the distributive property to simplify each expression. See Example 5.*

**51.** $6(b + 6)$ **52.** $5(a + 3)$ **53.** $7(x - 1)$

**54.** $4(y - 4)$ **55.** $3(7t + 1)$ **56.** $8(2c + 5)$

**57.** $^-2(5r + 3)$ **58.** $^-5(6z + 2)$ **59.** $^-9(k + 4)$

**60.** $^-3(p + 7)$ **61.** $50(m - 6)$ **62.** $25(n - 1)$

*Simplify each expression. See Example 6.*

**63.** $10 + 2(4y + 3)$ **64.** $4 + 7(x^2 + 3)$ **65.** $6(a^2 - 2) + 15$

**66.** $5(b - 4) + 25$ **67.** $2 + 9(m - 4)$ **68.** $6 + 3(n - 8)$

**69.** $^-5(k + 5) + 5k$ **70.** $^-7(p + 2) + 7p$ **71.** $4(6x - 3) + 12$

**72.** $6(3y - 3) + 18$ **73.** $5 + 2(3n + 4) - n$ **74.** $8 + 8(4z + 5) - z$

**75.** $^-p + 6(2p - 1) + 5$ **76.** $^-k + 3(4k - 1) + 2$

77. Explain the difference between *simplifying* an expression and *evaluating* an expression.

78. Simplify each expression. Are the answers equivalent? Explain why or why not.

$$5(3x + 2) \qquad 5(2 + 3x)$$

79. Explain what makes two terms *like* terms. Include several examples in your explanation.

80. Explain how to combine like terms. Include an example in your explanation.

81. Explain and correct the error made by a student who simplified this expression.

$$\underbrace{{}^-2x + 7x} + 8$$
$$5x^2 \quad\ \ + 8$$

82. Explain and correct the error made by a student who simplified this expression.

$$\underbrace{{}^-10a + 6a} - \underset{\downarrow}{7} + \underset{\downarrow}{2}$$
$$-4a \quad + \underbrace{{}^-7 + 2}$$
$$-4a \quad + \quad {}^-5$$
$$4a - 5$$

*Simplify.*

83. $^-4(3y) - 5 + 2(5y + 7)$

84. $6(^-3x) - 9 + 3(^-2x + 6)$

85. $^-10 + 4(^-3b + 3) + 2(6b - 1)$

86. $12 + 2(4a - 4) + 4(^-2a - 1)$

87. $^-5(^-x + 2) + 8(^-x) + 3(^-2x - 2) + 16$

88. $^-7(^-y) + 6(y - 1) + 3(^-2y) + 6 - y$

# 2.3 SOLVING EQUATIONS USING ADDITION

Now you are ready for a look at the "heart" of algebra, writing and solving **equations.** *Writing* an equation is a way to show the relationship between what you *know* about a problem and what you *don't* know. Then, *solving* the equation is a way to figure out the part that you didn't know and answer your question.

The questions you can answer by writing and solving equations are as varied as the careers people choose. A zoo keeper can solve an equation that answers the question of how long to incubate the egg of a particular tropical bird. An aerobics instructor can solve an equation that answers the question of how hard a certain person should exercise for maximum benefit.

**OBJECTIVES**

1. Determine whether a given number is a solution of an equation.
2. Solve equations, using the addition property of equality.
3. Simplify equations before using the addition property of equality.

**1** **Determine whether a given number is a solution of an equation.** Let's start with the example from the beginning of this chapter: ordering textbooks for math classes. The expression we used to order books was $c + 5$, where $c$ was the class limit (the maximum number of students allowed in the class). Suppose that 30 prealgebra books were ordered. What is the class limit for prealgebra? To answer this question, write an equation showing the relationship between what you know and what you don't know.

You don't know the class limit. ↓   You do know the total number of books ordered.

$$c + 5 = 30$$

You do know that 5 extra books were ordered.

> **NOTE**
>
> An equation has an equal sign. Notice the similarity in the words **equa**tion and **equa**l. An expression does *not* have an equal sign.

The equal sign in an equation is like the balance point on a playground teeter-totter, or seesaw. To have a true equation, the two sides must balance.

These equations balance, so we can use the $=$ sign.

$6 + 8 = 14$

$10 = 5 \cdot 2$

$3 \cdot 2 = 5 + 1$

These equations do *not* balance, so we write $\neq$ to mean "not equal to."

$6 + 8 \neq 15$

$10 \neq 4 \cdot 2$

$4 + 5 \neq 5 \cdot 4$

**1** (a) Which of these numbers, 95, 65, or 70, is the solution of the equation $c + 15 = 80$?

When an equation has a variable, we **solve the equation** by finding a number that can replace the variable and make the equation balance. For the example about ordering prealgebra textbooks:

$$c + 5 = 30$$

What number can replace $c$ so that the equation balances?

Try replacing $c$ with **15**.  $15 + 5 \neq 30$   Does *not* balance: $15 + 5$ is only 20.

Try replacing $c$ with **40**.  $40 + 5 \neq 30$   Does *not* balance: $40 + 5$ is more than 30.

Try replacing $c$ with **25**.  $25 + 5 = 30$   Balances: $25 + 5$ is 30.

The **solution** is 25 because 25 is the *only* number that makes the equation balance. By solving the equation, you have answered the question about the class limit for prealgebra. The class limit is 25.

**NOTE**

Most of the equations that you will solve in this book are linear equations that have only one solution, that is, one number that makes the equation balance. In Chapter 9, and in other algebra courses, you will solve equations that have two or more solutions.

(b) Which of these numbers, 20, 24, or 32, is the solution of the equation $28 = c - 4$?

**Example 1**  Identifying the Solution of an Equation

Which of these numbers, 70, 40, or 60, is the solution of the equation $c - 10 = 50$?

Replace $c$ with each of the numbers. The one that makes the equation balance is the solution.

$70 - 10 \neq 50$   $40 - 10 \neq 50$   $60 - 10 = 50$

Does *not* balance: $70 - 10$ is more than 50.

Does *not* balance: $40 - 10$ is only 30.

Balances: $60 - 10$ is 50.

The solution is 60 because, when $c$ is 60, the equation balances.

**Work Problem 1 at the Side.**

ANSWERS
1. (a) The solution is 65.
   (b) The solution is 32.

**2** **Solve equations, using the addition property of equality.** When solving the book ordering equation, $c + 5 = 30$, you could just look at the equation and think, "What number, plus 5, would balance with 30?" You could easily see that $c$ had to be 25. Not all equations can be solved this easily, so you'll need some tools for the harder ones. The first tool is called the *addition property of equality*.

### Addition Property of Equality

If $a = b$, then $a + c = b + c$.

In other words, you may add the same number to both sides of an equation and still keep it balanced.

Think of the teeter-totter. If there are 3 children of the same size on each side, it will balance. If 2 more children climb onto the left side, the only way to keep the balance is to have 2 more children of the same size climb onto the right side as well.

$$3 = 3$$

$$3 + 2 = 3 + 2$$

All the tools you will learn to use with equations have one goal.

### Goal in Solving an Equation

The goal is to end up with the variable (letter) on one side of the equal sign balancing a number on the other side.

We work on the original equation until we get:

variable = number    or    number = variable

Once we have arrived at that point, the number balancing the variable is the solution to the original equation.

### Example 2 — Using the Addition Property of Equality

Solve each equation and check the solution.

**(a)** $c + 5 = 30$

We want to get the variable, $c$, by itself on the left side of the equal sign. To do that, we add the *opposite* of 5, which is $^-5$. Then $5 + {^-5}$ will be 0.

$$\begin{array}{r} c + 5 = 30 \\ {^-5} \quad {^-5} \\ \hline c + 0 = 25 \end{array}$$

Add $^-5$ to the left side. To keep the balance, add $^-5$ to the right side also. $30 + {^-5}$ is 25. $5 + {^-5}$ is 0.

Recall that adding 0 to any number leaves the number unchanged, so $c + 0$ is $c$.

$$\underbrace{c + 0}_{c} = 25$$
$$c = 25$$

Because $c$ *balances* with 25, the *solution* is 25.

**Check the solution** by replacing $c$ with 25 in the *original equation*.

$$c + 5 = 30 \quad \text{Original equation}$$
$$\underbrace{25 + 5}_{30} = 30 \quad \text{Replace } c \text{ with 25.}$$
$$30 = 30 \quad \text{Balances}$$

Because the equation balances when we use 25 to replace the variable, we know that **25 is the correct solution**. If it had *not* balanced, we would need to rework the problem, find our error, and correct it.

**(b)** $^-5 = x - 3$

We want the variable, $x$, by itself on the right side of the equal sign. (Remember, it doesn't matter which side of the equal sign the variable is on, just so it ends up by itself.) To see what number to add, we change the subtraction to adding the opposite.

$$^-5 = x - 3$$
$$^-5 = x + {^-3} \quad \text{Change subtraction to adding the opposite.}$$
$$\underline{\phantom{-5 =} 3 \quad\quad 3} \quad \text{To get } x \text{ by itself on the right side, add the opposite of } ^-3, \text{ which is 3. Then } ^-3 + 3 \text{ is 0.}$$
$$^-2 = x + 0 \quad \text{Adding 0 to } x \text{ leaves } x \text{ unchanged.}$$
$$^-2 = x \quad x \text{ balances with } ^-2, \text{ so } ^-2 \text{ is the solution.}$$

To keep the balance, add 3 to the left side also; $^-5 + 3$ is $^-2$.

We check the solution by replacing $x$ with $^-2$ in the *original equation*. If the equation balances when we use $^-2$, we know that it is the correct solution. If the equation does *not* balance when we use $^-2$, we made an error and need to try solving the equation again.

**Continued on Next Page**

**Check**

$$-5 = x - 3 \quad \text{Original equation}$$
$$-5 = -2 - 3 \quad \text{Replace } x \text{ with } -2.$$
$$-5 = -2 + -3 \quad \text{Change subtraction to adding the opposite.}$$
$$-5 = -5 \quad \text{Balances; this shows } -2 \text{ is the correct solution.}$$

When $x$ is replaced with $-2$ the equation balances, so **$-2$ is the correct solution**.

> **CAUTION**
>
> When checking the solution to Example 2(b), we ended up with $-5 = -5$. Notice that $-5$ is **not** the solution. The solution is $-2$, the number used to replace $x$ in the original equation.

Work Problem ❷ at the Side.

**3** Simplify equations before using the addition property of equality. Sometimes you can simplify the expression on one or both sides of the equal sign. Doing so will make it easier to solve the equation.

### Example 3  Simplifying before Solving Equations

Solve each equation and check each solution.

**(a)** $y + 8 = 3 - 7$

You cannot simplify the left side because $y$ and 8 are *not* like terms.

$$y + 8 = 3 - 7 \quad \text{Simplify the right side by changing subtraction to adding the opposite.}$$
$$y + 8 = 3 + -7 \quad \text{Add } 3 + -7.$$
$$y + 8 = -4$$

To get $y$ by itself on the left side, add the opposite of 8, which is $-8$.
$8 + -8$ is 0.

$$y + 8 = -4$$
$$\quad\; -8 \quad\; -8 \quad \text{To keep the balance, add } -8 \text{ to the right side also.}$$
$$y + 0 = -12 \quad -4 + -8 \text{ is } -12.$$
$$y = -12$$

The solution is $-12$. Now check the solution.

**Check**

Add $-12 + 8$.

$$y + 8 = 3 - 7 \quad \text{Go back to the } \textit{original} \text{ equation and replace } y \text{ with } -12.$$
$$-12 + 8 = 3 - 7 \quad \text{Change } 3 - 7 \text{ to } 3 + -7.$$
$$-4 = 3 + -7 \quad \text{Add } 3 + -7.$$
$$-4 = -4 \quad \text{Balances; so } -12 \text{ is the correct solution.}$$

When $y$ is replaced with $-12$ the equation balances, so **$-12$ is the correct solution**.

*Continued on Next Page*

---

❷ Solve each equation and check each solution.

**(a)** $12 = y + 5$

**Check**

**(b)** $b - 2 = -6$

**Check**

**ANSWERS**

**2. (a)** $y = 7$
Check $\quad 12 = y + 5$
$\quad\quad\quad 12 = 7 + 5$
Balances $\;\; 12 = 12$

**(b)** $b = -4$
Check $\quad b - 2 = -6$
$\quad\quad\quad -4 + -2 = -6$
Balances $\quad -6 = -6$

## Chapter 2 Understanding Variables and Solving Equations

❸ Simplify each side of the equation when possible. Then solve the equation and check the solution.

(a) $2 - 8 = k - 2$

Check

(b) $4r + 1 - 3r = {^-8} + 11$

Check

---

(b) $^-2 + 2 = {^-4b} - 6 + 5b$

| | | |
|---|---|---|
| Simplify the left side by adding $^-2 + 2$. | $\underbrace{^-2 + 2} = {^-4b} - 6 + 5b$ | Simplify the right side by changing subtraction to adding the opposite. |
| | $0 = {^-4b} + {^-6} + 5b$ | Find like terms. |
| | $0 = \underbrace{^-4b + 5b} + {^-6}$ | Combine $^-4b + 5b$. |
| To keep the balance, add 6 to the left side also. | $0 = 1b + {^-6}$ <br> $\underline{\;\;6\;\;} \;\;\;\;\;\;\;\;\;\;\;\; \underline{\;\;6\;\;}$ | To get $1b$ by itself, add the opposite of $^-6$, which is 6. |
| | $6 = 1b + 0$ | |
| | $6 = 1b$ | $1b$ is equivalent to $b$. |
| | $6 = b$ | |

The solution is 6.

| Check | $^-2 + 2 = {^-4b} - 6 + 5b$ | Go back to the *original* equation and replace each $b$ with 6. |
|---|---|---|
| Add $^-2 + 2$. | $\underbrace{^-2 + 2} = {^-4 \cdot 6} - 6 + 5 \cdot 6$ | Do multiplications first. |
| | $0 = {^-24} + {^-6} + 30$ | Change subtraction to adding the opposite. |
| | $0 = \underbrace{^-30 + 30}$ | Add from left to right. |
| | $0 = 0$ | Balances |

When $b$ is replaced with 6 the equation balances, so **6 is the correct solution**.

---

**CAUTION**

When checking a solution, always go back to the *original* equation. That way you will catch any errors you made when simplifying each side of the equation.

---

**Work Problem ❸ at the Side.**

---

**ANSWERS**

3. (a) $k = {^-4}$
   Check  $2 - 8 = k - 2$
   $\underbrace{2 + {^-8}} = \underbrace{^-4 + {^-2}}$
   Balances  $^-6 = {^-6}$

(b) $r = 2$
   Check  $4r + 1 - 3r = {^-8} + 11$
   $\underbrace{4 \cdot 2} + 1 - \underbrace{3 \cdot 2} = \underbrace{^-8 + 11}$
   $8 + 1 - 6 = 3$
   $\underbrace{9 + {^-6}} = 3$
   Balances  $3 = 3$

## 2.3 Exercises

*In each list of numbers, find the one that is a solution of the given equation. See Example 1.*

1. $n - 50 = 8$

   58, 42, 60

2. $r - 20 = 5$

   15, 30, 25

3. $^-6 = y + 10$

   $^-4, ^-16, 16$

4. $^-4 = x + 13$

   $17, ^-17, ^-9$

5. $t + 12 = 0$

   $0, ^-12, ^-24$

6. $b - 8 = 0$

   $8, 0, ^-8$

*Solve each equation and check each solution. See Example 2.*

7. $p + 5 = 9$      **Check** $p + 5 = 9$

8. $a + 3 = 12$     **Check** $a + 3 = 12$

9. $8 = r - 2$     **Check** $8 = r - 2$

10. $3 = b - 5$     **Check** $3 = b - 5$

11. $^-5 = n + 3$     **Check**

12. $^-1 = a + 8$     **Check**

13. $^-4 + k = 14$     **Check**

14. $^-9 + y = 7$     **Check**

15. $y - 6 = 0$     **Check**

16. $k - 15 = 0$     **Check**

17. $7 = r + 13$          **Check**

18. $12 = z + 19$          **Check**

19. $x - 12 = {}^-1$          **Check**

20. $m - 3 = {}^-9$          **Check**

21. ${}^-5 = {}^-2 + t$          **Check**

22. ${}^-1 = {}^-10 + w$          **Check**

*A solution is given for each equation. Show how to check the solution. If the solution is correct, leave it. If the solution is not correct, solve the equation and check your new solution. See Example 2.*

23. $z - 5 = 3$          **Check** $z - 5 = 3$
    The solution is ${}^-2$.          $\downarrow$

24. $x - 9 = 4$          **Check** $x - 9 = 4$
    The solution is 13.          $\downarrow$

25. $7 + x = {}^-11$          **Check**
    The solution is ${}^-18$.

26. $2 + k = {}^-7$          **Check**
    The solution is ${}^-5$.

27. ${}^-10 = {}^-10 + b$          **Check**
    The solution is 10.

28. $0 = {}^-14 + a$          **Check**
    The solution is 0.

*Simplify each side of the equation when possible. Then solve the equation and check the solution. Show your work. See Example 3.*

**29.** $c - 4 = {}^-8 + 10$     **Check**

**30.** $b - 8 = 10 - 6$     **Check**

**31.** ${}^-1 + 4 = y - 2$     **Check**

**32.** $2 + 3 = k - 4$     **Check**

**33.** $10 + b = {}^-14 - 6$     **Check**

**34.** $1 + w = {}^-8 - 8$     **Check**

**35.** $t - 2 = 3 - 5$     **Check**

**36.** $p - 8 = {}^-10 + 2$     **Check**

**37.** $10z - 9z = {}^-15 + 8$     **Check**

**38.** $2r - r = 5 - 10$     **Check**

**39.** ${}^-5w + 2 + 6w = {}^-4 + 9$     **Check**

**40.** ${}^-2t + 4 + 3t = 6 - 7$     **Check**

Solve each equation. Show your work. See Examples 2 and 3.

**41.** $^-3 - 3 = 4 - 3x + 4x$

**42.** $^-5 - 5 = ^-2 - 6b + 7b$

**43.** $^-3 + 7 - 4 = ^-2a + 3a$

**44.** $6 - 11 + 5 = ^-8c + 9c$

**45.** $y - 75 = ^-100$

**46.** $a - 200 = ^-100$

**47.** $^-x + 3 + 2x = 18$

**48.** $^-s + 2s - 4 = 13$

**49.** $82 = ^-31 + k$

**50.** $^-5 = 72 + w$

**51.** $^-2 + 11 = 2b - 9 - b$

**52.** $^-6 + 7 = 2h - 1 - h$

**53.** $r - 6 = 7 - 10 - 8$

**54.** $m - 5 = 2 - 9 + 1$

**55.** $^-14 = n + 91$

**56.** $66 = x - 28$

**57.** $^-9 + 9 = 5 + h$

**58.** $18 - 18 = 6 + p$

**59.** A student did this work when solving an equation. Do you agree that the solution is $^-7$? Explain why or why not.

$$^-8 + 1 = x + 7$$
$$^-7 = x + 7$$
$$\underline{^-7} = \underline{^-7}$$
$$^-14 = x + 0$$
$$^-14 = x$$

**Check**
$$^-8 + 1 = x + 7$$
$$^-8 + 1 = ^-14 + 7$$
$$^-7 = ^-7$$
Balances, so $^-7$ is the solution.

**60.** A student did this work when solving an equation. Show how to check the solution. If the solution does not check, find and correct the errors.

$$^-3 - 6 = n - 5$$
$$^-3 + 6 = n - 5$$
$$3 = n - 5$$
$$\underline{^-5} \qquad \underline{^-5}$$
$$^-2 = n + 0$$
$$^-2 = n$$

**61.** West Community College always orders 10 extra robes for the graduation ceremony. The college ordered 305 robes this year. Solving the equation $g + 10 = 305$ will give you the number of graduates ($g$) this year. Solve the equation.

**62.** Refer to Exercise 61. The college ordered 278 robes last year. Solve the equation $g + 10 = 278$ to find the number of graduates last year.

**63.** The warmer the temperature, the faster a field cricket chirps. Solving the equation $92 = c + 40$ will give you the number of chirps (in 15 seconds) when the temperature is 92 degrees. Solve the equation.

**64.** Refer to Exercise 63. Solve the equation $77 = c + 40$ to find the number of times a field cricket chirps (in 15 seconds) when the temperature is 77 degrees.

65. During the summer months, Ernesto spends an average of only $45 per month on parking fees by riding his bike to work on nice days. This is $65 less per month than what he spends for parking in the winter. Solving the equation $p - 65 = 45$ will give you his monthly parking fees in the winter. Solve the equation.

66. By walking to work several times a week in the summer, Aimee spends an average of $56 less per month on parking fees. If she spends $98 per month on parking in the summer, solve the equation $p - 56 = 98$ to find her monthly parking fees in the winter.

*Solve each equation. Show your work.*

67. $^-17 - 1 + 26 - 38 = {^-3} - m - 8 + 2m$

68. $19 - 38 - 9 + 11 = {^-t} - 6 + 2t - 6$

69. $^-6x + 2x + 6 + 5x = |0 - 9| - |{^-6} + 5|$

70. $^-h - |{^-9} - 9| + 8h - 6h = {^-12} - |{^-5} + 0|$

**RELATING CONCEPTS (Exercises 71–72)** FOR INDIVIDUAL OR GROUP WORK

*Use what you have learned about solving equations to* **work Exercises 71 and 72 in order.**

71. **(a)** Write two *different* equations that have $^-2$ as the solution. Be sure that you have to use the *addition property of equality* to solve the equations. Show how to solve each equation. Use Exercises 29–34 as models.

    **(b)** Follow the directions in part (a), but this time write two equations that have 0 as the solution.

72. Not all equations have solutions that are integers. Try solving these equations.

    **(a)** $x + 1 = 1\frac{1}{2}$

    **(b)** $\frac{1}{4} = y - 1$

    **(c)** $\$2.50 + n = \$3.35$

    **(d)** Write two more equations that have fraction or decimal solutions.

# 2.4 SOLVING EQUATIONS USING DIVISION

**OBJECTIVES**

1. Solve equations, using the division property of equality.
2. Simplify equations before using the division property of equality.
3. Solve equations such as $^{-}x = 5$.

**1** **Solve equations, using the division property of equality.** In **Section 2.1** you worked with the expression for finding the perimeter of a square-shaped garden, that is, finding the total distance around all four sides of the garden:

$4s$, where $s$ is the length of one side of the square.

Suppose you know that 24 feet of fencing were used around a square-shaped garden. What was the length of one side of the garden? To answer this question, write an equation showing the relationship between what you know and what you don't know.

You don't know the length of one side.
You do know that there are 4 sides.
You do know the perimeter.

$$4\ s = 24$$

To solve the equation, what number can replace $s$ so that the equation balances? You can see that $s$ is 6 feet.

$$4 \cdot 6 = 24$$

Balances:
$4 \cdot 6$ is exactly 24.

The **solution** is **6 feet** because **6** is the *only* number that makes the equation balance. You have answered the question about the length of one side: The length is 6 feet.

There is a tool that you can use to solve equations such as $4s = 24$. It is the *division property of equality*.

### Division Property of Equality

If $a = b$, then $\dfrac{a}{c} = \dfrac{b}{c}$ as long as $c$ is not 0.

In other words, you may divide both sides of an equation by the same nonzero number and still keep it balanced.

In **Section 2.3**, you saw that *adding* the same number to both sides of an equation kept it balanced. We could also have *subtracted* the same number from both sides because subtraction is defined as adding the opposite. Now we're saying that you can *divide* both sides by the same number. In Chapter 4 we'll *multiply* both sides by the same number.

### Equality Principle for Solving an Equation

As long as you do the *same* thing to *both* sides of an equation, the balance is maintained and you still have a true equation. (The only exception is that you cannot divide by 0.)

**120** Chapter 2 Understanding Variables and Solving Equations

## ❶ Solve each equation and check each solution.

(a) $4s = 44$

Check

(b) $27 = {}^-9p$

Check

(c) ${}^-40 = {}^-5x$

Check

(d) $7t = {}^-70$

Check

---

**ANSWERS**

1. (a) $s = 11$
   Check $4s = 44$
   $4 \cdot 11 = 44$
   Balances $44 = 44$

(b) $p = {}^-3$
   Check $27 = {}^-9p$
   $27 = {}^-9 \cdot {}^-3$
   Balances $27 = 27$

(c) $x = 8$
   Check ${}^-40 = {}^-5x$
   ${}^-40 = {}^-5 \cdot 8$
   Balances ${}^-40 = {}^-40$

(d) $t = {}^-10$
   Check $7t = {}^-70$
   $7 \cdot {}^-10 = {}^-70$
   Balances ${}^-70 = {}^-70$

---

### Example 1  Using the Division Property of Equality

Solve each equation and check each solution.

**(a)** $4s = 24$

As with any equation, the goal is to get the variable by itself on one side of the equal sign. On the left side we have $4s$, which means $4 \cdot s$. The variable is multiplied by 4. Division is the opposite of multiplication, so dividing by 4 can be used to "undo" multiplying by 4.

The result of $4 \cdot s \div 4$ is just $s$.

To see how this method works, we replace $s$ with several specific values.

Evaluate $4 \cdot s \div 4$ when $s$ is ${}^-3$.

$4 \cdot {}^-3 \div 4$
${}^-12 \div 4$
${}^-3$

The result is the original value of $s$.

Evaluate $4 \cdot s \div 4$ when $s$ is $25$.

$4 \cdot 25 \div 4$
$100 \div 4$
$25$

The result is the original value of $s$.

In algebra we usually write division using a fraction bar.

Divide $4s$ by 4. The fraction bar indicates division: $4s \div 4$ is $s$.

$$\frac{4s}{4} = \frac{24}{4}$$

To keep the balance, divide the right side by 4 also. $24 \div 4$ is 6.

$s = 6$

So, as we already knew, 6 is the solution. We check the solution by replacing $s$ with 6 in the original equation.

Check
$4s = 24$  Original equation
$4 \cdot 6 = 24$  Replace $s$ with 6.
$24 = 24$  Balances

When $s$ is replaced with 6 the equation balances, so **6 is the correct solution.**

**(b)** $42 = {}^-6w$

On the right side of the equation, the variable is *multiplied* by ${}^-6$. To undo the multiplication, *divide* by ${}^-6$.

To keep the balance, divide by ${}^-6$ on the left side also.

$$\frac{42}{{}^-6} = \frac{{}^-6w}{{}^-6}$$

Use division to undo multiplication: ${}^-6 \cdot w \div {}^-6$ is $w$.

${}^-7 = w$

The solution is ${}^-7$.

Check
$42 = {}^-6w$  Original equation
$42 = {}^-6 \cdot {}^-7$  Replace $w$ with ${}^-7$.
$42 = 42$  Balances

When $w$ is replaced with ${}^-7$ the equation balances, so **${}^-7$ is the correct solution.**

**Work Problem ❶ at the Side.**

**Section 2.4  Solving Equations Using Division**  121

> **CAUTION**
>
> Be careful to divide both sides by the *same* number as the coefficient of the variable term. In Example 1(b), the coefficient of $^-6w$ is $^-6$, so divide both sides by $^-6$. (Do **not** divide by the *opposite* of $^-6$, which is 6. Use the opposite only when you're *adding* the same number to both sides.)

**2** Simplify equations before using the division property of equality. You can sometimes simplify the expression on one or both sides of the equal sign, as you did in **Section 2.3**.

**Example 2** Simplifying before Solving Equations

Solve each equation and check each solution.

**(a)** $4y - 7y = ^-12$

| | | |
|---|---|---|
| Simplify the left side by combining like terms. | $4y - 7y = ^-12$ | The right side cannot be simplified. |
| Change subtraction to adding the opposite. | $4y + {^-7y} = ^-12$ | |
| Divide by the coefficient, which is $^-3$. | $\dfrac{^-3y}{^-3} = \dfrac{^-12}{^-3}$ | To keep the balance, divide by $^-3$ on the right side also. |
| | $y = 4$ | $^-12 \div {^-3}$ is 4. |

The solution is 4.

**Check**

| | | |
|---|---|---|
| | $4y - 7y = ^-12$ | Go back to the *original* equation and replace each $y$ with 4. |
| Do multiplications first. | $4 \cdot 4 - 7 \cdot 4 = ^-12$ | |
| Change subtraction to adding the opposite. | $16 - 28 = ^-12$ | |
| | $16 + {^-28} = ^-12$ | |
| | $^-12 = ^-12$ | Balances |

When $y$ is replaced with 4 the equation balances, so **4 is the correct solution**.

**(b)** $3 - 10 + 7 = h + 7h$

| | | |
|---|---|---|
| Change subtraction to adding the opposite. | $3 - 10 + 7 = h + 7h$ | Write the understood 1 as the coefficient of $h$. |
| Add from left to right. | $3 + {^-10} + 7 = 1h + 7h$ | Combine like terms. |
| | $^-7 + 7 = 8h$ | |
| To keep the balance, divide by 8 on the left side also: $0 \div 8$ is 0. | $\dfrac{0}{8} = \dfrac{8h}{8}$ | Divide by the coefficient, which is 8. |
| | $0 = h$ | |

The solution is 0.

**Check**

$3 - 10 + 7 = h + 7h$
$3 + {^-10} + 7 = 0 + 7 \cdot 0$
$^-7 + 7 = 0 + 0$
$0 = 0$  Balances

Go back to the *original* equation and replace each $h$ with 0.

When $h$ is replaced with 0 the equation balances, so **0 is the correct solution**.

────────────── Work Problem **2** at the Side.

**2** Simplify each side of the equation when possible. Then solve the equation and check the solution.

**(a)** $^-28 = ^-6n + 10n$

**Check**

**(b)** $p - 14p = ^-2 + 18 - 3$

**Check**

**ANSWERS**

2. **(a)** $n = ^-7$
   Check  $^-28 = ^-6n + 10n$
   $^-28 = ^-6 \cdot {^-7} + 10 \cdot {^-7}$
   $^-28 = 42 + {^-70}$
   Balances  $^-28 = ^-28$

**(b)** $p = ^-1$
   Check  $p - 14p = ^-2 + 18 - 3$
   $^-1 - 14(^-1) = 16 - 3$
   $^-1 - {^-14} = 13$
   $^-1 + {^+14} = 13$
   Balances  $13 = 13$

**3** Solve each equation and check each solution.

(a) $^-k = ^-12$

Check

(b) $7 = ^-t$

Check

(c) $^-m = ^-20$

Check

**ANSWERS**
3. (a) $k = 12$
  Check $^-1k = ^-12$
  $^-1 \cdot 12 = ^-12$
  Balances $^-12 = ^-12$

(b) $t = ^-7$
  Check $7 = ^-1t$
  $7 = ^-1 \cdot ^-7$
  Balances $7 = 7$

(c) $m = 20$
  Check $^-1m = ^-20$
  $^-1 \cdot 20 = ^-20$
  Balances $^-20 = ^-20$

**3** Solve equations such as $^-x = 5$. When solving equations, do *not* leave a negative sign in front of the variable.

### Example 3 Solving an Equation of the Type $^-x = 5$

Solve $^-x = 5$ and check the solution.

It may look as if there is nothing more we can do to the equation $^-x = 5$, but $^-x$ is *not* the same as $x$. To see this, we write in the understood $^-1$ as the coefficient of $^-x$.

$$^-x = 5 \text{ can be written } ^-1x = 5$$

We want the coefficient of $x$ to be $^+1$, not $^-1$. To accomplish that, we can divide both sides by the coefficient of $x$, which is $^-1$.

$$\frac{^-1x}{^-1} = \frac{5}{^-1}$$ 
Divide both sides by $^-1$.
On the left side, $^-1 \div ^-1$ is 1.
On the right side, $5 \div ^-1$ is $^-5$.

$$1x = ^-5$$

Now $x$ is by itself on one side of the equal sign and has a coefficient of $^+1$. The solution is $^-5$.

Check   $^-x = 5$ ← Go back to the *original* equation.

$^-1x = 5$   Write in the understood $^-1$ as the coefficient of $^-x$.

$^-1 \cdot ^-5 = 5$   Replace $x$ with $^-5$.

$5 = 5$   Balances

When $x$ is replaced with $^-5$ the equation balances, so **$^-5$ is the correct solution**.

### CAUTION

As the last step in solving an equation, do *not* leave a negative sign in front of a variable. For example, do *not* leave $^-y = ^-8$. Write in the understood $^-1$ as the coefficient, so that

$^-y = ^-8$   is written as   $^-1y = ^-8$.

Then divide both sides by $^-1$ to get $y = 8$. The solution is 8.

Work Problem **3** at the Side.

## 2.4 Exercises

*Solve each equation and check each solution. See Example 1.*

1. $6z = 12$     **Check** $6z = 12$
2. $8k = 24$     **Check** $8k = 24$

3. $48 = 12r$     **Check**
4. $99 = 11m$     **Check**

5. $3y = 0$     **Check**
6. $5a = 0$     **Check**

7. $^-7k = 70$     **Check**
8. $^-6y = 36$     **Check**

9. $^-54 = {^-9}r$     **Check**
10. $^-36 = {^-4}p$     **Check**

11. $^-25 = 5b$     **Check**
12. $^-70 = 10x$     **Check**

*Simplify where possible. Then solve each equation and check each solution. See Example 2.*

13. $2r = {^-7} + 13$     **Check** $2r = \underbrace{{^-7} + 13}$
14. $6y = 28 - 4$     **Check** $6y = \underbrace{28 - 4}$

15. $^-12 = 5p - p$     **Check**
16. $20 = z - 11z$     **Check**

*Solve each equation. Show your work. See Examples 1 and 2.*

**17.** $3 - 28 = 5a$

**18.** $^-55 + 7 = 8n$

**19.** $x - 9x = 80$

**20.** $4c - c = {^-27}$

**21.** $13 - 13 = 2w - w$

**22.** $^-11 + 11 = 8t - 7t$

**23.** $3t + 9t = 20 - 10 + 26$

**24.** $6m + 6m = 40 + 20 - 12$

**25.** $0 = {^-9t}$

**26.** $^-10 = 10b$

**27.** $^-14m + 8m = 6 - 60$

**28.** $7w - 14w = 1 - 50$

**29.** $100 - 96 = 31y - 35y$

**30.** $150 - 139 = 20x - 9x$

*Use multiplication to simplify the side of the equation with the variable. Then solve each equation.*

**31.** $3(2z) = {}^-30$

**32.** $2(4k) = 16$

**33.** $50 = {}^-5(5p)$

**34.** $60 = 4({}^-3a)$

**35.** $^-2(^-4k) = 56$

**36.** $^-5(4r) = {}^-80$

**37.** $^-90 = {}^-10(^-3b)$

**38.** $^-90 = {}^-5(^-2y)$

*Solve each equation. See Example 3.*

**39.** $^-x = 32$

**40.** $^-c = 23$

**41.** $^-2 = {}^-w$

**42.** $^-75 = {}^-t$

**43.** $^-n = {}^-50$

**44.** $^-x = {}^-1$

**45.** $10 = {}^-p$

**46.** $100 = {}^-k$

**47.** Look again at the solutions to Exercises 39–46. Describe the pattern you see. Then write a rule for solving equations with a negative sign in front of the variable, such as $^-x = 5$.

**48.** Explain the division property of equality in your own words.

**49.** Explain and correct the error made by a student who solved this equation.

$$3x = \underbrace{16 - 1}$$
$$\frac{3x}{^-3} = \frac{15}{^-3}$$
$$x = {^-5}$$

**50.** Write two *different* equations that have $^-4$ as the solution. Be sure that you have to use the division property of equality to solve the equations. Show how to solve each equation.

**51.** The perimeter of a triangle with sides of equal length is 3 times the length of one side (*s*). If the perimeter is 45 ft, solving the equation $3s = 45$ will give the length of one side. Solve the equation.

**52.** Refer to Exercise 51. If the perimeter of the triangle is 63 inches, solve the equation $3s = 63$ to find the length of one side.

**53.** The perimeter of a pentagon with sides of equal length is 5 times the length of one side (*s*). If the perimeter is 120 meters, solve the equation $120 = 5s$ to find the length of one side.

**54.** Refer to Exercise 53. If the pentagon has a perimeter of 335 yards, solving the equation $335 = 5s$ will give the length of one side. Solve the equation.

*Solve each equation. Show your work.*

**55.** $89 - 116 = {^-4}(^-4y) - 9(2y) + y$

**56.** $58 - 208 = {^-b} + 8(^-3b) + 5(^-5b)$

**57.** $^-37(14x) + 28(21x) = |72 - 72| + |^-166 + 96|$

**58.** $6a - 10a - 3(2a) = |^-25 - 25| - 5(8)$

# 2.5 SOLVING EQUATIONS WITH SEVERAL STEPS

**OBJECTIVES**

1. Solve equations, using the addition and division properties of equality.
2. Solve equations, using the distributive, addition, and division properties.

**1** Solve equations, using the addition and division properties of equality.
To solve some equations, you need to use both the addition property of equality (see **Section 2.3**) and the division property of equality (see **Section 2.4**). Here are the steps.

### Solving an Equation by Using the Addition and Division Properties

*Step 1* *Add* the same amount to both sides of the equation so that the variable term (the variable and its coefficient) ends up by itself on one side of the equal sign.

*Step 2* *Divide* both sides by the coefficient of the variable term to find the solution.

*Step 3* *Check* the solution by going back to the *original* equation.

**Example 1** Solving an Equation with Several Steps

Solve this equation and check the solution: $5m + 1 = 16$.

*Step 1* Get the variable term by itself on one side of the equal sign.
The variable term is $5m$. Adding $^-1$ to the left side of the equation will leave $5m$ by itself. To keep the balance, add $^-1$ to the right side also.

$$5m + 1 = 16$$
$$\underline{\phantom{5m+} ^-1 \phantom{==} ^-1}$$
$$\underbrace{5m + 0}_{} = 15$$
$$5m = 15$$

*Step 2* Divide both sides by the coefficient of the variable term. In $5m$, the coefficient is 5, so divide both sides by 5.

$$\frac{5m}{5} = \frac{15}{5}$$
$$m = 3$$

*Step 3* Check the solution by going back to the *original* equation.

$$5m + 1 = 16 \quad \text{Use the original equation}$$
$$\downarrow \quad\quad\quad\quad \text{and replace } m \text{ with 3.}$$
$$5(3) + 1 = 16$$
$$\underbrace{15}_{} + 1 = 16$$
$$\underbrace{16}_{} = 16 \quad \text{Balances}$$

When $m$ is replaced with 3 the equation balances, so **3 is the correct solution**.

━━━━━━━━━━━━━━━━━━━━ Work Problem **1** at the Side.

**1** Solve each equation and check each solution.

(a) $2r + 7 = 13$

Check

(b) $20 = 6y - 4$

Check

(c) $^-10z - 9 = 11$

Check

**ANSWERS**
1. (a) $r = 3$
   Check  $\underbrace{2(3)}_{} + 7 = 13$
   $\phantom{Check}$ $6 + 7 = 13$
   Balances  $13 = 13$

(b) $y = 4$
   Check  $20 = \underbrace{6(4)}_{} - 4$
   $\phantom{Check}$ $20 = \underbrace{24}_{} - 4$
   Balances  $20 = 20$

(c) $z = ^-2$
   Check  $\underbrace{^-10(^-2)}_{} - 9 = 11$
   $\phantom{Check}$ $\underbrace{20}_{} - 9 = 11$
   Balances  $11 = 11$

So far, variable terms have appeared on just one side of the equal sign. But some equations start with variable terms on both sides. In that case, you can use the addition property of equality to add the same *variable term* to both sides of the equation, just as you have added the same *number* to both sides.

❷ Solve each equation *two ways*. First keep the variable term on the *left* side when you solve. Then solve again, keeping the variable term on the *right* side. Compare the solutions.

(a) $3y - 1 = 2y + 7$

$3y - 1 = 2y + 7$

(b) $3p - 2 = p - 6$

$3p - 2 = p - 6$

First decide whether to keep the variable term on the left side, or to keep the variable term on the right side. It doesn't matter which one you keep; just pick one side or the other. Then use the addition property to "get rid of" the variable term on the *other* side by adding its opposite.

### Example 2  Solving an Equation with Variable Terms on Both Sides

Solve this equation and check the solution: $2x - 2 = 5x - 11$.

First let's keep $2x$, the variable term on the *left* side. That means we need to "get rid of" $5x$ on the *right* side. We can do that by adding the opposite of $5x$, which is $^-5x$.

To keep the balance, add $^-5x$ to the left side also. Write $^-5x$ under $2x$, *not* under 2.

$$2x - 2 = 5x - 11$$
$$\underline{^-5x} \qquad \underline{^-5x}$$
$$^-3x - 2 = 0 - 11$$
$$^-3x + ^-2 = 0 + ^-11$$

Write $^-5x$ under $5x$, *not* under 11. $5x + ^-5x$ is $0x$, or 0.
Change subtractions to adding the opposite.

To get $^-3x$ by itself, add 2 to both sides.

$$\underline{\phantom{^-3x+}2\phantom{=0+}2}$$
$$^-3x + 0 = 0 + ^-9$$

Divide both sides by $^-3$, the coefficient of the variable term.

$$\frac{^-3x}{^-3} = \frac{^-9}{^-3}$$
$$x = 3$$

Suppose that, in the first step, we decided to keep $5x$ on the *right* side and "get rid of" $2x$ on the *left* side. Let's see what happens.

$$2x - 2 = 5x - 11$$

Add $^-2x$ to both sides.

$$\underline{^-2x} \qquad \underline{^-2x}$$
$$0 - 2 = 3x - 11$$
$$0 + ^-2 = 3x + ^-11$$

Change subtractions to adding the opposite.
To get $3x$ by itself, add 11 to both sides.

$$\underline{\phantom{0+}11\phantom{=3x+}11}$$
$$0 + 9 = 3x + 0$$

$$\frac{9}{3} = \frac{3x}{3}$$

Divide both sides by 3.

$$3 = x$$

The two solutions are the same. In both cases, $x$ balances with 3.

Notice that we used the addition principle *twice:* once to "get rid of" a variable term and once to "get rid of" a number. We could have done those steps in the reverse order without changing the result.

> **NOTE**
> More than one sequence of steps will work to solve complicated equations. The basic approach is the following:
> - Simplify each side of the equation, if possible.
> - Get the variable term by itself on one side of the equal sign and a number by itself on the other side.
> - Divide both sides by the coefficient of the variable term.

**Work Problem ❷ at the Side.**

**ANSWERS**
2. (a) $y = 8$ and $8 = y$
   (b) $p = ^-2$ and $^-2 = p$

Section 2.5 Solving Equations with Several Steps  129

**2** Solve equations, using the distributive, addition, and division properties.

If an equation contains parentheses, check to see whether you can use the distributive property to remove them.

**Example 3** Solving an Equation by Using the Distributive Property

Solve this equation and check the solution: $^-6 = 3(y - 2)$.

We can use the distributive property to simplify the right side of the equation. Recall from **Section 2.2** that

$3(y - 2)$ can be written as $3 \cdot y - 3 \cdot 2$
$\phantom{3(y - 2) \text{ can be written as }} 3y - 6$

So the original equation $^-6 = 3(y - 2)$ becomes $^-6 = 3y - 6$.

$^-6 = 3y - 6$    Change subtraction to adding the opposite.

$^-6 = 3y + {}^-6$    To get $3y$ by itself, add 6 to both sides.
$\underline{\phantom{-}6 \phantom{= 3y +}} \phantom{-} \underline{6\phantom{xx}}$

$\dfrac{0}{3} = \dfrac{3y}{3}$    Divide both sides by 3, the coefficient of $3y$.

$0 = y$

The solution is 0.

**Check**    $^-6 = 3(y - 2)$    Go back to the *original* equation and replace $y$ with 0.

$^-6 = 3(0 - 2)$    Follow the order of operations; work inside parentheses first.

$^-6 = 3(0 + {}^-2)$    Change subtraction to addition.

$^-6 = \phantom{xx}3({}^-2)$

$^-6 = \phantom{xx}{}^-6$    Balances

When $y$ is replaced with 0 the equation balances, so **0 is the correct solution**.

━━━━━━━━━━━━━━━━━━━━━━━ Work Problem **3** at the Side.

Here is a summary of all the steps you can use to solve an equation. Sometimes you will use only two or three steps, and sometimes you will need all five steps.

**Solving an Equation**

*Step 1*   If possible, use the **distributive property** to remove parentheses.

*Step 2*   **Combine** any like terms on the left side of the equation. Combine any like terms on the right side of the equation.

*Step 3*   **Add** the same amount to both sides of the equation so that the variable term ends up by itself on one side of the equal sign and a number is by itself on the other side. You may have to do this step more than once.

*Step 4*   **Divide** both sides by the coefficient of the variable term to find the solution.

*Step 5*   **Check** your solution by going back to the *original* equation. Replace the variable with your solution. Follow the order of operations to complete the calculations. If the two sides of the equation balance, your solution is correct.

**3** Solve each equation and check each solution.

(a) $^-12 = 4(y - 1)$

(b) $5(m + 4) = 20$

(c) $6(t - 2) = 18$

**ANSWERS**
3. (a) $y = {}^-2$   (b) $m = 0$   (c) $t = 5$

**4** Solve each equation and check each solution.

**(a)** $3(b + 7) = 2b - 1$

**(b)** $6 - 2n = 14 + 4(n - 5)$

---

**Example 4  Solving an Equation**

Solve this equation and check the solution: $8 + 5(m + 2) = 6 + 2m$.

*Step 1*  Use the distributive property on the left side.

$$8 + 5(m + 2) = 6 + 2m$$

*Step 2*  Combine like terms on the left side.

$$8 + 5m + 10 = 6 + 2m$$   No like terms on the right side.

$$5m + 18 = 6 + 2m$$

*Step 3*  Add $^-2m$ to both sides.

$$\underline{\phantom{3m + 18} \;\; -2m \phantom{xx} \underline{\;\;-2m\;\;}}$$
$$3m + 18 = 6 + 0$$
$$3m + 18 = 6$$

*Step 3*  To get $3m$ by itself, add $^-18$ to both sides.

$$\underline{\phantom{3m} \;\; -18 \phantom{xx} -18\;\;}$$
$$3m + 0 = \,^-12$$

*Step 4*  Divide both sides by 3, the coefficient of the variable term $3m$.

$$\frac{3m}{3} = \frac{^-12}{3}$$
$$m = -4$$

The solution is $^-4$.

*Step 5*  Check

$$8 + 5(m + 2) = 6 + 2m$$   Replace each $m$ with $^-4$.

$$8 + 5(^-4 + 2) = 6 + 2(^-4)$$
$$8 + 5(^-2) = 6 + \,^-8$$
$$8 + \,^-10 = \,^-2$$
$$^-2 = \,^-2$$   Balances

When $m$ is replaced with $^-4$ the equation balances, so **$^-4$ is the correct solution**.

**Work Problem 4 at the Side.**

---

ANSWERS

**4.** (a) $b = \,^-22$  (b) $n = 2$

## 2.5 Exercises

*Solve each equation and check each solution. See Example 1.*

1. $7p + 5 = 12$  **Check** $7p + 5 = 12$

2. $6k + 3 = 15$  **Check** $6k + 3 = 15$

3. $2 = 8y - 6$  **Check**

4. $10 = 11p - 12$  **Check**

5. $-3m + 1 = 1$  **Check**

6. $-4k + 5 = 5$  **Check**

7. $28 = -9a + 10$  **Check**

8. $75 = -10w + 25$  **Check**

9. $-5x - 4 = 16$  **Check**

10. $-12b - 3 = 21$  **Check**

Chapter 2 Understanding Variables and Solving Equations

*Solve each equation **two** ways. First keep the variable term on the left side when you solve it. Then solve it again, keeping the variable term on the right side. Finally, check your solution. See Example 2.*

**11.** $6p - 2 = 4p + 6$ $\qquad\qquad$ $6p - 2 = 4p + 6$ $\qquad\qquad$ **Check** $\quad 6p - 2 = 4p + 6$

**12.** $5y - 5 = 2y + 10$ $\qquad\qquad$ $5y - 5 = 2y + 10$ $\qquad\qquad$ **Check** $\quad 5y - 5 = 2y + 10$

**13.** $^-2k - 6 = 6k + 10$ $\qquad\qquad$ $^-2k - 6 = 6k + 10$ $\qquad\qquad$ **Check**

**14.** $5x + 4 = {^-3x} - 4$ $\qquad\qquad$ $5x + 4 = {^-3x} - 4$ $\qquad\qquad$ **Check**

**15.** $^-18 + 7a = 2a + 7$ $\qquad\qquad$ $^-18 + 7a = 2a + 7$ $\qquad\qquad$ **Check**

**16.** $^-9 + 2z = 9z + 12$     $^-9 + 2z = 9z + 12$     **Check**

*Use the distributive property to help you solve each equation. Show your work. See Example 3.*

**17.** $8(w - 2) = 32$     **18.** $9(b - 4) = 27$     **19.** $^-10 = 2(y + 4)$

**20.** $^-3 = 3(x + 6)$     **21.** $^-4(t + 2) = 12$     **22.** $^-5(k + 3) = 25$

**23.** $6(x - 5) = ^-30$     **24.** $7(r - 7) = ^-49$     **25.** $^-12 = 12(h - 2)$

**26.** $^-11 = 11(c - 3)$     **27.** $0 = ^-2(y + 2)$     **28.** $0 = ^-9(b + 1)$

*Solve each equation. Show your work. See Example 4.*

**29.** $6m + 18 = 0$

**30.** $8p - 40 = 0$

**31.** $6 = 9w - 12$

**32.** $8 = 8h + 24$

**33.** $5x = 3x + 10$

**34.** $7n = {}^-2n - 36$

**35.** $2a + 11 = 8a - 7$

**36.** $r - 10 = 10r + 8$

**37.** $7 - 5b = 28 + 2b$

**38.** $1 - 8t = {}^-9 - 3t$

**39.** ${}^-20 + 2k = k - 4k$

**40.** $6y - y = {}^-16 + y$

**41.** $10(c - 6) + 4 = 2 + c - 58$

**42.** $8(z + 7) - 6 = z + 60 - 10$

**43.** ${}^-18 + 13y + 3 = 3(5y - 1) - 2$

**44.** $3 + 5h - 9 = 4(3h + 4) - 1$

**45.** $6 - 4n + 3n = 20 - 35$

**46.** $^-19 + 8 = 6p - 7p - 5$

**47.** $6(c - 2) = 7(c - 6)$

**48.** $^-3(5 + x) = 4(x - 2)$

**49.** $^-5(2p + 2) - 7 = 3(2p + 5)$

**50.** $4(3m - 6) = 72 + 3(m - 8)$

**51.** $^-6b - 4b + 7b = 10 - b + 3b$

**52.** $w + 8 - 5w = {}^-w - 15w + 11w$

**53.** Solve $^-2t - 10 = 3t + 5$. Show each step you take while solving it. Next to each step, write a sentence that explains what you did in that step. Be sure to tell when you used the addition property of equality and when you used the division property of equality.

**54.** Explain the distributive property in your own words. Show two examples of using the distributive property to remove parentheses in an expression.

**55.** Here is one student's solution to an equation. Show how to check the solution. If the solution doesn't check, explain the error and correct it.

$$
\begin{aligned}
-8 + 4a &= 2a + 2 \\
\underline{-2a} & \quad \underline{-2a} \\
-10 + 4a &= 0 + 2 \\
-10 + 4a &= 2 \\
\underline{\phantom{-}10} & \quad \underline{\phantom{-}10} \\
0 + 4a &= 12 \\
\dfrac{4a}{4} &= \dfrac{12}{4} \\
a &= 3
\end{aligned}
$$

**56.** Here is one student's solution to an equation. Show how to check the solution. If the solution doesn't check, explain the error and correct it.

$$
\begin{aligned}
2(x + 4) &= -16 \\
2x + 4 &= -16 \\
\underline{\phantom{2x}-4} & \quad \underline{-4} \\
2x + 0 &= -20 \\
\dfrac{2x}{2} &= \dfrac{-20}{2} \\
x &= -10
\end{aligned}
$$

---

**RELATING CONCEPTS (Exercises 57–60)** FOR INDIVIDUAL OR GROUP WORK

*Work Exercises 57–60 in order.*

**57. (a)** Suppose that the sum of two numbers is negative and you know that one of the numbers is positive. What can you conclude about the other number?

**(b)** How can you tell, just by looking, that the solution to $x + 5 = -7$ cannot be a positive number? Recall your answer from part (a).

**58. (a)** Suppose that the sum of two numbers is positive, and you know that one of the numbers is negative. What can you conclude about the other number?

**(b)** How can you tell, just by looking, that the solution to $-8 + d = 2$ cannot be a negative number? Recall your answer from part (a).

**59. (a)** Suppose the product of two numbers is negative and you know that one of the numbers is negative. What can you conclude about the other number?

**(b)** How can you tell, just by looking, that the solution to $-15n = -255$ cannot be negative?

**60. (a)** Suppose the product of two numbers is positive and you know that one of the numbers is negative. What can you conclude about the other number?

**(b)** How can you tell, just by looking, that the solution to $437 = -23y$ cannot be positive?

# Chapter 2

## SUMMARY

### KEY TERMS

**2.1** 
| | | |
|---|---|---|
| | **variable** | A variable is a letter that represents a number that varies or changes, depending on the situation. |
| | **constant** | A constant is a number that is added or subtracted in an expression. It does not vary. For example, 5 is the constant in the expression $c + 5$. |
| | **expression** | An expression expresses, or tells, the rule for doing something. It is a combination of operations on variables and numbers. |
| | **evaluate the expression** | To evaluate an expression, replace each variable with specific values (numbers) and follow the order of operations. |
| | **coefficient** | The number part in a multiplication expression is the coefficient. For example, 4 is the coefficient in the expression $4s$. |
| **2.2** | **simplifying expressions** | To simplify an expression, write it in a simpler way by combining all the like terms. |
| | **term** | Each addend in an expression is a term. |
| | **variable term** | A variable term has a number part (called the coefficient) multiplied by a variable part (a letter). An example is $4s$. |
| | **like terms** | Like terms are terms with exactly the same variable parts (the same letters and exponents). The coefficients may be different. |
| **2.3** | **equations** | An equation has an equal sign. It shows the relationship between what is known about a problem and what isn't known. |
| | **solve the equation** | To solve an equation, find a number that can replace the variable and make the equation balance. |
| | **solution** | A solution of an equation is a number that can replace the variable and make the equation balance. |
| | **addition property of equality** | The addition property of equality states that adding the same quantity to both sides of an equation will keep it balanced. |
| | **check the solution** | To check the solution of an equation, go back to the *original* equation and replace the variable with the solution. If the equation balances, the solution is correct. |
| **2.4** | **division property of equality** | The division property of equality states that dividing both sides of an equation by the same nonzero number will keep it balanced. |

### TEST YOUR WORD POWER

*See how well you have learned the vocabulary in this chapter. Answers follow the Quick Review.*

1. **A variable**
   (a) can only be the letter $x$
   (b) is never an addend in an expression
   (c) is the solution of an equation
   (d) represents a number that varies.

2. Which expression has 2 as a **coefficient**?
   (a) $x^2$
   (b) $2x$
   (c) $x + 2$
   (d) $2 - x$

3. Which expression has 4 as a **constant** term?
   (a) $4y$
   (b) $y^4$
   (c) $4 + y$
   (d) $\dfrac{y}{4}$

4. Which expression has four **terms**?
   (a) $2 + 3x = {}^-6 + x$
   (b) $2 + 3x - 6 + x$
   (c) $(2)(3x)({}^-6)(x)$
   (d) $2(3x) = {}^-6(x)$

5. **Like terms**
   (a) can be multiplied but not added
   (b) have the same coefficients
   (c) have the same solutions
   (d) have the same variable parts.

6. To **simplify an expression,**
   (a) combine all like terms, using the distributive property
   (b) multiply the exponents
   (c) add all the numbers in the expression
   (d) add the same quantity to both sides.

## QUICK REVIEW

*Concepts*

### 2.1 Evaluating Expressions

Replace each variable with the specified value. Then follow the order of operations to simplify the expression.

*Examples*

The expression for ordering textbooks for two prealgebra classes is $2c + 10$, where $c$ is the class limit. Evaluate the expression when the class limit is 24.

$$2c + 10 \quad \text{Replace } c \text{ with 24.}$$
$$2 \cdot 24 + 10 \quad \text{Multiply first.}$$
$$48 + 10 \quad \text{Add last.}$$
$$58 \quad \text{Order 58 books.}$$

### 2.1 Using Exponents with Variables

An exponent next to a variable tells how many times to use the variable as a factor in multiplication.

Rewrite $^-6x^4$ without exponents.

$^-6x^4$ can be written as $^-6 \cdot x \cdot x \cdot x \cdot x$
Coefficient is $^-6$.    $x$ is used as a factor 4 times.

### 2.1 Evaluating Expressions with Exponents

Rewrite the expression without exponents, replace each variable with the specified value, and multiply all the factors.

Evaluate $x^3 y$ when $x$ is $^-4$ and $y$ is 5.

$x^3 y$ means $x \cdot x \cdot x \cdot y$    Replace $x$ with $^-4$ and $y$ with 5.
$^-4 \cdot ^-4 \cdot ^-4 \cdot 5$    Multiply two factors at a time.
$16 \cdot ^-4 \cdot 5$
$^-64 \cdot 5$
$^-320$

### 2.2 Identifying Like Terms

Like terms have *exactly* the same letters and exponents. The coefficients may be different.

List the like terms in this expression. Then identify the coefficients of the like terms.

$$^-3b + ^-3b^2 + 3ab + b + 3$$

The like terms are $^-3b$ and $b$. The coefficient of $^-3b$ is $^-3$, and the coefficient of $b$ is understood to be 1.

### 2.2 Combining Like Terms

*Step 1* If there are any variable terms with no coefficient, write in the understood 1.

*Step 2* Change any subtractions to adding the opposite.

*Step 3* Find like terms.

*Step 4* Add the coefficients of like terms, keeping the variable part the same.

Simplify $4x^2 - 10 + x^2 + 15$.

Write understood 1.

$4x^2 - 10 + 1x^2 + 15$    Change subtraction to adding the opposite.

$4x^2 + ^-10 + 1x^2 + 15$

$4x^2 + 1x^2 + ^-10 + 15$    Combine $4x^2 + 1x^2$. The variable part stays the same. Also combine $^-10 + 15$.

$(4 + 1)x^2 + 5$

$5x^2 + 5$

The simplified expression is $5x^2 + 5$.

## Concepts

### 2.2 Simplifying Multiplication Expressions

Use the associative property to rewrite the expression so that the two number parts can be multiplied. The variable part stays the same.

### 2.2 Using the Distributive Property

Multiplication distributes over addition and over subtraction. Be careful to multiply *every* term inside the parentheses by the number outside the parentheses.

### 2.3 Solving and Checking Equations Using the Addition Property of Equality

If possible, *simplify* the expression on one or both sides of the equal sign. Next, to get the variable by itself on one side of the equal sign, *add* the same number to both sides. Finally, *check* the solution by going back to the original equation and replacing the variable with the solution. If the equation balances, the solution is correct.

## Examples

Simplify: $-7(5k)$.

Use the associative property of multiplication.

$-7 \cdot (5 \cdot k)$ can be written as $(-7 \cdot 5) \cdot k$
$= -35 \cdot k$
$= -35k$

The simplified expression is $-35k$.

Simplify.

(a) $6(w - 4)$ can be written as $6 \cdot w - 6 \cdot 4$
$= 6w - 24$

The simplified expression is $6w - 24$.

(b) $-3(2b + 5)$ can be written as $-3 \cdot 2b + -3 \cdot 5$
$= -6b + -15$

Use the definition of subtraction "in reverse" to write $-6b + -15$ as $-6b - 15$.

The simplified expression is $-6b - 15$.

Solve this equation and check the solution.

$-5 + 8 = 9 + r$   Simplify the left side by adding $-5 + 8$.
$3 = 9 + r$
$\underline{-9} \quad \underline{-9}$   To get $r$ by itself, add the opposite of 9, which is $-9$, to both sides.
$-6 = 0 + r$
$-6 = r$

The solution is $-6$.

**Check**  $-5 + 8 = 9 + r$   Use the original equation and replace $r$ with $-6$.
$-5 + 8 = 9 + -6$
$3 = 3$   Balances

When $r$ is replaced with $-6$ the equation balances, so $-6$ is the correct solution.

## Concepts

### 2.4 Solving and Checking Equations Using the Division Property of Equality

If possible, *simplify* the expression on one or both sides of the equal sign. Next, to get the variable by itself on one side of the equal sign, *divide* both sides by the coefficient of the variable term. Finally, *check* the solution by going back to the original equation and replacing the variable with the solution. If the equation balances, the solution is correct.

## Examples

Solve this equation and check the solution.

Simplify the left side. Change subtraction to adding the opposite.

$$2h - 6h = 18 + 22$$

Simplify the right side. Add $18 + 22$.

$$2h + {}^-6h = 40$$

Divide by $^-4$, the coefficient of $^-4h$.

$$\frac{^-4h}{^-4} = \frac{40}{^-4}$$

Also divide 40 by $^-4$ to keep the balance.

$$h = {}^-10$$

The solution is $^-10$.

**Check**
$$2h - 6h = 18 + 22 \quad \text{Replace } h \text{ with } {}^-10.$$
$$2({}^-10) - 6({}^-10) = 40$$
$$^-20 - {}^-60 = 40$$
$$^-20 + {}^+60$$
$$40 = 40 \quad \text{Balances}$$

When $h$ is replaced with $^-10$ the equation balances, so $^-10$ is the correct solution.

### 2.4 Solving Equations Such as $^-x = 5$

As the last step in solving an equation, do *not* leave a negative sign in front of the variable, such as $^-x = 5$, because $^-x$ is *not* the same as $x$. Divide both sides by $^-1$, the understood coefficient of $^-x$.

Solve this equation and check the solution.

$$9 = {}^-n$$

Write the understood $^-1$ as the coefficient of $n$.

$$9 = {}^-n \quad \text{can be written as} \quad 9 = {}^-1n$$

Now divide both sides by $^-1$.

$$\frac{9}{^-1} = \frac{^-1n}{^-1}$$

$$^-9 = n$$

The solution is $^-9$.

**Check**
$$9 = {}^-n \quad \text{Original equation}$$
$$9 = {}^-1n \quad \text{Write understood } {}^-1.$$
$$9 = {}^-1({}^-9) \quad \text{Replace } n \text{ with } {}^-9.$$
$$9 = 9 \quad \text{Balances}$$

When $n$ is replaced with $^-9$ the equation balances, so $^-9$ is the correct solution.

## Chapter 2 Summary

### Concepts

**2.5 Solving Equations with Several Steps**

*Step 1* If possible, use the distributive property to remove parentheses.

*Step 2* Combine any like terms on the left side of the equal sign. Combine any like terms on the right side of the equal sign.

*Step 3* Add the same amount to both sides of the equation so that the variable term ends up by itself on one side of the equal sign, and a number is by itself on the other side. You may have to do this step more than once.

*Step 4* Divide both sides by the coefficient of the variable term to find the solution.

*Step 5* Check the solution by going back to the original equation. Replace the variable with the solution. If the equation balances, the solution is correct.

### Examples

Solve this equation and check the solution.

$$3 + 2(y + 8) = 5y + 4$$
$$3 + 2y + 16 = 5y + 4$$
$$2y + 19 = 5y + 4$$
$$\underline{-2y \qquad\quad -2y}$$
$$0 + 19 = 3y + 4$$
$$\underline{\phantom{0+}-4 \qquad\quad -4}$$
$$0 + 15 = 3y + 0$$
$$\frac{15}{3} = \frac{3y}{3}$$
$$5 = y$$

The solution is 5.

**Check**
$$3 + 2(y + 8) = 5y + 4$$
$$3 + 2(5 + 8) = 5(5) + 4$$
$$3 + 2(13) = 25 + 4$$
$$3 + 26 = 29$$
$$29 = 29$$

When $y$ is replaced with 5 the equation balances, so 5 is the correct solution.

---

**ANSWERS TO TEST YOUR WORD POWER**

**1. (d)** *Example:* In $c + 5$, the variable is $c$.  **2. (b)** *Example:* $2y^3$ and $2n$ also have 2 as a coefficient.
**3. (c)** *Example:* $-5a^2 + 4$ also has 4 as a constant term.  **4. (b)** *Example:* $3y^2 - 6y + 2y - 5$ also has four terms. Choices (a) and (d) are equations, not expressions; choice (c) has four *factors*.  **5. (d)** *Example:* $7n^2$ and $-3n^2$ are like terms.  **6. (a)** *Example:* To simplify $4a - 9 + 6a$, combine $4a$ and $6a$ by adding the coefficients. The simplified expression is $10a - 9$.

# Focus on Real-Data Applications

## Algebraic Expressions and Tuition Costs

Algebraic expressions are useful in real-life scenarios in which the same set of instructions are repeated for different choices of numbers. Below is the description of how tuition and fees are calculated for "Resident of District" students at North Harris Montgomery Community College District (NHMCCD) in Texas for 2000–2001. The information is given in the college's schedule and can be found at the Web site www.nhmccd.edu.*

**Fees Required at NHMCCD**

[*Residents of the district pay*] tuition at the rate of $26 per credit hour, a $4 per credit hour technology fee, and a registration fee of $12.

### For Group Discussion

1. Calculate the tuition and fees for a student who is a resident of the district and who enrolls at NHMCCD for the specified number of credit hours in parts **(a)**, **(b)**, and **(c)** below. Then, for part **(d)** let $x$ represent the number of credit hours. Pay attention to the *process* you used in your calculations so that you can write the algebraic expression for $x$ credit hours.

   **(a)** 3 credit hours: _____     **(b)** 9 credit hours: _____

   **(c)** 12 credit hours: _____    **(d)** $x$ credit hours: _____ dollars

   *Write the algebraic expression that represents the tuition and fees for each institution for one semester. Let $x$ represent the number of credit hours. If you have difficulty, first calculate the costs for 3 or 9 credit hours and focus on the process that you used to get the answer.*

2. American River College, California (nonresident student) www.arc.losrios.cc.ca.us*
   Enrollment: $11 per credit hour; parking: $30 per semester; an additional nonresident enrollment: $134 per credit hour; other fees: $8

3. Austin Community College, Texas (out-of-district student) www.austin.cc.tx.us*
   Tuition: $31 per credit hour; parking: $10 per semester; an additional out-of-district tuition: $75 per credit hour; student service fee: $3

4. Valdosta State University, Georgia (in-state student) www.valdosta.edu*
   Matriculation: $78 per credit hour; health fee: $66; student services fee: $78; athletics fees: $97; technology fee: $38; parking fee: $25; special fees also apply.

5. Your college tuition and fees

---

*Note that URLs sometimes change, although that is unlikely for academic institutions. If the Web address given does not work, use a search engine, such as www.yahoo.com, to find the new URL.

# Chapter 2 REVIEW EXERCISES

**[2.1]**

1. Identify the variable, the coefficient, and the constant term in this expression.
$$-3 + 4k$$

2. The expression for ordering test tubes for a chemistry lab is $4c + 10$, where $c$ is the class limit. Evaluate the expression when
   (a) the class limit is 15.
   (b) the class limit is 24.

3. Rewrite each expression without exponents.
   (a) $x^2 y^4$
   (b) $5ab^3$

4. Evaluate each expression when $m$ is 2, $n$ is $-3$, and $p$ is 4.
   (a) $n^2$
   (b) $n^3$
   (c) $-4mp^2$
   (d) $5m^4 n^2$

**[2.2]** Simplify.

5. $ab + ab^2 + 2ab$

6. $-3x + 2y - x - 7$

7. $-8(-2g^3)$

8. $4(3r^2 t)$

9. $5(k + 2)$

10. $-2(3b + 4)$

11. $3(2y - 4) + 12$

12. $-4 + 6(4x + 1) - 4x$

13. Write an expression with four terms that *cannot* be simplified.

**[2.3]** Solve each equation and check each solution.

14. $16 + n = 5$    **Check**

15. $-4 + 2 = 2a - 6 - a$   **Check**

**[2.4]** Solve each equation. Show your work.

16. $48 = -6m$

17. $k - 5k = -40$

18. $-17 + 11 + 6 = 7t$

**19.** $^-2p + 5p = 3 - 21$

**20.** $^-30 = 3(^-5r)$

**21.** $12 = {^-h}$

**[2.5]** *Solve each equation. Show your work.*

**22.** $12w - 4 = 8w + 12$

**23.** $0 = {^-4}(c + 2)$

**24.** Every Wednesday is "treat day" at the office. The person who brings the treats buys two treat items for each employee plus four extras. If Roger brings 34 doughnuts, solve the equation $34 = 2n + 4$ to find the number of employees.

## MIXED REVIEW EXERCISES

*Solve each equation. Show your work.*

**25.** $12 + 7a = 4a - 3$

**26.** $^-2(p - 3) = {^-14}$

**27.** $10y = 6y + 20$

**28.** $2m - 7m = 5 - 20$

**29.** $20 = 3x - 7$

**30.** $b + 6 = 3b - 8$

**31.** $z + 3 = 0$

**32.** $3(2n - 1) = 3(n + 3)$

**33.** $^-4 + 46 = 7(^-3t + 6)$

**34.** $6 + 10d - 19 = 2(3d + 4) - 1$

**35.** $^-4(3b + 9) = 24 + 3(2b - 8)$

# Chapter 2 TEST

*Study Skills Workbook*
**Activity 8**

1. Identify the parts of this expression: $^-7w + 6$.
   Choose from these labels: variable, constant term, coefficient.

   1. _____

2. The expression for buying hot dogs for the company picnic is $3a + 2c$, where $a$ is the number of adults and $c$ is the number of children. Evaluate the expression when there are 45 adults and 21 children.

   2. _____

*Rewrite each expression without exponents.*

3. $x^5 y^3$
4. $4ab^4$

   3. _____

   4. _____

5. Evaluate $^-2s^2 t$ when $s$ is $^-5$ and $t$ is 4.

   5. _____

*Simplify each expression.*

6. $3w^3 - 8w^3 + w^3$
7. $xy - xy$

   6. _____

   7. _____

8. $^-6c - 5 + 7c + 5$
9. $3m^2 - 3m + 3mn$

   8. _____

   9. _____

10. $^-10(4b^2)$
11. $^-5(^-3k)$

    10. _____

    11. _____

12. $7(3t + 4)$
13. $^-4(a + 6)$

    12. _____

    13. _____

14. $^-8 + 6(x - 2) + 5$
15. $^-9b - c - 3 + 9 + 2c$

    14. _____

    15. _____

*Solve each equation and check each solution.*

16. $^-4 = x - 9$    **Check**

17. $^-7w = 77$    **Check**

18. $^-p = 14$    **Check**

19. $^-15 = ^-3(a + 2)$    **Check**

*Solve each equation. Show your work.*

20. $6n + 8 - 5n = {}^-4 + 4$

21. $5 - 20 = 2m - 3m$

22. $^-2x + 2 = 5x + 9$

23. $3m - 5 = 7m - 13$

24. $2 + 7b - 44 = {}^-3b + 12 + 9b$

25. $3c - 24 = 6(c - 4)$

26. Write an equation that requires the *addition* property of equality to solve it and has $^-4$ as its solution. Then write a different equation that requires the *division* property of equality to solve it and has $^-4$ as its solution. Show how to solve each equation.

# Cumulative Review Exercises  Chapters 1–2

1. Write this number in words.
   306,000,004,210

2. Write this number, using digits.
   Eight hundred million, sixty-six thousand

3. Write < or > between each pair of numbers to make a true statement.
   (a) ⁻3 ___ ⁻10   (b) ⁻1 ___ 0

4. Name the property illustrated by each example.
   (a) ⁻6 + 2 = 2 + ⁻6
   (b) 0 · 25 = 0
   (c) 5(⁻6 + 4) = 5 · ⁻6 + 5 · 4

5. (a) Round 9047 to the nearest hundred.
   (b) Round 289,610 to the nearest thousand.

*Simplify*

6. 0 − 8

7. |⁻6| + |4|

8. ⁻3(⁻10)

9. (⁻5)²

10. $\dfrac{^{-}42}{^{-}6}$

11. ⁻19 + 19

12. (⁻4)³

13. $\dfrac{^{-}14}{0}$

14. ⁻5 · 12

15. ⁻20 − 20

16. $\dfrac{45}{^{-}5}$

17. ⁻50 + 25

18. ⁻10 + 6(4 − 7)

19. $\dfrac{^{-}20 - 3(^{-}5) + 16}{(^{-}4)^2 - 3^3}$

*First use front end rounding to estimate the answer to each application problem. Then find the exact answer.*

20. One Siberian tiger was tracked for 22 days while it was searching for food. It traveled 616 miles. What was the average distance it traveled each day?
    *Estimate:*
    
    *Exact:*

21. The temperature in Siberia got down to ⁻48 degrees one night. The next day the temperature rose 23 degrees. What was the daytime temperature?
    *Estimate:*
    
    *Exact:*

22. Doug owned 52 shares of Mathtronic stock that had a total value of $2132. Yesterday the value of each share dropped $8. What are his shares worth now?
    *Estimate:*
    
    *Exact:*

23. Ikuko's monthly rent is $552 plus $35 for parking. How much will she spend for rent and parking in one year?
    *Estimate:*
    
    *Exact:*

24. Rewrite $^-4ab^3c^2$ without exponents.

25. Evaluate $3xy^3$ when $x$ is $^-5$ and $y$ is $^-2$.

*Simplify.*

26. $3h - 7h + 5h$

27. $c^2d - c^2d$

28. $4n^2 - 4n + 6 - 8 + n^2$

29. $^-10(3b^2)$

30. $7(4p - 4)$

31. $3 + 5(^-2w^2 - 3) + w^2$

*Solve each equation and check each solution.*

32. $3x = x - 8$ **Check**

33. $^-44 = ^-2 + 7y$ **Check**

34. $2k - 5k = ^-21$ **Check**

35. $m - 6 = ^-2m + 6$ **Check**

*Solve each equation. Show your work.*

36. $4 - 4x = 18 + 10x$

37. $18 = ^-r$

38. $^-8b - 11 + 7b = b - 1$

39. $^-2(t + 1) = 4(1 - 2t)$

40. $5 + 6y - 23 = 5(2y + 8) - 10$

# Solving Application Problems

3.1 Problem Solving: Perimeter

3.2 Problem Solving: Area

3.3 Solving Application Problems with One Unknown Quantity

3.4 Solving Application Problems with Two Unknown Quantities

In 1900 there were only 8000 cars and 144 miles of paved roads in the United States. But 100 years later there are 185 million licensed drivers and 132 million registered cars. (*Source:* Federal Highway Administration.) A handy formula for drivers to use is the distance formula, $d = rt$. In Section 3.1, Exercises 51–54, you'll see how to use the formula to find your driving time, rate (speed), or distance traveled on long trips.

## 3.1 PROBLEM SOLVING: PERIMETER

**OBJECTIVES**

1. Use the formula for perimeter of a square to find the perimeter or the length of one side.
2. Use the formula for perimeter of a rectangle to find the perimeter, the length, or the width.
3. Find the perimeter of parallelograms, triangles, and irregular shapes.

**1** Use the formula for perimeter of a square to find the perimeter or the length of one side. If you have ever studied geometry, you probably used several different formulas such as $P = 2l + 2w$ and $A = lw$. A **formula** is just a shorthand way of writing a rule for solving a particular type of problem. A formula uses variables (letters) and it has an equal sign, so it is an equation. That means you can use the equation-solving techniques you learned in Chapter 2 to work with formulas.

But let's start at the beginning. Geometry was developed centuries ago when people needed a way to measure land. The name *geometry* comes from the Greek words *ge*, meaning earth, and *metron*, meaning measure. Today we still use geometry to measure land. It is also important in architecture, construction, navigation, art and design, physics, chemistry, and astronomy. You can use geometry at home when you buy carpet or wallpaper, hang a picture, or do home repairs. In this chapter you'll learn about two basic ideas, perimeter and area. Other geometry concepts will appear in later chapters.

In **Section 2.1**, you found the *perimeter* of a square garden.

> **Perimeter**
>
> The distance around the outside edges of any flat shape is called the **perimeter** of the shape.

To review, a **square** has four sides that are all the same length. Also, the sides meet to form 90° (90 degree) angles. This means that the sides form "square corners." (For more information on angles, see **Section 6.5**.) Two examples of squares are shown below.

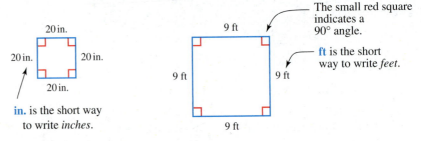

To find the *perimeter* of a square, we can "unfold" the shape so the four sides lie end-to-end, as shown below.

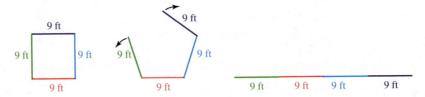

**Unfold the square so the sides lie end-to-end.**

Now we can see the total length of the four sides. The total length is the perimeter of the square. We can find the perimeter by adding.

$$\text{Perimeter} = 9 \text{ ft} + 9 \text{ ft} + 9 \text{ ft} + 9 \text{ ft} = 36 \text{ ft}$$

A shorter way is to multiply the length of one side times 4, because all 4 sides are the same length.

## Finding the Perimeter of a Square

Perimeter of a square = side + side + side + side

or,   P = 4 • side

P = 4s

**Example 1** Finding the Perimeter of a Square

Find the perimeter of the square on the previous page that measures 9 ft on each side.

Use the formula for perimeter of a square, $P = 4s$. You know that for this particular square, the value of $s$ is 9 ft.

$P = 4s$          Formula for perimeter of a square

$P = 4 \cdot 9 \text{ ft}$   Replace $s$ with 9 ft. Multiply 4 times 9 ft.

$P = 36 \text{ ft}$   Write 36 ft; ft is the unit of measure.

The perimeter of the square is 36 ft. Notice that this answer matches the result obtained from adding the four sides.

Work Problem ❶ at the Side.

**Example 2** Finding the Length of One Side of a Square

If the perimeter of a square is 40 cm, find the length of one side. (Note: **cm** is the short way to write *centimeters*.)

Use the formula for perimeter of a square, $P = 4s$. This time you know that the value of $P$ (the perimeter) is 40 cm.

$P = 4s$          Formula for perimeter of a square

$40 \text{ cm} = 4s$   Replace $P$ with 40 cm.

$\dfrac{40 \text{ cm}}{4} = \dfrac{4s}{4}$   To get the variable by itself on the right side, divide both sides by 4.

$10 \text{ cm} = s$

The length of one side of the square is 10 cm.

**Check** Check the solution by drawing a square with each side 10 cm. The perimeter is 10 cm + 10 cm + 10 cm + 10 cm = 40 cm. This result matches the perimeter given in the problem.

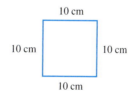

Work Problem ❷ at the Side.

---

❶ Find the perimeter of each square, using the appropriate formula.

(a) The 20 in. square shown on the previous page

(b) A square measuring 14 miles on each side (*Hint:* Draw a sketch of the square and label each side with its length.)

❷ Use the perimeter of each square and the appropriate formula to find the length of one side. Then check your solution by drawing a square, labeling each side, and finding the perimeter.

(a) Perimeter is 28 in.

(b) Perimeter is 100 ft.

(c) Perimeter is 64 cm.

**ANSWERS**
1. (a) $P = 80$ in.
   (b) $P = 56$ miles

2. (a) $s = 7$ in.

   $P = 7 \text{ in.} + 7 \text{ in.} + 7 \text{ in.} + 7 \text{ in.} = 28 \text{ in.}$

   (b) $s = 25$ ft

   $P = 25 \text{ ft} + 25 \text{ ft} + 25 \text{ ft} + 25 \text{ ft} = 100 \text{ ft}$

   (c) $s = 16$ cm

   $P = 16 \text{ cm} + 16 \text{ cm} + 16 \text{ cm} + 16 \text{ cm}$
   $= 64 \text{ cm}$

**152** Chapter 3 Solving Application Problems

❸ Find the perimeter of each rectangle by using the appropriate formula. Check your solutions by adding the lengths of the four sides.

(a)
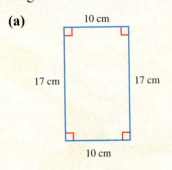

(b) 6 m wide and 11 m long (*Hint:* First draw a sketch of the rectangle and label the length of each side.)

(c)
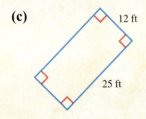

**ANSWERS**

3. (a) $P = 54$ cm
    Check 17 cm + 17 cm + 10 cm + 10 cm = 54 cm

(b)
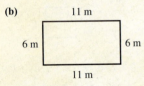

$P = 34$ m
Check 11 m + 11 m + 6 m + 6 m = 34 m

(c) $P = 74$ ft
Check 25 ft + 25 ft + 12 ft + 12 ft = 74 ft

**2** Use the formula for perimeter of a rectangle to find the perimeter, the length, or the width. A **rectangle** is a figure with four sides that intersect to form 90° angles. Each set of opposite sides is parallel and congruent (has the same length). Three examples of rectangles are shown below.

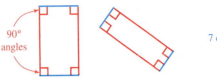 

Each longer side of a rectangle is called the length (*l*) and each shorter side is called the width (*w*).

Look at the rectangle above with the lengths of the sides labeled. To find the perimeter (distance around), you could unfold the shape so the sides lie end-to-end. Then add the lengths of the sides.

$$P = 12 \text{ cm} + 12 \text{ cm} + 7 \text{ cm} + 7 \text{ cm} = 38 \text{ cm}$$

Because the two long sides are both 12 cm, and the two short sides are both 7 cm, you can also use the formula below.

**Finding the Perimeter of a Rectangle**

Perimeter of a rectangle = length + length + width + width

$$P = (2 \cdot \text{length}) + (2 \cdot \text{width})$$
$$P = 2l + 2w$$

**Example 3** Finding the Perimeter of a Rectangle

Find the perimeter of this rectangle.

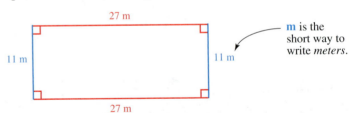

*m* is the short way to write *meters*.

The length is **27 m**, and the width is **11 m**.

$P = \quad 2l \quad + \quad 2w$     Replace *l* with 27 m and *w* with 11 m.

$P = 2 \cdot 27 \text{ m} + 2 \cdot 11 \text{ m}$     Do the multiplications first.

$P = \quad 54 \text{ m} \quad + \quad 22 \text{ m}$     Add last.

$P = 76 \text{ m}$

The perimeter of the rectangle (the distance you would walk around the outside edges of the rectangle) is 76 m.

**Check** To check the solution, add the lengths of the four sides.

$$P = 27 \text{ m} + 27 \text{ m} + 11 \text{ m} + 11 \text{ m}$$
$$P = 76 \text{ m} \quad \text{Matches the solution above}$$

**Work Problem** ❸ **at the Side.**

Section 3.1 Problem Solving: Perimeter   153

## Example 4 — Finding the Length or Width of a Rectangle

If the perimeter of a rectangle is 20 ft and the width is 3 ft, find the length.

First draw a sketch of the rectangle and label the widths as 3 ft.

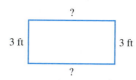

Then use the formula for perimeter of a rectangle, $P = 2l + 2w$. The value of $P$ is 20 ft and the value of $w$ is 3 ft.

$P = 2l + 2w$    Formula for perimeter of a rectangle

$20 \text{ ft} = 2l + 2 \cdot 3 \text{ ft}$    Replace $P$ with 20 ft and $w$ with 3 ft. Simplify the right side by multiplying $2 \cdot 3$ ft.

$20 \text{ ft} = 2l + 6 \text{ ft}$    To get $2l$ by itself, add $^-6$ ft to both sides: $6 + {^-6}$ is 0.

$\underline{^-6 \text{ ft} \phantom{= 2l + } ^-6 \text{ ft}}$

$14 \text{ ft} = 2l + 0$

$\dfrac{14 \text{ ft}}{2} = \dfrac{2l}{2}$    To get $l$ by itself, divide both sides by 2.

$7 \text{ ft} = l$

The length is 7 ft.

**Check** To check the solution, put the length measurements on your sketch. Then add the four measurements.

$P = 7 \text{ ft} + 7 \text{ ft} + 3 \text{ ft} + 3 \text{ ft}$

$P = 20 \text{ ft}$

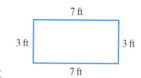

A perimeter of 20 ft matches the information in the original problem, so 7 ft is the correct length of the rectangle.

*Work Problem 4 at the Side.*

**3** Find the perimeter of parallelograms, triangles, and irregular shapes. A **parallelogram** is a four-sided figure with opposite sides parallel. Some examples are shown below. Notice that opposite sides have the same length.

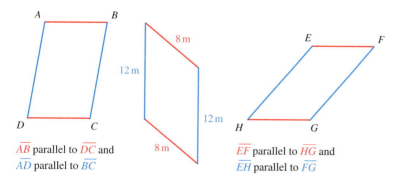

$\overline{AB}$ parallel to $\overline{DC}$ and $\overline{AD}$ parallel to $\overline{BC}$

$\overline{EF}$ parallel to $\overline{HG}$ and $\overline{EH}$ parallel to $\overline{FG}$

Perimeter is the distance around a flat shape, so the easiest way to find the perimeter of a parallelogram is to add the lengths of the four sides.

---

**4** Use the perimeter of each rectangle and the appropriate formula to find the length or width. Draw a sketch of each rectangle and use it to check your solution.

(a) The perimeter of a rectangle is 36 in. and the width is 8 in. Find the length.

(b) A rectangle has a width of 4 cm. The perimeter is 32 cm. Find the length.

(c) A rectangle with a perimeter of 14 ft has a length of 6 ft. Find the width.

**ANSWERS**

4. (a) $l = 10$ in.

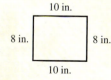

Check $10 \text{ in.} + 10 \text{ in.} + 8 \text{ in.} + 8 \text{ in.} = 36 \text{ in.}$

(b) $l = 12$ cm

    12 cm

4 cm [         ] 4 cm

    12 cm

Check $12 \text{ cm} + 12 \text{ cm} + 4 \text{ cm} + 4 \text{ cm} = 32 \text{ cm}$

(c) $w = 1$ ft

    6 ft

1 ft [         ] 1 ft

    6 ft

Check $6 \text{ ft} + 6 \text{ ft} + 1 \text{ ft} + 1 \text{ ft} = 14 \text{ ft}$

**154** Chapter 3 Solving Application Problems

**5** Find the perimeter of each parallelogram.

(a)
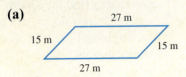

(b)

**6** Find the perimeter of each triangle.

(a)
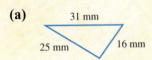

(b) A triangle with sides that each measure 5 in. Draw a sketch of the triangle and label the length of each side.

**7** How much fencing will be needed to go around a flower bed with the measurements shown below?

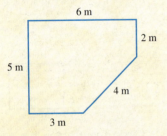

**ANSWERS**
5. (a) $P = 84$ m  (b) $P = 18$ ft
6. (a) $P = 72$ mm
   (b) $P = 15$ in.
7. 20 m of fencing are needed.

**Example 5** Finding the Perimeter of a Parallelogram

Find the perimeter of the middle parallelogram on the previous page.

$$P = 12 \text{ m} + 12 \text{ m} + 8 \text{ m} + 8 \text{ m}$$
$$P = 40 \text{ m}$$

**Work Problem 5 at the Side.**

A **triangle** is a figure with exactly three sides. Some examples are shown below.

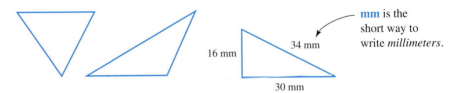

*mm is the short way to write millimeters.*

To find the perimeter of a triangle (the distance around the edges), add the lengths of the three sides.

**Example 6** Finding the Perimeter of a Triangle

Find the perimeter of the triangle above on the right.

To find the perimeter, add the lengths of the sides.

$$P = 16 \text{ mm} + 30 \text{ mm} + 34 \text{ mm}$$
$$P = 80 \text{ mm}$$

**Work Problem 6 at the Side.**

As with any other shape, you can find the perimeter (distance around) of an irregular shape by adding the lengths of the sides.

**Example 7** Finding the Perimeter of an Irregular Shape

The floor of a room has the shape shown below.

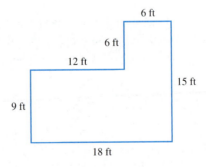

Suppose you want to put a new wallpaper border along the top of all the walls. How much material do you need?

Find the perimeter of the room by adding the lengths of the sides.

$$P = 9 \text{ ft} + 12 \text{ ft} + 6 \text{ ft} + 6 \text{ ft} + 15 \text{ ft} + 18 \text{ ft}$$
$$P = 66 \text{ ft}$$

You need 66 ft of wallpaper border.

**Work Problem 7 at the Side.**

# 3.1 Exercises

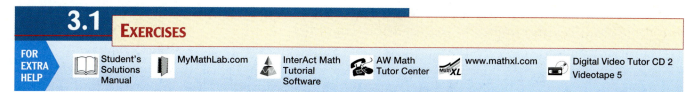

*Find the perimeter of each square, using the appropriate formula. See Example 1.*

1.
2.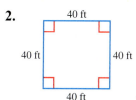
3.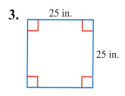
4.

*Draw a sketch of each square and label the lengths of the sides. Then find the perimeter. (Sketches may vary; show your sketches to your instructor.)*

5. A square park measuring 1 mile on each side
6. A square garden measuring 4 meters on each side
7. A 22 mm square postage stamp
8. A 10 in. square piece of cardboard

*For the given perimeter of each square, find the length of one side using the appropriate formula. See Example 2.*

9. The perimeter is 120 ft.
10. The perimeter is 52 cm.
11. The perimeter is 4 mm.
12. The perimeter is 20 miles.

13. A square parking lot with a perimeter of 92 yards
14. A square building with a perimeter of 144 meters
15. A square closet with a perimeter of 8 ft
16. A square bedroom with a perimeter of 44 ft

*Find the perimeter of each rectangle, using the appropriate formula. Check your solutions by adding the lengths of the four sides. See Example 3.*

17.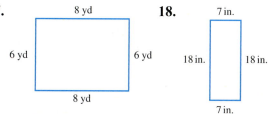
18. 
19.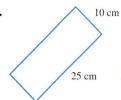
20.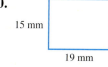

*Draw a sketch of each rectangle and label the lengths of the sides. Then find the perimeter by using the appropriate formula. (Sketches may vary; show your sketches to your instructor.)*

**21.** A rectangular living room 20 ft long by 16 ft wide

**22.** A rectangular placemat 45 cm long by 30 cm wide

**23.** An 8 in. by 5 in. rectangular piece of paper

**24.** A 2 ft by 3 ft rectangular window

*For each rectangle, you are given the perimeter and either the length or width. Find the unknown measurement by using the appropriate formula. Draw a sketch of each rectangle and use it to check your solution. See Example 4. (Show your sketches to your instructor.)*

**25.** The perimeter is 30 cm and the width is 6 cm.

**26.** The perimeter is 48 yards and the length is 14 yards.

**27.** The length is 4 miles and the perimeter is 10 miles.

**28.** The width is 8 meters and the perimeter is 34 meters.

**29.** A 6 ft long rectangular table has a perimeter of 16 ft.

**30.** A 13 in. wide rectangular picture frame has a perimeter of 56 in.

**31.** A rectangular door 1 meter wide has a perimeter of 6 meters.

**32.** A rectangular house 33 ft long has a perimeter of 118 ft.

*Find the perimeter of each shape. See Examples 5–7.*

**33.**

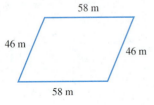

**34.**

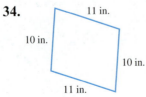

**35.** Parallelogram

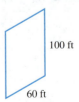

**36.** Parallelogram

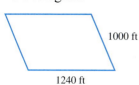

**37.**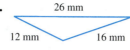

**38.**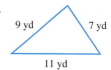

**39.**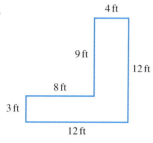

**40.**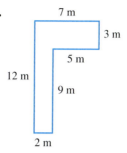

**41.**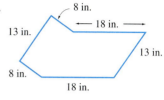

**42.**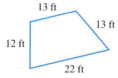

**43.**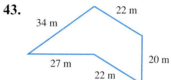

**44.**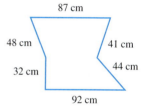

*For each shape, you are given the perimeter and the lengths of all sides except one. Find the length of the unlabeled side.*

**45.** The perimeter is 115 cm.

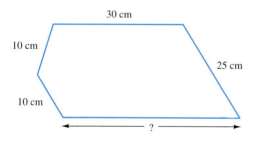

**46.** The perimeter is 63 in.

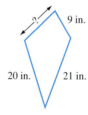

*For each irregular figure, first find the length of the unlabeled side. Then find the perimeter.*

**47.**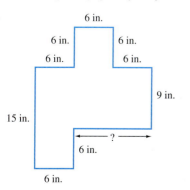

**48.**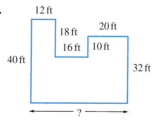

**49.** In an *equilateral* triangle, all sides have the same length.

(a) Draw sketches of four different equilateral triangles, label the lengths of the sides, and find the perimeter.

(b) Write a "shortcut" rule (a formula) for finding the perimeter of an equilateral triangle.

(c) Will your formula work for other kinds of triangles that are not equilateral? Explain why or why not.

**50.** Be sure that you have done Exercise 49 first.

(a) Draw a sketch of a figure with five sides of equal length. Write a "shortcut" rule (a formula) for finding the perimeter of this shape.

(b) Draw a sketch of a figure with six sides of equal length. Write a formula for finding the perimeter of the shape.

(c) Write a formula for finding the perimeter of a shape with 10 sides of equal length.

(d) Write a formula for finding the perimeter of a shape with *n* sides of equal length.

### RELATING CONCEPTS (Exercises 51–54) FOR INDIVIDUAL OR GROUP WORK

*A formula that has many uses for drivers is $d = rt$, called the distance formula. If you are driving a car, then*

$d$ is the *distance* you travel (how many miles)
$r$ is the *rate* (how fast you are driving in miles per hour)
$t$ is the *time* (how many hours you drive)

Use the distance formula as you **work Exercises 51–54 in order.**

**51.** Suppose you are driving on Interstate highways at a rate of 70 miles per hour. Use the distance formula to find out how far you will travel in (a) 2 hours; (b) 5 hours; (c) 8 hours.

**52.** If an ice storm slows your driving rate to 35 miles per hour, how far will you travel in (a) 2 hours, (b) 5 hours, (c) 8 hours? Show how to find each answer using the formula. (d) Then explain how to find each answer using the results from Exercise 51 instead of the formula.

**53.** Use the distance formula to find out how many hours you would have to drive to travel the 3000 miles from Boston to San Francisco if your average rate is (a) 60 miles per hour; (b) 50 miles per hour; (c) 20 miles per hour (which was the speed limit 100 years ago).

**54.** Use the distance formula to find the average driving rate (speed) on each of these trips. (Distances are from *World Almanac,* 2001.)

(a) It took 11 hours to drive 671 miles from Atlanta to Chicago

(b) Sam drove 1539 miles from New York City to Dallas in 27 hours.

(c) Carlita drove 16 hours to travel 1040 miles from Memphis to Denver.

## 3.2 PROBLEM SOLVING: AREA

**1** Use the formula for area of a rectangle to find the area, the length, or the width.

### OBJECTIVES

**1** Use the formula for area of a rectangle to find the area, the length, or the width.

**2** Use the formula for area of a square to find the area or the length of one side.

**3** Use the formula for area of a parallelogram to find the area, the base, or the height.

**4** Solve application problems involving perimeter and area of rectangles, squares, or parallelograms.

### Difference between Perimeter and Area

**Perimeter** is the *distance around the outside edges* of a flat shape.
**Area** is the amount of *surface inside* a flat shape.

The *perimeter* of a rectangle is the distance around the *outside edges.* Recall that we unfolded a shape and laid the sides end-to-end so we could see the total distance. The *area* of a rectangle is the amount of surface *inside* the rectangle. We measure area by finding the number of squares of a certain size needed to cover the surface inside the rectangle. Think of covering the floor of a rectangular living room with carpet. Carpet is measured in square yards, that is, square pieces that measure 1 yard along each side. Here is a drawing of a living room floor.

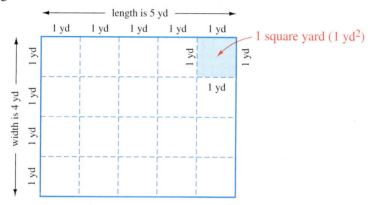

You can see from the drawing that it takes 20 squares to cover the floor. We say that the area of the floor is 20 *square yards.* A short way to write square yards is yd$^2$.

20 **square yards** can be written as 20 **yd$^2$**

To find the number of squares, you can count them, or you can multiply the number of squares in the length (5) times the number of squares in the width (4) to get 20. The formula is given below.

### Finding the Area of a Rectangle

Area of a rectangle = length • width

$A = lw$

Remember to use *square units* when measuring area.

Squares of many sizes can be used to measure area. For smaller areas, you might use the ones shown at the right.

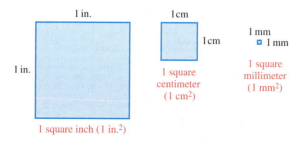

**Actual-size drawings**

① Find the area of each rectangle, using the appropriate formula.

(a)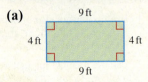

(b) A rectangle is 35 mi long and 20 mi wide. (First make a sketch of the rectangle and label the lengths of the sides.)

(c) A rectangular patio measures 3 m by 2 m. (First make a sketch of the patio and label the lengths of the sides.)

ANSWERS
1. (a) $A = 36 \text{ ft}^2$
(b) 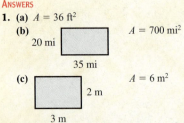 $A = 700 \text{ mi}^2$

(c) $A = 6 \text{ m}^2$

Other sizes of squares that are often used to measure area are listed here, but they are too large to draw on this page.

1 square meter (1 m²)        1 square foot (1 ft²)
1 square kilometer (1 km²)   1 square yard (1 yd²)
                             1 square mile (1 mi²)

### CAUTION

The raised 2 in $4^2$ means that you multiply $4 \cdot 4$ to get 16. The raised 2 in cm² or yd² is a short way to write the word "square." It means that you multiplied cm times cm to get cm², or yd times yd to get yd². Recall that a short way to write $x \cdot x$ is $x^2$. Similarly, cm · cm is cm². When you see 5 cm², say "five square centimeters." Do *not* multiply $5 \cdot 5$, because the exponent applies only to the *first* thing to its left. The exponent applies to cm, *not* to the number.

### Example 1  Finding the Area of Rectangles

Find the area of each rectangle.

(a)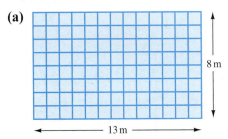

The length of this rectangle is 13 m and the width is 8 m. Use the formula $A = lw$.

$A = l \cdot w$         Replace $l$ with 13 m and $w$ with 8 m.

$A = \textbf{13 m} \cdot \textbf{8 m}$     Multiply 13 times 8 to get 104.

$A = 104 \text{ m}^2$    Multiply m times m to get m².

The area of the rectangle is 104 m². If you count the number of squares in the sketch, you will also get 104 m². (Each square in the sketch represents 1 m by 1 m, which is 1 square meter or 1 m².)

(b) A rectangle measuring 7 cm by 21 cm
First make a sketch of the rectangle. The length is 21 cm (the longer measurement) and the width is 7 cm. Then use the formula for area of a rectangle, $A = lw$.

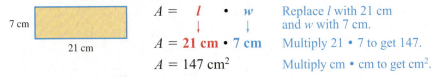

$A = l \cdot w$         Replace $l$ with 21 cm and $w$ with 7 cm.

$A = \textbf{21 cm} \cdot \textbf{7 cm}$   Multiply $21 \cdot 7$ to get 147.

$A = 147 \text{ cm}^2$   Multiply cm · cm to get cm².

The area of the rectangle is 147 cm².

**Work Problem ① at the Side.**

### CAUTION

The units for *area* will always be *square* units (cm², m², yd², mi², and so on). The units for *perimeter* will always be *linear* units (cm, m, yd, mi, and so on), *not* square units.

Section 3.2 Problem Solving: Area  **161**

## Example 2  Finding the Length or Width of a Rectangle

If the area of a rectangular rug is 12 yd² and the length is 4 yd, find the width.
First draw a sketch of the rug and label the length as 4 yd.

4 yd

Use the formula for area of a rectangle, $A = lw$.
The value of $A$ is 12 yd² and the value of $l$ is 4 yd.

$A = l \cdot w$    Replace $A$ with 12 yd² and replace $l$ with 4 yd.

$12 \text{ yd}^2 = 4 \text{ yd} \cdot w$    To get $w$ by itself, divide both sides by 4 yd.

$$\frac{12 \text{ yd} \cdot \text{yd}}{4 \text{ yd}} = \frac{4 \text{ yd} \cdot w}{4 \text{ yd}}$$    On the left side, rewrite yd² as yd · yd. Then $\frac{\text{yd}}{\text{yd}}$ is 1, so they "cancel out."

$3 \text{ yd} = w$    On the left side, 12 yd ÷ 4 is 3 yd.

The width of the rug is 3 yd.

**Check** To check the solution, put the width measurement on your sketch. Then use the area formula.

$A = l \cdot w$
$A = 4 \text{ yd} \cdot 3 \text{ yd}$
$A = 12 \text{ yd}^2$

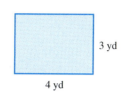
3 yd
4 yd

An area of 12 yd² matches the information in the original problem. So 3 yd is the correct width of the rug.

 Work Problem ❷ at the Side.

**2** Use the formula for area of a square to find the area or the length of one side. As with a rectangle, you can multiply length times width to find the area (surface inside) of a square. Because the length and the width are the same in a square, the formula is written as shown below.

### Finding the Area of a Square

Area of a square = side · side

$A = s \cdot s$

$A = s^2$

Remember to use *square units* when measuring area.

## Example 3  Finding the Area of a Square

Find the area of a square highway sign that is 4 ft on each side.
Use the formula for area of a square, $A = s^2$.

$A = s^2$    Remember that $s^2$ means $s \cdot s$.

$A = s \cdot s$    Replace $s$ with 4 ft.

$A = 4 \text{ ft} \cdot 4 \text{ ft}$    Multiply 4 · 4 to get 16.

$A = 16 \text{ ft}^2$    Multiply ft · ft to get ft².

The area of the sign is 16 ft².

 Continued on Next Page

❷ Use the area of each rectangle and the appropriate formula to find the length or width. Draw a sketch of each rectangle and use it to check your solution.

(a) The area of a microscope slide is 12 cm², and the length is 6 cm. Find the width.

(b) A child's play lot is 10 ft wide and has an area of 160 ft². Find the length.

(c) A hallway is 31 m long and has an area of 93 m². Find the width of the hall.

**ANSWERS**
2. (a) $w = 2$ cm

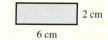
2 cm
6 cm

Check $A = 6 \text{ cm} \cdot 2 \text{ cm}$
$A = 12 \text{ cm}^2$
Matches original problem

(b) $l = 16$ ft

10 ft
16 ft

Check $A = 16 \text{ ft} \cdot 10 \text{ ft}$
$A = 160 \text{ ft}^2$
Matches original problem

(c) $w = 3$ m

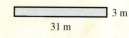
3 m
31 m

Check $A = 31 \text{ m} \cdot 3 \text{ m}$
$A = 93 \text{ m}^2$
Matches original problem

## 162  Chapter 3 Solving Application Problems

**3** Find the area of each square, using the appropriate formula. Make a sketch of each square.

(a) A 12 in. square piece of fabric

(b) A square township 7 mi on a side

(c) A square earring measuring 20 mm on each side

**4** Given the area of each square, find the length of one side by inspection.

(a) The area of a square-shaped nature center is 16 mi².

(b) A square floor has an area of 100 m².

(c) A square clock face has an area of 81 in.²

**ANSWERS**

3. (a) 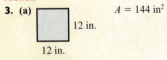 $A = 144$ in²

(b) 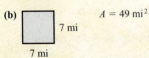 $A = 49$ mi²

(c) $A = 400$ mm²
20 mm
20 mm

4. (a) $s = 4$ mi   (b) $s = 10$ m   (c) $s = 9$ in.

**CAUTION**

Be careful! $s^2$ means $s \cdot s$. It does **not** mean $2 \cdot s$. In this example $s$ is 4 ft, so $(4 \text{ ft})^2$ is 4 ft · 4 ft = 16 ft². It is **not** 2 · 4 ft = 8 ft.

**Check** Check the solution by drawing a square with each side 4 ft. You can multiply length (4 ft) times width (4 ft), as for a rectangle. So the area is 4 ft · 4 ft, or 16 ft². This result matches the solution you got by using the formula $A = s^2$.

4 ft
4 ft

Work Problem **3** at the Side.

### Example 4  Finding the Length of One Side of a Square

If the area of a square township is 49 mi², what is the length of one side of the township?

Use the formula for area of a square, $A = s^2$. The value of $A$ is 49 mi².

$A = s^2$ — Replace $A$ with 49 mi².

$49 \text{ mi}^2 = s^2$ — To get $s$ by itself, we have to "undo" the squaring of $s$. This is called *finding the square root* (more on square roots in Chapter 5).

$49 \text{ mi}^2 = s \cdot s$ — For now, solve by inspection. Ask, what number times itself gives 49?

$49 \text{ mi}^2 = 7 \text{ mi} \cdot 7 \text{ mi}$ — 7 · 7 is 49, so 7 mi · 7 mi is 49 mi².

The value of $s$ is 7 mi, so the length of one side of the township is 7 mi. Notice how this result matches the information about the township in Margin Problem 3(b) at the left.

Work Problem **4** at the Side.

**3**  Use the formula for area of a parallelogram to find the area, the base, or the height. To find the area of a parallelogram, first draw a dashed line inside the figure, as shown here.

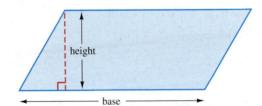

Try this yourself by tracing this parallelogram onto a piece of paper.

The length of the dashed line is the *height* of the parallelogram. It forms a 90° angle (a square corner) with the base. A 90° angle is also called a *right angle*. The height is the shortest distance between the base and the opposite side.

Now cut off the triangle created on the left side of the parallelogram and move it to the right side, as shown below.

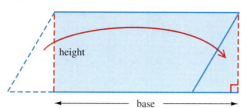

You have made the parallelogram into a rectangle. You can see that the area of the parallelogram and the rectangle are the same. The area of the rectangle is *length* times *width*. For the parallelogram, the area translates into *base* times *height*.

### Finding the Area of a Parallelogram

Area of a parallelogram = base • height

$$A = bh$$

Remember to use *square units* when measuring area.

### Example 5 — Finding the Area of Parallelograms

Find the area of each parallelogram.

(a) 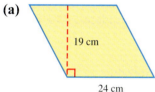  (b)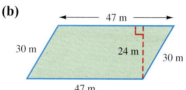

(a) The base is 24 cm and the height is 19 cm. The formula for the area is $A = bh$.

$A = \quad b \quad • \quad h$   Replace *b* with 24 cm and *h* with 19 cm.
$A = \textbf{24 cm} • \textbf{19 cm}$   Multiply 24 • 19 to get 456.
$A = 456 \text{ cm}^2$   Multiply cm • cm to get $cm^2$.

The area of the parallelogram is 456 $cm^2$.

(b) Use the formula for area of a parallelogram, $A = bh$.

$A = \quad b \quad • \quad h$   Replace *b* with 47 m and *h* with 24 m.
$A = \textbf{47 m} • \textbf{24 m}$   Multiply 47 • 24 to get 1128.
$A = 1128 \text{ m}^2$   Multiply m • m to get $m^2$.

Notice that the 30 m sides are *not* used in finding the area. But you would use them when finding the *perimeter* of the parallelogram.

= Work Problem ❺ at the Side.

### Example 6 — Finding the Base or Height of a Parallelogram

If the area of a parallelogram is 24 $ft^2$ and the base is 6 ft, find the height.
  First draw a sketch of the parallelogram and label the base as 6 ft.
  Use the formula for the area of a parallelogram, $A = bh$.

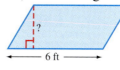

= Continued on Next Page

Section 3.2 Problem Solving: Area

❺ Find the area of each parallelogram.

(a)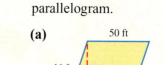

(b)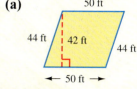

(c) A parallelogram with base 8 cm and height 1 cm.

**ANSWERS**
5. (a) $A = 2100 \text{ ft}^2$  (b) $A = 180 \text{ in.}^2$
   (c) $A = 8 \text{ cm}^2$

**164** Chapter 3 Solving Application Problems

**6** Use the area of each parallelogram and the appropriate formula to find the base or height. Draw a sketch of each parallelogram and use it to check your solution.

(a) The area of a parallelogram is 140 in.² and the base is 14 in. Find the height.

(b) A parallelogram has an area of 4 yd². The height is 1 yd. Find the base.

The value of $A$ is 24 ft², and the value of $b$ is 6 ft.

$A = b \cdot h$    Replace $A$ with 24 ft² and $b$ with 6 ft.

$24 \text{ ft}^2 = 6 \text{ ft} \cdot h$    To get $h$ by itself, divide both sides by 6 ft.

$$\frac{24 \text{ ft} \cdot \text{ft}}{6 \text{ ft}} = \frac{6 \text{ ft} \cdot h}{6 \text{ ft}}$$    On the left side, rewrite ft² as ft · ft. Then $\frac{\text{ft}}{\text{ft}}$ is 1, so they "cancel out."

$4 \text{ ft} = h$    On the left side, 24 ft ÷ 6 is 4 ft.

The height of the parallelogram is 4 ft.

**Check** To check the solution, put the height measurement on your sketch. Then use the area formula.

$A = b \cdot h$
$A = 6 \text{ ft} \cdot 4 \text{ ft}$
$A = 24 \text{ ft}^2$

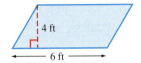

An area of 24 ft² matches the information in the original problem. So 4 ft is the correct height of the parallelogram.

**Work Problem 6 at the Side.**

---

**4** **Solve application problems involving perimeter and area of rectangles, squares, or parallelograms.** When you are solving problems, first decide whether you need to find the perimeter or the area.

**Example 7** Solving an Application Problem Involving Perimeter or Area

A group of neighbors is fixing up a playground area for their children. The rectangular lot is 22 yd by 16 yd. If chain-link fencing costs $6 per yard, how much will they spend to put a fence around the lot?

**7** If sod costs $3 per square yard, how much will the neighbors in Example 7 spend to cover the playground with grass?

First draw a sketch of the rectangular lot and label the lengths of the sides. The fence will go around the edges of the lot, so you need to find the *perimeter* of the lot.

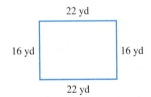

$P = 2l + 2w$    Formula for perimeter of a rectangle

$P = 2 \cdot 22 \text{ yd} + 2 \cdot 16 \text{ yd}$    Replace $l$ with 22 yd and $w$ with 16 yd.

$P = 44 \text{ yd} + 32 \text{ yd}$

$P = 76 \text{ yd}$

The perimeter of the lot is 76 yd, so the neighbors need to buy 76 yd of fencing. The cost of the fencing is $6 *per yard*, which means $6 *for 1 yard*. To find the cost for 76 yd, multiply $6 · 76. The neighbors will spend $456 on the fence.

**Work Problem 7 at the Side.**

---

**ANSWERS**

6. (a) $h = 10$ in.

Check $A = 14$ in. · 10 in.
$A = 140$ in.²
Matches original problem

(b) $b = 4$ yd

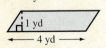

Check $A = 4$ yd · 1 yd
$A = 4$ yd²
Matches original problem

7. $1056 (Find the area by multiplying 22 yd · 16 yd to get 352 yd². Then multiply $3 · 352 to get $1056.)

## 3.2 Exercises

*Find the area of each rectangle, square, or parallelogram using the appropriate formula. See Examples 1, 3, and 5.*

1.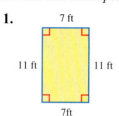
2.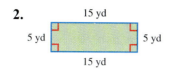
3.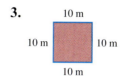
4.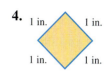

5.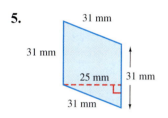
6.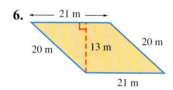
7.
8.

*In Exercises 9–16, first draw a sketch of the shape and label the lengths of the sides or base and height. Then find the area. (Sketches may vary; show your sketches to your instructor.)*

9. A rectangular calculator that measures 15 cm by 7 cm
10. A rectangular piece of plywood that is 8 ft long and 2 ft wide
11. A parallelogram with height of 9 ft and base of 8 ft
12. A parallelogram measuring 18 mm on the base and 3 mm on the height

13. A fire burned a square-shaped forest 25 mi on a side.
14. An 11 in. square pillow
15. A piece of window glass 1 m on each side
16. A table 12 ft long by 3 ft wide

*Use the area of each rectangle and either its length or width, and the appropriate formula, to find the other measurement. Draw a sketch of each rectangle and use it to check your solution. See Example 2. (Sketches may vary; show your sketches to your instructor.)*

17. The area of a desk is 18 ft$^2$, and the width is 3 ft. Find its length.
18. The area of a classroom is 630 ft$^2$, and the length is 30 ft. Find its width.
19. A parking lot is 90 yd long and has an area of 7200 yd$^2$. Find its width.

**20.** A playground is 60 yd wide and has an area of 6000 yd². Find its length.

**21.** A 154 in.² photo has a width of 11 in. Find its length.

**22.** A 15 in.² note card has a width of 3 in. Find its length.

*Given the area of each square, find the length of one side by inspection. See Example 4.*

**23.** A square floor has an area of 36 m².

**24.** A square stamp has an area of 9 cm².

**25.** The area of a square sign is 4 ft².

**26.** The area of a square piece of metal is 64 in.².

*Use the area of each parallelogram and either its base or height, and the appropriate formula, to find the other measurement. Draw a sketch of each parallelogram and use it to check your solution. See Example 6. (Sketches may vary; show your sketches to your instructor.)*

**27.** The area is 500 cm², and the base is 25 cm. Find the height.

**28.** The area is 1500 m², and the height is 30 m. Find the base.

**29.** The height is 13 in. and the area is 221 in.². Find the base.

**30.** The base is 19 cm, and the area is 114 cm². Find the height.

**31.** The base is 9 m, and the area is 9 m². Find the height.

**32.** The area is 25 mm², and the height is 5 mm. Find the base.

*Explain and correct the **two** errors made by students in Exercises 33 and 34.*

**33.**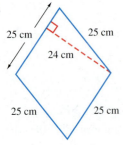

$P = 25 \text{ cm} + 24 \text{ cm} + 25 \text{ cm} + 25 \text{ cm} + 25 \text{ cm}$

$P = 124 \text{ cm}^2$

**34.**

7 ft

$A = s^2$

$A = 2 \cdot 7 \text{ ft}$

$A = 14 \text{ ft}$

*Find both the perimeter and the area of each rectangle, square, or parallelogram.*

**35.**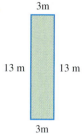

**36.**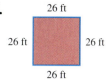

**37.**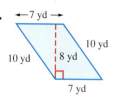

**38.**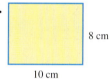

**39.**

**40.**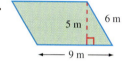

*Solve each application problem. You may need to find the perimeter, the area, or one of the side measurements. In Exercises 43–46, draw a sketch for each problem and label the sketch with the appropriate measurements. See Example 7. (Sketches may vary; show your sketches to your instructor.)*

**41.** The new Advanced Photo System (APS) cameras allow you to choose from three different print sizes each time you snap a photo. The choices are shown below. Find the perimeter and area of each size print. (*Source:* Kodak.)

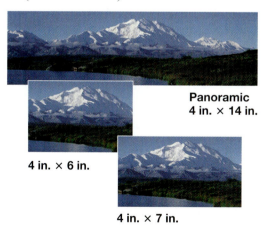

**42.** The Monterey Bay Aquarium in California lets visitors look into a million gallon tank through an acrylic panel that is 13 in. thick. The panel is 54 ft long and 15 ft high. What is the perimeter and the area of the panel? (*Source: AAA California Tour Book.*)

**43.** Tyra's kitchen is 4 m wide and 5 m long. She is pasting a decorative strip that costs $6 per meter around the top edge of all the walls. How much will she spend?

**44.** The Wang's family room measures 20 ft by 25 ft. They are covering the floor with square tiles that measure 1 ft on a side and cost $1 each. How much will they spend on tile?

**45.** Mr. and Mrs. Gomez are buying carpet for their square-shaped bedroom that is 5 yd wide. The carpet is $23 per square yard and padding and installation is another $6 per square yard. How much will they spend in all?

**46.** A page in this book measures about 27 cm from top to bottom and 21 cm from side to side. Find the perimeter and the area of the page.

**47.** A regulation football field is 100 yd long (excluding end zones) and has an area of 5300 yd². Find the width of the field. (*Source:* NFL.)

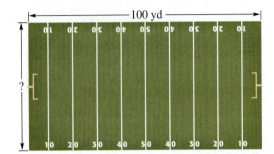

**48.** There are 14,790 ft² of ice in the rectangular playing area for a major league hockey game (excluding the area behind the goal lines). If the playing area is 85 ft wide, how long is it? (*Source:* NHL.)

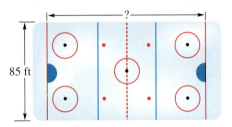

### RELATING CONCEPTS (Exercises 49–52)   FOR INDIVIDUAL OR GROUP WORK

*Use your knowledge of perimeter and area to* **work Exercises 49–52 in order.**

**49.** Suppose you have 12 ft of fencing to make a square or rectangular garden plot. Draw sketches of *all* the possible plots that use exactly 12 ft of fencing and label the lengths of the sides. Use only *whole number* lengths. (*Hint:* There are three possibilities.)

**50. (a)** Find the area of each plot in Exercise 49.

**(b)** Which plot has the greatest area?

**51.** Repeat Exercise 49 using 16 ft of fencing. Be sure to draw *all* possible plots that have whole number lengths for the sides.

**52. (a)** Find the area of each plot in Exercise 51.

**(b)** Compare your results to those from Exercise 50. What do you notice about the plots with the greatest area?

## 3.3 SOLVING APPLICATION PROBLEMS WITH ONE UNKNOWN QUANTITY

**OBJECTIVES**
1. Translate word phrases into algebraic expressions.
2. Translate sentences into equations.
3. Solve application problems with one unknown quantity.

**1** Translate word phrases into algebraic expressions. In **Sections 3.1** and **3.2** you worked with applications involving perimeter and area. You were able to use well-known rules (formulas) to set up equations that could be solved. However, you will encounter many problems for which no formula is available. Then you need to analyze the problem and translate the words into an equation that fits the particular situation. We'll start by translating word phrases into algebraic expressions.

### Example 1  Translating Word Phrases into Algebraic Expressions

Write each phrase as an algebraic expression. Use $x$ as the variable.

| Words | Algebraic Expression | |
|---|---|---|
| A number **plus** 2 | $x + 2$ or $2 + x$ | |
| The **sum** of 8 and a number | $8 + x$ or $x + 8$ | Two correct ways to write each addition expression. |
| 5 **more than** a number | $x + 5$ or $5 + x$ | |
| $^-35$ **added to** a number | $^-35 + x$ or $x + {}^-35$ | |
| A number **increased by** 6 | $x + 6$ or $6 + x$ | |
| 9 **less than** a number | $x - 9$ | |
| A number **subtracted from** 3 | $3 - x$ | Only one correct way to write each subtraction expression. |
| 3 **subtracted from** a number | $x - 3$ | |
| A number **decreased by** 4 | $x - 4$ | |
| 10 **minus** a number | $10 - x$ | |

**CAUTION**

Recall that addition can be done in any order, so $x + 2$ gives the same result as $2 + x$. This is *not* true in subtraction, so be careful. $10 - x$ does *not* give the same result as $x - 10$.

**Work Problem ❶ at the Side.**

### Example 2  Translating Word Phrases into Algebraic Expressions

Write each phrase as an algebraic expression. Use $x$ as the variable.

| Words | Algebraic Expression |
|---|---|
| 8 **times** a number | $8x$ |
| The **product** of 12 and a number | $12x$ |
| **Double** a number (meaning "2 times") | $2x$ |
| The **quotient** of $^-6$ and a number | $\dfrac{^-6}{x}$ |
| A number **divided by** 10 | $\dfrac{x}{10}$ |
| 15 **subtracted from** 4 **times** a number | $4x - 15$ |
| The result **is** | $=$ |

**Work Problem ❷ at the Side.**

❶ Write each phrase as an algebraic expression. Use $x$ as the variable.

(a) 15 less than a number

(b) 12 more than a number

(c) A number increased by 13

(d) A number minus 8

(e) 10 plus a number

(f) A number subtracted from 6

(g) 6 subtracted from a number

❷ Write each phrase as an algebraic expression. Use $x$ as the variable.

(a) Double a number

(b) The product of $^-8$ and a number

(c) The quotient of 15 and a number

(d) 5 times a number subtracted from 30

**ANSWERS**

1. (a) $x - 15$  (b) $x + 12$ or $12 + x$
   (c) $x + 13$ or $13 + x$  (d) $x - 8$
   (e) $10 + x$ or $x + 10$  (f) $6 - x$
   (g) $x - 6$

2. (a) $2x$  (b) $^-8x$  (c) $\dfrac{15}{x}$
   (d) $30 - 5x$ (*not* $5x - 30$)

**3** Translate each sentence into an equation and solve it. Check your solution by going back to the words in the original problem.

(a) If 3 times a number is added to 4, the result is 19. Find the number.

(b) If 7 is subtracted from 6 times a number, the result is −25. Find the number.

**ANSWERS**

3. (a) $3x + 4 = 19$
$x = 5$
Check $3 \cdot 5 + 4$ does equal 19
$15 + 4$
$19$

(b) $6x - 7 = {}^-25$
$x = {}^-3$
Check $6 \cdot {}^-3 - 7$ does equal ${}^-25$
${}^-18 + {}^-7$
${}^-25$

**2** **Translate sentences into equations.** The next example shows you how to translate a sentence into an equation that you can solve.

**Example 3** Translating a Sentence into an Equation

If 5 times a number is added to 11, the result is 26. Find the number.

Let $x$ represent the unknown number. Use the information in the problem to write an equation.

5 times a number added to 11 is 26.
$5x \quad\quad + \quad 11 = 26$

Next, solve the equation.

$5x + 11 = 26$  To get $5x$ by itself, add $^-11$ to both sides.
$\underline{\quad -11 \quad -11\quad}$
$5x + 0 = 15$
$\dfrac{5x}{5} = \dfrac{15}{5}$  To get $x$ by itself, divide both sides by 5.
$x = 3$

The number is 3.

**Check** Go back to the words of the original problem.

If 5 times a number is added to 11, the result is 26.
$5 \cdot 3 + 11 = 26$

Does $5 \cdot 3 + 11$ really equal 26? Yes, $5 \cdot 3 + 11 = 15 + 11 = 26$.

So 3 is the correct solution because it "works" when you put it back into the original problem.

**Work Problem 3 at the Side.**

**3** **Solve application problems with one unknown quantity.** Now you are ready to tackle application problems. The steps we will use are summarized below.

### Solving an Application Problem

*Step 1*  **Read** the problem once to see what it is about. Read it carefully a second time. As you read, make a sketch or write word phrases that identify the known and the unknown parts of the problem.

*Step 2(a)*  If there is one unknown quantity, **assign a variable** to represent it. Write down what your variable represents.

*Step 2(b)*  If there is more than one unknown quantity, **assign a variable** to represent "the thing you know the least about." Then write variable expression(s), using the same variable, to show the relationship of the other unknown quantities to the first one.

*Step 3*  **Write an equation,** using your sketch or word phrases as the guide.

*Step 4*  **Solve** the equation.

*Step 5*  **State the answer** to the question in the problem and label your answer.

*Step 6*  **Check** whether your answer fits all the facts given in the *original* statement of the problem. If it does, you are done. If it doesn't, start again at Step 1.

### Example 4 — Solving an Application Problem with One Unknown Quantity

Heather had put some money aside in an envelope for household expenses. Yesterday she took out $20 for groceries. Today a friend paid back a loan and Heather put the $34 in the envelope. Now she has $43 in the envelope. How much was in the envelope at the start?

**Step 1**  **Read** the problem once. It is about money in an envelope. Read it a second time and write word phrases.

> **Unknown:** amount of money in the envelope at the start
> **Known:** took out $20; put in $34; ended up with $43

**Step 2(a)**  There is only one unknown quantity, so **assign a variable** to represent it. Let $m$ represent the money at the start.

**Step 3**  **Write an equation,** using the phrases you wrote as a guide.

| money at the start | took out $20 | put in $34 | ended up with $43 |
|---|---|---|---|
| $m$ | $-\$20$ | $+\$34$ | $=\$43$ |

**Step 4**  **Solve** the equation.

$$m - 20 + 34 = 43 \quad \text{Change subtraction to adding the opposite.}$$
$$m + {}^-20 + 34 = 43 \quad \text{Simplify the left side.}$$
$$m + 14 = 43$$
$$\underline{\phantom{m+}{}^-14 \qquad {}^-14} \quad \text{To get } m \text{ by itself, add } {}^-14 \text{ to both sides.}$$
$$m + 0 = 29$$
$$m = 29$$

**Step 5**  **State the answer** to the question, "How much was in the envelope at the start?" There was $29 in the envelope.

**Step 6**  **Check** the solution by going back to the *original* problem and inserting the solution.

Started with $29 in the envelope
Took out $20, so $29 − $20 = $9 in the envelope
Put in $34, so $9 + $34 = $43
Now has $43 ← Matches

Because $29 "works" when you put it back into the original problem, you know it is the correct solution.

**Work Problem ④ at the Side.**

---

**④** Some people got on an empty bus at its first stop. At the second stop, 3 people got on. At the third stop, 5 more people got on. At the fourth stop, 10 people got off, but 4 people were still on the bus. How many people got on at the first stop? Show your work for each of the six problem-solving steps.

**ANSWERS**

**4.** *Step 1* **Read.**
Unknown: number of people who got on at first stop
Known: 3 got on; 5 got on; 10 got off; 4 people still on bus

*Step 2(a)* **Assign a variable.**
Let $p$ be people who got on at first stop. (You may use any letter you like as the variable.)

*Step 3* **Write an equation.**
$p + 3 + 5 - 10 = 4$

*Step 4* **Solve.**
$$p + 3 + 5 + {}^-10 = 4$$
$$p + {}^-2 = 4$$
$$\underline{\phantom{p+}{}^+2 \qquad {}^+2}$$
$$p + 0 = 6$$
$$p = 6$$

*Step 5* **State the answer.**
6 people got on at the first stop.

*Step 6* **Check.**
6 got on at first stop
3 got on at 2nd stop:   $6 + 3 = 9$
5 got on at 3rd stop:   $9 + 5 = 14$
10 got off at 4th stop: $14 - 10 = 4$
4 people are left. ← Matches

**Chapter 3** Solving Application Problems

**5** Five donors each gave the same amount of money to a college to use for scholarships. From the money, scholarships of $1250, $900, and $850 were given to students; $250 was left. How much money did each donor give to the college? Show your work for each of the six problem-solving steps.

**Example 5** Solving an Application Problem with One Unknown Quantity

Three friends each put in the same amount of money to buy a gift. After they spent $2 for a card and $31 for the gift, they had $6 left. How much money had each friend put in originally?

*Step 1*     **Read** the problem. It is about 3 friends buying a gift.

           **Unknown:** amount of money each friend contributed
           **Known:** 3 friends put in money; spent $2 and $31; had $6 left

*Step 2(a)*    There is only one unknown quantity. **Assign a variable,** $m$, to represent the amount of money each friend contributed.

*Step 3*     **Write an equation.**

| number of friends | | amount each friend put in | | spent on card | | spent on gift | | left over |
|---|---|---|---|---|---|---|---|---|
| 3 | • | $m$ | − | $2 | − | $31 | = | $6 |

To see why this is multiplication, think of an example. If each friend put in $10, how much money would there be? 3 • $10, or $30

*Step 4*     **Solve.**

$3m - 2 - 31 = 6$    Change subtractions to adding the opposite.
$3m + {}^-2 + {}^-31 = 6$    Simplify the left side.
$3m + {}^-33 = 6$
       $+33$    $+33$    To get $3m$ by itself, add 33 to both sides.
$3m + 0 = 39$
$\dfrac{3m}{3} = \dfrac{39}{3}$    To get $m$ by itself, divide both sides by 3.
$m = 13$

*Step 5*     **State the answer.** Each friend put in $13.

*Step 6*     **Check** the solution by putting it back into the *original* problem.

3 friends each put in $13, so 3 • $13 = $39.
Spent $2, spent $31, so $39 − $2 − $31 = $6.
Had $6 left ⟵——— Matches ———

$13 is the correct solution because it "works."

**Work Problem** **5** at the Side.

---

**ANSWERS**

**5.** *Step 1* **Read.**
Unknown: money given by each donor
Known: 5 donors; gave out $1250, $900, $850; $250 left

*Step 2(a)* **Assign a variable.**
Let $m$ be each donor's money.

*Step 3* **Write an equation.**
$5 \cdot m - \$1250 - \$900 - \$850 = \$250$

*Step 4* **Solve.**
$5m + {}^-1250 + {}^-900 + {}^-850 = 250$

$5m + {}^-3000 = 250$
      $+3000$    $+3000$
$5m + 0 = 3250$

$\dfrac{5m}{5} = \dfrac{3250}{5}$

$m = 650$

*Step 5* **State the answer.**
Each donor gave $650.

*Step 6* **Check.**
5 donors each gave $650, so

5 • $650 = $3250

Gave out $1250, $900, $850, so

$3250 − 1250 − 900 − 850 = $250

Had $250 left ⟵— Matches —

Section 3.3 Solving Application Problems with One Unknown Quantity  173

**Example 6** **Solving a More Complex Application Problem with One Unknown Quantity**

Michael has completed 5 less than three times as many lab experiments as David. If Michael has completed 13 experiments, how many experiments has David completed?

*Step 1* **Read** the problem. It is about the number of experiments done by two students.

**Unknown:** number of experiments David did
**Known:** Michael did 5 less than 3 times the number David did; Michael did 13.

*Step 2(a)* **Assign a variable.** Let $n$ represent the number of experiments David did.

*Step 3* **Write an equation.**

The number Michael did  is  5 less than 3 times David's number.

$$13 = 3n - 5$$

*Step 4* **Solve.**

$13 = 3n - 5$   Change subtraction to adding the opposite.

$13 = 3n + {}^-5$
$\underline{{}^+5 \qquad\qquad {}^+5}$   To get $3n$ by itself, add 5 to both sides.
$18 = 3n + 0$

$\dfrac{18}{3} = \dfrac{3n}{3}$   To get $n$ by itself, divide both sides by 3.

$6 = n$

*Step 5* **State the answer.** David did 6 experiments.

*Step 6* **Check** the solution by putting it back into the *original* problem.

3 times David's number      $3 \cdot 6 = 18$
Less 5                       $18 - 5 = 13$
Michael did 13. ←———— Matches ————┘

The correct solution is: David did 6 experiments.

**Work Problem** ❻ **at the Side.**

❻ Susan donated $10 more than twice what LuAnn donated. If Susan donated $22, how much did LuAnn donate?

Show your work for each of the six problem-solving steps.

**ANSWERS**

**6.** *Step 1* **Read.**
Unknown: LuAnn's donation
Known: Susan donated $10 more than twice what LuAnn donated; Susan donated $22.

*Step 2(a)* **Assign a variable.**
Let $d$ be LuAnn's donation.

*Step 3* **Write an equation.**
$2d + \$10 = \$22$
(or $\$10 + 2d = \$22$)

*Step 4* **Solve.**
$2d + 10 = 22$
$\phantom{2d +} \underline{{}^-10 \quad {}^-10}$
$2d + \phantom{1}0 = 12$

$\dfrac{2d}{2} = \dfrac{12}{2}$

$d = 6$

*Step 5* **State the answer.**
LuAnn donated $6.

*Step 6* **Check.**
$10 more than twice $6 is
$\$10 + 2 \cdot \$6 = \$22.$ ←┐
Susan donated $22. ←Matches ┘

# Focus on Real-Data Applications

## Formulas

In **Section 3.1**, you found the perimeter of a triangle by adding the lengths of the three sides. A formula for this is shown here.

$$P = a + b + c$$

where $P$ is the perimeter and $a$, $b$, and $c$ are the lengths of the sides.

1. Use the formula to find the length of the third side in this triangle. Replace $P$ with 33 ft, replace $a$ with 12 ft, and replace $b$ with 7 ft. Then solve the equation.

Use the formula to find the length of the third side in each of these triangles.

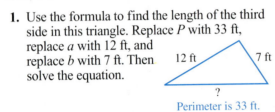

Perimeter is 24 cm.  Perimeter is 64 in.

4. Describe *in words* what you were doing as you found the length of the third side. What things were being added? What were you doing with the perimeter?

5. How can you simplify the equation when *two* sides of the triangle are the *same* length?

Use the simplified equation to find the unknown lengths in these triangles.

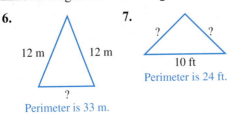

Perimeter is 33 m.  Perimeter is 24 ft.

In **Section 2.2** you worked with the expression $60A - 900$ for the vocabulary of a child. Now we can use this formula.

$$V = 60A - 900$$

where $V$ is the number of vocabulary words and $A$ is the age of the child in months.

Use the formula to find the *age* of each child.

8. The child's vocabulary is 180 words.

9. The child's vocabulary is 1440 words.

10. The child's vocabulary is 60 words.

11. Describe *in words* what you were doing as you found each child's age.

12. Find the child's age if the child's vocabulary is 0 words.

13. Refer to Problem 12 above. Are there any other ages that would also have a vocabulary of 0 words?

14. Is the formula useful for all ages of children? Explain your answer.

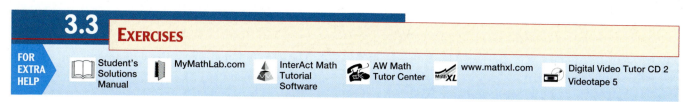

Write an algebraic expression, using x as the variable. See Examples 1 and 2.

1. 14 plus a number

2. The sum of a number and ⁻8

3. ⁻5 added to a number

4. 16 more than a number

5. 20 minus a number

6. A number decreased by 25

7. 9 less than a number

8. A number subtracted from ⁻7

9. Subtract 4 from a number.

10. 3 fewer than a number

11. ⁻6 times a number

12. The product of ⁻3 and a number

13. Double a number

14. A number times 10

15. A number divided by 2

16. 4 divided by a number

17. Twice a number added to 8

18. Five times a number plus 5

19. 10 fewer than seven times a number

20. 12 less than six times a number

21. The sum of twice a number and the number

22. Triple a number subtracted from the number

*Translate each sentence into an equation and solve it. Check your solution by going back to the words in the original problem. See Example 3.*

**23.** If four times a number is decreased by 2, the result is 26. Find the number.

**24.** The sum of 8 and five times a number is 53. Find the number.

**25.** If a number is added to twice the number, the result is −15. What is the number?

**26.** If a number is subtracted from three times the number, the result is −8. What is the number?

**27.** If the product of some number and 5 is increased by 12, the result is seven times the number. Find the number.

**28.** If eight times a number is subtracted from eleven times the number, the result is −9. Find the number.

**29.** When three times a number is subtracted from 30, the result is 2 plus the number. What is the number?

**30.** When twice a number is decreased by 8, the result is the number increased by 7. Find the number.

*Solve each application problem. Use the six problem-solving steps you learned in this section. See Examples 4–6.*

**31.** Ricardo gained 15 pounds over the winter. He went on a diet and lost 28 pounds. Then he regained 5 pounds and weighed 177 pounds. How much did he weigh originally?

**32.** Mr. Chee deposited $80 into his checking account. Then, after writing a $23 check for gas and a $90 check for his child's day care, the balance in his account was $67. How much was in his account before he made the deposit?

**33.** There were 18 cookies in Magan's cookie jar. While she was busy in another room, her children ate some of the cookies. Magan bought three dozen cookies and added them to the jar. At that point she had 49 cookies in the jar. How many cookies did her children eat?

**34.** The Greens had a 20-pound bag of bird seed in their garage. Mice got into the bag and ate some of it. They bought an 8-pound bag of seed and put all the seed in a metal container. They now have 24 pounds of seed. How much did the mice eat?

**35.** A college bookstore ordered six boxes of red pens. The store sold 32 red pens last week and 35 red pens this week. Five pens were left on the shelf. How many pens were in each box?

**36.** A local charity received a donation of eight cartons filled with cans of soup. The charity gave out 100 cans of soup yesterday and 92 cans today before running out. How many cans were in each carton?

**37.** The 14 music club members each paid the same amount for dues. The club also earned $340 selling magazine subscriptions. They spent $575 to organize a jazz festival. Now their bank account is overdrawn by $25. How much did each member pay in dues?

**38.** The manager of an apartment complex had 11 packages of light bulbs on hand. He replaced 29 burned out bulbs in hallway lights and 7 bulbs in the party room. Eight bulbs were left. How many bulbs were in each package?

**39.** When 75 is subtracted from four times Tamu's age, the result is Tamu's age. How old is Tamu?

**40.** If three times Linda's age is decreased by 36, the result is twice Linda's age. How old is Linda?

**41.** While shopping for clothes, Consuelo spent $3 less than twice what Brenda spent. Consuelo spent $81. How much did Brenda spend?

**42.** Dennis weighs 184 pounds. His weight is 2 pounds less than six times his child's weight. How much does his child weigh?

**43.** Paige bought five bags of candy for Halloween. Forty-eight children visited her home and she gave each child three pieces of candy. At the end of the night she still had one bag of candy. How many pieces of candy were in each bag?

**44.** A restaurant ordered four packages of paper napkins. Yesterday they used up one package, and today they used up 140 napkins. Two packages plus 60 napkins remain. How many napkins are in each package?

**45.** The recommended daily intake of iron for an adult female is 3 mg more than twice the recommended amount for a newborn infant. The amount for an adult female is 15 mg. How much should the infant receive? (*Source:* Food and Nutrition Board.)

**46.** A cheetah's sprinting speed is 61 miles per hour less than three times a zebra's running speed. A cheetah can sprint 68 miles per hour. Find the zebra's running speed. (*Source: Grolier Multimedia Encyclopedia*.)

## 3.4 SOLVING APPLICATION PROBLEMS WITH TWO UNKNOWN QUANTITIES

**1** **Solve application problems with two unknown quantities.** In the preceding section, the problems had only one unknown quantity. As a result, we used Step 2(a) rather than Step 2(b) in the problem-solving steps. For easy reference, we repeat the steps here.

**OBJECTIVES**

  Solve application problems with two unknown quantities.

### Solving an Application Problem

*Step 1*   **Read** the problem once to see what it is about. Read it carefully a second time. As you read, make a sketch or write word phrases that identify the known and the unknown parts of the problem.

*Step 2(a)*   If there is one unknown quantity, **assign a variable** to represent it. Write down what your variable represents.

*Step 2(b)*   If there is more than one unknown quantity, **assign a variable** to represent "the thing you know the least about." Then write variable expression(s), using the same variable, to show the relationship of the other unknown quantities to the first one.

*Step 3*   **Write an equation,** using your sketch or word phrases as the guide.

*Step 4*   **Solve** the equation.

*Step 5*   **State the answer** to the question in the problem and label your answer.

*Step 6*   **Check** whether your answer fits all the facts given in the *original* statement of the problem. If it does, you are done. If it doesn't, start again at Step 1.

Now you are ready to solve problems with two unknown quantities.

### Example 1   Solving an Application Problem with Two Unknown Quantities

Last month, Sheila worked 72 hours more than Russell. Together they worked a total of 232 hours. Find the number of hours each person worked last month.

*Step 1*   **Read** the problem. It is about the number of hours worked by Sheila and by Russell.

  **Unknowns:** hours worked by Sheila;
   hours worked by Russell
  **Known:** Sheila worked 72 hours more than Russell;
   232 hours total for Sheila and Russell

*Step 2(b)*   There are *two* unknowns so **assign a variable** to represent "the thing you know the least about." You know the *least* about the hours worked by Russell, so let $h$ represent Russell's hours.

Sheila worked 72 hours more than Russell, so her hours are $h + 72$, that is, Russell's hours ($h$) plus 72 more.

*Step 3*   **Write an equation.**

$$\underset{\text{hours worked by Russell}}{h} + \underset{\text{hours worked by Sheila}}{h + 72} = \underset{\text{total hours worked}}{232}$$

*Continued on Next Page*

❶ In a day of work, Keonda made $12 more than her daughter. Together they made $182. Find the amount that each person made. (*Hint:* Which amount do you know the *least* about, Keonda's or her daughter's? Let $m$ be that amount.) Use the six problem-solving steps.

*Step 4* **Solve.**

$$\underline{h + h} + 72 = 232 \qquad \text{Simplify the left side by combining like terms.}$$
$$2h + 72 = 232$$
$$\phantom{2h +} -72 \phantom{= 2}-72 \qquad \text{To get } 2h \text{ by itself, add } {-72} \text{ to both sides.}$$
$$2h + 0 = 160$$
$$\frac{2h}{2} = \frac{160}{2} \qquad \text{To get } h \text{ by itself, divide both sides by 2.}$$
$$h = 80$$

*Step 5* **State the answer.**
Because $h$ represents Russell's hours, and the solution of the equation is $h = 80$, Russell worked 80 hours.

$h$ + 72 represents Sheila's hours. Replace $h$ with 80.

$80 + 72 = 152$, so Sheila worked 152 hours.

The final answer is: Russell worked 80 hours and Sheila worked 152 hours.

*Step 6* **Check** the solution by putting both numbers back into the *original* problem.

**"Sheila worked 72 hours more than Russell."**
Sheila's 152 hours are 72 more than Russell's 80 hours, so the solution checks.

$$\begin{array}{r} 152 \\ -\ 72 \\ \hline 80 \end{array}$$

**"Together they worked a total of 232 hours."**
Sheila's 152 hours + Russell's 80 hours = 232 hours so the solution checks.

$$\begin{array}{r} 152 \\ +\ 80 \\ \hline 232 \end{array}$$

You've answered the question correctly because 80 hours and 152 hours fit all the facts given in the problem.

> **CAUTION**
>
> Check the solution to an application problem by putting the numbers back in the *original* problem. If they do *not* work, recheck your work or try solving the problem in a different way.

**Work Problem ❶ at the Side.**

### Example 2  Solving a Geometry Application with Two Unknown Quantities

The length of a rectangle is 2 cm more than the width. The perimeter is 68 cm. Find the length and width.

*Step 1* **Read** the problem. It is about a rectangle. Make a sketch of a rectangle.

**Unknowns:** length of the rectangle; width of the rectangle
**Known:** The length is 2 cm more than the width; the perimeter is 68 cm.

*Continued on Next Page*

**ANSWERS**
1. Daughter made $m$.
   Keonda made $m + 12$.
   $m + m + 12 = 182$
   Daughter made $85.
   Keonda made $97.

   **Check** $97 − $85 = $12
   and $97 + $85 = $182

*Step 2(b)* There are *two* unknowns so **assign a variable** to represent "the thing you know the least about."
You know the *least* about the width, so let **w** represent the **width**.

The length is 2 cm more than the width, so the **length** is *w* + 2.

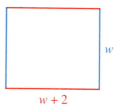

*Step 3* **Write an equation.**
Use the formula for perimeter of a rectangle, $P = 2l + 2w$, to help you write the equation.

$$P = 2 \cdot l + 2 \cdot w \quad \text{Replace } P \text{ with 68.}$$
$$\quad\quad\quad\quad\quad\quad\quad\quad \text{Replace } l \text{ with } (w + 2).$$
$$68 = 2(w + 2) + 2 \cdot w$$

*Step 4* **Solve.**

$$68 = 2(w + 2) + 2w \quad \text{Use the distributive property.}$$
$$68 = 2w + 4 + 2w \quad \text{Combine like terms.}$$
$$68 = 4w + 4$$
$$\underline{-4 \quad\quad\quad -4} \quad \text{To get } 4w \text{ by itself, add } {-4} \text{ to both sides.}$$
$$64 = 4w + 0$$
$$\frac{64}{4} = \frac{4w}{4} \quad \text{To get } w \text{ by itself, divide both sides by 4.}$$
$$16 = w$$

*Step 5* **State the answer.**

*w* represents the width, and *w* = 16, so the width is 16 cm.

*w* + 2 represents the length. Replace *w* with 16.

16 + 2 = 18, so the length is 18 cm. ← The label for both answers is cm.

The final answer is: The width is 16 cm and the length is 18 cm.

*Step 6* **Check** the solution by putting the measurements on your sketch and going back to the original problem.

**"The length of a rectangle is 2 cm more than the width."**

18 cm is 2 cm more than 16 cm, so the solution checks.

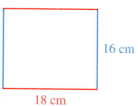

**"The perimeter is 68 cm."**

$$P = 2 \cdot 18 \text{ cm} + 2 \cdot 16 \text{ cm}$$
$$P = 36 \text{ cm} + 32 \text{ cm}$$
$$P = 68 \text{ cm} \longleftarrow \text{This matches the perimeter given in the original problem, so the solution checks.}$$

═══════ Work Problem ❷ at the Side.

❷ Make a sketch to help solve this problem.

The length of Ann's rectangular garden plot is 3 yd more than the width. She used 22 yd of fencing around the edge. Find the length and the width of the garden, using the six problem-solving steps.

**ANSWERS**

2.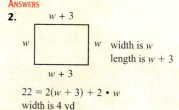
width is *w*
length is *w* + 3

$22 = 2(w + 3) + 2 \cdot w$
width is 4 yd
length is 7 yd
**Check** 7 yd is 3 yd more than 4 yd.
$P = 2 \cdot 7 \text{ yd} + 2 \cdot 4 \text{ yd} = 22 \text{ yd}$
Matches perimeter given in the original problem

# Focus on Real-Data Applications

## Connections: Arithmetic to Algebra

As you begin to learn algebra, you may be frustrated when asked to work problems algebraically that are simple enough to be solved using arithmetic. Why bother with algebra? This activity shows that the *process* you use when solving a problem using arithmetic becomes *automated* when you use an **algebraic equation.** As the applications become more complicated and you find it more and more difficult to solve a problem using just arithmetic, the algebraic process becomes a powerful tool. In the example, compare the steps of the arithmetic process to the steps involved in solving the algebraic equation. Note that the step-by step *operations* are identical.

Problem: Kevin rented a chain saw for a one-time $15 sharpening fee plus $18-a-day rental. His total bill was $105. For how many days did Kevin rent the chain saw?

| Arithmetic Solution | Algebraic Solution |
| --- | --- |
| To solve the problem, we have to first find the amount spent on rental fee alone. Then we have to find the number of days that gives that rental fee. The process is as follows. | First, let $n$ be the number of rental days. Then, the daily rental fee times the number of days, plus the sharpening fee, equals the total charge: $18n + 15 = 105$. Solve the equation. |
| 1. First subtract the sharpening fee from the total charge to find the rental fee: $105 - 15 = 90$. | 1. Add $-15$ to both sides. $$18n + 15 = 105$$ $$\underline{\phantom{18n+}-15 \quad -15}$$ $$18n \phantom{+15} = \phantom{10}90$$ |
| 2. Divide the rental fee by the daily rental charge to find the number of days: $90 \div 18 = 5$. | 2. Divide both sides by 18. $$\frac{18n}{18} = \frac{90}{18}$$ $$n = 5$$ |
| 3. *Answer:* Kevin rented the saw for 5 days. | 3. *Answer:* Kevin rented the saw for 5 days. |

In the following problems, the first one can be solved easily using both arithmetic and algebraic methods. Show that the processes are the same, in a manner similar to the above example. The second problem should be much easier to work using an algebraic equation. Show your work on separate paper, using a format similar to the one in the example.

1. Jose made a $300 down payment on a used car. His monthly payment was $150, and he paid $3900 in all. For how many months did he make payments?

| Arithmetic Solution | Algebraic Solution |
| --- | --- |
|  | Let $m$ be the number of months of payments. |

2. A 295 ft long rope is to be cut into three parts. Two parts are the same length, and the third part is 47 ft shorter than the other two parts. Find the length of each part.

| Arithmetic Solution | Algebraic Solution |
| --- | --- |
|  | Let $x$ be the length of Part 1 and Part 2. _____ is the length of Part 3. |

## 3.4 Exercises

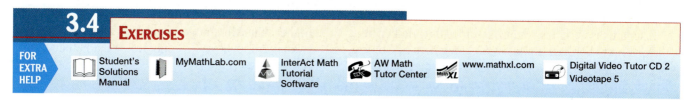

*Solve each application problem by using the six problem-solving steps you learned in this section. See Example 1.*

1. My sister is 9 years older than I am. The sum of our ages is 51. Find our ages.

2. Ed and Marge were candidates for city council. Marge won, with 93 more votes than Ed. The total number of votes cast in the election was 587. Find the number of votes received by each candidate.

3. Last year, Lien earned $1500 more than her husband. Together they earned $37,500. How much did each of them earn?

4. A $149,000 estate is to be divided between two charities so that one charity receives $18,000 less than the other. How much will each charity receive?

5. Jason paid five times as much for his computer as he did for his printer. He paid a total of $1320 for both items. What did each item cost?

6. The attendance at the Saturday night baseball game was three times the attendance at Sunday's game. In all, 56,000 fans attended the games. How many fans were at each game?

7. A board is 78 cm long. Rosa cut the board into two pieces, with one piece 10 cm longer than the other. Find the length of both pieces. (*Hint:* Make a sketch of the board. Which piece do you know the least about? Let $x$ represent the length of that piece.)

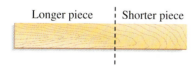

8. A rope is 21 yd long. Marcos cut it into two pieces so that one piece is 3 yd longer than the other. Find the length of each piece.

9. A wire is cut into two pieces, with one piece 7 ft shorter than the other. The wire was 31 ft long before it was cut. How long was each piece?

10. A 90 cm pipe is cut into two pieces so that one piece is 6 cm shorter than the other. Find the length of each piece.

11. In the United States Congress, the number of Representatives is 65 less than five times the number of Senators. There are a total of 535 members of Congress. Find the number of Senators and the number of Representatives. (*Source:* World Almanac, 2001.)

12. Florida's record low temperature is 68 degrees higher than Montana's record low. The sum of the two record lows is ⁻72 degrees. What is the record low for each state? (*Source:* National Climatic Data Center.)

13. A fence is 706 m long. It is to be cut into three parts. Two parts are the same length, and the third part is 25 m longer than each of the other two. Find the length of each part.

14. A wooden railing is 82 m long. It is to be divided into four pieces. Three pieces will be the same length, and the fourth piece will be 2 m longer than each of the other three. Find the length of each piece.

*In Exercises 15–20, use the formula for the perimeter of a rectangle, $P = 2l + 2w$. Make a sketch to help you solve each problem. See Example 2.*

15. The perimeter of a rectangle is 48 yd. The width is 5 yd. Find the length.

16. The length of a rectangle is 27 cm, and the perimeter is 74 cm. Find the width of the rectangle.

17. A rectangular dog pen is twice as long as it is wide. The perimeter of the pen is 36 ft. Find the length and the width of the pen.

18. A new city park is a rectangular shape. The length is triple the width. It will take 240 meters of fencing to go around the park. Find the length and width of the park.

19. The length of a rectangular jewelry box is 3 in. more than twice the width. The perimeter is 36 in. Find the length and the width.

20. The perimeter of a rectangular house is 122 ft. The width is 5 ft less than the length. Find the length and the width.

21. A photograph measures 8 in. by 10 in. Earl put it in a frame that is 2 in. wide. Find the area of the frame and its outside perimeter. Make a sketch.

22. Barb had a 16 in. by 20 in. photograph. She cropped 3 in. off each side of the photo. What are the perimeter and the area of the cropped photo? Make a sketch.

# Chapter 3

## SUMMARY

### KEY TERMS

**3.1** **formula** — Formulas are well-known rules for solving common types of problems. They are written in a shorthand form that uses variables.

**perimeter** — Perimeter is the distance around the outside edges of a flat shape. It is measured in linear units such as in., ft, yd, mm, cm, m, and so on.

**square** — A square is a figure with four sides that are all the same length and meet to form 90° angles (square corners). See example at the right.

**rectangle** — A rectangle is a four-sided figure with all sides meeting at 90° angles. The opposite sides are the same length. See example at the right.

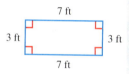

**parallelogram** — A parallelogram is a four-sided figure with both pairs of opposite sides parallel and equal in length. See example at the right.

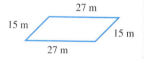

**triangle** — A triangle is a figure with exactly three sides. See example at the right.

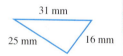

**3.2** **area** — Area is the surface inside a two-dimensional (flat) shape. It is measured by determining the number of squares of a certain size needed to cover the surface inside the shape. Some of the commonly used units for measuring area are square inches (in.$^2$), square feet (ft$^2$), square yards (yd$^2$), square centimeters (cm$^2$), and square meters (m$^2$).

### NEW FORMULAS

Perimeter of a square: $P = 4s$

Perimeter of a rectangle: $P = 2l + 2w$

Area of a square: $A = s^2$

Area of a rectangle: $A = lw$

Area of a parallelogram: $A = bh$

## Test Your Word Power

See how well you have learned the vocabulary in this chapter. Answers follow the Quick Review.

1. The **perimeter** of a flat shape is
   (a) measured in square units
   (b) found by adding the lengths of the sides
   (c) measured in cubic units
   (d) found by multiplying length times width.

2. The **area** of a flat shape is
   (a) found by adding length plus width
   (b) measured in linear units
   (c) found by squaring the height
   (d) measured in square units

3. When working with a **square shape**,
   (a) the perimeter formula is $P = s^2$
   (b) the area formula is $A = \frac{1}{2}bh$
   (c) all sides have the same length
   (d) all angles measure 180°.

4. When working with a **rectangular shape**,
   (a) the area formula is $A = lw$
   (b) opposite sides have different lengths
   (c) the perimeter formula is $P = bh$
   (d) the width is the longer measurement.

5. In *all* **triangles**
   (a) the sides meet at 90° angles
   (b) the sides have the same length
   (c) there are exactly three sides
   (d) the area formula is $A = s^2$.

6. In *all* **parallelograms**
   (a) there are exactly six sides
   (b) the height line is parallel to the base
   (c) all sides have the same length
   (d) both pairs of opposite sides are parallel.

## Quick Review

| Concepts | Examples |
|---|---|

### 3.1 Finding Perimeter

To find the perimeter of *any* shape, add the lengths of the sides. Perimeter is measured in linear units (cm, m, ft, yd, and so on).

Or, for some shapes, there are formulas that you can use.

Find the perimeter of each figure.

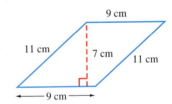

$P = 9 \text{ cm} + 11 \text{ cm} + 9 \text{ cm} + 11 \text{ cm}$

$P = 40 \text{ cm}$

Perimeter of a square: $P = 4s$

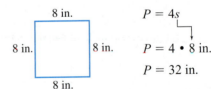

$P = 4s$
$P = 4 \cdot 8 \text{ in.}$
$P = 32 \text{ in.}$

Perimeter of a rectangle: $P = 2l + 2w$

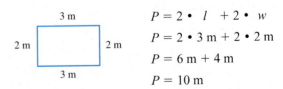

$P = 2 \cdot l + 2 \cdot w$
$P = 2 \cdot 3 \text{ m} + 2 \cdot 2 \text{ m}$
$P = 6 \text{ m} + 4 \text{ m}$
$P = 10 \text{ m}$

## Chapter 3 Summary

*Concepts*

**3.1 Finding the Length of One Side of a Square**
If you know the perimeter of a square, use the formula $P = 4s$. Replace $P$ with the value for the perimeter and solve the equation for $s$.

*Examples*

If the perimeter of a square room is 44 ft, find the length of one side.

$$P = 4s \qquad \text{Replace } P \text{ with 44 ft.}$$
$$44 \text{ ft} = 4s$$
$$\frac{44 \text{ ft}}{4} = \frac{4s}{4} \qquad \text{Divide both sides by 4.}$$
$$11 \text{ ft} = s$$

The length of one side is 11 ft.

**Check** Check the solution by drawing a sketch of the room. The perimeter is
11 ft + 11 ft + 11 ft + 11 ft = 44 ft. The result matches the perimeter given in the problem.

**3.1 Finding the Length or Width of a Rectangle**
If you know the perimeter of a rectangle and either its width or length, use the formula $P = 2l + 2w$. Replace $P$ and either $l$ or $w$ with the values that you know. Then solve the equation.

The width of a rectangular rug is 8 ft. The perimeter is 36 ft. Find the length.

$$P = 2l + 2w \qquad \text{Replace } P \text{ with 36 ft and } w \text{ with 8 ft.}$$
$$36 \text{ ft} = 2l + 2 \cdot 8 \text{ ft}$$
$$36 \text{ ft} = 2l + 16 \text{ ft}$$
$$\underline{\phantom{36 \text{ ft}} {-}16 \text{ ft} \qquad {-}16 \text{ ft}} \qquad \text{Add } {-}16 \text{ ft to both sides.}$$
$$20 \text{ ft} = 2l + 0$$
$$\frac{20 \text{ ft}}{2} = \frac{2l}{2} \qquad \text{Divide both sides by 2.}$$
$$10 \text{ ft} = l \qquad \text{The length is 10 ft.}$$

**Check** To check the solution, draw a sketch of the rectangle and label the lengths of the sides. Then add the four measurements.
10 ft + 10 ft + 8 ft + 8 ft = 36 ft
The result matches the perimeter given in the problem.

## Concepts

### 3.2 Finding Area
Use the appropriate formula. Remember to measure area in *square* units (cm$^2$, m$^2$, ft$^2$, yd$^2$, and so on).

Area of a rectangle: $A = lw$,
where $l$ is the length and $w$ is the width.

Area of a square: $A = s^2$, which means $s \cdot s$,
where $s$ is the length of one side.

Area of a parallelogram: $A = bh$,
where $b$ is the base and $h$ is the height.

### 3.2 Finding the Unknown Length in a Rectangle or Parallelogram
If you know the area of a rectangle or parallelogram and one of the measurements, use the appropriate area formula (see above). Replace $A$ and one of the other variables with the values that you know. Then solve the equation.

### 3.2 Finding the Length of One Side of a Square
If you know the area of a square, use the formula $A = s^2$. Replace $A$ with the value that you know. Then solve the equation by asking, "What number, times itself, gives the value of $A$?"

## Examples

Find the area of each figure.

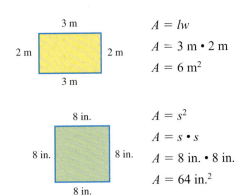

$A = lw$
$A = 3 \text{ m} \cdot 2 \text{ m}$
$A = 6 \text{ m}^2$

$A = s^2$
$A = s \cdot s$
$A = 8 \text{ in.} \cdot 8 \text{ in.}$
$A = 64 \text{ in.}^2$

$A = bh$
$A = 4 \text{ cm} \cdot 3 \text{ cm}$
$A = 12 \text{ cm}^2$

The area of a parallelogram is 72 yd$^2$, and its height is 9 yd. Find the base.

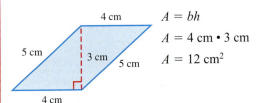

Replace $A$ with 72 yd$^2$ and $h$ with 9 yd.

Divide both sides by 9 yd.

$8 \text{ yd} = b$

The base of the parallelogram is 8 yd.

A square ceiling has an area of 100 ft$^2$. What is the length of each side of the ceiling?

$A = s^2$    Replace $A$ with 100 ft$^2$.

**100 ft$^2$** $= s^2$    Rewrite $s^2$ as $s \cdot s$.

100 ft$^2$ $= s \cdot s$    Ask, "What number times itself gives 100?"

100 ft$^2$ = 10 ft $\cdot$ 10 ft

The value of $s$ is 10 ft, so each side of the ceiling is 10 ft long.

## Chapter 3 Summary

| Concepts | Examples |
|---|---|
| | |

### 3.3 Translating Sentences into Equations

Translate word phrases into symbols using $x$ (or any other letter) as the variable. Then solve the equation. Check the solution by putting it back in the original problem.

If 10 is subtracted from three times a number, the result is 14.

Let $x$ represent the unknown number.

$$3x - 10 = 14$$
$$3x + {}^-10 = 14$$
$$\underline{\phantom{3x +}\ {}^+10\ \ \ {}^+10}\quad \text{Add 10 to both sides.}$$
$$3x + 0 = 24$$
$$\frac{3x}{3} = \frac{24}{3}\quad \text{Divide both sides by 3.}$$
$$x = 8\quad \text{The number is 8.}$$

**Check** If 10 is subtracted from three times 8, do you get 14? Yes, $3 \cdot 8 - 10 = 24 - 10 = 14$.

### 3.3 Solving Application Problems with One Unknown Quantity

Use the six problem-solving steps outlined in Section 3.3. They are listed below in abbreviated form.

Denise had some money in her purse this morning. She gave $15 to her daughter and paid $4 to park in the lot at work. At that point she still had $27. How much was in her purse this morning?

*Step 1* **Read** the problem and identify what is known and what is unknown.

*Step 1* Unknown: money in purse this morning
Known: took out $15; took out $4; still had $27.

*Step 2* **Assign a variable** to represent the unknown quantity.

*Step 2* Let $m$ represent the money in her purse this morning.

*Step 3* **Write an equation.**

*Step 3* $m - \$15 - \$4 = \$27$

*Step 4* **Solve** the equation.

*Step 4*
$$m - 15 - 4 = 27$$
$$m + {}^-15 + {}^-4 = 27$$
$$m + {}^-19 = 27$$
$$\underline{\phantom{m +}\ {}^+19\ \ \ {}^+19}$$
$$m + 0 = 46$$
$$m = 46$$

*Step 5* **State the answer.**

*Step 5* She had $46 in her purse this morning.

*Step 6* **Check** whether your answer fits all the facts given in the *original* statement of the problem.

*Step 6* Started with $46.
Took out $15, so $46 − $15 = $31
Took out $4, so $31 − $4 = $27
Had $27 left ←— Matches ——

| Concepts | Examples |
|---|---|
| **3.4 Solving Application Problems with Two Unknown Quantities** <br> Use the six problem-solving steps outlined in Section 3.3. | Last week, Brian earned $50 more than twice what Dan earned. How much did each person earn if the total for both of them was $254? <br><br> *Step 1*   Unknowns: Brian's earnings; Dan's earnings <br>   Known: Brian earned $50 more than twice what Dan earned; the sum of their earnings was $254. |
| In *Step 2*, there are *two* unknown quantities, so assign a variable to represent "the thing you know the least about." Then write variable expressions, using the same variable, to show the relationship of the other unknown quantities to the first one. | *Step 2*   You know the least about Dan's earnings, so let $m$ represent Dan's earnings. <br> Brian earned $50 more than twice what Dan earned, so Brian's earnings are $2m + \$50$. <br><br> *Step 3*   $\underbrace{m + 2m}_{} + 50 = 254$ <br><br> *Step 4*   $3m + 50 = 254$ <br> $\phantom{3m +}\; -50 \;\; -50$ <br> $\underbrace{3m}_{} + \;\; 0 = 204$ <br> $\dfrac{3m}{3} = \dfrac{204}{3}$ <br> $m = 68$ <br><br> *Step 5*   $m$ represents Brian's earnings, so Brian earned $68. <br><br> $2m + \$50$ represents Dan's earnings, <br> and $2 \cdot \$68 + \$50$ is $\$136 + \$50 = \$186$. <br> Brian earned $68; Dan earned $186. <br><br> *Step 6*   Is $186 actually $50 more than twice $68? Yes, the solution checks. <br> Does $68 + \$186 = \$254$? Yes, the solution checks. |

## ANSWERS TO TEST YOUR WORD POWER

1. **(b)** *Example:* If a triangle has sides measuring 4 ft, 10 ft, and 8 ft, then $P = 4 \text{ ft} + 10 \text{ ft} + 8 \text{ ft} = 22 \text{ ft}$.
2. **(d)** *Example:* Area is measured in square units such as in.$^2$, ft$^2$, yd$^2$, cm$^2$, and km$^2$.
3. **(c)** *Example:* If one side of a square is 5 in. long, all the other sides will also be 5 in. long.
4. **(a)** *Example:* If the length of a rectangle is 12 cm and the width is 8 cm, then $A = 12 \text{ cm} \cdot 8 \text{ cm} = 96 \text{ cm}^2$.
5. **(c)** *Examples:*

6. **(d)** *Example:* In parallelogram *ABCD*, sides *AB* and *DC* are parallel, and sides *AD* and *BC* are parallel.

# Chapter 3 Review Exercises

**[3.1]** *In Exercises 1–4, find the perimeter of each shape.*

1. Square

28 cm

2. Rectangle

8 mi, 3 mi

3. Parallelogram

7 yd, 5 yd, 14 yd

4.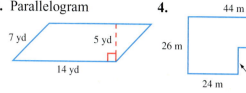

44 m, 14 m, 20 m, 13 m, 24 m, 26 m

5. A square card table has a perimeter of 12 ft. Find the length of one side of the table.

6. A rectangular playground has a perimeter of 128 yd. Find its length if it is 31 yd wide.

7. A rectangular watercolor painting that is 21 in. long has a perimeter of 72 in. What is the width of the painting?

**[3.2]** *In Exercises 8–10, draw a sketch of each shape and label the lengths of the sides or the base and height. Then find the area, using the appropriate formula. Show your sketches to your instructor.*

8. A tablecloth that measures 5 ft by 8 ft

9. A 25 m square dance floor

10. A parallelogram-shaped lot with a base of 16 yd and a height of 13 yd

11. A rectangular patio that is 14 ft long has an area of 126 ft². Find its width.

12. A parallelogram has an area of 88 cm². If the base is 11 cm, what is the height?

13. The area of a square piece of land is 100 mi². What is the length of one side?

**[3.3]** *Write each phrase as an algebraic expression. Use x as the variable.*

14. A number subtracted from 57

15. The sum of 15 and twice a number

16. The product of ⁻9 and a number

*Translate each sentence into an equation and solve it. Show your work.*

**17.** The sum of four times a number and 6 is $^-30$. What is the number?

**18.** When twice a number is subtracted from 10, the result is 4 plus the number. Find the number.

**[3.3–3.4]** *Use the six problem-solving steps to solve each problem.*

**19.** Grace wrote a $600 check for her rent. Then she deposited her $750 paycheck and a $75 tax refund into her account. The new balance was $309. How much was in her account before she wrote the rent check?

**20.** Yoku ordered four boxes of candles for his restaurant. One candle was put on each of the 25 tables. There were 23 candles left. How many candles were originally in each box?

**21.** $1000 in prize money in an essay contest is being split between Reggie and Donald. Donald should get $300 more than Reggie. How much will each man receive?

**22.** A rectangular photograph is twice as long as it is wide. The perimeter of the photograph is 84 cm. Find the length and the width of the photograph.

## MIXED REVIEW EXERCISES

*Use the information in the advertisement to work Exercises 23–26.*

**Build Your Own Dog Pen**
Kit #1 includes 20 feet of fencing.
Kit #2 includes 36 feet of fencing.

**23.** Anthony made a square dog pen using Kit #2.

  (a) What was the length of each side of the pen?

  (b) What was the area of the pen?

**24.** First draw sketches of two *different* rectangular dog pens that you could build using all the fencing in Kit #1. Label the lengths of the sides. Then find the area of each pen.

**25.** Timotha bought Kit #2. But she used some of the fencing around her garden, so she went back and bought Kit #1. Now she has 41 ft of fencing for a dog pen. How much fencing did she use around her garden? Use the six problem-solving steps.

**26.** Diana bought Kit #2. The pen she built had a length that was 2 ft more than the width. Find the length and width of the pen. Use the six problem-solving steps.

# Chapter 3 TEST

*Study Skills Workbook*
Activity 10

*Find the perimeter of each shape.*

1.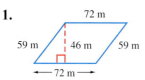

2.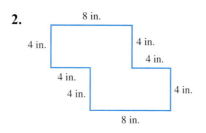

1. _____

2. _____

3. A square wetland 3 mi on a side

4. A rectangular mirror that measures 2 ft by 4 ft

3. _____

4. _____

5.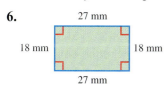

5. _____

*Find the area of each shape.*

6.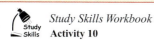

7.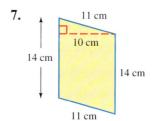

6. _____

8. A rectangular animal preserve is 55 mi wide and 68 mi long.

9. A square room measures 6 m on each side.

7. _____

8. _____

9. _____

*Solve each problem by using the appropriate formula. Show your work.*

10. A square table has a perimeter of 12 ft. Find the length of one side.

10. _____

11. The Mercado family has 34 ft of fencing to put around a garden plot. The plot is rectangular in shape. If it is 6 ft wide, find the length of the plot.

11. _____

12. The area of a parallelogram is 65 in.$^2$, and the base is 13 in. What is the height of the parallelogram?

12. _____

13. _____

**13.** A rectangular postage stamp has a length of 4 cm and an area of 12 cm². Find its width.

14. _____

**14.** A square bulletin board has an area of 16 ft². How long is each side of the bulletin board?

15. _____

**15.** Explain the difference between ft and ft². For which types of problems might you use each of these units?

*Translate each sentence into an equation and solve it. Show your work.*

16. _____

**16.** If 40 is added to four times a number, the result is zero. Find the number.

**17.** When 7 times a number is decreased by 23, the result is the number plus 7. What is the number?

17. _____

*Solve each application problem, using the six problem-solving steps.*

18. _____

**18.** Josephine had $43 in her wallet. Her son used some of it when he bought groceries. Josephine found $16 in her desk drawer and put it in her wallet. She counted $44 in the wallet. How much money did her son spend on groceries?

19. _____

**19.** Ray is 39 years old. His age is 4 years more than five times his daughter's age. How old is his daughter?

20. _____

**20.** A board is 118 cm long. Karin cut it into two pieces, with one piece 4 cm longer than the other. Find the length of both pieces.

21. _____

**21.** The perimeter of a rectangular building is 420 ft. The length is four times as long as the width. Find the length and the width. Draw a sketch to help you solve this problem.

22. _____

**22.** Marcella and her husband Tim spent a total of 19 hours redecorating their living room. Tim spent 3 hours less time than Marcella. How long did each person work on the room?

# Cumulative Review Exercises — Chapters 1–3

1. Write this number in words.
   4,000,206,300

2. Write this number, using digits.
   seventy million, five thousand, four hundred eighty-nine

3. Write $<$ or $>$ between each pair of numbers to make a true statement.
   $^-7$ ___ $^-1$     $0$ ___ $^-5$

4. Name the property illustrated by each equation.
   (a) $1 \cdot 97 = 97$
   (b) $^-10 + 0 = ^-10$
   (c) $(3 \cdot {}^-7) \cdot 6 = 3 \cdot ({}^-7 \cdot 6)$

5. (a) Round 3795 to the nearest ten.
   (b) Round 493,662 to the nearest ten thousand.

*Simplify.*

6. $^-12 - 12$

7. $^-3(^-9)$

8. $|7| - |^-10|$

9. $^-40 \div 2 \cdot 5$

10. $3 - 8 + 10$

11. $\dfrac{0}{^-6}$

12. $^-8 + 5(2 - 3)$

13. $(^-3)^2 + 4^2$

14. $4 - 3(^-6 \div 3) + 7(0 - 6)$

15. $\dfrac{4 - 2^3 + 5^2 - 3(^-2)}{^-1(3) - 6(^-2) - 9}$

16. Rewrite $10w^2xy^4$ without exponents.

17. Evaluate $^-6cd^3$ when $c$ is 5 and $d$ is $^-2$.

*Simplify.*

18. $^-4k + k + 5k$

19. $m^2 + 2m + 2m^2$

20. $xy^3 - xy^3$

21. $5(^-4a)$

22. $^-8 + x + 5 - 2x^2 - x$

23. $^-3(4n + 3) + 10$

*Solve each equation and check each solution.*

24. $6 - 20 = 2x - 9x$    **Check**

25. $^-5y = y + 6$    **Check**

*Solve each equation. Show your work.*

**26.** $3b - 9 = 19 - 4b$

**27.** $^-16 - h + 2 = h - 10$

**28.** $^-5(2x + 4) = 3x - 20$

**29.** $6 + 4(a + 8) = ^-8a + 5 + a$

*Find the perimeter and area of each shape.*

**30.**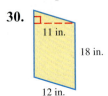

**31.** A square plaza is 15 m on a side.

**32.** A rectangular piece of plywood is 4 ft wide and 8 ft long.

*Translate each sentence into an equation and solve it. Show your work.*

**33.** When $^-50$ is added to five times a number, the result is 0. What is the number?

**34.** When three times a number is subtracted from 10, the result is two times the number. Find the number.

*Solve each application problem, using the six problem-solving steps.*

**35.** Some people were waiting in line at the bank. Three people made deposits and left. Six more people got in line. Then two more people were helped, but five people were still in line. How many were in the line originally?

**36.** Twelve soccer players each paid the same amount toward a team trip. Expenses for the trip were $2200, and now the team bank account is overdrawn by $40. How much did each player originally pay?

**37.** A group of 192 students was divided into two smaller groups. One smaller group was three times the size of the other. How many students were in each smaller group?

**38.** A rectangular swimming pool is 14 ft longer than it is wide. If the perimeter of the pool is 92 ft, find the length and the width.

# Rational Numbers: Positive and Negative Fractions

## 4

Americans spend over $230 billion each year buying building materials for their home construction, remodeling, and repair projects. (*Source:* U.S. Bureau of the Census.) One of the small but essential items for many home projects is nails. There are many kinds of nails. For example, you'll need common nails to build a new deck, drywall nails to attach wallboard in an unfinished basement, and finishing nails to fasten trim. You'll also need to buy the right size nail for each project. Nails are sized according to the "penny" system, which is a number followed by the abbreviation "d." (The "d" is the traditional British abbreviation for "penny.") An algebraic expression, and the ability to solve equations with fractions, will help you find the relationship between a nail's "penny size" and its length in inches. (See Section 4.7, Exercises 33–36.)

- 4.1 Introduction to Signed Fractions
- 4.2 Writing Fractions in Lowest Terms
- 4.3 Multiplying and Dividing Signed Fractions
- 4.4 Adding and Subtracting Signed Fractions
- 4.5 Problem Solving: Mixed Numbers and Estimating

   Summary Exercises on Fractions
- 4.6 Exponents, Order of Operations, and Complex Fractions
- 4.7 Problem Solving: Equations Containing Fractions
- 4.8 Geometry Applications: Area and Volume

# 4.1 Introduction to Signed Fractions

## Objectives

1. Use a fraction to name the part of a whole that is shaded.
2. Identify numerators, denominators, proper fractions, and improper fractions.
3. Graph positive and negative fractions on a number line.
4. Find the absolute value of a fraction.
5. Write equivalent fractions.

**1** Write fractions for the shaded portion and the unshaded portion of each figure.

(a)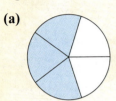

(b)

(c)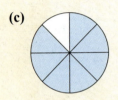

**1** **Use a fraction to name the part of a whole that is shaded.** In Chapters 1–3 you worked with integers. Recall that a list of integers can be written as follows.

$$\ldots, {}^-6, {}^-5, {}^-4, {}^-3, {}^-2, {}^-1, 0, 1, 2, 3, 4, 5, 6, \ldots$$

The dots show that the list goes on forever in both directions.

Now we will work with *fractions*.

### Fractions

A **fraction** is a number of the form $\dfrac{a}{b}$ where $a$ and $b$ are integers and $b$ is not 0.

One use for fractions is situations in which we need a number that is between two integers. Here is an example.

A recipe uses $\frac{2}{3}$ cup of milk.

$\frac{2}{3}$ is between 0 and 1.

$\frac{2}{3}$ is a fraction because it is of the form $\frac{a}{b}$ and 2 and 3 are integers.

The number $\frac{2}{3}$ is a fraction that represents 2 of 3 equal parts. We read $\frac{2}{3}$ as "two thirds."

### Example 1 Using Fractions to Represent Part of One Whole

Use fractions to represent the shaded portion and the unshaded portion of each figure.

(a)

The figure has 3 equal parts. The 2 shaded parts are represented by the fraction $\frac{2}{3}$. The *un*shaded part is $\frac{1}{3}$ of the figure.

(b) 

The 4 shaded parts of the 7-part figure are represented by the fraction $\frac{4}{7}$. The *un*shaded part is $\frac{3}{7}$ of the figure.

**Work Problem 1 at the Side.**

---

**Answers**

1. (a) $\frac{3}{5}$; $\frac{2}{5}$  (b) $\frac{1}{6}$; $\frac{5}{6}$  (c) $\frac{7}{8}$; $\frac{1}{8}$

Section 4.1 Introduction to Signed Fractions 199

Fractions can also be used to represent more than one whole object.

### Example 2 — Using Fractions to Represent More Than One Whole

Use a fraction to represent the shaded parts.

(a)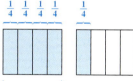

(b)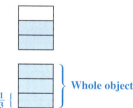

An area equal to 5 of the $\frac{1}{4}$ parts is shaded, so $\frac{5}{4}$ is shaded.

An area equal to 5 of the $\frac{1}{3}$ parts is shaded, so $\frac{5}{3}$ is shaded.

**Work Problem ❷ at the Side.**

**❷** Identify numerators, denominators, proper fractions, and improper fractions. In the fraction $\frac{2}{3}$, the number 2 is the *numerator* and 3 is the *denominator*. The bar between the numerator and the denominator is the *fraction bar*.

Fraction bar → $\frac{2}{3}$ ← Numerator
← Denominator

### Numerator and Denominator

The **denominator** of a fraction shows the number of equal parts in the whole, and the **numerator** shows how many parts are being considered.

### NOTE

Recall that a fraction bar, —, is a symbol for division and division by 0 is undefined. Therefore a fraction with a denominator of 0 is also undefined.

### Example 3 — Identifying Numerators and Denominators

Identify the numerator and denominator in each fraction.

(a) $\frac{5}{9}$

(b) $\frac{11}{7}$

$\frac{5}{9}$ ← Numerator
← Denominator

$\frac{11}{7}$ ← Numerator
← Denominator

**Work Problem ❸ at the Side.**

Fractions are sometimes called *proper* or *improper* fractions.

### Proper and Improper Fractions

If the numerator of a fraction is *smaller* than the denominator, the fraction is a **proper fraction**. A proper fraction is less than 1.

If the numerator is *greater than or equal to* the denominator, the fraction is an **improper fraction**. An improper fraction is greater than or equal to 1.

---

**❷** Write fractions for the shaded portions.

(a)

(b)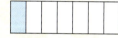

**❸** Identify the numerator and the denominator. Draw a picture with shaded parts to show each fraction. Your drawings may vary, but they should have the correct number of shaded parts.

(a) $\frac{2}{3}$

(b) $\frac{1}{4}$

(c) $\frac{8}{5}$

(d) $\frac{5}{2}$

**ANSWERS**

2. (a) $\frac{8}{7}$  (b) $\frac{7}{4}$

3. (a) N: 2; D: 3

(b) N: 1; D: 4

(c) N: 8; D: 5

(d) N: 5; D: 2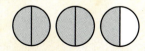

**4** From this group of fractions:

$$\frac{3}{4} \quad \frac{8}{7} \quad \frac{5}{7} \quad \frac{6}{6} \quad \frac{1}{2} \quad \frac{2}{1}$$

(a) list all proper fractions.

(b) list all improper fractions.

**Proper Fractions**

$$\frac{1}{2} \quad \frac{5}{11} \quad \frac{35}{36}$$

**Improper Fractions**

$$\frac{9}{7} \quad \frac{126}{125} \quad \frac{7}{7}$$

### Example 4 — Classifying Types of Fractions

(a) Identify all proper fractions in this list.

$$\frac{3}{4} \quad \frac{5}{9} \quad \frac{17}{5} \quad \frac{9}{7} \quad \frac{3}{3} \quad \frac{12}{25} \quad \frac{1}{9} \quad \frac{5}{3}$$

Proper fractions have a numerator that is *smaller* than the denominator. The proper fractions in the list are shown below.

$$\frac{3}{4} \leftarrow \text{3 is smaller than 4.} \qquad \frac{5}{9} \quad \frac{12}{25} \quad \frac{1}{9}$$

(b) Identify all improper fractions in the list in part (a).

Improper fractions have a numerator that is *equal to or greater* than the denominator. The improper fractions in the list are shown below.

$$\frac{17}{5} \leftarrow \text{17 is greater than 5.} \qquad \frac{9}{7} \quad \frac{3}{3} \quad \frac{5}{3}$$

**Work Problem** ❹ **at the Side.**

**3** Graph positive and negative fractions on a number line. Sometimes we need *negative* numbers that are between two integers. For example, $-\frac{3}{4}$ is between 0 and ⁻1. Graphing numbers on a number line helps us see the difference between $\frac{3}{4}$ and $-\frac{3}{4}$. Both represent 3 out of 4 equal parts, but they are in opposite directions from 0 on the number line. For $\frac{3}{4}$, divide the distance from 0 to 1 into 4 equal parts. Then start at 0, count over 3 parts, and make a dot. For $-\frac{3}{4}$, repeat the same process between 0 and ⁻1.

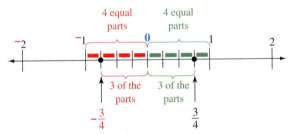

### CAUTION

In Chapters 1–3 we used a raised negative sign to help you avoid confusion between negative numbers and subtraction. Now you are ready to start writing the negative sign in the more traditional way. In this chapter, the negative sign will still be red, but it will be centered on the number instead of raised: for example, $-2$ instead of ⁻2. When the negative sign might be confused with the sign for subtraction, we will write parentheses around the negative number. Here is an example.

$$3 - (-2) \quad \text{means} \quad 3 \text{ minus (negative 2)}$$

For fractions, the negative sign will be written in front of the fraction bar: for example, $-\frac{3}{4}$. As with integers, the negative sign tells you that a fraction is *less than 0*; it is to the *left* of 0 on the number line. When there is *no* sign in front of a fraction, the fraction is assumed to be positive. For example, $\frac{3}{4}$ is assumed to be $+\frac{3}{4}$. It is to the *right* of 0 on the number line.

**ANSWERS**

4. (a) $\frac{3}{4}, \frac{5}{7}, \frac{1}{2}$   (b) $\frac{8}{7}, \frac{6}{6}, \frac{2}{1}$

Section 4.1 Introduction to Signed Fractions 201

> **Example 5** Graphing Positive and Negative Fractions

Graph each fraction on the number line.

(a) $\dfrac{2}{5}$

There is *no* sign in front of $\dfrac{2}{5}$, so it is *positive*. Because $\dfrac{2}{5}$ is between 0 and 1, we divide that space into 5 equal parts. Then we start at 0 and count to the right 2 parts.

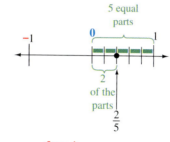

(b) $-\dfrac{4}{5}$

The fraction is *negative,* so it is between 0 and $-1$. We divide that space into 5 equal parts. Then we start at 0 and count to the left 4 parts.

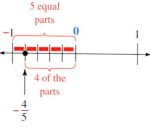

**Work Problem** ❺ **at the Side.**

**4** **Find the absolute value of a fraction.** In **Section 1.2** we said that the *absolute value* of a number was its distance from 0 on the number line. Two vertical bars indicate absolute value, as shown below.

$\left|-\dfrac{3}{4}\right|$ is read "the absolute value of negative three-fourths."

As with integers, the absolute value of fractions will *always* be positive (or 0) because it is the *distance* from 0 on the number line.

> **Example 6** Finding the Absolute Value of Fractions

Find each absolute value: $\left|\dfrac{1}{2}\right|$ and $\left|-\dfrac{1}{2}\right|$.

The distance from 0 to $\dfrac{1}{2}$ on the number line is $\dfrac{1}{2}$ space, so $\left|\dfrac{1}{2}\right| = \dfrac{1}{2}$.

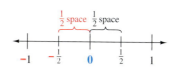

The distance from 0 to $-\dfrac{1}{2}$ is also $\dfrac{1}{2}$ space, so $\left|-\dfrac{1}{2}\right| = \dfrac{1}{2}$.

**Work Problem** ❻ **at the Side.**

❺ Graph each fraction on the number line.

(a) $\dfrac{2}{4}$

(b) $\dfrac{1}{2}$

(c) $-\dfrac{2}{3}$

❻ Find each absolute value.

(a) $\left|-\dfrac{3}{4}\right|$

(b) $\left|\dfrac{5}{8}\right|$

(c) $|0|$

**ANSWERS**

5. (a)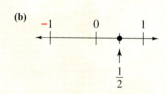

(b)

(c)

6. (a) $\dfrac{3}{4}$  (b) $\dfrac{5}{8}$  (c) 0

**5** **Write equivalent fractions.** You may have noticed in Margin Problems 5(a) and 5(b) that $\frac{2}{4}$ and $\frac{1}{2}$ were at the same point on the number line. Both of them were halfway between 0 and 1. There are actually *many* different names for this point. We illustrate some of them here.

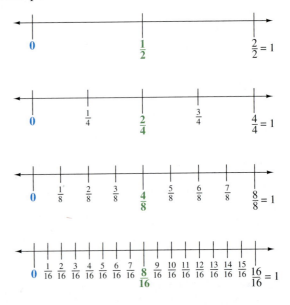

That is, $\frac{8}{16} = \frac{4}{8} = \frac{2}{4} = \frac{1}{2}$. If you have ever used a standard ruler with inches divided into sixteenths, you have probably already noticed that these distances are the same. Although the fractions look different, they all name the same point that is halfway between 0 and 1. In other words, they have the same value. We say that they are *equivalent fractions*.

### Equivalent Fractions

Fractions that represent the same number (the same point on a number line) are **equivalent fractions**.

Drawing number lines is tedious, so we usually find equivalent fractions by multiplying or dividing both the numerator and denominator by the same number. We can use some of the fractions that we just graphed to illustrate this method.

$$\frac{1}{2} = \frac{1 \cdot 2}{2 \cdot 2} = \frac{2}{4} \qquad\qquad \frac{8}{16} = \frac{8 \div 4}{16 \div 4} = \frac{2}{4}$$

Multiply both numerator and denominator by 2.   Divide both numerator and denominator by 4.

### Writing Equivalent Fractions

If $a$, $b$, and $c$ are numbers (and $b$ and $c$ are not 0), then

$$\frac{a}{b} = \frac{a \cdot c}{b \cdot c} \quad \text{or} \quad \frac{a}{b} = \frac{a \div c}{b \div c}.$$

In other words, if the numerator and denominator of a fraction are multiplied or divided by the *same* nonzero number, the result is an *equivalent* fraction.

Section 4.1 Introduction to Signed Fractions

### Example 7 — Writing Equivalent Fractions

**(a)** Write $-\dfrac{1}{2}$ as an equivalent fraction with a denominator of 16.

In other words, $-\dfrac{1}{2} = -\dfrac{?}{16}$.

The original denominator is 2. *Multiplying* 2 times 8 gives 16, the new denominator. To write an equivalent fraction, multiply *both* the numerator and denominator by 8.

$$-\dfrac{1}{2} = -\dfrac{1 \cdot 8}{2 \cdot 8} = -\dfrac{8}{16}$$

Keep the negative sign.

So, $-\dfrac{1}{2}$ is equivalent to $-\dfrac{8}{16}$.

**(b)** Write $\dfrac{12}{15}$ as an equivalent fraction with a denominator of 5.

In other words, $\dfrac{12}{15} = \dfrac{?}{5}$.

The original denominator is 15. *Dividing* 15 by 3 gives 5, the new denominator. To write an equivalent fraction, divide *both* the numerator and denominator by 3.

$$\dfrac{12}{15} = \dfrac{12 \div 3}{15 \div 3} = \dfrac{4}{5}$$

So, $\dfrac{12}{15}$ is equivalent to $\dfrac{4}{5}$.

= Work Problem ❼ at the Side.

Look back at the set of four number lines on the previous page. Notice that there are many different names for 1.

$$\dfrac{2}{2} = 1 \qquad \dfrac{4}{4} = 1 \qquad \dfrac{8}{8} = 1 \qquad \dfrac{16}{16} = 1$$

Because a fraction bar is a symbol for division, you can think of $\dfrac{2}{2}$ as $2 \div 2$, which equals 1. Similarly, $\dfrac{4}{4}$ is $4 \div 4$, which also is 1, and so on. These examples illustrate one of the division properties from **Section 1.7**.

### Division Properties

If $a$ is any number (except 0), then $\dfrac{a}{a} = 1$. In other words, when a nonzero number is divided by itself, the result is 1.

For example, $\dfrac{6}{6} = 1$ and $\dfrac{-4}{-4} = 1$.

Also recall that when any number is divided by 1, the result is the number. That is, $\dfrac{a}{1} = a$.

For example, $\dfrac{6}{1} = 6$ and $-\dfrac{12}{1} = -12$.

---

❼ **(a)** Write $\dfrac{2}{5}$ as an equivalent fraction with a denominator of 20.

**(b)** Write $-\dfrac{21}{28}$ as an equivalent fraction with a denominator of 4.

**Answers**

7. **(a)** $\dfrac{2}{5} = \dfrac{2 \cdot 4}{5 \cdot 4} = \dfrac{8}{20}$

**(b)** $-\dfrac{21}{28} = -\dfrac{21 \div 7}{28 \div 7} = -\dfrac{3}{4}$

**8** Simplify each fraction by dividing the numerator by the denominator.

(a) $\dfrac{10}{10}$

(b) $-\dfrac{3}{1}$

(c) $\dfrac{8}{2}$

(d) $-\dfrac{25}{5}$

### Example 8 Using Division to Simplify Fractions

Simplify each fraction by dividing the numerator by the denominator.

(a) $\dfrac{5}{5}$    Think of $\dfrac{5}{5}$ as $5 \div 5$. The result is 1, so $\dfrac{5}{5} = 1$.

(b) $\dfrac{12}{4}$    Think of $\dfrac{12}{4}$ as $12 \div 4$. The result is 3, so $\dfrac{12}{4} = 3$.

(c) $-\dfrac{6}{1}$    Think of $-\dfrac{6}{1}$ as $-6 \div 1$. The result is $-6$, so $-\dfrac{6}{1} = -6$.

Keep the negative sign.

**Work Problem 8 at the Side.**

> **NOTE**
>
> The title of this chapter is *Rational Numbers: Positive and Negative Fractions*. Rational numbers are numbers that can be written in the form $\dfrac{a}{b}$, where $a$ and $b$ are integers and $b$ is not 0. In Example 8(c) above, you saw that an integer can be written in the form $\dfrac{a}{b}$ ($-6$ can be written as $-\dfrac{6}{1}$). So rational numbers include all the integers and all the fractions. In Chapter 5 you'll work with rational numbers that are in decimal form.

**ANSWERS**

**8.** (a) 1   (b) $-3$   (c) 4   (d) $-5$

## 4.1 EXERCISES

FOR EXTRA HELP: Student's Solutions Manual, MyMathLab.com, InterAct Math Tutorial Software, AW Math Tutor Center, www.mathxl.com, Digital Video Tutor CD 2 Videotape 6

*Write the fractions that represent the shaded and unshaded portions of each figure. See Examples 1 and 2.*

1.

2.

3.

4.

5.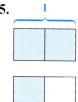

6.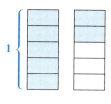

7.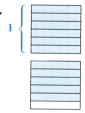

8.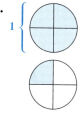

9. What fraction of these 11 coins are dimes? What fraction are pennies? What fraction are nickels?

10. What fraction of these eight recording artists are men? What fraction are women? What fraction are wearing white shirts?

11. In an American Sign Language (A.S.L.) class of 25 students, eight are deaf. What fraction of the students are deaf? What fraction are not deaf?

12. Of 35 motorcycles in the parking lot, 17 are Harley Davidsons. What fraction of the motorcycles are *not* Harley Davidsons? What fraction are Harley Davidsons?

13. Of 71 cars making up a freight train, 58 are boxcars. What fraction of the cars are *not* boxcars? What fraction are boxcars?

14. A college cheerleading squad has 12 members. If five of the cheerleaders are sophomores and the rest are freshmen, find the fraction of the members that are sophomores and the fraction that are freshmen.

The circle graph shows the results of a survey on where women would like to have flowers delivered on Valentine's Day. Use the graph to answer Exercises 15–18.

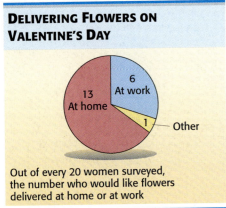

Source: FTD, Inc.

**15.** What fraction of the women would like flowers delivered at work?

**16.** What fraction of the women would like flowers delivered at home?

**17.** What fraction would like flowers delivered either at home or at work?

**18.** What fraction picked a location other than home or work?

Identify the numerator and denominator in each fraction. See Example 3.

**19.** $\frac{3}{4}$    **20.** $\frac{5}{8}$    **21.** $\frac{12}{7}$    **22.** $\frac{8}{3}$

List the proper and improper fractions in each group of numbers. See Example 4.

|  | Proper | Improper |
|---|---|---|
| **23.** $\frac{8}{5}, \frac{1}{3}, \frac{5}{8}, \frac{6}{6}, \frac{12}{2}, \frac{7}{16}$ | _____ | _____ |
| **24.** $\frac{1}{6}, \frac{5}{8}, \frac{15}{14}, \frac{11}{9}, \frac{7}{7}, \frac{3}{4}$ | _____ | _____ |
| **25.** $\frac{3}{4}, \frac{3}{2}, \frac{5}{5}, \frac{9}{11}, \frac{7}{15}, \frac{19}{18}$ | _____ | _____ |
| **26.** $\frac{12}{12}, \frac{15}{11}, \frac{13}{12}, \frac{11}{8}, \frac{17}{17}, \frac{19}{12}$ | _____ | _____ |

**27.** Write a fraction of your own choice. Label the *three* parts of the fraction and write a sentence describing what each part represents. Draw a figure with shaded parts to illustrate your fraction.

**28.** Give one example of a proper fraction and one example of an improper fraction. What determines whether a fraction is proper or improper? Draw figures with shaded parts to illustrate your fractions.

*Graph each pair of fractions on the number line. See Example 5.*

**29.** $\frac{1}{4}, -\frac{1}{4}$

**30.** $-\frac{1}{3}, \frac{1}{3}$

**31.** $-\frac{3}{5}, \frac{3}{5}$

**32.** $\frac{5}{6}, -\frac{5}{6}$

**33.** $\frac{7}{8}, -\frac{7}{8}$

**34.** $-\frac{3}{4}, \frac{3}{4}$

*Write a positive or negative fraction to describe each situation.*

**35.** The baby lost $\frac{3}{4}$ pound in weight while she was sick.

**36.** Greta needed $\frac{1}{3}$ cup of brown sugar for the cookie recipe.

**37.** The oil level in my car is $\frac{1}{2}$ quart below normal.

**38.** The mice who were on an experimental diet lost an average of $\frac{1}{4}$ ounce.

**39.** The Brown's driveway is $\frac{3}{10}$ mile long.

**40.** Marcel cut $\frac{5}{8}$ in. from the bottom of the door so that it wouldn't scrape the floor.

*Find each absolute value. See Example 6.*

**41.** $\left|-\frac{2}{5}\right|$

**42.** $\left|-\frac{2}{3}\right|$

**43.** $\left|\frac{9}{10}\right|$

**44.** $|0|$

**45.** $\left|-\frac{13}{6}\right|$

**46.** $\left|-\frac{4}{4}\right|$

**47.** Rewrite each fraction as an equivalent fraction with a denominator of 24. See Example 7.

(a) $\dfrac{1}{2} = \dfrac{}{24}$  (b) $\dfrac{1}{3} = \dfrac{}{}$  (c) $\dfrac{2}{3} = \dfrac{}{}$  (d) $\dfrac{1}{4} = \dfrac{}{}$  (e) $\dfrac{3}{4} = \dfrac{}{}$

(f) $\dfrac{1}{6} = \dfrac{}{}$  (g) $\dfrac{5}{6} = \dfrac{}{}$  (h) $\dfrac{1}{8} = \dfrac{}{}$  (i) $\dfrac{3}{8} = \dfrac{}{}$  (j) $\dfrac{5}{8} = \dfrac{}{}$

**48.** Rewrite each fraction as an equivalent fraction with a denominator of 36. See Example 7.

(a) $\dfrac{1}{2} = \dfrac{}{36}$  (b) $\dfrac{1}{3} = \dfrac{}{}$  (c) $\dfrac{2}{3} = \dfrac{}{}$  (d) $\dfrac{1}{4} = \dfrac{}{}$  (e) $\dfrac{3}{4} = \dfrac{}{}$

(f) $\dfrac{1}{6} = \dfrac{}{}$  (g) $\dfrac{5}{6} = \dfrac{}{}$  (h) $\dfrac{1}{9} = \dfrac{}{}$  (i) $\dfrac{4}{9} = \dfrac{}{}$  (j) $\dfrac{8}{9} = \dfrac{}{}$

**49.** Rewrite each fraction as an equivalent fraction with a denominator of 3. See Example 7.

(a) $-\dfrac{2}{6} = -\dfrac{}{3}$  (b) $-\dfrac{4}{6} = \dfrac{}{}$  (c) $-\dfrac{12}{18} = \dfrac{}{}$  (d) $-\dfrac{6}{18} = \dfrac{}{}$  (e) $-\dfrac{200}{300} = \dfrac{}{}$

(f) Write two more fractions that are equivalent to $-\dfrac{1}{3}$ and two more fractions equivalent to $-\dfrac{2}{3}$.

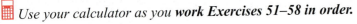

**50.** Rewrite each fraction as an equivalent fraction with a denominator of 4. See Example 7.

(a) $-\dfrac{2}{8} = -\dfrac{}{4}$  (b) $-\dfrac{6}{8} = \dfrac{}{}$  (c) $-\dfrac{15}{20} = \dfrac{}{}$  (d) $-\dfrac{50}{200} = \dfrac{}{}$  (e) $-\dfrac{150}{200} = \dfrac{}{}$

(f) Write two more fractions that are equivalent to $-\dfrac{1}{4}$ and two more fractions equivalent to $-\dfrac{3}{4}$.

---

**RELATING CONCEPTS (Exercises 51–58)**    **FOR INDIVIDUAL OR GROUP WORK**

*Use your calculator as you work Exercises 51–58 in order.*

**51.** (a) Write $\tfrac{3}{8}$ as an equivalent fraction with a denominator of 3912.

(b) Explain how you solved part (a).

**52.** (a) Write $\tfrac{7}{9}$ as an equivalent fraction with a denominator of 5472.

(b) Explain how you solved part (a).

**53.** (a) Is $-\tfrac{697}{3485}$ equivalent to $-\tfrac{1}{2}$, $-\tfrac{1}{3}$, or $-\tfrac{1}{5}$?

(b) Explain how you solved part (a).

**54.** (a) Is $-\tfrac{817}{4902}$ equivalent to $-\tfrac{1}{4}$, $-\tfrac{1}{6}$, or $-\tfrac{1}{8}$?

(b) Explain how you solved part (a).

*(continued)*

Section 4.1   **209**

 *Find a number to replace the ? that will make the two fractions equivalent.*

**55.** $\dfrac{1183}{2028} = \dfrac{?}{12}$     **56.** $\dfrac{2775}{6105} = \dfrac{?}{11}$     **57.** $\dfrac{13}{?} = \dfrac{1157}{1335}$     **58.** $\dfrac{9}{?} = \dfrac{891}{1584}$

**59.** Explain how to write equivalent fractions. Show one example in which the new denominator is larger than the original denominator, and one example in which the new denominator is smaller.

**60.** Explain how to find the absolute value of a fraction. Draw a number line to illustrate your explanation and include a positive fraction and a negative fraction as examples.

**61.** Can you write $\tfrac{3}{5}$ as an equivalent fraction with a denominator of 18? Explain why or why not. If not, what denominators could you use instead of 18?

**62.** Can you write $\tfrac{3}{4}$ as an equivalent fraction with a denominator of 0? Explain why or why not.

*Simplify each fraction by dividing the numerator by the denominator. See Example 8.*

**63.** $\dfrac{10}{1}$     **64.** $\dfrac{9}{9}$     **65.** $-\dfrac{16}{16}$     **66.** $-\dfrac{7}{1}$

**67.** $-\dfrac{18}{3}$     **68.** $-\dfrac{40}{4}$     **69.** $\dfrac{24}{8}$     **70.** $\dfrac{42}{6}$

**71.** $\dfrac{12}{12}$     **72.** $-\dfrac{5}{5}$     **73.** $\dfrac{14}{7}$     **74.** $\dfrac{8}{2}$

*There are many correct ways to draw the answers for Exercises 75–92, so ask your instructor to check your work.*

**75.** Shade $\tfrac{3}{5}$ of this figure. What fraction is unshaded?

**76.** Shade $\tfrac{5}{6}$ of this figure. What fraction is unshaded?

**77.** Shade $\tfrac{3}{8}$ of this figure. What fraction is unshaded?

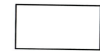

**78.** Shade $\tfrac{1}{3}$ of this figure. What fraction is unshaded?

79. Shade $\frac{7}{4}$ of this figure. What fraction is unshaded?

80. Shade $\frac{6}{5}$ of this figure. What fraction is unshaded?

81. Shade $\frac{4}{3}$ of this figure. What fraction is unshaded?

82. Shade $\frac{11}{8}$ of this figure. What fraction is unshaded?

83. Shade $\frac{6}{6}$ of this figure. What fraction is unshaded?

84. Shade $\frac{10}{10}$ of this figure. What fraction is unshaded?

85. Shade $\frac{10}{5}$ of this figure.

86. Shade $\frac{8}{4}$ of this figure.

87. Draw a group of seven faces. Draw some of the faces smiling and some sad. What fraction of your faces are smiling? What fraction are sad?

88. Draw a group of eight apples. Draw some of the apples with a stem and some of them with a bite taken out. What fraction of your apples have stems? What fraction have a bite taken out?

89. Draw a group of figures. Make $\frac{1}{10}$ of the figures circles, $\frac{6}{10}$ of the figures squares, and $\frac{3}{10}$ of the figures triangles. Then shade $\frac{1}{6}$ of the squares and $\frac{2}{3}$ of the triangles.

90. Write a group of capital letters. Make $\frac{4}{9}$ of the letters A's, $\frac{2}{9}$ of the letters B's, and $\frac{3}{9}$ of the letters C's. Then draw a line under $\frac{3}{4}$ of the A's and $\frac{1}{2}$ of the B's.

91. Draw a group of punctuation marks. Make $\frac{5}{12}$ of them exclamation points, $\frac{1}{12}$ of them commas, $\frac{3}{12}$ of them periods, and add enough question marks to make a full $\frac{12}{12}$ in all. Then circle $\frac{2}{5}$ of the exclamation points.

92. Draw a group of symbols. Make $\frac{2}{15}$ of them addition signs, $\frac{4}{15}$ of them subtraction signs, $\frac{2}{15}$ of them division signs, and add enough equal signs to make a full $\frac{15}{15}$ in all. Then circle $\frac{3}{4}$ of the subtraction signs.

## 4.2 WRITING FRACTIONS IN LOWEST TERMS

**1** **Identify fractions written in lowest terms.** You can see from these drawings that $\frac{1}{2}$ and $\frac{4}{8}$ are different names for the same amount of pizza.

$\frac{1}{2}$ of the pizza has pepperoni on it.

$\frac{4}{8}$ of the pizza has pepperoni on it.

### OBJECTIVES

**1** Identify fractions written in lowest terms.
**2** Write a fraction in lowest terms using common factors.
**3** Write a number as a product of prime factors.
**4** Write a fraction in lowest terms, using prime factorization.
**5** Write a fraction with variables in lowest terms.

You saw in the last section that $\frac{1}{2}$ and $\frac{4}{8}$ are equivalent fractions. But we say that the fraction $\frac{1}{2}$ is in *lowest terms* because the numerator and denominator have no *common factor* other than 1. That means that 1 is the only number that divides evenly into both 1 and 2. However, the fraction $\frac{4}{8}$ is *not* in lowest terms because its numerator and denominator have a common factor of 4. That means 4 will divide evenly into both 4 and 8.

### NOTE

Recall that *factors* are numbers being multiplied to give a product. For example,

1 • 4 = 4, so 1 and 4 are factors of 4.

2 • 4 = 8, so 2 and 4 are factors of 8.

4 is a factor of both 4 and 8, so 4 is a *common factor* of those numbers.

**❶** Are the following fractions in lowest terms? If not, find a common factor of the numerator and denominator (other than 1).

(a) $\frac{2}{3}$

(b) $-\frac{8}{10}$

(c) $-\frac{9}{11}$

(d) $\frac{15}{20}$

### Writing a Fraction in Lowest Terms

A fraction is written in **lowest terms** when the numerator and denominator have no common factor other than 1. Examples are $\frac{1}{3}, \frac{3}{4}, \frac{2}{5},$ and $\frac{7}{10}$. When you work with fractions, always write the final answer in lowest terms.

**Example 1** Identifying Fractions Written in Lowest Terms

Are the following fractions in lowest terms?

(a) $\frac{3}{8}$

The numerator and denominator have no common factor other than 1, so the fraction is in lowest terms.

(b) $\frac{21}{36}$

The numerator and denominator have a common factor of 3, so the fraction is *not* in lowest terms.

━━━ Work Problem **❶** at the Side.

**ANSWERS**
1. (a) yes  (b) No; 2 is a common factor.
   (c) yes  (d) No; 5 is a common factor.

❷ Write in lowest terms.

(a) $\dfrac{5}{10}$

(b) $\dfrac{9}{12}$

(c) $-\dfrac{24}{30}$

(d) $\dfrac{15}{40}$

(e) $-\dfrac{50}{90}$

**Answers**

2. (a) $\dfrac{1}{2}$  (b) $\dfrac{3}{4}$  (c) $-\dfrac{4}{5}$  (d) $\dfrac{3}{8}$  (e) $-\dfrac{5}{9}$

❷ **Write a fraction in lowest terms using common factors.** We will show you two methods for writing a fraction in lowest terms. The first method, dividing by a common factor, works best when the numerator and denominator are small numbers.

**Example 2** Using Common Factors to Write Fractions in Lowest Terms

Write each fraction in lowest terms.

(a) $\dfrac{20}{24}$

The *largest* common factor of 20 and 24 is 4. Divide both numerator and denominator by **4**.

$$\dfrac{20}{24} = \dfrac{20 \div 4}{24 \div 4} = \dfrac{5}{6}$$

(b) $\dfrac{30}{50} = \dfrac{30 \div 10}{50 \div 10} = \dfrac{3}{5}$   Divide both numerator and denominator by 10.

(c) $-\dfrac{24}{42} = -\dfrac{24 \div 6}{42 \div 6} = -\dfrac{4}{7}$   Divide both numerator and denominator by 6. Keep the negative sign.

(d) $\dfrac{60}{72}$

Suppose we made an error and thought that 4 was the largest common factor of 60 and 72. Dividing by 4 gives the following.

$$\dfrac{60}{72} = \dfrac{60 \div 4}{72 \div 4} = \dfrac{15}{18}$$

But $\dfrac{15}{18}$ is *not* in lowest terms because 15 and 18 have a common factor of 3. Therefore, divide by 3.

$$\dfrac{15}{18} = \dfrac{15 \div 3}{18 \div 3} = \dfrac{5}{6} \leftarrow \text{Lowest terms}$$

The fraction $\dfrac{60}{72}$ could have been written in lowest terms in one step by dividing by 12, the *largest* common factor of 60 and 72.

$$\dfrac{60}{72} = \dfrac{60 \div 12}{72 \div 12} = \dfrac{5}{6} \quad \left\{ \begin{array}{l} \text{Same answer} \\ \text{as above} \end{array} \right.$$

Either way works. Just keep dividing until the fraction is in lowest terms.

This method of writing a fraction in lowest terms by dividing by a common factor is summarized below.

**Dividing by a Common Factor to Write a Fraction in Lowest Terms**

*Step 1* Find the *largest* number that will divide evenly into both the numerator and denominator. This number is a **common factor**.

*Step 2* **Divide** both numerator and denominator by the common factor.

*Step 3* **Check** to see if the new fraction has any common factors (besides 1). If it does, repeat Steps 2 and 3. If the only common factor is 1, the fraction is in lowest terms.

**Work Problem ❷ at the Side.**

**3** **Write a number as a product of prime factors.** In Example 2(d) on the previous page, the largest common factor of 60 and 72 was difficult to see quickly. You can handle a problem like that by writing the numerator and denominator as a product of *prime numbers*.

## Prime Numbers

A **prime number** is a whole number that has exactly *two different* factors, itself and 1.

The number 3 is a prime number because it can be divided evenly only by itself and 1. The number 8 is *not* a prime number. The number 8 is a *composite number* because it can be divided evenly by 2 and 4, as well as by itself and 1.

## Composite Numbers

A number with a factor other than itself or 1 is called a **composite number.**

### CAUTION

A prime number has *only two* different factors, itself and 1. The number 1 is *not* a prime number because it does not have *two different* factors; the only factor of 1 is 1. Also, 0 is *not* a prime number. Therefore, 0 and 1 are neither prime nor composite numbers.

**Example 3** Finding Prime Numbers

Which of the following numbers are prime?

2  5  8  11  15

The number 8 can be divided by 4 and 2, so it is *not* prime. Also, 15 is *not* prime because 15 can be divided by 5 and 3. The other numbers in the list, 2, 5, and 11, are prime. Each of these numbers is divisible only by itself and 1.

**Work Problem ❸ at the Side.**

For reference, here are the prime numbers smaller than 100.

| | | | | |
|---|---|---|---|---|
| 2 | 3 | 5 | 7 | 11 |
| 13 | 17 | 19 | 23 | 29 |
| 31 | 37 | 41 | 43 | 47 |
| 53 | 59 | 61 | 67 | 71 |
| 73 | 79 | 83 | 89 | 97 |

### CAUTION

All prime numbers are odd numbers except the number 2. Be careful though, because *not all odd numbers are prime numbers*. For example, 9, 15, and 21 are odd numbers but they are *not* prime numbers.

The *prime factorization* of a number can be especially useful when working with fractions.

## Prime Factorization

A **prime factorization** of a number is a factorization in which every factor is a prime number.

❸ Which of the following are prime numbers?

1, 2, 3, 4, 7, 9, 13, 19, 25, 29

---

ANSWERS
**3.** 2, 3, 7, 13, 19, 29

**4** Find the prime factorization of each number.

(a) 8

(b) 42

(c) 90

(d) 100

(e) 81

**ANSWERS**
4. (a) 2 • 2 • 2   (b) 2 • 3 • 7
   (c) 2 • 3 • 3 • 5   (d) 2 • 2 • 5 • 5
   (e) 3 • 3 • 3 • 3

### Example 4  Factoring by Using the Division Method

(a) Find the prime factorization of 48.

All prime factors:
$2\overline{)48}$ ← Divide 48 by 2 (first prime).
$2\overline{)24}$ ← Divide 24 by 2.
$2\overline{)12}$ ← Divide 12 by 2.
$2\overline{)6}$ ← Divide 6 by 2.
$3\overline{)3}$ ← Divide 3 by 3.
1 ← Continue to divide until the quotient is 1.

Because all factors (divisors) are prime, the prime factorization of 48 is

2 • 2 • 2 • 2 • 3

Check by multiplying the factors to see if the product is 48.
Yes, 2 • 2 • 2 • 2 • 3 does equal 48.

**NOTE**

You may write the factors in any order because multiplication is commutative and associative. So you could write the factorization of 48 as 3 • 2 • 2 • 2 • 2. We will show the factors from smallest to largest in our examples.

(b) Find the prime factorization of 225.

All prime factors:
$3\overline{)225}$ ← 225 is not divisible by 2; use 3.
$3\overline{)75}$ ← Divide 75 by 3.
$5\overline{)25}$ ← 25 is not divisible by 3; use 5.
$5\overline{)5}$ ← Divide 5 by 5.
1 ← Quotient is 1.

So, 225 = 3 • 3 • 5 • 5.

**CAUTION**

When you're using the division method of factoring, the last quotient is 1. Do *not* list 1 as a prime factor because 1 is not a prime number.

**Work Problem ❹ at the Side.**

Another method of factoring uses what is called a *factor tree*.

### Example 5  Factoring by Using a Factor Tree

Find the prime factorization of each number.

(a) 60
   Try to divide 60 by the first prime number, 2. Write the factors under the 60. Circle the 2, because it is a prime.

**Continued on Next Page**

Try dividing 30 by 2. Write the factors under the 30.

Because 15 cannot be evenly divided by 2, try dividing 15 by the next prime number, 3.

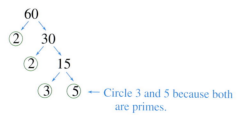

← Circle 3 and 5 because both are primes.

No uncircled factors remain, so you have found the prime factorization (the circled factors).

$$60 = 2 \cdot 2 \cdot 3 \cdot 5$$

**(b)** 72

Divide by 2, the first prime number.

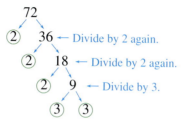

So, $72 = 2 \cdot 2 \cdot 2 \cdot 3 \cdot 3$.

**(c)** 45

Because 45 cannot be divided evenly by 2, try dividing by the next prime, 3.

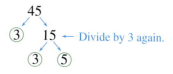

So, $45 = 3 \cdot 3 \cdot 5$.

## NOTE

Here is a reminder about the quick way to see whether a number can be divided evenly by 2, 3, or 5.

A number is divisible by 2 if the ones digit is 0, 2, 4, 6, or 8. For example, 30, 512, 76, and 3018 are all divisible by 2.

A number is divisible by 3 if the *sum* of the digits is divisible by 3. For example, 129 is divisible by 3 because $1 + 2 + 9 = 12$ and 12 is divisible by 3.

A number is divisible by 5 if it has 0 or 5 in the ones place. For example, 85, 610, and 1725 are all divisible by 5.

See **Section R.4** for more information.

Work Problem ❺ at the Side.

❺ Complete each factor tree and write the prime factorization.

**(a)**

**(b)** 35

**(c)** 90

**ANSWERS**

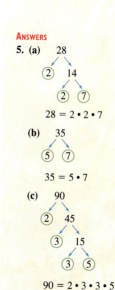

5. (a) $28 = 2 \cdot 2 \cdot 7$
(b) $35 = 5 \cdot 7$
(c) $90 = 2 \cdot 3 \cdot 3 \cdot 5$

**Calculator Tip**  You can use your calculator to find the prime factorization of a number. Here is an example that uses 539.

Try dividing 539 by the first prime number, 2.

$$539 \div 2 = 269.5 \quad \text{Does not divide evenly}$$

Try dividing by the next prime number, 3.

$$539 \div 3 = 179.6666667 \quad \text{Does not divide evenly}$$

Keep trying the next prime numbers until you find one that divides evenly.

$$539 \div 5 = 107.8 \quad \text{Does not divide evenly}$$
$$539 \div 7 = 77 \quad \text{Divides evenly}$$

Once you have found that 7 works, try using it again.

$$77 \div 7 = 11 \quad \text{Divides evenly}$$

Because 11 is prime, you're finished. The prime factorization of 539 is $7 \cdot 7 \cdot 11$.

Now try factoring 2431 using your calculator. (The answer is at the bottom left of this page.)

**4** **Write a fraction in lowest terms, using prime factorization.**  Now you can use prime factorization to write fractions in lowest terms. This is a good method to use when the numerator and denominator are larger numbers.

**Example 6** **Using Prime Factorization to Write Fractions in Lowest Terms**

**(a)** Write $\frac{20}{35}$ in lowest terms.

20 can be written as $2 \cdot 2 \cdot 5$ ← Prime factors
35 can be written as $5 \cdot 7$ ← Prime factors

$$\frac{20}{35} = \frac{2 \cdot 2 \cdot 5}{5 \cdot 7}$$

The numerator and denominator have 5 as a common factor. Dividing both numerator and denominator by 5 will give an equivalent fraction.

$$\frac{20}{35} = \frac{2 \cdot 2 \cdot 5}{5 \cdot 7} = \frac{2 \cdot 2 \cdot 5}{7 \cdot 5} = \frac{2 \cdot 2 \cdot \boxed{5 \div 5}}{7 \cdot \boxed{5 \div 5}} = \frac{2 \cdot 2 \cdot 1}{7 \cdot 1} = \frac{4}{7}$$

Multiplication is commutative.  Any number divided by itself is 1.

$\frac{20}{35}$ is written in lowest terms as $\frac{4}{7}$.

To shorten the work, you may use slashes to indicate the divisions. For example, the work on $\frac{20}{35}$ can be shown as follows.

$$\frac{20}{35} = \frac{2 \cdot 2 \cdot \cancel{5}^{1}}{\cancel{5}_{1} \cdot 7}$$

Slashes indicate $5 \div 5$, and the result is 1.

**CALCULATOR TIP ANSWER**
$2431 = 11 \cdot 13 \cdot 17$

Continued on Next Page

**(b)** Write $\frac{60}{72}$ in lowest terms.

Use the prime factorizations of 60 and 72 from Examples 5(a) and 5(b).

$$\frac{60}{72} = \frac{2 \cdot 2 \cdot 3 \cdot 5}{2 \cdot 2 \cdot 2 \cdot 3 \cdot 3}$$

This time there are three common factors. Use slashes to show the three divisions.

$$\frac{60}{72} = \frac{\cancel{2} \cdot \cancel{2} \cdot \cancel{3} \cdot 5}{\cancel{2} \cdot \cancel{2} \cdot 2 \cdot \cancel{3} \cdot 3} = \frac{5}{6}$$ ← Multiply 1 · 1 · 1 · 5 to get 5.
← Multiply 1 · 1 · 2 · 1 · 3 to get 6.

2 ÷ 2 is 1.
2 ÷ 2 is 1.
3 ÷ 3 is 1.

**(c)** $\frac{18}{90}$

$$\frac{18}{90} = \frac{\cancel{2} \cdot \cancel{3} \cdot \cancel{3}}{\cancel{2} \cdot \cancel{3} \cdot \cancel{3} \cdot 5} = \frac{1}{5}$$ ← Multiply 1 · 1 · 1 to get 1.
← Multiply 1 · 1 · 1 · 5 to get 5.

**6** Use the method of prime factorization to write each fraction in lowest terms.

(a) $\frac{16}{48}$

(b) $\frac{28}{60}$

(c) $\frac{74}{111}$

(d) $\frac{124}{340}$

**CAUTION**

In Example 6(c), all factors of the numerator divided out. But 1 · 1 · 1 is still 1, so the final answer is $\frac{1}{5}$ (**not** 5).

This method of writing a fraction in lowest terms is summarized as follows.

**Using Prime Factorization to Write a Fraction in Lowest Terms**

*Step 1* Write the **prime factorization** of both numerator and denominator.

*Step 2* Use slashes to show where you are **dividing** the numerator and denominator by any common factors.

*Step 3* **Multiply** the remaining factors in the numerator and in the denominator.

Work Problem **6** at the Side.

**5** Write a fraction with variables in lowest terms. Fractions may have variables in the numerator or denominator. Examples are shown below.

$$\frac{6}{2x} \qquad \frac{3xy}{9xy} \qquad \frac{4b^3}{8ab} \qquad \frac{7ab^2}{n^2}$$

You can use prime factorization to write these fractions in lowest terms.

**ANSWERS**

**6.** (a) $\frac{1}{3}$ (b) $\frac{7}{15}$ (c) $\frac{2}{3}$ (d) $\frac{31}{85}$

**Chapter 4** Rational Numbers: Positive and Negative Fractions

**7** Write each fraction in lowest terms.

(a) $\dfrac{5c}{15}$

(b) $\dfrac{10x^2}{8x^2}$

(c) $\dfrac{9a^3}{11b^3}$

(d) $\dfrac{6m^2n}{9n^2}$

**Example 7** Writing Fractions with Variables in Lowest Terms

Write each fraction in lowest terms.

(a) $\dfrac{6}{2x}$ ← Prime factors of 6 are 2 • 3.
← 2x means 2 • x.

$$\dfrac{6}{2x} = \dfrac{\overset{1}{\cancel{2}} \cdot 3}{\underset{1}{\cancel{2}} \cdot x} = \dfrac{3}{x} \quad \begin{array}{l}\leftarrow 1 \cdot 3 \text{ is } 3.\\ \leftarrow 1 \cdot x \text{ is } x.\end{array}$$

3xy means 3 • x • y.

(b) $\dfrac{3xy}{9xy} = \dfrac{3 \cdot x \cdot y}{3 \cdot 3 \cdot x \cdot y} = \dfrac{\overset{1}{\cancel{3}} \cdot \overset{1}{\cancel{x}} \cdot \overset{1}{\cancel{y}}}{\underset{1}{\cancel{3}} \cdot 3 \cdot \underset{1}{\cancel{x}} \cdot \underset{1}{\cancel{y}}} = \dfrac{1}{3}$

The prime factors of 9 are 3 • 3.

$b^3$ means $b \cdot b \cdot b$.

(c) $\dfrac{4b^3}{8ab} = \dfrac{2 \cdot 2 \cdot b \cdot b \cdot b}{2 \cdot 2 \cdot 2 \cdot a \cdot b} = \dfrac{\overset{1}{\cancel{2}} \cdot \overset{1}{\cancel{2}} \cdot \overset{1}{\cancel{b}} \cdot b \cdot b}{\underset{1}{\cancel{2}} \cdot \underset{1}{\cancel{2}} \cdot 2 \cdot a \cdot \underset{1}{\cancel{b}}} = \dfrac{b^2}{2a} \quad \begin{array}{l}\leftarrow b \cdot b \text{ is } b^2.\\ \leftarrow 2 \cdot a \text{ is } 2a.\end{array}$

The prime factors of 8 are 2 • 2 • 2.

(d) $\dfrac{7ab^2}{n^2} = \dfrac{7 \cdot a \cdot b \cdot b}{n \cdot n}$ There are no common factors.

$\dfrac{7ab^2}{n^2}$ is already in lowest terms.

**Work Problem 7 at the Side.**

**ANSWERS**

7. (a) $\dfrac{c}{3}$ (b) $\dfrac{5}{4}$ (c) already in lowest terms
(d) $\dfrac{2m^2}{3n}$

## 4.2 Exercises

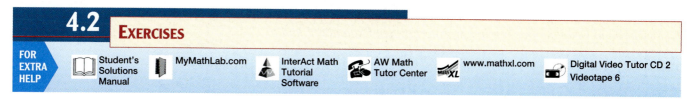

*Label each number as* prime *or* composite. *See Example 3.*

1. 9    2    8    5    11    10    21

2. 12    3    7    6    15    13    25

*Find the prime factorization of each number. See Examples 4 and 5.*

3. 6        4. 12        5. 20        6. 30

7. 25       8. 18        9. 36        10. 56

11. 44      12. 68       13. 88       14. 64

15. 75                   16. 80

17. Write definitions of a composite number and a prime number. Give three examples of each. Which whole numbers are neither prime nor composite?

18. With the exception of the number 2, all prime numbers are odd numbers. Nevertheless, not all odd numbers are prime numbers. Explain why these statements are true.

*Write each numerator and denominator as a product of prime factors. Then use the prime factorization to write the fraction in lowest terms. See Examples 1 and 6.*

19. $\dfrac{8}{16}$
20. $\dfrac{6}{8}$
21. $\dfrac{32}{48}$

22. $\dfrac{9}{27}$
23. $\dfrac{14}{21}$
24. $\dfrac{20}{32}$

25. $\dfrac{36}{42}$
26. $\dfrac{22}{33}$
27. $\dfrac{63}{70}$

28. $\dfrac{72}{80}$
29. $\dfrac{27}{45}$
30. $\dfrac{36}{63}$

31. $\dfrac{12}{18}$
32. $\dfrac{63}{90}$
33. $\dfrac{35}{40}$

34. $\dfrac{36}{48}$
35. $\dfrac{90}{180}$
36. $\dfrac{16}{64}$

37. $\dfrac{210}{315}$
38. $\dfrac{96}{192}$

39. $\dfrac{429}{495}$
40. $\dfrac{135}{182}$

*Write your answers to Exercises 41–46 in lowest terms.*

**41.** There are 60 minutes in an hour. What fraction of an hour is
   (a) 15 minutes?  (b) 30 minutes?
   (c) 6 minutes?   (d) 60 minutes?

**42.** There are 24 hours in a day. What fraction of a day is
   (a) 8 hours?   (b) 18 hours?
   (c) 12 hours?  (d) 3 hours?

**43.** SueLynn's monthly income is $1500.
   (a) She spends $500 on rent. What fraction of her income is spent on rent?
   (b) She spends $300 on food. What fraction of her income is spent on food?
   (c) What fraction of her income is left for other expenses?

**44.** There are 10,000 students at Minneapolis Community and Technical College.
   (a) 7500 of the students receive some form of financial aid. What fraction of the students receive financial aid?
   (b) 6000 of the students are women. What fraction are women?
   (c) What fraction of the students are men?

**45.** What fraction of the time spent on household chores is done by (a) husbands, (b) wives, (c) children?

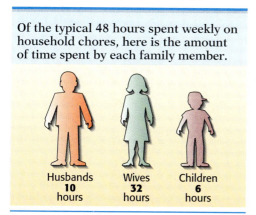

Of the typical 48 hours spent weekly on household chores, here is the amount of time spent by each family member.

Husbands 10 hours   Wives 32 hours   Children 6 hours

*Source: Journal of Marriage and the Family.*

**46.** A survey asked people whether certain types of advertising were believable. Out of every 100 people in the survey, here is the number who said the advertising was believable.

| Type of Advertising | Number Who Said Advertising Was Believable |
|---|---|
| Computer software | 35 |
| Pharmaceutical companies | 28 |
| Auto manufacturers | 18 |
| Insurance companies | 15 |

*Source: Porter Novella.*

What fraction of the people said each type of advertising was believable?

**47.** Explain the error in each of these problems and correct it.

   (a) $\dfrac{9}{36} = \dfrac{\cancel{3} \cdot \cancel{3}}{2 \cdot 2 \cdot \cancel{3} \cdot \cancel{3}} = 4$

   (b) $\dfrac{9}{16} = \dfrac{9 \div 3}{16 \div 4} = \dfrac{3}{4}$

**48.** (a) Explain how you could use your calculator to find the prime factorization of 437. Then find the prime factorization.

   (b) The text lists all the prime numbers less than 100. Use the divisibility rules and your calculator to find at least five prime numbers between 100 and 150.

*Write each fraction in lowest terms. See Example 7.*

49. $\dfrac{16c}{40}$  50. $\dfrac{36}{54a}$  51. $\dfrac{20x}{35x}$

52. $\dfrac{21n}{28n}$  53. $\dfrac{18r^2}{15rs}$  54. $\dfrac{18ab}{48b^2}$

55. $\dfrac{6m}{42mn^2}$  56. $\dfrac{10g^2}{90g^2h}$  57. $\dfrac{9x^2}{16y^2}$

58. $\dfrac{5rst}{8st}$  59. $\dfrac{7xz}{9xyz}$  60. $\dfrac{6a^3}{23b^3}$

61. $\dfrac{21k^3}{6k^2}$  62. $\dfrac{16x^3}{12x^4}$  63. $\dfrac{13a^2bc^3}{39a^2bc^3}$

64. $\dfrac{22m^3n^4}{55m^3n^4}$  65. $\dfrac{14c^2d}{14cd^2}$  66. $\dfrac{19rs}{19s^3}$

67. $\dfrac{210ab^3c}{35b^2c^2}$  68. $\dfrac{81w^4xy^2}{300wy^4}$  69. $\dfrac{25m^3rt^2}{36n^2s^3w^2}$

70. $\dfrac{42a^5b^4c^3}{7a^4b^3c^2}$  71. $\dfrac{33e^2fg^3}{11efg}$  72. $\dfrac{21xy^2z^3}{17ab^2c^3}$

## 4.3 Multiplying and Dividing Signed Fractions

**OBJECTIVES**
1. Multiply signed fractions.
2. Multiply fractions that involve variables.
3. Divide signed fractions.
4. Divide fractions that involve variables.
5. Solve application problems involving multiplying and dividing fractions.

**1 Multiply signed fractions.** Suppose that you give $\frac{1}{3}$ of your candy bar to your friend Ann. Then Ann gives $\frac{1}{2}$ of her share to Tim. How much of the bar does Tim get to eat?

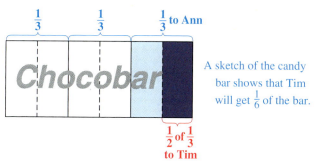

A sketch of the candy bar shows that Tim will get $\frac{1}{6}$ of the bar.

Tim's share is $\frac{1}{2}$ **of** $\frac{1}{3}$ candy bar. When used with fractions, the word **of** indicates multiplication.

$$\frac{1}{2} \text{ of } \frac{1}{3} \text{ means } \frac{1}{2} \cdot \frac{1}{3}$$

Tim's share is $\frac{1}{6}$ bar, so $\frac{1}{2} \cdot \frac{1}{3} = \frac{1}{6}$.

This example illustrates the rule for multiplying fractions.

### Multiplying Fractions

If $a$, $b$, $c$, and $d$ are numbers (but $b$ and $d$ are not 0), then

$$\frac{a}{b} \cdot \frac{c}{d} = \frac{a \cdot c}{b \cdot d}$$

In other words, multiply the numerators and multiply the denominators.

When we apply this rule to find Tim's part of the bar, we get

$$\frac{1}{2} \cdot \frac{1}{3} = \frac{1 \cdot 1}{2 \cdot 3} = \frac{1}{6} \quad \begin{array}{l}\leftarrow \text{Multiply numerators.} \\ \leftarrow \text{Multiply denominators.}\end{array}$$

### Example 1 Multiplying Signed Fractions

Multiply.

(a) $-\frac{5}{8} \cdot -\frac{3}{4}$    Recall that the product of two negative numbers is a positive number.

Multiply the numerators and multiply the denominators.

$$-\frac{5}{8} \cdot -\frac{3}{4} = \frac{5 \cdot 3}{8 \cdot 4} = \frac{15}{32} \leftarrow \text{Lowest terms}$$

The product of two negative numbers is positive.

The answer is in lowest terms because 15 and 32 have no common factor other than 1.

*Continued on Next Page*

**Chapter 4** Rational Numbers: Positive and Negative Fractions

① Multiply.

(a) $-\dfrac{3}{4} \cdot \dfrac{1}{2}$

(b) $\left(-\dfrac{2}{5}\right)\left(-\dfrac{2}{3}\right)$

(c) $\dfrac{3}{4}\left(\dfrac{3}{8}\right)$

(b) $\left(\dfrac{4}{7}\right)\left(-\dfrac{2}{5}\right) = -\dfrac{4 \cdot 2}{7 \cdot 5} = -\dfrac{8}{35}$ — Recall that the product of a negative number and a positive number is negative.

**Work Problem ① at the Side.**

Sometimes the result won't be in lowest terms. For example, find $\dfrac{3}{10}$ of $\dfrac{5}{6}$.

$$\dfrac{3}{10} \text{ of } \dfrac{5}{6} \text{ means } \dfrac{3}{10} \cdot \dfrac{5}{6} = \dfrac{3 \cdot 5}{10 \cdot 6} = \dfrac{15}{60} \quad \{\text{Not in lowest terms}$$

Now write $\dfrac{15}{60}$ in lowest terms.

$$\dfrac{15}{60} = \dfrac{\overset{1}{\cancel{3}} \cdot \overset{1}{\cancel{5}}}{2 \cdot 2 \cdot \underset{1}{\cancel{3}} \cdot \underset{1}{\cancel{5}}} = \dfrac{1}{4} \leftarrow \text{Lowest terms}$$

You used prime factorization in **Section 4.2** to write fractions in lowest terms. You can also use it when multiplying fractions. Writing the prime factors of the original fractions and dividing out common factors *before* multiplying usually saves time. If you divide out *all* the common factors, the result will automatically be in lowest terms. Let's see how that works when finding $\dfrac{3}{10}$ of $\dfrac{5}{6}$.

3 and 5 are already prime.

$$\dfrac{3}{10} \cdot \dfrac{5}{6} = \dfrac{3 \cdot 5}{2 \cdot 5 \cdot 2 \cdot 3} = \dfrac{\overset{1}{\cancel{3}} \cdot \overset{1}{\cancel{5}}}{2 \cdot \cancel{5} \cdot 2 \cdot \cancel{3}} = \dfrac{1}{4} \quad \{\text{Same result as above}$$

Write 10 as 2 · 5.     Write 6 as 2 · 3.     Divide common factors.

**CAUTION**

When you are working with fractions, always write the final result in lowest terms. Visualizing $\dfrac{15}{60}$ is hard to do. But when $\dfrac{15}{60}$ is written as $\dfrac{1}{4}$, working with it is much easier. Lowest terms is the simplest way to write a fraction.

**Example 2** Using Prime Factorization to Multiply Fractions

(a) Multiply $-\dfrac{8}{5}\left(\dfrac{5}{12}\right)$.

Multiplying a negative number times a positive number gives a negative product.

Write 8 as 2 · 2 · 2.

$$-\dfrac{8}{5}\left(\dfrac{5}{12}\right) = -\dfrac{2 \cdot 2 \cdot 2 \cdot 5}{5 \cdot 2 \cdot 2 \cdot 3} = -\dfrac{\overset{1}{\cancel{2}} \cdot \overset{1}{\cancel{2}} \cdot 2 \cdot \overset{1}{\cancel{5}}}{\cancel{5} \cdot \cancel{2} \cdot \cancel{2} \cdot 3} = -\dfrac{2}{3} \quad \{\text{Lowest terms}$$

5 is already prime.     Write 12 as 2 · 2 · 3.     Negative product

**Continued on Next Page**

---

**ANSWERS**

1. (a) $-\dfrac{3}{8}$  (b) $\dfrac{4}{15}$  (c) $\dfrac{9}{32}$

**(b)** Find $\dfrac{2}{9}$ of $\dfrac{15}{16}$.

Recall that, when used with fractions, *of* indicates multiplication.

2 is already prime. Write 15 as 3 • 5.

$$\dfrac{2}{9} \cdot \dfrac{15}{16} = \dfrac{2 \cdot 3 \cdot 5}{3 \cdot 3 \cdot 2 \cdot 2 \cdot 2 \cdot 2} = \dfrac{\overset{1}{\cancel{2}} \cdot \overset{1}{\cancel{3}} \cdot 5}{\cancel{3} \cdot 3 \cdot \cancel{2} \cdot 2 \cdot 2 \cdot 2} = \dfrac{5}{24} \quad \begin{cases} \text{Lowest} \\ \text{terms} \end{cases}$$

Write 9 as 3 • 3. Write 16 as 2 • 2 • 2 • 2.

**Work Problem ❷ at the Side.**

❷ Use prime factorization to multiply these fractions.

**(a)** $\dfrac{15}{28} \cdot -\dfrac{6}{5}$

**(b)** $\dfrac{12}{7}\left(\dfrac{7}{24}\right)$

**(c)** $\left(-\dfrac{11}{18}\right)\left(-\dfrac{9}{20}\right)$

🖩 **Calculator Tip** If your *scientific* calculator has a fraction key $\boxed{a^{b/c}}$, you can do calculations with fractions. (Most *graphing* calculators do not have a fraction key.) You'll also need the change of sign key $\boxed{+/-}$ to enter negative fractions, as you did to enter negative integers in **Section 1.4**.

Start by entering several different fractions. Clear your calculator after each one.

To enter $\tfrac{3}{4}$, press 3 $\boxed{a^{b/c}}$ 4. The display will show $\boxed{3\lrcorner 4}$.
                                                         ↑
                                                    Fraction bar

To enter $-\tfrac{9}{10}$, press 9 $\boxed{a^{b/c}}$ 10 $\boxed{+/-}$. The display will show $\boxed{-9\lrcorner 10}$.

Try entering a fraction that is *not* in lowest terms. As soon as you press an operation key, such as $\boxed{\times}$ or $\boxed{\div}$, most calculators will automatically show the fraction in lowest terms. Suppose that you start to enter the multiplication problem $\tfrac{4}{16} \cdot \tfrac{2}{3}$. Press 4 $\boxed{a^{b/c}}$ 16 $\boxed{\times}$. The display shows $\boxed{1\lrcorner 4}$, or $\tfrac{1}{4}$, which is $\tfrac{4}{16}$ in lowest terms. The calculator will always show fractions in lowest terms.

Let's check the result of Example 2(a): Multiply $-\tfrac{8}{5}\left(\tfrac{5}{12}\right)$ by pressing

8 $\boxed{a^{b/c}}$ 5 $\boxed{+/-}$ $\boxed{\times}$ 5 $\boxed{a^{b/c}}$ 12 $\boxed{=}$.   The display shows $\boxed{-2\lrcorner 3}$.
  $\underbrace{\qquad\qquad\qquad}_{-\tfrac{8}{5}}$        $\underbrace{\qquad\qquad}_{\tfrac{5}{12}}$                                    $\underbrace{\qquad}_{-\tfrac{2}{3}}$

Now try Example 2(b): Find $\tfrac{2}{9}$ of $\tfrac{15}{16}$. (Did you get $\tfrac{5}{24}$?)

There are some limitations to the calculations that you can do using the fraction key.

Try entering the fraction $\tfrac{9}{1000}$. What happens?

(You can't enter denominators >999.)

Try doing this multiplication: $\tfrac{7}{10} \cdot \tfrac{3}{100}$. The result should be $\tfrac{21}{1000}$. What happens?

(The answer is given in decimal form because the denominator is >999.)

**CAUTION**

The fraction key is useful for *checking* your work. But knowing the rules for fraction computation is important because you'll need them when fractions involve variables. You *cannot* enter fractions such as $\tfrac{3x}{5}$ or $\tfrac{9}{m^2}$ on your calculator (see Example 4, coming up shortly).

**Answers**

2. **(a)** $-\dfrac{3 \cdot \overset{1}{\cancel{5}} \cdot \overset{1}{\cancel{2}} \cdot 3}{2 \cdot \cancel{2} \cdot 7 \cdot \cancel{5}} = -\dfrac{9}{14}$

**(b)** $\dfrac{\overset{1}{\cancel{2}} \cdot \overset{1}{\cancel{2}} \cdot \overset{1}{\cancel{3}} \cdot \overset{1}{\cancel{7}}}{\cancel{7} \cdot \cancel{2} \cdot \cancel{2} \cdot 2 \cdot \cancel{3}} = \dfrac{1}{2}$

**(c)** $\dfrac{11 \cdot \overset{1}{\cancel{3}} \cdot \overset{1}{\cancel{3}}}{2 \cdot \cancel{3} \cdot \cancel{3} \cdot 2 \cdot 2 \cdot 5} = \dfrac{11}{40}$

## Chapter 4 Rational Numbers: Positive and Negative Fractions

**❸** Use prime factorization to find these products.

(a) $\dfrac{3}{4}$ of 36

(b) $-10 \cdot \dfrac{2}{5}$

(c) $\left(-\dfrac{7}{8}\right)(-24)$

**❹** Use prime factorization to find these products.

(a) $\dfrac{2c}{5} \cdot \dfrac{c}{4}$

(b) $\left(\dfrac{m}{6}\right)\left(\dfrac{9}{m^2}\right)$

(c) $\dfrac{w^2}{y} \cdot \dfrac{x^2 y}{w}$

### Example 3  Multiplying a Fraction and an Integer

Find $\dfrac{2}{3}$ of 6.

We can write 6 in fraction form as $\dfrac{6}{1}$. Recall that $\dfrac{6}{1}$ means $6 \div 1$, which is 6. So we can write any integer $a$ as $\dfrac{a}{1}$.

$$\dfrac{2}{3} \text{ of } 6 \text{ means } \dfrac{2}{3} \cdot \dfrac{6}{1} = \dfrac{2 \cdot 2 \cdot \cancel{3}}{\cancel{3} \cdot 1} = \dfrac{4}{1} = 4$$

4 : 1 is 4.

**Work Problem ❸ at the Side.**

**2** **Multiply fractions that involve variables.** The multiplication method involving the use of prime factors also works when there are variables in the numerators and/or denominators of the fractions.

### Example 4  Multiplying Fractions with Variables

Multiply.

(a) $\dfrac{3x}{5} \cdot \dfrac{2}{9x}$

$3x$ means $3 \cdot x$.

$$\dfrac{3x}{5} \cdot \dfrac{2}{9x} = \dfrac{3 \cdot x \cdot 2}{5 \cdot 3 \cdot 3 \cdot x} = \dfrac{\cancel{3} \cdot \cancel{x} \cdot 2}{5 \cdot \cancel{3} \cdot 3 \cdot \cancel{x}} = \dfrac{2}{15}$$

$\dfrac{x}{x}$ is 1.

The prime factors of 9 are $3 \cdot 3$, so $9x$ is $3 \cdot 3 \cdot x$.

(b) $\left(\dfrac{3y}{4x}\right)\left(\dfrac{2x^2}{y}\right)$

$2x^2$ means $2 \cdot x \cdot x$.

$$\left(\dfrac{3y}{4x}\right)\left(\dfrac{2x^2}{y}\right) = \dfrac{3 \cdot y \cdot 2 \cdot x \cdot x}{2 \cdot 2 \cdot x \cdot y} = \dfrac{3 \cdot \cancel{y} \cdot 2 \cdot \cancel{x} \cdot x}{2 \cdot 2 \cdot \cancel{x} \cdot \cancel{y}} = \dfrac{3x}{2}$$

The prime factors of 4 are $2 \cdot 2$, so $4x$ is $2 \cdot 2 \cdot x$.

**Work Problem ❹ at the Side.**

**3** **Divide signed fractions.** To divide fractions, we will rewrite division problems as multiplication problems. For division, you will leave the first number (the dividend) as it is but change the second number (the divisor) to its *reciprocal*.

**Answers**

3. (a) $\dfrac{3 \cdot \cancel{2} \cdot \cancel{2} \cdot 3 \cdot 3}{\cancel{2} \cdot \cancel{2} \cdot 1} = \dfrac{27}{1} = 27$

(b) $-\dfrac{2 \cdot \cancel{5} \cdot 2}{1 \cdot \cancel{5}} = -\dfrac{4}{1} = -4$

(c) $\dfrac{7 \cdot \cancel{2} \cdot \cancel{2} \cdot \cancel{2} \cdot 3}{\cancel{2} \cdot \cancel{2} \cdot \cancel{2}} = \dfrac{21}{1} = 21$

4. (a) $\dfrac{\cancel{2} \cdot c \cdot c}{5 \cdot \cancel{2} \cdot 2} = \dfrac{c^2}{10}$

(b) $\dfrac{\cancel{m} \cdot \cancel{3} \cdot 3}{2 \cdot \cancel{3} \cdot \cancel{m} \cdot m} = \dfrac{3}{2m}$

(c) $\dfrac{\cancel{w} \cdot w \cdot x \cdot x \cdot \cancel{y}}{\cancel{y} \cdot \cancel{w}} = \dfrac{wx^2}{1} = wx^2$

## Reciprocal of a Fraction

Two numbers are **reciprocals** of each other if their product is 1. The reciprocal of the fraction $\frac{a}{b}$ is $\frac{b}{a}$ because

$$\frac{a}{b} \cdot \frac{b}{a} = \frac{\cancel{a} \cdot \cancel{b}}{\cancel{b} \cdot \cancel{a}} = \frac{1}{1} = 1$$

Notice that you "flip" or "invert" a fraction to find its reciprocal. Here are some examples.

| Number | Reciprocal | | Reason |
|---|---|---|---|
| $\frac{1}{6}$ | $\frac{6}{1}$ | Because | $\frac{1}{6} \cdot \frac{6}{1} = \frac{6}{6} = 1$ |
| $-\frac{2}{5}$ | $-\frac{5}{2}$ | Because | $\left(-\frac{2}{5}\right)\left(-\frac{5}{2}\right) = \frac{10}{10} = 1$ |
| 4 <br> Think of 4 as $\frac{4}{1}$. | $\frac{1}{4}$ | Because | $4 \cdot \frac{1}{4} = \frac{4}{1} \cdot \frac{1}{4} = \frac{4}{4} = 1$ |

> **NOTE**
>
> Every number has a reciprocal except 0. Why not 0? Recall that a number times its reciprocal equals 1. But that doesn't work for 0.
>
> $0 \cdot$ (reciprocal) $= 1$
>
> Put any number here. When you multiply it by 0, you get 0, never 1.

## Dividing Fractions

If $a$, $b$, $c$, and $d$ are numbers (but $b$, $c$, and $d$ are not 0), then we have the following.

$$\frac{a}{b} \div \frac{c}{d} = \frac{a}{b} \cdot \frac{d}{c}$$

Reciprocals

In other words, change division to multiplying by the reciprocal of the divisor.

Use this method to find the quotient for $\frac{2}{3} \div \frac{1}{6}$. Rewrite it as a multiplication problem and then use the steps for multiplying fractions.

$$\frac{2}{3} \div \frac{1}{6} = \frac{2}{3} \cdot \frac{6}{1} = \frac{2 \cdot 2 \cdot \cancel{3}}{\cancel{3} \cdot 1} = \frac{4}{1} = 4$$

Change division to multiplication.

The reciprocal of $\frac{1}{6}$ is $\frac{6}{1}$.

Does it make sense that $\frac{2}{3} \div \frac{1}{6} = 4$? Let's compare dividing fractions to dividing whole numbers.

$15 \div 3$ is asking, "How many 3s are in 15?"

$\frac{2}{3} \div \frac{1}{6}$ is asking, "How many $\frac{1}{6}$s are in $\frac{2}{3}$?"

The figure below illustrates $\frac{2}{3} \div \frac{1}{6}$.

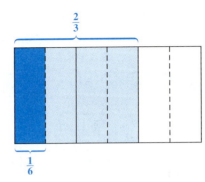

How many $\frac{1}{6}$s are in $\frac{2}{3}$?
There are 4 of the $\frac{1}{6}$ pieces in $\frac{2}{3}$.
So $\frac{2}{3} \div \frac{1}{6} = 4$.

As a final check on this method of dividing, try changing $15 \div 3$ into a multiplication problem. You know that the quotient should be 5.

$$15 \div 3 = 15 \cdot \frac{1}{3} = \frac{15}{1} \cdot \frac{1}{3} = \frac{\overset{1}{\cancel{3}} \cdot 5 \cdot 1}{1 \cdot \underset{1}{\cancel{3}}} = \frac{5}{1} = 5 \quad \left\{ \text{The quotient we expected} \right.$$

Reciprocals

### Example 5  Dividing Signed Fractions

Rewrite each division problem as a multiplication problem. Then multiply.

(a) $\dfrac{3}{10} \div \dfrac{4}{5} = \dfrac{3}{10} \cdot \dfrac{5}{4} = \dfrac{3 \cdot \overset{1}{\cancel{5}}}{2 \cdot \underset{1}{\cancel{5}} \cdot 2 \cdot 2} = \dfrac{3}{8}$

Reciprocals

**NOTE**

When multiplying fractions, you don't always have to factor the numerator and denominator completely into prime numbers. In part (a) above, if you notice that 5 is a common factor of the numerator and denominator, you can write

$$\frac{3}{10} \cdot \frac{5}{4} = \frac{3 \cdot \overset{1}{\cancel{5}}}{2 \cdot \underset{1}{\cancel{5}} \cdot 4} = \frac{3}{8}.$$

Factor 10 into $2 \cdot 5$.

Leave 4 as it is.

If no common factors are obvious to you, then write out the complete prime factorization to help find the common factors.

Continued on Next Page

**(b)** $2 \div \left(-\dfrac{1}{3}\right)$

First notice that the numbers have different signs. In a division problem, different signs mean that the quotient is negative. Then write 2 in fraction form as $\dfrac{2}{1}$.

$$2 \div \left(-\dfrac{1}{3}\right) = \dfrac{2}{1} \cdot \left(-\dfrac{3}{1}\right) = -\dfrac{2 \cdot 3}{1 \cdot 1} = -\dfrac{6}{1} = -6$$

Reciprocals ; Negative product

**(c)** $-\dfrac{3}{4} \div (-8) = -\dfrac{3}{4} \cdot \left(-\dfrac{1}{8}\right) = \dfrac{3 \cdot 1}{4 \cdot 8} = \dfrac{3}{32}$

Reciprocals ; No common factor to divide out

The quotient is positive because both numbers in the problem were negative. When signs match, the quotient is positive.

**(d)** $\dfrac{9}{16} \div 0$

This is *undefined,* as dividing by 0 was undefined for integers (see **Section 1.7**). Recall that 0 does *not* have a reciprocal, so you can't change the division to multiplying by the reciprocal of the divisor.

**(e)** $0 \div \dfrac{9}{16} = 0 \cdot \dfrac{16}{9} = 0$  Recall that 0 divided by any nonzero number gives a result of 0.

Reciprocals

*Work Problem* ❺ *at the Side.*

**4** **Divide fractions that involve variables.** The method for dividing fractions also works when there are variables in the numerators and/or denominators of the fractions.

### Example 6  Dividing Fractions with Variables

Divide.

**(a)** $\dfrac{x^2}{y} \div \dfrac{x}{3y} = \dfrac{x^2}{y} \cdot \dfrac{3y}{x} = \dfrac{x \cdot x \cdot 3 \cdot y}{y \cdot x} = \dfrac{3x}{1} = 3x$

Reciprocals

**(b)** $\dfrac{8b}{5} \div b^2 = \dfrac{8b}{5} \cdot \dfrac{1}{b^2} = \dfrac{8 \cdot b \cdot 1}{5 \cdot b \cdot b} = \dfrac{8}{5b}$

Write $b^2$ as $\dfrac{b^2}{1}$.
The reciprocal is $\dfrac{1}{b^2}$.

*Work Problem* ❻ *at the Side.*

---

❺ Rewrite each division problem as a multiplication problem. Then multiply.

**(a)** $-\dfrac{3}{4} \div \dfrac{5}{8}$

**(b)** $0 \div \left(-\dfrac{7}{12}\right)$

**(c)** $\dfrac{5}{6} \div 10$

**(d)** $-9 \div \left(-\dfrac{9}{16}\right)$

**(e)** $\dfrac{2}{5} \div 0$

❻ Divide.

**(a)** $\dfrac{c^2 d^2}{4} \div \dfrac{c^2 d}{4}$

**(b)** $\dfrac{20}{7h} \div \dfrac{5h}{7}$

**(c)** $\dfrac{n}{8} \div mn$

**ANSWERS**

5. **(a)** $-\dfrac{3}{4} \cdot \dfrac{8}{5} = -\dfrac{3 \cdot 2 \cdot 4}{4 \cdot 5} = -\dfrac{6}{5}$

**(b)** $0 \cdot \left(-\dfrac{12}{7}\right) = 0$

**(c)** $\dfrac{5}{6} \cdot \dfrac{1}{10} = \dfrac{5 \cdot 1}{6 \cdot 2 \cdot 5} = \dfrac{1}{12}$

**(d)** $-\dfrac{9}{1} \cdot \left(-\dfrac{16}{9}\right) = \dfrac{9 \cdot 16}{1 \cdot 9} = \dfrac{16}{1} = 16$

**(e)** undefined; can't be written as multiplication because 0 doesn't have a reciprocal

6. **(a)** $d$   **(b)** $\dfrac{4}{h^2}$   **(c)** $\dfrac{1}{8m}$

**230** Chapter 4 Rational Numbers: Positive and Negative Fractions

**7** Look for indicator words or draw sketches to help you with these problems.

(a) How many times can a $\frac{2}{3}$ quart spray bottle be filled before 18 quarts of window cleaner are used up?

(b) A retiring police officer will receive $\frac{5}{8}$ of her highest annual salary as retirement income. If her highest annual salary is $48,000, how much will she receive as retirement income?

**ANSWERS**

7. (a) $18 \div \frac{2}{3} = \frac{18}{1} \cdot \frac{3}{2} = \frac{\overset{1}{\cancel{2}} \cdot 9 \cdot 3}{1 \cdot \cancel{2}} = \frac{27}{1} = 27$

The bottle can be filled 27 times.

(b) $\frac{5}{8} \cdot \frac{48,000}{1} = \frac{5 \cdot \cancel{8} \cdot 6000}{\cancel{8} \cdot 1} = \frac{30,000}{1} = 30,000$

The officer will receive $30,000.

---

**5** Solve application problems involving multiplying and dividing fractions.

When you're solving application problems, some indicator words are used to suggest multiplication and some are used to suggest division.

| Indicator Words for Multiplication | Indicator Words for Division |
|---|---|
| product | per |
| double | each |
| triple | goes into |
| times | divided by |
| twice | divided into |
| of (when *of* follows a fraction) | divided equally |

Look for these indicator words in the following examples. However, you won't always find an indicator word. Then, you need to think through the problem to decide what to do. Sometimes, drawing a sketch of the situation described in the problem will help you decide which operation to use.

**Example 7** Using Indicator Words and Sketches to Solve Application Problems

(a) Lois gives $\frac{1}{10}$ **of** her income to her church. Last month she earned $1980. How much of that did she give to her church?

Notice the word **of**. Because the word **of** *follows the fraction* $\frac{1}{10}$, it indicates multiplication.

$$\frac{1}{10} \text{ of } 1980 = \frac{1}{10} \cdot \frac{1980}{1} = \frac{1 \cdot \overset{1}{\cancel{10}} \cdot 198}{\cancel{10} \cdot 1} = \frac{198}{1} = 198$$

Lois gave $198 to her church.

(b) The apparel design class is making infant snowsuits to give to a local shelter. A fabric store donated 12 yd of fabric for the project. If one snowsuit needs $\frac{2}{3}$ yd of fabric, how many suits can the class make?

The word **of** appears in the second sentence: "A fabric store donated 12 yd **of** fabric." But there is *no fraction* before the word **of**, so it is *not* an indicator to multiply. Let's try a sketch. There is a 12 yd piece of fabric. One snowsuit will use $\frac{2}{3}$ yd. The question is, how many $\frac{2}{3}$ yd pieces can be cut from the 12 yards?

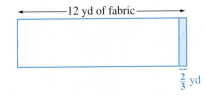

Cutting 12 yards into equal size pieces indicates division. How many $\frac{2}{3}$s are in 12?

$$12 \div \frac{2}{3} = \frac{12}{1} \cdot \frac{3}{2} = \frac{\overset{1}{\cancel{2}} \cdot 6 \cdot 3}{1 \cdot \cancel{2}} = \frac{18}{1} = 18$$

Reciprocals

The class can make 18 snowsuits.

**Work Problem 7 at the Side.**

## 4.3 Exercises

*Multiply. Write the products in lowest terms. See Examples 1–4.*

1. $-\dfrac{3}{8} \cdot \dfrac{1}{2}$

2. $\left(\dfrac{2}{3}\right)\left(-\dfrac{5}{7}\right)$

3. $\left(-\dfrac{3}{8}\right)\left(-\dfrac{12}{5}\right)$

4. $\dfrac{4}{9} \cdot \dfrac{12}{7}$

5. $\dfrac{21}{30}\left(\dfrac{5}{7}\right)$

6. $\left(-\dfrac{6}{11}\right)\left(-\dfrac{22}{15}\right)$

7. $10\left(-\dfrac{3}{5}\right)$

8. $-20\left(\dfrac{3}{4}\right)$

9. $\dfrac{4}{9}$ of 81

10. $\dfrac{2}{3}$ of 48

11. $\dfrac{3x}{4} \cdot \dfrac{5}{xy}$

12. $\dfrac{2}{5a^2} \cdot \dfrac{a}{8}$

*Divide. Write the quotients in lowest terms. See Examples 5 and 6.*

13. $\dfrac{1}{6} \div \dfrac{1}{3}$

14. $-\dfrac{1}{2} \div \dfrac{2}{3}$

15. $-\dfrac{3}{4} \div \left(-\dfrac{5}{8}\right)$

16. $\dfrac{7}{10} \div \dfrac{2}{5}$

17. $6 \div \left(-\dfrac{2}{3}\right)$

18. $-7 \div \left(-\dfrac{1}{4}\right)$

19. $-\dfrac{2}{3} \div 4$

20. $\dfrac{5}{6} \div (-15)$

21. $\dfrac{11c}{5d} \div 3c$

22. $8x^2 \div \dfrac{4x}{7}$

23. $\dfrac{ab^2}{c} \div \dfrac{ab}{c}$

24. $\dfrac{mn}{6} \div \dfrac{n}{3m}$

25. Explain and correct the error in each of these calculations.

   (a) $\dfrac{3}{14} \cdot \dfrac{7}{9} = \dfrac{\cancel{3} \cdot \cancel{7}}{2 \cdot \cancel{7} \cdot \cancel{3} \cdot 3} = 6$

   (b) $8 \cdot \dfrac{2}{3} = \dfrac{8}{1} \cdot \dfrac{3}{2} = \dfrac{\cancel{2} \cdot 4 \cdot 3}{1 \cdot \cancel{2}} = \dfrac{12}{1} = 12$

26. Explain and correct the error in each of these calculations.

   (a) $\dfrac{3}{4} \cdot \dfrac{8}{9} = \dfrac{3 \cdot 8}{4 \cdot 9} = \dfrac{24}{36}$

   (b) $\dfrac{2}{5} \cdot \dfrac{3}{8} = \dfrac{\cancel{2}}{5} \cdot \dfrac{3}{\cancel{8}} = \dfrac{3}{10}$

**27.** Explain and correct the error in each of these calculations.

(a) $\dfrac{2}{3} \div 4 = \dfrac{2}{3} \cdot \dfrac{4}{1} = \dfrac{2 \cdot 4}{3 \cdot 1} = \dfrac{8}{3}$

(b) $\dfrac{5}{6} \div \dfrac{10}{9} = \dfrac{6}{5} \cdot \dfrac{10}{9} = \dfrac{2 \cdot \cancel{3} \cdot 2 \cdot \cancel{5}}{\cancel{5} \cdot \cancel{3} \cdot 3} = \dfrac{4}{3}$

**28.** Explain and correct the error in each of these calculations.

(a) $\dfrac{1}{2} \div 0 = 0$

(b) $\dfrac{\cancel{2}^{1}}{10} \div \dfrac{1}{\cancel{2}_{1}} = \dfrac{1}{10}$

**29.** Your friend missed class and is confused about how to divide fractions. Write a short explanation for your friend.

**30.** Mary spilled coffee on her math homework, and part of one problem is covered up.

$\dfrac{3}{\blacksquare} \div \dfrac{4}{5}$

She knows the answer given in the back of the book is $\dfrac{3}{4}$. Describe how to find the missing number.

*Find each product or quotient. Write all answers in lowest terms. See Examples 1–6.*

**31.** $\dfrac{4}{5} \div 3$

**32.** $\left(-\dfrac{20}{21}\right)\left(-\dfrac{14}{15}\right)$

**33.** $-\dfrac{3}{8}\left(\dfrac{3}{4}\right)$

**34.** $-\dfrac{8}{17} \div \dfrac{4}{5}$

**35.** $\dfrac{3}{5}$ of 35

**36.** $\dfrac{2}{3} \div (-6)$

**37.** $-9 \div \left(-\dfrac{3}{5}\right)$

**38.** $\dfrac{7}{8} \cdot \dfrac{25}{21}$

**39.** $\dfrac{12}{7} \div 0$

**40.** $\dfrac{5}{8}$ of $(-48)$

**41.** $\left(\dfrac{11}{2}\right)\left(-\dfrac{5}{6}\right)$

**42.** $\dfrac{3}{4} \div \dfrac{3}{16}$

**43.** $\dfrac{4}{7}$ of $14b$

**44.** $\dfrac{ab}{6} \div \dfrac{b}{9}$

**45.** $\dfrac{12}{5} \div 4d$

**46.** $\dfrac{18}{7} \div 2t$

**47.** $\dfrac{x^2}{y} \div \dfrac{w}{2y}$

**48.** $\dfrac{5}{6}$ of $18w$

*Solve each application problem. See Example 7.*

**49.** Al is helping Tim make a mahogany lamp table for Jill's birthday. Find the area of the rectangular top of the table if it is $\frac{4}{5}$ yd long by $\frac{3}{8}$ yd wide.

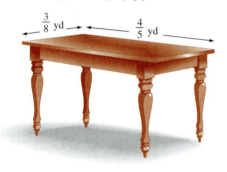

**50.** A dog bed is $\frac{7}{8}$ yd by $\frac{10}{9}$ yd. Find its area.

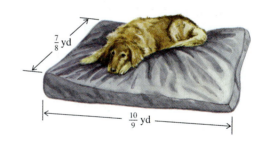

**51.** How many $\frac{1}{8}$-ounce eye drop dispensers can be filled with 10 ounces of eye drops?

**52.** Ms. Shaffer has a piece of property with an area that is $\frac{9}{10}$ acre. She wishes to divide it into three equal parts for her children. How many acres of land will each child get?

**53.** Todd estimates that it will cost him $12,400 to attend a community college for one year. He thinks he can earn $\frac{3}{4}$ of the cost and borrow the balance. Find the amount he must earn and the amount he must borrow.

**54.** Joyce Chen wants to make vests to sell at a craft fair. Each vest requires $\frac{3}{4}$ yd of material. She has 36 yd of material. Find the number of vests she can make.

**55.** Pam Trizlia has a small pickup truck that can carry $\frac{2}{3}$ cord of firewood. Find the number of trips needed to deliver 6 cords of wood.

**56.** At the Garlic Festival Fun Run, $\frac{5}{12}$ of the runners are women. If there are 780 runners, how many are women? How many are men?

**57.** One-third of the players elected to the Baseball Hall of Fame were pitchers during their playing careers. If 183 players are in the Hall of Fame, how many were pitchers? (*Source:* National Baseball Hall of Fame.)

**58.** Parking lot A is $\frac{1}{4}$ mile long and $\frac{3}{16}$ mile wide, and parking lot B is $\frac{3}{8}$ mile long and $\frac{1}{8}$ mile wide. Which parking lot has the larger area?

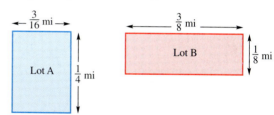

**59.** An adult male alligator may weigh 400 pounds. A newly hatched alligator weighs only $\frac{1}{8}$ pound. The adult weighs how many times the hatchling? (*Source:* St. Marks NWR.)

**60.** A female alligator lays about 35 eggs in a nest of marsh grass and mud. But $\frac{4}{5}$ of the eggs or hatchlings will fall prey to raccoons, wading birds, or larger alligators. How many of the 35 eggs will hatch and survive? (*Source:* St. Marks NWR.)

*The table below shows the earnings for the Gomez family last year and the circle graph shows how they spent their earnings. Use this information to work Exercises 61–64.*

| Month | Earnings | Month | Earnings |
|---|---|---|---|
| January | $3050 | July | $3160 |
| February | $2875 | August | $2355 |
| March | $3325 | September | $2780 |
| April | $3020 | October | $3675 |
| May | $2880 | November | $3310 |
| June | $3265 | December | $4305 |

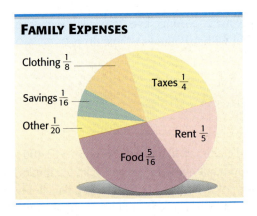

**FAMILY EXPENSES**

Clothing $\frac{1}{8}$, Taxes $\frac{1}{4}$, Savings $\frac{1}{16}$, Other $\frac{1}{20}$, Rent $\frac{1}{5}$, Food $\frac{5}{16}$

**61.** **(a)** What was the family's total income for the year?

**(b)** Find the amount of the family's rent for the year.

**62.** **(a)** How much did the family pay in taxes during the year?

**(b)** How much more did the family spend on taxes than on rent?

**63.** How much did the family spend for food and clothing last year?

**64.** Find the amount the family saved during the year.

*There are a total of about 150 million companion pets in the United States. This table shows the fraction of pets that are dogs, cats, birds, and horses. Use the table to answer Exercises 65–68.*

**COMPANION PETS IN THE UNITED STATES**

| Type of Pet | Fraction of All Pets |
|---|---|
| Dog | $\frac{2}{5}$ |
| Cat | $\frac{23}{50}$ |
| Bird | $\frac{1}{10}$ |
| Horse | $\frac{3}{100}$ |

*Source:* American Veterinary Medical Association.

**65.** How many U.S. pets are birds?

**66.** How many U.S. pets are dogs?

**67.** How many dogs and cats are pets?

**68.** How many birds and cats are pets?

Section 4.4 Adding and Subtracting Signed Fractions    235

## 4.4 ADDING AND SUBTRACTING SIGNED FRACTIONS

**1** **Add and subtract like fractions.** You probably remember learning something about "common denominators" in other math classes. When fractions have the *same* denominator, we say that they have a *common* denominator, which makes them **like fractions.** When fractions have different denominators, they are called **unlike fractions.** Here are some examples.

**OBJECTIVES**

**1** Add and subtract like fractions.

**2** Find the lowest common denominator for unlike fractions.

**3** Add and subtract unlike fractions.

**4** Add and subtract unlike fractions that contain variables.

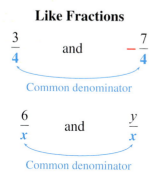

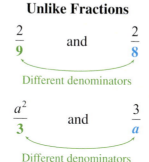

You can add or subtract fractions *only* when they have a common denominator. To see why, let's look at more pizzas.

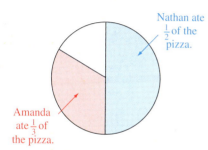

What fraction of the pizza has been eaten? We can't write a fraction until the pizza is cut into pieces of *equal* size. That's what the denominator of a fraction tell us: the number of *equal* size pieces in the pizza.

Now the pizza is cut into 6 *equal* pieces, and we can find out how much was eaten.

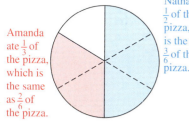

Nathan ate $\frac{1}{2}$ of the pizza, which is the same as $\frac{3}{6}$ of the pizza.

$$\frac{1}{2} + \frac{1}{3}$$
$$\downarrow \quad \downarrow$$
$$\frac{3}{6} + \frac{2}{6}$$

$$= \frac{5}{6} \leftarrow \text{Number of pieces eaten}$$
$$\phantom{= \frac{5}{6}} \leftarrow \text{Number of equal pieces in the pizza}$$

### Adding and Subtracting Like Fractions

You can add or subtract fractions *only* when they have a common denominator. If $a$, $b$, and $c$ are numbers (and $b$ is not 0), then

$$\frac{a}{b} + \frac{c}{b} = \frac{a+c}{b} \quad \text{and} \quad \frac{a}{b} - \frac{c}{b} = \frac{a-c}{b}$$

In other words, add or subtract the numerators and write the result over the common denominator. Then check to be sure that the answer is in lowest terms.

**1** Write each sum or difference in lowest terms.

(a) $\dfrac{1}{6} + \dfrac{5}{6}$

(b) $-\dfrac{11}{12} + \dfrac{5}{12}$

(c) $-\dfrac{2}{9} - \dfrac{3}{9}$

(d) $\dfrac{8}{ab} + \dfrac{3}{ab}$

**ANSWERS**

1. (a) $\dfrac{6}{6} = 1$  (b) $-\dfrac{1}{2}$
   (c) $-\dfrac{5}{9}$  (d) $\dfrac{11}{ab}$

## Example 1  Adding and Subtracting Like Fractions

Find each sum or difference.

(a) $\dfrac{1}{8} + \dfrac{3}{8}$

These are *like* fractions because they have a *common denominator of 8*. So they are ready to be added. Add the numerators and write the sum over the common denominator.

$$\dfrac{1}{8} + \dfrac{3}{8} = \dfrac{1+3}{8} = \dfrac{4}{8}$$

Common denominator

Now write $\dfrac{4}{8}$ in lowest terms.

$$\dfrac{4}{8} = \dfrac{\cancel{2} \cdot \cancel{2}}{\cancel{2} \cdot \cancel{2} \cdot 2} = \dfrac{1}{2}$$

The sum, in lowest terms, is $\dfrac{1}{2}$.

**CAUTION**

Add *only* the numerators. **Do not add the denominators.** In part (a) above we *kept the common denominator.*

$$\dfrac{1}{8} + \dfrac{3}{8} = \dfrac{1+3}{8} \quad \text{not} \quad \dfrac{1}{8} + \dfrac{3}{8} = \dfrac{1+3}{8+8} = \dfrac{4}{16}$$
Incorrect

To help you understand why we add *only* the numerators, think of $\dfrac{1}{8}$ as $1\left(\dfrac{1}{8}\right)$ and $\dfrac{3}{8}$ as $3\left(\dfrac{1}{8}\right)$. Then we use the distributive property.

$$\dfrac{1}{8} + \dfrac{3}{8} = 1\left(\dfrac{1}{8}\right) + 3\left(\dfrac{1}{8}\right) = (1+3)\left(\dfrac{1}{8}\right) = 4\left(\dfrac{1}{8}\right) = \dfrac{4}{8} = \dfrac{1}{2}$$

Use the distributive property.

Or think about a pie cut into 8 equal pieces. If you eat 1 piece and your friend eats 3 pieces, together you've eaten 4 of the 8 pieces or $\dfrac{4}{8}$ of the pie (*not* $\dfrac{4}{16}$ of the pie, which would be 4 out of 16 pieces).

(b) $-\dfrac{3}{5} + \dfrac{4}{5} = \dfrac{-3+4}{5} = \dfrac{1}{5}$ ← Lowest terms

Common denominator

(c) $\dfrac{3}{10} - \dfrac{7}{10} = \dfrac{3-7}{10} = \dfrac{3+(-7)}{10} = \dfrac{-4}{10}$ or $-\dfrac{4}{10}$

Rewrite subtraction as adding the opposite.

Common denominator

Now write $-\dfrac{4}{10}$ in lowest terms.

$$-\dfrac{4}{10} = -\dfrac{\cancel{2} \cdot 2}{\cancel{2} \cdot 5} = -\dfrac{2}{5}$$ ← Lowest terms

(d) $\dfrac{5}{x^2} - \dfrac{2}{x^2} = \dfrac{5-2}{x^2} = \dfrac{3}{x^2}$

Common denominator

**Work Problem 1 at the Side.**

**2** **Find the lowest common denominator for unlike fractions.** When we first tried to add the pizza eaten by Nathan and Amanda, we could *not* do so because the pizza was *not cut into pieces of the same size*. So we rewrote $\frac{1}{2}$ and $\frac{1}{3}$ as equivalent fractions that both had 6 as the common denominator.

$$\frac{1}{2} = \frac{1 \cdot 3}{2 \cdot 3} = \frac{3}{6}$$
$$\frac{1}{3} = \frac{1 \cdot 2}{3 \cdot 2} = \frac{2}{6}$$ Common denominator

Then we could add the fractions, because the pizza was cut into pieces of *equal* size.

$$\frac{1}{2} + \frac{1}{3} = \frac{3}{6} + \frac{2}{6} = \frac{3+2}{6} = \frac{5}{6} \leftarrow \text{Lowest terms}$$

In other words, when you want to add or subtract unlike fractions, the first thing you must do is rewrite them so that they have a common denominator.

### A Common Denominator for Unlike Fractions

To find a common denominator for two unlike fractions, find a number that is divisible by *both* of the original denominators.
For example, the common denominator for $\frac{1}{2}$ and $\frac{1}{3}$ is 6 because 2 goes into 6 evenly and 3 goes into 6 evenly.

Notice that 12 is also a common denominator for $\frac{1}{2}$ and $\frac{1}{3}$ because 2 and 3 both go into 12 evenly.

$$\frac{1}{2} = \frac{1 \cdot 6}{2 \cdot 6} = \frac{6}{12}$$
$$\frac{1}{3} = \frac{1 \cdot 4}{3 \cdot 4} = \frac{4}{12}$$ Common denominator

Now that the fractions have a common denominator, we can add them.

$$\frac{1}{2} + \frac{1}{3} = \frac{6}{12} + \frac{4}{12} = \frac{6+4}{12} = \frac{10}{12} \begin{cases} \text{Not in} \\ \text{lowest} \\ \text{terms} \end{cases} \quad \text{but} \quad \frac{10}{12} = \frac{\overset{1}{\cancel{2}} \cdot 5}{\underset{1}{\cancel{2}} \cdot 6} = \frac{5}{6} \begin{cases} \text{Same} \\ \text{result} \\ \text{as above} \end{cases}$$

Both 6 and 12 worked as common denominators for adding $\frac{1}{2}$ and $\frac{1}{3}$, but using the smaller number saved some work. You should always try to find the smallest common denominator. If you don't for some reason, you can still work the problem—but it may take you longer. You'll have to divide out some common factors at the end in order to write the answer in lowest terms.

### Least Common Denominator

The **least common denominator** (LCD) for two fractions is the *smallest* positive number divisible by both denominators of the original fractions. For example, both 6 and 12 are common denominators for $\frac{1}{2}$ and $\frac{1}{3}$, but 6 is smaller, so it is the LCD.

There are several ways to find the LCD. When the original denominators are small numbers, you can often find the LCD by inspection. *Hint:* Always check to see if the larger denominator will work as the LCD.

**238** Chapter 4 Rational Numbers: Positive and Negative Fractions

**② Find the LCD for each pair of fractions.**

(a) $\dfrac{3}{5}$ and $\dfrac{3}{10}$

(b) $\dfrac{1}{2}$ and $\dfrac{2}{5}$

(c) $\dfrac{3}{4}$ and $\dfrac{1}{6}$

(d) $\dfrac{5}{6}$ and $\dfrac{7}{18}$

**③ Use prime factorization to find the LCD for each pair of fractions.**

(a) $\dfrac{1}{10}$ and $\dfrac{13}{14}$

(b) $\dfrac{5}{12}$ and $\dfrac{17}{20}$

(c) $\dfrac{7}{15}$ and $\dfrac{7}{9}$

**ANSWERS**

2. (a) 10  (b) 10  (c) 12  (d) 18

3. (a) $10 = 2 \cdot 5$
$14 = 2 \cdot 7$  LCD $= 2 \cdot 5 \cdot 7 = 70$

(b) $12 = 2 \cdot 2 \cdot 3$
$20 = 2 \cdot 2 \cdot 5$  LCD $= 2 \cdot 2 \cdot 3 \cdot 5 = 60$

(c) $15 = 3 \cdot 5$
$9 = 3 \cdot 3$  LCD $= 3 \cdot 3 \cdot 5 = 45$

---

**Example 2  Finding the LCD by Inspection**

(a) Find the LCD for $\dfrac{2}{3}$ and $\dfrac{1}{9}$.

Check to see if 9 (the larger denominator) will work as the LCD. Is 9 divisible by 3 (the other denominator)? Yes, so 9 is the LCD for $\dfrac{2}{3}$ and $\dfrac{1}{9}$.

(b) Find the LCD for $\dfrac{5}{8}$ and $\dfrac{5}{6}$.

Check to see if 8 (the larger denominator) will work. No, 8 is not divisible by 6. So start checking numbers that are multiples of 8, that is, 16, 24, and 32. Notice that 24 will work because it is divisible by 8 and by 6.

The LCD for $\dfrac{5}{8}$ and $\dfrac{5}{6}$ is 24.

**Work Problem ② at the Side.**

For larger denominators, you can use prime factorization to find the LCD. Factor each denominator completely into prime numbers. Then use the factors to build the LCD.

**Example 3  Using Prime Factors to Find the LCD**

(a) What is the LCD for $\dfrac{7}{12}$ and $\dfrac{13}{18}$?

Write 12 and 18 as the product of prime factors. Then use enough prime factors in the LCD to "cover" both 12 and 18.

$12 = 2 \cdot 2 \cdot 3$
$18 = 2 \cdot 3 \cdot 3$

LCD $= 2 \cdot 2 \cdot 3 \cdot 3 = 36$

(Factors of 12 and Factors of 18)

Check whether 36 is divisible by 12 (yes) and by 18 (yes). So 36 is the LCD for $\dfrac{7}{12}$ and $\dfrac{13}{18}$.

**CAUTION**

When finding the LCD, notice that we did *not* have to repeat the factors that 12 and 18 have in common. If we had used *all* the 2s and 3s, we would get a common denominator, but not the *smallest* one.

(b) What is the LCD for $\dfrac{11}{15}$ and $\dfrac{9}{70}$?

$15 = 3 \cdot 5$
$70 = 2 \cdot 5 \cdot 7$

LCD $= 3 \cdot 5 \cdot 2 \cdot 7 = 210$

(Factors of 15 and Factors of 70)

Check whether 210 is divisible by 15 (yes) and divisible by 70 (yes). So 210 is the LCD for $\dfrac{11}{15}$ and $\dfrac{9}{70}$.

**Work Problem ③ at the Side.**

### Section 4.4 Adding and Subtracting Signed Fractions   239

**3** **Add and subtract unlike fractions.** Here are the steps for adding or subtracting unlike fractions. The key idea is that you must rewrite the fractions so that they have a common denominator before you can add or subtract them.

> **Adding and Subtracting Unlike Fractions**
>
> *Step 1* Find the LCD, the smallest number divisible by both denominators in the problem.
>
> *Step 2* Rewrite each original fraction as an equivalent fraction whose denominator is the LCD.
>
> *Step 3* Add or subtract the numerators of the like fractions. Keep the common denominator.
>
> *Step 4* Write the sum or difference in lowest terms.

**Example 4** **Adding and Subtracting Unlike Fractions**

Find each sum or difference.

(a) $\dfrac{1}{5} + \dfrac{3}{10}$

*Step 1*  The larger denominator (10) is the LCD.

*Step 2*  $\dfrac{1}{5} = \dfrac{1 \cdot 2}{5 \cdot 2} = \dfrac{2}{10}$ ← LCD    and    $\dfrac{3}{10}$ already has the LCD.

*Step 3*  Add the numerators. Write the sum over the common denominator.

$$\dfrac{1}{5} + \dfrac{3}{10} = \dfrac{2}{10} + \dfrac{3}{10} = \dfrac{2+3}{10} = \dfrac{5}{10}$$

*Step 4*  Write $\dfrac{5}{10}$ in lowest terms.

$$\dfrac{5}{10} = \dfrac{\cancel{5}^{\,1}}{2 \cdot \cancel{5}_{\,1}} = \dfrac{1}{2} \;\; \leftarrow \text{Lowest terms}$$

(b) $\dfrac{3}{4} - \dfrac{5}{6}$

*Step 1*  The LCD is 12.

*Step 2*  $\dfrac{3}{4} = \dfrac{3 \cdot 3}{4 \cdot 3} = \dfrac{9}{12}$ ← LCD    and    $\dfrac{5}{6} = \dfrac{5 \cdot 2}{6 \cdot 2} = \dfrac{10}{12}$ ← LCD

*Step 3*  Subtract the numerators. Write the difference over the common denominator.

$9 + (-10)$ is $-1$.

$$\dfrac{3}{4} - \dfrac{5}{6} = \dfrac{9}{12} - \dfrac{10}{12} = \dfrac{9 - 10}{12} = \dfrac{-1}{12} \;\; \text{or} \;\; -\dfrac{1}{12}$$

*Step 4*  $-\dfrac{1}{12}$ is in lowest terms.

*Continued on Next Page*

Chapter 4 Rational Numbers: Positive and Negative Fractions

**4** Find each sum or difference. Write all answers in lowest terms.

(a) $-\dfrac{2}{3} + \dfrac{1}{6}$

(b) $\dfrac{1}{12} - \dfrac{5}{6}$

(c) $3 - \dfrac{4}{5}$

(d) $\dfrac{9}{16} + \left(-\dfrac{5}{12}\right)$

**ANSWERS**

4. (a) $-\dfrac{1}{2}$  (b) $-\dfrac{3}{4}$  (c) $\dfrac{11}{5}$  (d) $\dfrac{7}{48}$

---

(c) $-\dfrac{5}{12} + \dfrac{5}{9}$

*Step 1* Use prime factorization to find the LCD.

$12 = 2 \cdot 2 \cdot 3$
$9 = 3 \cdot 3$

Factors of 12
$LCD = 2 \cdot 2 \cdot 3 \cdot 3 = 36$
Factors of 9

*Step 2* $-\dfrac{5}{12} = -\dfrac{5 \cdot 3}{12 \cdot 3} = -\dfrac{15}{36}$  and  $\dfrac{5}{9} = \dfrac{5 \cdot 4}{9 \cdot 4} = \dfrac{20}{36}$

*Step 3* Add the numerators. Keep the common denominator.

$$-\dfrac{5}{12} + \dfrac{5}{9} = -\dfrac{15}{36} + \dfrac{20}{36} = \dfrac{-15 + 20}{36} = \dfrac{5}{36}$$

*Step 4* $\dfrac{5}{36}$ is in lowest terms.

(d) $4 - \dfrac{2}{3}$

*Step 1* Think of 4 as $\dfrac{4}{1}$. The LCD for $\dfrac{4}{1}$ and $\dfrac{2}{3}$ is 3, the larger denominator.

*Step 2* $\dfrac{4}{1} = \dfrac{4 \cdot 3}{1 \cdot 3} = \dfrac{12}{3}$  and  $\dfrac{2}{3}$ already has the LCD.

*Step 3* Subtract the numerators. Keep the common denominator.

$$\dfrac{4}{1} - \dfrac{2}{3} = \dfrac{12}{3} - \dfrac{2}{3} = \dfrac{12 - 2}{3} = \dfrac{10}{3}$$

*Step 4* $\dfrac{10}{3}$ is in lowest terms.

**Work Problem** **4** **at the Side.**

**4** **Add and subtract unlike fractions that contain variables.** We use the same steps to add or subtract unlike fractions with variables in the numerators or denominators.

**Example 5** Adding and Subtracting Unlike Fractions with Variables

Find each sum or difference.

(a) $\dfrac{1}{4} + \dfrac{b}{5}$

*Step 1* The LCD is 20.

*Step 2* $\dfrac{1}{4} = \dfrac{1 \cdot 5}{4 \cdot 5} = \dfrac{5}{20}$  and  $\dfrac{b}{5} = \dfrac{b \cdot 4}{5 \cdot 4} = \dfrac{4b}{20}$

*Step 3* $\dfrac{1}{4} + \dfrac{b}{5} = \dfrac{5}{20} + \dfrac{4b}{20} = \dfrac{5 + 4b}{20}$ ← Add the numerators.
← Keep the common denominator.

*Step 4* $\dfrac{5 + 4b}{20}$ is in lowest terms.

**Continued on Next Page**

> **CAUTION**
> 
> In part (a), we could *not* add $5 + 4b$ in the numerator of the answer because 5 and $4b$ are *not* like terms. We *could* add $5b + 4b$ but *not* $5 + 4b$.
> 
> Variable parts match.

**(b)** $\dfrac{2}{3} - \dfrac{6}{x}$

*Step 1*    The LCD is $3 \cdot x$, or $3x$.

*Step 2*    $\dfrac{2}{3} = \dfrac{2 \cdot x}{3 \cdot x} = \dfrac{2x}{3x}$    and    $\dfrac{6}{x} = \dfrac{6 \cdot 3}{x \cdot 3} = \dfrac{18}{3x}$

*Step 3*    $\dfrac{2}{3} - \dfrac{6}{x} = \dfrac{2x}{3x} - \dfrac{18}{3x} = \dfrac{2x - 18}{3x}$ ← Keep the common denominator.

*Step 4*    $\dfrac{2x - 18}{3x}$ is in lowest terms.

> **NOTE**
> 
> Notice that we found the LCD for $\frac{2}{3} - \frac{6}{x}$ by multiplying the two denominators. The LCD is $3 \cdot x$ or $3x$.
> 
> Multiplying the two denominators will *always* give you a common denominator, but it may not be the *smallest* common denominator. Here are more examples.
> 
> $\dfrac{1}{3} - \dfrac{2}{5}$    If you multiply the denominators, $3 \cdot 5 = 15$ and 15 is the LCD.
> 
> $\dfrac{5}{6} + \dfrac{3}{4}$    If you multiply the denominators, $6 \cdot 4 = 24$ and 24 will work. But you'll save some time by using the *smallest* common denominator, which is 12.
> 
> $\dfrac{7}{y} + \dfrac{a}{4}$    If you multiply the denominators, $y \cdot 4 = 4y$ and $4y$ is the LCD.

━━ **Work Problem ❺ at the Side.**

---

❺ Find each sum or difference.

**(a)** $\dfrac{5}{6} - \dfrac{h}{2}$

**(b)** $\dfrac{7}{t} + \dfrac{3}{5}$

**(c)** $\dfrac{4}{x} - \dfrac{8}{3}$

---

**ANSWERS**

**5.** **(a)** $\dfrac{5 - 3h}{6}$    **(b)** $\dfrac{35 + 3t}{5t}$    **(c)** $\dfrac{12 - 8x}{3x}$

# Focus on Real-Data Applications

## Music

The time signature at the beginning of a piece of music looks like a fraction. Commonly used time signatures are $\frac{2}{4}$, $\frac{3}{4}$, $\frac{4}{4}$, and $\frac{6}{8}$. Musicians use the time signature to tell how long to hold each note. The values of different notes can be written as fractions:

$\mathbf{o} = 1 \qquad \mathbf{\circ} = \frac{1}{2} \qquad \mathbf{\bullet} = \frac{1}{4} \qquad \mathbf{\flat} = \frac{1}{8} \qquad \mathbf{\flat} = \frac{1}{16}$

Music is divided into measures. In $\frac{4}{4}$ time, each measure contains notes that add up to $\frac{4}{4}$ (or 1). In $\frac{2}{4}$ time the notes in each measure add up to $\frac{2}{4}$ $\left(\text{or } \frac{1}{2}\right)$, and so on for $\frac{3}{4}$ time and $\frac{6}{8}$ time.

Write one or more notes in each measure to make it add up to its time signature. Use as many different kinds of notes as possible.

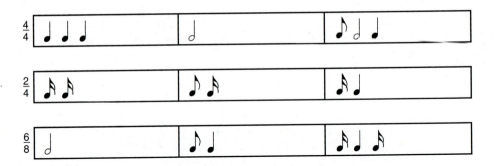

Below are excerpts from "Jingle Bells" and "The Star-Spangled Banner." Divide each line of music into measures based on the time signature.

## 4.4 Exercises

*Find each sum or difference. Write all answers in lowest terms. See Examples 1–5.*

1. $\dfrac{3}{4} + \dfrac{1}{8}$

2. $\dfrac{1}{3} + \dfrac{1}{2}$

3. $-\dfrac{1}{14} + \left(-\dfrac{3}{7}\right)$

4. $-\dfrac{2}{9} + \dfrac{2}{3}$

5. $\dfrac{2}{3} - \dfrac{1}{6}$

6. $\dfrac{5}{12} - \dfrac{1}{4}$

7. $\dfrac{3}{8} - \dfrac{3}{5}$

8. $\dfrac{1}{3} - \dfrac{3}{5}$

9. $-\dfrac{5}{8} + \dfrac{1}{12}$

10. $-\dfrac{13}{16} + \dfrac{13}{16}$

11. $-\dfrac{7}{20} - \dfrac{5}{20}$

12. $-\dfrac{7}{9} - \dfrac{5}{6}$

13. $0 - \dfrac{7}{18}$

14. $-\dfrac{7}{8} + 3$

15. $2 - \dfrac{6}{7}$

16. $5 - \dfrac{2}{5}$

17. $-\dfrac{1}{2} + \dfrac{3}{24}$

18. $\dfrac{7}{10} + \dfrac{7}{15}$

19. $\dfrac{1}{5} + \dfrac{c}{3}$

20. $\dfrac{x}{4} + \dfrac{2}{3}$

21. $\dfrac{5}{m} - \dfrac{1}{2}$

22. $\dfrac{2}{9} - \dfrac{4}{y}$

23. $\dfrac{3}{b^2} + \dfrac{5}{b^2}$

24. $\dfrac{10}{xy} - \dfrac{7}{xy}$

**25.** $\dfrac{c}{7} + \dfrac{3}{b}$

**26.** $\dfrac{2}{x} - \dfrac{y}{5}$

**27.** $-\dfrac{4}{c^2} - \dfrac{d}{c}$

**28.** $-\dfrac{1}{n} + \dfrac{m}{n^2}$

**29.** $-\dfrac{11}{42} - \dfrac{11}{70}$

**30.** $\dfrac{7}{45} - \dfrac{7}{20}$

**31.** A key step in adding or subtracting unlike fractions is to rewrite the fractions so that they have a common denominator. Explain why this step is necessary.

**32.** Explain how to write a fraction with an indicated denominator. As part of your explanation, show how to change $\tfrac{3}{4}$ to an equivalent fraction having 12 as a denominator.

**33.** Explain the error in each calculation and correct it.

(a) $\dfrac{3}{4} + \dfrac{2}{5} = \dfrac{3+2}{4+5} = \dfrac{5}{9}$

(b) $\dfrac{5}{6} - \dfrac{4}{9} = \dfrac{5}{18} - \dfrac{4}{18} = \dfrac{5-4}{18} = \dfrac{1}{18}$

**34.** Explain the error in each calculation and correct it.

(a) $-\dfrac{1}{4} + \dfrac{7}{12} = -\dfrac{3}{12} + \dfrac{7}{12} = \dfrac{-3+7}{12} = \dfrac{4}{12}$

(b) $\dfrac{3}{10} - \dfrac{1}{4} = \dfrac{3-1}{10-4} = \dfrac{2}{6} = \dfrac{1}{3}$

## Relating Concepts (Exercises 35–36)  For Individual or Group Work

*As you **work Exercises 35 and 36 in order,** think about the properties you learned when working with integers. Then explain what each pair of problems illustrates. (For a review of the properties, see the Chapter 1 Summary.)*

**35.** (a) $-\dfrac{2}{3} + \dfrac{3}{4} = $ _____   $\dfrac{3}{4} + \left(-\dfrac{2}{3}\right) = $ _____

(b) $\dfrac{5}{6} - \dfrac{1}{2} = $ _____   $\dfrac{1}{2} - \dfrac{5}{6} = $ _____

(c) $\left(-\dfrac{2}{3}\right)\left(\dfrac{9}{10}\right) = $ _____   $\left(\dfrac{9}{10}\right)\left(-\dfrac{2}{3}\right) = $ _____

(d) $\dfrac{2}{5} \div \dfrac{1}{15} = $ _____   $\dfrac{1}{15} \div \dfrac{2}{5} = $ _____

**36.** (a) $-\dfrac{7}{12} + \dfrac{7}{12} = $ _____   $\dfrac{3}{5} + \left(-\dfrac{3}{5}\right) = $ _____

(b) $-\dfrac{13}{16} \div \left(-\dfrac{13}{16}\right) = $ _____   $\dfrac{1}{8} \div \dfrac{1}{8} = $ _____

(c) $\dfrac{5}{6} \cdot 1 = $ _____   $1\left(-\dfrac{17}{20}\right) = $ _____

(d) $\left(-\dfrac{4}{5}\right)\left(-\dfrac{5}{4}\right) = $ _____   $7 \cdot \dfrac{1}{7} = $ _____

*Solve each application problem. Write all answers in lowest terms.*

**37.** When installing cabinets, Pam Phelps must be certain that the proper type and size of mounting hardware is used. Find the total length of the bolt shown.

**38.** How much fencing will be needed to enclose this rectangular wildflower preserve?

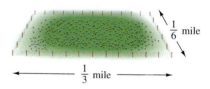

**39.** The owner of Racy's Feed Store ordered $\frac{1}{3}$ cubic yard of corn, $\frac{3}{8}$ cubic yard of oats, and $\frac{1}{4}$ cubic yard of washed medium mesh gravel. How many cubic yards of material were ordered?

**40.** A flower grower purchased $\frac{9}{10}$ acre of land one year and $\frac{3}{10}$ acre the next year. She then sold $\frac{7}{10}$ acre of land. How much land does she now have?

**41.** A forester planted $\frac{5}{12}$ acre in seedlings in the morning and $\frac{11}{12}$ acre in the afternoon. The next day, $\frac{7}{12}$ acre of seedlings were destroyed by a brush fire. How many acres of seedlings remained?

**42.** Adrian Ortega drives a tanker for the British Petroleum Company. He leaves the refinery with his tanker filled to $\frac{7}{8}$ of capacity. If he delivers $\frac{1}{4}$ of the tanker's capacity at the first stop and $\frac{1}{3}$ of the tanker's capacity at the second stop, find the fraction of the tanker's capacity remaining.

*The circle graph shows the fraction of U.S. workers who used various methods to learn about computers. Use the graph to answer Exercises 43–46.*

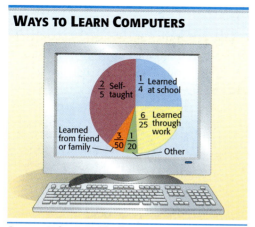

*Source:* John J. Heldrich Center for Workforce Development.

**43.** What fraction of workers are self-taught or learned from friends or family?

**44.** What fraction of workers learned at school or through work?

**45.** What is the difference in the fraction of workers who are self-taught and those who learned at work?

**46.** What is the difference in the fraction of workers who are self-taught and those who learned at school?

*Refer to the circle graph to answer Exercises 47–52.*

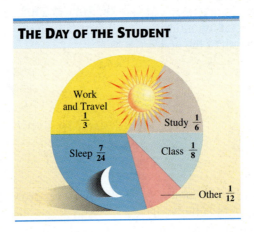

47. What fraction of the day was spent in class and study? How many hours is that?

48. What fraction of the day was spent in work and travel and sleep? How many hours is that?

49. In which activity was the greatest amount of time spent? How many hours did this activity take?

50. In which activity was the least amount of time spent? How many hours did this activity take?

51. How much more of the day was spent sleeping than studying? Write your answer as a fraction of the day.

52. How much more of the day was spent working and traveling than in class? Write your answer as a fraction of the day.

53. A hazardous waste dump site will require $\frac{7}{8}$ mile of security fencing. The site has four sides with three of the sides measuring $\frac{1}{4}$ mile, $\frac{1}{6}$ mile, and $\frac{3}{8}$ mile. Find the length of the fourth side.

54. Chakotay is fitting a turquoise stone into a bear claw pendant. Find the diameter of the hole in the pendant. (The diameter is the distance across the center of the hole.)

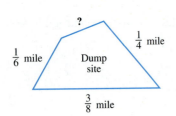

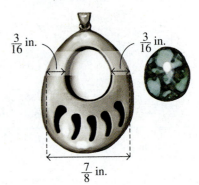

# 4.5 Problem Solving: Mixed Numbers and Estimating

**OBJECTIVES**

1. Identify mixed numbers and graph them on a number line.
2. Rewrite mixed numbers as improper fractions, or the reverse.
3. Estimate the answer and multiply or divide mixed numbers.
4. Estimate the answer and add or subtract mixed numbers.
5. Solve application problems containing mixed numbers.

**1** **Identify mixed numbers and graph them on a number line.** When a fraction and a whole number are written together, the result is a **mixed number**. For example, the mixed number

$$3\tfrac{1}{2} \quad \text{represents} \quad 3 + \tfrac{1}{2}$$

or 3 wholes and $\tfrac{1}{2}$ of a whole. Read $3\tfrac{1}{2}$ as "three and one half."

One common use of mixed numbers is to measure things. Examples are shown below.

Juan worked $5\tfrac{1}{2}$ hours.   The box weighs $2\tfrac{3}{4}$ pounds.

The park is $1\tfrac{7}{10}$ miles long.   Add $1\tfrac{2}{3}$ cups of flour.

### Example 1   Illustrating a Mixed Number with a Diagram and a Number Line

As this diagram shows, the mixed number $3\tfrac{1}{2}$ is equivalent to the improper fraction $\tfrac{7}{2}$.

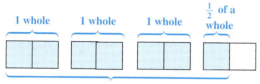

We can also use a number line to show mixed numbers, as in this graph of $3\tfrac{1}{2}$ and $-3\tfrac{1}{2}$.

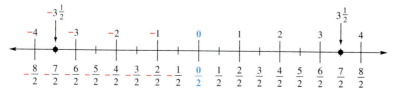

The number line shows the following.

$3\tfrac{1}{2}$   is equivalent to   $\tfrac{7}{2}$

$-3\tfrac{1}{2}$   is equivalent to   $-\tfrac{7}{2}$

**NOTE**

$3\tfrac{1}{2}$ represents $3 + \tfrac{1}{2}$.

$-3\tfrac{1}{2}$ represents $-3 + (-\tfrac{1}{2})$, which can also be written as $-3 - \tfrac{1}{2}$.

In algebra we usually work with the improper fraction form of mixed numbers, especially for negative mixed numbers. However, positive mixed numbers are frequently used in daily life, so it's important to know how to work with them. For example, we usually say $3\tfrac{1}{2}$ inches rather than $\tfrac{7}{2}$ inches.

**Work Problem 1 at the Side.**

**1** (a) Use these diagrams to write $1\tfrac{2}{3}$ as an improper fraction.

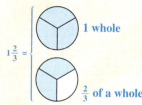

(b) Now graph $1\tfrac{2}{3}$ and $-1\tfrac{2}{3}$ on this number line.

(c) Use these diagrams to write $2\tfrac{1}{4}$ as an improper fraction.

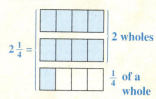

(d) Now graph $2\tfrac{1}{4}$ and $-2\tfrac{1}{4}$ on this number line.

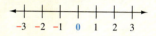

**ANSWERS**

1. (a) $\tfrac{5}{3}$    (b)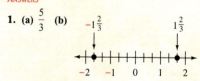

(c) $\tfrac{9}{4}$    (d)

**Chapter 4** Rational Numbers: Positive and Negative Fractions

❷ Write each mixed number as an equivalent improper fraction.

(a) $3\frac{2}{3}$

(b) $4\frac{7}{10}$

(c) $5\frac{3}{4}$

(d) $8\frac{5}{6}$

❷ **Rewrite mixed numbers as improper fractions, or the reverse.** You can use the following steps to write $3\frac{1}{2}$ as an improper fraction without drawing a diagram or a number line.

*Step 1* Multiply 2 times 3 and add 1 to the product.

$$3\frac{1}{2} \quad 2 \cdot 3 = 6 \quad \text{Then } 6 + 1 = 7.$$

*Step 2* Use 7 (from Step 1) as the numerator and 2 as the denominator.

$$3\frac{1}{2} = \frac{7}{2} \leftarrow (2 \cdot 3) + 1$$
— Same denominator

To see why this method works, recall that $3\frac{1}{2}$ represents $3 + \frac{1}{2}$. Let's add $3 + \frac{1}{2}$.

$$3 + \frac{1}{2} = \frac{3}{1} + \frac{1}{2} = \frac{6}{2} + \frac{1}{2} = \frac{6+1}{2} = \frac{7}{2} \quad \left\{\begin{array}{l}\text{Same result}\\\text{as above}\end{array}\right.$$

Common denominator

In summary, use the following steps to *write a mixed number as an improper fraction*.

---

### Writing a Mixed Number as an Improper Fraction

*Step 1* **Multiply** the denominator of the fraction times the whole number and **add** the numerator of the fraction to the product.

*Step 2* Write the result of Step 1 as the **numerator** and keep the original **denominator**.

---

**Example 2** Writing a Mixed Number as an Improper Fraction

Write $7\frac{2}{3}$ as an improper fraction (numerator greater than denominator).

*Step 1* $7\frac{2}{3} \quad 3 \cdot 7 = 21 \quad \text{Then } 21 + 2 = 23.$

*Step 2* $7\frac{2}{3} = \frac{23}{3} \leftarrow (3 \cdot 7) + 2$
— Same denominator

**Work Problem** ❷ **at the Side.**

---

**ANSWERS**

2. (a) $\frac{11}{3}$ (b) $\frac{47}{10}$ (c) $\frac{23}{4}$ (d) $\frac{53}{6}$

We used *multiplication* for the first step in writing a mixed number as an improper fraction. To work in *reverse*, writing an improper fraction as a mixed number, we will use *division*. Recall that the fraction bar is a symbol for division.

### Writing an Improper Fraction as a Mixed Number

To write an *improper fraction* as a mixed number, divide the numerator by the denominator. The quotient is the whole number part (of the mixed number), the remainder is the numerator of the fraction part, and the denominator remains the same.

Always check to be sure that the fraction part of the mixed number is in lowest terms. Then the mixed number is in *simplest form*.

**Example 3** Writing Improper Fractions as Mixed Numbers

Write each improper fraction as an equivalent mixed number.

(a) $\dfrac{17}{5}$

Divide 17 by 5.

$$\begin{array}{r} 3 \\ 5\overline{)17} \\ 15 \\ \hline 2 \end{array}$$ ← Whole number part

← Remainder

The quotient **3** is the whole number part of the mixed number. The remainder **2** is the numerator of the fraction, and the denominator stays as **5**.

$$\dfrac{17}{5} = 3\dfrac{2}{5}$$ ← Remainder

Same denominator

Let's look at a drawing of $\frac{17}{5}$ to check our work.

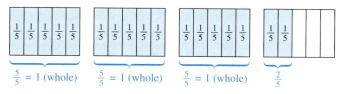

$\frac{5}{5} = 1$ (whole)  $\frac{5}{5} = 1$ (whole)  $\frac{5}{5} = 1$ (whole)  $\frac{2}{5}$

**Continued on Next Page**

❸ Write each improper fraction as an equivalent mixed number in simplest form.

(a) $\dfrac{5}{2}$

(b) $\dfrac{14}{4}$

(c) $\dfrac{33}{5}$

(d) $\dfrac{58}{10}$

❹ Round each mixed number to the nearest whole number.

(a) $2\dfrac{3}{4}$

(b) $6\dfrac{3}{8}$

(c) $4\dfrac{2}{3}$

(d) $1\dfrac{7}{10}$

(e) $3\dfrac{1}{2}$

(f) $5\dfrac{4}{9}$

In other words,

$$\dfrac{17}{5} = \dfrac{5}{5} + \dfrac{5}{5} + \dfrac{5}{5} + \dfrac{2}{5}$$
$$= 1 + 1 + 1 + \dfrac{2}{5}$$
$$= 3 + \dfrac{2}{5}$$
$$= 3\dfrac{2}{5}$$

(b) $\dfrac{26}{4}$

Divide 26 by 4.

$$\begin{array}{r} 6 \\ 4\overline{)26} \\ 24 \\ \hline 2 \end{array}$$ so $\dfrac{26}{4} = 6\dfrac{2}{4} = 6\dfrac{1}{2}$ {Simplest form}

Write $\dfrac{2}{4}$ in lowest terms.

You could write $\dfrac{26}{4}$ in lowest terms first.

$$\dfrac{26}{4} = \dfrac{\overset{1}{\cancel{2}} \cdot 13}{\underset{1}{\cancel{2}} \cdot 2} = \dfrac{13}{2}$$ Then $\begin{array}{r} 6 \\ 2\overline{)13} \\ 12 \\ \hline 1 \end{array}$ so $\dfrac{13}{2} = 6\dfrac{1}{2}$ {Same result as above}

**Work Problem ❸ at the Side.**

**3** Estimate the answer and multiply or divide mixed numbers. Once you have rewritten mixed numbers as improper fractions, you can use the steps you learned in **Section 4.3** to multiply and divide. However, it's a good idea to estimate the answer before you start any other work.

**Example 4** Rounding Mixed Numbers

To estimate answers, first round each mixed number to the *nearest whole number*. If the numerator is *half* of the denominator or *more*, round up the whole number part. If the numerator is *less* than half the denominator, leave the whole number as it is.

(a) Round $1\dfrac{5}{8}$ ← 5 is more than 4.  $1\dfrac{5}{8}$ rounds up to 2
← Half of 8 is 4.

(b) Round $3\dfrac{2}{5}$ ← 2 is less than $2\dfrac{1}{2}$.  $3\dfrac{2}{5}$ rounds to 3
← Half of 5 is $2\dfrac{1}{2}$.

**Work Problem ❹ at the Side.**

---

ANSWERS

3. (a) $2\dfrac{1}{2}$  (b) $3\dfrac{1}{2}$  (c) $6\dfrac{3}{5}$  (d) $5\dfrac{4}{5}$

4. (a) 3  (b) 6  (c) 5  (d) 2  (e) 4  (f) 5

Section 4.5 Problem Solving: Mixed Numbers and Estimating   251

### Multiplying and Dividing Mixed Numbers

*Step 1* **Rewrite** each mixed number as an improper fraction.

*Step 2* **Multiply** or **divide** the improper fractions.

*Step 3* Write the answer in lowest terms and change it to a mixed number or whole number where possible. This step gives you an answer that is in **simplest form**.

**Example 5** Estimating the Answer and Multiplying Mixed Numbers

First, round the numbers and estimate each answer. Then find the exact answer. Write exact answers in simplest form.

(a) $2\frac{1}{2} \cdot 3\frac{1}{5}$

Estimate the answer by rounding the mixed numbers.

$2\frac{1}{2}$ rounds to 3   and   $3\frac{1}{5}$ rounds to 3

$3 \cdot 3 = 9$ ← Estimated answer

To find the exact answer, first rewrite each mixed number as an improper fraction.

Step 1   $2\frac{1}{2} = \frac{5}{2}$   and   $3\frac{1}{5} = \frac{16}{5}$

Next, multiply.

$$2\frac{1}{2} \cdot 3\frac{1}{5} = \frac{5}{2} \cdot \frac{16}{5} = \frac{5 \cdot 2 \cdot 8}{2 \cdot 5} = \frac{8}{1} = 8 \quad \begin{cases}\text{Simplest}\\ \text{form}\end{cases}$$

The estimate was 9, so an exact answer of 8 is reasonable.

(b) $\left(3\frac{5}{8}\right)\left(4\frac{4}{5}\right)$

First, round each mixed number and estimate the answer.

$3\frac{5}{8}$ rounds to 4   and   $4\frac{4}{5}$ rounds to 5

$4 \cdot 5 = 20$ ← Estimated answer

Now find the exact answer.

$$\left(3\frac{5}{8}\right)\left(4\frac{4}{5}\right) = \left(\frac{29}{8}\right)\left(\frac{24}{5}\right) = \frac{29 \cdot 3 \cdot 8}{8 \cdot 5} = \frac{87}{5} = 17\frac{2}{5} \quad \begin{cases}\text{Simplest}\\ \text{form}\end{cases}$$

The exact answer of $17\frac{2}{5}$ is close to the estimated answer of 20.

**Work Problem ❺ at the Side.**

❺ First, round the numbers and estimate each answer. Then find the exact answer. Write exact answers in simplest form.

(a) $2\frac{1}{4} \cdot 7\frac{1}{3}$

_____ • _____ = _____ estimate

(b) $\left(4\frac{1}{2}\right)\left(1\frac{2}{3}\right)$

_____ • _____ = _____ estimate

(c) $3\frac{3}{5} \cdot 4\frac{4}{9}$

_____ • _____ = _____ estimate

(d) $3\frac{1}{5} \cdot 5\frac{3}{8}$

_____ • _____ = _____ estimate

**ANSWERS**

5. (a) *Estimate:* $2 \cdot 7 = 14$; *Exact:* $16\frac{1}{2}$

   (b) *Estimate:* $5 \cdot 2 = 10$; *Exact:* $7\frac{1}{2}$

   (c) *Estimate:* $4 \cdot 4 = 16$; *Exact:* 16

   (d) *Estimate:* $3 \cdot 5 = 15$; *Exact:* $17\frac{1}{5}$

**252** Chapter 4 Rational Numbers: Positive and Negative Fractions

**6** First, round the numbers and estimate each answer. Then find the exact answer. Write exact answers in simplest form.

(a) $6\frac{1}{4} \div 3\frac{1}{3}$

$\underline{\phantom{00}} \div \underline{\phantom{00}} = \underline{\phantom{00}}$ estimate

(b) $3\frac{3}{8} \div 2\frac{4}{7}$

$\underline{\phantom{00}} \div \underline{\phantom{00}} = \underline{\phantom{00}}$ estimate

(c) $8 \div 5\frac{1}{3}$

$\underline{\phantom{00}} \div \underline{\phantom{00}} = \underline{\phantom{00}}$ estimate

(d) $4\frac{1}{2} \div 6$

$\underline{\phantom{00}} \div \underline{\phantom{00}} = \underline{\phantom{00}}$ estimate

**ANSWERS**

6. (a) Estimate: $6 \div 3 = 2$; Exact: $1\frac{7}{8}$

(b) Estimate: $3 \div 3 = 1$; Exact: $1\frac{5}{16}$

(c) Estimate: $8 \div 5 = 1\frac{3}{5}$; Exact: $1\frac{1}{2}$

(d) Estimate: $5 \div 6 = \frac{5}{6}$; Exact: $\frac{3}{4}$

### Example 6 Estimating the Answer and Dividing Mixed Numbers

First, round the numbers and estimate each answer. Then find the exact answer. Write exact answers in simplest form.

(a) $3\frac{3}{5} \div 1\frac{1}{2}$

To estimate the answer, round each mixed number to the nearest whole number.

$3\frac{3}{5} \div 1\frac{1}{2}$

Rounded

$4 \div 2 = 2$ ← Estimate

To find the exact answer, first rewrite each mixed number as an improper fraction.

$3\frac{3}{5} \div 1\frac{1}{2} = \frac{18}{5} \div \frac{3}{2}$

Now rewrite the problem as multiplying by the reciprocal of $\frac{3}{2}$.

$\frac{18}{5} \div \frac{3}{2} = \frac{18}{5} \cdot \frac{2}{3} = \frac{\cancel{3} \cdot 6 \cdot 2}{5 \cdot \cancel{3}} = \frac{12}{5} = 2\frac{2}{5}$ { Simplest form

Reciprocals

The estimate was 2, so an exact answer of $2\frac{2}{5}$ is reasonable.

(b) $4\frac{3}{8} \div 5$

First, round the numbers and estimate the answer.

$4\frac{3}{8} \div 5$

Rounded

$4 \div 5$    Write $4 \div 5$ using a fraction bar. } $\frac{4}{5}$ ← Estimate

Now find the exact answer.

Write 5 as $\frac{5}{1}$.

$4\frac{3}{8} \div 5 = \frac{35}{8} \div \frac{5}{1} = \frac{35}{8} \cdot \frac{1}{5} = \frac{\cancel{5} \cdot 7 \cdot 1}{8 \cdot \cancel{5}} = \frac{7}{8}$ { Simplest form

Reciprocals

The estimate was $\frac{4}{5}$, so an exact answer of $\frac{7}{8}$ is reasonable. They are both less than 1.

**Work Problem** 6 **at the Side.**

Section 4.5 Problem Solving: Mixed Numbers and Estimating   253

**4** **Estimate the answer and add or subtract mixed numbers.** The steps you learned for adding and subtracting fractions in **Section 4.4** will also work for mixed numbers: Just rewrite the mixed numbers as equivalent improper fractions. Again, it is a good idea to estimate the answer before you start any other work.

### Example 7  Estimating the Answer and Adding or Subtracting Mixed Numbers

First, estimate each answer. Then add or subtract to find the exact answer. Write exact answers in simplest form.

(a) $2\frac{3}{8} + 3\frac{3}{4}$

To estimate the answer, round each mixed number to the nearest whole number.

$$2\frac{3}{8} + 3\frac{3}{4}$$
$$\downarrow \quad \downarrow$$
$$2 + 4 = 6 \leftarrow \text{Estimate}$$

To find the exact answer, first rewrite each mixed number as an equivalent improper fraction.

$$2\frac{3}{8} + 3\frac{3}{4} = \frac{19}{8} + \frac{15}{4}$$

You can't add fractions until they have a common denominator. The LCD for $\frac{19}{8}$ and $\frac{15}{4}$ is 8. Rewrite $\frac{15}{4}$ as an equivalent fraction with a denominator of 8.

$$\frac{19}{8} + \frac{15}{4} = \frac{19}{8} + \frac{30}{8} = \frac{19 + 30}{8} = \frac{49}{8} = 6\frac{1}{8} \quad \begin{cases} \text{Simplest} \\ \text{form} \end{cases}$$

Common denominator

The estimate was 6, so an exact answer of $6\frac{1}{8}$ is reasonable.

(b) $4\frac{2}{3} - 2\frac{4}{5}$

Round each number and estimate the answer.

$$4\frac{2}{3} - 2\frac{4}{5}$$
$$\downarrow \quad \downarrow$$
$$5 - 3 = 2 \leftarrow \text{Estimate}$$

To find the exact answer, rewrite the mixed numbers as improper fractions and subtract.

$$4\frac{2}{3} - 2\frac{4}{5} = \frac{14}{3} - \frac{14}{5} = \frac{70}{15} - \frac{42}{15} = \frac{70 - 42}{15} = \frac{28}{15} = 1\frac{13}{15} \quad \begin{cases} \text{Simplest} \\ \text{form} \end{cases}$$

LCD is 15.

The estimate was 2, so an exact answer of $1\frac{13}{15}$ is reasonable.

— Continued on Next Page

**7** First, round the numbers and estimate each answer. Then add or subtract to find the exact answer.

(a) $5\dfrac{1}{3} - 2\dfrac{5}{6}$

$\underline{\phantom{0}} - \underline{\phantom{0}} = \underline{\phantom{0}}$ estimate

(b) $\dfrac{3}{4} + 3\dfrac{1}{8}$

$\underline{\phantom{0}} + \underline{\phantom{0}} = \underline{\phantom{0}}$ estimate

(c) $6 - 3\dfrac{4}{5}$

$\underline{\phantom{0}} - \underline{\phantom{0}} = \underline{\phantom{0}}$ estimate

**ANSWERS**

7. (a) $5 - 3 = 2; \; 2\dfrac{1}{2}$

  (b) $1 + 3 = 4; \; 3\dfrac{7}{8}$

  (c) $6 - 4 = 2; \; 2\dfrac{1}{5}$

---

(c) $5 - 1\dfrac{3}{8}$

$5 - 1\dfrac{3}{8}$

$5 - 1 = 4$ ← Estimate

Write 5 as $\dfrac{5}{1}$.

$5 - 1\dfrac{3}{8} = \dfrac{5}{1} - \dfrac{11}{8} = \dfrac{40}{8} - \dfrac{11}{8} = \dfrac{29}{8} = 3\dfrac{5}{8}$ {Simplest form

LCD is 8.

The estimate was 4, so an exact answer of $3\dfrac{5}{8}$ is reasonable.

**Work Problem 7 at the Side.**

**NOTE**

In some situations the method of rewriting mixed numbers as improper fractions may result in very large numerators. Consider this example.

Last year Hue's child was $48\dfrac{3}{8}$ in. tall. This year the child is $51\dfrac{1}{4}$ in. tall. How much has the child grown?

First, estimate the answer by rounding each mixed number to the nearest whole number.

$51\dfrac{1}{4} - 48\dfrac{3}{8}$

$51 - 48 = 3$ in. ← Estimate

To find the exact answer, rewrite the mixed numbers as improper fractions.

Rewrite $\dfrac{205}{4}$ as $\dfrac{410}{8}$.

$51\dfrac{1}{4} - 48\dfrac{3}{8} = \dfrac{205}{4} - \dfrac{387}{8} = \dfrac{410}{8} - \dfrac{387}{8} = \dfrac{410 - 387}{8} = \dfrac{23}{8} = 2\dfrac{7}{8}$ in.

LCD is 8.

You can also use the fraction key $\boxed{a^{b/c}}$ on your *scientific* calculator to solve this problem.

To enter $51\dfrac{1}{4}$, press

51 $\boxed{a^{b/c}}$ 1 $\boxed{a^{b/c}}$ 4 $\boxed{a^{b/c}}$. The display shows $\boxed{51\_1\_4}$.

↑ Whole number  ↑ Numerator  ↑ Denominator

Then press $\boxed{-}$ 48 $\boxed{a^{b/c}}$ 3 $\boxed{a^{b/c}}$ 8 $\boxed{=}$. The display shows $\boxed{2\_7\_8}$.

↑ Subtract.   $48\dfrac{3}{8}$   $2\dfrac{7}{8}$

Either way the exact answer is $2\dfrac{7}{8}$ in., which is close to the estimate of 3 in.

Another efficient method for handling large mixed numbers is to rewrite them in decimal form. You will learn how to do that in Chapter 5.

Section 4.5 Problem Solving: Mixed Numbers and Estimating 255

**5** Solve application problems containing mixed numbers. Rounding mixed numbers to the nearest whole number can also help you decide whether to solve an application problem by adding, subtracting, multiplying, or dividing.

### Example 8 Solving Application Problems with Mixed Numbers

First, estimate the answer to each application problem. Then find the exact answer.

**(a)** Gary needs to haul $15\frac{3}{4}$ **tons** of sand to a construction site. His truck can carry $2\frac{1}{4}$ **tons**. How many trips will he need to make?

First, round each mixed number to the nearest whole number.

$$15\frac{3}{4} \text{ rounds to } 16 \quad \text{and} \quad 2\frac{1}{4} \text{ rounds to } 2$$

Now read the problem again, *using the rounded numbers.*

Gary needs to haul **16 tons** of sand to a construction site. His truck can carry **2 tons**. How many trips will he need to make?

Using the rounded numbers in the problem makes it easier to see that you need to *divide*.

$$16 \div 2 = 8 \text{ trips} \leftarrow \text{Estimate}$$

To find the exact answer, use the original mixed numbers and divide.

$$15\frac{3}{4} \div 2\frac{1}{4} = \frac{63}{4} \div \frac{9}{4} = \frac{63}{4} \cdot \frac{4}{9} = \frac{7 \cdot 9 \cdot 4}{4 \cdot 9} = \frac{7}{1} = 7 \quad \{\text{Simplest form}$$

Reciprocals

Gary needs to make 7 trips to haul all the sand. This result is close to the estimate of 8 trips.

**(b)** Zenitia worked $3\frac{5}{6}$ **hours** on Monday and $6\frac{1}{2}$ **hours** on Tuesday. How much longer did she work on Tuesday than on Monday?

First, round each mixed number to the nearest whole number.

$$3\frac{5}{6} \text{ rounds to } 4 \quad \text{and} \quad 6\frac{1}{2} \text{ rounds to } 7$$

Now read the problem again, *using the rounded numbers.*

Zenitia worked **4 hours** on Monday and **7 hours** on Tuesday. How much longer did she work on Tuesday than on Monday?

Using the rounded numbers in the problem makes it easier to see that you need to *subtract*.

$$7 - 4 = 3 \text{ hours} \leftarrow \text{Estimate}$$

To find the exact answer, use the original mixed numbers and subtract.

Write answer in simplest form.

$$6\frac{1}{2} - 3\frac{5}{6} = \frac{13}{2} - \frac{23}{6} = \frac{39}{6} - \frac{23}{6} = \frac{39-23}{6} = \frac{16}{6} = 2\frac{4}{6} = 2\frac{2}{3}$$

LCD is 6.

Zenitia worked $2\frac{2}{3}$ hours longer on Tuesday. This result is close to the estimate of 3 hours.

═══════ Work Problem **8** at the Side.

**8** First, round the numbers and estimate the answer to each problem. Then find the exact answer.

**(a)** Richard's son grew $3\frac{5}{8}$ inches last year and $2\frac{1}{4}$ inches this year. How much has his height increased over the two years?

*Estimate:*

*Exact:*

**(b)** Ernestine used $2\frac{1}{2}$ packages of chocolate chips in her cookie recipe. Each package has $5\frac{1}{2}$ ounces of chips. How many ounces of chips did she use in the recipe?

*Estimate:*

*Exact:*

**Answers**
**8. (a)** $4 + 2 = 6$ inches; $5\frac{7}{8}$ inches

**(b)** $3 \cdot 6 = 18$ ounces; $13\frac{3}{4}$ ounces

# Focus on Real-Data Applications

## Recipes

The side of the corn starch box shown below has useful recipes for Fun-Time Dough and for Great Gravy. Suppose you have made the recipes before and know from experience that 1 pound of Fun-Time Dough is enough for three children.

**ARGO CORN STARCH**
**FAVORITE RECIPES**

**Fun-Time Dough**

$1\frac{1}{2}$ cups Argo Corn Starch
$\frac{1}{2}$ cup flour
2 cups water
2 tsp cream of tartar
1 cup salt
1 T. vegetable oil

Mix all ingredients together in saucepan. Cook over medium heat, stirring constantly, until mixture gathers on the stirring spoon and forms dough. This will take about 6 minutes. Dump onto waxed paper until cool enough to handle and knead to form a pliable mass. Store in covered container or plastic bag. Food coloring may be added to make different colors.
Makes about 2 lbs. of Fun-Time Dough.

**Great Gravy**

3 T. bacon fat or meat drippings
2 T. Argo Corn Starch
$1\frac{1}{2}$ cups water
$\frac{3}{4}$ tsp. salt
$\frac{1}{8}$ tsp. pepper

Blend fat and Argo Corn Starch over low heat until it is a rich brown color, stirring constantly. Gradually add water, salt, and pepper. Heat to boiling over direct heat and then boil gently 2 minutes, stirring constantly.
Makes $1\frac{1}{2}$ cups.

—— Satisfaction Guaranteed ——

1. If you make the recipe as written, you will have enough Fun-Time Dough for how many children?

2. Suppose you work in a day care center. How many pounds of Fun-Time Dough will you need for nine children?

3. If you double the recipe, you would have 4 pounds of dough. By what number should you multiply each ingredient amount to make 3 pounds of dough?

4. Fill in the blanks with the ingredient amounts needed to make 3 pounds of Fun-Time Dough. Show your work.

   Corn starch: _____   Flour: _____

   Water: _____   Cream of tartar: _____

   Salt: _____   Vegetable oil: _____

You decide to make Great Gravy for Thanksgiving dinner.

5. How much gravy does the recipe make?

6. Suppose you decide to double the recipe. By what factor will you change the ingredient amounts?

7. You did not make enough! Suppose you decide to halve the recipe to make more gravy. Now, by what factor will you change the ingredient amounts?

8. If you need 4 cups of gravy, by what factor must you multiply each ingredient amount? Explain how you determined your answer.

## 4.5 Exercises

*Graph the mixed numbers or improper fractions on the number line. See Example 1.*

1. Graph $2\frac{1}{3}$ and $-2\frac{1}{3}$.

2. Graph $1\frac{3}{4}$ and $-1\frac{3}{4}$.

3. Graph $\frac{3}{2}$ and $-\frac{3}{2}$.

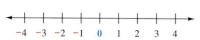

4. Graph $\frac{11}{3}$ and $-\frac{11}{3}$.

*Write each mixed number as an improper fraction. See Example 2.*

5. $4\frac{1}{2}$
6. $2\frac{1}{4}$
7. $-1\frac{3}{5}$
8. $-1\frac{5}{6}$
9. $2\frac{3}{8}$
10. $3\frac{4}{9}$
11. $-5\frac{7}{10}$
12. $-4\frac{5}{7}$
13. $10\frac{11}{15}$
14. $12\frac{9}{11}$

*Write each improper fraction as a mixed number in simplest form. See Example 3.*

15. $\frac{13}{3}$
16. $\frac{11}{2}$
17. $-\frac{10}{4}$
18. $-\frac{14}{5}$
19. $\frac{22}{6}$
20. $\frac{28}{8}$
21. $-\frac{51}{9}$
22. $-\frac{44}{10}$
23. $\frac{188}{16}$
24. $\frac{200}{15}$

*First, round the mixed numbers to the nearest whole number and estimate each answer. Then find the exact answer. Write exact answers in simplest form. See Examples 4–7.*

25. *Exact:*
$2\frac{1}{4} \cdot 3\frac{1}{2}$
*Estimate:*
___ • ___ = ___

26. *Exact:*
$\left(1\frac{1}{2}\right)\left(3\frac{3}{4}\right)$
*Estimate:*
___ • ___ = ___

27. *Exact:*
$3\frac{1}{4} \div 2\frac{5}{8}$
*Estimate:*
___ ÷ ___ = ___

28. *Exact:*
$2\frac{1}{4} \div 1\frac{1}{8}$
*Estimate:*
___ ÷ ___ = ___

29. *Exact:*
$3\frac{2}{3} + 1\frac{5}{6}$
*Estimate:*
___ + ___ = ___

30. *Exact:*
$4\frac{4}{5} + 2\frac{1}{3}$
*Estimate:*
___ + ___ = ___

**31.** *Exact:*

$$4\frac{1}{4} - \frac{7}{12}$$

*Estimate:*

____ − ____ = ____

**32.** *Exact:*

$$10\frac{1}{3} - 6\frac{5}{6}$$

*Estimate:*

____ − ____ = ____

**33.** *Exact:*

$$5\frac{2}{3} \div 6$$

*Estimate:*

____ ÷ ____ = ____

**34.** *Exact:*

$$1\frac{7}{8} \div 6\frac{1}{4}$$

*Estimate:*

____ ÷ ____ = ____

**35.** *Exact:*

$$8 - 1\frac{4}{5}$$

*Estimate:*

____ − ____ = ____

**36.** *Exact:*

$$7 - 3\frac{3}{10}$$

*Estimate:*

____ − ____ = ____

*Find the perimeter and the area of each square or rectangle. Write all answers in simplest form.*

**37.**

$1\frac{3}{4}$ in.

$1\frac{3}{4}$ in.

$1\frac{3}{4}$ in.

**38.**

$2\frac{1}{3}$ ft

$2\frac{1}{3}$ ft

**39.**

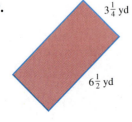

$3\frac{1}{4}$ yd

$6\frac{1}{2}$ yd

**40.**

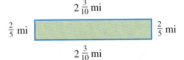

$2\frac{3}{10}$ mi

$\frac{2}{5}$ mi

$\frac{2}{5}$ mi

$2\frac{3}{10}$ mi

*First, estimate the answer to each application problem. Then find the exact answer. Write all answers in simplest form. See Example 8.*

**41.** A carpenter has two pieces of oak trim. One piece of trim is $12\frac{1}{2}$ ft long and the other is $8\frac{2}{3}$ ft long. How many feet of oak trim does he have in all?

*Estimate:*

*Exact:*

**42.** On Monday, $5\frac{3}{4}$ tons of cans were recycled, and, on Tuesday, $9\frac{3}{5}$ tons were recycled. How many tons were recycled on these two days?

*Estimate:*

*Exact:*

**43.** The directions for mixing an insect spray say to use $1\frac{3}{4}$ ounces of chemical in each gallon of water. How many ounces of chemical should be mixed with $12\frac{1}{2}$ gallons of water?

*Estimate:*

*Exact:*

**44.** Shirley Cicero wants to make 16 holiday wreaths to sell at the craft fair. Each wreath requires $2\frac{1}{4}$ yd of ribbon. How many yards does she need?

*Estimate:*

*Exact:*

**45.** The Boy Scout troop has volunteered to pick up trash along a 4-mile stretch of highway. So far they have done $1\frac{7}{10}$ miles. How much do they have left to do?

*Estimate:*

*Exact:*

**46.** The gas tank on a Jeep Cherokee has a capacity of $21\frac{3}{8}$ gallons. Scott started with a full tank and then used $8\frac{1}{2}$ gallons of gasoline. Find the number of gallons that remain.

*Estimate:*

*Exact:*

**47.** Suppose that a dress requires $2\frac{3}{4}$ yd of material. How much material would be needed for 7 dresses?

*Estimate:*

*Exact:*

**48.** A cookie recipe uses $\frac{2}{3}$ cup brown sugar. How much brown sugar is needed to make $2\frac{1}{2}$ times the original recipe?

*Estimate:*

*Exact:*

**49.** A landscaper has $9\frac{5}{8}$ cubic yards of peat moss in a truck. If she unloads $1\frac{1}{2}$ cubic yards at the first stop and 3 cubic yards at the second stop, how much peat moss remains in the truck?

*Estimate:*

*Exact:*

**50.** Marv bought 10 yd of Italian silk fabric. He used $3\frac{7}{8}$ yd to make a jacket. How much fabric is left for other sewing projects?

*Estimate:*

*Exact:*

**51.** Melissa worked $18\frac{3}{4}$ hours over the last five days. If she worked the same amount each day, how long was she at work each day?

*Estimate:*

*Exact:*

**52.** Michael is cutting a $10\frac{1}{2}$ ft board into shelves for a bookcase. Each shelf will be $1\frac{3}{4}$ ft long. How many shelves can he cut?

*Estimate:*

*Exact:*

**53.** Find the length of the arrow shaft.

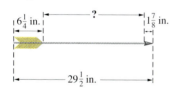

*Estimate:*

*Exact:*

**54.** Find the length of the indented section on this board.

*Estimate:*

*Exact:*

*First, estimate the answer to each application problem. Then use your calculator to find the exact answer.*

**55.** A craftsperson must attach a lead strip around all four sides of a stained glass window before it is installed. Find the length of lead stripping needed for the window shown.

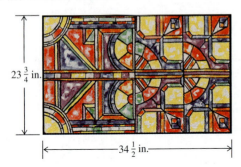

Estimate:

Exact:

**56.** To complete a custom order, Zak Morten of Home Depot must find the number of inches of brass trim needed to go around the four sides of the lamp base plate shown. Find the length of brass trim needed.

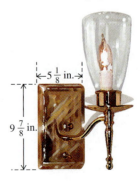

Estimate:

Exact:

**57.** A fishing boat anchor requires $10\frac{3}{8}$ pounds of steel. Find the number of anchors that can be manufactured with 25,730 pounds of steel.

Estimate:

Exact:

**58.** Each apartment requires $62\frac{1}{2}$ square yards of carpet. Find the number of apartments that can be carpeted with 6750 square yards of carpet.

Estimate:

Exact:

**59.** Claire and Deb create custom hat bands that people can put around the crowns of their hats. The finished bands are the lengths shown in the table. The strip of fabric for each band must include the finished length plus an extra $\frac{3}{4}$ in. for the seam.

| Band Size | Finished Length |
|---|---|
| Small | $21\frac{7}{8}$ in. |
| Medium | $22\frac{5}{8}$ in. |
| Large | $23\frac{1}{2}$ in. |

What length of fabric strip is needed to make 4 small bands, 5 medium bands, and 3 large bands, including the seam allowance?

Estimate:

Exact:

**60.** Three sides of a parking lot are $108\frac{1}{4}$ ft, $162\frac{3}{8}$ ft, and $143\frac{1}{2}$ ft. If the distance around the lot is $518\frac{3}{4}$ ft, find the length of the fourth side.

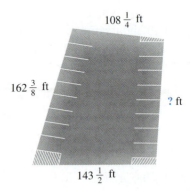

Estimate:

Exact:

# Summary Exercises on FRACTIONS

1. Write fractions that represent the shaded and unshaded portion of each figure.

   (a) 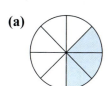   (b)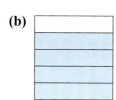

2. Graph $\frac{2}{3}$ and $-\frac{2}{3}$ on the number line.

3. Rewrite each fraction with the indicated denominator.

   (a) $-\dfrac{4}{5} = -\dfrac{\phantom{30}}{30}$   (b) $\dfrac{2}{7} = \dfrac{\phantom{14}}{14}$

4. Simplify.

   (a) $\dfrac{15}{15}$   (b) $-\dfrac{24}{6}$   (c) $\dfrac{9}{1}$

5. Write the prime factorization of each number.

   (a) 72   (b) 105

6. Write each fraction in lowest terms.

   (a) $\dfrac{24}{30}$   (b) $\dfrac{175}{200}$

*Simplify.*

7. $\left(-\dfrac{3}{4}\right)\left(-\dfrac{2}{3}\right)$

8. $-\dfrac{7}{8} + \dfrac{2}{3}$

9. $\dfrac{7}{16} + \dfrac{5}{8}$

10. $\dfrac{5}{8} \div \dfrac{3}{4}$

11. $\dfrac{2}{3} - \dfrac{4}{5}$

12. $\dfrac{7}{12}\left(-\dfrac{9}{14}\right)$

13. $-21 \div \left(-\dfrac{3}{8}\right)$

14. $\dfrac{7}{8} - \dfrac{5}{12}$

15. $-\dfrac{35}{45} \div \dfrac{10}{15}$

16. $-\dfrac{5}{6} - \dfrac{3}{4}$

17. $\dfrac{7}{12} + \dfrac{5}{6} + \dfrac{2}{3}$

18. $\dfrac{5}{8}$ of 56

*First, round the numbers and estimate each answer. Then find the exact answer.*

19. *Exact:*

    $4\dfrac{3}{4} + 2\dfrac{5}{6}$

    *Estimate:*

    ___ + ___ = ___

20. *Exact:*

    $2\dfrac{2}{9} \cdot 5\dfrac{1}{7}$

    *Estimate:*

    ___ • ___ = ___

21. *Exact:*

    $6 - 2\dfrac{7}{10}$

    *Estimate:*

    ___ − ___ = ___

22. *Exact:*

    $1\dfrac{3}{5} \div 3\dfrac{1}{2}$

    *Estimate:*

    ___ ÷ ___ = ___

23. *Exact:*

    $4\dfrac{2}{3} \div 1\dfrac{1}{6}$

    *Estimate:*

    ___ ÷ ___ = ___

24. *Exact:*

    $3\dfrac{5}{12} - \dfrac{3}{4}$

    *Estimate:*

    ___ − ___ = ___

*Solve each application problem. Write all answers in simplest form.*

25. When installing cabinets, Cecil Feathers must be certain that the proper type and size of mounting screw is used.

    (a) Find the total length of the screw shown.

    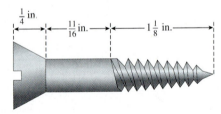

    (b) If the screw is put into a board that is $1\frac{3}{4}$ in. thick, how much of the screw will stick out the back of the board?

26. Find the perimeter and the area of this postage stamp.

    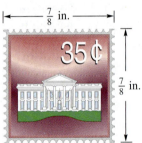

27. A batch of cookies requires $\frac{3}{4}$ pound of chocolate chips. If you have nine pounds of chocolate chips, how many batches of cookies can you make?

28. The Municipal Utility District says that the cost of operating a hair dryer is $\frac{1}{5}$¢ per minute. Find the cost of operating the hair dryer for a half hour.

29. A survey asked 1500 adults if UFOs (unidentified flying objects) are real. The circle graph shows the fraction of the adults who gave each answer. How many adults gave each answer?

    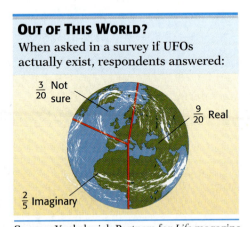

    **Source:** Yankelovich Partners for *Life* magazine.

30. Find the diameter of the hole in the mounting bracket shown. (The diameter is the distance across the center of the hole.) Then find the perimeter of the bracket.

    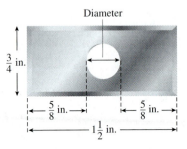

31. A bottle of contact lens daily cleaning solution holds $\frac{2}{3}$ fluid ounce. (*Source:* Alcon Laboratories, Inc.) How many bottles can be filled with $15\frac{1}{3}$ fluid ounces?

32. A home builder bought two parcels of land that were $5\frac{7}{8}$ acres and $10\frac{3}{4}$ acres. She is setting aside $2\frac{7}{8}$ acres for a park and using the rest for $1\frac{1}{4}$ acre home lots. How many lots will be in the development?

## 4.6 EXPONENTS, ORDER OF OPERATIONS, AND COMPLEX FRACTIONS

**OBJECTIVES**
1. Use exponents with fractions.
2. Use the order of operations with fractions.
3. Simplify complex fractions.

**1** **Use exponents with fractions.** We have used exponents as a quick way to write repeated multiplication of integers and variables. Here are two examples as a review.

$$(-3)^2 = \underbrace{(-3) \cdot (-3)}_{\text{Two factors of } -3} = 9 \quad \text{and} \quad x^3 = \underbrace{x \cdot x \cdot x}_{\text{Three factors of } x}$$

Base — Exponent

The meaning of an exponent remains the same when a fraction is the base.

### Example 1   Using Exponents with Fractions

Simplify.

**(a)** $\left(-\dfrac{1}{2}\right)^3$

The exponent indicates that there are three factors of $-\dfrac{1}{2}$.

$$\left(-\dfrac{1}{2}\right)^3 = \underbrace{\left(-\dfrac{1}{2}\right)\left(-\dfrac{1}{2}\right)\left(-\dfrac{1}{2}\right)}_{\text{Three factors of } -\frac{1}{2}}$$

Watch the signs carefully. Multiply $\left(-\dfrac{1}{2}\right)\left(-\dfrac{1}{2}\right)$ to get $\dfrac{1}{4}$.

$$= \underbrace{\dfrac{1}{4} \left(-\dfrac{1}{2}\right)}$$

Now multiply $\dfrac{1}{4}\left(-\dfrac{1}{2}\right)$ to get $-\dfrac{1}{8}$.

$$= -\dfrac{1}{8}$$

The product is negative.

**(b)** $\left(\dfrac{3}{4}\right)^2 \cdot \left(\dfrac{2}{3}\right)^3$

$$\left(\dfrac{3}{4}\right)^2 \cdot \left(\dfrac{2}{3}\right)^3 = \underbrace{\left(\dfrac{3}{4} \cdot \dfrac{3}{4}\right)}_{\text{Two factors of } \frac{3}{4}} \cdot \underbrace{\left(\dfrac{2}{3} \cdot \dfrac{2}{3} \cdot \dfrac{2}{3}\right)}_{\text{Three factors of } \frac{2}{3}}$$

$$= \dfrac{\cancel{3} \cdot \cancel{3} \cdot \cancel{2} \cdot \cancel{2} \cdot \cancel{2}}{\cancel{2} \cdot \cancel{2} \cdot \cancel{2} \cdot 2 \cdot \cancel{3} \cdot \cancel{3} \cdot 3}$$

Divide out all common factors.

$$= \dfrac{1}{6}$$

**Work Problem 1 at the Side.**

1. Simplify.

(a) $\left(-\dfrac{3}{5}\right)^2$

(b) $\left(\dfrac{1}{3}\right)^4$

(c) $\left(-\dfrac{2}{3}\right)^3 \left(\dfrac{1}{2}\right)^2$

(d) $\left(-\dfrac{1}{2}\right)^2 \left(\dfrac{1}{4}\right)^2$

**2** **Use the order of operations with fractions.** The order of operations that you used in **Section 1.8** for integers also applies to fractions.

**ANSWERS**
1. (a) $\dfrac{9}{25}$   (b) $\dfrac{1}{81}$   (c) $-\dfrac{2}{27}$   (d) $\dfrac{1}{64}$

## Chapter 4 Rational Numbers: Positive and Negative Fractions

**2** Simplify.

(a) $\dfrac{1}{3} - \dfrac{5}{9}\left(\dfrac{3}{4}\right)$

(b) $-\dfrac{3}{4} + \left(-\dfrac{1}{2}\right)^2 \div \dfrac{2}{3}$

(c) $\dfrac{12}{5} - \dfrac{1}{6}\left(3 - \dfrac{3}{5}\right)$

### Order of Operations

*Step 1* Work inside **parentheses** or **other grouping symbols**.

*Step 2* Simplify expressions with **exponents**.

*Step 3* Do the remaining **multiplications and divisions** as they occur from left to right.

*Step 4* Do the remaining **additions and subtractions** as they occur from left to right.

### Example 2  Using the Order of Operations with Fractions

Simplify.

(a) $-\dfrac{1}{3} + \dfrac{1}{2}\left(\dfrac{4}{5}\right)$

There is no work to be done inside the parentheses. There are no exponents, so start with Step 3, multiplying and dividing.

$-\dfrac{1}{3} + \dfrac{1 \cdot 4}{2 \cdot 5}$  Multiply.

$-\dfrac{1}{3} + \dfrac{4}{10}$  Now add. The LCD is 30.
$-\dfrac{1}{3} = -\dfrac{10}{30}$ and $\dfrac{4}{10} = \dfrac{12}{30}$

$\dfrac{-10 + 12}{30}$  Add the numerators.

$\dfrac{2}{30}$  Write $\dfrac{2}{30}$ in lowest terms: $\dfrac{\cancel{2}^{1}}{\cancel{2}_{1} \cdot 15} = \dfrac{1}{15}$

$\dfrac{1}{15}$  The answer is in lowest terms.

(b) $-2 + \left(\dfrac{1}{4} - \dfrac{3}{2}\right)^2$  Work inside parentheses. The LCD for $\dfrac{1}{4}$ and $\dfrac{3}{2}$ is 4. Rewrite $\dfrac{3}{2}$ as $\dfrac{6}{4}$ and subtract.

$-2 + \left(\dfrac{1 - 6}{4}\right)^2$

$-2 + \left(\dfrac{-5}{4}\right)^2$  Simplify the term with the exponent. Multiply $\left(-\dfrac{5}{4}\right)\left(-\dfrac{5}{4}\right)$. Signs match, so the product is positive.

$-2 + \left(\dfrac{25}{16}\right)$  Add last. Write $-2$ as $-\dfrac{2}{1}$.

$-\dfrac{2}{1} + \dfrac{25}{16}$  The LCD is 16. Rewrite $-\dfrac{2}{1}$ as $-\dfrac{32}{16}$.

$\dfrac{-32 + 25}{16}$  Add the numerators.

$-\dfrac{7}{16}$  The answer is in lowest terms.

**Work Problem** ❷ **at the Side.**

**ANSWERS**

**2.** (a) $-\dfrac{1}{12}$  (b) $-\dfrac{3}{8}$  (c) 2

## Section 4.6 Exponents, Order of Operations, and Complex Fractions

**3** **Simplify complex fractions.** We have used both the symbol ÷ and a fraction bar to indicate division. For example,

Indicates division → $\dfrac{6}{2}$ can be written as 6 ÷ 2 ← Indicates division

That means we could write $-\dfrac{4}{5} \div \left(-\dfrac{3}{10}\right)$ using a fraction bar instead of ÷.

$-\dfrac{4}{5} \div \left(-\dfrac{3}{10}\right)$ can be written as $\dfrac{-\frac{4}{5}}{-\frac{3}{10}}$ ← Indicates division

↑ Indicates division

The result looks a bit complicated, and its name reflects that fact. We call it a *complex fraction*.

### Complex Fractions

A **complex fraction** is a fraction in which the numerator and/or denominator contain one or more fractions.

**Example 3** Simplifying a Complex Fraction

Simplify: $\dfrac{-\frac{4}{5}}{-\frac{3}{10}}$

Rewrite the complex fraction using the ÷ symbol for division. Then follow the steps for dividing fractions.

$$\dfrac{-\frac{4}{5}}{-\frac{3}{10}} = -\dfrac{4}{5} \div -\dfrac{3}{10} = -\dfrac{4}{5} \cdot -\dfrac{10}{3} = \dfrac{4 \cdot 2 \cdot \cancel{5}}{\cancel{5} \cdot 3} = \dfrac{8}{3} \text{ or } 2\dfrac{2}{3}$$

↑ Reciprocals

The quotient is positive because the numbers in the problem had matching signs (both were negative).

━━━ Work Problem **3** at the Side.

**3** Simplify.

(a) $\dfrac{-\frac{3}{5}}{\frac{9}{10}}$

(b) $\dfrac{6}{\frac{3}{4}}$ ← *Hint:* Write 6 as $\frac{6}{1}$.

(c) $\dfrac{-\frac{15}{16}}{-5}$

---

ANSWERS

**3.** (a) $-\dfrac{2}{3}$  (b) 8  (c) $\dfrac{3}{16}$

# Focus on Real-Data Applications

## Heart-Rate Training Zone

Performing aerobic exercise is beneficial both for improving aerobic fitness and for burning fat. For best results, you should keep your heart rate within the training zone for a minimum of 12 minutes. If you train at the higher end of the training zone, you will burn glycogen and improve aerobic fitness. Training for longer periods at the lower end of the training zone results in your body using fat reserves for energy.

*Example:* The training zone (TZ) is based on your heart rate (HR) for one minute. To see if you are in the training zone, measure your heart rate for 15 seconds. Compare it to the 15-second training zone. Find the exact answer, and then round to the nearest whole number.

| Instruction | Calculation | Example (age 22) |
|---|---|---|
| Calculate maximum heart rate (MHR) | 220 − your age | $220 - 22 = 198$ |
| Calculate lower limit of training zone (TZ) | $\frac{3}{5} \times$ (MHR) | $\frac{3}{5} \times (198) = \frac{594}{5} = 118\frac{4}{5}$ |
| Calculate upper limit of training zone (TZ) | $\frac{4}{5} \times$ (MHR) | $\frac{4}{5} \times (198) = \frac{792}{5} = 158\frac{2}{5}$ |
| Calculate the exact 15-second training zone. Round the results to the nearest whole number. | $\left(\frac{1}{4} \times \text{lower TZ}, \frac{1}{4} \times \text{Upper TZ}\right)$ | $\frac{1}{4} \times \frac{594}{5} = 29\frac{7}{10}; \frac{1}{4} \times \frac{792}{5} = 39\frac{3}{5}$ $29\frac{7}{10} < \text{HR} < 39\frac{3}{5}$ $30 < \text{HR} < 40$ |

| Age | MHR | Lower Limit of TZ | Upper Limit of TZ | 15-Second TZ (exact) | 15-Second TZ (rounded) |
|---|---|---|---|---|---|
| 18 | | | | | |
| 25 | | | | | |
| 30 | | | | | |
| 40 | | | | | |
| 50 | | | | | |
| 60 | | | | | |

1. Suppose you work in a physical fitness center and decide to design a poster to remind the clients of the training zone for their age. Compute the exact 15-second training zone for people of each of the following ages. Write fractions in lowest terms. Then round the answers to the nearest whole.

2. Explain why the lower and upper training zones (TZ) are multiplied by $\frac{1}{4}$.

3. Explain why the 15-second training zone is lower for a person aged 50 in comparison to a person aged 20.

4. Based on the chart, what would you tell a 45-year-old person about their 15-second training zone?

Section 4.6  **267**

## 4.6 EXERCISES

*Simplify. See Example 1.*

1. $\left(-\dfrac{3}{4}\right)^2$

2. $\left(-\dfrac{4}{5}\right)^2$

3. $\left(\dfrac{2}{5}\right)^3$

4. $\left(\dfrac{1}{4}\right)^3$

5. $\left(-\dfrac{1}{3}\right)^3$

6. $\left(-\dfrac{3}{5}\right)^3$

7. $\left(\dfrac{1}{2}\right)^5$

8. $\left(\dfrac{1}{3}\right)^4$

9. $\left(\dfrac{7}{10}\right)^2$

10. $\left(\dfrac{8}{9}\right)^2$

11. $\left(-\dfrac{6}{5}\right)^2$

12. $\left(-\dfrac{8}{7}\right)^2$

13. $\dfrac{15}{16}\left(\dfrac{4}{5}\right)^3$

14. $-8\left(-\dfrac{3}{8}\right)^2$

15. $\left(\dfrac{1}{3}\right)^4 \left(\dfrac{9}{10}\right)^2$

16. $\left(\dfrac{4}{5}\right)^2 \left(\dfrac{1}{2}\right)^6$

17. $\left(-\dfrac{3}{2}\right)^3 \left(-\dfrac{2}{3}\right)^2$

18. $\left(\dfrac{5}{6}\right)^2 \left(-\dfrac{2}{5}\right)^3$

## Relating Concepts (Exercises 19–20) For Individual or Group Work

*Use your knowledge of exponents as you **work Exercises 19 and 20 in order.***

**19. (a)** Evaluate this series of examples.

$\left(-\dfrac{1}{2}\right)^2 =$ _____     $\left(-\dfrac{1}{2}\right)^6 =$ _____

$\left(-\dfrac{1}{2}\right)^3 =$ _____     $\left(-\dfrac{1}{2}\right)^7 =$ _____

$\left(-\dfrac{1}{2}\right)^4 =$ _____     $\left(-\dfrac{1}{2}\right)^8 =$ _____

$\left(-\dfrac{1}{2}\right)^5 =$ _____     $\left(-\dfrac{1}{2}\right)^9 =$ _____

**20.** Several drops of ketchup fell on Ron's homework. Explain how he can figure out what real number is covered by each drop. Be careful. More than one number may work, or there may not be any real number that works.

**(a)** $(\blacksquare)^2 = \dfrac{4}{9}$     **(b)** $(\blacksquare)^3 = -\dfrac{1}{27}$

**(c)** $(\blacksquare)^4 = \dfrac{1}{16}$     **(d)** $(\blacksquare)^2 = -\dfrac{9}{16}$

**(e)** $(\blacksquare)^2 (\blacksquare)^2 = \dfrac{1}{36}$

**(b)** Explain the pattern in the sign of the answers.

---

*Simplify. See Example 2.*

**21.** $\dfrac{1}{5} - \dfrac{7}{10}(6)$

**22.** $\dfrac{2}{9} - 4\left(\dfrac{5}{6}\right)$

**23.** $\left(\dfrac{4}{3} \div \dfrac{8}{3}\right) + \left(-\dfrac{3}{4} \cdot \dfrac{1}{4}\right)$

**24.** $\left(-\dfrac{1}{3} \cdot \dfrac{3}{5}\right) + \left(\dfrac{3}{4} \div \dfrac{1}{4}\right)$

**25.** $-\dfrac{3}{10} \div \dfrac{3}{5}\left(-\dfrac{2}{3}\right)$

**26.** $5 \div \left(-\dfrac{10}{3}\right)\left(-\dfrac{4}{9}\right)$

Section 4.6    **269**

**27.** $\dfrac{8}{3}\left(\dfrac{1}{4} - \dfrac{1}{2}\right)^2$

**28.** $\dfrac{1}{3}\left(\dfrac{4}{5} - \dfrac{3}{10}\right)^3$

**29.** $-\dfrac{3}{8} + \dfrac{2}{3}\left(-\dfrac{2}{3} + \dfrac{1}{6}\right)$

**30.** $\dfrac{1}{6} + 4\left(\dfrac{2}{5} - \dfrac{7}{10}\right)$

**31.** $2\left(\dfrac{1}{3}\right)^3 - \dfrac{2}{9}$

**32.** $8\left(-\dfrac{3}{4}\right)^2 + \dfrac{3}{2}$

**33.** $\left(-\dfrac{2}{3}\right)^3\left(\dfrac{1}{8} - \dfrac{1}{2}\right) - \dfrac{2}{3}\left(\dfrac{1}{8}\right)$

**34.** $\left(\dfrac{3}{5}\right)^2\left(\dfrac{5}{9} - \dfrac{2}{3}\right) \div \left(-\dfrac{1}{5}\right)^2$

**35.** A square operation key on a calculator is $\tfrac{3}{8}$ in. on each side. (*Source:* Texas Instruments.) What is the area of the key? Use the formula $A = s^2$.

**36.** A square lot for sale in the country is $\tfrac{3}{10}$ mile on a side. Find the area of the lot by using the formula $A = s^2$.

**37.** A rectangular parking lot at the megamall is $\tfrac{7}{10}$ mile long and $\tfrac{1}{4}$ mile wide. How much fencing is needed to enclose the lot? Use the formula $P = 2l + 2w$.

**38.** A computer chip in a rectangular shape is $\tfrac{7}{8}$ in. long and $\tfrac{5}{16}$ in. wide. An insulating strip must be put around all sides of the chip. Find the length of the strip by using the formula $P = 2l + 2w$.

*Simplify. See Example 3.*

39. $\dfrac{-\dfrac{7}{9}}{-\dfrac{7}{36}}$

40. $\dfrac{\dfrac{15}{32}}{-\dfrac{5}{64}}$

41. $\dfrac{-15}{\dfrac{6}{5}}$

42. $\dfrac{-6}{-\dfrac{5}{8}}$

43. $\dfrac{\dfrac{4}{7}}{8}$

44. $\dfrac{-\dfrac{11}{5}}{3}$

45. $\dfrac{\left(\dfrac{2}{5}\right)^2}{\left(-\dfrac{4}{3}\right)^2}$

46. $\dfrac{\left(-\dfrac{5}{6}\right)^2}{\left(\dfrac{1}{2}\right)^3}$

47. $\dfrac{\dfrac{5}{6}+\dfrac{2}{3}}{2\dfrac{2}{5}}$

48. $\dfrac{\dfrac{1}{2}+\dfrac{3}{4}}{3\dfrac{1}{3}}$

49. $\dfrac{4\dfrac{1}{2}}{\dfrac{1}{2}-\dfrac{3}{4}}$

50. $\dfrac{1\dfrac{2}{3}}{\dfrac{3}{10}-\dfrac{4}{5}}$

## 4.7 PROBLEM SOLVING: EQUATIONS CONTAINING FRACTIONS

**1** **Use the multiplication property of equality to solve equations containing fractions.** In **Section 2.4** you used the division property of equality to solve an equation such as $4s = 24$. The division property says that you may divide both sides of an equation by the same nonzero number and that the equation will still be balanced. Now that you have some experience with fractions, let's look again at how the division property works.

**OBJECTIVES**

**1** Use the multiplication property of equality to solve equations containing fractions.

**2** Use both the addition and multiplication properties of equality to solve equations containing fractions.

**3** Solve application problems using equations containing fractions.

$$4s = 24$$

4s means 4 • s

$$\frac{4 \cdot s}{4} = \frac{24}{4} \quad \text{Divide both sides by 4.}$$

Divide out the common factor of 4.

$$\frac{\overset{1}{\cancel{4}} \cdot s}{\underset{1}{\cancel{4}}} = 6$$

$$s = 6$$

Because multiplication and division are related to each other, we can also *multiply* both sides of an equation by the same nonzero number and keep it balanced.

### Multiplication and Division Properties of Equality

If $a = b$, then $a \cdot c = b \cdot c$. This is the **multiplication property of equality.**

Also $\frac{a}{c} = \frac{b}{c}$ as long as $c$ is not 0 (**division property of equality**).

In other words, you may multiply or divide both sides of an equation by the same nonzero number and it will still be balanced.

**Example 1** Using the Multiplication Property of Equality

Solve each equation and check each solution.

(a) $\frac{1}{2}b = 5$

As in Chapter 2, you want the variable by itself on one side of the equal sign. In this example, you want $1b$, not $\frac{1}{2}b$, on the left side. (Recall that $1b$ is equivalent to $b$.)

In **Section 4.3** you learned that the product of a number and its reciprocal is 1. Thus, multiplying $\frac{1}{2}$ by $\frac{2}{1}$ will give the desired coefficient of 1.

$$\frac{1}{2}b = 5$$

$$\frac{2}{1}\left(\frac{1}{2}b\right) = \frac{2}{1}(5) \quad \text{Multiply both sides by } \frac{2}{1} \text{ (the reciprocal of } \frac{1}{2}\text{)}.$$

On the left side, use the associative property to regroup the factors.

$$\left(\frac{2}{1} \cdot \frac{1}{2}\right)b = \frac{2}{1}\left(\frac{5}{1}\right) \quad \text{On the right side, 5 is equivalent to } \frac{5}{1}.$$

$$\left(\frac{\overset{1}{\cancel{2}}}{1} \cdot \frac{1}{\cancel{2}}\right)b = \frac{10}{1}$$

$1b$ is equivalent to $b$.

$$1b = 10$$
$$b = 10$$

**Continued on Next Page**

Once you understand the process, you don't have to show every step. Here is a shorthand solution of the same problem.

$$\frac{1}{2}b = 5$$

$$\frac{\cancel{2}}{1}\left(\frac{1}{\cancel{2}}b\right) = \frac{2}{1}(5)$$

$$b = 10$$

The solution is 10. Check the solution by going back to the *original* equation.

**Check**  $\frac{1}{2}b = 5$    Replace $b$ with 10 in the original equation.

$\frac{1}{2}(10) = 5$    Multiply on the left side: $\frac{1}{2}(10)$ is $\frac{1}{2} \cdot \frac{10}{1}$, or $\frac{1}{2} \cdot \frac{2 \cdot 5}{1}$.

$\frac{1 \cdot \cancel{2} \cdot 5}{\cancel{2} \cdot 1} = 5$

$5 = 5$    Balances

When $b$ is 10 the equation balances, so 10 is the correct solution.

**(b)** $12 = -\frac{3}{4}x$

$12 = -\frac{3}{4}x$    Multiply both sides by $-\frac{4}{3}$ (the reciprocal of $-\frac{3}{4}$). The reciprocal of a negative number is also negative.

$-\frac{4}{3}(12) = -\frac{\cancel{4}}{\cancel{3}}\left(-\frac{\cancel{3}}{\cancel{4}}x\right)$

On the left side $-\frac{4}{3}(12)$ is $-\frac{4}{3} \cdot \frac{12}{1}$, or $-\frac{4 \cdot \cancel{3} \cdot 4}{\cancel{3} \cdot 1} = -16$

$-16 = x$

The solution is $-16$. Check the solution by going back to the *original* equation.

**Check**  $12 = -\frac{3}{4}x$    Replace $x$ with $-16$ in the original equation.

$12 = -\frac{3}{4}(-16)$    The product of two negative numbers is positive. $-\frac{3}{4}(-16)$ is $\frac{3}{4} \cdot \frac{16}{1}$, or $\frac{3}{4} \cdot \frac{4 \cdot 4}{1}$.

$12 = \frac{3 \cdot \cancel{4} \cdot 4}{\cancel{4} \cdot 1}$

$12 = 12$    Balances

When $x$ is $-16$ the equation balances, so $-16$ is the correct solution.

**Continued on Next Page**

## Section 4.7 Problem Solving: Equations Containing Fractions

**(c)** $-\dfrac{2}{5}n = -\dfrac{1}{3}$

$-\dfrac{\cancel{5}}{\cancel{2}}\left(-\dfrac{\cancel{2}}{\cancel{5}}n\right) = \left(-\dfrac{5}{2}\right)\left(-\dfrac{1}{3}\right)$    Multiply both sides by $-\dfrac{5}{2}$ (the reciprocal of $-\dfrac{2}{5}$).

$n = \dfrac{5 \cdot 1}{2 \cdot 3}$    The product of two negative numbers is positive.

$n = \dfrac{5}{6}$

**Check**

$-\dfrac{2}{5}n = -\dfrac{1}{3}$    Original equation

$\left(-\dfrac{2}{5}\right)\left(\dfrac{5}{6}\right) = -\dfrac{1}{3}$    Replace $n$ with $\dfrac{5}{6}$.

$-\dfrac{\cancel{2} \cdot \cancel{5}}{\cancel{5} \cdot \cancel{2} \cdot 3} = -\dfrac{1}{3}$    Multiply on the left side.

$-\dfrac{1}{3} = -\dfrac{1}{3}$    Balances

When $n$ is $\dfrac{5}{6}$ the equation balances, so $\dfrac{5}{6}$ is the correct solution.

**Work Problem ❶ at the Side.**

**❷** **Use both the addition and multiplication properties of equality to solve equations containing fractions.** In **Section 2.5** you used both the addition and *division* properties of equality to solve equations. Now you can use both the addition and *multiplication* properties.

### Example 2   Using the Addition and Multiplication Properties of Equality

Solve each equation and check each solution.

**(a)** $\dfrac{1}{3}c + 5 = 7$

The first step is to get the variable term $\dfrac{1}{3}c$ by itself on the left side of the equal sign. Recall that to "get rid of" the 5 on the left side, add the opposite of 5, which is $-5$, to both sides.

$\dfrac{1}{3}c + 5 = 7$

$\underline{\phantom{xx}-5 = -5}$    Add $-5$ to both sides.

$\dfrac{1}{3}c + 0 = 2$

$\dfrac{1}{3}c = 2$

$\dfrac{\cancel{3}}{1}\left(\dfrac{1}{\cancel{3}}c\right) = \dfrac{3}{1}\left(\dfrac{2}{1}\right)$    Multiply both sides by $\dfrac{3}{1}$ (the reciprocal of $\dfrac{1}{3}$).

$c = 6$

The solution is 6. Check the solution by going back to the *original* equation.

*Continued on Next Page*

---

**❶** Solve each equation. Check each solution.

**(a)** $\dfrac{1}{6}m = 3$    Check

**(b)** $\dfrac{3}{2}a = -9$    Check

**(c)** $\dfrac{3}{14} = -\dfrac{2}{7}x$    Check

**ANSWERS**

**1. (a)** $m = 18$    Check    $\dfrac{1}{6}m = 3$
$\dfrac{1}{6}(18) = 3$
Balances    $3 = 3$

**(b)** $a = -6$    Check    $\dfrac{3}{2}a = -9$
$\dfrac{3}{2}(-6) = -9$
Balances    $-9 = -9$

**(c)** $x = -\dfrac{3}{4}$    Check    $\dfrac{3}{14} = -\dfrac{2}{7}x$
$\dfrac{3}{14} = -\dfrac{2}{7}\left(-\dfrac{3}{4}\right)$
Balances    $\dfrac{3}{14} = \dfrac{3}{14}$

## 274  Chapter 4  Rational Numbers: Positive and Negative Fractions

**❷** Solve each equation. Check each solution.

(a) $18 = \dfrac{4}{5}x + 2$   Check

(b) $\dfrac{1}{4}h - 5 = 1$   Check

(c) $\dfrac{4}{3}r + 4 = -8$   Check

**Check**  $\dfrac{1}{3}c + 5 = 7$   Original equation

$\dfrac{1}{3}(6) + 5 = 7$   Replace $c$ with 6.

$\underbrace{2 + 5}_{7} = 7$

$7 = 7$   Balances

When $c$ is 6 the equation balances, so 6 is the correct solution.

(b) $-3 = \dfrac{2}{3}y + 7$

To get the variable term $\tfrac{2}{3}y$ by itself on the right side, add $-7$ to both sides.

$$-3 = \dfrac{2}{3}y + 7$$
$$\underline{\phantom{-3 =} -7 \qquad\quad -7}\quad \text{Add } -7 \text{ to both sides.}$$
$$-10 = \dfrac{2}{3}y + 0$$

$\dfrac{3}{2}(-10) = \dfrac{\cancel{3}}{\cancel{2}}\left(\dfrac{\cancel{2}}{\cancel{3}}y\right)$   Multiply both sides by $\tfrac{3}{2}$ (the reciprocal of $\tfrac{2}{3}$).

$-15 = y$

**Check**  $-3 = \dfrac{2}{3}y + 7$   Original equation

$-3 = \dfrac{2}{3}(-15) + 7$   Replace $y$ with $-15$.

$-3 = \underbrace{-10 + 7}$

$-3 = -3$   Balances

When $y$ is $-15$ the equation balances, so $-15$ is the correct solution.

**Work Problem ❷ at the Side.**

**3**   Solve application problems using equations containing fractions.   Use the six problem-solving steps from **Section 3.3** to solve application problems.

**ANSWERS**

2. (a) $x = 20$   Check   $18 = \dfrac{4}{5}x + 2$
$18 = \dfrac{4}{5}(20) + 2$
$18 = 16 + 2$
Balances   $18 = 18$

(b) $h = 24$   Check   $\dfrac{1}{4}h - 5 = 1$
$\dfrac{1}{4}(24) - 5 = 1$
$6 - 5 = 1$
Balances   $1 = 1$

(c) $r = -9$   Check   $\dfrac{4}{3}r + 4 = -8$
$\dfrac{4}{3}(-9) + 4 = -8$
$-12 + 4 = -8$
Balances   $-8 = -8$

## Example 3 Solving an Application Problem Using an Equation with Fractions

The expression for finding a person's approximate systolic blood pressure is $100 + \frac{\text{age}}{2}$. Suppose your friend's systolic blood pressure is 116 (and he has normal blood pressure). Find his age.

*Step 1* **Read** the problem. It is about blood pressure and age.

**Unknown:** friend's age

**Known:** Blood pressure expression is $100 + \frac{\text{age}}{2}$; friend's pressure is 116.

*Step 2* **Assign a variable.** Let $a$ represent the friend's age.

*Step 3* **Write an equation.**

$$100 + \frac{\text{age}}{2} \text{ is blood pressure}$$

$$100 + \frac{a}{2} = 116$$

*Step 4* **Solve.**

$$100 + \frac{a}{2} = 116$$

$$\underline{-100 \qquad -100} \qquad \text{Add } -100 \text{ to both sides.}$$

$$0 + \frac{a}{2} = 16$$

$$\frac{a}{2} = 16$$

$\frac{1}{2}a$ is equivalent to $\frac{a}{2}$ because $\frac{1}{2}a$ is $\frac{1}{2} \cdot \frac{a}{1}$

$$\frac{1}{2}a = 16$$

$$\frac{\overset{1}{\cancel{2}}}{1}\left(\frac{1}{\cancel{2}}a\right) = \frac{2}{1}(16) \qquad \text{Multiply both sides by } \frac{2}{1}$$
$$\text{(the reciprocal of } \frac{1}{2}\text{).}$$

$$a = 32$$

*Step 5* **State the answer.** Your friend is 32 years old.

*Step 6* **Check** the solution by putting it back into the *original* problem.

Approximate systolic blood pressure is $100 + \frac{\text{age}}{2}$.

If the age is 32, then $100 + \frac{32}{2} = 100 + 16 = 116$.

Friend's blood pressure is 116. ← Matches

Age 32 is the correct solution because it "works" when you put it back into the original problem.

— Work Problem ③ at the Side.

---

③ A woman's systolic blood pressure is 111. Find her age, using the expression for systolic blood pressure from Example 3 and the six problem-solving steps. (Assume that the woman has normal blood pressure.)

---

**ANSWERS**

**3.** Age is $a$.

$100 + \frac{a}{2} = 111$   The woman is 22 years old.

**Check**   $100 + \frac{22}{2} = 100 + 11 = 111$

# Focus on Real-Data Applications

## Hotel Expenses

Mathematics teachers attending conferences in New Orleans, Louisiana, and San Jose, California, found the following information about hotel rates on the organizations' Internet Web sites.

| Hotel | Single | Double | Triple | Quad | Suites |
|---|---|---|---|---|---|
| **New Orleans (January 2001)** | | | | | |
| Marriott | $116 | $121 | $121 | $121 | $603 |
| Sheraton (Club) | $157 | $169 | $194 | $219 | $598 |
| **San Jose (February 2001)** | | | | | |
| Crowne Plaza | $157 | $157 | $167 | $177 | — |
| Hilton Towers | $167 | $187 | $207 | $227 | — |
| Hyatt Sainte Claire | $156 | $176 | $196 | $216 | — |

1. The double rate is for two people sharing a room. What fractional part does each person pay?

2. (a) Multiply the double rate at the Hilton Towers in San Jose by $\frac{1}{2}$. What is the result?

   (b) Divide the double rate at the Hilton Towers in San Jose by 2. What is the result?

   (c) Explain what happened. How much money would one person owe if he or she shared a double room at the Hilton Towers in San Jose?

3. The triple rate is for three people sharing a room. What fractional part does each person pay?

4. How much money would one person owe if he or she shared a triple room at the Hyatt Sainte Claire in San Jose? How much money would each person save if they could book a triple room at the Crowne Plaza instead of the Hyatt Sainte Claire? Find your answer using two different methods, based on your observations in Problem 2.

5. The quad rate is for four people sharing a room. What fractional part does each person pay?

6. How much money would one person owe if he or she shared a quad room at the Sheraton in New Orleans? Find your answer using two different methods, based on your observations in Problem 2.

7. How many people would have to share a suite at the Sheraton in New Orleans for the cost per person to be less than sharing a quad room at the same hotel? Round the answer to the nearest whole number. (*Hint:* Estimate the cost per person for a quad room first. Then estimate the number of people needed to share the cost of the suite. Check your work using actual values.)

8. Suppose you have a travel allotment of $500 that can be spent on transportation, hotel, food, and registration fees. You and a colleague decide to attend the 3-day New Orleans conference and plan to share a room at the Marriott. Registration costs $150; the flight costs $129 round-trip; the taxi ride to the hotel costs $20 per person each way; and you budget $35 per day for meals. How much out-of-pocket expense will you have to pay? How much would you save if you could recruit a third person to share the room?

## 4.7 Exercises

*Solve each equation and check each solution. See Examples 1 and 2.*

1. $\dfrac{1}{3}a = 10$     **Check**

2. $7 = \dfrac{1}{5}y$     **Check**

3. $-20 = \dfrac{5}{6}b$     **Check**

4. $-\dfrac{4}{9}w = 16$     **Check**

5. $-\dfrac{7}{2}c = -21$     **Check**

6. $-25 = \dfrac{5}{3}x$     **Check**

7. $\dfrac{9}{16} = \dfrac{3}{4}m$     **Check**

8. $\dfrac{5}{12}k = \dfrac{15}{16}$     **Check**

9. $\dfrac{3}{10} = -\dfrac{1}{4}d$     **Check**

10. $-\dfrac{7}{8}h = -\dfrac{1}{6}$     **Check**

11. $\dfrac{1}{6}n + 7 = 9$     **Check**

12. $3 + \dfrac{1}{4}p = 5$     **Check**

13. $-10 = \frac{5}{3}r + 5$   Check

14. $0 = 6 + \frac{3}{2}t$   Check

15. $\frac{3}{8}x - 9 = 0$   Check

16. $\frac{1}{3}s - 10 = -5$   Check

*Solve each equation. Show your work.*

17. $7 - 2 = \frac{1}{5}y + 2 - 6$

18. $0 - 8 = \frac{1}{10}k - 8 + 5$

19. $-3 + 7 + \frac{2}{3}n = -10 + 2$

20. $-\frac{2}{5}m - 3 = 6 - 10 - 5$

21. $3x + \frac{1}{2} = \frac{3}{4}$

22. $4y + \frac{1}{3} = \frac{7}{9}$

23. $\frac{3}{10} = -4b - \frac{1}{5}$

24. $\frac{5}{6} = -3c - \frac{2}{3}$

**25.** Check the solution given for each equation. If a solution doesn't check, show how to find the correct solution.

(a) $\frac{1}{6}x + 1 = -2$

$x = 18$

(b) $-\frac{3}{2} = \frac{9}{4}k$

$k = -\frac{2}{3}$

**26.** Check the solution given for each equation. If a solution doesn't check, show how to find the correct solution.

(a) $-\frac{3}{4}y = -\frac{5}{8}$

$y = \frac{5}{6}$

(b) $16 = -\frac{7}{3}w + 2$

$w = 6$

**27.** Write two different equations that have 8 as a solution. Write your equations with a fraction as the coefficient of the variable term.

**28.** Write two different equations that have $-12$ as the solution. Write your equations with a fraction as the coefficient of the variable term.

*In Exercises 29–32, find each person's age using the six problem-solving steps and this expression for approximate systolic blood pressure:* $100 + \frac{\text{age}}{2}$. *Assume that all the people have normal blood pressure. See Example 3.*

**29.** A man has systolic blood pressure of 109. How old is he?

**30.** A man has systolic blood pressure of 118. How old is he?

**31.** A woman has systolic blood pressure of 122. How old is she?

**32.** A woman has systolic blood pressure of 113. How old is she?

*As you read at the start of this chapter, nails for construction projects are classified by the "penny" system, which is a number that indicates the nail's length. The expression for finding the length of a nail, in inches, is $\dfrac{\text{penny size}}{4} + \dfrac{1}{2}$ inch. (Source: Season by Season Home Maintenance.) In Exercises 33–36, find the penny size for each nail using this expression and the six problem-solving steps.*

**33.** The length of a common nail is 3 in. What is its penny size?

**34.** The length of a drywall nail is 2 in. Find the nail's penny size.

**35.** The length of a box nail is $2\frac{1}{2}$ in. What penny size would you ask for when buying these nails?

**36.** The length of a finishing nail is $1\frac{1}{2}$ in. What is its penny size?

*An expression for the recommended weight of an adult is $\dfrac{11}{2}$ (height in inches) $- 220$.*

*In Exercises 37–40, find each person's height using this expression and the six problem-solving steps. Assume that all the people are at their recommended weight.*

**37.** A man weighs 209 pounds. What is his height in inches?

**38.** A woman weighs 110 pounds. What is her height in inches?

**39.** A woman weighs 132 pounds. What is her height in inches?

**40.** A man weighs 176 pounds. What is his height in inches?

# 4.8 GEOMETRY APPLICATIONS: AREA AND VOLUME

**1** **Find the area of a triangle.** In **Section 3.1** you worked with triangles, which are flat shapes that have exactly three sides. You found the perimeter of a triangle by adding the lengths of the three sides. Now you are ready to find the area of a triangle (the amount of surface inside the triangle).

You can find the *height* of a triangle by measuring the distance from one corner of the triangle to the opposite side (the base). The height must be *perpendicular* to the base, that is, it must form a right angle with the base. Sometimes you have to extend the base before you can draw the height perpendicular to it.

### OBJECTIVES

**1** Find the area of a triangle.
**2** Find the volume of a rectangular solid.
**3** Find the volume of a pyramid.

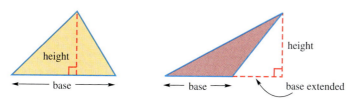

If you cut out two identical triangles and turn one upside down, you can fit them together to form a parallelogram.

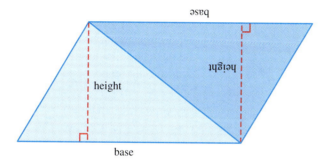

Recall from **Section 3.2** that the area of the parallelogram is *base* times *height*. Because each triangle is *half* of the parallelogram, the area of one triangle is

$\frac{1}{2}$ of base times height.

Use the following formula to find the area of a triangle.

### Finding the Area of a Triangle

$$\text{Area of triangle} = \frac{1}{2} \cdot \text{base} \cdot \text{height}$$

$$A = \frac{1}{2}bh$$

Remember to use *square units* when measuring area.

**1** Find the area of each triangle.

(a)
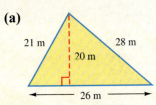

(b)
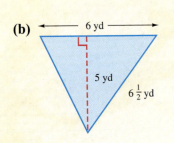

(c)
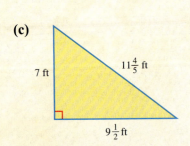

**ANSWERS**
1. (a) 260 m² (b) 15 yd²
(c) $\frac{133}{4}$ ft² or $33\frac{1}{4}$ ft²

### Example 1  Finding the Area of Triangles

Find the area of each triangle.

(a)
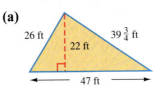

The base is 47 ft and the height is 22 ft. You do *not* need the 26 ft or $39\frac{3}{4}$ ft sides to find the area.

$A = \frac{1}{2} \cdot b \cdot h$ — Replace $b$ with 47 ft and $h$ with 22 ft.

$A = \frac{1}{2} \cdot 47 \text{ ft} \cdot 22 \text{ ft}$

$A = \frac{1}{2} \cdot \frac{47 \text{ ft}}{1} \cdot \frac{22 \text{ ft}}{1}$ — Divide out the common factor of 2.

$A = \frac{1 \cdot 47 \text{ ft} \cdot \cancel{2} \cdot 11 \text{ ft}}{\cancel{2} \cdot 1 \cdot 1}$ — Multiply $47 \cdot 11$ to get 517.

$A = 517 \text{ ft}^2$ — Multiply ft · ft to get ft².

(b)
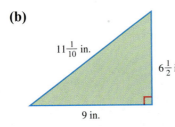

Because two sides of the triangle are perpendicular to each other, use those sides as the base and the height. (Remember that the height must be perpendicular to the base.)

$A = \frac{1}{2} bh$ — Formula for area of a triangle

$A = \frac{1}{2} \cdot 9 \text{ in.} \cdot 6\frac{1}{2} \text{ in.}$ — Replace $b$ with 9 in. and $h$ with $6\frac{1}{2}$ in.

$A = \frac{1}{2} \cdot \frac{9 \text{ in.}}{1} \cdot \frac{13 \text{ in.}}{2}$ — Write 9 in. and $6\frac{1}{2}$ in. as improper fractions.

$A = \frac{1 \cdot 9 \text{ in.} \cdot 13 \text{ in.}}{2 \cdot 1 \cdot 2}$ — Multiply 9 · 13 to get 117 and in. · in. to get in.².

$A = \frac{117}{4} \text{ in.}^2$ or $29\frac{1}{4} \text{ in.}^2$

**Work Problem 1 at the Side.**

### Example 2  Using the Concept of Area

Find the area of the shaded part in this figure.

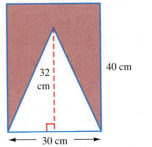

The *entire* figure is a rectangle.

$A = lw$
$A = 30 \text{ cm} \cdot 40 \text{ cm} = 1200 \text{ cm}^2$

**Continued on Next Page**

The *un*shaded part is a triangle.

$$A = \frac{1}{2}bh$$

$$A = \frac{1}{2} \cdot \frac{30 \text{ cm}}{1} \cdot \frac{32 \text{ cm}}{1}$$

$$A = \frac{1 \cdot \overset{1}{\cancel{2}} \cdot 15 \text{ cm} \cdot 32 \text{ cm}}{\underset{1}{\cancel{2}} \cdot 1 \cdot 1}$$

$$A = 480 \text{ cm}^2$$

Subtract to find the area of the shaded part.

$$A = \overbrace{1200 \text{ cm}^2}^{\text{Entire area}} - \overbrace{480 \text{ cm}^2}^{\text{Unshaded part}} = \overbrace{720 \text{ cm}^2}^{\text{Shaded part}}$$

────────────── Work Problem ❷ at the Side.

❷ Find the area of the shaded part in this figure.

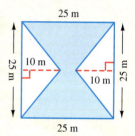

## 2 ▸ Find the volume of a rectangular solid.

A shoe box and a cereal box are examples of three-dimensional (or solid) figures. The three dimensions are length, width, and height. (A rectangle or square is a two-dimensional figure. The two dimensions are length and width.) If we want to know how much the shoe box will hold, we find its *volume*. We measure volume by seeing how many cubes of a certain size will fill the space inside the box. Three sizes of *cubic units* are shown here. Notice that all the edges of a cube have the same length.

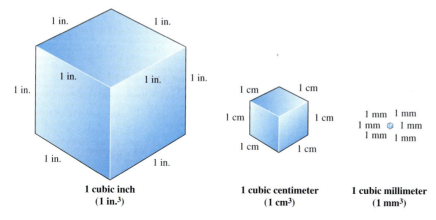

1 cubic inch (1 in.³)    1 cubic centimeter (1 cm³)    1 cubic millimeter (1 mm³)

Some other sizes of cubes that are used to measure volume are 1 cubic foot (1 ft³), 1 cubic yard (1 yd³), and 1 cubic meter (1 m³).

### CAUTION

The raised 3 in $4^3$ means that you multiply $4 \cdot 4 \cdot 4$ to get 64. The raised 3 in cm³ or ft³ is a short way to write the word "cubic." It means that you multiplied cm times cm times cm to get cm³, or ft times ft times ft to get ft³. Recall that a short way to write $x \cdot x \cdot x$ is $x^3$. Similarly, cm · cm · cm is cm³. When you see 5 cm³, say "five cubic centimeters." Do *not* multiply $5 \cdot 5 \cdot 5$ because the exponent applies only to the *first* thing to its left. The exponent applies to cm, *not* to the 5.

### Volume

**Volume** is a measure of the space inside a solid shape. The volume of a solid is the number of cubic units it takes to fill the solid.

**ANSWERS**

**2.** $A = 625 \text{ m}^2 - 125 \text{ m}^2 - 125 \text{ m}^2 = 375 \text{ m}^2$

**3** Find the volume of each rectangular solid.

(a)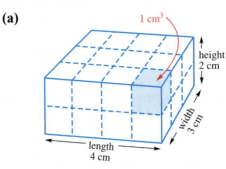

Use the following formula to find the volume of *rectangular solids* (box-like shapes).

### Finding the Volume of Rectangular Solids

Volume of rectangular solid = length • width • height

$V = lwh$

Remember to use *cubic units* when measuring volume.

**Example 3** Finding the Volume of Rectangular Solids

Find the volume of each box.

(a)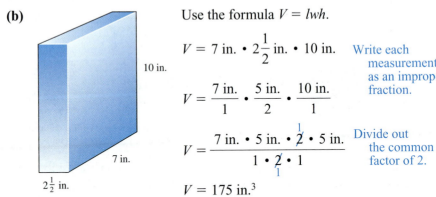

Each cube that fits in the box is 1 cubic centimeter (1 cm³). To find the volume, you can count the number of cubes.

Bottom layer has 12 cubes.  
Top layer has 12 cubes.  
} Total of 24 cubes (24 cm³)

Or you can use the formula for rectangular solids.

$V = l \cdot w \cdot h$
$V = 4 \text{ cm} \cdot 3 \text{ cm} \cdot 2 \text{ cm}$   Multiply 4 • 3 • 2 to get 24.
$V = 24 \text{ cm}^3$   Multiply cm • cm • cm to get cm³.

(b) Length $6\frac{1}{4}$ ft, width $3\frac{1}{2}$ ft, height 2 ft

(b)

Use the formula $V = lwh$.

$V = 7 \text{ in.} \cdot 2\frac{1}{2} \text{ in.} \cdot 10 \text{ in.}$   Write each measurement as an improper fraction.

$V = \frac{7 \text{ in.}}{1} \cdot \frac{5 \text{ in.}}{2} \cdot \frac{10 \text{ in.}}{1}$

$V = \frac{7 \text{ in.} \cdot 5 \text{ in.} \cdot \overset{1}{\cancel{2}} \cdot 5 \text{ in.}}{1 \cdot \underset{1}{\cancel{2}} \cdot 1}$   Divide out the common factor of 2.

$V = 175 \text{ in.}^3$

**Work Problem 3 at the Side.**

**3** **Find the volume of a pyramid.** A pyramid is a solid shape like the one shown below. The base of a pyramid may be a square or a rectangle.

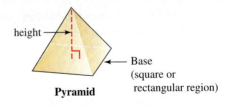

**Pyramid** — Base (square or rectangular region)

The height is the distance from the base to the highest point of the pyramid. The height must be perpendicular to the base.

ANSWERS

**3.** (a) 72 m³   (b) $\frac{175}{4}$ ft³ or $43\frac{3}{4}$ ft³

Use this formula to find the volume of a pyramid.

### Finding the Volume of a Pyramid

$$\text{Volume of pyramid} = \frac{1}{3} \cdot B \cdot h$$

$$V = \frac{1}{3}Bh$$

where $B$ is the area of the square or rectangular base of the pyramid and $h$ is the height of the pyramid.

Remember to use *cubic units* when measuring volume.

**Example 4** Finding the Volume of a Pyramid

Find the volume of the pyramid.

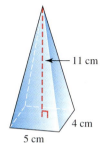

First find the value of $B$ in the formula, which is the area of the rectangular base. Recall that the area of a rectangle is found by multiplying length times width.

$$B = 5 \text{ cm} \cdot 4 \text{ cm}$$
$$\mathbf{B = 20 \text{ cm}^2}$$

Next, find the volume.

$V = \frac{1}{3}Bh$      Formula for volume of pyramid

$V = \frac{1}{3} \cdot 20 \text{ cm}^2 \cdot 11 \text{ cm}$      Replace $B$ with 20 cm² and $h$ with 11 cm.

$V = \frac{1}{3} \cdot \frac{20 \text{ cm}^2}{1} \cdot \frac{11 \text{ cm}}{1}$

$V = \frac{1 \cdot 20 \text{ cm}^2 \cdot 11 \text{ cm}}{3 \cdot 1 \cdot 1}$      There are no common factors to divide out.

$V = \frac{220}{3} \text{ cm}^3 \quad \text{or} \quad 73\frac{1}{3} \text{ cm}^3$

================== Work Problem ❹ at the Side.

❹ Find the volume of a pyramid with a square base measuring 10 ft by 10 ft and a height of 6 ft.

**ANSWERS**
4. $V = 200 \text{ ft}^3$

## Focus on Real-Data Applications

## Quilt Patterns

People who make quilts often base their designs on a block that is cut into a grid of 4, 9, 16, or 25 squares. The quilter chooses various colors for the pieces. Each quilt design shown was selected from an Archive of American Quilt Designs.

1. Identify the makeup of the block as 4, 9, 16, etc. Each color is what fractional part of the block?

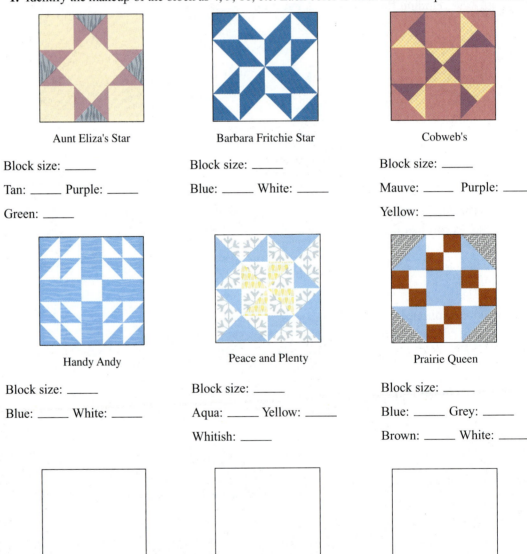

Aunt Eliza's Star

Block size: _____
Tan: _____ Purple: _____
Green: _____

Barbara Fritchie Star

Block size: _____
Blue: _____ White: _____

Cobweb's

Block size: _____
Mauve: _____ Purple: _____
Yellow: _____

Handy Andy

Block size: _____
Blue: _____ White: _____

Peace and Plenty

Block size: _____
Aqua: _____ Yellow: _____
Whitish: _____

Prairie Queen

Block size: _____
Blue: _____ Grey: _____
Brown: _____ White: _____

2. Use the blocks to design and color your own quilt patterns. Tell the fractional part of the block that is represented by each color.

3. Find the next two numbers in this pattern: 4, 9, 16, 25, _____, _____.

4. Explain how the pattern works.

## 4.8 Exercises

*Find the perimeter and area of each triangle. See Example 1.*

1.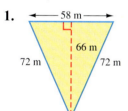
2.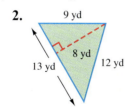
3.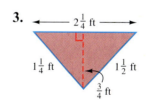
4.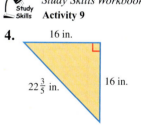

5.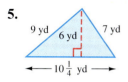
6.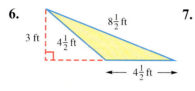
7.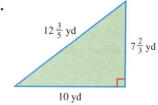
8.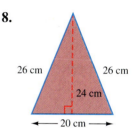

*Find the shaded area in each figure. See Example 2.*

9.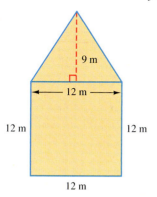
10.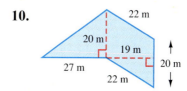

11.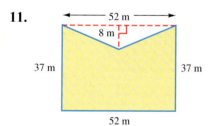
12.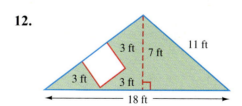

13. Explain the difference between perimeter, area, and volume.

14. Explain where the $\frac{1}{2}$ comes from in the formula for area of a triangle.

*Solve each application problem.*

15. A triangular tent flap measures $3\frac{1}{2}$ ft along the base and has a height of $4\frac{1}{2}$ ft. How much canvas is needed to make the flap?

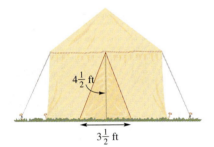

16. A wooden sign in the shape of a right triangle has perpendicular sides measuring $1\frac{1}{2}$ yd and 1 yd. How much surface area does the sign have?

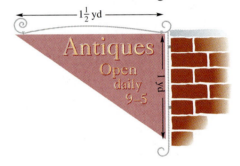

17. A triangular space between three streets has the measurements shown. How much new curbing will be needed to go around the space? How much sod will be needed to cover the space?

18. Each gable end of a new house has a span of 36 ft and a rise of $9\frac{1}{2}$ ft. What is the total area of both gable ends of the house?

*Find the volume of each figure. See Examples 3 and 4.*

**19.**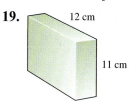

**20.**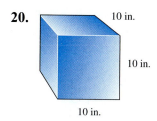

**21.**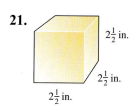

**22.**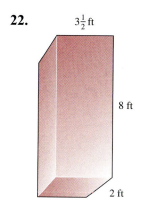

**23.**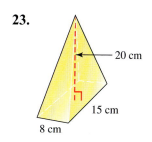

**24.**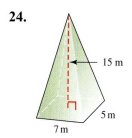

**25.** A box to hold pencils measures 3 in. by 8 in. by $\frac{3}{4}$ in. high. Find the volume of the box. (*Source:* Faber Castell.)

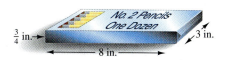

**26.** A train is being loaded with shipping crates. Each crate is 12 ft long, 8 ft wide, and $2\frac{1}{4}$ ft high. How much space will each crate take?

**27.** One of the ancient stone pyramids in Egypt has a square base that measures 145 m on each side. The height is 93 m. What is the volume of the pyramid? (*Source: Columbia Encyclopedia.*)

**28.** A cardboard model of an ancient stone pyramid has a square base that is $10\frac{3}{8}$ in. on each side. The height is $6\frac{1}{2}$ in. Find the volume of the model.

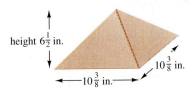

**29.** Find the volume of the object.

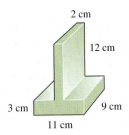

**30.** Find the volume of the shaded part. (*Hint:* Notice the hole that goes through the center of the shape.)

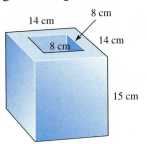

*The following are some answers from a student's test. The number part of each answer is correct, but some of the units are not. Find the errors, explain what is wrong, and correct the errors.*

**31.** There are *two* errors in this list of answers.

$$A = 135^2 \text{ ft}$$
$$V = 76 \text{ yd}^3$$
$$P = 5\frac{1}{2} \text{ in.}^2$$

**32.** There are *two* errors in this list of answers.

$$P = 215 \text{ cm}$$
$$A = 8\frac{1}{2} \text{ m}$$
$$V = 98^3 \text{ in.}$$

# Chapter 4

## SUMMARY

### KEY TERMS

| | | |
|---|---|---|
| 4.1 | fraction | A fraction is a number of the form $\frac{a}{b}$, where $a$ and $b$ are integers and $b$ is not 0. |
| | numerator | The top number in a fraction is the numerator. It shows how many of the equal parts are being considered. |
| | denominator | The bottom number in a fraction is the denominator. It shows how many equal parts are in the whole. |
| | proper fraction | In a proper fraction, the numerator is smaller than the denominator. The fraction is less than 1. |
| | improper fraction | In an improper fraction, the numerator is greater than or equal to the denominator. The fraction is greater than or equal to 1. |
| | equivalent fractions | Equivalent fractions have the same value even though they look different. When graphed on a number line, they are names for the same point. |
| 4.2 | lowest terms | A fraction is written in lowest terms when its numerator and denominator have no common factor other than 1. |
| | prime number | A prime number is a whole number that has exactly two different factors, itself and 1. The first few prime numbers are 2, 3, 5, 7, 11, 13, and 17. |
| | composite number | A composite number has at least one factor other than itself or 1. Examples are 4, 6, 9, and 10. The numbers 0 and 1 are neither prime nor composite. |
| | prime factorization | In a prime factorization, every factor is a prime number. For example, the prime factorization of 24 is 2 • 2 • 2 • 3. |
| 4.3 | reciprocal | Two numbers are reciprocals of each other if their product is 1. The reciprocal of $\frac{a}{b}$ is $\frac{b}{a}$ because $\frac{a}{b} \cdot \frac{b}{a} = 1$. |
| 4.4 | like fractions | Like fractions have the same denominator. |
| | unlike fractions | Unlike fractions have different denominators. |
| | least common denominator | The least common denominator (LCD) for two fractions is the smallest positive number that can be divided evenly by both denominators. |
| 4.5 | mixed number | A mixed number is a fraction and a whole number written together. It represents the sum of the whole number and the fraction. For example, $3 + \frac{1}{2}$ is written as $3\frac{1}{2}$. |
| 4.6 | complex fraction | A complex fraction is a fraction in which the numerator and/or denominator contain one or more fractions. |
| 4.7 | multiplication property of equality | The multiplication property of equality states that you may multiply both sides of an equation by the same nonzero number and still keep it balanced. |
| 4.8 | volume | Volume is a measure of the space inside a solid shape. Volume is measured in cubic units, such as in.$^3$, ft$^3$, yd$^3$, mm$^3$, cm$^3$, and so on. |

### NEW FORMULAS

**Area of a triangle:** $\quad A = \dfrac{1}{2}bh$

**Volume of a rectangular solid:** $\quad V = lwh$

**Volume of a pyramid:** $\quad V = \dfrac{1}{3}Bh$

## Chapter 4 Rational Numbers: Positive and Negative Fractions

### TEST YOUR WORD POWER

*See how well you have learned the vocabulary in this chapter. Answers follow the Quick Review.*

1. A **fraction** is in lowest terms if
   (a) it has a value less than 1
   (b) its numerator and denominator have no common factor other than 1
   (c) its numerator and denominator are composite numbers
   (d) it is rewritten as a mixed number.

2. The **denominator** of a fraction
   (a) is written above the fraction bar
   (b) is a prime number
   (c) shows how many equal parts are in the whole
   (d) is the smallest number divisible by the numerator.

3. The **LCD** of two fractions is the
   (a) smallest number divisible by both denominators
   (b) largest factor common to both denominators
   (c) smallest number divisible by both numerators
   (d) smallest prime number that divides evenly into both denominators.

4. Two numbers are **reciprocals** of each other if
   (a) they have the same prime factorizations
   (b) their sum is 0
   (c) they are written in lowest terms
   (d) their product is 1.

5. **Volume** is
   (a) measured in square units
   (b) the space inside a solid shape
   (c) the sum of the lengths of the sides of a shape
   (d) found by multiplying base times height.

6. A **mixed number**
   (a) has a value equal to 1
   (b) is the reciprocal of an improper fraction
   (c) is the sum of a whole number and a fraction
   (d) has a value less than 1.

7. A whole number is **prime** if
   (a) it is divisible by itself and 1
   (b) it has only composite factors
   (c) it cannot be divided
   (d) it has exactly two factors, itself and 1.

8. **Equivalent fractions**
   (a) have the same denominators
   (b) are written in lowest terms
   (c) name the same point on a number line
   (d) are reciprocals of each other.

### QUICK REVIEW

| Concepts | Examples |
|---|---|
| **4.1 Understanding Fraction Terminology** <br> The *numerator* is the top number. The *denominator* is the bottom number. In a *proper fraction* the numerator is smaller than the denominator. In an *improper fraction* the numerator is greater than or equal to the denominator. | **Proper fractions** $\quad \dfrac{2}{3}, \dfrac{3}{4}, \dfrac{15}{16}, \dfrac{1}{8} \quad \leftarrow$ Numerator <br> $\qquad\qquad\qquad\qquad\qquad\qquad\qquad\quad \leftarrow$ Denominator <br><br> **Improper fractions** $\quad \dfrac{17}{8}, \dfrac{19}{12}, \dfrac{11}{2}, \dfrac{5}{3}, \dfrac{7}{7}$ |
| **4.1 Writing Equivalent Fractions** <br> Multiply or divide the numerator and denominator by the same nonzero number. The result is an equivalent fraction. | $\dfrac{1}{2} = \dfrac{1 \cdot 8}{2 \cdot 8} = \dfrac{8}{16} \quad \leftarrow$ Equivalent to $\dfrac{1}{2}$ <br><br> $-\dfrac{12}{15} = -\dfrac{12 \div 3}{15 \div 3} = -\dfrac{4}{5} \quad \leftarrow$ Equivalent to $-\dfrac{12}{15}$ |

## Chapter 4 Summary

| Concepts | Examples |
|---|---|
| **4.2 Finding Prime Factorizations**<br>A prime factorization of a number shows the number as the product of prime numbers. The first few prime numbers are 2, 3, 5, 7, 11, 13, and 17. You can use a division method or a factor tree to find the prime factorization. | Find the prime factorization of 24.<br>**Division Method:**<br>$2\overline{)24}$ ← Divide 24 by 2, the first prime.<br>$2\overline{)12}$ ← Divide 12 by 2.<br>$2\overline{)6}$ ← Divide 6 by 2.<br>$3\overline{)3}$ ← Divide 3 by 3.<br>$1$ ← Continue to divide until the quotient is 1.<br>$24 = 2 \cdot 2 \cdot 2 \cdot 3$<br><br>**Factor Tree Method:**<br>Circle each prime number.<br>$24 = 2 \cdot 2 \cdot 2 \cdot 3$ |
| **4.2 Writing Fractions in Lowest Terms**<br>Write the prime factorization of both numerator and denominator. Divide out all common factors, using slashes to show the division. Multiply any remaining factors in the numerator and in the denominator. | $\dfrac{18}{90} = \dfrac{2 \cdot 3 \cdot 3}{2 \cdot 3 \cdot 3 \cdot 5} = \dfrac{1}{5}$<br><br>$\dfrac{2b^3}{8ab} = \dfrac{2 \cdot b \cdot b \cdot b}{2 \cdot 2 \cdot 2 \cdot a \cdot b} = \dfrac{b^2}{4a}$ |
| **4.3 Multiplying Fractions**<br>Multiply the numerators and multiply the denominators. The product must be written in lowest terms. One way to do this is to write each original numerator and each original denominator as the product of primes and divide out any common factors before multiplying. | $\left(-\dfrac{7}{10}\right)\left(\dfrac{5}{6}\right) = -\dfrac{7 \cdot 5}{2 \cdot 5 \cdot 2 \cdot 3} = -\dfrac{7}{12}$<br><br>$\dfrac{3x^2}{5} \cdot \dfrac{2}{9x} = \dfrac{3 \cdot x \cdot x \cdot 2}{5 \cdot 3 \cdot 3 \cdot x} = \dfrac{2x}{15}$ |
| **4.3 Dividing Fractions**<br>Rewrite the division problem as multiplying by the reciprocal of the divisor. In other words, the first number (dividend) stays the same and the second number (divisor) is changed to its reciprocal. Then use the steps for multiplying fractions. The quotient must be in lowest terms. Division by 0 is undefined. | $2 \div \left(-\dfrac{1}{3}\right) = \dfrac{2}{1} \cdot \left(-\dfrac{3}{1}\right) = -\dfrac{2 \cdot 3}{1 \cdot 1} = -6$<br>Reciprocals<br><br>$\dfrac{x^2}{y^2} \div \dfrac{x}{3y} = \dfrac{x^2}{y^2} \cdot \dfrac{3y}{x} = \dfrac{x \cdot x \cdot 3 \cdot y}{y \cdot y \cdot x} = \dfrac{3x}{y}$<br>Reciprocals |

| Concepts | Examples |
|---|---|
| **4.4 Adding and Subtracting Like Fractions**<br>You can add or subtract fractions *only* when they have the *same* denominator. Add or subtract the numerators and write the result over the common denominator. Be sure that the final result is in lowest terms. | $\dfrac{3}{10} - \dfrac{7}{10} = \dfrac{3-7}{10} = \dfrac{-4}{10}$ or $-\dfrac{4}{10}$<br><br>Write $-\dfrac{4}{10}$ in lowest terms. $\quad -\dfrac{4}{10} = -\dfrac{\overset{1}{\cancel{2}} \cdot 2}{\underset{1}{\cancel{2}} \cdot 5} = -\dfrac{2}{5}$ {Lowest terms<br><br>$\dfrac{5}{a} + \dfrac{7}{a} = \dfrac{5+7}{a} = \dfrac{12}{a}$ {Lowest terms |
| **4.4 Finding the Lowest Common Denominator (LCD)**<br>Write the prime factorization of each denominator. Then use enough prime factors in the LCD to "cover" both denominators. | What is the LCD for $\dfrac{5}{12}$ and $\dfrac{5}{18}$?<br><br>$12 = 2 \cdot 2 \cdot 3$<br>$18 = 2 \cdot 3 \cdot 3$<br>$\text{LCD} = 2 \cdot 2 \cdot 3 \cdot 3 = 36$<br><br>The LCD for $\dfrac{5}{12}$ and $\dfrac{5}{18}$ is 36. |
| **4.4 Adding and Subtracting Unlike Fractions**<br>Find the LCD. Rewrite each original fraction as an equivalent fraction whose denominator is the LCD. Then add or subtract the numerators and keep the common denominator. Be sure that the final result is in lowest terms. | $-\dfrac{5}{12} + \dfrac{7}{9}$<br>The LCD is 36.<br>Rewrite: $\quad -\dfrac{5}{12} = -\dfrac{5 \cdot 3}{12 \cdot 3} = -\dfrac{15}{36}$<br>Rewrite: $\quad \dfrac{7}{9} = \dfrac{7 \cdot 4}{9 \cdot 4} = \dfrac{28}{36}$<br>Add: $\quad -\dfrac{15}{36} + \dfrac{28}{36} = \dfrac{-15+28}{36} = \dfrac{13}{36}$ {Lowest terms<br><br>$\dfrac{2}{3} - \dfrac{6}{x}$<br>The LCD is $3 \cdot x$ or $3x$.<br>Rewrite: $\quad \dfrac{2}{3} = \dfrac{2 \cdot x}{3 \cdot x} = \dfrac{2x}{3x}$<br>Rewrite: $\quad \dfrac{6}{x} = \dfrac{6 \cdot 3}{x \cdot 3} = \dfrac{18}{3x}$<br>Subtract: $\quad \dfrac{2x}{3x} - \dfrac{18}{3x} = \dfrac{2x - 18}{3x}$ {Lowest terms |

| Concepts | Examples |
|---|---|
| **4.5 Mixed Numbers and Improper Fractions** <br> **Changing Mixed Numbers to Improper Fractions** <br> Multiply denominator by whole number, add numerator, and place over denominator. <br><br> **Changing Improper Fractions to Mixed Numbers** <br> Divide numerator by denominator and place remainder over denominator. | Mixed to improper: $7\frac{2}{3} = \frac{23}{3}$ ← $(3 \cdot 7) + 2$ <br> Same denominator <br><br> Improper to mixed: $\frac{17}{5} = 3\frac{2}{5}$ <br> Same denominator |
| **4.5 Multiplying Mixed Numbers** <br> First, round the numbers and estimate the answer. Then follow these steps to find the exact answer. <br> *Step 1* Rewrite each mixed number as an improper fraction. <br> *Step 2* Multiply. <br> *Step 3* Write the answer in lowest terms and change the answer to a mixed number if desired. Then the answer is in simplest form. | Estimate: $1\frac{3}{5} \cdot 3\frac{1}{3}$ Rounded $2 \cdot 3 = 6$ <br><br> Exact: $1\frac{3}{5} \cdot 3\frac{1}{3} = \frac{8}{5} \cdot \frac{10}{3}$ <br> $= \frac{8 \cdot 2 \cdot \cancel{5}}{\cancel{5} \cdot 3}$ <br> $= \frac{16}{3} = 5\frac{1}{3}$ ← Close to estimate |
| **4.5 Dividing Mixed Numbers** <br> First, round the numbers and estimate the answer. Then follow these steps to find the exact answer. <br> *Step 1* Rewrite each mixed number as an improper fraction. <br> *Step 2* Divide. (Rewrite as multiplication by the reciprocal of the divisor.) <br> *Step 3* Write the answer in lowest terms and change the answer to a mixed number if desired. Then the answer is in simplest form. | Estimate: $3\frac{3}{4} \div 2\frac{2}{5}$ Rounded $4 \div 2 = 2$ <br><br> Exact: $3\frac{3}{4} \div 2\frac{2}{5} = \frac{15}{4} \div \frac{12}{5}$ Reciprocal of $\frac{12}{5}$ is $\frac{5}{12}$. <br> $= \frac{15}{4} \cdot \frac{5}{12}$ <br> $= \frac{\cancel{3} \cdot 5 \cdot 5}{4 \cdot \cancel{3} \cdot 4}$ <br> $= \frac{25}{16} = 1\frac{9}{16}$ ← Close to estimate |
| **4.5 Adding and Subtracting Mixed Numbers** <br> First round the numbers and estimate the answer. Then rewrite the mixed numbers as improper fractions and follow the steps for adding and subtracting fractions. Write the answer in simplest form. | $2\frac{3}{8} + 3\frac{3}{4}$ <br> $2 + 4 = 6$ ← Estimate <br><br> $2\frac{3}{8} + 3\frac{3}{4} = \frac{19}{8} + \frac{15}{4} = \frac{19}{8} + \frac{30}{8} = \frac{19 + 30}{8}$ <br> $= \frac{49}{8} = 6\frac{1}{8}$ ← Close to estimate |

| Concepts | Examples |
|---|---|

### 4.6 Exponents and Order of Operations

The meaning of an exponent is the same for fractions as it is for integers. An exponent is a way to write repeated multiplication.

$$\left(-\frac{2}{3}\right)^2 \text{ means } \left(-\frac{2}{3}\right)\left(-\frac{2}{3}\right) = \frac{2 \cdot 2}{3 \cdot 3} = \frac{4}{9}$$

The product of two negative numbers is positive.

The order of operations is also the same for fractions as for integers.

1. Work inside *parentheses* or *other grouping symbols*.
2. Simplify expressions with *exponents*.
3. Do the remaining *multiplications and divisions* as they occur from left to right.
4. Do the remaining *additions and subtractions* as they occur from left to right.

Simplify.

$-\dfrac{2}{3} + 3\left(\dfrac{1}{4}\right)^2$    Cannot work inside parentheses. Use exponent: $\frac{1}{4} \cdot \frac{1}{4}$ is $\frac{1}{16}$.

$-\dfrac{2}{3} + 3\left(\dfrac{1}{16}\right)$    Multiply next: $3\left(\frac{1}{16}\right)$ is $\frac{3}{1} \cdot \frac{1}{16} = \frac{3}{16}$.

$-\dfrac{2}{3} + \dfrac{3}{16}$    Add last. The LCD is 48.

$-\dfrac{32}{48} + \dfrac{9}{48}$    Rewrite $-\frac{2}{3}$ as $-\frac{32}{48}$. Rewrite $\frac{3}{16}$ as $\frac{9}{48}$.

$\dfrac{-32 + 9}{48}$    Add the numerators. Keep the common denominator.

$-\dfrac{23}{48}$    The answer is in lowest terms.

### 4.6 Simplifying Complex Fractions

Recall that the fraction bar indicates division. Rewrite the complex fraction using the ÷ symbol for division. Then follow the steps for dividing fractions.

Simplify. $\dfrac{-\frac{4}{5}}{10}$    Rewrite as $-\dfrac{4}{5} \div 10$.

$$-\frac{4}{5} \div 10 = -\frac{4}{5} \cdot \frac{1}{10} = -\frac{\overset{1}{\cancel{2}} \cdot 2 \cdot 1}{5 \cdot \underset{1}{\cancel{2}} \cdot 5} = -\frac{2}{25}$$

Reciprocals

## Concepts

### 4.7 Solving Equations Containing Fractions

1. If necessary, add the same number to both sides of the equation so that the variable term is by itself on one side of the equal sign.
2. Multiply both sides by the reciprocal of the coefficient of the variable term.
3. To check your solution, go back to the original equation and replace the variable with your solution. If the equation balances, your solution is correct. If it does not balance, rework the problem.

## Examples

Solve the equation. Check the solution.

$$\frac{1}{3}b + 6 = 10$$
$$\underline{\phantom{\frac{1}{3}b + 0} \ -6 \ -6} \quad \text{Add } -6 \text{ to both sides.}$$
$$\frac{1}{3}b + 0 = 4$$
$$\frac{1}{3}b = 4$$

$$\frac{\cancel{3}}{1}\left(\frac{1}{\cancel{3}}b\right) = \frac{3}{1}(4) \quad \text{Multiply both sides by } \tfrac{3}{1} \text{ (the reciprocal of } \tfrac{1}{3}\text{).}$$

$$b = 12$$

**Check** $\quad \frac{1}{3} b + 6 = 10 \quad$ Original equation

$\quad\quad\quad \frac{1}{3}(12) + 6 = 10 \quad$ Replace $b$ with 12.

$\quad\quad\quad\quad 4 + 6 = 10$

$\quad\quad\quad\quad\quad 10 = 10 \quad$ Balances, so 12 is the correct solution.

### 4.8 Finding the Area of a Triangle

Use this formula to find the area of a triangle.

$$\text{Area} = \frac{1}{2} \cdot \text{base} \cdot \text{height}$$

$$A = \frac{1}{2}bh$$

Remember that area is measured in **square units**.

Find the area of this triangle.

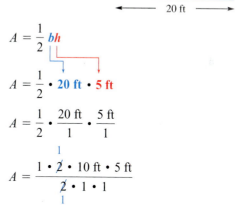

$$A = \frac{1}{2}bh$$

$$A = \frac{1}{2} \cdot 20 \text{ ft} \cdot 5 \text{ ft}$$

$$A = \frac{1}{2} \cdot \frac{20 \text{ ft}}{1} \cdot \frac{5 \text{ ft}}{1}$$

$$A = \frac{1 \cdot \cancel{2} \cdot 10 \text{ ft} \cdot 5 \text{ ft}}{\cancel{2} \cdot 1 \cdot 1}$$

$$A = 50 \text{ ft}^2 \quad \text{Measure area in square units.}$$

### 4.8 Finding the Volume of a Rectangular Solid

Use this formula to find the volume of box-like solids.

$$\text{Volume} = \text{length} \cdot \text{width} \cdot \text{height}$$

$$V = lwh$$

Volume is measured in **cubic units**.

Find the volume of this box.

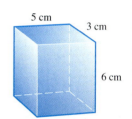

$$V = l \cdot w \cdot h$$

$$V = 5 \text{ cm} \cdot 3 \text{ cm} \cdot 6 \text{ cm}$$

$$V = 90 \text{ cm}^3$$

Measure volume in cubic units.

| Concepts | Examples |
|---|---|
| **4.8 Finding the Volume of a Pyramid**<br>Use this formula to find the volume of a pyramid.<br>$$\text{Volume} = \frac{1}{3} \cdot B \cdot h$$<br>$$V = \frac{1}{3} Bh$$<br>where $B$ is the area of the square or rectangular base and $h$ is the height of the pyramid.<br>Volume is measured in **cubic units**. | Find the volume of a pyramid with a square base 2 cm by 2 cm and a height of 6 cm.<br>Area of square base = 2 cm • 2 cm = **4 cm²**<br>$$V = \frac{1}{3} \cdot \textbf{B} \cdot \textbf{h}$$<br>$$V = \frac{1}{3} \cdot \textbf{4 cm}^2 \cdot \textbf{6 cm}$$<br>$$V = \frac{1}{3} \cdot \frac{4 \text{ cm}^2}{1} \cdot \frac{6 \text{ cm}}{1}$$<br>$$V = \frac{1 \cdot 4 \text{ cm}^2 \cdot \cancel{3}^1 \cdot 2 \text{ cm}}{\cancel{3}_1 \cdot 1 \cdot 1}$$<br>$V = 8$ **cm³**    Measure volume in cubic units. |

**ANSWERS TO TEST YOUR WORD POWER**

1. **(b)** *Example:* $\frac{2}{5}$ is in lowest terms, but $\frac{4}{10}$ is not.
2. **(c)** *Example:* In $\frac{3}{4}$ the denominator, 4, shows that the whole is divided into 4 equal parts.
3. **(a)** *Example:* The LCD of $\frac{1}{4}$ and $\frac{5}{6}$ is 12, because 12 is the smallest number that can be divided evenly by 4 and by 6.
4. **(d)** *Example:* The reciprocal of $\frac{3}{8}$ is $\frac{8}{3}$ because $\frac{3}{8} \cdot \frac{8}{3} = \frac{24}{24} = 1$.
5. **(b)** *Example:* The volume of a rectangular solid (box-like shape) is the space inside the box, measured in cubic units.
6. **(c)** *Example:* The mixed number $3\frac{1}{2}$ is $3 + \frac{1}{2}$.
7. **(d)** *Example:* 7 is prime because it has exactly two factors, 7 and 1.
8. **(c)** *Example:* $\frac{1}{2}$ and $\frac{2}{4}$ are equivalent fractions because they both name a point halfway between 0 and 1.

# Chapter 4 Review Exercises

**[4.1]** 1. What fraction of these figures are squares? What fraction are triangles?

2. Write fractions to represent the shaded and unshaded portions of this figure.

3. Graph $-\frac{1}{2}$ and $1\frac{1}{2}$ on the number line.

4. Simplify each fraction.

(a) $-\frac{20}{5}$ (b) $\frac{8}{1}$ (c) $-\frac{3}{3}$

**[4.2]** *Write each fraction in lowest terms.*

5. $\frac{28}{32}$

6. $\frac{54}{90}$

7. $\frac{16}{25}$

8. $\frac{15x^2}{40x}$

9. $\frac{7a^3}{35a^3b}$

10. $\frac{12mn^2}{21m^3n}$

**[4.3]** *Multiply or divide. Write all answers in lowest terms.*

11. $-\frac{3}{8} \div (-6)$

12. $\frac{2}{5}$ of $(-30)$

13. $\frac{4}{9}\left(\frac{2}{3}\right)$

14. $\frac{7}{3x^3} \cdot \frac{x^2}{14}$

15. $\frac{ab}{5} \div \frac{b}{10a}$

16. $\frac{18}{7} \div 3k$

**[4.4]** *Add or subtract. Write all answers in lowest terms.*

17. $-\frac{5}{12} + \frac{5}{8}$

18. $\frac{2}{3} - \frac{4}{5}$

19. $4 - \frac{5}{6}$

20. $\frac{7}{9} + \frac{13}{18}$

21. $\frac{n}{5} + \frac{3}{4}$

22. $\frac{3}{10} - \frac{7}{y}$

299

**300** Chapter 4 Rational Numbers: Positive and Negative Fractions

**[4.5]** *First, round the mixed numbers to the nearest whole number and estimate each answer. Then find the exact answer.*

**23.** *Exact*

$$2\frac{1}{4} \div 1\frac{5}{8}$$

*Estimate*

___ ÷ ___ = ___

**24.** *Exact*

$$7\frac{1}{3} - 4\frac{5}{6}$$

*Estimate*

___ − ___ = ___

**25.** *Exact*

$$1\frac{3}{4} + 2\frac{3}{10}$$

*Estimate*

___ + ___ = ___

**[4.6]** *Simplify.*

**26.** $\left(-\dfrac{3}{4}\right)^3$

**27.** $\left(\dfrac{2}{3}\right)^2 \left(-\dfrac{1}{2}\right)^4$

**28.** $\dfrac{2}{5} + \dfrac{3}{10}(-4)$

**29.** $-\dfrac{5}{8} \div \left(-\dfrac{1}{2}\right)\left(\dfrac{14}{15}\right)$

**30.** $\dfrac{\frac{5}{8}}{\frac{1}{16}}$

**31.** $\dfrac{\frac{8}{9}}{-6}$

**[4.7]** *Solve each equation. Show your work.*

**32.** $-12 = -\dfrac{3}{5}w$

**33.** $18 + \dfrac{6}{5}r = 0$

**34.** $3x - \dfrac{2}{3} = \dfrac{5}{6}$

**[4.8]** *Find the area of the triangle and the volume of each solid.*

**35.**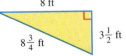
8 ft, $8\frac{3}{4}$ ft, $3\frac{1}{2}$ ft

**36.**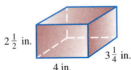
$2\frac{1}{2}$ in., 4 in., $3\frac{1}{4}$ in.

**37.**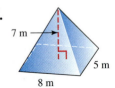
7 m, 5 m, 8 m

### MIXED REVIEW EXERCISES

*Solve each application problem.*

**38.** A chili recipe that makes 10 servings uses $2\frac{1}{2}$ pounds of meat. How much meat will be in each serving? How much meat would be needed to make 30 servings?

**39.** Yanli worked as a math tutor for $4\frac{1}{2}$ hours on Monday, $2\frac{3}{4}$ hours on Tuesday, and $3\frac{2}{3}$ hours on Friday. How much longer did she work on Monday than on Friday? How many hours did she work in all?

**40.** There are 60 children in the day care center. If $\frac{1}{5}$ of the children are preschoolers, $\frac{2}{3}$ of the children are toddlers, and the rest are infants, find the number of children in each age group.

**41.** A rectangular city park is $\frac{3}{4}$ mile long and $\frac{3}{10}$ mile wide. Find the perimeter and area of the park.

# Chapter 4 Test

*Study Skills Workbook*
Activity 11

1. Write fractions to represent the shaded and unshaded portions of the figure.

2. Graph $-\frac{2}{3}$ and $2\frac{1}{3}$ on the number line at the right.

*Write each fraction in lowest terms.*

3. $\dfrac{21}{84}$

4. $\dfrac{25}{54}$

5. $\dfrac{6a^2b}{9b^2}$

*Add, subtract, multiply, or divide, as indicated. Write all answers in lowest terms.*

6. $\dfrac{1}{6} + \dfrac{7}{10}$

7. $-\dfrac{3}{4} \div \dfrac{3}{8}$

8. $\dfrac{5}{8} - \dfrac{4}{5}$

9. $(-20)\left(-\dfrac{7}{10}\right)$

10. $\dfrac{\frac{4}{9}}{-6}$

11. $4 - \dfrac{7}{8}$

12. $-\dfrac{2}{9} + \dfrac{2}{3}$

13. $\dfrac{21}{24}\left(\dfrac{9}{14}\right)$

14. $\dfrac{12x}{7y} \div 3x$

15. $\dfrac{6}{n} - \dfrac{1}{4}$

16. $\dfrac{2}{3} + \dfrac{a}{5}$

17. $\dfrac{5}{9b^2} \cdot \dfrac{b}{10}$

18. Simplify.

    $\left(-\dfrac{1}{2}\right)^3\left(\dfrac{2}{3}\right)^2$

19. Simplify.

    $\dfrac{1}{6} + 4\left(\dfrac{2}{5} - \dfrac{7}{10}\right)$

1. _____

2.

3. _____

4. _____

5. _____

6. _____

7. _____

8. _____

9. _____

10. _____

11. _____

12. _____

13. _____

14. _____

15. _____

16. _____

17. _____

18. _____

19. _____

301

*First, round the numbers and estimate each answer. Then find the exact answer. Write exact answers in simplest form.*

20. $4\frac{4}{5} \div 1\frac{1}{8}$

21. $3\frac{2}{5} \cdot 1\frac{9}{10}$

*Solve each equation. Show your work.*

22. $7 = \frac{1}{5}d$

23. $-\frac{3}{10}t = \frac{9}{14}$

24. $0 = \frac{1}{4}b - 2$

25. $\frac{4}{3}x + 7 = -13$

*Find the area of each triangle.*

26.

27.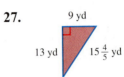

*Find the volume of each solid.*

28.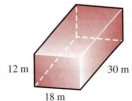

29.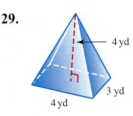

*Solve each application problem.*

30. Ann-Marie Sargent is training for an upcoming wheelchair race. She rides $4\frac{5}{6}$ hours on Monday, $6\frac{2}{3}$ hours on Tuesday, and $3\frac{1}{4}$ hours on Wednesday. How many hours did she spend in all? How many more hours did she train on Tuesday than on Monday?

31. A new vaccine is synthesized at the rate of $2\frac{1}{2}$ ounces per day. How long will it take to synthesize $8\frac{3}{4}$ ounces?

32. There are 8448 students at the Metro Community College campus. If $\frac{7}{8}$ of the students work either full time or part time, find the total number of students who work.

# Cumulative Review Exercises — Chapters 1–4

1. Write this number in words:
   505,008,238

2. Write this number using digits:
   thirty-five billion, six hundred million, nine hundred sixteen.

3. (a) Round 60,719 to the nearest hundred.

   (b) Round 99,505 to the nearest thousand.

   (c) Round 3206 to the nearest ten.

4. Name the property illustrated by each example.
   (a) $-3 \cdot 6 = 6 \cdot (-3)$

   (b) $(7 + 18) + 2 = 7 + (18 + 2)$

   (c) $5(-10 + 7) = 5 \cdot (-10) + 5 \cdot 7$

*Simplify.*

5. $9(-6)$

6. $-10 - 10$

7. $\dfrac{-14}{0}$

8. $6 + 3(2 - 7)^2$

9. $|-8| + |2|$

10. $-8 + 24 \div 2$

11. $(-4)^2 - 2^5$

12. $\dfrac{-45 \div (-5) \cdot 3}{-5 - 4(0 - 8)}$

13. $-3 \div \dfrac{3}{8}$

14. $-\dfrac{5}{6}(-42)$

15. $\dfrac{5a^2}{12} \cdot \dfrac{18}{a}$

16. $\dfrac{7}{x^2} \div \dfrac{7y^2}{3x}$

17. $\dfrac{3}{10} - \dfrac{5}{6}$

18. $-\dfrac{3}{8} + \dfrac{11}{16}$

19. $\dfrac{2}{3} - \dfrac{b}{7}$

20. $\dfrac{8}{5} + \dfrac{3}{n}$

21. $3\dfrac{1}{4} \div 2\dfrac{1}{4}$

22. $2\dfrac{2}{5} - 1\dfrac{3}{4}$

23. $\left(\dfrac{1}{2}\right)^3 (-2)^3$

24. $\dfrac{\dfrac{7}{12}}{-\dfrac{14}{15}}$

*Solve each equation. Show your work.*

**25.** $-5a + 2a = 12 - 15$

**26.** $y = -7y - 40$

**27.** $-10 = 6 + \dfrac{4}{9}k$

**28.** $20 + 5x = -2x - 8$

**29.** $3(-2m + 5) = -4m + 9 + m$

*Find the perimeter and area of each figure.*

**30.** 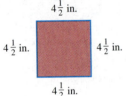 $4\frac{1}{2}$ in. on all sides

**31.** 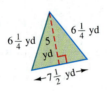 $6\frac{1}{4}$ yd, $6\frac{1}{4}$ yd, 5 yd, $7\frac{1}{2}$ yd

**32.** 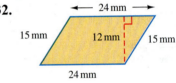 24 mm, 15 mm, 12 mm, 15 mm, 24 mm

*Solve each application problem by using the six problem-solving steps.*

**33.** Three new bags of birdseed each weighed the same amount. Sixteen pounds of seed were used from one bag and 25 pounds from another. There were still 79 pounds of seed left. How much did each bag weigh originally?

**34.** A rectangular parking lot is twice as long as it is wide. If the perimeter of the lot is 102 yd, find the length and width of the parking lot.

# Rational Numbers: Positive and Negative Decimals

**5**

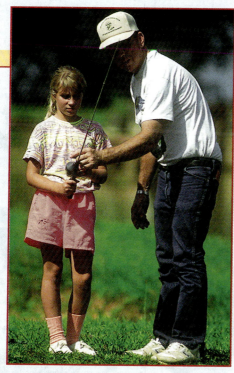

**F**reshwater fishing is America's third most popular recreational activity, and saltwater fishing is ninth. (*Source:* Sporting Goods Manufacturers Association.) In Section 5.3, Exercises 57–60, this father will use decimal numbers when he pays for his daughter's equipment. But will decimals help him and his daughter to catch their limit? (See also Section 5.1, Exercises 59–62, and Section 5.6, Exercises 65–68.)

- **5.1** Reading and Writing Decimal Numbers
- **5.2** Rounding Decimal Numbers
- **5.3** Adding and Subtracting Signed Decimal Numbers
- **5.4** Multiplying Signed Decimal Numbers
- **5.5** Dividing Signed Decimal Numbers
- **5.6** Fractions and Decimals
- **5.7** Problem Solving with Statistics: Mean, Median, Mode, and Variability
- **5.8** Geometry Applications: Pythagorean Theorem and Square Roots
- **5.9** Problem Solving: Equations Containing Decimals
- **5.10** Geometry Applications: Circles, Cylinders, and Surface Area

# 5.1 READING AND WRITING DECIMAL NUMBERS

**OBJECTIVES**

1. Write parts of a whole using decimals.
2. Identify the place value of a digit.
3. Read decimals.
4. Write decimals as fractions or mixed numbers.

In Chapter 4 you worked with rational numbers written in fraction form to represent parts of a whole. In this chapter, we will use rational numbers written as **decimals** to show parts of a whole. For example, our money system is based on decimals. One dollar is divided into 100 equivalent parts. One cent ($0.01) is one of the parts, and a dime ($0.10) is 10 of the parts. Metric measurement (see Chapter 8) is also based on decimals.

 **Write parts of a whole using decimals.**

Decimals are used when a whole is divided into 10 equivalent parts or into 100 or 1000 or 10,000 equivalent parts. In other words, decimals are fractions with denominators that are a power of 10. For example, the square at the right is cut into 10 equivalent parts. Written as a fraction, each part is $\frac{1}{10}$ of the whole. Written as a decimal, each part is **0.1**. Both $\frac{1}{10}$ and 0.1 are read as "*one tenth*."

One-tenth of the square is shaded.

The dot in 0.1 is called the **decimal point**.

0.1
↑
Decimal point

The square at the right has 7 of its 10 parts shaded.

Written as a *fraction*, $\frac{7}{10}$ of the square is shaded.

Written as a *decimal*, **0.7** of the square is shaded.

Both $\frac{7}{10}$ and 0.7 are read as "*seven tenths*."

Seven-tenths of the square is shaded.

**Work Problem 1 at the Side.**

① There are 10 dimes in one dollar. Each dime is $\frac{1}{10}$ of a dollar. Write a fraction, a decimal, and the words that name the yellow shaded portion of each dollar.

(a)

(b)

(c)

Each square below is cut into 100 equivalent parts. Written as a fraction, each part is $\frac{1}{100}$ of the whole.
Written as a decimal, each part is **0.01** of the whole.
Both $\frac{1}{100}$ and 0.01 are read as "*one hundredth*."

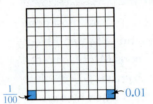 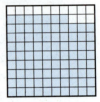

Eighty-seven hundredths of the square is shaded.

The square above on the right has 87 of its 100 parts shaded.

Written as a fraction, $\frac{87}{100}$ of the total area is shaded.

Written as a decimal, **0.87** of the total area is shaded.

Both $\frac{87}{100}$ and 0.87 are read as "*eighty-seven hundredths*."

**ANSWERS**

1. (a) $\frac{1}{10}$; 0.1; one tenth
   (b) $\frac{3}{10}$; 0.3; three tenths
   (c) $\frac{9}{10}$; 0.9; nine tenths

Section 5.1 Reading and Writing Decimal Numbers  **307**

Work Problem ❷ at the Side.

The example below shows several numbers written as fractions, as decimals, and in words.

**Example 1** Using the Decimal Forms of Fractions

| | Fraction | Decimal | Read As |
|---|---|---|---|
| (a) | $\frac{4}{10}$ | 0.4 | four tenths |
| (b) | $-\frac{9}{100}$ | $-0.09$ | negative nine hundredths |
| (c) | $\frac{71}{100}$ | 0.71 | seventy-one hundredths |
| (d) | $\frac{8}{1000}$ | 0.008 | eight thousandths |
| (e) | $-\frac{45}{1000}$ | $-0.045$ | negative forty-five thousandths |
| (f) | $\frac{832}{1000}$ | 0.832 | eight hundred thirty-two thousandths |

Work Problem ❸ at the Side.

**2** **Identify the place value of a digit.** The decimal point separates the *whole number part* from the *fractional part* in a decimal number. In the chart below, you see that the **place value names** for fractional parts are similar to those on the whole number side but end in "*ths*."

**Decimal Place Value Chart**

hundred thousands | ten thousands | thousands | hundreds | tens | **ones** | tenths | hundredths | thousandths | ten-thousandths | hundred-thousandths
100,000 | 10,000 | 1000 | 100 | 10 | 1 | $\frac{1}{10}$ | $\frac{1}{100}$ | $\frac{1}{1000}$ | $\frac{1}{10,000}$ | $\frac{1}{100,000}$

Whole number part | Decimal point (Read "and") | Fractional part

Notice that the ones place is at the center. (There is no "oneths" place.) Also notice that each place is 10 times the value of the place to its right.

❷ Write the portion of each square that is shaded as a fraction, as a decimal, and in words.

(a)
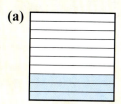

(b)
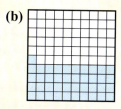

❸ Write each decimal as a fraction.

(a) $-0.7$

(b) $0.2$

(c) $-0.03$

(d) $0.69$

(e) $0.047$

(f) $-0.351$

**ANSWERS**

2. (a) $\frac{3}{10}$; 0.3; three tenths

   (b) $\frac{41}{100}$; 0.41; forty-one hundredths

3. (a) $-\frac{7}{10}$ (b) $\frac{2}{10}$ (c) $-\frac{3}{100}$

   (d) $\frac{69}{100}$ (e) $\frac{47}{1000}$ (f) $-\frac{351}{1000}$

## Chapter 5 Rational Numbers: Positive and Negative Decimals

**4** Identify the place value of each digit.

(a) 971.54

(b) 0.4

(c) 5.60

(d) 0.0835

**5** Tell how to read each decimal in words.

(a) 0.6

(b) 0.46

(c) 0.05

(d) 0.409

(e) 0.0003

(f) 0.0703

(g) 0.088

**ANSWERS**

4. (a) 9 7 1 . 5 4 — hundreds, tens, ones, tenths, hundredths  (b) 0 . 4 — ones, tenths
   (c) 5 . 6 0 — ones, tenths, hundredths  (d) 0 . 0 8 3 5 — ones, tenths, hundredths, thousandths, ten-thousandths

5. (a) six tenths
   (b) forty-six hundredths
   (c) five hundredths
   (d) four hundred nine thousandths
   (e) three ten-thousandths
   (f) seven hundred three ten-thousandths
   (g) eighty-eight thousandths

### CAUTION

In this chapter, if a number does *not* have a decimal point, it is an *integer*. An integer has no fractional part. If you want to show the decimal point in an integer, it is just to the *right* of the digit in the ones place. For example:

$$8 = 8. \qquad 306 = 306. \qquad -42 = -42.$$

↑ Decimal point    ↑ Decimal point    ↑ Decimal point

### Example 2  Identifying the Place Value of a Digit

Identify the place value of each digit.

(a) 178.36

1 7 8 . 3 6 — hundreds, tens, ones, tenths, hundredths

(b) 0.00935

0 . 0 0 9 3 5 — ones, tenths, hundredths, thousandths, ten-thousandths, hundred-thousandths

Notice in Example 2(b) that we do *not* use commas on the right side of the decimal point.

**Work Problem 4 at the Side.**

**3** Read decimals.  A decimal is read according to its form as a fraction.

0. 9 — ones, tenths

We read 0.9 as "nine tenths" because 0.9 is the same as $\frac{9}{10}$. Notice that 0.9 ends in the tenths place.

0. 0 2 — ones, tenths, hundredths

We read 0.02 as "two hundredths" because 0.02 is the same as $\frac{2}{100}$. Notice that 0.02 ends in the hundredths place.

### Example 3  Reading Decimal Numbers

Tell how to read each decimal in words.

(a) 0.3

Because $0.3 = \frac{3}{10}$, read the decimal as: three ten**ths**.

(b) 0.49   Read it as:   forty-nine hundred**ths**.

(c) 0.08   Read it as:   eight hundred**ths**.

(d) 0.918   Read it as:   nine hundred eighteen thousand**ths**.

(e) 0.0106   Read it as:   one hundred six ten-thousand**ths**.

**Work Problem 5 at the Side.**

## Reading a Decimal Number

*Step 1*  Read any whole number part to the *left* of the decimal point as you normally would.

*Step 2*  Read the decimal point as "*and*."

*Step 3*  Read the part of the number to the *right* of the decimal point as if it were an ordinary whole number.

*Step 4*  Finish with the place value name of the rightmost digit; these names all end in "*ths*."

**NOTE**

If there is *no whole number part*, you will use only Steps 3 and 4.

### Example 4  Reading Decimal Numbers

Read each decimal.

(a)
9 is in tenths place.

16.9

sixteen **and** nine **tenths**

16.9 is read "sixteen and nine tenths."

(b)
5 is in hundredths place.

482.35

four hundred eighty-two **and** thirty-five **hundredths**

482.35 is read "four hundred eighty-two and thirty-five hundredths."

3 is in thousandths place.

(c) 0.063 is "sixty-three **thousandths**." (No whole number part.)

(d) 11.1085 is "eleven **and** one thousand eighty-five **ten-thousandths**."

**CAUTION**

Use "and" *only* when reading a decimal point. A common mistake is to read the whole number 405 as "four hundred *and* five." But there is *no decimal point* shown in 405, so it is read "four hundred five."

Work Problem ❻ at the Side.

**4**  **Write decimals as fractions or mixed numbers.**  Knowing how to read decimals will help you when writing decimals as fractions or mixed numbers.

## Writing a Decimal as a Fraction or Mixed Number

*Step 1*  The digits to the right of the decimal point are the numerator of the fraction.

*Step 2*  The denominator is 10 for tenths, 100 for hundredths, 1000 for thousandths, 10,000 for ten-thousandths, and so on.

*Step 3*  If the decimal has a whole number part, the fraction will be a mixed number with the same whole number part.

❻ Tell how to read each decimal in words.

(a) 3.8

(b) 15.001

(c) 0.0073

(d) 64.309

**Answers**

6. (a) three and eight tenths
   (b) fifteen and one thousandth
   (c) seventy-three ten-thousandths
   (d) sixty-four and three hundred nine thousandths

**7** Write each decimal as a fraction or mixed number.

(a) 0.7

(b) 12.21

(c) 0.101

(d) 0.007

(e) 1.3717

**8** Write each decimal as a fraction or mixed number in lowest terms.

(a) 0.5

(b) 12.6

(c) 0.85

(d) 3.05

(e) 0.225

(f) 420.0802

**ANSWERS**

7. (a) $\frac{7}{10}$  (b) $12\frac{21}{100}$  (c) $\frac{101}{1000}$
   (d) $\frac{7}{1000}$  (e) $1\frac{3717}{10,000}$

8. (a) $\frac{1}{2}$  (b) $12\frac{3}{5}$  (c) $\frac{17}{20}$  (d) $3\frac{1}{20}$
   (e) $\frac{9}{40}$  (f) $420\frac{401}{5000}$

## Example 5 — Writing Decimals as Fractions or Mixed Numbers

Write each decimal as a fraction or mixed number.

(a) 0.19

The digits to the right of the decimal point, 19, are the numerator of the fraction. The denominator is 100 for hundredths because the rightmost digit is in the hundredths place.

$$0.1\underset{\uparrow}{9} = \frac{19}{100} \leftarrow \text{100 for hundredths}$$

Hundredths place

(b) 0.863

$$0.86\underset{\uparrow}{3} = \frac{863}{1000} \leftarrow \text{1000 for thousandths}$$

Thousandths place

(c) 4.0099

The whole number part stays the same.

$$4.009\underset{\uparrow}{9} = 4\frac{99}{10{,}000} \leftarrow \text{10,000 for ten-thousandths}$$

Ten-thousandths place

**Work Problem 7 at the Side.**

**CAUTION**

After you write a decimal as a fraction or a mixed number, check to see if the fraction is in lowest terms.

## Example 6 — Writing Decimals as Fractions or Mixed Numbers

Write each decimal as a fraction or mixed number in lowest terms.

(a) $0.4 = \frac{4}{10} \leftarrow$ 10 for tenths

Write $\frac{4}{10}$ in lowest terms. $\quad \frac{4}{10} = \frac{4 \div 2}{10 \div 2} = \frac{2}{5} \leftarrow$ Lowest terms

(b) $0.75 = \frac{75}{100} = \frac{75 \div 25}{100 \div 25} = \frac{3}{4} \leftarrow$ Lowest terms

(c) $18.105 = 18\frac{105}{1000} = 18\frac{105 \div 5}{1000 \div 5} = 18\frac{21}{200} \leftarrow$ Lowest terms

(d) $42.8085 = 42\frac{8085}{10{,}000} = 42\frac{8085 \div 5}{10{,}000 \div 5} = 42\frac{1617}{2000} \leftarrow$ Lowest terms

**Work Problem 8 at the Side.**

**Calculator Tip** In this book we will write 0.45 instead of just .45, to emphasize that there is no whole number. Your *scientific* calculator shows these zeros also. Enter 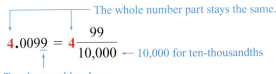 and notice that the display automatically shows 0.45 even though you did not press 0. For comparison, enter the whole number 45 by pressing (4) (5) (+) and notice where the decimal point is shown in the display. It automatically appears to the *right* of the 5. (*Graphing* calculators do not automatically show a zero in the ones place.)

## 5.1 Exercises

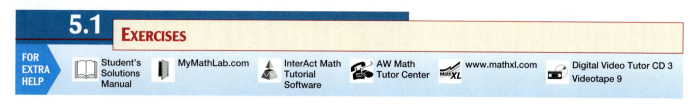

*Identify the digit that has the given place value. See Example 2.*

1. 70.489
   tens
   ones
   tenths

2. 135.296
   ones
   tenths
   tens

3. 0.2518
   hundredths
   thousandths
   ten-thousandths

4. 0.9347
   hundredths
   thousandths
   ten-thousandths

5. 93.01472
   thousandths
   ten-thousandths
   tenths

6. 0.51968
   tenths
   ten-thousandths
   hundredths

7. 314.658
   tens
   tenths
   hundreds

8. 51.325
   tens
   tenths
   hundredths

9. 149.0832
   hundreds
   hundredths
   ones

10. 3458.712
    hundreds
    hundredths
    tenths

11. 6285.7125
    thousands
    thousandths
    hundredths

12. 5417.6832
    thousands
    thousandths
    ones

*Write the decimal number that has the specified place values. See Example 2.*

13. 0 ones, 5 hundredths, 1 ten, 4 hundreds, 2 tenths

14. 7 tens, 9 tenths, 3 ones, 6 hundredths, 8 hundreds

15. 3 thousandths, 4 hundredths, 6 ones, 2 ten-thousandths, 5 tenths

16. 8 ten-thousandths, 4 hundredths, 0 ones, 2 tenths, 6 thousandths

17. 4 hundredths, 4 hundreds, 0 tens, 0 tenths, 5 thousandths, 5 thousands, 6 ones

18. 7 tens, 7 tenths, 6 thousands, 6 thousandths, 3 hundreds, 3 hundredths, 2 ones

*Write each decimal as a fraction or mixed number in lowest terms. See Examples 1, 5, and 6.*

**19.** 0.7  **20.** 0.1  **21.** 13.4  **22.** 9.8  **23.** 0.35

**24.** 0.85  **25.** 0.66  **26.** 0.33  **27.** 10.17  **28.** 31.99

**29.** 0.06  **30.** 0.08  **31.** 0.205  **32.** 0.805

**33.** 5.002  **34.** 4.008  **35.** 0.686  **36.** 0.492

*Tell how to read each decimal in words. See Examples 1, 3, and 4.*

**37.** 0.5  **38.** 0.2

**39.** 0.78  **40.** 0.55

**41.** 0.105  **42.** 0.609

**43.** 12.04  **44.** 86.09

**45.** 1.075  **46.** 4.025

*Write each decimal in numbers. See Examples 3 and 4.*

**47.** Six and seven tenths

**48.** Eight and twelve hundredths

**49.** Thirty-two hundredths

**50.** One hundred eleven thousandths

**51.** Four hundred twenty and eight thousandths

**52.** Two hundred and twenty-four thousandths

**53.** Seven hundred three ten-thousandths

**54.** Eight hundred and six hundredths

**55.** Seventy-five and thirty thousandths

**56.** Sixty and fifty hundredths

**57.** Anne read the number 4302 as "four thousand three hundred and two." Explain what is wrong with the way Anne read the number.

**58.** Jerry read the number 9.0106 as "nine and one hundred and six ten-thousandths." Explain the error he made.

*The dad on the first page of this chapter needs to select the correct fishing line for his daughter's reel. Fishing line is sold according to how many pounds of "pull" the line can withstand before breaking. Use the table to answer Exercises 59–62. Write all fractions in lowest terms. (Note: The diameter of the fishing line is its thickness.)*

**FISHING LINE**

| Test Strength (pounds) | Average Diameter (inches) |
| --- | --- |
| 4 | 0.008 |
| 8 | 0.010 |
| 12 | 0.013 |
| 14 | 0.014 |
| 17 | 0.015 |
| 20 | 0.016 |

*Source:* Berkley Outdoor Technologies Group.

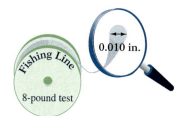

The diameter is the distance across the end of the line, or its thickness.

**59.** Write the diameter of 8-pound test line in words and as a fraction.

**60.** Write the diameter of 17-pound test line in words and as a fraction.

**61.** What is the test strength of the line with a diameter of $\frac{13}{1000}$ inch?

**62.** What is the test strength of the line with a diameter of sixteen thousandths inch?

Suppose your job is to take phone orders for precision parts. Use the table below. In Exercises 63–68, write the correct part number that matches what you hear the customer say over the phone. In Exercises 67–68, write the words you would say to the customer.

| Part Number | Size in Centimeters |
|---|---|
| 3-A | 0.06 |
| 3-B | 0.26 |
| 3-C | 0.6 |
| 3-D | 0.86 |
| 4-A | 1.006 |
| 4-B | 1.026 |
| 4-C | 1.06 |
| 4-D | 1.6 |
| 4-E | 1.602 |

63. "Please send the six-tenths centimeter bolt."

    Part number _____

64. "The part missing from our order was the one and six hundredths size."

    Part number _____

65. "The size we need is one and six thousandths centimeters."

    Part number _____

66. "Do you still stock the twenty-six hundredths centimeter bolt?"

    Part number _____

67. "What size is part number 4-E?" Write your answer in words.

68. "What size is part number 4-B?" Write your answer in words.

### Relating Concepts (Exercises 69–76) For Individual or Group Work

*Use your knowledge of place value to* **work Exercises 69–76 in order.**

69. Look back at the Decimal Place Value Chart in this section. What do you think would be the names of the next four places to the *right* of hundred-thousandths? What information did you use to come up with these names?

70. A common mistake is to think that the first place to the right of the decimal point is "oneths" and the second place is "tenths." Why might someone make that mistake? How would you explain why there is no "oneths" place?

71. Use your answer from Exercise 69 to write 0.72436955 in words.

72. Use your answer from Exercise 69 to write 0.000678554 in words.

73. Write 8006.500001 in words.

74. Write 20,060.000505 in words.

75. Write this decimal using digits: three hundred two thousand forty ten-millionths.

76. Write this decimal using digits: nine billion, eight hundred seventy-six million, five hundred forty-three thousand, two hundred ten and one hundred million two hundred thousand three hundred billionths.

## 5.2 ROUNDING DECIMAL NUMBERS

**Section 1.5** showed how to round integers. For example, 89 rounded to the nearest ten is 90, and 8512 rounded to the nearest hundred is 8500.

**1** **Learn the rules for rounding decimals.** It is also important to be able to **round** decimals. For example, a store is selling 2 candy mints for $0.75 but you want only one mint. The price of each mint is $0.75 ÷ 2, which is $0.375, but you cannot pay part of a cent. Is $0.375 closer to $0.37 or to $0.38? Actually, it's exactly halfway between. When this happens in everyday situations, the rule is to round *up*. The store will charge you $0.38 for the mint.

### OBJECTIVES

1. Learn the rules for rounding decimals.
2. Round decimals to any given place.
3. Round money amounts to the nearest cent or nearest dollar.

### Rounding a Decimal Number

*Step 1*   Find the place to which the rounding is being done. Draw a "cut-off" line *after* that place to show that you are cutting off and dropping the rest of the digits.

*Step 2*   Look **only** at the **first** digit you are cutting off.

*Step 3A*  If this digit is **4 or less,** the part of the number you are keeping *stays the same.*

*Step 3B*  If this digit is **5 or more,** you must **round up** the part of the number you are keeping.

*Step 4*   You can use the ≈ symbol or the ≐ symbol to indicate that the rounded number is now an approximation (close, but not exact). Both symbols mean "is approximately equal to." (In this book we will use the ≈ symbol.)

**CAUTION**

Do *not* move the decimal point when rounding.

**2** **Round decimals to any given place.** These examples show you how to round decimals.

### Example 1   Rounding a Decimal Number

Round 14.39652 to the nearest thousandth. Is it closer to 14.396 or to 14.397?

*Step 1*   Draw a "cut-off" line after the thousandths place.

$$14.396 \,|\, 52$$

Thousandths

You are cutting off the 5 and 2. They will be dropped.

*Step 2*   Look *only* at the *first* digit you are cutting off. Ignore the other digits you are cutting off.

$$14.396 \,|\, 52$$

Look *only* at the 5.
Ignore the 2.

*Continued on Next Page*

**1** Round to the nearest thousandth.

(a) 0.33492

(b) 8.00851

(c) 265.42068

(d) 10.70180

*Step 3* If the first digit you are cutting off is 5 or more, round up the part of the number you are keeping.

$$14.396\,|\,52$$
$$+\;\;\;0.001$$
$$\overline{14.397}$$

First digit cut is *5 or more,* so round up by adding 1 thousandth to the part you are keeping.

So, 14.39652 rounded to the nearest thousandth is 14.397.
You can write 14.39652 ≈ 14.397.

**CAUTION**

When rounding integers in **Section 1.5,** you kept all the digits but changed some to zeros. With decimals, you cut off and *drop the extra digits.* In the example above, 14.39652 rounds to 14.397 ***not*** 14.39700.

**Work Problem** ❶ **at the Side.**

In Example 1, the rounded number 14.397 had *three decimal places.* **Decimal places** are the number of digits to the *right* of the decimal point. The first decimal place is tenths, the second is hundredths, the third is thousandths, and so on.

**Example 2** Rounding Decimals to Different Places

Round to the place indicated.

**(a)** Round 5.3496 to the nearest tenth. (Is it closer to 5.3 or to 5.4?)

*Step 1* Draw a cut-off line after the tenths place.

$$5\,.\,3\,|\,4\,9\,6$$
Tenths ⬏     You are cutting off the 4, 9, and 6.

*Step 2*     $5\,.\,3\,|\,4\,9\,6$
Look *only* at the 4.
Ignore these digits.

*Step 3*     $5\,.\,3\,|\,4\,9\,6$
First digit cut is *4 or less,* so the part you are keeping stays the same.

5.3 ⬅ Stays the same

So, 5.3496 rounded to the nearest tenth is 5.3 (one decimal place for tenths). You can write 5.3496 ≈ 5.3. Notice that it does ***not*** round to 5.3000 (which would be ten-thousandths instead of tenths).

**(b)** Round 0.69738 to the nearest hundredth. (Is it closer to 0.69 or to 0.70?)

*Step 1*     $0\,.\,6\,9\,|\,7\,3\,8$     Draw a cut-off line after the hundredths place.
Hundredths

*Step 2*     $0\,.\,6\,9\,|\,7\,3\,8$     Look *only* at the 7.

Continued on Next Page

ANSWERS
**1.** (a) 0.335  (b) 8.009  (c) 265.421  (d) 10.702

Step 3    0.69|738 ── First digit cut is *5 or more,* so round up by adding 1 hundredth to the part you are keeping.

$$\begin{array}{r} \overset{1}{0.6\,9} \\ +\,0.0\,1 \\ \hline 0.7\,0 \end{array}$$ ← Keep this part.
← To round up, add 1 hundredth.
← 9 + 1 is 10; write 0 and carry 1 to the tenths place.

So, 0.69738 rounded to the nearest hundredth is 0.70. Hundredths is *two* decimal places so you *must* write the 0 in the hundredths place.
You can write 0.69738 ≈ 0.70.

**(c)** Round 0.01806 to the nearest thousandth. (Is it closer to 0.018 or to 0.019?)

0.018|06 ── First digit cut is *4 or less,* so the part you are keeping stays the same.
0.018

So, 0.01806 rounded to the nearest thousandth is 0.018 (three decimal places for thousandths).
You can write 0.01806 ≈ 0.018.

**(d)** Round 57.976 to the nearest tenth. (Is it closer to 57.9 or to 58.0?)

57.9|76 ── First digit cut is *5 or more,* so round up by adding 1 tenth to the part you are keeping.

$$\begin{array}{r} \overset{1}{57.9} \\ +\,0.1 \\ \hline 58.0 \end{array}$$ ← 9 + 1 is 10; write the 0 and carry 1 to the ones place.

So, 57.976 rounded to the nearest tenth is 58.0. You can write 57.976 ≈ 58.0.
You *must* write the 0 in the tenths place to show that the number was rounded to the nearest tenth.

**CAUTION**

Check that your rounded answer shows *exactly* the number of decimal places called for, even if a 0 is in that place. Be sure your answer shows *one decimal place* if you rounded to *tenths,* *two decimal places* for *hundredths,* or *three decimal places* for *thousandths.*

Work Problem ❷ at the Side.

**3** Round money amounts to the nearest cent or nearest dollar. When you are shopping in a store, money amounts are usually rounded to the nearest cent. There are 100 cents in a dollar.

Each cent is $\frac{1}{100}$ of a dollar.

Another way to write $\frac{1}{100}$ is 0.01. So rounding to the *nearest cent* is the same as rounding to the *nearest hundredth of a dollar.*

❷ Round to the place indicated.

(a) 0.8988 to the nearest hundredth

(b) 5.8903 to the nearest hundredth

(c) 11.0299 to the nearest thousandth

(d) 0.545 to the nearest tenth

**ANSWERS**
2. (a) 0.90 (b) 5.89 (c) 11.030 (d) 0.5

**3** Round each money amount to the nearest cent.

(a) $14.595

You pay _____

(b) $578.0663

You pay _____

(c) $0.849

You pay _____

(d) $0.0548

You pay _____

---

**Example 3** Rounding to the Nearest Cent

Round each money amount to the nearest cent.

(a) $2.4238 (Is it closer to $2.42 or to $2.43?)

$2.42|38 ← First digit cut is *4 or less,* so the part you are keeping stays the same.

$2.42 ← You pay $2.42.

$2.4238 rounded to the nearest cent is $2.42.

(b) $0.695 (Is it closer to $0.69 or to $0.70?)

$0.69|5 ← 5 or more; round up

$0.69
+ $0.01 ← To round up, add 1 hundredth (1 cent).
$0.70 ← You pay $0.70.

$0.695 rounded to the nearest cent is $0.70.

**Work Problem 3 at the Side.**

> **NOTE**
>
> A few stores round *all* money amounts up to the next higher cent, even if the next digit is *less* than 5. In Example 3(a) above, some stores would round $2.4238 up to $2.43, even though it is closer to $2.42.

It is also common to round money amounts to the nearest dollar. For example, you can do that on your federal and state income tax returns to make the calculations easier.

**Example 4** Rounding to the Nearest Dollar

Round to the nearest dollar.

(a) $48.69 (Is it closer to $48 or to $49?)

$48.|69 ← First digit cut is *5 or more,* so round up by adding $1.

$48
+   1
$49

$48.69 rounded to the nearest dollar is $49.

> **CAUTION**
>
> $48.69 rounded to the nearest dollar is $49. Be careful to write the answer as $49 to show that the rounding is to the *nearest dollar.* Writing $49.00 would show rounding to the nearest *cent.*

**Continued on Next Page**

---

**ANSWERS**

**3.** (a) $14.60  (b) $578.07  (c) $0.85
(d) $0.05

**(b)** $594.36 (Is it closer to $594 or to $595?)

$594.|36    ← First digit cut is *4 or less*, so the part you keep stays the same.

$594

$594.36 rounded to the nearest dollar is $594.

**(c)** $349.88 (Is it closer to $349 or to $350?)

$349.|88    ← 5 or more; round up by adding $1.

$349
+    1
─────
$350

$349.88 rounded to the nearest dollar is $350.

**(d)** $2689.50 rounded to the nearest dollar is $2690.

**(e)** $0.61 rounded to the nearest dollar is $1.

**Calculator Tip**  Accountants and other people who work with money amounts often set their calculators to automatically round to 2 decimal places (nearest cent) or to round to 0 decimal places (nearest dollar). Your calculator may have this feature.

Work Problem ④ at the Side.

---

④ Round to the nearest dollar.

**(a)** $29.10

**(b)** $136.49

**(c)** $990.91

**(d)** $5949.88

**(e)** $49.60

**(f)** $0.55

**(g)** $1.08

---

**ANSWERS**
**4. (a)** $29  **(b)** $136  **(c)** $991
    **(d)** $5950  **(e)** $50  **(f)** $1  **(g)** $1

# Focus on Real-Data Applications

## Lawn Fertilizer

### Gotta Be Green

A lot's being said about personal responsibility these days, and the idea seems to be ending up on the front lawn—literally! Each spring, homeowners across the country gear up to green up their lawns, and the increased use of fertilizer has a lot of environmentalists concerned about the potential effects of chemical runoff into nearby rivers and streams.

Each year, according to a study conducted by the University of Minnesota's Department of Agriculture, each household in the Minneapolis/St. Paul metro area uses an average of 36 pounds of lawn fertilizer. That adds up to 25,529,295 pounds, or 12,765 tons. Add to that another 193,000 pounds of weed killer and you're looking at the total picture for keeping it green in the Twin Cities.

*Source: Minneapolis Star Tribune.*

1. According to the article,
   (a) How many pounds of lawn fertilizer are used each year in the *entire metro area?*
   (b) Do a division on your calculator to find the number of *households* in the metro area.
   (c) Would it make sense to round your answer to part (b)? If so, how would you round it?
   (d) How many pounds of *weed killer* are used each year in the entire metro area?
   (e) Do a division on your calculator to find the number of pounds of *weed killer* used by each household in the metro area. Round your answer to the nearest hundredth.

2. There are 2000 pounds in one ton.
   (a) Find the number of tons equivalent to 25,529,295 pounds of fertilizer.
   (b) Does your answer match the figure given in the article? If not, what did the author of the article do to get 12,765 tons?
   (c) Is the author's figure accurate? Why or why not?
   (d) Find the number of tons equivalent to 193,000 pounds of weed killer.
   (e) How can you write the division problem 193,000 ÷ 2,000 in a simpler way? Explain what you did.

3. According to the article,
   (a) "each household in the Minneapolis/St. Paul metro area uses an average of 36 pounds of lawn fertilizer" each year. What mathematical operation do you do to find an average?

   (b) When the calculations were done to find the average, the answer was probably not *exactly* 36 pounds. List six different values that are *less than* 36 that would round to 36. List two values with one decimal place; two values with two decimal places, and two values with three decimal places.

   (c) List six different values that are *greater than* 36 that would round to 36. List two values each with one, two, and three decimal places.

   (d) What is the *smallest* number that can be rounded to 36? What is the *largest* number?

320

## 5.2 Exercises

*Round each number to the place indicated. See Examples 1 and 2.*

1. 16.8974 to the nearest tenth
2. 193.845 to the nearest hundredth
3. 0.95647 to the nearest thousandth
4. 96.81584 to the nearest ten-thousandth
5. 0.799 to the nearest hundredth
6. 0.952 to the nearest tenth
7. 3.66062 to the nearest thousandth
8. 1.5074 to the nearest hundredth
9. 793.988 to the nearest tenth
10. 476.1196 to the nearest thousandth
11. 0.09804 to the nearest ten-thousandth
12. 176.004 to the nearest tenth
13. 48.512 to the nearest one
14. 3.385 to the nearest one
15. 9.0906 to the nearest hundredth
16. 30.1290 to the nearest thousandth
17. 82.000151 to the nearest ten-thousandth
18. 0.400594 to the nearest ten-thousandth

*Nardos is grocery shopping. The store will round the amount she pays for each item to the nearest cent. Write the rounded amounts. See Example 3.*

19. Soup is three cans for $2.45, so one can is $0.81666. Nardos pays _____.
20. Orange juice is two cartons for $2.69, so one carton is $1.345. Nardos pays _____.
21. Facial tissue is four boxes for $4.89, so one box is $1.2225. Nardos pays _____.
22. Muffin mix is three packages for $1.75, so one package is $0.58333. Nardos pays _____.
23. Candy bars are six for $2.99, so one bar is $0.4983. Nardos pays _____.
24. Boxes of spaghetti are four for $3.59, so one box is $0.8975. Nardos pays _____.

*As she gets ready to do her income tax return, Ms. Chen rounds each amount to the nearest dollar. Write the rounded amounts. See Example 4.*

25. Income from job, $48,649.60
26. Income from interest on bank account, $69.58
27. Union dues, $310.08
28. Federal withholding, $6064.49
29. Donations to charity, $848.91
30. Medical expenses, $609.38

*Round each money amount as indicated.*

**31.** $499.98 to the nearest dollar

**32.** $9899.59 to the nearest dollar

**33.** $0.996 to the nearest cent

**34.** $0.09929 to the nearest cent

**35.** $999.73 to the nearest dollar

**36.** $9999.80 to the nearest dollar

*The table lists speed records for various types of transportation. Use the table to answer Exercises 37–40.*

| Record | Speed (miles per hour) |
|---|---|
| Land speed record (specially built car) | 763.04 |
| Motorcycle speed record (specially adapted motorcycle) | 322.15 |
| Fastest train (regular passenger service) | 162.7 |
| Fastest X-15 (military jet) | 4520 |
| Boeing 737-300 airplane (regular passenger service) | 495 |
| Indianapolis 500 Auto Race (fastest average winning speed) | 185.984 |
| Daytona 500 Auto Race (fastest average winning speed) | 177.602 |

*Source: The Top 10 of Everything 2000.*

**37.** Round these speed records to the nearest tenth.

(a) Indianapolis 500 average winning speed

(b) Land speed record

**38.** Round these speed records to the nearest hundredth.

(a) Daytona 500 average winning speed

(b) Indianapolis 500 average winning speed

**39.** Round these speed records to the nearest whole number.

(a) Motorcycle

(b) Train

**40.** Round these speed records to the nearest hundred.

(a) X-15 military jet

(b) Boeing 737-300 airplane

### RELATING CONCEPTS (Exercises 41–44)  FOR INDIVIDUAL OR GROUP WORK

*Use your knowledge about rounding money amounts to* **work Exercises 41–44 in order.**

**41.** Explain what happens when you round $0.499 to the nearest dollar. Why does this happen?

**42.** Look again at Exercise 41. How else could you round $0.499 that would be more helpful? What kind of guideline does this suggest about rounding to the nearest dollar?

**43.** Explain what happens when you round $0.0015 to the nearest cent. Why does this happen?

**44.** Suppose you want to know which of these amounts is less, so you round them both to the nearest cent.

$0.5968     $0.6014

Explain what happens. Describe what you could do instead of rounding to the nearest cent.

## 5.3 Adding and Subtracting Signed Decimal Numbers

**1** **Add and subtract positive decimals.** When adding or subtracting *whole* numbers, you line up the numbers in columns so that you are adding ones to ones, tens to tens, and so on. A similar idea applies to adding or subtracting *decimal* numbers. With decimals, you line up the decimal points to be sure that you are adding tenths to tenths, hundredths to hundredths, and so on.

### Adding and Subtracting Decimal Numbers

*Step 1* Write the numbers in columns with the decimal points lined up.

*Step 2* If necessary, write in zeros so both numbers have the same number of decimal places. Then add or subtract as if they were whole numbers.

*Step 3* Line up the decimal point in the answer directly below the decimal points in the problem.

### Example 1  Adding Decimal Numbers

Add.

**(a)** 16.92 and 48.34

*Step 1* Write the numbers in columns with the decimal points lined up.

```
   tens
   ones
    tenths
    hundredths
   1 6 . 9 2
 + 4 8 . 3 4
```
⎯ Decimal points are lined up.

*Step 2* Add as if these were whole numbers.

```
    1 1
   1 6 . 9 2
 + 4 8 . 3 4
   ─────────
   6 5 . 2 6
```
*Step 3* ⎯ Decimal point in answer is lined up under decimal points in problem.

**(b)** 5.897 + 4.632 + 12.174

Write the numbers in columns with the decimal points lined up. Then add.

```
    1 1   2 1
      5 . 8 9 7
      4 . 6 3 2
 +  1 2 . 1 7 4
    ───────────
    2 2 . 7 0 3
```
⎯ Decimal points are lined up.

**Work Problem** ❶ **at the Side.**

In Example 1(a) above, both numbers had *two* decimal places (two digits to the right of the decimal point). In Example 1(b), all the numbers had *three decimal places* (three digits to the right of the decimal point). That made it easy to add tenths to tenths, hundredths to hundredths, and so on.

---

**OBJECTIVES**

**1** Add and subtract positive decimals.

**2** Add and subtract negative decimals.

**3** Estimate the answer when adding or subtracting decimals.

❶ Find each sum.

**(a)** 2.86 + 7.09

**(b)** 13.761 + 8.325

**(c)** 0.319 + 56.007 + 8.252

**(d)** 39.4 + 0.4 + 177.2

**ANSWERS**

**1.** (a) 9.95  (b) 22.086  (c) 64.578
(d) 217.0

**Chapter 5** Rational Numbers: Positive and Negative Decimals

❷ Find each sum.

(a) $6.54 + 9.8$

If the number of decimal places does *not* match, you can write in zeros as placeholders to make them match. This is shown in Example 2.

### Example 2  Writing Zeros as Placeholders Before Adding

Add.

(a) $7.3 + 0.85$

There are two decimal places in 0.85 (tenths and hundredths), so write a 0 in the hundredths place in 7.3 so that it has two decimal places also.

$$\begin{array}{r} 7.30 \\ + 0.85 \\ \hline 8.15 \end{array} \leftarrow \text{One 0 is written in.}$$

$7.30$ is equivalent to $7.3$ because $7\frac{30}{100}$ in lowest terms is $7\frac{3}{10}$.

(b) **0.831 + 222.2 + 10**

(b) $6.42 + 9 + 2.576$

Write in zeros so that all the addends have three decimal places. Notice how the whole number 9 is written with the decimal point on the *right* side. (If you put the decimal point on the *left* side of the 9, you would turn it into the decimal fraction 0.9.)

$$\begin{array}{r} 6.420 \\ 9.000 \\ + 2.576 \\ \hline 17.996 \end{array}$$

← One 0 is written in.
← 9 is a whole number; decimal point and three zeros are written in.
← No zeros are needed.

(c) $8.64 + 39.115 + 3.0076$

**NOTE**

Writing zeros to the right of a *decimal* number does *not* change the value of the number, as shown in Example 2(a) above.

**Work Problem ❷ at the Side.**

### Example 3  Subtracting Decimal Numbers

Subtract.

(a) 15.82 from 28.93

(d) $5 + 429.823 + 0.76$

*Step 1*

$$\begin{array}{r} 28.93 \\ - 15.82 \end{array}$$

Line up decimal points. Then you will be subtracting hundredths from hundredths and tenths from tenths.

*Step 2*

$$\begin{array}{r} 28.93 \\ - 15.82 \\ \hline 13\;11 \end{array}$$

Both numbers have two decimal places; no need to write in zeros.

Subtract as if they were whole numbers.

**Continued on Next Page**

**ANSWERS**
2. (a) 16.34  (b) 233.031  (c) 50.7626
   (d) 435.583

*Step 3*    28.93
           − 15.82
           ───────
            13.11
              ↑ Decimal point in answer is lined up.

**(b)** 146.35 minus 58.98
Borrowing is needed here.

Line up decimal points.

```
  0 13 15  12 15
  1̷ 4̷ 6̷ . 3̷ 5̷
−     5 8 . 9 8
  ─────────────
      8 7 . 3 7
```

═══════════════════ Work Problem ❸ at the Side.

**Example 4** **Writing Zeros as Placeholders Before Subtracting**

Subtract.

**(a)** 16.5 from 28.362
Use the same steps as in Example 3. Remember to write in zeros so both numbers have three decimal places.

Line up decimal points.
```
  28.362
− 16.500    ← Write two zeros.
  ──────
  11.862    ← Subtract as usual.
```

**(b)** 59.7 − 38.914
```
  59.700    ← Write two zeros.
− 38.914
  ──────
  20.786    ← Subtract as usual.
```

**(c)** 12 less 5.83
```
  12.00     ← Write a decimal point and two zeros.
−  5.83
  ─────
   6.17     ← Subtract as usual.
```

═══════════════════ Work Problem ❹ at the Side.

**2** **Add and subtract negative decimals.** The rules that you used to add integers in **Section 1.3** will also work for positive and negative decimal numbers.

### Adding Signed Numbers

To add two numbers with the *same* sign, add the absolute values of the numbers. Use the common sign as the sign of the sum.

To add two numbers with *unlike* signs, subtract the smaller absolute value from the larger absolute value. Use the sign of the number with the larger absolute value as the sign of the sum.

❸ Subtract.

**(a)** 22.7 from 72.9

**(b)** 6.425 from 11.813

**(c)** $20.15 − $19.67

❹ Subtract.

**(a)** 18.651 from 25.3

**(b)** 5.816 − 4.98

**(c)** 40 less 3.66

**(d)** 1 − 0.325

**ANSWERS**
**3.** (a) 50.2   (b) 5.388   (c) 0.48
**4.** (a) 6.649   (b) 0.836   (c) 36.34
     (d) 0.675

**5** Add.

(a) $13.245 + (-18)$

(b) $-0.7 + (-0.33)$

(c) $-6.02 + 100.5$

**ANSWERS**

**5.** (a) $-4.755$ (b) $-1.03$ (c) $94.48$

---

**Example 5** Adding Positive and Negative Decimal Numbers

Add.

(a) $-3.7 + (-16)$

Both addends are negative, so the sum will be negative. To begin, $|-3.7|$ is 3.7 and $|-16|$ is 16. Then add the absolute values.

$$\begin{array}{r} 3.7 \\ +16.0 \\ \hline 19.7 \end{array}$$ ← Write a decimal point and one 0.

$$-3.7 + (-16) = -19.7$$

Both negative     Negative sum

**NOTE**

In Chapter 4 the negative sign was **red** to help you distinguish it from the subtraction symbol. From now on it will be **black**. We will continue to write parentheses around negative numbers when the negative sign might be confused with other symbols. Thus in part (a) above

$$-3.7 \; \textbf{+} \; (-16) \quad \text{means} \quad \textbf{negative} \; 3.7 \; \textbf{plus negative} \; 16.$$

(b) $-5.23 + 0.792$

The addends have different signs. To begin, $|-5.23|$ is 5.23 and $|0.792|$ is 0.792. Then subtract the smaller absolute value from the larger.

$$\begin{array}{r} 5.230 \\ -\;0.792 \\ \hline 4.438 \end{array}$$ ← Write one 0.

$$-5.23 + 0.792 = -4.438$$

Number with larger     Answer is
absolute value          negative.
is negative.

**Work Problem 5 at the Side.**

In **Section 1.4** you rewrote subtraction of integers as addition of the first number to the opposite of the second number. This same strategy works with positive and negative decimal numbers.

**Example 6** Subtracting Positive and Negative Decimal Numbers

Subtract.

(a) $4.3 - 12.73$

Rewrite subtraction as adding the opposite.

$$4.3 \; \textbf{–} \; \textbf{12.73}$$
$$4.3 \; + \; (\textbf{–12.73})$$

The opposite of 12.73 is $-12.73$.

$-12.73$ has the larger absolute value, so the answer will be negative.

$$4.3 + (-12.73) = -8.43$$

Answer is negative.

Subtract the absolute values:
$$\begin{array}{r} 12.73 \\ -\;4.30 \\ \hline 8.43 \end{array}$$

**Continued on Next Page**

**(b)** $-3.65 - (-4.8)$

Rewrite subtraction as adding the opposite.

$-3.65 - (\mathbf{-4.8})$    The opposite of $-4.8$
    ↓    ↓       is $4.8$.
$-3.65 + \mathbf{4.8}$

$4.8$ has the larger absolute value, so the answer will be positive.

$-3.65 + 4.8 = 1.15$    Subtract the absolute values:
    ↑                   $4.80$
  Answer is      $-\ 3.65$
  positive.          $1.15$

**(c)** $14.2 - \underbrace{(1.69 + 0.48)}$    Work inside parentheses first.

$14.2 -\ \ \ (2.17)$    Change subtraction to adding the opposite.

$14.2 + (-2.17)$    $14.2$ has the larger absolute value, so the answer will be positive.
$\underbrace{\phantom{14.2 + (-2.17)}}$
    $12.03$

**Work Problem 6 at the Side.**

**6** Subtract.

**(a)** $-0.37 - (-6)$

**(b)** $5.8 - 10.03$

**(c)** $-312.72 - 65.7$

**(d)** $0.8 - (6 - 7.2)$

### 3   Estimate the answer when adding or subtracting decimals.

A common error when working decimal problems by hand is to misplace the decimal point in the answer. Or, when using a calculator, you may accidentally press the wrong key. Using *front end rounding* to estimate the answer will help you avoid these mistakes. Start by rounding each number to the highest possible place (as you did in **Section 1.5**). Here are several examples. Notice that in the rounded numbers only the leftmost digit is something other than 0.

| | | | | | |
|---|---|---|---|---|---|
| $3.25$ | rounds to | $3$ | $6.812$ | rounds to | $7$ |
| $532.6$ | rounds to | $500$ | $26.397$ | rounds to | $30$ |
| $7094.2$ | rounds to | $7000$ | $351.24$ | rounds to | $400$ |

### Example 7   Estimating Decimal Answers

First, use front end rounding to round each number. Then add or subtract the rounded numbers to get an estimated answer. Finally, find the exact answer.

**(a)** Add $194.2$ and $6.825$.

     *Estimate:*            *Exact:*
        $200$ ← Rounds to    $194.200$
     $+\ \ \ \ 7$ ← Rounds to    $+\ \ \ \ 6.825$
        $207$                  $201.025$

The estimate goes out to the hundreds place (three places to the *left* of the decimal point), and so does the exact answer. Therefore, the decimal point is probably in the correct place in the exact answer.

**(b)** $\$69.42 - \$13.78$

     *Estimate:*            *Exact:*
      $\$70$ ← Rounds to    $\$69.42$
   $-\ \ \ 10$ ← Rounds to    $-\ \ 13.78$
      $\$60$                 $\$55.64$ ← Answer is close to estimate, so the problem is probably set up correctly.

**Continued on Next Page**

**ANSWERS**
**6.** **(a)** $5.63$   **(b)** $-4.23$   **(c)** $-378.42$
    **(d)** $2$

**7** Use front end rounding to estimate each answer. Then find the exact answer.

(a) $2.83 + 5.009 + 76.1$

  *Estimate:*

  *Exact:*

(b) $19.28 less $1.53

  *Estimate:*

  *Exact:*

(c) $11.365 - 38$

  *Estimate:*

  *Exact:*

(d) $-214.6 + 300.72$

  *Estimate:*

  *Exact:*

(c) $-1.861 - 7.3$

Rewrite subtraction as adding the opposite. Then round each number.

$$-1.861 \quad - \quad 7.3$$
$$\downarrow \qquad\qquad \downarrow$$
$$-1.861 \quad + \quad (-7.3)$$
$$\downarrow \qquad\qquad \downarrow$$
Rounded $\quad -2 \quad + \quad (-7) = -9 \quad$ Estimate

To find the exact answer, add the absolute values.

$\quad\quad 1.861 \leftarrow$ Absolute value of $-1.861$
$+\ 7.300 \leftarrow$ Absolute value of $-7.3$ with two zeros written in
$\quad\quad 9.161$

The answer will be negative because both numbers are negative.

$$-1.861 + (-7.3) = -9.161 \quad \text{Exact}$$

**Work Problem 7 at the Side.**

**Calculator Tip** If you are *adding* decimal numbers, you can enter them in any order on your calculator. Try these; jot down the answers.

$\quad 9.82\ \boxed{+}\ 1.86\ \boxed{=}$ _____ $\quad\quad 1.86\ \boxed{+}\ 9.82\ \boxed{=}$ _____

The answers are the same because addition is *commutative* (see **Section 1.3**). But subtraction is *not* commutative. It *does* matter which number you enter first. Try these:

$\quad 9.82\ \boxed{-}\ 1.86\ \boxed{=}$ _____ $\quad\quad 1.86\ \boxed{-}\ 9.82\ \boxed{=}$ _____

The answers are 7.96 and −7.96. As you know, positive numbers are *greater* than 0, but negative numbers are *less* than 0. So it is important to do subtraction in the correct order, particularly if it is in your checkbook!

**ANSWERS**

7. (a) $3 + 5 + 80 = 88$; 83.939
   (b) $20 − $2 = $18; $17.75
   (c) $10 + (-40) = -30$; $-26.635$
   (d) $-200 + 300 = 100$; 86.12

## 5.3 Exercises

*Find each sum. See Examples 1 and 2.*

1.  5.69
    0.24
    + 11.79

2.  372.1
    33.7
    + 42.3

3.  0.38
    7
    + 4.6

4.  3.7
    0.812
    + 55

5. $14.23 + 8 + 74.63 + 18.715 + 0.286$

6. $197.4 + 0.72 + 17.43 + 25 + 1.4$

7. $27.65 + 18.714 + 9.749 + 3.21$

8. $58.546 + 19.2 + 8.735 + 14.58$

9. Explain and correct the error that a student made when he added $0.72 + 6 + 39.5$ this way:

    0.72
    6
    + 39.50
    ———
    40.28

10. Explain and correct the error that a student made when she added $7.21 + 65 + 13.15$ this way:

    7.21
    .65
    + 13.15
    ———
    21.01

11. Show why 0.3 is equivalent to 0.3000.

12. Explain why 7 may be written as 7.0 but not as 0.7.

*Find each difference. See Examples 3 and 4.*

13. $90.5 - 0.8$

14. $303.72 - 0.68$

15. $0.4 - 0.291$

16. $0.35 - 0.088$

17. $6 - 5.09$

18. $80 - 16.3$

19. $15 - 8.339$

20. $44 - 0.08$

21. Explain and correct the error that a student made when he subtracted 7.45 from 15.32 this way:

    7.45
    − 15.32
    ———
    12.13

22. Explain the difference between saying "subtract 2.9 from 8" and saying "2.9 minus 8."

*This drawing of a human skeleton shows the average length of the longest bones, in inches. Use the drawing to answer Exercises 23–26.*

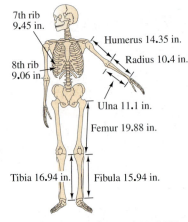

7th rib 9.45 in.
8th rib 9.06 in.
Humerus 14.35 in.
Radius 10.4 in.
Ulna 11.1 in.
Femur 19.88 in.
Tibia 16.94 in.
Fibula 15.94 in.

**Source:** *The Top 10 of Everything 2000.*

23. (a) What is the combined length of the humerus and radius bones?

    (b) What is the difference in the lengths of these two bones?

24. (a) What is the total length of the femur and tibia bones?

    (b) How much longer is the femur than the tibia?

25. (a) Find the sum of the lengths of the humerus, ulna, femur, and tibia.

    (b) How much shorter is the 8th rib than the 7th rib?

26. (a) What is the difference in the lengths of the two bones in the lower arm?

    (b) What is the difference in the lengths of the two bones in the lower leg?

*Find each sum or difference. See Examples 5 and 6.*

27. $24.008 + (-0.995)$

28. $0.77 - 3.06$

29. $-6.05 + (-39.7)$

30. $-6.409 + 8.224$

31. $0.9 - 7.59$

32. $-489.7 - 38$

33. $-2 - 4.99$

34. $2.068 - (-32.7)$

35. $-5.009 + 0.73$

36. $-0.33 - 65$

37. $-1.7035 - (5 - 6.7)$

38. $60 + (-0.9345 + 1.4)$

39. $8000 - (8002.63 - 8)$

40. $-210 - (-0.7306 + 0.5)$

*Use your estimation skills to pick the most reasonable answer for each example. Do **not** solve the problems. Circle your choice. See Example 7.*

**41.** 12 − 11.725

2.75    0.275    27.5

**42.** 20 − 1.37

0.1863    1.863    18.63

**43.** 6.5 + 0.007

6.507    0.6507    65.07

**44.** 9.67 + 0.09

0.976    9.76    0.00976

**45.** 456.71 − 454.9

18.1    181    1.81

**46.** 803.25 − 0.6

802.65    0.80265    8.0265

**47.** 6004.003 + 52.7172

605.67202    60,567.202    6056.7202

**48.** 128.35 + 97.0093

2253.593    225.3593    22.53593

*Use front end rounding to estimate each sum or difference. Then find the exact answer to each application problem.*

**49.** Tom has agreed to work 42.5 hours a week as a car wash attendant. So far this week he has worked 16.35 hours. How many more hours must he work?

*Estimate:*

*Exact:*

**50.** The U.S. population was about 281.42 million in 2000. The Census Bureau estimates that it will be 393.9 million in 2050. The increase in population during that 50-year period is how many millions of people? (*Source:* U.S. Bureau of the Census.)

*Estimate:*

*Exact:*

**51.** Mrs. Little Owl put two checks in the deposit envelope at the automated teller machine. There was a $310.14 paycheck and a $0.95 refund check. How much did she deposit in her account?

*Estimate:*

*Exact:*

**52.** Rodney Green's paycheck stub showed wages of $274.19 at the regular rate of pay and $72.94 at the overtime rate. What were his total wages?

*Estimate:*

*Exact:*

**53.** The tallest known land mammal is a prehistoric ancestor of the rhino, measuring 6.4 m. Compare the rhino's height to the combined heights of these NBA basketball stars: Kevin Garnett at 2.1 m, Karl Malone at 2.06 m, and David Robinson at 2.16 m. Is their combined height greater or less than the prehistoric rhino? By how much? (*Source: Harper's Index* and NBA.)

6.4 m

*Estimate:*

*Exact:*

**54.** Sammy works in a veterinarian's office. He weighed two kittens. One was 3.9 ounces and the other was 4.05 ounces. What was the difference in the weight of the two kittens?

*Estimate:*

*Exact:*

**55.** Steven One Feather gave the cashier a $20 bill to pay for $9.12 worth of groceries. How much change did he get?

*Estimate:*

*Exact:*

**56.** The cost of Julie's tennis racket, with tax, is $41.09. She gave the clerk two $20 bills and a $10 bill. What amount of change did Julie receive?

*Estimate:*

*Exact:*

*The dad on the first page of this chapter bought fishing equipment for his daughter during a sale. Use the information in the store's sale ad to answer Exercises 57–60. When estimating, round prices to the nearest dollar.*

Source: Wal-Mart.

57. What is the difference in price between the most expensive and least expensive spinning reel?

    Estimate:

    Exact:

58. How much more would 330 yd of line cost than three bobbers?

    Estimate:

    Exact:

59. What is the total cost of the middle-priced rod, the second most expensive reel, a one-tray tackle box, 110 yd of line, and a package of environmentally safe split shot?

    Estimate:

    Exact:

60. Dad also bought his daughter a cap for $8.49, SPF-45 sunscreen for $6.97, and a child-size flotation vest for $19.99. How much did he spend on these items?

    Estimate:

    Exact:

*Olivia Sanchez kept track of her expenses for one month. Use her list to answer Exercises 61–64.*

61. What were Olivia's total expenses for the month?

62. How much did Olivia pay for telephone, cable TV, and Internet access?

| Monthly Expenses | |
|---|---|
| Rent | $994 |
| Car payment | $190.78 |
| Car repairs, gas | $205 |
| Cable TV | $39.95 |
| Internet access | $19.95 |
| Electricity | $40.80 |
| Telephone | $57.32 |
| Groceries | $186.81 |
| Entertainment | $97.75 |
| Clothing, laundry | $107 |

63. What was the difference in the amounts spent for groceries and for the car payment?

64. How much more did Olivia spend on rent than on all her car expenses?

*Find the length of the dashed line in each rectangle or circle.*

65.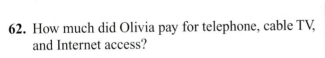

66.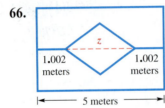

67.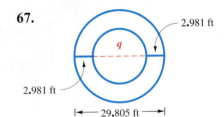

# 5.4 MULTIPLYING SIGNED DECIMAL NUMBERS

**1** **Multiply positive and negative decimals.** The decimals 0.3 and 0.07 can be multiplied by writing them as fractions.

$$0.3 \times 0.07 = \frac{3}{10} \times \frac{7}{100} = \frac{3 \times 7}{10 \times 100} = \frac{21}{1000} = 0.021$$

1 decimal place + 2 decimal places → 3 decimal places

**OBJECTIVES**

**1** Multiply positive and negative decimals.

**2** Estimate the answer when multiplying decimals.

Can you see a way to multiply decimals without writing them as fractions? Try these steps. Remember that each number in a multiplication problem is called a *factor*, and the answer is called the *product*.

### Multiplying Two Decimal Numbers

*Step 1* Multiply the factors (the numbers being multiplied) as if they were whole numbers.

*Step 2* Find the *total* number of decimal places in *both* factors.

*Step 3* Write the decimal point in the product (the answer) so it has the same number of decimal places as the total from Step 2. You may need to write in extra zeros on the left side of the product in order to get the correct number of decimal places.

*Step 4* If two factors have the *same sign*, the product is *positive*. If two factors have *different signs*, the product is *negative*.

**NOTE**

When multiplying decimals, you do *not* need to line up decimal points. (You *do* need to line up decimal points when adding or subtracting.)

### Example 1  Multiplying Decimal Numbers

Multiply 8.34 times (−4.2).

*Step 1* Multiply the numbers as if they were whole numbers.

$$\begin{array}{r} 8.34 \\ \times\ 4.2 \\ \hline 1668 \\ 3336\phantom{0} \\ \hline 35028 \end{array}$$

*Step 2* Count the total number of decimal places in both factors.

$$\begin{array}{r} 8.3\,4 \leftarrow \text{2 decimal places} \\ \times\ \ 4.2 \leftarrow \text{1 decimal place} \\ \hline 1\,6\,6\,8 \quad \text{3 total decimal places} \\ 3\,3\,3\,6\phantom{0} \\ \hline 3\,5\,0\,2\,8 \end{array}$$

— Continued on Next Page

**Chapter 5** Rational Numbers: Positive and Negative Decimals

**❶ Multiply.**

(a) $-2.6(0.4)$

(b) $(45.2)(0.25)$

(c) $\quad 0.104 \leftarrow$ 3 decimal places
$\quad \underline{\times \quad\quad 7} \leftarrow$ 0 decimal places
$\quad\quad\quad\quad\quad \leftarrow$ 3 decimal places in the product

(d) $(-3.18)^2$
*Hint:* Recall that squaring a number means multiplying the number times itself, so this is $(-3.18)(-3.18)$.

**❷ Multiply.**

(a) $0.04(-0.09)$

(b) $(0.2)(0.008)$

(c) $(-0.063)(-0.04)$

(d) $(0.003)^2$

**ANSWERS**
1. (a) $-1.04$  (b) 11.300 or 11.3  (c) 0.728
   (d) 10.1124
2. (a) $-0.0036$  (b) 0.0016  (c) 0.00252
   (d) 0.000009

*Step 3* Count over 3 places in the product and write the decimal point. Count from *right to left*.

$\quad\quad 8.3\,4 \leftarrow$ 2 decimal places
$\underline{\times\quad 4.2} \leftarrow$ 1 decimal place
$\quad\quad 1\,6\,6\,8 \quad\quad$ 3 total decimal places
$\underline{3\,3\,3\,6\quad\quad}$
$\quad 3\,5.0\,2\,8 \leftarrow$ 3 decimal places in product
Count over 3 places from right to left to position the decimal point.

*Step 4* The factors have *different* signs, so the product is *negative*:
8.34 times $(-4.2) = -35.028$.

**Work Problem ❶ at the Side.**

**Example 2** Writing Zeros as Placeholders in the Product

Multiply $(-0.042)(-0.03)$.

Start by multiplying, then count decimal places.

$\quad\quad 0.0\,4\,2 \leftarrow$ 3 decimal places
$\underline{\times\quad 0.0\,3} \leftarrow$ 2 decimal places
$\quad\quad 1\,2\,6 \leftarrow$ 5 decimal places needed in product

After multiplying, the answer has only three decimal places, but five are needed. So write two zeros on the *left* side of the answer.

$\quad\quad 0.0\,4\,2 \quad\quad\quad\quad\quad 0.0\,4\,2 \leftarrow$ 3 decimal places
$\underline{\times\quad 0.0\,3} \quad\quad\quad\quad \underline{\times\quad 0.0\,3} \leftarrow$ 2 decimal places
$\quad\mathbf{0\,0}\,1\,2\,6 \quad\quad\quad\quad .0\,0\,1\,2\,6 \leftarrow$ 5 decimal places
$\quad\uparrow\uparrow$
Write two zeros $\quad\quad\quad\quad$ Now count over 5 places
on *left* side of answer. $\quad\quad$ and write in the decimal point.

The final product is 0.00126, which has five decimal places. The product is *positive* because the factors have the *same* sign (both factors are negative).

**Work Problem ❷ at the Side.**

🖩 **Calculator Tip** When working with money amounts, you may need to write a 0 in your answer. For example, try multiplying $\$3.54 \times 5$ on your calculator. Write down the result.

$$3.54 \;\otimes\; 5 \;\ominus\; \underline{\quad\quad\quad}$$

Notice that the result is 17.7, which is *not* the way to write a money amount. You have to write the 0 in the hundredths place: $\$17.7\mathbf{0}$ is correct. The calculator does not show the "extra" zero because:

$$17.70 \text{ or } 17\frac{70}{100} \quad \text{reduces to} \quad 17\frac{7}{10} \text{ or } 17.7.$$

So keep an eye on your calculator—it doesn't know when you're working with money amounts.

**2** **Estimate the answer when multiplying decimals.** If you are doing multiplication problems by hand, estimating the answer helps you check that the decimal point is in the right place. When you are using a calculator, estimating helps you catch an error like pressing the ÷ key instead of the × key.

**Example 3** **Estimating Before Multiplying**

First, use front end rounding to estimate $(76.34)(12.5)$. Then find the exact answer.

*Estimate:*

$$\begin{array}{r} 80 \\ \times\ 10 \\ \hline 800 \end{array}$$

*Exact:*

$$\begin{array}{r} 76.34 \\ \times\ \ 12.5 \\ \hline 3817\ 0 \\ 1526\ 8\ \ \\ 7\ 6\ 3\ 4\ \ \ \ \\ \hline 954.250 \end{array}$$

← 2 decimal places
← 1 decimal place

3 decimal places are in the product.

Both the estimate and the exact answer go out to the hundreds place, so the decimal point in 954.250 is probably in the correct place.

**Work Problem** ❸ **at the Side.**

❸ First, use front end rounding to estimate the answer. Then find the exact answer.

(a) $(11.62)(4.01)$

(b) $(-5.986)(-33)$

(c) $8(\$4.35)$

(d) $58.6(-17.4)$

**Answers**

**3.** (a) $(10)(4) = 40$; $46.5962$
(b) $(-6)(-30) = 180$; $197.538$
(c) $8(\$4) = \$32$; $\$34.80$
(d) $60(-20) = -1200$; $-1019.64$

# Focus on Real-Data Applications

## Life Insurance Benefits

Employees often must choose benefit options for programs such as medical, dental, and life insurance. Payment is made through payroll deductions. For example, BP Amoco offers Group Universal Life Insurance coverage for an employee or spouse, based on the employee's eligible pay. The rates charged to the employee depend on the age of the employee or spouse at the time enrolled in the plan, the level of coverage, and the use of tobacco. Each employee may choose coverage that is a whole number multiple of his or her eligible pay, rounded up to the next thousand dollars. For example, if an employee chooses life insurance that is triple the eligible pay, then the increment is 3.

The formula for computing the monthly cost of insurance for an employee or spouse is

$$\underbrace{\$\_\_\_\_\_}_{\text{(eligible pay)}} \times \underbrace{\_\_\_\_\_}_{\substack{\text{(increment of} \\ \text{eligible pay)}}} = \underbrace{\$\_\_\_\_\_*}_{\substack{\text{(coverage} \\ \text{amount)}}} \div \$1000 \times \underbrace{\$\_\_\_\_\_}_{\substack{\text{(monthly} \\ \text{rate)}}} = \underbrace{\$\_\_\_\_\_}_{\substack{\text{(monthly cost} \\ \text{of insurance)}}}$$

*Round the coverage amount up to the next $1000 if it is not a whole multiple of $1000. (For example, $34,200 would be rounded up to $35,000.)

The following table shows the monthly rates available.

| | The monthly rate per $1000 of coverage is . . . | |
|---|---|---|
| If your age is . . . | Non-Tobacco User | Tobacco User |
| Under 25 | $0.040 | $0.050 |
| 25–39 | $0.047 | $0.056 |
| 40–44 | $0.092 | $0.112 |
| 45–49 | $0.147 | $0.178 |
| 50–54 | $0.239 | $0.290 |
| 55–59 | $0.378 | $0.458 |
| 60–64 | $0.645 | $0.781 |
| 65–69 | $0.992 | $1.201 |
| 70–74 | $1.726 | $2.090 |

Source: *Employee Benefits Handbook,* BP Amoco, December 2000.

Use the formula and the table to answer the problems.

1. Suppose you work in BP Amoco's human resources department and must advise new employees of their benefit options. Calculate the monthly cost of insurance, to the nearest cent, for each of these employees.

| Employee | Age | Smoker? | Eligible Pay | Increment | Monthly Cost of Insurance |
|---|---|---|---|---|---|
| Rebecca C. | 25 | Yes | $23,600 | 2 | |
| Roger J. | 58 | Yes | $85,750 | 3 | |
| Stan S. | 49 | No | $45,850 | 3 | |
| Diana H. | 42 | No | $53,270 | 2 | |
| Hulon M. | 63 | No | $58,100 | 2 | |

2. Eric W., aged 55, is a nonsmoker who earns $42,700 in eligible pay. He wants to limit his monthly cost for life insurance to less than $50.00 per month. How much total insurance can he purchase and what will be his monthly costs?

3. Sarah F. and Ellen S. each earn $64,250 and purchase insurance equivalent to twice her eligible salary. Sarah F. is a smoker aged 72, whereas Ellen S. is a nonsmoker aged 30. Calculate the premium cost for each person. Explain why you believe the difference in costs between the two policies is or is not justified.

## 5.4 Exercises

*Multiply. See Example 1.*

1. 0.042
   × 3.2

2. 0.571
   × 2.9

3. −21.5(7.4)

4. −85.4(−3.5)

5. (−23.4)(−0.666)

6. 0.896(−0.799)

7. $51.88
   × 665

8. $736.75
   × 118

*Use the fact that (72)(6) = 432 to solve Exercises 9–16 by simply counting decimal places and writing the decimal point in the correct location. Be sure to indicate the sign of the product.*

9. 72(−0.6) =    4 3 2

10. 7.2(−6) =    4 3 2

11. (7.2)(0.06) =    4 3 2

12. (0.72)(0.6) =    4 3 2

13. −0.72(−0.06) =    4 3 2

14. −72(−0.0006) =    4 3 2

15. (0.0072)(0.6) =    4 3 2

16. (0.072)(0.006) =    4 3 2

*Multiply. See Example 2.*

17. (0.006)(0.0052)

18. (0.0052)(0.009)

19. (−0.003)$^2$

20. (0.0004)$^2$

### Relating Concepts (Exercises 21–22) For Individual or Group Work

*Look for patterns as you **work Exercises 21 and 22 in order**.*

21. Do these multiplications:

    (5.96)(10) = _____    (3.2)(10) = _____
    (0.476)(10) = _____   (80.35)(10) = _____
    (722.6)(10) = _____   (0.9)(10) = _____

    What pattern do you see? Write a "rule" for multiplying by 10. What do you think the rule is for multiplying by 100? by 1000? Write the rules and try them out on the numbers above.

22. Do these multiplications:

    (59.6)(0.1) = _____   (3.2)(0.1) = _____
    (0.476)(0.1) = _____  (80.35)(0.1) = _____
    (65)(0.1) = _____     (523)(0.1) = _____

    What pattern do you see? Write a "rule" for multiplying by 0.1. What do you think the rule is for multiplying by 0.01? by 0.001? Write the rules and try them out on the numbers above.

*First, use front end rounding to estimate the answer. Then multiply to find the exact answer. See Example 3.*

**23.** Estimate:  Exact:
  Rounds to  39.6
  × _____ ← Rounds to  × 4.8

**24.** Estimate:  Exact:
    18.7
  × _____    × 2.3

**25.** Estimate:  Exact:
    37.1
  × _____    × 42

**26.** Estimate:  Exact:
    5.08
  × _____    × 71

**27.** Estimate:  Exact:
    6.53
  × _____    × 4.6

**28.** Estimate:  Exact:
    7.51
  × _____    × 8.2

**29.** Estimate:  Exact:
    2.809
  × _____    × 6.85

**30.** Estimate:  Exact:
    73.52
  × _____    × 22.34

*Even with most of the problem missing, you can tell whether or not these answers are reasonable. Circle* reasonable *or* unreasonable. *If the answer is unreasonable, move the decimal point or insert a decimal point to make the answer reasonable.*

**31.** How much was his car payment?   $18.90
  reasonable
  unreasonable, should be _____

**32.** How many hours did she work today?   25 hours
  reasonable
  unreasonable, should be _____

**33.** How tall is her son?   60.5 inches
  reasonable
  unreasonable, should be _____

**34.** How much does he pay for rent now?   $6.92
  reasonable
  unreasonable, should be _____

**35.** What is the price of one gallon of milk?   $319
  reasonable
  unreasonable, should be _____

**36.** How long is the living room?   16.8 feet
  reasonable
  unreasonable, should be _____

**37.** How much did the baby weigh?   0.095 pounds
  reasonable
  unreasonable, should be _____

**38.** What was the sale price of the jacket?   $1.49
  reasonable
  unreasonable, should be _____

*Solve each application problem. Round money answers to the nearest cent when necessary.*

**39.** LaTasha worked 50.5 hours over the last two weeks. She earns $18.73 per hour. How much did she make?

**40.** Michael's time card shows 42.2 hours at $10.03 per hour. What are his gross earnings?

41. Sid needs 0.6 meter of canvas material to make a carry-all bag that fits on his wheelchair. If canvas is $4.09 per meter, how much will Sid spend? (*Note:* $4.09 *per* meter means $4.09 for *one* meter.)

42. How much will Mrs. Nguyen pay for 3.5 yards of lace trim that costs $0.87 per yard?

43. Michelle filled the tank of her SUV with regular unleaded gas. Use the information shown on the pump to find how much she paid for gas.

**Source:** Holiday.

44. Ground beef and spicy chicken wings are on sale. Juma bought 1.7 pounds of wings. Use the information in the ad to find the amount she paid.

45. Ms. Rolack is a real estate broker who helps people sell their homes. Her fee is 0.07 times the price of the home. What was her fee for selling a $175,300 home?

46. Alex Rodriguez, shortstop for the Seattle Mariners, had a 2000 batting average of 0.316. If he went to bat 554 times, how many hits did he make? (*Hint:* Multiply his batting average by the number of times at bat.) Round to the nearest whole number. (*Source: World Almanac*, 2001.)

47. Judy Lewis pays $28.96 per month for basic cable TV. How much will she pay for cable over one year? How much would she pay in a year for the deluxe cable package that costs $59.95 per month?

48. Chuck's car payment is $220.27 per month for three years. How much will he pay altogether?

49. Paper for the copy machine at the library costs $0.015 per sheet. How much will the library pay for 5100 sheets?

50. A student group collected 2200 pounds of plastic as a fund raiser. How much will they make if the recycling center pays $0.142 per pound?

51. The National Aquarium in Baltimore charges $11.95 for adults, $10.50 for seniors, and $7.50 for children. How much will a mother with four children spend for her family and three senior relatives? (*Source:* Lyon Group.)

52. (Complete Exercise 51 first.) How much *less* would the same family spend at the Texas State Aquarium, which charges $8 for adults, $5.75 for seniors, and $4.50 for children? (*Source:* Lyon Group.)

53. Ms. Sanchez paid $29.95 a day to rent a car, plus $0.29 per mile. Find the cost of her rental for a four-day trip of 926 miles.

54. The Bell family rented a motor home for $375 per week plus $0.35 per mile. What was the rental cost for their three-week vacation trip of 2650 miles?

55. Barry bought 16.5 meters of rope at $0.47 per meter and three meters of wire at $1.05 per meter. How much change did he get from three $5 bills?

56. Susan bought a VCR that cost $229.88. She paid $45 down and $37.98 per month for six months. How much could she have saved by paying cash?

*Use the information below from the Look Smart mail order catalog to answer Exercises 57–60.*

| Knit Shirt Ordering Information | | |
|---|---|---|
| 43–2A | short sleeve, solid colors | $14.75 each |
| 43–2B | short sleeve, stripes | $16.75 each |
| 43–3A | long sleeve, solid colors | $18.95 each |
| 43–3B | long sleeve, stripes | $21.95 each |
| Extra-large size, add $2 per shirt. | | |
| Monogram, $4.95 each. Gift box, $5 each. | | |

| Total Price of Items (excluding monograms and gift boxes) | Shipping, Packing, and Handling |
|---|---|
| $0–25.00 | $3.50 |
| $25.01–75.00 | $5.95 |
| $75.01–125.00 | $7.95 |
| $125.01+ | $9.95 |
| Shipping to each additional address add $4.25. | |

57. Find the total cost of ordering four long-sleeve, solid-color shirts and two short-sleeve, striped shirts, all in the extra-large size, and all shipped to your home.

58. What is the total cost of eight long-sleeve shirts, five in solid colors and three striped? Include the cost of shipping the solid shirts to your home and the striped shirts to your brother's home.

59. (a) What is the total cost, including shipping, of sending three short-sleeve solid-color shirts, with monograms, in a gift box to your aunt for her birthday?

   (b) How much did the monograms, gift box, and shipping add to the cost of your gift?

60. (a) Suppose you order one of each type of shirt for yourself, adding a monogram on each of the solid-color shirts. At the same time, you order three long-sleeved striped shirts, in the extra-large size, shipped to your dad in a gift box. Find the total cost of your order.

   (b) What is the difference in total cost (excluding shipping) between the shirts for yourself and the gift for your dad?

## 5.5 DIVIDING SIGNED DECIMAL NUMBERS

There are two kinds of decimal division problems; those in which a decimal is divided by an integer, and those in which a decimal is divided by a decimal. First recall the parts of a division problem.

Divisor → 2)16 ← Dividend, with 8 as Quotient

$16 \div 2 = 8$ (Dividend ÷ Divisor = Quotient)

$\dfrac{16}{2} = 8$ (Dividend over Divisor = Quotient)

**OBJECTIVES**

1. Divide a decimal by an integer.
2. Divide a decimal by a decimal.
3. Estimate the answer when dividing decimals.
4. Use the order of operations with decimals.

**1** **Divide a decimal by an integer.** When the divisor is an integer, use these steps.

### Dividing a Decimal Number by an Integer

*Step 1* Write the decimal point in the quotient (answer) directly above the decimal point in the dividend.

*Step 2* Divide as if both numbers were whole numbers.

*Step 3* If both numbers have the *same sign*, the quotient is *positive*. If they have *different signs*, the quotient is *negative*.

### Example 1  Dividing Decimals by Integers

Divide.

**(a)** $21.93 \div (-3)$

Dividend: 21.93; Divisor: −3

First consider $21.93 \div 3$.    3)21.93

*Step 1* Write the decimal point in the quotient directly above the decimal point in the dividend.

3)21.93  — Decimal points lined up

*Step 2* Divide as if the numbers were whole numbers.

$\phantom{3)}\ 7.31$
$3)21.93$

Check by multiplying the quotient times the divisor.

**Check**
$\phantom{\times}7.31$
$\times\ \ \ 3$
$\overline{\phantom{0}21.93}$   Matches

*Step 3* The quotient is $-7.31$ because the numbers had *different* signs.

$21.93 \div (-3) = -7.31$

Different signs → Negative quotient

*Continued on Next Page*

**1** Divide. Check your answers by multiplying.

(a) $4\overline{)93.6}$

(b) $6\overline{)6.804}$

(c) $\dfrac{278.3}{11}$

(d) $-0.51835 \div 5$

(e) $-213.45 \div (-15)$

**ANSWERS**
1. (a) 23.4; (23.4)(4) = 93.6
   (b) 1.134; (1.134)(6) = 6.804
   (c) 25.3; (25.3)(11) = 278.3
   (d) −0.10367; (−0.10367)(5) = −0.51835
   (e) 14.23; (14.23)(−15) = −213.45

---

(b)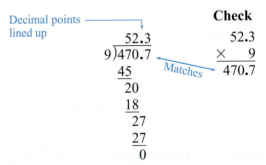

Write the decimal point in the quotient directly above the decimal point in the dividend. Then divide as if they were whole numbers.

```
              52.3           Check
         9)470.7              52.3
            45       Matches ×   9
            20               470.7
            18
            27
            27
             0
```
Decimal points lined up

The quotient is 52.3 and is *positive* because both numbers had the *same* sign.

**Work Problem 1 at the Side.**

### Example 2  Writing Extra Zeros to Complete a Division

Divide 1.5 by 8.

Keep dividing until the remainder is 0, or until the digits in the quotient begin to repeat in a pattern. In Example 1(b) above, you ended up with a remainder of 0. But sometimes you run out of digits in the dividend before that happens. If so, write extra zeros on the right side of the dividend so you can continue dividing.

```
       0.1
    8)1.5    ← All digits have been used.
      8
      7     ← Remainder is not yet 0.
```

Write a 0 after the 5 in the dividend so you can continue dividing. Keep writing more zeros in the dividend if needed. Recall that writing zeros to the *right* of a decimal number does **not** change its value.

```
      0.1 8 7 5              Check
    8)1.5 0 0 0              0.1875
      8                   ×       8
      7 0                    1.5000
      6 4
        6 0
        5 6
          4 0
          4 0
             0
```
← Three zeros needed to complete the division
← Stop dividing when the remainder is 0.
← Matches dividend, so 0.1875 is correct

**CAUTION**

Notice that in decimals the dividend may *not* be the larger number, as it was in whole numbers. In Example 2 the dividend is 1.5, which is *smaller* than 8.

Section 5.5 Dividing Signed Decimal Numbers    **343**

**Calculator Tip** When *multiplying* numbers, you can enter them in any order because multiplication is commutative (see **Section 1.6**). But division is *not* commutative. It *does* matter which number you enter first. Try Example 2 both ways; jot down your answers.

$1.5 \div 8 =$ _____    $8 \div 1.5 =$ _____

Notice that the first answer, 0.1875, matches the result from Example 2. But the second answer is much different: 5.333333333. Be careful to enter the dividend first.

Work Problem ❷ at the Side.

In the next example the remainder is never 0 even if we keep dividing.

**Example 3** Rounding a Decimal Quotient

Divide 4.7 by 3. Round the quotient to the nearest thousandth.

Write extra zeros in the dividend so that you can continue dividing.

```
      1.5 6 6 6
   3)4.7 0 0 0   ← Three zeros added so far
      3
      ─
      1 7
      1 5
      ───
        2 0
        1 8
        ───
          2 0
          1 8
          ───
            2 0
            1 8
            ───
              2   ← Remainder is still not 0.
```

Notice that the digit 6 in the quotient is repeating. It will continue to do so. The remainder will never be 0. There are two ways to show that an answer is a **repeating decimal** that goes on forever. You can write three dots after the answer, or you can write a bar above the digits that repeat (in this case, the 6).

$1.5\underbrace{666}_{\text{Three dots}}\ldots$    or    $1.5\overline{6}$ ← Bar above repeating digit

**CAUTION**

Do not use *both* the dots *and* the bar at the same time. Use three dots *or* the bar.

When repeating decimals occur, round the answer according to the directions in the problem. In this example, to round to thousandths, divide out one *more* place, to ten-thousandths.

$4.7 \div 3 = 1.5666\ldots$ rounds to $1.567$

Check the answer by multiplying 1.567 by 3. Because 1.567 is a rounded answer, the check will *not* give exactly 4.7, but it should be very close.

$(1.567)(3) = 4.701$ ← Does not equal exactly 4.7 because 1.567 was rounded

❷ Divide. Check your answers by multiplying.

(a) $\dfrac{6.4}{5}$

(b) $30.87 \div (-14)$

(c) $\dfrac{-259.5}{-30}$

(d) $0.3 \div 8$

**Answers**
2. (a) 1.28; (1.28)(5) = 6.40 or 6.4
   (b) −2.205; (−2.205)(−14) = 30.870 or 30.87
   (c) 8.65; (8.65)(−30) = −259.50 or −259.5
   (d) 0.0375; (0.0375)(8) = 0.3000 or 0.3

**3** Divide. Round answers to the nearest thousandth. If it is a repeating decimal, also write the answer using a bar. Check your answers by multiplying.

(a) $13\overline{)267.01}$

(b) $6\overline{)20.5}$

(c) $\dfrac{10.22}{9}$

(d) $16.15 \div 3$

(e) $116.3 \div 11$

**ANSWERS**

3. (a) 20.539; no repeating digits visible on calculator;
   $(20.539)(13) = 267.007$
   (b) 3.417; $3.41\overline{6}$; $(3.417)(6) = 20.502$
   (c) 1.136; $1.13\overline{5}$; $(1.136)(9) = 10.224$
   (d) 5.383; $5.38\overline{3}$; $(5.383)(3) = 16.149$
   (e) 10.573; $10.5\overline{72}$; $(10.573)(11) = 116.303$

**CAUTION**

When you're checking quotients that you've rounded, the check will *not* match the dividend exactly, but it should be very close.

Work Problem **3** at the Side.

**2** **Divide a decimal by a decimal.** To divide by a *decimal* divisor, first change the divisor to a whole number. Then divide as before. To see how this is done, write the problem in fraction form. Here is an example.

$$1.2\overline{)6.36} \quad \text{can be written} \quad \dfrac{6.36}{1.2}$$

In **Section 4.1** you learned that multiplying the numerator and denominator by the same number gives an equivalent fraction. We want the divisor (1.2) to be a whole number. Multiplying by 10 will accomplish that.

Decimal divisor → $\dfrac{6.36}{1.2} = \dfrac{(6.36)(10)}{(1.2)(10)} = \dfrac{63.6}{12}$ ← Whole number divisor

The short way to multiply by 10 is to move the decimal point *one place* to the *right* in both the divisor and the dividend.

$$1.2\overline{)6.36} \quad \text{is equivalent to} \quad 12\overline{)63.6}$$

**NOTE**

Moving the decimal points the *same* number of places in **both** the divisor and dividend will *not* change the answer.

**Dividing by a Decimal Number**

*Step 1* Count the number of decimal places in the divisor and move the decimal point that many places to the *right*. (This changes the divisor to a whole number.)

*Step 2* Move the decimal point in the dividend the *same* number of places to the *right*. (Write in extra zeros if needed.)

*Step 3* Write the decimal point in the quotient directly above the decimal point in the dividend. Then divide as usual.

*Step 4* If both numbers have the *same sign*, the quotient is *positive*. If they have *different signs*, the quotient is *negative*.

**Example 4** Dividing by Decimal Numbers

(a) $\dfrac{27.69}{0.003}$

Move the decimal point in the divisor *three* places to the *right* so that 0.003 becomes the whole number 3. In order to move the decimal point in the dividend the same number of places, write in an extra 0.

*Continued on Next Page*

## Section 5.5 Dividing Signed Decimal Numbers

$$0.003\overline{)27.690}$$

Moving decimal point three places is the same as multiplying by 1000.

Move decimal points in divisor and dividend. Then line up decimal point in answer.

$$3\overline{)27690.} = 9230.$$

Divide as usual.

**(b)** Divide $-5$ by $-4.2$ and round the quotient to the nearest hundredth.

First consider $5 \div 4.2$. Move the decimal point in the divisor one place to the right so that 4.2 becomes the whole number 42. The decimal point in the dividend starts on the right side of 5 and is also moved one place to the right.

```
      1.1 9 0
4.2 )5.0 0 0 0
     4 2
     ‾‾‾
       8 0
       4 2
       ‾‾‾
         3 8 0
         3 7 8
         ‾‾‾‾‾
             2 0
```

← In order to round to hundredths, divide out one *more* place, to thousandths.

Rounding the quotient to the nearest hundredth gives 1.19. The quotient is *positive* because both the divisor and dividend had the *same* sign (both were negative).

$$-5 \div (-4.2) \approx 1.19$$

Same sign — Positive quotient

= Work Problem **4** at the Side.

**3** **Estimate the answer when dividing decimals.** Estimating the answer to a division problem helps you catch errors. Compare the estimate to your exact answer. If they are very different, do the division again.

### Example 5 — Estimating Before Dividing

First, use front end rounding to estimate the answer. Then divide to find the exact answer.

$$580.44 \div 2.8$$

Here is how one student solved this problem. She rounded 580.44 to 600 and 2.8 to 3 to estimate the answer.

*Estimate:*   *Exact:*

```
   200              2 7.3
3 )600         2.8 )5 8 0.4 4
                   5 6
                   ‾‾‾
                     2 0 4
                     1 9 6
                     ‾‾‾‾‾
                         8 4
                         8 4
                         ‾‾‾
                           0
```

Very different; need to rework the problem

Notice that the estimate, which is in the hundreds, is very different from the exact answer, which is only in the tens. This tells the student that she needs to rework the problem. Can you find the error? (The exact answer should be 207.3, which fits with the estimate of 200.)

---

**4** Divide. If the quotient does not come out even, round to the nearest hundredth.

**(a)** $0.2\overline{)1.04}$

**(b)** $0.06\overline{)1.8072}$

**(c)** $0.005\overline{)32}$

**(d)** $-8.1 \div 0.025$

**(e)** $\dfrac{7}{1.3}$

**(f)** $-5.3091 \div (-6.2)$

**ANSWERS**
**4.** **(a)** 5.2   **(b)** 30.12   **(c)** 6400   **(d)** $-324$
   **(e)** 5.38 (rounded)   **(f)** 0.86 (rounded)

**5** Decide whether each answer is reasonable by using front end rounding to estimate the answer. If the exact answer is *not* reasonable, find and correct the error.

(a) $42.75 \div 3.8 = 1.125$

Estimate:

(b) $807.1 \div 1.76 = 458.580$ to nearest thousandth

Estimate:

(c) $48.63 \div 52 = 93.519$ to nearest thousandth

Estimate:

(d) $9.0584 \div 2.68 = 0.338$

Estimate:

Work Problem **5** at the Side.

**4** **Use the order of operations with decimals.** Use the order of operations from **Section 1.8** when a decimal problem involves more than one operation.

### Order of Operations

*Step 1* Work inside *parentheses* or *other grouping symbols*.

*Step 2* Simplify expressions with *exponents*.

*Step 3* Do the remaining *multiplications and divisions* as they occur from left to right.

*Step 4* Do the remaining *additions and subtractions* as they occur from left to right.

**Example 6** Using the Order of Operations

Simplify by using the order of operations.

(a) $2.5 + (-6.3)^2 + 9.62$   Use the exponent: $(-6.3)(-6.3)$ is $39.69$.

   $2.5 + 39.69 + 9.62$   Add from left to right.

   $42.19 + 9.62$

   $51.81$

(b) $1.82 + (5.2 - 6.7)(5.8)$   Work inside parentheses.

   $1.82 + (-1.5)(5.8)$   Multiply next.

   $1.82 + (-8.7)$   Add last.

   $-6.88$

(c) $3.7^2 - 1.8 \div 5(1.5)$   Use the exponent first.

   $13.69 - 1.8 \div 5(1.5)$   Multiply and divide from left to right.

   $13.69 - 0.36(1.5)$

   $13.69 - 0.54$   Subtract last.

   $13.15$

---

ANSWERS

**5.** (a) Estimate is $40 \div 4 = 10$; exact answer is not reasonable, should be $11.25$.
(b) Estimate is $800 \div 2 = 400$; exact answer is reasonable.
(c) Estimate is $50 \div 50 = 1$; exact answer is not reasonable, should be $0.935$.
(d) Estimate is $9 \div 3 = 3$; exact answer is not reasonable, should be $3.38$.

Section 5.5 Dividing Signed Decimal Numbers    **347**

Work Problem ❻ at the Side.

**Calculator Tip**  Most calculators that have parentheses keys $($ $)$ can handle calculations like those in Example 6 just by entering the numbers in the order given. For example, the keystrokes for Example 6(b) are:

$$1.82 \; [+] \; [(] \; 5.2 \; [-] \; 6.7 \; [)] \; [\times] \; 5.8 \; [=] \quad \text{Answer is } -6.88.$$

(Parentheses)

Standard, four-function calculators generally do *not* give the correct answer if you enter the numbers in the order given. Check the instruction manual that came with your calculator for information on "order of calculations" to see if your machine has the rules for order of operations built into it. For a quick check, try entering this problem:

$$2 \; [+] \; 2 \; [\times] \; 2 \; [=]$$

If the result is 6, the calculator follows the order of operations. If the result is 8, it does *not* have the rules built into it. Use the space below to explain how this test works.

Answer: The test works because a calculator that follows the order of operations will automatically do the multiplication first. If the calculator does *not* have the rules built into it, it will work from left to right.

Following Order of Operations    Working from Left to Right
$$2 + \underbrace{2 \times 2}$$          $$\underbrace{2 + 2} \times 2$$
$$\underbrace{2 + \;\; 4}$$              $$\underbrace{4 \;\; \times 2}$$
$$\quad 6 \leftarrow \text{Correct}$$    $$\quad 8 \leftarrow \text{Incorrect}$$

❻ Simplify by using the order of operations.

(a) $-4.6 - 0.79 + 1.5^2$

(b) $3.64 \div 1.3(3.6)$

(c) $0.08 + 0.6(2.99 - 3)$

(d) $10.85 - 2.3(5.2) \div 3.2$

**ANSWERS**
6. (a) $-3.14$  (b) $10.08$  (c) $0.074$
   (d) $7.1125$

# Focus on Real-Data Applications

## Dollar-Cost Averaging

To gain the most benefit from investing in the stock market, you should purchase shares when stock prices are low and sell stocks when prices are high. Unfortunately, it is very difficult to predict whether prices will rise or fall from one day to the next. Financial advisors recommend using *dollar-cost averaging* instead of trying to guess how the market will change. By investing the same amount of money at regular intervals, such as the beginning of each month, you will be buying more shares when the stock prices are low and fewer shares when stock prices are high. Of course, *dollar-cost averaging* does not guarantee that you will make a profit. However, over time, your shares should cost less than the market average.

Many financial Web sites report current and historical data on stock performance. Suppose you invested $100 at the first of each month in 2000 and purchased shares of Microsoft at its closing price. The table shows data on Microsoft closing prices during 2000.

|  | Amount Invested | Price per Share | Number of Shares Bought |
|---|---|---|---|
| January | $100 | 97.875 | ($100 ÷ 97.8750) = 1.0217 shares |
| February | $100 | 89.375 |  |
| March | $100 | 106.25 |  |
| April | $100 | 69.75 |  |
| May | $100 | 62.5625 |  |
| June | $100 | 80 |  |
| July | $100 | 69.8125 |  |
| August | $100 | 69.8125 |  |
| September | $100 | 60.3125 |  |
| October | $100 | 68.875 |  |
| November | $100 | 57.375 |  |
| December | $100 | 43.375 |  |
| Total Investment |  |  |  |

*Source:* www.biz.yahoo

1. Calculate the total amount invested and enter the value in the table.
2. Calculate the number of shares bought each month. Round the number of shares to the nearest ten-thousandth. The calculation for January is shown as an example.
3. Calculate the average market price per share. (*Hint:* Add the monthly prices per share and divide by 12.)
4. Calculate the average price based on *dollar-cost averaging*. (*Hint:* Divide the total amount invested by the total number of shares.)
5. Rank the following scenarios in order of which was the best investment (most profit or least loss) at the end of December 2000. Show the value of each investment as a basis for your answer.

    (a) $1200 invested in Microsoft in January 2000.

    (b) $1200 invested in Microsoft using *dollar-cost averaging*.

    (c) $1200 invested in Microsoft in December 2000.

## 5.5 Exercises

**Divide.** See Examples 1, 2, and 4.

1. $27.3 \div (-7)$
2. $-50.4 \div 8$
3. $\dfrac{4.23}{9}$
4. $\dfrac{1.62}{6}$

5. $-20.01 \div (-0.05)$
6. $-16.04 \div (-0.08)$
7. $1.5\overline{)54}$
8. $2.4\overline{)132}$

Use the fact that $108 \div 18 = 6$ to work Exercises 9–16 simply by moving decimal points. See Examples 1, 2, and 4.

9. $1.8\overline{)0.108}$
10. $18\overline{)10.8}$
11. $0.018\overline{)108}$
12. $0.18\overline{)1.08}$
13. $0.18\overline{)10.8}$
14. $0.18\overline{)108}$
15. $18\overline{)0.0108}$
16. $1.8\overline{)0.0108}$

**Divide.** Round quotients to the nearest hundredth when necessary. See Examples 3 and 4.

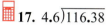
17. $4.6\overline{)116.38}$
18. $2.6\overline{)4.992}$
19. $\dfrac{-3.1}{-0.006}$
20. $\dfrac{-1.7}{0.09}$

**Divide.** Round quotients to the nearest thousandth. See Examples 3 and 4.

21. $-240 \div 9.88$
22. $-7643 \div (-5.36)$
23. $0.034\overline{)342.81}$
24. $0.043\overline{)1748.4}$

### Relating Concepts (Exercises 25–26) For Individual or Group Work

First, look back at your work in Section 5.4, Exercises 21 and 22. Then look for patterns as you **work Exercises 25 and 26 in order.**

25. Do these division problems.

    $3.77 \div 10 =$ _____   $9.1 \div 10 =$ _____
    $0.886 \div 10 =$ _____   $30.19 \div 10 =$ _____
    $406.5 \div 10 =$ _____   $6625.7 \div 10 =$ _____

    **(a)** What pattern do you see? Write a "rule" for dividing by 10. What do you think the rule is for dividing by 100? by 1000? Write the rules and try them out on the numbers above.

    **(b)** Compare your rules to the ones you wrote in Section 5.4, Exercise 21.

26. Do these division problems.

    $40.2 \div 0.1 =$ _____   $7.1 \div 0.1 =$ _____
    $0.339 \div 0.1 =$ _____   $15.77 \div 0.1 =$ _____
    $46 \div 0.1 =$ _____   $873 \div 0.1 =$ _____

    **(a)** What pattern do you see? Write a "rule" for dividing by 0.1. What do you think the rule is for dividing by 0.01? by 0.001? Write the rules and try them out on the numbers above.

    **(b)** Compare your rules to the ones you wrote in Section 5.4, Exercise 22.

*Decide whether each answer is* **reasonable** *or* **unreasonable** *by using front end rounding to estimate the answer. If the exact answer is not reasonable, find and correct the error. See Example 5.*

27. $37.8 \div 8 = 47.25$
    Estimate:

28. $345.6 \div 3 = 11.52$
    Estimate:

29. $54.6 \div 48.1 \approx 1.135$
    Estimate:

30. $2428.8 \div 4.8 = 50.6$
    Estimate:

31. $307.02 \div 5.1 = 6.2$
    Estimate:

32. $395.415 \div 5.05 = 78.3$
    Estimate:

33. $9.3 \div 1.25 = 0.744$
    Estimate:

34. $78 \div 14.2 = 0.182$
    Estimate:

*Solve each application problem. Round money answers to the nearest cent when necessary.*

35. Alfred has discovered that Batman's favorite brand of superhero tights are on sale. He's been told to buy only one pair for Robin. How much will he pay for one pair?

**Special Purchase!**
**Tights**
**6 pairs for $23.98**
Stock up now!

36. The bookstore has a special price on notepads. How much did Randall pay for one notepad?

*Notepads 4 for $1.69*

37. It will take 21 months for Aimee to pay off her charge account balance of $408.66. How much is she paying each month?

38. Marcella Anderson bought 2.6 meters of suede fabric for $18.19. How much did she pay per meter?

39. Adrian Webb bought 619 bricks to build a barbecue pit, paying $185.70. Find the cost per brick. (*Hint:* Cost *per* brick means the cost for *one* brick.)

40. Lupe Wilson is a newspaper distributor. Last week she paid the newspaper $130.51 for 842 copies. Find the cost per copy.

**41.** Darren Jackson earned $356.80 for 40 hours of work. Find his earnings per hour.

**42.** At a CD manufacturing company, 400 CDs cost $289. Find the cost per CD.

**43.** It took 16.35 gallons of gas to fill Kim's car gas tank. She had driven 346.2 miles since her last fill-up. How many miles per gallon did she get? Round to the nearest tenth.

**44.** Mr. Rodriquez pays $53.19 each month to Household Finance. How many months will it take him to pay off $1436.13?

*Use the table of longest long jumps (through the year 2000) to answer Exercises 45–50. To find an average, add up the values you are interested in and then divide the sum by the number of values. Round your answer to the nearest hundredth. Some of the other exercises may require subtraction or multiplication.*

| Athlete | Country | Year | Length (meters) |
|---|---|---|---|
| M. Powell | U.S. | 1991 | 8.95 |
| B. Beamon | U.S. | 1968 | 8.90 |
| C. Lewis | U.S. | 1991 | 8.87 |
| R. Emmiyan | USSR | 1987 | 8.86 |
| L. Myricks | U.S. | 1988 | 8.74 |
| E. Walder | U.S. | 1994 | 8.74 |
| I. Pedroso | Cuba | 1995 | 8.71 |
| K. Streete-Thompson | U.S. | 1994 | 8.63 |
| J. Beckford | Jamaica | 1997 | 8.62 |

*Source:* www.Olympics.com

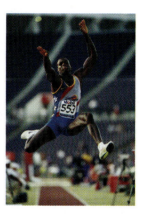

**45.** Find the average length of the long jumps made by U.S. athletes.

**46.** Find the average length of all the long jumps listed in the table.

**47.** How much longer was the second-place jump than the third-place jump?

**48.** If the first-place athlete made six jumps of the same length, what was the total distance jumped?

**49.** What was the total length jumped by the top three athletes?

**50.** How much less was the last-place jump than the next-to-last jump?

*Simplify by using the order of operations. See Example 6.*

**51.** $7.2 - 5.2 + 3.5^2$

**52.** $6.2 + 4.3^2 - 9.72$

**53.** $38.6 + 11.6(10.4 - 13.4)$

**54.** $2.25 - 1.06(0.85 - 3.95)$

**55.** $-8.68 - 4.6(10.4) \div 6.4$

**56.** $25.1 + 11.4 \div 7.5(-3.75)$

**57.** $33 - 3.2(0.68 + 9) + (-1.3)^2$

**58.** $0.6 + (-1.89 + 0.11) \div 0.004(0.5)$

*Solve each application problem.*

**59.** Soup is on sale at six cans for $3.25, or you can purchase individual cans for $0.57. How much will you save per can if you buy six cans? Round to the nearest cent.

**60.** Nadia's diet says she can eat 3.5 ounces of chicken nuggets. The package weighs 10.5 ounces and contains 15 nuggets. How many nuggets can Nadia eat?

**61.** In 1998, the U.S. Treasury spent about $386,500,000 to print 9,200,000,000 pieces of paper money. How much did it cost to print each piece, to the nearest cent? (*Source:* U.S. Department of the Treasury.)

9,200,000,000 pieces of paper money printed in 1998

**62.** Mach 1 is the speed of sound. Dividing a vehicle's speed by the speed of sound gives its speed on the Mach scale. In 1997, a specially built car with two 110,000-horsepower engines broke the world land speed record by traveling 763.035 miles per hour. The speed of sound changes slightly with the weather. That day it was 748.11 miles per hour. What was the car's Mach speed, to the nearest hundredth? (*Source:* Associated Press.)

*General Mills will give a school 10¢ for each box top logo from its cereals and other products. A school can earn up to $10,000 per year. Use this information to answer Exercises 63–66. (Source: General Mills.)*

**63.** How many box tops would a school need to collect in one year to earn the maximum amount?

**64.** (Complete Exercise 63 first.) If a school has 550 children, how many box tops would each child need to collect to reach the maximum?

**65.** How many box tops would need to be collected during each of the 38 weeks in the school year to reach the maximum amount?

**66.** How many box tops would each of the 550 children need to collect during each of the 38 weeks of school to reach the maximum amount?

# 5.6 FRACTIONS AND DECIMALS

Writing fractions as equivalent decimals can help you do calculations or compare the size of two numbers more easily.

**OBJECTIVES**
1. Write fractions as equivalent decimals.
2. Compare the size of fractions and decimals.

**1** Write fractions as equivalent decimals. Recall that a fraction is one way to show division (see **Section 1.7**). For example, $\frac{3}{4}$ means $3 \div 4$. If you are doing the division by hand, write it as $4\overline{)3}$. When you do the division, the result is $0.75$, the decimal equivalent of $\frac{3}{4}$.

### Writing a Fraction as a Decimal

*Step 1* Divide the numerator of the fraction by the denominator.

*Step 2* If necessary, round the answer to the place indicated.

Work Problem **1** at the Side.

**Example 1** Writing Fractions or Mixed Numbers as Decimals

**(a)** Write $\frac{1}{8}$ as a decimal.

$\frac{1}{8}$ means $1 \div 8$. Write it as $8\overline{)1}$. The decimal point in the dividend is on the right side of the 1. Write extra zeros in the dividend so you can continue dividing until the remainder is 0.

$$\frac{1}{8} \Rightarrow 1 \div 8 \Rightarrow 8\overline{)1} \Rightarrow 8\overline{)1.000}$$

Decimal points lined up. Three extra zeros needed. Remainder is 0.

Therefore, $\frac{1}{8} = 0.125$.

To check, write $0.125$ as a fraction, then write it in lowest terms.

$0.125 = \frac{125}{1000}$  In lowest terms: $\frac{125 \div 125}{1000 \div 125} = \frac{1}{8}$ {Original fraction}

**📱 Calculator Tip** When using your calculator to write fractions as decimals, enter the numbers from the top down. Remember that the *order* in which you enter the numbers *does* matter in division. Example 1(a) works like this:

$\frac{1}{8}$ Top down   Enter 1 ÷ 8 =   Answer is $0.125$.

What happens if you enter 8 ÷ 1 =? Do you see why that cannot possibly be correct? (Answer: $8 \div 1 = 8$, and a proper fraction like $\frac{1}{8}$ cannot be equivalent to a whole number.)

*Continued on Next Page*

**1** Rewrite each fraction so you could do the division by hand. Do *not* complete the division.

(a) $\frac{1}{9}$ is written $9\overline{)\phantom{0}}$

(b) $\frac{2}{3}$ is written $\overline{)\phantom{0}}$

(c) $\frac{5}{4}$ is written $\overline{)\phantom{0}}$

(d) $\frac{3}{10}$ is written $\overline{)\phantom{0}}$

(e) $\frac{21}{16}$ is written $\overline{)\phantom{0}}$

(f) $\frac{1}{50}$ is written $\overline{)\phantom{0}}$

**ANSWERS**
1. (a) $9\overline{)1}$  (b) $3\overline{)2}$  (c) $4\overline{)5}$
   (d) $10\overline{)3}$  (e) $16\overline{)21}$  (f) $50\overline{)1}$

Chapter 5 Rational Numbers: Positive and Negative Decimals

**❷ Write each fraction or mixed number as a decimal.**

(a) $\dfrac{1}{4}$

(b) $2\dfrac{1}{2}$

(c) $\dfrac{5}{8}$

(d) $4\dfrac{3}{5}$

(e) $\dfrac{7}{8}$

**ANSWERS**
2. (a) 0.25  (b) 2.5  (c) 0.625
   (d) 4.6  (e) 0.875

---

**(b)** Write $2\dfrac{3}{4}$ as a decimal.

One method is to divide 3 by 4 to get 0.75 for the fraction part. Then add the whole number part to 0.75.

$$\dfrac{3}{4} \Rightarrow \begin{array}{r} 0.75 \\ 4\overline{)3.00} \\ \underline{2\ 8} \\ 20 \\ \underline{20} \\ 0 \end{array}$$

Fraction part ⟶

Whole number part ⟶ 2.00
$+\ 0.75$
$\overline{2.75}$

So, $2\dfrac{3}{4} = 2.75$

*Whole number parts match.*

**Check**  $2.75 = 2\dfrac{75}{100} = 2\dfrac{3}{4}$ ⟵ Lowest terms

A second method is to write $2\dfrac{3}{4}$ as an improper fraction before dividing numerator by denominator.

$$2\dfrac{3}{4} = \dfrac{11}{4}$$

$$\dfrac{11}{4} \Rightarrow 11 \div 4 \Rightarrow 4\overline{)11} \Rightarrow \begin{array}{r} 2.75 \\ 4\overline{)11.00} \\ \underline{8} \\ 3\ 0 \\ \underline{2\ 8} \\ 20 \\ \underline{20} \\ 0 \end{array}$$ ⟵ Two extra zeros needed.

*Whole number parts match.*

So, $2\dfrac{3}{4} = 2.75$

$\dfrac{3}{4}$ is equivalent to $\dfrac{75}{100}$ or 0.75.

**Work Problem ❷ at the Side.**

---

**Example 2** **Writing a Fraction as a Decimal with Rounding**

Write $\dfrac{2}{3}$ as a decimal and round to the nearest thousandth.

$\dfrac{2}{3}$ means $2 \div 3$. To round to thousandths, divide out one *more* place, to ten-thousandths.

$$\dfrac{2}{3} \Rightarrow 2 \div 3 \Rightarrow 3\overline{)2} \Rightarrow \begin{array}{r} 0.6666 \\ 3\overline{)2.0000} \\ \underline{1\ 8} \\ 20 \\ \underline{18} \\ 20 \\ \underline{18} \\ 20 \\ \underline{18} \\ 2 \end{array}$$ ⟵ Four zeros needed for ten-thousandths

Written as a repeating decimal, $\dfrac{2}{3} = 0.\overline{6}$.
Rounded to the nearest thousandth, $\dfrac{2}{3} \approx 0.667$.

**Calculator Tip** Try Example 2 on your calculator. Enter 2 ÷ 3. Which answer do you get?

          0.6666666667    or    0.666666666

Most *scientific* and *graphing* calculators will show a 7 as the last digit. Because the 6s keep on repeating forever, the calculator automatically rounds in the last decimal place it has room to show. If you have a 10-digit display space, the calculator is rounding as shown below.

      0.6666666666 (11 digits)  rounds to  0.666666667

          Next digit is 5 or more, so 6 rounds to 7.

Other calculators, especially standard, four-function ones, may *not* round. They just cut off, or *truncate*, the extra digits. Such a calculator would show 0.6666666 in the display.

Would this difference in calculators show up when changing $\frac{1}{3}$ to a decimal? Why not? (Answer: The repeating digit is a 3, which is less than 5, so it stays as a 3 whether it's rounded or not.)

                                  Work Problem ❸ at the Side.

**2** **Compare the size of fractions and decimals.** You can use a number line to compare fractions and decimals. For example, the number line below shows the space between 0 and 1. The locations of some commonly used fractions are marked, along with their decimal equivalents.

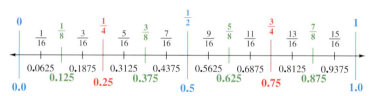

The next number line shows the locations of some commonly used fractions between 0 and 1 that are equivalent to repeating decimals. The decimal equivalents use a bar above repeating digits.

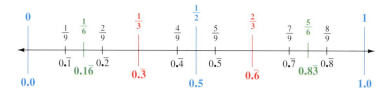

**Example 3** **Using a Number Line to Compare Numbers**

Use the number lines above to decide whether to write >, <, or = in the blank between each pair of numbers.

**(a)** 0.6875 _____ 0.625

You learned in **Section 1.2** that the number farther to the right on the number line is the greater number. On the first number line, 0.6875 is to the *right* of 0.625, so use the > symbol.

      0.6875  **is greater than**  0.625      0.6875 > 0.625

*Continued on Next Page*

❸ Write as decimals. Round to the nearest thousandth.

(a) $\frac{1}{3}$

(b) $2\frac{7}{9}$

(c) $\frac{10}{11}$

(d) $\frac{3}{7}$

(e) $3\frac{5}{6}$

**ANSWERS**

3. (a) $\frac{1}{3} \approx 0.333$  (b) $2\frac{7}{9} \approx 2.778$

   (c) $\frac{10}{11} \approx 0.909$  (d) $\frac{3}{7} \approx 0.429$

   (e) $3\frac{5}{6} \approx 3.833$

**4** Use the number lines in the text to help you decide whether to write <, >, or = in each blank.

(a) 0.4375 _____ 0.5

(b) 0.75 _____ 0.6875

(c) 0.625 _____ 0.0625

(d) $\dfrac{2}{8}$ _____ 0.375

(e) $0.8\overline{3}$ _____ $\dfrac{5}{6}$

(f) $\dfrac{1}{2}$ _____ $0.\overline{5}$

(g) $0.\overline{1}$ _____ $0.1\overline{6}$

(h) $\dfrac{8}{9}$ _____ $0.\overline{8}$

(i) $0.\overline{7}$ _____ $\dfrac{4}{6}$

(j) $\dfrac{1}{4}$ _____ 0.25

**5** Arrange each group in order from smallest to largest.

(a) 0.7, 0.703, 0.7029

(b) 6.39, 6.309, 6.4, 6.401

(c) 1.085, $1\dfrac{3}{4}$, 0.9

(d) $\dfrac{1}{4}, \dfrac{2}{5}, \dfrac{3}{7}$, 0.428

**ANSWERS**
4. (a) < (b) > (c) > (d) < (e) =
 (f) < (g) < (h) = (i) > (j) =
5. (a) 0.7, 0.7029, 0.703
 (b) 6.309, 6.39, 6.4, 6.401
 (c) 0.9, 1.085, $1\dfrac{3}{4}$
 (d) $\dfrac{1}{4}, \dfrac{2}{5}$, 0.428, $\dfrac{3}{7}$

---

(b) $\dfrac{3}{4}$ _____ 0.75

On the first number line, $\dfrac{3}{4}$ and 0.75 are at the same point on the number line. They are equivalent, so use the = symbol.

$$\dfrac{3}{4} = 0.75$$

(c) 0.5 _____ $0.\overline{5}$

On the second number line, 0.5 is to the *left* of $0.\overline{5}$ (which is actually 0.555 . . .), so use the < symbol.

0.5  is less than  $0.\overline{5}$     $0.5 < 0.\overline{5}$

(d) $\dfrac{2}{6}$ _____ $0.\overline{3}$

Write $\dfrac{2}{6}$ in lowest terms as $\dfrac{1}{3}$.
On the second number line you can see that $\dfrac{1}{3}$ and $0.\overline{3}$ are equivalent.

$$\dfrac{1}{3} = 0.\overline{3}.$$

**Work Problem 4 at the Side.**

You can also compare fractions by first writing each one as a decimal. You can then compare the decimals by writing each one with the same number of decimal places.

### Example 4  Arranging Numbers in Order

Write each group of numbers in order, from smallest to largest.

(a) 0.49   0.487   0.4903

It is easier to compare decimals if they are all tenths, or all hundredths, and so on. Because 0.4903 has four decimal places (ten-thousandths), write zeros to the right of 0.49 and 0.487 so they also have four decimal places. Writing zeros to the right of a decimal number does *not* change its value (see **Section 5.3**). Then find the smallest and largest number of ten-thousandths.

0.49 = 0.4900 = **4900** ten-thousandths  ← 4900 is in the middle.
0.487 = 0.4870 = **4870** ten-thousandths  ← 4870 is the smallest.
0.4903 = **4903** ten-thousandths  ← 4903 is the largest.

From smallest to largest, the correct order is shown below.

0.487   0.49   0.4903

(b) $2\dfrac{5}{8}$   2.63   2.6

Write $2\dfrac{5}{8}$ as $\dfrac{21}{8}$ and divide $8\overline{)21}$ to get the decimal form, 2.625. Then, because 2.625 has three decimal places, write zeros so all the numbers have three decimal places.

$2\dfrac{5}{8}$ = 2.625 = 2 and **625** thousandths ← 625 is in the middle.

2.63 = 2.630 = 2 and **630** thousandths ← 630 is largest.

2.6 = 2.600 = 2 and **600** thousandths ← 600 is smallest.

From smallest to largest, the correct order is shown below.

2.6   $2\dfrac{5}{8}$   2.63

**Work Problem 5 at the Side.**

## 5.6 Exercises

*Write each fraction or mixed number as a decimal. Round to the nearest thousandth when necessary. See Examples 1 and 2.*

1. $\dfrac{1}{2}$
2. $\dfrac{1}{4}$
3. $\dfrac{3}{4}$
4. $\dfrac{1}{10}$
5. $\dfrac{3}{10}$
6. $\dfrac{7}{10}$
7. $\dfrac{9}{10}$
8. $\dfrac{4}{5}$
9. $\dfrac{3}{5}$
10. $\dfrac{2}{5}$
11. $\dfrac{7}{8}$
12. $\dfrac{3}{8}$
13. $2\dfrac{1}{4}$
14. $1\dfrac{1}{2}$
15. $14\dfrac{7}{10}$
16. $23\dfrac{3}{5}$
17. $3\dfrac{5}{8}$
18. $2\dfrac{7}{8}$
19. $\dfrac{1}{3}$
20. $\dfrac{2}{3}$
21. $\dfrac{5}{6}$
22. $\dfrac{1}{6}$
23. $1\dfrac{8}{9}$
24. $5\dfrac{4}{7}$

### Relating Concepts (Exercises 25–28) For Individual or Group Work

*Use your knowledge of fractions and decimals as you **work Exercises 25–28 in order.***

25. **(a)** Explain how you can tell that Keith made an error *just by looking at his final answer*. Here is his work.

$$\dfrac{5}{9} = 5\overline{)9.0}\ \ \ \text{so}\ \ \ \dfrac{5}{9} = 1.8$$

with the long division showing 1.8, 5, 40, 40, 0.

**(b)** Show the correct way to change $\dfrac{5}{9}$ to a decimal. Explain why your answer makes sense.

26. **(a)** How can you prove to Sandra that $2\dfrac{7}{20}$ is *not* equivalent to 2.035? Here is her work.

$$2\dfrac{7}{20} = 20\overline{)7.00}\ \ \ \text{so}\ \ \ 2\dfrac{7}{20} = 2.035$$

with long division showing 0.35, 6 0, 1 00, 1 00, 0.

**(b)** What is the correct answer? Show how to prove that it is correct.

27. Ving knows that $\dfrac{3}{8} = 0.375$. How can he write $1\dfrac{3}{8}$ as a decimal *without* having to do a division? How can he write $3\dfrac{3}{8}$ as a decimal? $295\dfrac{3}{8}$? Explain your answer.

28. Iris has found a shortcut for writing mixed numbers as decimals.

$$2\dfrac{7}{10} = 2.7 \qquad 1\dfrac{13}{100} = 1.13$$

Does her shortcut work for all mixed numbers? Explain when it works and why it works.

*Find the decimal or fraction equivalent for each number. Write fractions in lowest terms.*

| Fraction | Decimal | Fraction | Decimal |
|---|---|---|---|
| 29. _____ | 0.4 | 30. _____ | 0.75 |
| 31. _____ | 0.625 | 32. _____ | 0.111 |
| 33. _____ | 0.35 | 34. _____ | 0.9 |
| 35. $\frac{7}{20}$ | _____ | 36. $\frac{1}{40}$ | _____ |
| 37. _____ | 0.04 | 38. _____ | 0.52 |
| 39. _____ | 0.15 | 40. _____ | 0.85 |
| 41. $\frac{1}{5}$ | _____ | 42. $\frac{1}{8}$ | _____ |
| 43. _____ | 0.09 | 44. _____ | 0.02 |

*Solve each application problem.*

**45.** The average length of a newborn baby is 20.8 inches. Charlene's baby is 20.08 inches long. Is her baby longer or shorter than the average? By how much?

**46.** The patient in room 830 is supposed to get 8.3 milligrams of medicine. She was actually given 8.03 milligrams. Did she get too much or too little medicine? What was the difference?

**47.** The label on the bottle of vitamins says that each capsule contains 0.5 gram of calcium. When checked, each capsule had 0.505 gram of calcium. Was there too much or too little calcium? What was the difference?

**48.** The glass mirror of the Hubble telescope had to be repaired in space in 1993 because it would not focus properly. The problem was that the mirror's outer edge had a thickness of 0.6248 cm when it was supposed to be 0.625 cm. Was the edge too thick or too thin? By how much? (*Source:* NASA.)

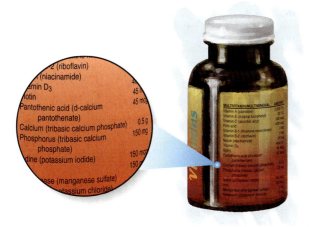

**49.** Precision Medical Parts makes an artificial heart valve that must measure between 0.998 centimeter and 1.002 centimeters. Circle the lengths that are acceptable:

1.01 cm, 0.9991 cm, 1.0007 cm, 0.99 cm.

**50.** The mice in a medical experiment must start out weighing between 2.95 ounces and 3.05 ounces. Circle the weights that can be used:

3.0 ounces, 2.995 ounces, 3.005 ounces,

3.055 ounces,

**51.** Ginny Brown hoped her crops would get $3\frac{3}{4}$ inches of rain this month. The newspaper said the area received 3.8 inches of rain. Was that more or less than Ginny had hoped for? By how much?

**52.** The mice in the experiment in Exercise 50 gained $\frac{3}{8}$ ounce. They were expected to gain 0.3 ounce. Was their actual gain more or less than expected? By how much?

*Arrange each group of numbers in order from smallest to largest. See Example 4.*

**53.** 0.54, 0.5455, 0.5399

**54.** 0.76, 0.7, 0.7006

**55.** 5.8, 5.79, 5.0079, 5.804

**56.** 12.99, 12.5, 13.0001, 12.77

**57.** 0.628, 0.62812, 0.609, 0.6009

**58.** 0.27, 0.281, 0.296, 0.3

**59.** 5.8751, 4.876, 2.8902, 3.88

**60.** 0.98, 0.89, 0.904, 0.9

**61.** 0.043, 0.051, 0.006, $\frac{1}{20}$

**62.** 0.629, $\frac{5}{8}$, 0.65, $\frac{7}{10}$

**63.** $\frac{3}{8}$, $\frac{2}{5}$, 0.37, 0.4001

**64.** 0.1501, 0.25, $\frac{1}{10}$, $\frac{1}{5}$

*The dad on the first page of this chapter found four boxes of fishing line in a sale bin. He knows that the thicker the line, the stronger it is. The diameter of the fishing line is its thickness. Use the information on the boxes to answer Exercises 65–68.*

**65.** Which color box has the strongest line?

**66.** Which color box has the line with the least strength?

**67.** What is the difference in line diameter between the weakest and strongest line?

**68.** What is the difference in line diameter between the blue and purple boxes?

*Some rulers for technical drawings show each inch divided into tenths. Use this scale drawing for Exercises 69–74. Change the measurements on the drawing to decimals and round them to the nearest tenth of an inch.*

**69.** Length (a) is _____

**70.** Length (b) is _____

**71.** Length (c) is _____

**72.** Length (d) is _____

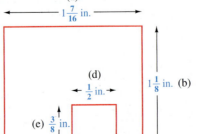

**73.** Length (e) is _____

**74.** Length (f) is _____

# 5.7 Problem Solving with Statistics: Mean, Median, Mode, and Variability

The word *statistics* originally came from words that mean *state numbers*. State numbers refer to numerical information, or *data*, gathered by the government such as the number of births, deaths, or marriages in a population. Today the word *statistics* has a much broader meaning; data from the fields of economics, social science, science, and business can all be organized and studied under the branch of mathematics called *statistics*.

**OBJECTIVES**
1. Find the mean of a list of numbers.
2. Find a weighted mean.
3. Find the median.
4. Find the mode.
5. Evaluate the variability of a set of data by finding the range of values.

**1** **Find the mean of a list of numbers.** Making sense of a long list of numbers can be difficult. So when you analyze data, one of the first things to look for is a *measure of central tendency*—a single number that you can use to represent the entire list of numbers. One such measure is the *average* or **mean**. The mean can be found with the following formula.

### Finding the Mean (Average)

$$\text{mean} = \frac{\text{sum of all values}}{\text{number of values}}$$

### Example 1 Finding the Mean (Average)

David had test scores of 84, 90, 95, 98, and 88. Find his mean (average) score.

Use the formula for finding the mean. Add up all the test scores and then divide the sum by the number of tests.

$$\text{mean} = \frac{84 + 90 + 95 + 98 + 88}{5} \quad \begin{array}{l} \leftarrow \text{Sum of test scores} \\ \leftarrow \text{Number of tests} \end{array}$$

$$\text{mean} = \frac{455}{5} \quad \text{Divide.}$$

$$\text{mean} = 91$$

David has a mean (average) score of 91.

➡ Work Problem **1** at the Side.

### Example 2 Applying the Mean (Average)

The sales of photo albums at Sarah's Card Shop for each day last week were $86, $103, $118, $117, $126, $158, and $149. Find the mean daily sales of photo albums.

To find the mean, add all the daily sales amounts and then divide the sum by the number of days (7).

$$\text{mean} = \frac{\$86 + \$103 + \$118 + \$117 + \$126 + \$158 + \$149}{7} \quad \begin{array}{l} \leftarrow \text{Sum of sales} \\ \leftarrow \text{Number of days} \end{array}$$

$$\text{mean} = \frac{\$857}{7}$$

$$\text{mean} \approx \$122.43 \quad \text{Rounded to nearest cent}$$

➡ Work Problem **2** at the Side.

**1** Tanya had test scores of 96, 98, 84, 88, 82, and 92. Find her mean (average) score.

**2** Find the mean for each list of numbers.

(a) Monthly long distance phone bills of $25.12, $42.58, $76.19, $32, $81.11, $26.41, $19.76, $59.32, $71.18, and $21.03

(b) The sales for one year at eight different office supply stores: $749,820; $765,480; $643,744; $824,222; $485,886; $668,178; $702,294; $525,800

**ANSWERS**
1. 90
2. (a) $\dfrac{\$454.70}{10} = \$45.47$

   (b) $\dfrac{\$5,365,424}{8} = \$670,678$

**3** Alison Nakano works downtown. Some days she can park in cheap lots that charge $6 or $7. Other days she has to park in lots that charge $9 or $10. Last month she kept track of the amount she spent each day for parking and the number of days she spent that amount. Find her average daily parking cost.

| Parking Fee | Frequency |
|---|---|
| $ 6 | 2 |
| $ 7 | 6 |
| $ 8 | 3 |
| $ 9 | 4 |
| $10 | 6 |

**2** **Find a weighted mean.** Some items in a list of data might appear more than once. In this case, we find a **weighted mean,** in which each value is "weighted" by multiplying it by the number of times it occurs.

### Example 3   Finding a Weighted Mean

The following table shows the amount of contribution and the number of times the amount was given (frequency) to a food pantry. Find the weighted mean.

| Contribution Value | Frequency |
|---|---|
| $ 3 | 4 ← 4 people each contributed $3. |
| $ 5 | 2 |
| $ 7 | 1 |
| $ 8 | 5 |
| $ 9 | 3 |
| $10 | 2 |
| $12 | 1 |
| $13 | 2 |

The same amount was given by more than one person: for example, $3 was given by four people, and $8 was given by five people. Other amounts, such as $12, were given by only one person.

To find the mean, multiply each contribution value by its frequency. Then add the products. Next, add the numbers in the *frequency* column to find the total number of values, that is, the total number of people who contributed money.

| Value | Frequency | Product |
|---|---|---|
| $ 3 | 4 | ($3 • 4) = $12 |
| $ 5 | 2 | ($5 • 2) = $10 |
| $ 7 | 1 | ($7 • 1) = $ 7 |
| $ 8 | 5 | ($8 • 5) = $40 |
| $ 9 | 3 | ($9 • 3) = $27 |
| $10 | 2 | ($10 • 2) = $20 |
| $12 | 1 | ($12 • 1) = $12 |
| $13 | 2 | ($13 • 2) = $26 |
| **Totals** | **20** | **$154** |

Finally, divide the totals.

$$\text{mean} = \frac{\$154}{20} = \$7.70$$

The mean contribution to the food pantry was $7.70.

**Work Problem 3 at the Side.**

**ANSWERS**
**3.** average ≈ $8.29 (to nearest cent)

Section 5.7 Problem Solving with Statistics: Mean, Median, Mode, and Variability

A common use of the weighted mean is to find a student's *grade point average (GPA)*, as shown by the next example.

**Example 4**  Applying the Weighted Mean

Find the GPA (grade point average) for a student who earned the following grades last semester. Assume A = 4, B = 3, C = 2, D = 1, and F = 0. The number of credits determines how many times the grade is counted (the frequency).

| Course | Credits | Grade | Credits • Grade |
|---|---|---|---|
| Mathematics | 4 | A (= 4) | 4 • 4 = 16 |
| Speech | 3 | C (= 2) | 3 • 2 = 6 |
| English | 3 | B (= 3) | 3 • 3 = 9 |
| Computer Science | 2 | A (= 4) | 2 • 4 = 8 |
| Theater | 2 | D (= 1) | 2 • 1 = 2 |
| **Totals** | **14** | | **41** |

It is common to round grade point averages to the nearest hundredth. So the grade point average for this student is rounded to 2.93.

$$\text{GPA} = \frac{41}{14} \approx 2.93$$

━━━━━━━━━━━ Work Problem ❹ at the Side.

**❹** Find the GPA (grade point average) for a student who earned the following grades. Round to the nearest hundredth.

| Course | Credits | Grade |
|---|---|---|
| Mathematics | 5 | A (= 4) |
| English | 3 | C (= 2) |
| Biology | 4 | B (= 3) |
| History | 3 | B (= 3) |

**3** **Find the median.** Because it can be affected by extremely high or low numbers, the mean is often a poor indicator of central tendency for a list of numbers. In cases like this, another measure of central tendency, called the *median*, can be used. The **median** divides a group of numbers in half; half the numbers lie above the median, and half lie below the median.

Find the median by listing the numbers *in order* from *smallest* to *largest*. If the list contains an *odd* number of items, the median is the *middle number*.

**Example 5**  Finding the Median

Find the median for this list of prices.

$7, $23, $15, $6, $18, $12, $24

First arrange the numbers in numerical order from smallest to largest.

Smallest → 6, 7, 12, 15, 18, 23, 24 ← Largest

Next, find the middle number in the list.

6, 7, 12, 15, 18, 23, 24

Three are below.   Three are above.
Middle number

The median price is $15.

━━━━━━━━━━━ Work Problem ❺ at the Side.

**❺** Find the median for the following number of customers helped each hour at the order desk.

35, 33, 27, 31, 39, 50, 59, 25, 30

If a list contains an *even* number of items, there is no single middle number. In this case, the median is defined as the mean (average) of the *middle two* numbers.

**ANSWERS**

**4.** GPA = $\frac{47}{15} \approx 3.13$

**5.** 33 customers (the middle number when the numbers are arranged from smallest to largest)

**6** Find the median for this list of measurements.

178 ft, 261 ft, 126 ft, 189 ft, 121 ft, 195 ft

**7** Find the mode for each list of numbers.

(a) Ages of part-time employees (in years): 28, 16, 22, 28, 34, 22, 28

(b) Total points on a screening exam: 312, 219, 782, 312, 219, 426, 507, 600

(c) Monthly commissions of sales people: $1706, $1289, $1653, $1892, $1301, $1782

**ANSWERS**

6. $\dfrac{178 + 189}{2} = 183.5$ ft

7. (a) 28 years
   (b) bimodal, 219 points and 312 points (this list has two modes)
   (c) no mode (no number occurs more than once)

### Example 6 — Finding the Median

Find the median for this list of ages, in years.

74, 7, 15, 13, 25, 28, 47, 59, 32, 68

First arrange the numbers in numerical order from smallest to largest. Then, because the list has an even number of ages, find the middle *two* numbers.

Smallest → 7, 13, 15, 25, **28, 32,** 47, 59, 68, 74 ← Largest
Middle two numbers

The median age is the mean of the two middle numbers.

$$\text{median} = \frac{28 + 32}{2} = \frac{60}{2} = 30 \text{ years}$$

**Work Problem 6 at the Side.**

**4 Find the mode.** Another statistical measure is the **mode**, which is the number that occurs *most often* in a list of numbers. For example, if the test scores for ten students were

74, 81, 39, 74, 82, 80, 100, 92, 74, and 85,

then the mode is 74. Three students earned a score of 74, so 74 appears more times on the list than any other score. It is *not* necessary to place the numbers in numerical order when looking for the mode, although that may help you find it more easily.

A list can have two modes; such a list is sometimes called *bimodal*. If no number occurs more frequently than any other number in a list, the list has *no mode.*

### Example 7 — Finding the Mode

Find the mode for each list of numbers.

(a) 51, 32, 49, 51, 49, 90, 49, 60, 17, 60
The number 49 occurs three times, which is more often than any other number. Therefore, 49 is the mode.

(b) 482, 485, 483, 485, 487, 487, 489, 486
Because both 485 and 487 occur twice, each is a mode. This list is *bimodal.*

(c) 10,708; 11,519; 10,972; 12,546; 13,905; 12,182
No number occurs more than once. This list has *no mode.*

### Measures of Central Tendency

The **mean** is the sum of all the values divided by the number of values. It is the mathematical *average*.

The **median** is the middle number in a group of values that are listed from smallest to largest. It divides a group of numbers in half.

The **mode** is the value that occurs most often in a group of values.

**Work Problem 7 at the Side.**

## Section 5.7 Problem Solving with Statistics: Mean, Median, Mode, and Variability

**5** ▌ **Evaluate the variability of a set of data by finding the range of values.** If two students each have a mean (average) score of 60 on their math tests, you might think they have done work of equal quality. However, a closer look at the test scores below gives a different impression.

|  | Test 1 | Test 2 | Test 3 | Test 4 | Test 5 | Mean | Median |
|---|---|---|---|---|---|---|---|
| Student Y | 55 | 60 | 60 | 60 | 65 | 60 | 60 |
| Student Z | 20 | 40 | 60 | 80 | 100 | 60 | 60 |

The means and medians are all 60. But Student Y's scores are all clustered around 60. Student Z's scores, however, have greater *variability* because they are spread over a wider range, from 20 to 100. The **variability** of a set of data is the spread of the data around the mean. A quick way to evaluate the variability is to look at the *range of values*. To find the range, subtract the lowest value from the highest value.

### Example 8  Finding the Range

(a) Find the range of test scores for Student Y, shown above.

Subtract Student Y's lowest test score of 55 from the highest test score of 65.

$$65 - 55 = 10 \quad \leftarrow \text{Range of scores for Student Y}$$

(b) Find the range of test scores for Student Z, shown above.

Highest score ⟶
$$100 - 20 = 80 \quad \leftarrow \text{Range of scores for Student Z}$$
Lowest score ⟶

**Work Problem ❽ at the Side.**

Generally, the greater the *range* in a set of values, the greater the *variability*. When analyzing data, look at the mean and/or median and also look at the variability. For example, if you were the teacher of Students Y and Z in Example 8 above, you would analyze their test data quite differently. Student Y has consistently gotten fairly low scores and may need some extra help. Student Z's scores, on the other hand, have improved dramatically, and he or she will probably do well on future tests.

### Example 9  Evaluating the Variability of Data

(a) Which set of data shows greater variability in the points scored in a basketball game? Show your work.

|  | Game 1 | Game 2 | Game 3 | Game 4 | Game 5 | Mean |
|---|---|---|---|---|---|---|
| Player A | 18 | 9 | 27 | 3 | 25 | 16.4 |
| Player B | 18 | 16 | 17 | 17 | 15 | 16.6 |

Range for Player A = 27 − 3 = 24
Range for Player B = 18 − 15 = 3
Player A's data has a greater range and greater variability.

*Continued on Next Page*

❽ Find the mean, median, and range of values for each set of data.

(a) Ages of students in Classroom B: 21, 23, 22, 21, 22, 23

Ages of students in Classroom C: 31, 18, 25, 17, 23, 21

(b) Expenses in December: $225, $350, $100, $325, $700

Expenses in January: $300, $325, $350, $325, $350

**ANSWERS**
8. (a) Classroom B: mean = 22, median = 22, range = 2
Classroom C: mean = 22.5, median = 22, range = 14
(b) December: mean = $340, median = $325, range = $600
January: mean = $330, median = $325, range = $50

**9** (a) Which set of temperature data, in degrees, has less variability? Both cities have a mean high of 72 degrees. Show your work.

City A had daily highs of 69, 73, 70, 75, 71, 74.

City B had daily highs of 76, 55, 63, 97, 92, 49.

(b) Which city would you rather visit on vacation? Explain why.

(c) Which set of weekly sales figures has greater variability? Show your work.

Salesperson Y: $8000; $7500; $7800; $8200; $8300

Salesperson Z: $8000; $12,000; $3000; $14,500; $2500

(d) Which salesperson would you like to hire? Explain why.

(b) If you are a coach, which player would you like on your team, Player A or Player B? Explain why.

Answers may vary. Some things to consider are: Player A had some very good games, and the means for the two players are nearly the same. But Player B is more consistent and perhaps more reliable. You would want to know more details about why Player A scored so few points in Games 2 and 4.

**Work Problem 9 at the Side.**

**ANSWERS**

9. (a) City A: $75 - 69 = 6$;
City B: $97 - 49 = 48$

City A's temperatures have less variability.

(b) Answers will vary; be sure you explained why you chose a particular city.

(c) Salesperson Y: $8300 - $7500 = $800;
Salesperson Z: $14,500 - $2500 = $12,000

Salesperson Z's figures have greater variability.

(d) Answers will vary; be sure you explained why you chose a particular salesperson.

## 5.7 Exercises

*Find the mean for each list of numbers. Round answers to the nearest tenth when necessary. See Example 1.*

1. Ages of infants at the child care center (in months): 4, 9, 6, 4, 7, 10, 9

2. Monthly electric bills: $53, $77, $38, $29, $49, $48

3. Final exam scores: 92, 51, 59, 86, 68, 73, 49, 80

4. Quiz scores: 18, 25, 21, 8, 16, 13, 23, 19

5. Annual salaries: $31,900; $32,850; $34,930; $39,712; $38,340, $60,000

6. Numbers of people attending baseball games: 27,500; 18,250; 17,357; 14,298; 33,110

*Solve each application problem. See Examples 2 and 3.*

7. The Athletic Shoe Store sold shoes at the following prices: $75.52, $36.15, $58.24, $21.86, $47.68, $106.57, $82.72, $52.14, $28.60, $72.92. Find the mean shoe sales amount.

8. In one evening, a waitress collected the following checks from her dinner customers: $30.10, $42.80, $91.60, $51.20, $88.30, $21.90, $43.70, $51.20. Find the mean dinner check amount.

9. The table below shows the face value (policy amount) of life insurance policies sold and the number of policies sold for each amount by the New World Life Company during one week. Find the weighted mean amount for the policies sold.

| Policy Amount | Number of Policies Sold |
| --- | --- |
| $ 10,000 | 6 |
| $ 20,000 | 24 |
| $ 25,000 | 12 |
| $ 30,000 | 8 |
| $ 50,000 | 5 |
| $100,000 | 3 |
| $250,000 | 2 |

10. Detroit Metro-Sales Company prepared the table below showing the gasoline mileage obtained by each of the cars in their automobile fleet. Find the weighted mean to determine the miles per gallon for the fleet of cars.

| Miles per Gallon | Number of Autos |
| --- | --- |
| 15 | 5 |
| 20 | 6 |
| 24 | 10 |
| 30 | 14 |
| 32 | 5 |
| 35 | 6 |
| 40 | 4 |

*Find the weighted mean. Round answers to the nearest tenth when necessary. See Example 3.*

11.
| Quiz Scores | Frequency |
|---|---|
| 3 | 4 |
| 5 | 2 |
| 6 | 5 |
| 8 | 5 |
| 9 | 2 |

12.
| Credits per Student | Frequency |
|---|---|
| 9 | 3 |
| 12 | 5 |
| 13 | 2 |
| 15 | 6 |
| 18 | 1 |

13.
| Hours Worked | Frequency |
|---|---|
| 12 | 4 |
| 13 | 2 |
| 15 | 5 |
| 19 | 3 |
| 22 | 1 |
| 23 | 5 |

14.
| Students per Class | Frequency |
|---|---|
| 25 | 1 |
| 26 | 2 |
| 29 | 5 |
| 30 | 4 |
| 32 | 3 |
| 33 | 5 |

*Find the GPA (grade point average) for students earning the following grades. Assume* A = 4, B = 3, C = 2, D = 1, *and* F = 0. *Round answers to the nearest hundredth. See Example 4.*

15.
| Course | Credits | Grade |
|---|---|---|
| Biology | 4 | B |
| Biology Lab | 2 | A |
| Mathematics | 5 | C |
| Health | 1 | F |
| Psychology | 3 | B |

16.
| Course | Credits | Grade |
|---|---|---|
| Chemistry | 3 | A |
| English | 3 | B |
| Mathematics | 4 | B |
| Theater | 2 | C |
| Astronomy | 3 | C |

17. Look again at the grades in Exercise 15. Find the student's GPA in each of these situations.
    (a) The student earned a B instead of an F in the 1-credit class.
    (b) The student earned a B instead of a C in the 5-credit class.
    (c) Both (a) and (b) happened.

18. List the credits for the courses you're taking at this time. List the lowest grades you think you will earn in each class and find your GPA. Then list the highest grades you think you will earn and find your GPA.

*Find the median for each list of numbers. See Examples 5 and 6.*

19. Number of e-mail messages received:
    9, 12, 14, 15, 23, 24, 28

20. Deliveries by a newspaper distributor:
    99, 108, 109, 123, 126, 129, 146, 168, 170

21. Students enrolled in algebra each semester:
    328, 549, 420, 592, 715, 483

22. Number of cars in the parking lot each day:
    520, 523, 513, 1283, 338, 509, 290, 420

23. Number of computer service calls taken each day:
    51, 48, 96, 40, 47, 23, 95, 56, 34, 48

24. Number of gallons of paint sold per week:
    1072, 1068, 1093, 1042, 1056, 205, 1009, 1081

*The table lists the cruising speed and distance flown without refueling for several types of larger airplanes used to carry passengers. Use the table to answer Exercises 25–28.*

| Type of Airplane | Cruising Speed (miles per hour) | Distance without Refueling (miles) |
|---|---|---|
| 747-400 | 565 | 7650 |
| 747-200 | 558 | 6450 |
| DC-9 | 505 | 1100 |
| DC-10 | 550 | 5225 |
| 727 | 530 | 1550 |
| 757 | 530 | 2875 |

*Source:* Northwest Airlines *WorldTraveler*.

25. What is the average distance flown without refueling, to the nearest mile?

26. Find the average cruising speed.

27. (a) Find the median distance.

    (b) Is the median similar to the average distance from Exercise 25? Explain why or why not.

28. (a) Find the median cruising speed.

    (b) Is the median similar to the average speed from Exercise 26? Explain why or why not.

*Find the mode or modes for each list of numbers. See Example 7.*

29. Number of samples taken each hour:
    3, 8, 5, 1, 7, 6, 8, 4, 5, 8

30. Monthly water bills:
    $21, $32, $46, $32, $49, $32, $49, $25, $32

31. Ages of retirees (in years):
    74, 68, 68, 68, 75, 75, 74, 74, 70, 77

32. Patients admitted to the hospital each week:
    30, 19, 25, 78, 36, 20, 45, 85, 38

33. The number of boxes of candy sold by each child:
    5, 9, 17, 3, 2, 8, 19, 1, 4, 20, 10, 6

34. The weights of soccer players (in pounds):
    158, 161, 165, 162, 165, 157, 163, 162

*The table lists monthly normal temperatures from November through April for some of the coldest and warmest U.S. cities. Use the table to answer Exercises 35–40. Round answers to the nearest whole degree. For help in determining variability, see Examples 8 and 9.*

| Normal Monthly Temperatures (in Degrees Fahrenheit) | | | | | | |
|---|---|---|---|---|---|---|
| City | Nov. | Dec. | Jan. | Feb. | Mar. | Apr. |
| Barrow, Alaska | −2 | −11 | −13 | −18 | −15 | −2 |
| Fairbanks, Alaska | 3 | −7 | −10 | −4 | 11 | 31 |
| Honolulu, Hawaii | 77 | 74 | 73 | 73 | 74 | 76 |
| Miami, Florida | 74 | 69 | 67 | 69 | 72 | 75 |

*Source:* National Climate Data Center.

35. (a) Find Barrow's mean temperature and Fairbanks' mean temperature for the six-month period.

    (b) How much warmer is Fairbanks' mean than Barrow's mean?

36. (a) Find Barrow's median temperature and Fairbanks' median temperature for the six-month period.

    (b) How much cooler is Barrow's median than Fairbanks' median?

**37.** Which set of temperatures has greater variability, Barrow or Fairbanks? Show how to calculate the range for each city.

**38. (a)** Find the mean and median for Honolulu's temperatures during the six-month period.

**(b)** Why are the mean and median so similar?

**39. (a)** Find the mean and median for Miami's temperatures during the six-month period.

**(b)** What do you notice about the mean and median? Explain how this is possible.

**40.** Which set of temperatures has less variability, Honolulu or Miami? Show how to calculate the range for each city.

---

**RELATING CONCEPTS (Exercises 41–44)**  **FOR INDIVIDUAL OR GROUP WORK**

*Use your knowledge of statistics as you* **work Exercises 41–44 in order.**

**41. (a)** Find the mean, median, and range for each student's test data. Round to the nearest tenth when necessary.
Student P: 92, 80, 61, 49, 82, 53
Student Q: 70, 76, 77, 60, 67, 72

**42. (a)** Find the mean, median, and range for each golfer's scores. Round to the nearest tenth when necessary.
Golfer G: 84, 87, 83, 89, 88
Golfer H: 94, 88, 76, 89, 85

**(b)** Which student's data has greater variability?

**(c)** If you were the teacher, what advice would you give to each student?

**(b)** Which golfer had scores with less variability?

**(c)** If you were the coach, what advice would you give to each golfer?

**43. (a)** Find the set of data on ages of tenants with the least variability. Show your work.
Building A: 63, 70, 66, 62, 74, 65, 74
Building B: 23, 20, 26, 18, 21, 23, 19
Building C: 45, 32, 69, 25, 50, 37, 41

**44. (a)** Find the set of data on monthly salaries of computer technicians that has the greatest variability. Show your work.
Company P: $4900, $5500, $5000, $7200, $6100
Company Q: $5800, $5900, $5500, $6000, $5900
Company R: $6200, $6400, $5800, $5200, $5600

**(b)** Based on your work in part (a), which building would you like to live in? Explain why.

**(b)** Based on your work in part (a), which company would you like to work for? Explain why.

## 5.8 Geometry Applications: Pythagorean Theorem and Square Roots

In **Section 3.2** you used this formula for area of a square, $A = s^2$. The blue square below has an area of 25 cm² because (5 cm)(5 cm) = 25 cm².

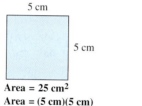

Area = 25 cm²
Area = (5 cm)(5 cm)

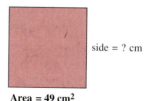

Area = 49 cm²
Area = (? cm)(? cm)

The red square above has an area of 49 cm². To find the length of a side, ask yourself, "What number can be multiplied by itself to give 49?" Because (7)(7) = 49, the length of each side is 7 cm. Also, because (7)(7) = 49 we say that 7 is the *square root* of 49, or $\sqrt{49} = 7$.

### Square Root

The positive **square root** of a positive number is one of two identical positive factors of that number.
For example, $\sqrt{36} = 6$ because (6)(6) = 36.

**NOTE**

There is another square root for 36. We know that

$$(6)(6) = 36 \quad \text{and} \quad (-6)(-6) = 36$$

so the *positive* square root of 36 is 6 and the *negative* square root of 36 is −6. In this section we will work only with positive square roots. You will learn more about negative square roots in other math courses.

Work Problem ❶ at the Side.

A number that has a whole number as its square root is called a *perfect square*. For example, 9 is a perfect square because $\sqrt{9} = 3$, and 3 is a whole number. The first few perfect squares are listed below.

### The First Twelve Perfect Squares

| | | | |
|---|---|---|---|
| $\sqrt{1} = 1$ | $\sqrt{16} = 4$ | $\sqrt{49} = 7$ | $\sqrt{100} = 10$ |
| $\sqrt{4} = 2$ | $\sqrt{25} = 5$ | $\sqrt{64} = 8$ | $\sqrt{121} = 11$ |
| $\sqrt{9} = 3$ | $\sqrt{36} = 6$ | $\sqrt{81} = 9$ | $\sqrt{144} = 12$ |

❶ **Find square roots using the square root key on a calculator.** If a number is *not* a perfect square, then you can find its *approximate* square root by using a calculator with a square root key.

### Objectives

1. Find square roots using the square root key on a calculator.
2. Find the unknown length in a right triangle.
3. Solve application problems involving right triangles.

❶ Find each square root.

(a) $\sqrt{36}$

(b) $\sqrt{25}$

(c) $\sqrt{9}$

(d) $\sqrt{100}$

(e) $\sqrt{121}$

**Answers**
1. (a) 6  (b) 5  (c) 3  (d) 10  (e) 11

**2** Use a calculator with a square root key to find each square root. Round to the nearest thousandth when necessary.

(a) $\sqrt{11}$

(b) $\sqrt{40}$

(c) $\sqrt{56}$

(d) $\sqrt{196}$

(e) $\sqrt{147}$

**Calculator Tip** To find a square root on a *scientific* calculator, use the $\sqrt{\phantom{x}}$ or the $\sqrt{x}$ key. (On some models you may have to press the 2nd key to access the square root function.) You do *not* need to use the = key. Try these.

To find $\sqrt{16}$ press:   16 $\sqrt{x}$   Answer is 4
To find $\sqrt{7}$ press:   7 $\sqrt{x}$   Answer is 2.645751311

For $\sqrt{7}$, your calculator shows 2.645751311, which is an *approximate* answer. We will round to the nearest thousandth, so $\sqrt{7} \approx 2.646$. To check, multiply 2.646 times 2.646. Do you get 7 as the result? No, you get 7.001316, which is very close to 7. The difference is due to rounding.

### Example 1   Finding the Square Root of Numbers

Use a calculator to find each square root. Round to the nearest thousandth.

(a) $\sqrt{35}$   Calculator shows 5.916079783; round to 5.916.

(b) $\sqrt{124}$   Calculator shows 11.13552873; round to 11.136.

**Work Problem 2 at the Side.**

**2** **Find the unknown length in a right triangle.** One place you will use square roots is when working with the *Pythagorean Theorem*. This theorem applies only to *right* triangles (triangles with a 90° angle). The longest side of a right triangle is called the **hypotenuse.** It is opposite the right angle. The other two sides are called *legs*. The legs form the right angle. Here are some right triangles.

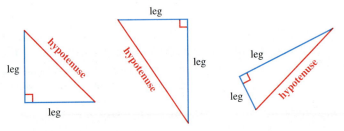

Examples of right triangles

### Pythagorean Theorem

$$(\text{hypotenuse})^2 = (\text{leg})^2 + (\text{leg})^2$$

In other words, square the length of each side. After you have squared all the sides, the sum of the squares of the two legs will equal the square of the hypotenuse.

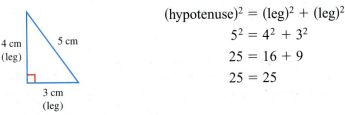

$$(\text{hypotenuse})^2 = (\text{leg})^2 + (\text{leg})^2$$
$$5^2 = 4^2 + 3^2$$
$$25 = 16 + 9$$
$$25 = 25$$

The theorem is named after Pythagoras, a Greek mathematician who lived about 2500 years ago. He and his followers may have used floor tiles to prove the theorem, as shown on the next page.

**Answers**
2. (a) $\sqrt{11} \approx 3.317$   (b) $\sqrt{40} \approx 6.325$
   (c) $\sqrt{56} \approx 7.483$   (d) $\sqrt{196} = 14$
   (e) $\sqrt{147} \approx 12.124$

Section 5.8 Geometry Applications: Pythagorean Theorem and Square Roots  373

Right triangle

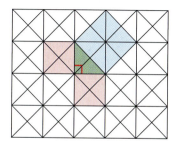

The green right triangle in the center of the floor tiles has sides *a*, *b*, and *c*. The pink square drawn on side *a* contains four triangular tiles. The pink square on side *b* contains four tiles. The blue square on side *c* contains eight tiles. The number of tiles in the square on side *c* equals the sum of the number of tiles in the squares on sides *a* and *b*, that is, 8 tiles = 4 tiles + 4 tiles. As a result, you often see the Pythagorean Theorem written as $c^2 = a^2 + b^2$.

If you know the lengths of any two sides in a right triangle, you can use the Pythagorean Theorem to find the length of the third side.

### Formulas Based on the Pythagorean Theorem

To find the hypotenuse:  hypotenuse $= \sqrt{(\text{leg})^2 + (\text{leg})^2}$

To find a leg:  leg $= \sqrt{(\text{hypotenuse})^2 - (\text{leg})^2}$

**Example 2**  Finding the Unknown Length in Right Triangles

Find the unknown length in each right triangle. Round to the nearest tenth when necessary.

(a)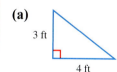

The unknown length is the side opposite the right angle, which is the hypotenuse. Use the formula for finding the hypotenuse.

$$\begin{aligned}\text{hypotenuse} &= \sqrt{(\text{leg})^2 + (\text{leg})^2} &&\text{Find the hypotenuse.}\\ \text{hypotenuse} &= \sqrt{(3)^2 + (4)^2} &&\text{Legs are 3 and 4.}\\ &= \sqrt{9 + 16} &&\text{(3)(3) is 9 and (4)(4) is 16.}\\ &= \sqrt{25}\\ &= 5\end{aligned}$$

The hypotenuse is 5 ft long.

(b)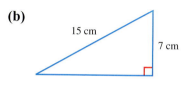

You *do* know the length of the hypotenuse (15 cm), so it is the length of one of the legs that is unknown. Use the formula for finding a leg.

$$\begin{aligned}\text{leg} &= \sqrt{(\text{hypotenuse})^2 - (\text{leg})^2} &&\text{Find a leg.}\\ \text{leg} &= \sqrt{(15)^2 - (7)^2} &&\text{Hypotenuse is 15; one leg is 7.}\\ &= \sqrt{225 - 49} &&\text{(15)(15) is 225 and (7)(7) is 49.}\\ &= \sqrt{176} &&\text{Use a calculator to find } \sqrt{176}.\\ &\approx 13.3 &&\text{Round 13.26649916 to 13.3.}\end{aligned}$$

The length of the leg is approximately 13.3 cm.

**Work Problem ❸ at the Side.**

❸ Find the unknown length in each right triangle. Round your answers to the nearest tenth when necessary.

(a)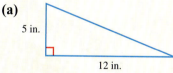

(b)

(c)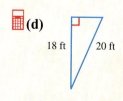

(d)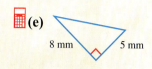

(e) 
8 mm   5 mm

**ANSWERS**

3. (a) $\sqrt{169} = 13$ in.  (b) $\sqrt{576} = 24$ cm
(c) $\sqrt{458} \approx 21.4$ m  (d) $\sqrt{76} \approx 8.7$ ft
(e) $\sqrt{89} \approx 9.4$ mm

**4**  These problems show ladders leaning against buildings. Find the unknown lengths. Round answers to the nearest tenth of a foot when necessary.

(a)

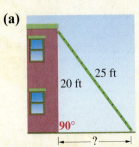

How far away from the building is the bottom of the ladder?

(b)

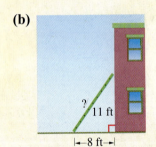

How long is the ladder?

(c) A 17 ft ladder is leaning against a building. The bottom of the ladder is 10 ft from the building. How high up on the building will the ladder reach? (*Hint:* Start by drawing a sketch of the building and the ladder.)

**CAUTION**

*Remember:* A small square drawn in one angle of a triangle indicates a right angle (90°). You can use the Pythagorean Theorem *only* on triangles that have a right angle.

**3** Solve application problems involving right triangles. The next example shows an application of the Pythagorean Theorem.

**Example 3** Using the Pythagorean Theorem

A television antenna is on the roof of a house, as shown. Find the length of the support wire. Round your answer to the nearest tenth of a meter if necessary.

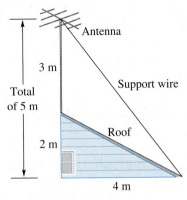

A right triangle is formed. The total length of the leg on the left is 5 m.

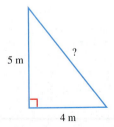

The support wire is the hypotenuse of the right triangle.

| | | |
|---|---|---|
| hypotenuse = $\sqrt{(\text{leg})^2 + (\text{leg})^2}$ | | Find the hypotenuse. |
| hypotenuse = $\sqrt{(5)^2 + (4)^2}$ | | Legs are 5 and 4. |
| $= \sqrt{25 + 16}$ | | $5^2$ is 25 and $4^2$ is 16. |
| $= \sqrt{41}$ | | Use a calculator to find $\sqrt{41}$. |
| $\approx 6.4$ | | Round 6.403124237 to 6.4. |

The length of the support wire is approximately 6.4 m.

Work Problem **4** at the Side.

**CAUTION**

You use the Pythagorean Theorem to find the *length* of one side, *not* the area of the triangle. Your answer will be in linear units, such as ft, yd, cm, m, and so on (*not* ft$^2$, yd$^2$, cm$^2$, m$^2$).

**ANSWERS**

**4.** (a) $\sqrt{225} = 15$ ft  (b) $\sqrt{185} \approx 13.6$ ft
  (c) $\sqrt{189} \approx 13.7$ ft

## 5.8 Exercises

*Find each square root. Starting with Exercise 5, find the square root using a calculator. Round your answers to the nearest thousandth when necessary. See Example 1.*

1. $\sqrt{16}$
2. $\sqrt{4}$
3. $\sqrt{64}$
4. $\sqrt{81}$

5. $\sqrt{11}$
6. $\sqrt{23}$
7. $\sqrt{5}$
8. $\sqrt{2}$

9. $\sqrt{73}$
10. $\sqrt{80}$
11. $\sqrt{101}$
12. $\sqrt{125}$

13. $\sqrt{361}$
14. $\sqrt{729}$
15. $\sqrt{1000}$
16. $\sqrt{2000}$

17. You know that $\sqrt{25} = 5$ and $\sqrt{36} = 6$. Using just that information (no calculator), describe how you could estimate $\sqrt{30}$. How would you estimate $\sqrt{26}$ or $\sqrt{35}$? Now check your estimates using a calculator.

18. Explain the relationship between *squaring* a number and finding the *square root* of a number. Include two examples to illustrate your explanation.

*Find the unknown length in each right triangle. Use a calculator to find square roots. Round your answers to the nearest tenth when necessary. See Example 2.*

19.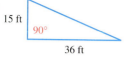

20.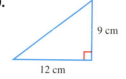

21.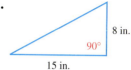

22.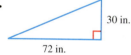

23.

24.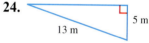

**25.**  3 in., 8 in.

**26.** 5 cm, 11 cm

**27.** 7 yd, 4 yd, 90°

**28.** 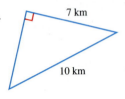 7 km, 10 km

**29.** 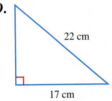 22 cm, 17 cm

**30.** 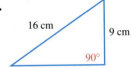 16 cm, 9 cm, 90°

**31.** 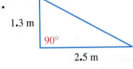 1.3 m, 90°, 2.5 m

**32.** 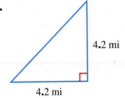 4.2 mi, 4.2 mi

**33.**  11.5 cm, 8.2 cm

**34.** 9.1 mm, 10.8 mm

**35.**  13.2 km, 90°, 21.6 km

**36.** 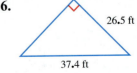 26.5 ft, 37.4 ft

Solve each application problem. Round your answers to the nearest tenth when necessary. See Example 3.

**37.** Find the length of this loading ramp.

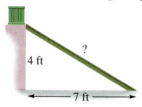

**38.** Find the unknown length in this roof plan.

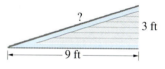

**39.** How high is the airplane above the ground?

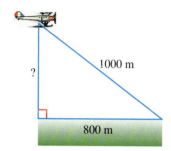

**40.** Find the height of this farm silo.

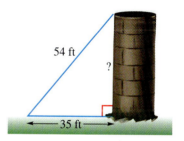

**41.** To reach his lady-love, a knight placed a 12 ft ladder against the castle wall. If the base of the ladder is 3 ft from the building, how high on the castle will the top of the ladder reach? Draw a sketch of the castle and ladder, and solve the problem.

**42.** William drove his car 15 miles north, then made a right turn and drove 7 miles east. How far is he, in a straight line, from his starting point? Draw a sketch to illustrate the problem, and solve the problem.

**43.** Explain the *two* errors made by a student in solving this problem. Also find the correct answer. Round to the nearest tenth.

$? = \sqrt{(13)^2 + (20)^2}$
$= \sqrt{169 + 400}$
$= \sqrt{569} \approx 23.9 \text{ m}^2$

**44.** Explain the *two* errors made by a student in solving this problem. Also find the correct answer. Round to the nearest tenth.

$? = \sqrt{(9)^2 + (7)^2}$
$= \sqrt{18 + 14}$
$= \sqrt{32} \approx 5.657 \text{ in.}$

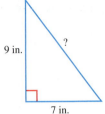

---

### RELATING CONCEPTS (Exercises 45–48)   FOR INDIVIDUAL OR GROUP WORK

*Use your knowledge of the Pythagorean Theorem to* **work Exercises 45–48 in order.** *Round answers to the nearest tenth.*

**45.** A major league baseball diamond is a square shape measuring 90 ft on each side. If the catcher throws a ball from home plate to second base, how far is he throwing the ball? (*Source:* American League of Professional Baseball Clubs.)

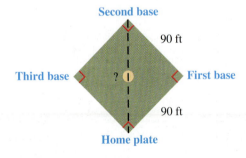

**46.** A softball diamond is only 60 ft on each side. (*Source:* Amateur Softball Association.)

(a) Draw a sketch of the diamond and label the sides and bases.

(b) How far is it to throw a ball from home plate to second base?

**47.** Look back at your answer to Exercise 45. Explain how you can tell the distance from third base to first base without doing any further calculations.

**48.** (a) Look back at your answer to Exercise 46. Suppose you measured the distance from home plate to second base on a softball diamond and found it was 80 ft. What would this tell you about the length of each side of the diamond? (Assume the diamond is still a square.)

(b) Bonus question: Find the length of each side, to the nearest tenth.

## 5.9 PROBLEM SOLVING: EQUATIONS CONTAINING DECIMALS

**OBJECTIVES**

1. Solve equations containing decimals by using the addition property of equality.
2. Solve equations containing decimals by using the division property of equality.
3. Solve equations containing decimals by using both properties of equality.
4. Solve application problems involving equations with decimals.

**1** Solve equations containing decimals by using the addition property of equality. In **Section 2.3** you used the addition property of equality to solve an equation like $c + 5 = 30$. The addition property says that you can add the *same* number to *both* sides of an equation and still keep it balanced. You can also use this property when an equation contains decimal numbers.

**Example 1**  Using the Addition Property of Equality

Solve each equation and check each solution.
(a) $w + 2.9 = -0.6$

The first step is to get the variable term ($w$) by itself on the left side of the equal sign. Use the addition property to "get rid of" the 2.9 on the left side by adding its opposite, $-2.9$, to both sides.

$$w + 2.9 = -0.6$$
$$\underline{\phantom{w+}-2.9 \quad -2.9} \quad \text{Add } -2.9 \text{ to both sides.}$$
$$w + 0 = -3.5$$
$$w = -3.5$$

The solution is $-3.5$. To check the solution, go back to the *original* equation.

**Check**
$w + 2.9 = -0.6$   Original equation
$-3.5 + 2.9 = -0.6$   Replace $w$ with $-3.5$.
$-0.6 = -0.6$   Balances, so $-3.5$ is the correct solution.

(b) $7 = -4.3 + x$

To get $x$ by itself on the right side of the equal sign, add 4.3 to both sides.

$$7 = -4.3 + x$$
$$\underline{+4.3 \quad\quad +4.3} \quad \text{Add } 4.3 \text{ to both sides.}$$
$$11.3 = 0 + x$$
$$11.3 = x$$

The solution is 11.3. To check the solution, go back to the original equation.

**Check**
$7 = -4.3 + x$   Original equation
$7 = -4.3 + 11.3$   Replace $x$ with 11.3.
$7 = 7$   Balances, so 11.3 is the correct solution.

━━━━━━━━━━━━━━━ Work Problem **1** at the Side.

**1** Solve each equation and check each solution.

(a) $8.1 = h + 9$   Check

(b) $-0.75 + y = 0$   Check

(c) $c - 6.8 = -4.8$   Check

**ANSWERS**

1. (a) $h = -0.9$   Check $8.1 = h + 9$
$8.1 = -0.9 + 9$
Balances $8.1 = 8.1$

(b) $y = 0.75$   Check $-0.75 + y = 0$
$-0.75 + 0.75 = 0$
Balances $0 = 0$

(c) $c = 2$   Check $c - 6.8 = -4.8$
$2 - 6.8 = -4.8$
Balances $-4.8 = -4.8$

**Chapter 5** Rational Numbers: Positive and Negative Decimals

❷ Solve each equation and check each solution.

(a) $-3y = -0.63$     Check

**2** Solve equations containing decimals by using the division property of equality. You can also use the division property of equality (from **Section 2.4**) when an equation contains decimals.

### Example 2 — Using the Division Property of Equality

Solve each equation and check each solution.

(a) $5x = 12.4$

On the left side of the equation, the variable is multiplied by 5. To undo the multiplication, divide both sides by 5.

$5x$ means $5 \cdot x$.

$$5x = 12.4$$

$$\frac{5 \cdot x}{5} = \frac{12.4}{5}$$  Divide both sides by 5.

On the left side, divide out the common factor of 5.

$$\frac{\overset{1}{\cancel{5}} \cdot x}{\underset{1}{\cancel{5}}} = 2.48$$  On the right side, $12.4 \div 5$ is 2.48.

$$x = 2.48$$

The solution is 2.48. To check the solution, go back to the original equation.

**Check**
$$5x = 12.4 \quad \text{Original equation}$$
$$5(\mathbf{2.48}) = 12.4 \quad \text{Replace } x \text{ with 2.48}$$
$$12.4 = 12.4 \quad \text{Balances, so 2.48 is the correct solution.}$$

(b) $2.25r = -18$     Check

(b) $-9.3 = 1.5t$

Signs are different, so quotient is negative.

$$\frac{-9.3}{1.5} = \frac{\cancel{1.5}t}{\cancel{1.5}} \quad \text{Divide both sides by the coefficient of the variable term, 1.5.}$$

$$-6.2 = t$$

The solution is $-6.2$. To check the solution, go back to the original equation.

**Check**
$$-9.3 = 1.5\,t \quad \text{Original equation}$$
$$-9.3 = 1.5\,(\mathbf{-6.2}) \quad \text{Replace } t \text{ with } -6.2.$$
$$-9.3 = -9.3 \quad \text{Balances, so } -6.2 \text{ is the correct solution.}$$

(c) $1.7 = 0.5n$     Check

**Work Problem** ❷ **at the Side.**

**3** Solve equations containing decimals by using both properties of equality. Sometimes you need to use both the addition and division properties to solve an equation, as shown in Example 3.

**ANSWERS**

2. (a) $y = 0.21$   Check   $-3y = -0.63$
    $-3(0.21) = -0.63$
    Balances   $-0.63 = -0.63$

(b) $r = -8$   Check   $2.25r = -18$
    $2.25(-8) = -18$
    Balances   $-18 = -18$

(c) $n = 3.4$   Check   $1.7 = 0.5n$
    $1.7 = 0.5(3.4)$
    Balances   $1.7 = 1.7$

## Example 3 Solving Equations with Several Steps

**(a)** $2.5b + 0.35 = -2.65$

The first step is to get the variable term $2.5b$ by itself on the left side of the equal sign.

$$\begin{aligned} 2.5b + 0.35 &= -2.65 \\ -0.35 &\phantom{=} -0.35 \\ \overline{2.5b + 0} &= -3.00 \end{aligned}$$ Add $-0.35$ to both sides.

$$2.5b = -3$$

The next step is to divide both sides by the coefficient of the variable term. In $2.5b$, the coefficient is $2.5$.

$$\frac{2.5b}{2.5} = \frac{-3}{2.5}$$ On the right side, signs do *not* match, so the quotient is *negative*.

$$b = -1.2$$

The solution is $-1.2$. To check the solution, go back to the original equation.

**Check**
$$2.5b + 0.35 = -2.65 \quad \text{Original equation}$$
$$2.5(-1.2) + 0.35 = -2.65 \quad \text{Replace } b \text{ with } -1.2.$$
$$-3 + 0.35 = -2.65$$
$$-2.65 = -2.65 \quad \text{Balances, so } -1.2 \text{ is the correct solution.}$$

**(b)** $5x - 0.98 = 2x + 0.4$

There is a variable term on both sides of the equation. You can choose to keep the variable term on the left side, or to keep the variable term on the right side. Either way will work. Just pick the left side or the right side.

Suppose that you decide to keep the variable term $5x$ on the left side. Use the addition property to "get rid of" $2x$ on the right side by adding its opposite, $-2x$, to both sides.

$$\begin{aligned} 5x - 0.98 &= 2x + 0.4 \\ -2x &\phantom{=} -2x \\ \overline{3x - 0.98} &= 0 + 0.4 \end{aligned}$$ Add $-2x$ to both sides.

Change subtraction to adding the opposite.

$$\begin{aligned} 3x + (-0.98) &= 0.4 \\ +0.98 &\phantom{=} +0.98 \\ \overline{3x + 0} &= 1.38 \end{aligned}$$ Add $0.98$ to both sides.

$$\frac{3x}{3} = \frac{1.38}{3} \quad \text{Divide both sides by 3.}$$

$$x = 0.46$$

The solution is $0.46$. To check the solution, go back to the original equation.

**Check**
$$5x - 0.98 = 2x + 0.4 \quad \text{Original equation}$$
$$5(0.46) - 0.98 = 2(0.46) + 0.4 \quad \text{Replace } x \text{ with } 0.46.$$
$$2.3 - 0.98 = 0.92 + 0.4$$
$$1.32 = 1.32 \quad \text{Balances, so } 0.46 \text{ is the correct solution.}$$

Work Problem ❸ at the Side.

---

❸ Solve each equation and check each solution.

**(a)** $4 = 0.2c - 2.6$     Check

**(b)** $3.1k - 4 = 0.5k + 13.42$

Check

**(c)** $-2y + 3 = 3y - 6$

Check

**ANSWERS**

**3. (a)** $c = 33$  Check $4 = 0.2c - 2.6$
$4 = 0.2(33) - 2.6$
$4 = 6.6 - 2.6$
Balances $4 = 4$

**(b)** $k = 6.7$  Check
$3.1k - 4 = 0.5k + 13.42$
$3.1(6.7) - 4 = 0.5(6.7) + 13.42$
$20.77 - 4 = 3.35 + 13.42$
Balances $16.77 = 16.77$

**(c)** $y = 1.8$  Check
$-2y + 3 = 3y - 6$
$-2(1.8) + 3 = 3(1.8) - 6$
$-3.6 + 3 = 5.4 - 6$
Balances $-0.6 = -0.6$

**4** During April, a special rate was offered on air-to-ground cell phone calls. The connection fee was $1.34 and the cost per minute was $2.69. Maureen made a call that cost $39. How long did the call last? Use the six problem-solving steps.

**4** Solve application problems involving equations with decimals. Use the six problem-solving steps from **Section 3.3**.

### Example 4 Solving an Application Problem

Many larger airplanes have special phones that can be used to call people on the ground. In January 2001, the cost of using the air-to-ground cell phone was $3.28 per minute plus a $2.99 connection charge. (*Source:* AT&T.) Hernando was billed $19.39 for one call. How many minutes did the call last?

*Step 1* **Read** the problem. It is about the cost of a telephone call.

**Unknown:** number of minutes the call lasted
**Known:** Costs are $3.28 per minute plus $2.99; total cost was $19.39.

*Step 2* **Assign a variable.** There is only one unknown, so let $m$ be the number of minutes.

*Step 3* **Write an equation.**

| cost per minute | | number of minutes | | connection charge | | total cost |
|---|---|---|---|---|---|---|
| 3.28 | · | $m$ | + | 2.99 | = | 19.39 |

*Step 4* **Solve** the equation.

$$3.28m + 2.99 = 19.39$$
$$\phantom{3.28m + } -2.99 \phantom{=} -2.99 \quad \text{Add } -2.99 \text{ to both sides.}$$
$$3.28m + 0 = 16.40$$

$$\frac{3.28m}{3.28} = \frac{16.40}{3.28} \quad \text{Divide both sides by 3.28.}$$

$$m = 5$$

*Step 5* **State the answer.** The call lasted 5 minutes.

*Step 6* **Check** the solution by putting it back into the original problem.

$3.28 per minute times 5 minutes = $16.40
$16.40 plus $2.99 connection charge = $19.39
Hernando was billed $19.39. ← Matches

Because 5 minutes "works" when put back into the original problem, it is the correct solution.

**Work Problem 4 at the Side.**

---

**ANSWERS**
**4.** Let $m$ be the number of minutes.
$1.34 + 2.69m = 39$
The call lasted 14 minutes.

## 5.9 Exercises

*Solve each equation and check each solution. See Examples 1–3.*

1. $h + 0.63 = 5.1$   **Check**
2. $-0.2 = k - 0.7$   **Check**

3. $-20.6 + n = -22$   **Check**
4. $g - 5 = 6.03$   **Check**

5. $0 = b - 0.008$   **Check**
6. $0.18 + m = -4.5$   **Check**

7. $2.03 = 7a$   **Check**
8. $-6.2c = 0$   **Check**

9. $0.8p = -96$   **Check**
10. $-10.16 = -4r$   **Check**

11. $-3.3t = -2.31$   **Check**
12. $8.3w = -49.8$   **Check**

13. $7.5x + 0.15 = -6$   **Check**
14. $0.8 = 0.2y + 3.4$   **Check**

15. $-7.38 = 2.05z - 7.38$   **Check**
16. $6.2h - 0.4 = 2.7$   **Check**

*Solve each equation. See Example 3(b).*

17. $3c + 10 = 6c + 8.65$
18. $2.1b + 5 = 1.6b + 10$
19. $8w - 6.4 = -6.4 + 5w$

**20.** $7r + 9.64 = -2.32 + 5r$  **21.** $-10.9 + 0.5p = 0.9p + 5.3$  **22.** $0.7x - 4.38 = x - 2.16$

*Solve each application problem using the six problem-solving steps. See Example 4.*

**23.** Most adult medication doses are for a person weighing 150 pounds. For a 45-pound child, the adult dose should be multiplied by 0.3. If the child's dose of a decongestant is 9 milligrams, what is the adult dose?

**24.** For a 30-pound child, an adult dose of medication should be multiplied by 0.2. If the child's dose of a cough suppressant is 3 milliliters, find the adult dose.

**25.** A storm blew down many trees. Several neighbors rented a chain saw for $275.80 and helped each other cut up and stack the wood. The rental company charges $65.95 per day plus a $12 sharpening fee. (*Source:* Central Rental.) How many days was the saw rented?

**26.** A 20-inch chain saw can be rented for $29.95 for the first two hours, and $9 for each additional hour. (*Source:* Central Rental.) Steve's rental charge was $56.95. How many hours did he rent the saw?

---

**RELATING CONCEPTS (Exercises 27–30)** **FOR INDIVIDUAL OR GROUP WORK**

*When doing aerobic exercises, it is important to increase your heart rate (the number of beats per minute) so that you get the maximum benefit from the exercise. But you don't want your heart rate to be so fast that it is dangerous. Here is an expression for finding a safe maximum heart rate for a healthy person with no heart disease:* $0.7(220 - a)$ *where a is the person's age.* **Work Exercises 27–30 in order.** *Write an equation and solve it to find each person's age. Assume all the people are healthy.*

**27.** How old is a person who has a maximum safe heart rate of 140 beats per minute? *Hint:* Use the distributive property to simplify $0.7(220 - a)$.

**28.** How old is a person who has a maximum safe heart rate of 126 beats per minute?

**29.** If a person's maximum safe heart rate is 134 (rounded to nearest whole number), how old is the person, to the nearest whole year?

**30.** If a person's maximum safe heart rate is 117 (rounded to the nearest whole number), how old is the person, to the nearest whole year?

# 5.10 GEOMETRY APPLICATIONS: CIRCLES, CYLINDERS, AND SURFACE AREA

**1** **Find the radius and diameter of a circle.** Suppose you start with one dot on a piece of paper. Then you draw many dots that are each 2 cm away from the first dot. If you draw enough dots (points) you'll end up with a circle. Each point on the circle is exactly 2 cm away from the *center* of the circle. The 2 cm distance is called the *radius*, *r*, of the circle. The distance across the circle (passing through the center) is called the *diameter*, *d*, of the circle. In this circle, the diameter is 4 cm.

**OBJECTIVES**

1. Find the radius and diameter of a circle.
2. Find the circumference of a circle.
3. Find the area of a circle.
4. Find the volume of a cylinder.
5. Find the surface area of a rectangular solid.
6. Find the surface area of a cylinder.

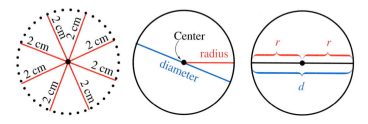

### Circle, Radius, and Diameter

A **circle** is a two-dimensional (flat) figure with all points the same distance from a fixed center point.

The **radius** (*r*) is the distance from the center of the circle to any point on the circle.

The **diameter** (*d*) is the distance across the circle passing through the center.

Using the circle above on the right as a model, you can see some relationships between the radius and diameter.

### Finding the Diameter and Radius of a Circle

$$\text{diameter} = 2 \cdot \text{radius}$$
$$d = 2r$$
$$\text{and} \quad r = \frac{d}{2}$$

---

**Example 1** Finding the Diameter and Radius of Circles

Find the unknown length of the diameter or radius in each circle.

(a)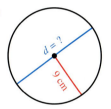

Because the radius is 9 cm, the diameter is twice as long.

$$d = 2 \cdot r$$
$$d = 2 \cdot 9 \text{ cm}$$
$$d = 18 \text{ cm}$$

*Continued on Next Page*

**1** Find the unknown length of the diameter or radius in each circle.

(a)

40 ft

(b)
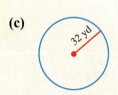
11 cm

(c)

32 yd

(d)
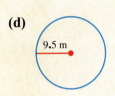
9.5 m

(b)

The radius is half the diameter.

$r = \dfrac{d}{2}$ so $r = \dfrac{17 \text{ m}}{2}$

$r = 8.5$ m or $8\dfrac{1}{2}$ m

**Work Problem ❶ at the Side.**

**2** **Find the circumference of a circle.** The perimeter of a circle is called its **circumference.** Circumference is the distance around the edge of a circle.

The diameter of the can in the drawing is about 10.6 cm, and the circumference of the can is about 33.3 cm. Dividing the circumference of the circle by the diameter gives an interesting result.

$$\dfrac{\text{Circumference}}{\text{diameter}} = \dfrac{33.3}{10.6} \approx 3.14 \quad \text{Rounded to the nearest hundredth}$$

Dividing the circumference of *any* circle by its diameter *always* gives an answer close to 3.14. This means that going around the edge of any circle is a little more than 3 times as far as going straight across the circle.

This ratio of circumference to diameter is called $\pi$ (the Greek letter **pi**, pronounced PIE). There is no decimal that is exactly equal to $\pi$, but here is the *approximate* value.

$$\pi \approx 3.14159265359$$

### Rounding the Value of *Pi* ($\pi$)

We usually round $\pi$ to 3.14. Therefore, calculations involving $\pi$ will give approximate answers and should be written using the $\approx$ symbol.

Use the following formulas to find the *circumference* of a circle.

### Finding the Circumference (Distance Around a Circle)

$$\text{Circumference} = \pi \cdot \text{diameter}$$
$$C = \pi d$$

or, because $d = 2r$ then $C = \pi \cdot 2r$ usually written $C = 2\pi r$

Remember to use linear units when measuring circumference.

**Answers**
**1.** (a) $r = 20$ ft (b) $r = 5.5$ cm (c) $d = 64$ yd (d) $d = 19$ m

Section 5.10 Geometry Applications: Circles, Cylinders, and Surface Area   387

## Example 2  Finding the Circumference of Circles

Find the circumference of each circle. Use 3.14 as the approximate value for $\pi$. Round answers to the nearest tenth.

(a)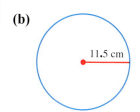

The *diameter* is 38 m, so use the formula with $d$ in it.

$C = \pi \cdot d$
$C \approx 3.14 \cdot 38 \text{ m}$
$C \approx 119.3 \text{ m}$   —Rounded

(b)

11.5 cm

In this example, the length of the *radius* is labeled, so it is easier to use the formula with $r$ in it.

$C = 2 \cdot \pi \cdot r$
$C \approx 2 \cdot 3.14 \cdot 11.5 \text{ cm}$
$C \approx 72.2 \text{ cm}$   Rounded

**Calculator Tip**   Most *scientific* calculators have a $\pi$ key. Try pressing it. With a 10-digit display, you'll see the value of $\pi$ to the nearest billionth.

3.141592654

But this is still an approximate value, although it is more precise than rounding $\pi$ to 3.14. Try finding the circumference in Example 2(a) above using the $\pi$ key.

$\pi$ × 38 = 119.3805208   Rounds to 119.4

When you used 3.14 as the approximate value of $\pi$, the result rounded to 119.3, so the answers are slightly different. In this book we will use 3.14. Our measurements of radius and diameter are given as whole numbers or with tenths, so it is acceptable to round $\pi$ to hundredths. And you may be using a calculator without a $\pi$ key.

Work Problem ❷ at the Side.

**3** Find the area of a circle.   To find the formula for the area of a circle, start by cutting two circles into many pie-shaped pieces.

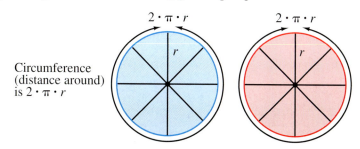

Circumference (distance around) is $2 \cdot \pi \cdot r$

Unfold the circles, much as you might "unfold" a peeled orange, and put them together as shown here.

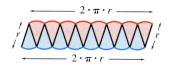

❷ Find the circumference of each circle. Use 3.14 as the approximate value for $\pi$. Round answers to the nearest tenth.

(a)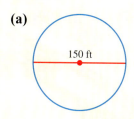
150 ft

(b)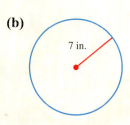
7 in.

(c) diameter 0.9 km

(d) radius 4.6 m

**Answers**
**2.** (a) $C \approx 471$ ft  (b) $C \approx 44.0$ in.
(c) $C \approx 2.8$ km  (d) $C \approx 28.9$ m

**388** Chapter 5 Rational Numbers: Positive and Negative Decimals

❸ Find the area of each circle. Use 3.14 for $\pi$. Round your answers to the nearest tenth.

(a)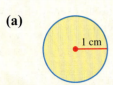

(b)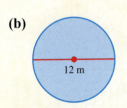

(*Hint:* The diameter is 12 m, so $r = $ _____ m)

(c)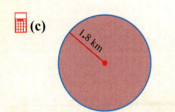

(d)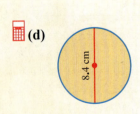

The figure is approximately a rectangle with width $r$ (the radius of the original circle) and length $2 \cdot \pi \cdot r$ (the circumference of the original circle). The area of the "rectangle" is length times width.

$$\text{Area of "rectangle"} = l \cdot w$$
$$\text{Area of "rectangle"} = 2 \cdot \pi \cdot r \cdot r$$
$$\text{Area of "rectangle"} = 2 \cdot \pi \cdot r^2 \quad \leftarrow \text{Recall that } r \cdot r \text{ is } r^2$$

Because the "rectangle" was formed from *two* circles, the area of *one* circle is half as much.

$$\frac{1}{2} \cdot 2 \cdot \pi \cdot r^2 = 1 \cdot \pi \cdot r^2 \quad \text{or simply} \quad \pi r^2$$

### Finding the Area of a Circle

Area of a circle = $\pi \cdot$ radius $\cdot$ radius
$$A = \pi r^2$$

Remember to use *square units* when measuring area.

**Example 3** Finding the Area of Circles

Find the area of each circle. Use 3.14 for $\pi$. Round your answers to the nearest tenth.

(a) A circle with a radius of 8.2 cm

Use the formula $A = \pi r^2$, which means $\pi \cdot r \cdot r$.

$$A = \pi \cdot r \cdot r$$
$$A \approx 3.14 \cdot 8.2 \text{ cm} \cdot 8.2 \text{ cm}$$
$$A \approx 211.1 \text{ cm}^2 \quad \text{Square units for area}$$

(b)

To use the formula, you need to know the radius ($r$). In this circle, the *diameter* is 10 ft. First find the radius.

$$r = \frac{d}{2}$$
$$r = \frac{10 \text{ ft}}{2} = 5 \text{ ft}$$

Now find the area.

$$A \approx 3.14 \cdot 5 \text{ ft} \cdot 5 \text{ ft}$$
$$A \approx 78.5 \text{ ft}^2 \quad \text{Square units for area}$$

**CAUTION**

When finding *circumference*, you can start with either the radius or the diameter. When finding *area*, you must use the *radius*. If you are given the diameter, divide it by 2 to find the radius. Then find the area.

**ANSWERS**
3. (a) $A \approx 3.1 \text{ cm}^2$   (b) $A \approx 113.0 \text{ m}^2$
   (c) $A \approx 10.2 \text{ km}^2$   (d) $A \approx 55.4 \text{ cm}^2$

Work Problem ❸ at the Side.

Section 5.10 Geometry Applications: Circles, Cylinders, and Surface Area  **389**

📱 **Calculator Tip**  You can find the area of the circle in Example 3(a) on your calculator. The first method works on all types of calculators.

$$3.14 \;\boxed{\times}\; 8.2 \;\boxed{\times}\; 8.2 \;\boxed{=}\; 211.1336$$

You round the answer to 211.1 (nearest tenth).

On a *scientific* or *graphing* calculator you can also use the $\boxed{x^2}$ key, which automatically squares the number you enter (that is, multiplies the number times itself).

$$3.14 \;\boxed{\times}\; 8.2 \;\boxed{x^2}\; \boxed{=}\; 211.1336$$

In the next example we will find the area of a *semicircle,* which is half the area of a circle.

**Example 4**  Finding the Area of a Semicircle

Find the area of the semicircle. Use 3.14 for $\pi$. Round your answer to the nearest tenth.

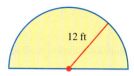

First, find the area of an entire circle with a radius of 12 ft.

$A = \pi \cdot r \cdot r$

$A \approx 3.14 \cdot 12 \text{ ft} \cdot 12 \text{ ft}$

$A \approx 452.16 \text{ ft}^2$   Do not round yet.

Divide the area of the whole circle by 2 to find the area of the semicircle.

$$\frac{452.16 \text{ ft}^2}{2} = 226.08 \text{ ft}^2$$

The *last* step is rounding 226.08 to the nearest tenth.

Area of semicircle $\approx 226.1 \text{ ft}^2$   Rounded

Work Problem ❹ at the Side.

**Example 5**  Applying the Concept of Circumference

A circular rug is 8 feet in diameter. The cost of fringe for the edge is $2.25 per foot. What will it cost to add fringe to the rug? Use 3.14 for $\pi$.

Circumference $= \pi \cdot d$

$C \approx 3.14 \cdot 8 \text{ ft}$

$C \approx 25.12 \text{ ft}$

cost = cost per foot • Circumference

$$\text{cost} = \frac{\$2.25}{1 \text{ ft}} \cdot \frac{25.12 \text{ ft}}{1}$$

cost = $56.52

Work Problem ❺ at the Side.

❹ Find the area of each semicircle. Use 3.14 for $\pi$. Round your answers to the nearest tenth.

(a)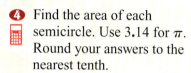

(b)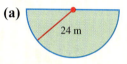

(c)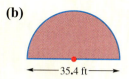

❺ Find the cost of binding around the edge of a circular rug that is 3 meters in diameter. The binder charges $4.50 per meter. Use 3.14 for $\pi$.

**ANSWERS**
4. (a) $A \approx 904.3 \text{ m}^2$  (b) $A \approx 491.9 \text{ ft}^2$
    (c) $A \approx 150.8 \text{ m}^2$
5. $42.39

**6** Find the cost of covering the underside of the rug in Problem 5 with a nonslip rubber backing. The rubber backing costs $2 per square meter.

### Example 6  Applying the Concept of Area

Find the cost of covering the rug in Example 5 with a plastic cover. The material for the cover costs $1.50 per square foot. Use 3.14 for $\pi$.

First find the radius.

$$r = \frac{d}{2} = \frac{8 \text{ ft}}{2} = 4 \text{ ft}$$

Then,

$$A = \pi \cdot r^2$$
$$A \approx 3.14 \cdot 4 \text{ ft} \cdot 4 \text{ ft}$$
$$A \approx 50.24 \text{ ft}^2$$

$$\text{cost} = \frac{\$1.50}{1 \text{ ft}^2} \cdot \frac{50.24 \text{ ft}^2}{1} = \$75.36$$

Work Problem **6** at the Side.

**4** Find the volume of a cylinder. Several *cylinders* are shown here.

These are called *right circular cylinders* because the top and bottom are circles, and the side makes a right angle with the top and bottom. Examples of cylinders are a soup can, a home water heater, and a piece of pipe.

Use the following formula to find the *volume* of a cylinder. Notice that the first part of the formula, $\pi \cdot r \cdot r$, is the *area* of the circular base.

### Finding the Volume of a Cylinder

Volume of cylinder = $\pi \cdot r \cdot r \cdot h$

$$V = \pi r^2 h$$

Remember to use *cubic units* when measuring volume.

**7** Find the volume of each cylinder. Use 3.14 for $\pi$. Round your answers to the nearest tenth.

(a)
12 ft, 4 ft

(b)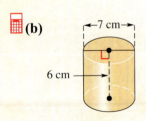
7 cm, 6 cm

(c) radius 14.5 yd, height 3.2 yd

### Example 7  Finding the Volume of Cylinders

Find the volume of each cylinder. Use 3.14 as the approximate value of $\pi$. Round your answers to the nearest tenth, if necessary.

(a)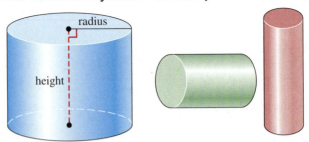
20 m, 9 m

The diameter is 20 m, so the radius is $\frac{20 \text{ m}}{2} = 10$ m. The height is 9 m. Use the formula to find the volume.

$$V = \pi \cdot r \cdot r \cdot h$$
$$V \approx 3.14 \cdot 10 \text{ m} \cdot 10 \text{ m} \cdot 9 \text{ m}$$
$$V \approx 2826 \text{ m}^3 \quad \text{Cubic units for volume}$$

(b)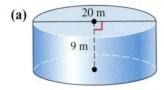
6.2 cm, 38.4 cm

$$V \approx 3.14 \cdot 6.2 \text{ cm} \cdot 6.2 \text{ cm} \cdot 38.4 \text{ cm}$$
$$V \approx 4634.94144 \quad \text{Now round to tenths.}$$
$$V \approx 4634.9 \text{ cm}^3 \quad \text{Cubic units for volume}$$

Work Problem **7** at the Side.

**ANSWERS**
6. $14.13
7. (a) $V \approx 602.9$ ft³  (b) $V \approx 230.8$ cm³
   (c) $V \approx 2112.6$ yd³

Section 5.10 Geometry Applications: Circles, Cylinders, and Surface Area   391

**5** **Find the surface area of a rectangular solid.** You have just learned how to find the *volume* of a cylinder. In **Section 4.8** you found the *volume* of a rectangular solid. For example, the volume of the cereal box shown below is $V = lwh = (7 \text{ in.})(2 \text{ in.})(10 \text{ in.}) = 140 \text{ in.}^3$ But if your company makes cereal boxes, you also need to know how much cardboard is needed for each box. You need to find the *surface area* of the box.

**8** Find the volume and surface area of each rectangular solid.

(a)
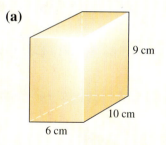

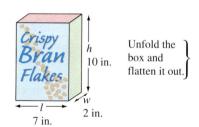 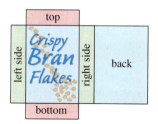

The unfolded box is made up of six rectangles: front, back, top, bottom, left side, right side.

**Surface area** is the area on the surface of a three-dimensional object (a solid). For a rectangular solid like the cereal box, the surface area is the sum of the areas of the six rectangular sides. Notice that the top and bottom have the same area, the front and back have the same area, and the left and right sides have the same area.

Surface Area = [top] $l \cdot w$ + [bottom] $l \cdot w$ + [front] $l \cdot h$ + [back] $l \cdot h$ + [left side] $w \cdot h$ + [right side] $w \cdot h$

$SA = l \cdot w + l \cdot w + l \cdot h + l \cdot h + w \cdot h + w \cdot h$

$SA = \underbrace{2lw} + \underbrace{2lh} + \underbrace{2wh}$

(b)
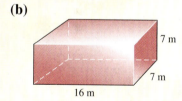

### Finding the Surface Area of a Rectangular Solid

$$\text{Surface Area} = 2 \cdot l \cdot w + 2 \cdot l \cdot h + 2 \cdot w \cdot h$$
$$SA = 2lw + 2lh + 2wh$$

Remember that area is measured in square units, so use *square units* when measuring *surface* area.

**Example 8** Finding the Volume and Surface Area of a Rectangular Solid

Find the volume and surface area of this shipping carton.

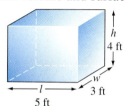

First find the volume.

$V = lwh$
$V = 5 \text{ ft} \cdot 3 \text{ ft} \cdot 4 \text{ ft}$
$V = 60 \text{ ft}^3$ ← Cubic units for volume

Next find the surface area.

$SA = \quad 2lw \quad + \quad 2lh \quad + \quad 2wh$
$SA = (2 \cdot 5 \text{ ft} \cdot 3 \text{ ft}) + (2 \cdot 5 \text{ ft} \cdot 4 \text{ ft}) + (2 \cdot 3 \text{ ft} \cdot 4 \text{ ft})$
$SA = \quad 30 \text{ ft}^2 \quad + \quad 40 \text{ ft}^2 \quad + \quad 24 \text{ ft}^2$
$SA = 94 \text{ ft}^2$ ← Square units for area

**ANSWERS**
8. (a) $V = 540 \text{ cm}^3$ (cubic cm for volume)
   $SA = 408 \text{ cm}^2$ (square cm for area)
   (b) $V = 784 \text{ m}^3$
   $SA = 546 \text{ m}^2$

Work Problem **8** at the Side.

**9** Find the volume and surface area of each cylinder. Use 3.14 for $\pi$. Round your answers to the nearest tenth.

(a)

**6** **Find the surface area of a cylinder.** You can use the same idea of "unfolding" a shape to find the surface area of a cylinder, such as the soup can shown below. Finding the surface area will tell you how much aluminum you need to make the can.

The unfolded soup can is made up of a rectangular side, a circular top, and a circular bottom.

Remember that the formula for the area of a circle is $\pi r^2$.

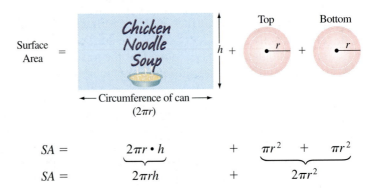

$$SA = 2\pi r \cdot h + \pi r^2 + \pi r^2$$
$$SA = 2\pi rh + 2\pi r^2$$

### Finding the Surface Area of a Right Circular Cylinder

$$\text{Surface Area} = 2 \cdot \pi \cdot r \cdot h + 2 \cdot \pi \cdot r \cdot r$$
$$SA = 2\pi rh + 2\pi r^2$$

Remember that area is measured in square units, so use *square units* when measuring *surface* area.

(b)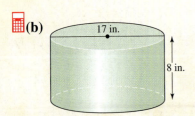

(*Hint:* You are given the *diameter* of the cylinder. Start by finding the radius.)

### Example 9  Finding the Volume and Surface Area of a Right Circular Cylinder

Find the volume and surface area of this water tank. Use 3.14 as the approximate value for $\pi$. Round your answers to the nearest tenth when necessary.

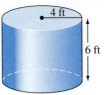

First find the volume.

$$V = \pi r^2 h$$
$V \approx 3.14 \cdot 4 \text{ ft} \cdot 4 \text{ ft} \cdot 6 \text{ ft}$
$V \approx 301.44 \text{ ft}^3$ ← Now round to tenths.
$V \approx 301.4 \text{ ft}^3$ ← Cubic units for volume

Now find the surface area.

$SA = 2\pi rh + 2\pi r^2$
$SA \approx (2 \cdot 3.14 \cdot 4 \text{ ft} \cdot 6 \text{ ft}) + (2 \cdot 3.14 \cdot 4 \text{ ft} \cdot 4 \text{ ft})$
$SA \approx 150.72 \text{ ft}^2 + 100.48 \text{ ft}^2$
$SA \approx 251.2 \text{ ft}^2$ ← Square units for area

**Work Problem** **9** **at the Side.**

**ANSWERS**
9. (a) $V \approx 1177.5$ cm³ (cubic units for volume)
    $SA \approx 628$ cm² (square units for area)
    (b) $V \approx 1814.9$ in.³
    $SA \approx 880.8$ in.²

Section 5.10  393

## 5.10 EXERCISES

FOR EXTRA HELP: Student's Solutions Manual | MyMathLab.com | InterAct Math Tutorial Software | AW Math Tutor Center | www.mathxl.com | Digital Video Tutor CD 4 Videotape 11

*Find the unknown length in each circle. See Example 1.*

1.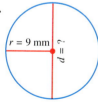

2.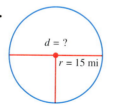

3.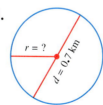

4.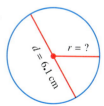

*Find the circumference and area of each circle. Use 3.14 as the approximate value for $\pi$. Round your answers to the nearest tenth. See Examples 2 and 3.*

5.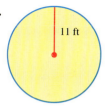

6.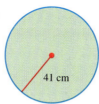

7.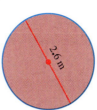

8.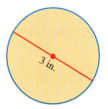

*Find the circumference and area of circles having the following diameters. Use 3.14 for $\pi$. Round your answers to the nearest tenth. See Examples 2 and 3.*

9. $d = 15$ cm

10. $d = 39$ ft

11. $d = 7\frac{1}{2}$ ft

12. $d = 4\frac{1}{2}$ yd

13. $d = 8.65$ km

14. $d = 19.5$ mm

*Find each shaded area. Note that Exercises 15 and 18 contain semicircles. Use 3.14 as the approximate value of $\pi$. Round your answers to the nearest tenth when necessary. See Example 4.*

15.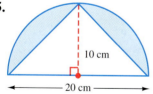

16.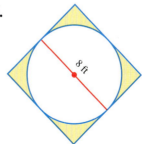

**17.**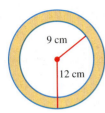

**18.**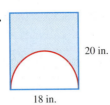

**19.** How would you explain π to a friend who is not in your math class? Write an explanation. Then make up a test question that requires the use of π, and show how to solve it.

**20.** Explain how circumference and perimeter are alike. How are they different? Make up two problems, one involving perimeter, the other circumference. Then show how to solve your problems.

*Solve each application problem. Use 3.14 as the approximate value of π. Round answers to the nearest tenth. See Examples 5 and 6.*

**21.** How far does a point on the tread of a tire move in one turn, if the diameter of the tire is 70 cm?

**22.** If you swing a ball held at the end of a string 2 m long, how far will the ball travel on each turn?

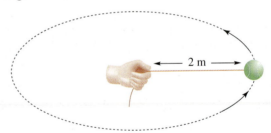

**23.** A wave energy extraction device is a huge undersea dome used to harness the power of ocean waves. The base of the dome has a radius of 125 ft. Find its circumference.

**24.** Find the area of the base of the dome in Exercise 23.

Solve each application problem. For Exercises 25–28, draw a circle, label the radius or diameter, and then solve the problem. *(Sketches may vary. Show your sketches to your instructor.)*

**25.** A radio station can be heard 150 miles in all directions during evening hours. How many square miles are in the station's broadcast area?

**26.** An earthquake was felt by people 900 km away in all directions from the epicenter (the source of the earthquake). How much area was affected by the quake?

**27.** The diameter of Diana Hestwood's wristwatch is 1 in., and the radius of the clock face on her kitchen wall is 3 in. (*Source:* Author's watch and clock.) Find the circumference and the area of each clock face.

**28.** The diameter of the largest known ball of twine is 12 ft 9 in. (*Source: Guinness Book of Records.*) The sign posted near the ball says it has a circumference of 40 ft. Is the sign correct? (*Hint:* First change 9 in. to feet and add it to 12 ft.)

**29.** A forester measures the circumference of a living tree at chest height, then calculates the diameter. From this information he can determine the approximate age of the tree. (*Source: Shoreview Press.*) If the circumference of one tree is 144 cm, find its diameter.

**30.** In Atlanta, Interstate 285 circles the city and is known as the "perimeter." If the circumference of the circle made by the highway is 62.8 miles, find

  **(a)** the diameter of the circle, and

  **(b)** the area inside the circle.

(*Source: Greater Atlanta Newcomer's Guide.*)

**31.** The Mormons traveled west to Utah by covered wagon in 1847. They tied a rag to a wagon wheel to keep track of the distance they traveled. The radius of the wheel was 2.33 ft. How far did the rag travel each time the wheel made a complete revolution? (*Source:* Trail of Hope.) Bonus question: How many wheel revolutions equaled one mile?

**32.** The National Audubon Society holds an annual end-of-year bird count. Volunteers count all the birds they can find in a circular area during a 24-hour period. Each circle has a diameter of 15 miles. In December 2000, about 1700 of these circular areas were counted throughout all 50 states. (*Source:* Patuxent Wildlife Research Center.)

  **(a)** What is the area of each circle?

  **(b)** Find the total area in the December 2000 count.

**33.** Find the cost of sod, at $1.76 per square foot, for this playing field that has a semicircle on each end.

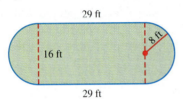

**34.** Find the area of this skating rink.

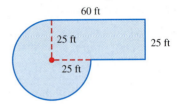

---

**RELATING CONCEPTS (Exercises 35–40)**  **FOR INDIVIDUAL OR GROUP WORK**

 *Use the table below to* **work Exercises 35–40 in order.**

*Find the best buy for each type of pizza. The best buy is the lowest cost per square inch of pizza. All the pizzas are circular in shape, and the measurement given on the menu board is the diameter of the pizza in inches. Use 3.14 as the approximate value of $\pi$. Round the area to the nearest tenth. Round cost per square inch to the nearest thousandth.*

| Pizza Menu | Small $7\frac{1}{2}"$ | Medium 13" | Large 16" |
|---|---|---|---|
| Cheese only | $2.80 | $ 6.50 | $ 9.30 |
| "The Works" | $3.70 | $ 8.95 | $14.30 |
| Deep-dish combo | $4.35 | $10.95 | $15.65 |

**35.** Find the area of a small pizza.

**36.** Find the area of a medium pizza.

**37.** Find the area of a large pizza.

**38.** What is the cost per square inch for each size of cheese pizza? Which size is the best buy?

**39.** What is the cost per square inch for each size of "The Works" pizza? Which size is the best buy?

**40.** You have a coupon for 95¢ off any small pizza. What is the cost per square inch for each size of deep dish combo pizza? Which size is the best buy?

Section 5.10    397

Find the volume and surface area of each figure. Use 3.14 as the approximate value of π. Round your answers to the nearest tenth when necessary. See Examples 7–9.

41.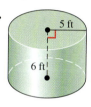

42.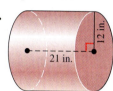

43.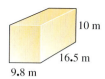

44.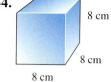

45.

46.

47.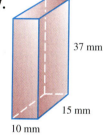

48.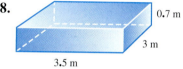

**398** Chapter 5 Rational Numbers: Positive and Negative Decimals

**49.** Explain the *two* errors made by a student in finding the volume of a cylinder with a diameter of 7 cm and a height of 5 cm. Find the correct answer.

$$V \approx 3.14 \cdot 7 \cdot 7 \cdot 5$$
$$V \approx 769.3 \text{ cm}^2$$

**50.** Look again at Exercise 44. The figure is a *cube*.
 (a) What is special about the measurements of a cube?

 (b) Find a shortcut you can use to calculate the surface area of a cube.

*Solve each application problem. Use 3.14 as the approximate value of $\pi$. Round your answers to the nearest tenth when necessary.*

**51.** A city sewer pipe has a diameter of 5 ft and a length of 200 ft. Find the volume of the pipe.

**52.** A cylindrical woven basket made by a Northwest Coast tribe is 8 cm high and has a diameter of 11 cm. What is the volume of the basket?

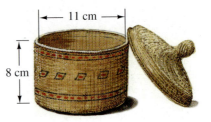

**53.** A box for graham crackers measures 5.5 in. by 2.8 in. by 8 in. high. Find the amount of cardboard needed to make the box.

**54.** A soda can is 12.5 cm tall and has a diameter of 6.7 cm. How much aluminum is needed to make the can?

# Chapter 5

## Summary

### Key Terms

| | | |
|---|---|---|
| **5.1** | **decimals** | Decimals, like fractions, are used to show parts of a whole. |
| | **decimal point** | A decimal point is the dot that is used to separate the whole number part from the fractional part of a decimal number. |
| | **place value names** | Place value is the value assigned to each place to the right or left of the decimal point. Whole numbers, such as ones and tens, are to the *left* of the decimal point. Fractional parts, such as tenths and hundredths, are to the *right* of the decimal point. |
| **5.2** | **round** | To round is to "cut off" a number after a certain place, such as to round to the nearest hundredth. The rounded number is less accurate than the original number. You can use the symbol "≈" to mean "approximately equal to." |
| | **decimal places** | Decimal places are the number of digits to the *right* of the decimal point; for example, 6.37 has two decimal places, 4.706 has three decimal places. |
| **5.5** | **repeating decimal** | A repeating decimal like the 6 in 0.166 . . . is a decimal with one or more digits that repeat forever. Use three dots to indicate that it is a repeating decimal; it never terminates (ends). Or you can write the number with a bar above the repeating digits, as in $0.1\overline{6}$. |
| **5.7** | **mean** | The mean is the sum of all the values divided by the number of values. It is often called the *average*. |
| | **weighted mean** | The weighted mean is a mean calculated so that each value is multiplied by its frequency. |
| | **median** | The median is the middle number in a group of values that are listed from smallest to largest. It divides a group of values in half. If there are an even number of values, the median is the mean (average) of the two middle values. |
| | **mode** | The mode is the value that occurs most often in a group of values. |
| | **variability** | The variability of a set of data is the spread of the data around the mean. A quick way to evaluate variability is to find the range of values. |
| **5.8** | **square root** | A positive square root of a positive number is one of two equal positive factors of the number. |
| | **hypotenuse** | The hypotenuse is the side of a right triangle opposite the 90° angle; it is the longest side.<br>*Example:* See the red side in the triangle at the right. |
| **5.10** | **circle** | A circle is a two-dimensional (flat) figure with all points the same distance from a fixed center point.<br>*Example:* See figure at the right. |
| | **radius** | Radius is the distance from the center of a circle to any point on the circle.<br>*Example:* See the red radius in the circle at the right. |
| | **diameter** | Diameter is the distance across a circle, passing through the center.<br>*Example:* See the blue diameter in the circle at the right. |
| | **circumference** | Circumference is the distance around a circle. |
| | **π (pi)** | π is the ratio of the circumference to the diameter of any circle. It is approximately equal to 3.14. |
| | **surface area** | Surface area is the area on the surface of a three-dimensional object (a solid). Surface area is measured in square units. |

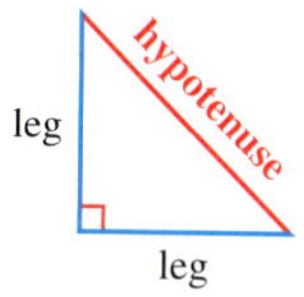

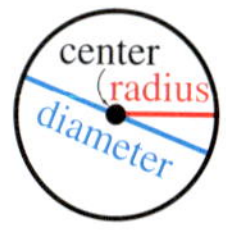

## New Formulas

$$\text{mean} = \frac{\text{sum of all values}}{\text{number of values}}$$

$$\text{hypotenuse} = \sqrt{(\text{leg})^2 + (\text{leg})^2}$$

$$\text{leg} = \sqrt{(\text{hypotenuse})^2 - (\text{leg})^2}$$

diameter of a circle: $d = 2r$

radius of a circle: $r = \dfrac{d}{2}$

circumference of a circle: $C = \pi d$ or $C = 2\pi r$

area of a circle: $A = \pi r^2$

volume of a cylinder: $V = \pi r^2 h$

surface area of a rectangular solid: $SA = 2lw + 2lh + 2wh$

surface area of a cylinder: $SA = 2\pi rh + 2\pi r^2$

## Test Your Word Power

*See how well you have learned the vocabulary in this chapter. Answers follow the Quick Review.*

1. **Decimal numbers** are like fractions in that they both
   (a) must be written in lowest terms
   (b) need common denominators
   (c) have decimal points
   (d) represent parts of a whole.

2. **Decimal places** refer to
   (a) the digits from 0 to 9
   (b) digits to the left of the decimal point
   (c) digits to the right of the decimal point
   (d) the number of zeros in a decimal number.

3. The **hypotenuse** is
   (a) the long base in a rectangle
   (b) the height in a parallelogram
   (c) the longest side in a right triangle
   (d) the distance across a circle, passing through the center.

4. The **decimal point**
   (a) separates the whole number part from the fractional part
   (b) is always moved when finding a quotient
   (c) separates tenths from hundredths
   (d) is at the far left side of a whole number.

5. The number $0.\overline{3}$ is an example of
   (a) an estimate
   (b) a repeating decimal
   (c) a rounded number
   (d) a truncated number.

6. $\pi$ is the ratio of
   (a) the diameter to the radius of a circle
   (b) the circumference to the diameter of a circle
   (c) the circumference to the radius of a circle
   (d) the diameter to the circumference of a circle.

7. The **median** for a set of values is
   (a) the mathematical average
   (b) the value that occurs most often
   (c) the sum of each value times its frequency
   (d) the middle value when the values are listed from smallest to largest.

8. The **circumference** of a circle is
   (a) the perimeter of the circle
   (b) found using the expression $\pi r^2$
   (c) the distance across the circle
   (d) measured in square units.

# Quick Review

## Concepts

### 5.1 Reading and Writing Decimals

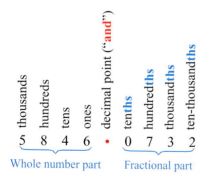

### 5.1 Writing Decimals as Fractions

The digits to the right of the decimal point are the numerator. The place value of the rightmost digit determines the denominator. Always write the fraction in lowest terms.

### 5.2 Rounding Decimals

Find the place to which you are rounding. Draw a cut-off line to the right of that place; the rest of the digits will be dropped. Look *only* at the first digit being cut. If it is *4 or less*, the part you are keeping stays the same. If it is *5 or more*, the part you are keeping rounds up. Do not move the decimal point when rounding. Use the sign "≈" to mean "approximately equal to."

### 5.3 Adding Positive and Negative Decimals

Estimate the answer by using front end rounding: round each number to the highest possible place.

To find the exact answer, line up the decimal points. If needed, write in zeros as placeholders. Add or subtract the absolute values as if they were whole numbers. Line up the decimal point in the answer.

If the numbers have the same sign, use the common sign as the sign of the sum. If the numbers have different signs, the sign of the sum is the sign of the number with the larger absolute value.

### 5.3 Subtracting Positive and Negative Decimals

Rewrite subtracting as adding the opposite of the second number. Then follow the rules for adding positive and negative decimals.

## Examples

Write each decimal in words.

15.38 — fifteen **and** thirty-eight **hundredths** (8 is in hundredths place.)

0.0103 — one hundred three **ten-thousandths** (3 is in ten-thousandths place.)

Write 0.45 as a fraction in lowest terms.

The numerator is 45. The rightmost digit, 5, is in the hundredths place, so the denominator is 100. Then write the fraction in lowest terms.

$$\frac{45}{100} = \frac{45 \div 5}{100 \div 5} = \frac{9}{20} \leftarrow \text{Lowest terms}$$

Round 0.17952 to the nearest thousandth. Is it closer to 0.179 or to 0.180?

```
0.179|52    First digit cut is 5 or more
            so round up by adding
            1 thousandth to the
            part you are keeping.
  0.179
+ 0.001  ← To round up, add 1 thousandth.
  0.180
```

0.17952 rounds to 0.180. Write 0.17952 ≈ 0.180.

Add 5.68 + 785.3 + 12 + 2.007.

```
Estimate:        Exact:
      6            5.680     Use zeros as
    800          785.300     placeholders so
     10           12.000     all numbers have
+     2         +  2.007     three decimal
    818          804.987     places.
                             Line up decimal
                             points.
```

All addends are positive, so the sign of the sum is positive. The estimate and exact answer are both in hundreds, so the decimal point is probably in the correct place.

Subtract.   4.2 − **12.91**

           4.2 + (**−12.91**)

| 4.2 | is 4.2 and | −12.91 | is 12.91.
Because −12.91 has the larger absolute value, the answer will be negative.

4.2 + (−12.91) = −8.71

| Concepts | Examples |
|---|---|

### 5.4 Multiplying Positive and Negative Decimals

**Step 1** Multiply as you would for whole numbers.

**Step 2** Count the total number of decimal places in both factors.

**Step 3** Write the decimal point in the answer so it has the same number of decimal places as the total from Step 2. You may need to write extra zeros on the left side of the product in order to get enough decimal places in the answer.

**Step 4** If two factors have the *same sign*, the product is *positive*. If two factors have *different signs*, the product is *negative*.

*Example:*

Multiply $(0.169)(-0.21)$.

$$\begin{array}{r} 0.169 \\ \times\ 0.21 \\ \hline 169 \\ 338\phantom{0} \\ \hline .03549 \end{array}$$

← 3 decimal places
← 2 decimal places
5 total decimal places

← 5 decimal places in product

Write in a 0 so you can count over 5 decimal places.
The factors have different signs, so the product is negative.

$$(0.169)(-0.21) = -0.03549$$

### 5.5 Dividing by a Decimal

**Step 1** Change the divisor to a whole number by moving the decimal point to the right.

**Step 2** Move the decimal point in the dividend the same number of places to the right.

**Step 3** Write the decimal point in the quotient directly above the decimal point in the dividend. Then divide as with whole numbers.

**Step 4** If the numbers have the *same sign*, the quotient is *positive*. If they have *different signs*, the quotient is *negative*.

*Example:*

Divide $-52.8$ by $-0.75$.

First consider $52.8 \div 0.75$.

$$\begin{array}{r} 70.4\phantom{0} \\ 0.75{\overline{\smash{\big)}\,52.800}} \\ \underline{525\phantom{00}} \\ 300 \\ \underline{300} \\ 0 \end{array}$$

Move decimal point two places to the right in divisor and dividend.
Write zeros in the dividend so you can move the decimal point and continue dividing until the remainder is 0.

The quotient is positive because both the divisor and dividend were negative (same signs means positive quotient).

### 5.6 Writing Fractions as Decimals

Divide the numerator by the denominator. If necessary, round to the place indicated.

*Example:*

Write $\frac{1}{8}$ as a decimal.

$\frac{1}{8}$ means $1 \div 8$. Write it as $8{\overline{\smash{\big)}\,1}}$.
The decimal point is on the right side of 1.

$$\begin{array}{r} 0.125 \\ 8{\overline{\smash{\big)}\,1.000}} \\ \underline{8\phantom{.000}} \\ 20\phantom{0} \\ \underline{16}\phantom{0} \\ 40 \\ \underline{40} \\ 0 \end{array}$$

← Write the decimal point and three zeros so you can continue dividing.

Therefore, $\frac{1}{8}$ is equivalent to $0.125$.

### 5.6 Comparing the Size of Fractions and Decimals

**Step 1** Write any fractions as decimals.

**Step 2** Write zeros so that all the numbers being compared have the same number of decimal places.

**Step 3** Use $<$ to mean "is less than," $>$ to mean "is greater than," or list the numbers from smallest to largest.

*Example:*

Arrange in order from smallest to largest.

$$0.505 \quad \frac{1}{2} \quad 0.55$$

$0.505 = 505$ thousandths

$\frac{1}{2} = 0.5 = 0.5\mathbf{00} = 500$ thousandths ← 500 is smallest.

$0.55 = 0.55\mathbf{0} = 550$ thousandths ← 550 is largest.

(smallest) $\frac{1}{2}$ $\quad 0.505 \quad 0.55$ (largest)

| Concepts | Examples |
|---|---|
| **5.7 Finding the Mean (Average) of a Set of Numbers** | Here are Heather Hall's test scores in her math course. |
| Step 1   Add all values to obtain a total. | 93  76  83  93 |
| Step 2   Divide the total by the number of values. | 78  82  87  85 |
| | Find Heather's mean score to the nearest tenth. |
| | $$\text{mean} = \frac{93 + 76 + 83 + 93 + 78 + 82 + 87 + 85}{8}$$ |
| | $$= \frac{677}{8} \approx 84.6$$ |
| **5.7 Finding the Median of a Set of Numbers** | Find the median for Heather Hall's scores from the previous example. |
| Step 1   Arrange the data from smallest to largest. | List the scores from smallest to largest. |
| Step 2   If there is an odd number of values, select the middle value. If there is an even number of values, find the average of the two middle values. | 76  78  82  **83  85**  87  93  93 |
| | Middle values |
| | The middle two values are 83 and 85. The average of these two values is |
| | $$\frac{83 + 85}{2} = 84 \quad \leftarrow \text{Median}$$ |
| **5.7 Determining the Mode of a Set of Values** | Find the mode for Heather's scores in the previous example. |
| Find the value that appears most often in the list of values. This is the mode. | The most frequently occurring score is 93 (it occurs twice). Therefore, the mode is 93. |
| If no value appears more than once, there is no mode. If two different values appear the same number of times, the list is bimodal. | |
| **5.7 Evaluating the Variability of Data** | Which set of test scores has greater variability? |
| The variability of a set of data is the spread of the data around the mean. A quick way to evaluate the variability of data is to look at the range of values. Subtract the lowest value from the highest value. Generally, the greater the range in a set of values, the greater the variability. | Student M:  78, 85, 82, 84, 90, 86 |
| | Student N:  87, 89, 65, 95, 58, 88 |
| | Range for Student M = 90 − 78 = 12 |
| | Range for Student N = 95 − 65 = 30 |
| | Student N's scores have greater range and greater variability. |
| **5.8 Finding the Square Root of a Number** | $\sqrt{64} = 8$ |
| Use the square root key on a calculator, √ or √x . Round to the nearest thousandth when necessary. | $\sqrt{43} \approx 6.557$    6.557438524 is rounded to the nearest thousandth. |

| Concepts | Examples |
|---|---|
| **5.8 Finding the Unknown Length in a Right Triangle**<br>To find the hypotenuse, use:<br>$$\text{hypotenuse} = \sqrt{(\text{leg})^2 + (\text{leg})^2}$$<br>The hypotenuse is the side opposite the right angle. It is the longest side in a right triangle. | Find the length of the hypotenuse. Round to the nearest tenth.<br>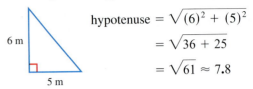<br>$$\begin{aligned}\text{hypotenuse} &= \sqrt{(6)^2 + (5)^2}\\ &= \sqrt{36 + 25}\\ &= \sqrt{61} \approx 7.8\end{aligned}$$<br>The hypotenuse is about 7.8 m long. |
| To find a leg, use:<br>$$\text{leg} = \sqrt{(\text{hypotenuse})^2 - (\text{leg})^2}$$<br>The legs are the sides that form the right angle. | Find the unknown length in this right triangle. Round to the nearest tenth.<br>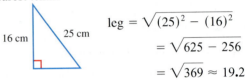<br>$$\begin{aligned}\text{leg} &= \sqrt{(25)^2 - (16)^2}\\ &= \sqrt{625 - 256}\\ &= \sqrt{369} \approx 19.2\end{aligned}$$<br>The leg is about 19.2 cm long. |
| **5.9 Solving Equations Containing Decimals**<br><br>*Step 1* Use the addition property of equality to get the variable term by itself on one side of the equal sign.<br><br>*Step 2* Divide both sides by the coefficient of the variable term to find the solution.<br><br><br><br>*Step 3* Check the solution by going back to the original equation. Replace the variable with the solution. If the equation balances, the solution is correct. | Solve the equation and check the solution.<br>$$\begin{aligned}4.5x + 0.7 &= -5.15\\ \underline{\phantom{4.5x}\;-0.7\phantom{x}} &\;\;\underline{-0.7} \quad \text{Add } -0.7 \text{ to both sides.}\\ 4.5x + 0 &= -5.85\\ \frac{4.5x}{4.5} &= \frac{-5.85}{4.5} \quad \text{Divide both sides by 4.5.}\\ x &= -1.3\end{aligned}$$<br>The solution is $-1.3$. To check the solution, go back to the original equation.<br>$$\begin{aligned}\text{Check}\quad 4.5x + 0.7 &= -5.15 \quad \text{Original equation}\\ 4.5(-1.3) + 0.7 &= -5.15 \quad \text{Replace } x \text{ with } -1.3.\\ -5.85 + 0.7 &= -5.15\\ -5.15 &= -5.15\end{aligned}$$<br>The equation balances, so $-1.3$ is the correct solution. |
| **5.10 Circles**<br>Use this formula to find the diameter of a circle, given the radius.<br>$$\text{diameter} = 2 \cdot \text{radius} \quad \text{or} \quad d = 2r$$ | Find the diameter of a circle if the radius is 3 ft.<br>$$d = 2 \cdot r = 2 \cdot 3 \text{ ft} = 6 \text{ ft}$$ |
| Use this formula to find the radius of a circle, given the diameter.<br>$$\text{radius} = \frac{\text{diameter}}{2} \quad \text{or} \quad r = \frac{d}{2}$$ | Find the radius of a circle if the diameter is 5 cm.<br>$$r = \frac{d}{2} = \frac{5 \text{ cm}}{2} = 2.5 \text{ cm}$$ |

| Concepts | Examples |
|---|---|

### 5.10 Circles (continued)

Use these formulas to find the circumference of a circle.

When you know the *radius*, use this formula.

$$C = 2 \cdot \pi \cdot \text{radius}$$
$$\text{or} \quad C = 2\pi r$$

When you know the *diameter*, use this formula.

$$C = \pi \cdot \text{diameter}$$
$$\text{or} \quad C = \pi d$$

Use 3.14 as the approximate value for $\pi$.

Find the circumference of a circle with a radius of 7 yd. Round your answer to the nearest tenth.

$$\text{Circumference} = 2 \cdot \pi \cdot r$$
$$C \approx 2 \cdot 3.14 \cdot 7 \text{ yd}$$
$$C \approx 44.0 \text{ yd} \quad \text{Rounded}$$

---

Use this formula to find the area of a circle.

$$A = \pi \cdot r \cdot r$$
$$\text{or} \quad A = \pi r^2$$

Use 3.14 as the approximate value of $\pi$.
Area is measured in **square units**.

Find the area of the circle. Round your answer to the nearest tenth.

$$\text{Area} = \pi \cdot r \cdot r$$
$$A \approx 3.14 \cdot 3 \text{ cm} \cdot 3 \text{ cm}$$
$$A \approx 28.3 \text{ cm}^2 \quad \text{Rounded}$$

### 5.10 Volume of a Cylinder

Use this formula to find the volume of a cylinder.

$$\text{Volume} = \pi \cdot r \cdot r \cdot h$$
$$\text{or} \quad V = \pi r^2 h$$

where $r$ is the radius of the circular base and $h$ is the height of the cylinder.

Volume is measured in **cubic units**.

Find the volume of a cylinder that is 10 m high with a diameter of 8 m.

Find the radius. $\quad r = \dfrac{8 \text{ m}}{2} = 4 \text{ m}$

$$V = \pi \cdot r \cdot r \cdot h$$
$$V \approx 3.14 \cdot 4 \text{ m} \cdot 4 \text{ m} \cdot 10 \text{ m}$$
$$V \approx 502.4 \text{ m}^3$$

### 5.10 Surface Area of a Rectangular Solid

Use this formula to find the surface area of a rectangular solid.

$$\text{Surface Area} = 2 \cdot l \cdot w + 2 \cdot l \cdot h + 2 \cdot w \cdot h$$
$$\text{or} \quad SA = 2lw + 2lh + 2wh$$

where $l$ is the length, $w$ is the width, and $h$ is the height of the solid.

Surface area is measured in **square units**.

Find the surface area of this packing crate.

$$SA = 2lw + 2lh + 2wh$$
$$SA = 2 \cdot 5 \text{ m} \cdot 3 \text{ m} + 2 \cdot 5 \text{ m} \cdot 6 \text{ m} + 2 \cdot 3 \text{ m} \cdot 6 \text{ m}$$
$$SA = 30 \text{ m}^2 + 60 \text{ m}^2 + 36 \text{ m}^2$$
$$SA = 126 \text{ m}^2$$

| Concepts | Examples |
|---|---|
| **5.10** *Surface Area of a Cylinder*<br>Use this formula to find the surface area of a cylinder.<br>$$\text{Surface Area} = 2 \cdot \pi \cdot r \cdot h + 2 \cdot \pi \cdot r \cdot r$$<br>or $\quad SA = 2\pi rh + 2\pi r^2$<br>where $r$ is the radius of the circular base and $h$ is the height of the cylinder. Use 3.14 as the approximate value of $\pi$. Surface area is measured in **square units**. | Find the surface area of a hot water tank with a height of 4.5 ft and a diameter of 1.8 ft. Round your answer to the nearest tenth.<br><br>Find the radius. $\quad r = \dfrac{1.8 \text{ ft}}{2} = 0.9 \text{ ft}$<br>$SA = 2\pi rh + 2\pi r^2$<br>$SA \approx 2 \cdot 3.14 \cdot 0.9 \text{ ft} \cdot 4.5 \text{ ft} + 2 \cdot 3.14 \cdot 0.9 \text{ ft} \cdot 0.9 \text{ ft}$<br>$SA \approx 25.434 \text{ ft}^2 + 5.0868 \text{ ft}^2$<br>$SA \approx 30.5208 \text{ ft}^2 \quad$ Now round to tenths.<br>$SA \approx 30.5 \text{ ft}^2$ |

**ANSWERS TO TEST YOUR WORD POWER**

1. **(d)** *Example:* For 0.7, the whole is cut into ten parts, and you are interested in 7 of the parts.
2. **(c)** *Examples:* The number 6.87 has two decimal places; 0.309 has three decimal places.
3. **(c)** *Example:* In triangle *ABC*, side *AC* is the hypotenuse.

4. **(a)** *Example:* In 5.42, the decimal point separates the whole number part, 5 ones, from the decimal part, 42 hundredths.
5. **(b)** *Example:* The bar above the 3 in $0.\overline{3}$ indicates that the 3 repeats forever.
6. **(b)** *Example:* The ratio of a circumference of 12.57 cm to a diameter of 4 cm is $\dfrac{12.57}{4} \approx 3.14$ (rounded).
7. **(d)** *Example:* For this set of prices—$4, $3, $7, $2, $5, $6, $7—the median is $5.
8. **(a)** *Example:* If the radius of a circle is 4 ft, then $C \approx 2 \cdot 3.14 \cdot 4 \approx 25.1$ ft.

# Chapter 5

## REVIEW EXERCISES

**[5.1]** *Name the digit that has the given place value.*

1. 243.059
   tenths
   hundredths

2. 0.6817
   ones
   tenths

3. $5824.39
   hundreds
   hundredths

4. 896.503
   tenths
   tens

5. 20.73861
   tenths
   ten-thousandths

*Write each decimal as a fraction or mixed number in lowest terms.*

6. 0.5

7. 0.75

8. 4.05

9. 0.875

10. 0.027

11. 27.8

*Write each decimal in words.*

12. 0.8

13. 400.29

14. 12.007

15. 0.0306

*Write each decimal in numbers.*

16. Eight and three tenths

17. Two hundred five thousandths

18. Seventy and sixty-six ten-thousandths

19. Thirty hundredths

**[5.2]** *Round to the place indicated.*

20. 275.635 to the nearest tenth

21. 72.789 to the nearest hundredth

22. 0.1604 to the nearest thousandth

23. 0.0905 to the nearest thousandth

24. 0.98 to the nearest tenth

*Round each money amount to the nearest cent.*

25. $15.8333

26. $0.698

27. $17,625.7906

*Round each income or expense item to the nearest dollar.*

28. The income from the pancake breakfast was $350.48.

29. Each member paid $129.50 in dues.

30. The refreshments cost $99.61.

31. The bank charges were $29.37.

**[5.3]** *Find each sum or difference.*

32. $0.4 - 6.07$

33. $-20 + 19.97$

34. $-1.35 + 7.229$

35. $0.005 + (3 - 9.44)$

*First, use front end rounding to estimate each answer. Then find the exact answer.*

36. American's favorite recreational activity in 1999 was swimming. About 95.1 million people went swimming at least once during the year. Bicycling was second with 56.3 million people. How many more people went swimming than bicycling? (*Source:* Sporting Goods Manufacturers Association.)

    *Estimate:*

    *Exact:*

37. Today, Jasmin wrote a check to the day care center for $215.53 and a check for $44.47 at the grocery store. What was the total of the two checks?

    *Estimate:*

    *Exact:*

38. Joey spent $1.59 for toothpaste, $5.33 for a gift, and $18.94 for a toaster. He gave the clerk three $10 bills. How much change did he get?

    *Estimate:*

    *Exact:*

39. Roseanne is training for a wheelchair race. She raced 2.3 kilometers on Monday, 4 kilometers on Wednesday, and 5.25 kilometers on Friday. How far did she race altogether?

    *Estimate:*

    *Exact:*

**[5.4]** *First, use front end rounding to estimate each answer. Then multiply to find the exact answer.*

40. *Estimate:*    *Exact:*
    
    $\phantom{0000}$ 6.138
    $\underline{\times \phantom{000}}$   $\underline{\times \phantom{00} 3.7}$

41. *Estimate:*    *Exact:*

    $\phantom{0000}$ 42.9
    $\underline{\times \phantom{000}}$   $\underline{\times \phantom{00} 3.3}$

*Multiply.*

42. $(-5.6)(-0.002)$

43. $(0.071)(-0.005)$

**[5.5]** *Decide if each answer is reasonable by rounding the numbers and estimating the answer. If the exact answer is not reasonable, find and correct the error.*

**44.** $706.2 \div 12 = 58.85$
Estimate:

**45.** $26.6 \div 2.8 = 0.95$
Estimate:

*Divide. Round to the nearest thousandth when necessary.*

**46.** $3\overline{)43.4}$

**47.** $\dfrac{-72}{-0.06}$

**48.** $-0.00048 \div 0.0012$

**[5.4–5.5]** *Solve each application problem.*

**49.** Adrienne worked 46.5 hours this week. Her hourly wage is $14.24 for the first 40 hours and 1.5 times that rate over 40 hours. Find her total earnings to the nearest dollar.

**50.** A book of 12 tickets costs $35.89 at the State Fair midway. What is the cost per ticket, to the nearest cent?

**51.** Stock in MathTronic sells for $3.75 per share. Kenneth is thinking of investing $500. How many whole shares could he buy?

**52.** Hamburger meat is on sale at $0.89 per pound. How much will Ms. Lee pay for 3.5 pounds of hamburger, to the nearest cent?

*Simplify by using the order of operations.*

**53.** $3.5^2 + 8.7(-1.95)$

**54.** $11 - 3.06 \div (3.95 - 0.35)$

**[5.6]** *Write each fraction as a decimal. Round to the nearest thousandth when necessary.*

**55.** $3\dfrac{4}{5}$

**56.** $\dfrac{16}{25}$

**57.** $1\dfrac{7}{8}$

**58.** $\dfrac{1}{9}$

*Arrange each group of numbers in order from smallest to largest.*

**59.** 3.68, 3.806, 3.6008

**60.** 0.215, 0.22, 0.209, 0.2102

**61.** $0.17, \dfrac{3}{20}, \dfrac{1}{8}, 0.159$

**[5.7]** *Find the mean and the median for each set of data.*

**62.** Digital cameras sold:
18, 12, 15, 24, 9, 42, 54, 87, 21, 3

**63.** Number of insurance claims processed:
54, 28, 35, 43, 17, 37, 68, 75, 39

**64.** Find the weighted mean.

| Dollar Value | Frequency |
|---|---|
| $42 | 3 |
| $47 | 7 |
| $53 | 2 |
| $55 | 3 |
| $59 | 5 |

**65. (a)** Find the mode or modes for each set of data.
Hiking boots at Store J priced at $107, $69, $139, $107, $160, $84, $160

Hiking boots at Store K priced at $119, $136, $99, $119, $139, $119, $95

**(b)** Which store's prices have greater variability? Show your work.

**[5.8]** *Find the unknown length in each right triangle. Use a calculator to find square roots. Round your answers to the nearest tenth when necessary.*

**66.**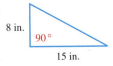

**67.**

**68.**

**69.**

**70.**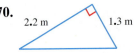

**71.**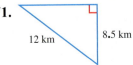

**[5.9]** *Solve each equation.*

**72.** $-0.1 = b - 0.35$

**73.** $-3.8x = 0$

**74.** $6.8 + 0.4n = 1.6$

**75.** $-0.375 + 1.75a = 2a$

**76.** $0.3y - 5.4 = 2.7 + 0.8y$

## Chapter 5 Review Exercises    411

**[5.10]** *Find the unknown radius or diameter.*

**77.** The radius of a circular irrigation field is 68.9 m. What is the diameter of the field?

**78.** The diameter of a juice can is 3 in. What is the radius of the can?

*Find the circumference and area of each circle. Use 3.14 as the approximate value for π. Round your answers to the nearest tenth.*

**79.**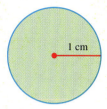

**80.**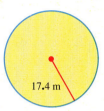

**81.**

*Find the volume and surface area of each solid. Use 3.14 as the approximate value for π. Round your answers to the nearest tenth when necessary.*

**82.**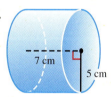

**83.**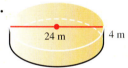

**84.** A rectangular cooler that measures 3.5 ft by 1.5 ft and is 1.5 ft high.

### MIXED REVIEW EXERCISES

*Simplify.*

**85.** $89.19 + 0.075 + 310.6 + 5$

**86.** $72.8(-3.5)$

**87.** $1648.3 \div 0.46$   Round to thousandths.

**88.** $30 - 0.9102$

**89.** $(4.38)(0.007)$

**90.** $0.005\overline{)0.047}$

**91.** $72.105 + 8.2 - 95.37$

**92.** $\dfrac{-81.36}{-9}$

**93.** $(0.6 - 1.22) + 4.8(-3.15)$

**94.** $0.455(18)$

**95.** $(-1.6)(-0.58)$

**96.** $0.218 \overline{)7.63}$

**97.** $-21.059 - 20.8$

**98.** $18.3 - 3^2 \div 0.5$

*Use the information in the ad to solve Exercises 99–103. Round money answers to the nearest cent. (Disregard any sales tax.)*

**99.** How much would one pair of men's socks cost?

**100.** How much more would one pair of men's socks cost than one pair of children's socks?

**101.** How much would Fernando pay for a dozen pair of men's socks?

**102.** How much would Akiko pay for five pairs of teen jeans and four pairs of women's jeans?

**103.** What is the difference between the cheapest sale price for athletic shoes and the highest regular price?

*Solve each equation.*

**104.** $4.62 = -6.6y$

**105.** $1.05x - 2.5 = 0.8x + 5$

*Solve each application problem. Round answers to the nearest tenth when necessary.*

**106.** A circular table has a diameter of 5 ft. How much rubber striping is needed to go around the edge of the table? What is the area of the table top?

**107.** Jerry missed one math test, so his test scores are 82, 0, 78, 93, 85. What is his average score?

**108.** LaRae drove 16 miles south, then made a 90° right turn and drove 12 miles west. How far is she, in a straight line, from her starting point?

**109.** A juice can that is 7 in. tall has a diameter of 4 in. What is the volume of the can?

# Chapter 5 TEST

*Write each decimal as a fraction or mixed number in lowest terms.*

**1.** 18.4

**2.** 0.075

*Write each decimal in words.*

**3.** 60.007

**4.** 0.0208

*First, use front end rounding to estimate each answer. Then find the exact answer.*

**5.** 7.6 + 82.0128 + 39.59

**6.** −5.79(1.2)

**7.** −79.1 − 3.602

**8.** −20.04 ÷ (−4.8)

*Find the exact answer.*

**9.** 670 − 0.996

**10.** 0.15)̄72

**11.** (−0.006)(−0.007)

**12.** Pat bought 3.4 meters of fabric. She paid $15.47. What was the cost per meter?

**13.** Davida ran a race in 3.059 minutes. Angela ran the race in 3.5 minutes. Who won? By how much?

**14.** Mr. Yamamoto bought 1.85 pounds of cheese at $2.89 per pound. How much did he pay for the cheese, to the nearest cent?

*Solve each equation. Show your work.*

**15.** −5.9 = y + 0.25

**16.** −4.2x = 1.47

**17.** 3a − 22.7 = 10

**18.** −0.8n + 1.88 = 2n − 6.1

1. _____

2. _____

3. _____

4. _____

5. Estimate: _____
   Exact: _____

6. Estimate: _____
   Exact: _____

7. Estimate: _____
   Exact: _____

8. Estimate: _____
   Exact: _____

9. _____

10. _____

11. _____

12. _____

13. _____

14. _____

15. _____

16. _____

17. _____

18. _____

*Arrange in order from smallest to largest.*

**19.** $0.44, 0.451, \dfrac{9}{20}, 0.4506$

*Use the order of operations to simplify.*

**20.** $6.3^2 - 5.9 + 3.4(-0.5)$

**21.** Find the mean number of books loaned: 52, 61, 68, 69, 73, 75, 79, 84, 91, 98.

**22.** Find the mode for hot tub temperatures (Fahrenheit) of 96°, 104°, 103°, 104°, 103°, 104°, 91°, 74°, 103°.

**23.** Find the weighted mean.

| Cost | Frequency |
|---|---|
| $ 6 | 7 |
| $10 | 3 |
| $11 | 4 |
| $14 | 2 |
| $19 | 3 |
| $24 | 1 |

**24.** (a) Find the median cost of a math textbook: $54.50, $48, $39.75, $89, $56.25, $49.30, $46.90, $51.80.

(b) Find the median cost of a biology textbook: $75, $50.45, $49.80, $30.95, $63, $97.55, $47.30.

(c) Which set of costs has less variability? Show your work.

*Find the unknown lengths. Round your answers to the nearest tenth when necessary.*

**25.**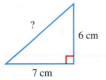

**26.**

*Use 3.14 as the approximate value for $\pi$ and round your answers to the nearest tenth.*

**27.** Find the radius.

**28.** Find the circumference.

**29.** Find the area.

**30.** Find the volume.

**31.** Find the surface area of the solid in Exercise 30.

# Cumulative Review Exercises   CHAPTERS 1–5

1. Write these numbers in words.

   (a) 45.0203

   (b) 30,000,650,008

2. Write these numbers using digits.

   (a) One hundred sixty million, five hundred

   (b) Seventy-five thousandths

*Round each number as indicated.*

3. 46,908 to the nearest hundred

4. 6.197 to the nearest hundredth

5. 0.66148 to the nearest thousandth

6. 9951 to the nearest hundred

*Simplify.*

7. $-5 - 8$

8. $-0.003(0.02)$

9. $\dfrac{-7}{0}$

10. $8 + 4(2 - 5)^2$

11. $-\dfrac{3}{8}(-48)$

12. $|4| - |-10|$

13. $0.721 + 55.9$

14. $3\dfrac{1}{3} - 1\dfrac{5}{6}$

15. $\dfrac{3}{b^2} \cdot \dfrac{b}{8}$

16. $12 - 0.853$

17. $\dfrac{3}{10} - \dfrac{3}{4}$

18. $\dfrac{\frac{5}{16}}{-10}$

19. $-3.75 \div (-2.9)$ Round answer to nearest tenth.

20. $\dfrac{x}{2} + \dfrac{3}{5}$

21. $5 \div \left(-\dfrac{5}{8}\right)$

22. $2^5 - 4^3$

23. $\dfrac{-36 \div (-2)}{-6 - 4(0 - 6)}$

24. $(0.8)^2 - 3.2 + 4(-0.8)$

25. $\dfrac{3}{4} \div \dfrac{3}{10}\left(\dfrac{1}{4} + \dfrac{2}{3}\right)$

*The ages (in years) of the students in a math class are: 19, 23, 24, 19, 20, 29, 26, 35, 20, 22, 26, 23, 25, 26, 20, 30. Use this data for Exercises 26–28.*

26. Find the mean age, to the nearest tenth.

27. Find the median age.

28. Find the mode.

*Solve each equation. Show your work.*

**29.** $3h - 4h = 16 - 12$

**30.** $-2x = x - 15$

**31.** $20 = 6r - 45.4$

**32.** $-3(y + 4) = 7y + 8$

**33.** $-0.8 + 1.4n = 2 + 0.7n$

*Find the unknown length, perimeter, area, or volume. Use 3.14 as the approximate value of π and round answers to the nearest tenth.*

**34.** Find the perimeter and the area.
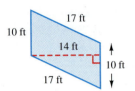

**35.** Find the circumference and the area.

**36.** Find the length of the third side and the area.

**37.** Find the volume and surface area.
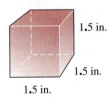

*First, use front end rounding to estimate each answer. Then find the exact answer.*

**38.** Lameck had two $10 bills. He spent $7.96 on gasoline and $0.87 for a candy bar at the convenience store. How much money does he have left?

*Estimate:*

*Exact:*

**39.** Toshihiro bought $2\frac{1}{3}$ yards of cotton fabric and $3\frac{7}{8}$ yards of wool fabric. How many yards did he buy in all?

*Estimate:*

*Exact:*

**40.** Paulette bought 2.7 pounds of grapes for $2.56. What was the cost per pound, to the nearest cent?

*Estimate:*

*Exact:*

**41.** Carter Community College received a $78,000 grant from a local computer company to help students pay tuition for computer classes. How much money could be given to each of 107 students? Round to the nearest dollar.

*Estimate:*

*Exact:*

*Solve each application problem using the six problem-solving steps.*

**42.** A $30,000 scholarship is being divided between two students so that one student receives three times as much as the other. How much will each student receive?

**43.** The perimeter of a rectangular game board is 48 in. The length is 4 in. more than the width. Find the length and width of the game board.

# Ratio, Proportion, and Line/Angle/Triangle Relationships

6.1  Ratios
6.2  Rates
6.3  Proportions
6.4  Problem Solving with Proportions
6.5  Geometry: Lines and Angles
6.6  Geometry Applications: Congruent and Similar Triangles

Over one-half of all adults in the United States own a cell phone. (*Source:* Gallup Poll results from www.gallup.com) Everyone likes to talk, but no one likes to pay the bills! Now you can get the best possible deal on cellular phone service by finding unit rates. (See Section 6.2, Exercises 47–50.)

## 6.1 Ratios

**OBJECTIVES**

1. Write ratios as fractions.
2. Solve ratio problems involving decimals or mixed numbers.
3. Solve ratio problems after converting units.

A **ratio** compares two quantities. You can compare two numbers, such as 8 and 4, or two measurements that have the *same* type of units, such as 3 days and 12 days. (*Rates* compare measurements with different types of units and are covered in the next section.)

Ratios can help you see important relationships. For example, if the ratio of your monthly expenses to your monthly income is 10 to 9, then you are spending $10 for every $9 you earn and going deeper into debt.

**1** **Write ratios as fractions.** A ratio can be written in three ways.

### Writing a Ratio

The ratio of $7 **to** $3 can be written as follows.

$$7 \text{ to } 3 \quad \text{or} \quad 7{:}3 \quad \text{or} \quad \frac{7}{3} \leftarrow \text{Fraction bar indicates "to."}$$

"**:**" indicates "**to**."

Writing a ratio as a fraction is the most common method, and the one we will use here. All three ways are read, "the ratio of 7 **to** 3." The word **to** separates the quantities being compared.

### Writing a Ratio as a Fraction

Order is important when you're writing a ratio. The quantity mentioned **first** is the **numerator**. The quantity mentioned **second** is the **denominator**. For example:

The ratio of **5** to **12** is written $\frac{5}{12}$.

### Example 1 — Writing Ratios

The Anasazi, ancestors of the Pueblo Indians, built multistory apartment towns in New Mexico about 1100 years ago. A room might measure 14 feet long, 11 feet wide, and 15 feet high.

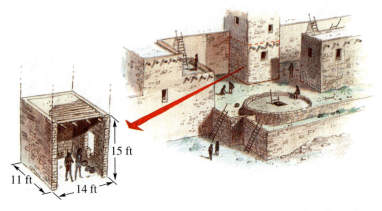

Continued on Next Page

Write each ratio as a fraction, using the room measurements.

**(a)** Ratio of length to width

The ratio of **length to width** is $\dfrac{14 \text{ feet}}{11 \text{ feet}} = \dfrac{14}{11}$.

Numerator (mentioned first)
Denominator (mentioned second)

You can divide out common *units* just as you divided out common *factors* when writing fractions in lowest terms. (See **Section 4.2**.) However, do *not* rewrite the fraction as a mixed number. Keep it as the ratio of 14 to 11.

**(b)** Ratio of width to height

The ratio of width **to** height is $\dfrac{11 \text{ feet}}{15 \text{ feet}} = \dfrac{11}{15}$.

### CAUTION

Remember, the *order* of the numbers is important in a ratio. Look for the words "ratio of *a* to *b*." Write the ratio as $\dfrac{a}{b}$, **not** $\dfrac{b}{a}$. The quantity mentioned first is the numerator.

**❶** Shane spent \$14 on meat, \$5 on milk, and \$7 on fresh fruit. Write the following ratios as fractions in lowest terms.

**(a)** The ratio of amount spent on fruit to amount spent on milk

**(b)** The ratio of amount spent on milk to amount spent on meat

Work Problem ❶ at the Side.

Any ratio can be written as a fraction. Therefore, you can write a ratio in *lowest terms,* just as you do with any fraction.

### Example 2  Writing Ratios in Lowest Terms

Write each ratio as a fraction in lowest terms.

**(a)** 60 days to 20 days

The ratio is $\dfrac{60}{20}$. Write this ratio in lowest terms by dividing the numerator and the denominator by 20.

$$\dfrac{60}{20} = \dfrac{60 \div 20}{20 \div 20} = \dfrac{3}{1} \quad \leftarrow \text{Ratio in lowest terms}$$

### CAUTION

In the fractions chapter you would have rewritten $\dfrac{3}{1}$ as 3. But a *ratio* compares *two* quantities, so you need to keep both parts of the ratio and write it as $\dfrac{3}{1}$.

**(c)** The ratio of amount spent on meat to amount spent on milk

**(b)** 50 ounces of medicine to 120 ounces of medicine

The ratio is $\dfrac{50}{120}$. Divide the numerator and the denominator by 10.

$$\dfrac{50}{120} = \dfrac{50 \div 10}{120 \div 10} = \dfrac{5}{12} \quad \leftarrow \text{Ratio in lowest terms}$$

── Continued on Next Page

**ANSWERS**

**1. (a)** $\dfrac{7}{5}$  **(b)** $\dfrac{5}{14}$  **(c)** $\dfrac{14}{5}$

**❷ Write each ratio as a fraction in lowest terms.**

(a) 9 hours to 12 hours

(b) 100 meters to 50 meters

(c) The ratio of width to length for this rectangle

Length 48 ft
Width 24 ft

**❸ Write each ratio as a ratio of whole numbers in lowest terms.**

(a) The price of Tamar's favorite brand of lipstick increased from $5.50 to $7.00. Find the ratio of the increase in price to the original price.

(b) Last week, Lance worked 4.5 hours each day. This week he cut back to 3 hours each day. Find the ratio of the decrease in hours to the original number of hours.

**ANSWERS**

2. (a) $\frac{3}{4}$  (b) $\frac{2}{1}$  (c) $\frac{1}{2}$

3. (a) $\frac{(1.50)(100)}{(5.50)(100)} = \frac{150 \div 50}{550 \div 50} = \frac{3}{11}$

(b) $\frac{(1.5)(10)}{(4.5)(10)} = \frac{15 \div 15}{45 \div 15} = \frac{1}{3}$

---

(c) 15 people in a large van to 6 people in a small van

The ratio is $\frac{15}{6} = \frac{15 \div 3}{6 \div 3} = \frac{5}{2}$ ← { Ratio in lowest terms

**NOTE**

Although $\frac{5}{2} = 2\frac{1}{2}$, ratios are *not* written as mixed numbers. Nevertheless, in Example 2(c), the ratio $\frac{5}{2}$ does mean the large van holds $2\frac{1}{2}$ times as many people as the small van.

**Work Problem ❷ at the Side.**

**OBJECTIVE 2 Solve ratio problems involving decimals or mixed numbers.** Sometimes a ratio compares two decimal numbers or two fractions. It is easier to understand if we rewrite the ratio as a ratio of two whole numbers.

**Example 3** Using Decimal Numbers in a Ratio

The price of a Sunday newspaper increased from $1.50 to $1.75. Find the ratio of the increase in price to the original price.

The words increase in price are mentioned first, so the increase will be the numerator. How much did the price go up? Use subtraction.

new price − original price = increase
$1.75 − $1.50 = $0.25

The words original price are mentioned second, so the original price of $1.50 is the denominator.

The ratio of increase in price to original price is

$\frac{0.25}{1.50}$ ← increase
← original price

Now rewrite the ratio as a ratio of whole numbers. Recall that if you multiply both the numerator and denominator of a fraction by the same number, you get an equivalent fraction. The decimals in this example are hundredths, so multiply by 100 to get whole numbers. (If the decimals are tenths, multiply by 10. If thousandths, multiply by 1000.) Then write the ratio in lowest terms.

$\frac{0.25}{1.50} = \frac{(0.25)(100)}{(1.50)(100)} = \frac{25}{150} = \frac{25 \div 25}{150 \div 25} = \frac{1}{6}$ ← { Ratio in lowest terms

Ratio as two whole numbers

**Work Problem ❸ at the Side.**

**Example 4** Using Mixed Numbers in Ratios

Write each ratio as a comparison of whole numbers in lowest terms.

(a) 2 days to $2\frac{1}{4}$ days

Write the ratio as follows. Divide out the common units.

$\frac{2 \text{ days}}{2\frac{1}{4} \text{ days}} = \frac{2}{2\frac{1}{4}}$

**Continued on Next Page**

Next, write 2 as $\frac{2}{1}$ and $2\frac{1}{4}$ as the improper fraction $\frac{9}{4}$.

$$\frac{2}{2\frac{1}{4}} = \frac{\frac{2}{1}}{\frac{9}{4}}$$

Now rewrite the problem in horizontal format, using the "÷" symbol for division. Finally, multiply by the reciprocal of the divisor, as you did in **Section 4.3**.

$$\frac{\frac{2}{1}}{\frac{9}{4}} = \frac{2}{1} \div \frac{9}{4} = \frac{2}{1} \cdot \frac{4}{9} = \frac{8}{9}$$

Reciprocals

The ratio, in lowest terms, is $\frac{8}{9}$.

**(b)** $3\frac{1}{4}$ to $1\frac{1}{2}$

Write the ratio as $\frac{3\frac{1}{4}}{1\frac{1}{2}}$. Then write $3\frac{1}{4}$ and $1\frac{1}{2}$ as improper fractions.

$$3\frac{1}{4} = \frac{13}{4} \quad \text{and} \quad 1\frac{1}{2} = \frac{3}{2}$$

The ratio is shown below.

$$\frac{3\frac{1}{4}}{1\frac{1}{2}} = \frac{\frac{13}{4}}{\frac{3}{2}}$$

Rewrite as a division problem in horizontal format, using the "÷" symbol. Then multiply by the reciprocal of the divisor.

$$\frac{13}{4} \div \frac{3}{2} = \frac{13}{4} \cdot \frac{2}{3} = \frac{13 \cdot \overset{1}{\cancel{2}}}{\cancel{2} \cdot 2 \cdot 3} = \frac{13}{6} \quad \leftarrow \begin{cases} \text{Ratio in} \\ \text{lowest terms} \end{cases}$$

Reciprocals

**NOTE**

We can also work Examples 4(a) and 4(b) by using decimals.
**(a)** $2\frac{1}{4}$ is equivalent to 2.25, so we have the ratio shown below.

$$\frac{2}{2\frac{1}{4}} = \frac{2}{2.25} = \frac{(2)(100)}{(2.25)(100)} = \frac{200}{225} = \frac{200 \div 25}{225 \div 25} = \frac{8}{9} \quad \leftarrow \text{Same result}$$

**(b)** $3\frac{1}{4}$ is equivalent to 3.25 and $1\frac{1}{2}$ is equivalent to 1.5.

$$\frac{3\frac{1}{4}}{1\frac{1}{2}} = \frac{3.25}{1.5} = \frac{(3.25)(100)}{(1.5)(100)} = \frac{325}{150} = \frac{325 \div 25}{150 \div 25} = \frac{13}{6} \quad \leftarrow \text{Same result}$$

This method would *not* work for fractions that are repeating decimals, such as $\frac{1}{3}$ or $\frac{5}{6}$.

Work Problem ❹ at the Side.

---

❹ Write each ratio as a ratio of whole numbers in lowest terms.

**(a)** $3\frac{1}{2}$ to 4

**(b)** $5\frac{5}{8}$ pounds to $3\frac{3}{4}$ pounds

**(c)** $3\frac{1}{2}$ inches to $\frac{7}{8}$ inch

**ANSWERS**

**4.** **(a)** $\frac{7}{8}$ **(b)** $\frac{3}{2}$ **(c)** $\frac{4}{1}$

**3** **Solve ratio problems after converting units.** When a ratio compares measurements, both measurements must be in the *same* units. For example, *feet* must be compared to *feet, hours* to *hours, pints* to *pints,* and *inches* to *inches.*

### Example 5  Ratio Applications Using Measurement

**(a)** Write the ratio of the length of the board on the left to the length of the board on the right. Compare in inches.

2 ft    30 in.

First, express 2 feet in inches. Because 1 foot has 12 inches, 2 feet is

$$2 \cdot \mathbf{12\ inches} = 24\ \text{inches.}$$

The length of the board on the left is 24 inches, so the ratio of the lengths is shown below. The common units divide out.

$$\frac{2\ \text{ft}}{30\ \text{in.}} = \frac{24\ \cancel{\text{inches}}}{30\ \cancel{\text{inches}}} = \frac{24}{30}$$

Write the ratio in lowest terms.

$$\frac{24}{30} = \frac{24 \div 6}{30 \div 6} = \frac{4}{5} \quad \left\{ \begin{array}{l} \text{Ratio in} \\ \text{lowest terms} \end{array} \right.$$

The shorter board on the left is $\frac{4}{5}$ the length of the longer board on the right.

**NOTE**

Notice that we wrote the ratio using the smaller unit (inches are smaller than feet). Using the smaller unit will help you avoid working with fractions. If we wrote the ratio using feet, then:

$$30\ \text{inches} = 2\frac{1}{2}\ \text{feet.}$$

The ratio in feet is shown below.

$$\frac{2\ \cancel{\text{feet}}}{2\frac{1}{2}\ \cancel{\text{feet}}} = \frac{2}{1} \div \frac{5}{2} = \frac{2}{1} \cdot \frac{2}{5} = \frac{4}{5} \quad \leftarrow \text{Same result}$$

The ratio is the same, but it takes more steps to get the answer. Using the smaller unit is usually easier.

**(b)** Write the ratio of 28 days to 3 weeks.

Since it is usually easier to write the ratio using the smaller measurement, compare in *days* because days are shorter than weeks.

First express 3 weeks in days. Because 1 week has 7 days, 3 weeks is

$$3 \cdot \mathbf{7\ days} = 21\ \text{days.}$$

So the ratio in days is shown below.

$$\frac{28\ \text{days}}{3\ \text{weeks}} = \frac{28\ \cancel{\text{days}}}{21\ \cancel{\text{days}}} = \frac{28}{21} = \frac{28 \div 7}{21 \div 7} = \frac{4}{3} \quad \leftarrow \text{Lowest terms}$$

The following table will help you set up ratios that compare measurements. You will work with these measurements again in Chapter 8.

## Measurement Comparisons

**Length**
1 foot = 12 inches
1 yard = 3 feet
1 mile = 5280 feet

**Weight**
1 pound = 16 ounces
1 ton = 2000 pounds

**Capacity (Volume)**
1 pint = 2 cups
1 quart = 2 pints
1 gallon = 4 quarts

**Time**
1 minute = 60 seconds
1 hour = 60 minutes
1 day = 24 hours
1 week = 7 days

Work Problem ❺ at the Side.

❺ Write each ratio as a fraction in lowest terms. (*Hint:* Recall that it is usually easier to write the ratio using the smaller measurement unit.)

**(a)** 9 inches to 6 feet

**(b)** 2 days to 8 hours

**(c)** 7 yards to 14 feet

**(d)** 3 quarts to 3 gallons

**(e)** 25 minutes to 2 hours

**(f)** 4 pounds to 12 ounces

ANSWERS

5. (a) $\frac{1}{8}$ (b) $\frac{6}{1}$ (c) $\frac{3}{2}$ (d) $\frac{1}{4}$
(e) $\frac{5}{24}$ (f) $\frac{16}{3}$

# Focus on Real-Data Applications

## Historical Ratios

The table below lists average prices and earnings from 1960, 1970, 1980, and 1990. (*Source:* Pages of Time.)

| Item | 1960 | 1970 | 1980 | 1990 | Today |
|---|---|---|---|---|---|
| First class postage stamp | $0.04 | $0.06 | $0.15 | $0.25 | |
| Movie ticket | $1.00 | $1.50 | $2.25 | $4.00 | |
| Gallon of gas | $0.26 | $0.36 | $1.19 | $1.34 | |
| New Car (to nearest hundred) | $2600 | $4000 | $7200 | $16,000 | |
| Yearly Income (to nearest hundred) | $5200 | $9400 | $19,200 | $28,900 | |
| Hourly Minimum Wage | $1.00 | $1.60 | $3.10 | $3.80 | |

1. Fill in the amount for each item in the Today column of the table. Use local prices in your city or neighborhood for a movie ticket and a gallon of gas. Use *The World Almanac* or an on-line search engine to find the average income and new car price.

2. Use the data on the price of a movie ticket to find these ratios. Write each ratio as a ratio of whole numbers in lowest terms.
   - (a) 1970 price to 1960 price
   - (b) 1980 price to 1970 price
   - (c) 1990 price to 1980 price
   - (d) Today price to 1990 price

3. What pattern do you see in the movie ticket ratios?

4. Use the data on postage stamp prices to find each ratio involving the increase in price. Write each ratio as a ratio of whole numbers in lowest terms.
   - (a) increase from 1960 to 1970 compared to 1960 price
   - (b) increase from 1970 to 1980 compared to 1970 price
   - (c) increase from 1980 to 1990 compared to 1980 price
   - (d) increase from 1990 to Today compared to 1990 price

5. What pattern do you see in the poastage stamp price increases?

6. Write a ratio in lowest terms comparing the cost of a new car to the average yearly income for each year. Then use your calculator to change each fractional ratio into a decimal. What pattern do you see?
   - (a) 1960
   - (b) 1970
   - (c) 1980
   - (d) 1990
   - (e) Today

7. How many gallons of gas could you buy with one hour of minimum wage earnings in each year? Round to the nearest tenth. What pattern do you see?
   - (a) 1960
   - (b) 1970
   - (c) 1980
   - (d) 1990
   - (d) Today

## 6.1 Exercises

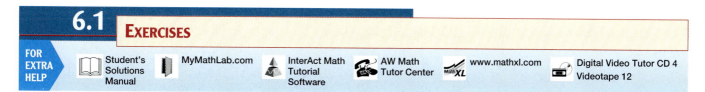

*Write each ratio as a fraction in lowest terms. See Examples 1 and 2.*

1. 8 to 9
2. 11 to 15
3. $100 to $50
4. 35¢ to 7¢

5. 30 minutes to 90 minutes
6. 9 pounds to 36 pounds

7. 80 miles to 50 miles
8. 300 people to 450 people

9. 6 hours to 16 hours
10. 45 books to 35 books

*Write each ratio as a ratio of whole numbers in lowest terms. See Examples 3 and 4.*

11. $4.50 to $3.50
12. $0.08 to $0.06
13. 15 to $2\frac{1}{2}$

14. 5 to $1\frac{1}{4}$
15. $1\frac{1}{4}$ to $1\frac{1}{2}$
16. $2\frac{1}{3}$ to $2\frac{2}{3}$

*Write each ratio as a fraction in lowest terms. For help, use the table of measurement relationships on page 423. See Example 5.*

17. 4 feet to 30 inches
18. 8 feet to 4 yards

19. 5 minutes to 1 hour
20. 8 quarts to 5 pints

21. 15 hours to 2 days
22. 3 pounds to 6 ounces

23. 5 gallons to 5 quarts
24. 3 cups to 3 pints

*The table shows the number of greeting cards that Americans buy for various occasions. Use the information to answer Exercises 25–30. Write each ratio as a fraction in lowest terms.*

| Holiday/Event | Cards Sold |
|---|---|
| Valentine's Day | 900 million |
| Mother's Day | 150 million |
| Father's Day | 95 million |
| Graduation | 60 million |
| Thanksgiving | 30 million |
| Halloween | 25 million |

*Source:* Hallmark Cards.

**25.** Find the ratio of Thanksgiving cards to graduation cards.

**26.** Find the ratio of Halloween cards to Mother's Day cards.

**27.** Find the ratio of Valentine's Day cards to Halloween cards.

**28.** Find the ratio of Mother's Day cards to Father's Day cards.

**29.** Explain how you might use the information in the table if you owned a shop selling gifts and greeting cards.

**30.** Why is the ratio of Valentine's Day cards to graduation cards $\frac{15}{1}$? Give two possible reasons.

*The bar graph shows the number of Americans who play various instruments. Use the graph to complete Exercises 31–34. Write each ratio as a fraction in lowest terms.*

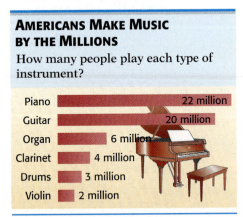

**AMERICANS MAKE MUSIC BY THE MILLIONS**
How many people play each type of instrument?

- Piano — 22 million
- Guitar — 20 million
- Organ — 6 million
- Clarinet — 4 million
- Drums — 3 million
- Violin — 2 million

*Source:* America by the Numbers.

**31.** Write six ratios that compare the least popular instrument to each of the other instruments.

**32.** Which two instruments give each of these ratios: **(a)** $\frac{5}{1}$; **(b)** $\frac{2}{1}$. There may be more than one correct answer.

**33.** Why might the ratio of guitar players to drum players be $\frac{20}{3}$? Give two possible explanations.

**34.** If you ran a music school, explain how you might use the information in the graph to decide how many teachers to hire.

*Use the circle graph of one family's monthly budget to complete Exercises 35–38. Write each ratio as a fraction in lowest terms.*

**35.** Find the ratio of taxes to transportation.

**36.** Find the ratio of rent to food.

**37.** Find the ratio of
  **(a)** rent to total income.
  **(b)** rent and utilities to taxes and miscellaneous.

**38.** Find the ratio of
  **(a)** utilities to total income.
  **(b)** food and taxes to transportation.

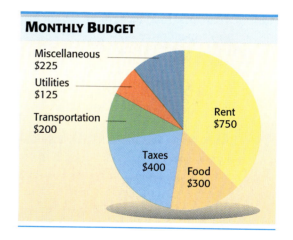

*For each figure, find the ratio of the length of the longest side to the length of the shortest side. Write each ratio as a fraction in lowest terms. See Examples 2–4.*

**39.**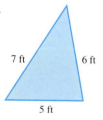

**40.**

**41.**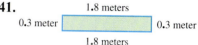

**42.**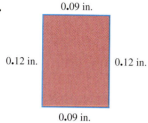

**43.**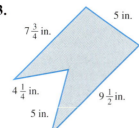

**44.**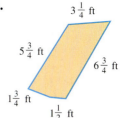

*Write each ratio as a fraction in lowest terms.*

**45.** The price of oil recently went from $10.80 to $13.50 per case of 12 quarts. Find the ratio of the increase in price to the original price.

**46.** The price of an antibiotic decreased from $8.80 to $5.60 for a bottle of 100 tablets. Find the ratio of the decrease in price to the original price.

**47.** The first time a movie was made in Minnesota, the cast and crew spent $59\frac{1}{2}$ days filming winter scenes. The next year, another movie was filmed in $8\frac{3}{4}$ weeks. Find the ratio of the first movie's filming time to the second movie's time. Compare in weeks.

**48.** The percheron, a large draft horse, measures about $5\frac{3}{4}$ feet at the shoulder. The prehistoric ancestor of the horse measured only $15\frac{3}{4}$ inches at the shoulder. Find the ratio of the percheron's height to its prehistoric ancestor's height. Compare in inches. (*Source: Eyewitness Books: Horse.*)

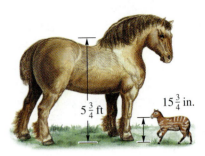

## RELATING CONCEPTS (Exercises 49–52)   FOR INDIVIDUAL OR GROUP WORK

*Use your knowledge of ratios to work Exercises 49–52 in order.*

**49.** In this painting, what is the ratio of the length of the longest side to the length of the shortest side? What other measurements could the painting have and still maintain the same ratio?

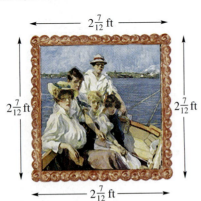

**50.** The ratio of my son's age to my daughter's age is 4 to 5. One possibility is that my son is 4 years old and my daughter is 5 years old. Find six other possibilities that fit the 4 to 5 ratio.

**51.** Amelia said that the ratio of her age to her mother's age is 5 to 3. Is this possible? Explain your answer.

**52.** Would you prefer that the ratio of your income to your friend's income be 1 to 3 or 3 to 1? Explain your answer.

## 6.2 RATES

A *ratio* compares two measurements with the same type of units, such as 9 feet to 12 feet (both length measurements). But many of the comparisons we make use measurements with different types of units, such as shown here.

$$160 \text{ dollars for } 8 \text{ hours} \quad \text{(money to time)}$$
$$450 \text{ miles on } 15 \text{ gallons} \quad \text{(distance to capacity)}$$

This type of comparison is called a **rate**.

### OBJECTIVES
1. Write rates as fractions.
2. Find unit rates.
3. Find the best buy based on cost per unit.

**1 Write rates as fractions.** Suppose you hiked 18 miles in 4 hours. The *rate* at which you hiked can be written as a fraction in lowest terms.

$$\frac{18 \text{ miles}}{4 \text{ hours}} = \frac{18 \text{ miles} \div 2}{4 \text{ hours} \div 2} = \frac{9 \text{ miles}}{2 \text{ hours}} \leftarrow \text{Lowest terms}$$

In a rate, you often find one of these words separating the quantities you are comparing.

**in    for    on    per    from**

### CAUTION
When writing a rate, always include the units, such as miles, hours, dollars, and so on. Because the units in a rate are different, the units do *not* divide out.

**① Write each rate as a fraction in lowest terms.**

(a) $6 for 30 packages

(b) 500 miles in 10 hours

### Example 1  Writing Rates in Lowest Terms

Write each rate as a fraction in lowest terms.

(a) 5 gallons of chemical for $60

$$\frac{5 \text{ gallons} \div 5}{60 \text{ dollars} \div 5} = \frac{1 \text{ gallon}}{12 \text{ dollars}} \quad \text{Write the units: gallons and dollars.}$$

(b) $1500 wages in 10 weeks

$$\frac{1500 \text{ dollars} \div 10}{10 \text{ weeks} \div 10} = \frac{150 \text{ dollars}}{1 \text{ week}}$$

(c) 2225 miles on 75 gallons of gas

$$\frac{2225 \text{ miles} \div 25}{75 \text{ gallons} \div 25} = \frac{89 \text{ miles}}{3 \text{ gallons}}$$

(c) 4 teachers for 90 students

(d) 1270 bushels on 30 acres

**Work Problem ① at the Side.**

**2 Find unit rates.** When the *denominator* of a rate is 1, it is called a **unit rate.** We use unit rates frequently. For example, you earn $12.75 for *1 hour* of work. This unit rate is written:

$$\$12.75 \text{ per hour} \quad \text{or} \quad \$12.75/\text{hour.}$$

You drive 28 miles on *1 gallon* of gas. This unit rate is written

$$28 \text{ miles per gallon} \quad \text{or} \quad 28 \text{ miles/gallon.}$$

Use **per** or a **/** mark when writing unit rates.

### ANSWERS
1. (a) $\dfrac{\$1}{5 \text{ packages}}$  (b) $\dfrac{50 \text{ miles}}{1 \text{ hour}}$
   (c) $\dfrac{2 \text{ teachers}}{45 \text{ students}}$  (d) $\dfrac{127 \text{ bushels}}{3 \text{ acres}}$

## 2 Find each unit rate.

(a) $4.35 for 3 pounds of cheese

(b) 304 miles on 9.5 gallons of gas

(c) $850 in 5 days

(d) 24-pound turkey for 15 people

**ANSWERS**
2. (a) $1.45/pound  (b) 32 miles/gallon
   (c) $170/day  (d) 1.6 pounds/person

### Example 2 — Finding Unit Rates

Find each unit rate.

(a) 337.5 miles on 13.5 gallons of gas
Write the rate as a fraction.

$$\frac{337.5 \text{ miles}}{13.5 \text{ gallons}}$$ ← The fraction bar indicates division.

Divide 337.5 by 13.5 to find the unit rate.

$$13.5\overline{)337.5} \quad 25.$$

$$\frac{337.5 \text{ miles} \div 13.5}{13.5 \text{ gallons} \div 13.5} = \frac{25 \text{ miles}}{1 \text{ gallon}}$$

The unit rate is 25 miles **per** gallon, or 25 miles/gallon.

(b) 549 miles in 18 hours

$$\frac{549 \text{ miles}}{18 \text{ hours}} \quad \text{Divide.} \quad 18\overline{)549.0} \quad 30.5$$

The unit rate is 30.5 miles per hour, or 30.5 miles/hour.

(c) $810 in 6 days

$$\frac{810 \text{ dollars}}{6 \text{ days}} \quad \text{Divide.} \quad 6\overline{)810} \quad 135$$

The unit rate is $135 per day, or $135/day.

**Work Problem 2 at the Side.**

### 3 ▬ Find the best buy based on cost per unit.
When shopping for groceries, household supplies, and health and beauty items, you will find many different brands and package sizes. You can save money by finding the lowest *cost per unit*.

#### Cost per Unit

**Cost per unit** is a rate that tells how much you pay for *one* item or *one* unit. Examples are $1.65 per gallon, $47 per shirt, and $2.98 per pound.

### Example 3 — Determining the Best Buy

The local store charges the following prices for pancake syrup. Find the best buy.

Continued on Next Page

The best buy is the container with the *lowest* cost per unit. All the containers are measured in *ounces* (oz), so you first need to find the *cost per ounce* for each one. Divide the price of the container by the number of ounces in it. Round to the nearest thousandth if necessary.

| Size | Cost per Unit (Rounded) |
|---|---|
| 12 ounces | $\dfrac{\$1.28}{12 \text{ ounces}} \approx \$0.107$ per ounce (highest) |
| 24 ounces | $\dfrac{\$1.81}{24 \text{ ounces}} \approx \$0.075$ per ounce (lowest) |
| 36 ounces | $\dfrac{\$2.73}{36 \text{ ounces}} \approx \$0.076$ per ounce |

The lowest cost per ounce is $0.075, so the 24-ounce container is the best buy.

**NOTE**

Earlier we rounded money amounts to the nearest hundredth (nearest cent). But when comparing unit costs, rounding to the nearest thousandth will help you see the difference between very similar unit costs. Notice that the 24-ounce and 36-ounce syrup containers would both have rounded to $0.08 per ounce if we had rounded to hundredths.

**③** Find the best buy (lowest cost per unit) for each purchase.

(a) 2 quarts for $3.25
3 quarts for $4.95
4 quarts for $6.48

Work Problem **③** at the Side.

**Calculator Tip** When using a calculator to find unit prices, remember that division is *not* commutative. In Example 3 you wanted to find cost per ounce. Let the *order* of the *words* help you enter the numbers in the correct order.

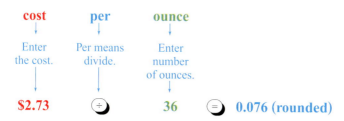

If you reversed the order and entered 36 ÷ 2.73 =, the result is the number of *ounces* per *dollar*. How could you use that information to find the best buy? (*Answer:* The best buy would be the greatest number of ounces per dollar.)

Finding the best buy is sometimes a complicated process. Things that affect the cost per unit can include "cents off" coupons and differences in how much use you'll get out of each unit.

(b) 6 cans of cola for $1.99
12 cans of cola for $3.49
24 cans of cola for $7

**ANSWERS**

**3.** (a) 4 quarts, at $1.62 per quart
(b) 12 cans, at about $0.291 per can

**4** Solve each problem.

(a) Some batteries claim to last longer than others. If you believe these claims, which brand is the best buy?

　　Four-pack of AA size batteries for $2.79

　　One AA size battery for $1.19; lasts twice as long

(b) Which tube of toothpaste is the best buy? You have a coupon for 85¢ off Brand C and a coupon for 20¢ off Brand D.

　　Brand C is $3.89 for 6 ounces.

　　Brand D is $1.59 for 2.5 ounces.

**Answers**

4. (a) One battery that lasts twice as long (like getting two) is the best buy. The cost per unit is $0.595 per battery. The four-pack is about $0.698 per battery.
   (b) Brand C with the 85¢ coupon is the best buy at about $0.507 per ounce. Brand D with the 20¢ coupon is $0.556 per ounce.

---

### Example 4 Solving Best Buy Applications

Solve each application problem.

(a) There are many brands of liquid laundry detergent. If you feel they all do a good job of cleaning your clothes, you can base your purchase on cost per unit. But some brands are "concentrated" so you can use less detergent for each load of clothes. Which of the choices shown below is the best buy?

try **SUDZY** to clean your clothes!
50 fluid ounces for $3.99
Does same number of washloads as the old 64-ounce bottle!

**WHITE-O** gets out ALL the stains!
One gallon (128 ounces) for $9.89
Does twice the washloads of the old gallon bottle!

To find Sudzy's unit cost, divide $3.99 by 64 ounces, not 50 ounces. You're getting as many clothes washed as if you bought 64 ounces. Similarly, to find White-O's unit cost, divide $9.89 by 256 ounces (twice 128 ounces, or 2 • 128 ounces = 256 ounces).

$$\text{Sudzy} \quad \frac{\$3.99}{64 \text{ ounces}} \approx \$0.062 \text{ per ounce}$$

$$\text{White-O} \quad \frac{\$9.89}{256 \text{ ounces}} \approx \$0.039 \text{ per ounce}$$

White-O has the lower cost per ounce and is the best buy. (However, if you try it and it really doesn't get out all the stains, Sudzy may be worth the extra cost.)

(b) "Cents-off" coupons also affect the best buy. Suppose you are looking at these choices for "extra strength" aspirin.

　　Brand X is $2.29 for 50 tablets.

　　Brand Y is $10.75 for 200 tablets.

You have a 40¢ coupon for Brand X and a 75¢ coupon for Brand Y. Which choice is the best buy?

To find the best buy, first subtract the coupon amounts, then divide to find the lowest cost per ounce.

$$\text{Brand X costs} \quad \$2.29 - \$0.40 = \$1.89$$

$$\frac{\$1.89}{50 \text{ tablets}} \approx \$0.038 \text{ per tablet}$$

$$\text{Brand Y costs} \quad \$10.75 - \$0.75 = \$10.00$$

$$\frac{\$10.00}{200 \text{ tablets}} = \$0.05 \text{ per tablet}$$

Brand X has the lower cost per tablet and is the best buy.

**Work Problem 4 at the Side.**

## 6.2 Exercises

*Write each rate as a fraction in lowest terms. See Example 1.*

1. 10 cups for 6 people
2. $12 for 30 pens
3. 15 feet in 35 seconds
4. 100 miles in 30 hours
5. 14 people for 28 dresses
6. 12 wagons for 48 horses
7. 25 letters in 5 minutes
8. 68 pills for 17 people
9. $63 for 6 visits
10. 25 doctors for 310 patients
11. 72 miles on 4 gallons
12. 132 miles on 8 gallons

*Find each unit rate. See Example 2.*

13. $60 in 5 hours
14. $2500 in 20 days
15. 50 eggs from 10 chickens
16. 36 children from 12 families
17. 7.5 pounds for 6 people
18. 44 bushels from 8 trees
19. $413.20 for 4 days
20. $74.25 for 9 hours

*Earl kept the following record of the gas he bought for his car. For each entry, find the number of miles he traveled and the unit rate. Round your answers to the nearest tenth.*

|     | Date | Odometer at Start | Odometer at End | Miles Traveled | Gallons Purchased | Miles per Gallon |
|-----|------|-------------------|------------------|----------------|-------------------|------------------|
| 21. | 2/4  | 27,432.3          | 27,758.2         |                | 15.5              |                  |
| 22. | 2/9  | 27,758.2          | 28,058.1         |                | 13.4              |                  |
| 23. | 2/16 | 28,058.1          | 28,396.7         |                | 16.2              |                  |
| 24. | 2/20 | 28,396.7          | 28,704.5         |                | 13.3              |                  |

*Source:* Author's car records.

*Find the best buy (based on the cost per unit) for each item. See Example 3. (Source: Piggly Wiggly.)*

**25.** Black pepper

**26.** Shampoo

**27.** Cereal
13 ounces for $2.80
15 ounces for $3.15
18 ounces for $3.98

**28.** Soup
2 cans for $0.95
3 cans for $1.45
5 cans for $2.29

**29.** Chunky peanut butter
12 ounces for $1.29
18 ounces for $1.79
28 ounces for $3.39
40 ounces for $4.39

**30.** Pork and beans
8 ounces for $0.37
16 ounces for $0.77
21 ounces for $0.99
31 ounces for $1.50

**31.** Suppose you are choosing between two brands of chicken noodle soup. Brand A is $0.48 per can and Brand B is $0.58 per can. But Brand B has more chunks of chicken in it. Which soup is the best buy? Explain your choice.

**32.** A small bag of potatoes costs $0.19 per pound. A large bag costs $0.15 per pound. But there are only two people in your family, so half the large bag would probably rot before you use it up. Which bag is the best buy? Explain.

*Solve each application problem. See Examples 2–4.*

**33.** Makesha lost 10.5 pounds in six weeks. What was her rate of loss in pounds per week?

**34.** Enrique's taco recipe uses four pounds of meat to feed 10 people. Give the rate in pounds per person.

**35.** Russ works 7 hours to earn $85.82. What is his pay rate per hour?

**36.** Find the cost of 1 gallon of gas if 18 gallons cost $26.28.

*The table lists data on the top three individual scoring basketball games of all time. Use the data to answer Exercises 37 and 38. Round answers to the nearest tenth.*

| Player | Date | Points | Min |
|---|---|---|---|
| Wilt Chamberlain | 3/2/62 | 100 | 48 |
| David Thompson | 4/9/78 | 73 | 43 |
| David Robinson | 4/24/94 | 71 | 44 |

*Source:* Elias Sports Bureau.

Wilt Chamberlain

**37.** What was Wilt Chamberlain's scoring rate in points per minute and in minutes per point?

**38.** Find David Robinson's scoring rate in points per minute and minutes per point.

**39.** In the 2000 Olympics, Michael Johnson ran the 400-meter event in a record time of approximately 44 seconds (actually 43.84 seconds). Give his rate in seconds per meter and in meters per second. Use 44 seconds as the time. (*Source:* www.Olympics.com)

**40.** Sofia can clean and adjust five hearing aids in four hours. Give her rate in hearing aids per hour and in hours per hearing aid.

**41.** A long-distance phone service advertised that all calls were 5¢ a minute, with a 50¢ minimum per completed call. (*Source:* VarTech Telecom, Inc.) Find the actual cost per minute for: **(a)** a three-minute call, **(b)** a four-minute call, and **(c)** a six-minute call. Round your answers to the nearest tenth of a cent.

**42.** Another long-distance phone service advertised a rate of 7¢ a minute, with no minimum per call, plus a $5.95 monthly fee. (*Source:* AT&T.) Find the actual cost per minute during one month if you make long-distance calls totaling **(a)** 10 minutes, **(b)** 30 minutes, or **(c)** 60 minutes. Round your answers to the nearest cent.

**43.** If you believe the claims that some batteries last longer, which is the best buy?

**44.** Which is the best buy, assuming these laundry detergents both clean equally well?

**45.** Three brands of cornflakes are available. Brand G is priced at $2.39 for 10 ounces. Brand K is $3.99 for 20.3 ounces and Brand P is $3.39 for 16.5 ounces. You have a coupon for 50¢ off Brand P and a coupon for 60¢ off Brand G. Which cereal is the best buy based on cost per unit?

**46.** Two brands of facial tissue are available. Brand K is on special at three boxes of 175 tissues each for $5. Brand S is priced at $1.29 per box of 125 tissues. You have a coupon for 20¢ off one box of Brand S and a coupon for 45¢ off one box of Brand K. How can you get the best buy on one box of tissue?

### RELATING CONCEPTS (Exercises 47–50)  FOR INDIVIDUAL OR GROUP WORK

*On the first page of this chapter, we said that unit rates can help you get the best deal on cell phone service. Use the information in the table to* **work Exercises 47–50 in order.**

| Company | Anytime Minutes | Weekend Minutes | Monthly Charge | Other Costs |
|---|---|---|---|---|
| Verizon | 500 | 1000 | $35 | $6.95 monthly access fee |
| Qwest | 400 | 1000 | $39.99 | $25 one-time activation fee |
| VoiceStream | 75 | 250 | $19.99 | None |
| Sprint PCS | 250 | 250 | $29.95 | None |

*Source:* Advertisements appearing in *Minneapolis Star Tribune*.

**Notes:**
1. All companies require that you sign up for 12 months of service and charge a $150 fee if you quit early.
2. Weekend minutes can only be used from midnight Friday to midnight Sunday.
3. Unused minutes cannot be carried over to the next month.

**47.** If you sign up for one year of service with Qwest, how much will the activation fee increase your average monthly charge?

**48.** Find the actual average cost per minute for each company, including "other costs." Assume you use all the minutes and no more. Decide how to round your answers so you can find the best buy.

**49.** How many hours is 1000 minutes? Over the course of a year, what is the average number of "weekend days" per month? How many hours would you have to talk on each "weekend day" to use up 1000 minutes per month? (Round answers to nearest tenth.)

**50.** Suppose that after two months you canceled your service because you found that you only used 100 minutes per month. Under those conditions, find the actual cost per minute for each company, to the nearest cent.

## 6.3 Proportions

**OBJECTIVES**
1. Write proportions.
2. Determine whether proportions are true or false.
3. Find the unknown number in a proportion.

**1** Write proportions. A **proportion** states that two ratios (or rates) are equivalent. For example,

$$\frac{\$20}{4 \text{ hours}} = \frac{\$40}{8 \text{ hours}}$$

is a proportion that says the rate $\frac{\$20}{4 \text{ hours}}$ is equivalent to the rate $\frac{\$40}{8 \text{ hours}}$. As the amount of money doubles, the number of hours also doubles. This proportion is read:

20 dollars **is to** 4 hours   **as**   40 dollars **is to** 8 hours.

### Example 1  Writing Proportions

Write each proportion.

**(a)** 6 feet is to 11 feet **as** 18 feet is to 33 feet.

$$\frac{6 \text{ feet}}{11 \text{ feet}} = \frac{18 \text{ feet}}{33 \text{ feet}} \quad \text{so} \quad \frac{6}{11} = \frac{18}{33} \quad \text{The common units (feet) divide out and are not written.}$$

**(b)** $9 is to 6 liters **as** $3 is to 2 liters.

$$\frac{\$9}{6 \text{ liters}} = \frac{\$3}{2 \text{ liters}} \quad \text{Units must be written.}$$

=== Work Problem **1** at the Side.

**1** Write each proportion.

(a) $7 is to 3 cans as $28 is to 12 cans.

(b) 9 meters is to 16 meters as 18 meters is to 32 meters.

(c) 5 is to 7 as 35 is to 49.

(d) 10 is to 30 as 60 is to 180.

**2** Determine whether proportions are true or false. There are two ways to see whether a proportion is true. One way is to *write both of the ratios in lowest terms*.

### Example 2  Writing Both Ratios in Lowest Terms

Are the following proportions true?

**(a)** $\frac{5}{9} = \frac{18}{27}$

Write each ratio in lowest terms.

$\frac{5}{9}$ ← Already in lowest terms    $\frac{18 \div 9}{27 \div 9} = \frac{2}{3}$ ← Lowest terms

Because $\frac{5}{9}$ is *not* equivalent to $\frac{2}{3}$, the proportion is *false*. The ratios are *not* proportional.

**(b)** $\frac{16}{12} = \frac{28}{21}$

Write each ratio in lowest terms.

$$\frac{16 \div 4}{12 \div 4} = \frac{4}{3} \quad \text{and} \quad \frac{28 \div 7}{21 \div 7} = \frac{4}{3}$$

Both ratios are equivalent to $\frac{4}{3}$, so the proportion is *true*. The ratios are proportional.

**2** Determine whether each proportion is true or false by writing both ratios in lowest terms.

(a) $\frac{6}{12} = \frac{15}{30}$

(b) $\frac{20}{24} = \frac{3}{4}$

(c) $\frac{25}{40} = \frac{30}{48}$

(d) $\frac{35}{45} = \frac{12}{18}$

(e) $\frac{21}{45} = \frac{56}{120}$

**ANSWERS**

1. (a) $\frac{\$7}{3 \text{ cans}} = \frac{\$28}{12 \text{ cans}}$  (b) $\frac{9}{16} = \frac{18}{32}$
   (c) $\frac{5}{7} = \frac{35}{49}$  (d) $\frac{10}{30} = \frac{60}{180}$
2. (a) true  (b) false  (c) true
   (d) false  (e) true

=== Work Problem **2** at the Side.

A second way to see whether a proportion is true is to find *cross products*.

### Using Cross Products to Determine Whether a Proportion Is True

To see whether a proportion is true, first multiply along one diagonal, then multiply along the other diagonal, as shown here.

$$5 \cdot 4 = 20$$
$$\frac{2}{5} = \frac{4}{10}$$
$$2 \cdot 10 = 20$$

Cross products are equal.

In this case the **cross products** are both 20. When cross products are *equal*, the proportion is *true*. If the cross products are *unequal*, the proportion is *false*.

**NOTE**

Why does the cross products test work? It is based on rewriting both fractions with the common denominator of 5 · 10 or 50. (We do not search for the *lowest* common denominator. We simply use the product of the two given denominators.)

$$\frac{2 \cdot 10}{5 \cdot 10} = \frac{20}{50} \quad \text{and} \quad \frac{4 \cdot 5}{10 \cdot 5} = \frac{20}{50}$$

We see that $\frac{2}{5}$ and $\frac{4}{10}$ are equivalent because both can be rewritten as $\frac{20}{50}$. The cross products test takes a shortcut by comparing only the two numerators (20 = 20).

**Example 3** Using Cross Products

Use cross products to see whether each proportion is true or false.

**(a)** $\frac{3}{5} = \frac{12}{20}$  Multiply along one diagonal and then along the other diagonal.

$$5 \cdot 12 = 60$$
$$\frac{3}{5} = \frac{12}{20}$$
$$3 \cdot 20 = 60$$

Equal

The cross products are *equal*, so the proportion is *true*.

*Continued on Next Page*

**(b)** $\dfrac{2\frac{1}{3}}{3\frac{1}{3}} = \dfrac{9}{16}$   Find the cross products.

$$\dfrac{2\frac{1}{3}}{3\frac{1}{3}} \:\times\: \dfrac{9}{16}$$

Changed to improper fractions

$$3\frac{1}{3} \cdot 9 = \dfrac{10}{\cancel{3}} \cdot \dfrac{\cancel{9}^{3}}{1} = \dfrac{30}{1} = 30$$

$$2\frac{1}{3} \cdot 16 = \dfrac{7}{3} \cdot \dfrac{16}{1} = \dfrac{112}{3} = 37\frac{1}{3}$$

Unequal

The cross products are *unequal,* so the proportion is *false.*

### NOTE

The numbers in a proportion do *not* have to be whole numbers. They may be fractions, mixed numbers, decimal numbers, and so on.

**Work Problem ❸ at the Side.**

**❸ Find the unknown number in a proportion.** Four numbers are used in a proportion. If any three of these numbers are known, the fourth can be found. For example, find the unknown number that will make this proportion true.

$$\dfrac{3}{5} = \dfrac{x}{40}$$

The variable *x* represents the unknown number. Start by finding the cross products.

$$\dfrac{3}{5} \:\times\: \dfrac{x}{40} \quad \begin{array}{l} 5 \cdot x \\ 3 \cdot 40 \end{array} \right] \text{Cross products}$$

To make the proportion true, the cross products must be equal. This gives us the following equation.

$$\underbrace{5 \cdot x}_{5x} = \underbrace{3 \cdot 40}_{120}$$

Recall from **Section 2.4** that we can solve an equation of this type by dividing both sides by the coefficient of the variable term. In this case, the coefficient of $5x$ is 5.

$$\dfrac{5x}{5} = \dfrac{120}{5} \quad \leftarrow \text{Divide both sides by 5.}$$

Divide out the common factor of 5.

$$\dfrac{\cancel{5}^{1} \cdot x}{\cancel{5}_{1}} = 24 \quad \leftarrow \text{Divide 120 by 5 to get 24.}$$

---

**❸** Find the cross products to see whether each proportion is true or false.

**(a)** $\dfrac{5}{9} = \dfrac{10}{18}$

**(b)** $\dfrac{32}{15} = \dfrac{16}{8}$

**(c)** $\dfrac{10}{17} = \dfrac{20}{34}$

**(d)** $\dfrac{2.4}{6} \:\times\: \dfrac{5}{12}$ $\quad\begin{array}{l}(6)(5) = \\ (2.4)(12) =\end{array}$

**(e)** $\dfrac{3}{4.25} = \dfrac{24}{34}$

**(f)** $\dfrac{1\frac{1}{6}}{2\frac{1}{3}} = \dfrac{4}{8}$

**ANSWERS**

**3.** **(a)** true  **(b)** false  **(c)** true
  **(d)** $(6)(5) = 30$; $(2.4)(12) = 28.8$; false
  **(e)** true  **(f)** true

Multiplying by 1 does *not* change a number, so in the numerator on the left side, 1 • *x* is the same as *x*.

$$\frac{x}{1} = 24$$

Dividing by 1 does *not* change a number, so $\frac{x}{1}$ is the same as *x*.

$$x = 24$$

The unknown number in the proportion is 24. The complete proportion is shown below.

$$\frac{3}{5} = \frac{24}{40} \leftarrow x \text{ is } 24.$$

*Check* by finding the cross products. If they are equal, you solved the problem correctly. If they are unequal, rework the problem.

$$\frac{3}{5} = \frac{24}{40}$$

5 • 24 = **120**
3 • 40 = **120**

Equal; proportion is true.

The cross products are equal, so the solution, *x* = 24, is correct.

### CAUTION

The solution is 24, which is the unknown number in the proportion. 120 is *not* the solution; it is the cross product you get when *checking* the solution.

*Solve* a proportion for an unknown number by using the following steps.

### Solving a Proportion to Find an Unknown Number

*Step 1* Find the cross products.

*Step 2* Show that the cross products are equivalent.

*Step 3* Divide both sides of the equation by the coefficient of the variable term.

*Step 4* Check by writing the solution in the proportion and finding the cross products.

**Example 4** Solving Proportions

Find the unknown number in each proportion. Round answers to the nearest hundredth when necessary.

(a) $\frac{16}{x} = \frac{32}{20}$

Recall that ratios can be rewritten in lowest terms. If desired, you can do that *before* finding the cross products. In this example, write $\frac{32}{20}$ in lowest terms as $\frac{8}{5}$, which gives the proportion $\frac{16}{x} = \frac{8}{5}$.

*Step 1*

$$\frac{16}{x} = \frac{8}{5}$$

*x* • 8
16 • 5

Find the cross products.

*Continued on Next Page*

*Step 2*  $x \cdot 8 = \underline{16 \cdot 5}$ ← Show that the cross products are equivalent.
$x \cdot 8 = 80$

*Step 3*  $\dfrac{x \cdot \cancel{8}}{\cancel{8}} = \dfrac{80}{8}$  ← Divide both sides by 8.

$x = 10$  ← Find $x$. (No rounding is necessary.)

*Step 4*  Write the solution in the proportion and check by finding the cross products.

$x$ is 10. →  $\dfrac{16}{10} = \dfrac{8}{5}$

$10 \cdot 8 = 80$
$16 \cdot 5 = 80$  } Equal; proportion is true.

The cross products are equal, so 10 is the correct solution.

**NOTE**

It is not necessary to write the ratios in lowest terms before solving. However, if you do, you will have smaller numbers to work with.

**(b)** $\dfrac{7}{12} = \dfrac{15}{x}$

$\dfrac{7}{12} = \dfrac{15}{x}$   $\begin{array}{l} 12 \cdot 15 = 180 \\ 7 \cdot x \end{array}$ } Find the cross products.

Show that the cross products are equivalent.

$7 \cdot x = 180$

Divide both sides by 7.

$\dfrac{\cancel{7} \cdot x}{\cancel{7}} = \dfrac{180}{7}$

$x \approx 25.71$ ← Rounded to nearest hundredth

When the division does not come out even, check for directions on how to round your answer. Divide out one more place, then round.

$\begin{array}{r} 25.714 \\ 7\overline{)180.000} \end{array}$ ← Divide out to thousandths. Round to hundredths.

Write the solution in the proportion and check by finding the cross products.

$\dfrac{7}{12} \approx \dfrac{15}{25.71}$   $\begin{array}{l} (12)(15) = 180 \\ (7)(25.71) = 179.97 \end{array}$ } Very close, but not equal

The cross products are slightly different because you rounded the value of $x$. However, they are close enough to see that the problem was done correctly and 25.71 is the approximate solution.

═══ Work Problem ❹ at the Side.

❹ Find the unknown numbers. Round to hundredths when necessary. Check your answers by finding the cross products.

(a) $\dfrac{1}{2} = \dfrac{x}{12}$

(b) $\dfrac{6}{10} = \dfrac{15}{x}$

(c) $\dfrac{28}{x} = \dfrac{21}{9}$

(d) $\dfrac{x}{8} = \dfrac{3}{5}$

(e) $\dfrac{14}{11} = \dfrac{x}{3}$

**ANSWERS**
4. (a) $x = 6$  (b) $x = 25$
   (c) $x = 12$  (d) $x = 4.8$
   (e) $x \approx 3.82$ (rounded to nearest hundredth)

The next example shows how to solve for the unknown number in a proportion with fractions or decimals.

### Example 5 — Solving Proportions with Mixed Numbers and Decimals

Find the unknown number in each proportion.

(a) $\dfrac{2\frac{1}{5}}{6} = \dfrac{x}{10}$

Find the cross products.

$6 \cdot x$

$2\frac{1}{5} \cdot 10$

Find $2\frac{1}{5} \cdot 10$.

$$2\frac{1}{5} \cdot 10 = \frac{11}{5} \cdot \frac{10}{1} = \frac{11 \cdot 2 \cdot \cancel{5}}{\cancel{5} \cdot 1} = \frac{22}{1} = 22$$

Changed to improper fraction

Show that the cross products are equivalent.

$6 \cdot x = 22$

Divide both sides by 6.

$$\dfrac{\cancel{6} \cdot x}{\cancel{6}} = \dfrac{22}{6}$$

Write the answer as a mixed number in lowest terms.

$$x = \dfrac{22 \div 2}{6 \div 2} = \dfrac{11}{3} = 3\frac{2}{3}$$

Write the solution in the proportion and check by finding the cross products.

$$6 \cdot 3\frac{2}{3} = \dfrac{2 \cdot \cancel{3}}{1} \cdot \dfrac{11}{\cancel{3}} = \dfrac{22}{1} = 22$$

$\dfrac{2\frac{1}{5}}{6} = \dfrac{3\frac{2}{3}}{10}$   Equal

$$2\frac{1}{5} \cdot 10 = \dfrac{11}{\cancel{5}} \cdot \dfrac{2 \cdot \cancel{5}}{1} = \dfrac{22}{1} = 22$$

The cross products are equal, so $3\frac{2}{3}$ is the correct solution.

**Continued on Next Page**

## NOTE

You can use decimal numbers and your calculator to solve Example 5(a). $2\frac{1}{5}$ is equivalent to 2.2, so the cross products are

$$6 \cdot x = (2.2)(10)$$

$$\frac{\overset{1}{\cancel{6}} \cdot x}{\underset{1}{\cancel{6}}} = \frac{22}{6}$$

When you divide 22 by 6 on your calculator, it shows 3.666666667. Write the answer using a bar to show the repeating digit: $3.\overline{6}$. Or round the answer to 3.67 (nearest hundredth).

**(b)** $\dfrac{1.5}{0.6} = \dfrac{2}{x}$

Show that cross products are equivalent.

$$(1.5)(x) = (0.6)(2)$$
$$(1.5)(x) = 1.2$$

Divide both sides by 1.5.

$$\frac{\overset{1}{\cancel{(1.5)}} (x)}{\underset{1}{\cancel{1.5}}} = \frac{1.2}{1.5}$$

$$x = \frac{1.2}{1.5} \qquad \text{Complete the division.}$$

$$x = 0.8 \qquad \quad 1.5\overline{)1.20}^{\,.8}$$

So the unknown number is 0.8. Check by finding the cross products.

$$\frac{1.5}{0.6} = \frac{2}{0.8} \quad \begin{matrix} (0.6)(2) = \mathbf{1.2} \\ \\ (1.5)(0.8) = \mathbf{1.2} \end{matrix} \Bigg\} \text{Equal}$$

The cross products are equal, so 0.8 is the correct solution.

**Work Problem ❺ at the Side.**

---

❺ Find the unknown numbers. Round to hundredths on the decimal problems when necessary. Check your answers by finding the cross products.

**(a)** $\dfrac{3\frac{1}{4}}{2} = \dfrac{x}{8}$

**(b)** $\dfrac{x}{3} = \dfrac{1\frac{2}{3}}{5}$

**(c)** $\dfrac{0.06}{x} = \dfrac{0.3}{0.4}$

**(d)** $\dfrac{2.2}{5} = \dfrac{13}{x}$

**(e)** $\dfrac{x}{6} = \dfrac{0.5}{1.2}$

**(f)** $\dfrac{0}{2} = \dfrac{x}{7.092}$

**ANSWERS**

5. **(a)** $x = 13$ **(b)** $x = 1$ **(c)** $x = 0.08$
   **(d)** $x \approx 29.55$ (rounded to nearest hundredth)
   **(e)** $x = 2.5$ **(f)** $x = 0$

# Focus on Real-Data Applications

## Tour of the West

Visits to Mount Rushmore, Devil's Tower, Yellowstone National Park, the Grand Tetons, Bryce Canyon, Zion National Park, the Painted Desert, and the Grand Canyon are highlights of a tour advertised to British citizens. (*Source:* www.archersdirect.co.uk) The itinerary is shown on the map to the right. (*Source:* www.mapquest.com)

The travel distances and times between the daily stopping points are estimates, based on information from the Web site www.mapquest.com. The table below gives the daily route, the travel distances, and the travel times between the locations.

*Source:* ©2000 MapQuest.com, Inc.; ©2000 AND Data Solutions B.V.

1. Calculate the average speed in miles per hour (mph) for each segment of the trip, rounded to the nearest whole number. Notice that you are working with rates that compare distance to time (miles to hours). (*Hint:* Divide the distance traveled by the time elapsed. You must first rewrite the time in decimal form. For example, on Day 4, 1 hr 40 min, is $1 + \frac{40}{60}$ or $1.6666666$, which rounds to $1.67$. The average speed is $56 \div 1.67 \approx 33.5$ or 34 mph.)

| Day | Location | Distance | Time | Average Speed (mph) |
|---|---|---|---|---|
| 1 | London, England to Denver, Colorado (CO) | 4693 mi | 14 hr flight | |
| 2 | Denver, CO | | | |
| 3 | Denver, CO to Custer, South Dakota (SD) | 365 mi | 8 hr | |
| 4 | Custer, SD to Lead, SD | 56 mi | 1 hr, 40 min | 34 mph |
| 5 | Lead, SD to Cody, Wyoming (WY) | 361 mi | 7 hr, 50 min | |
| 6 | Cody, WY to Yellowstone National Park, WY | 110 mi | 3 hr, 10 min | |
| 7 | Yellowstone, WY to Jackson, WY | 131 mi | 3 hr, 45 min | |
| 8 | Jackson, WY to Salt Lake City, Utah (UT) | 269 mi | 7 hr | |
| 9 | Salt Lake City, UT | | | |
| 10 | Salt Lake City, UT to Cedar City, UT | 251 mi | 4 hr, 20 min | |
| 11 | Cedar City, UT to Page, Arizona (AZ) | 155 mi | 4 hr, 30 min | |
| 12 | Page, AZ to Grand Canyon, AZ | 138 mi | 4 hr | |
| 13 | Grand Canyon, AZ to Las Vegas, Nevada (NV) | 279 mi | 6 hr, 30 min | |
| 14 | Las Vegas, NV | | | |
| 15 | Las Vegas, NV to San Francisco, CA | 420 mi | 1 hr 40 min flight | |
| | San Francisco, CA to London, England | 5376 mi | 17 hr flight | |
| | Totals (driving portion of the tour) | | | |

2. Calculate the total driving distance and total driving time during this tour. Find the overall average speed.

3. Why do you think the average speeds for this trip are not closer to 55 mph?

4. If a friend from Scotland asked your opinion about how interesting and feasible these tourist attractions would be, what advice would you give?

## 6.3 Exercises

*Write each proportion. See Example 1.*

1. $9 is to 12 cans as $18 is to 24 cans.

2. 28 people is to 7 cars as 16 people is to 4 cars.

3. 200 adults is to 450 children as 4 adults is to 9 children.

4. 150 trees is to 1 acre as 1500 trees is to 10 acres.

5. 120 feet is to 150 feet as 8 feet is to 10 feet.

6. $6 is to $9 as $10 is to $15.

7. 2.2 hours is to 3.3 hours as 3.2 hours is to 4.8 hours.

8. 4 meters is to 4.75 meters as 6 meters is to 7.125 meters.

*Determine whether each proportion is* true *or* false *by writing the ratios in lowest terms. See Example 2.*

9. $\dfrac{6}{10} = \dfrac{3}{5}$

10. $\dfrac{1}{4} = \dfrac{9}{36}$

11. $\dfrac{5}{8} = \dfrac{25}{40}$

12. $\dfrac{2}{3} = \dfrac{20}{27}$

13. $\dfrac{150}{200} = \dfrac{200}{300}$

14. $\dfrac{100}{120} = \dfrac{75}{100}$

*Use cross products to determine whether each proportion is* true *or* false. *Circle the correct answer. See Example 3.*

15. $\dfrac{2}{9} = \dfrac{6}{27}$

True    False

16. $\dfrac{20}{25} = \dfrac{4}{5}$

True    False

17. $\dfrac{20}{28} = \dfrac{12}{16}$

True    False

18. $\dfrac{16}{40} = \dfrac{22}{55}$

    True    False

19. $\dfrac{110}{18} = \dfrac{160}{27}$

    True    False

20. $\dfrac{600}{420} = \dfrac{20}{14}$

    True    False

21. $\dfrac{3.5}{4} = \dfrac{7}{8}$

    True    False

22. $\dfrac{36}{23} = \dfrac{9}{5.75}$

    True    False

23. $\dfrac{18}{16} = \dfrac{2.8}{2.5}$

    True    False

24. $\dfrac{0.26}{0.39} = \dfrac{1.3}{1.9}$

    True    False

25. $\dfrac{6}{3\frac{2}{3}} = \dfrac{18}{11}$

    True    False

26. $\dfrac{16}{13} = \dfrac{2}{1\frac{5}{8}}$

    True    False

27. Suppose Jerome Walton of the Atlanta Braves had 16 hits in 50 times at bat, and Mariano Duncan of the New York Yankees was at bat 400 times and got 128 hits. Paul is trying to convince Jamie that the two men hit equally well. Show how you could use a proportion and cross products to see whether Paul is correct.

28. Jay worked 3.5 hours and packed 91 cartons. Craig packed 126 cartons in 5.25 hours. To see if the men worked equally fast, Barry set up this proportion:

    $$\dfrac{3.5}{91} = \dfrac{126}{5.25}$$

    Explain what is wrong with Barry's proportion and write a correct one. Is the correct proportion true or false?

Solve each proportion to find the unknown number. Round your answers to hundredths when necessary. Check your answers by finding cross products. See Examples 4 and 5.

**29.** $\dfrac{1}{3} = \dfrac{x}{12}$

**30.** $\dfrac{x}{6} = \dfrac{15}{18}$

**31.** $\dfrac{15}{10} = \dfrac{3}{x}$

**32.** $\dfrac{5}{x} = \dfrac{20}{8}$

**33.** $\dfrac{x}{11} = \dfrac{32}{4}$

**34.** $\dfrac{12}{9} = \dfrac{8}{x}$

**35.** $\dfrac{42}{x} = \dfrac{18}{39}$

**36.** $\dfrac{49}{x} = \dfrac{14}{18}$

**37.** $\dfrac{x}{25} = \dfrac{4}{20}$

**38.** $\dfrac{6}{x} = \dfrac{4}{8}$

**39.** $\dfrac{8}{x} = \dfrac{24}{30}$

**40.** $\dfrac{32}{5} = \dfrac{x}{10}$

**41.** $\dfrac{99}{55} = \dfrac{44}{x}$

**42.** $\dfrac{x}{12} = \dfrac{101}{147}$

**43.** $\dfrac{0.7}{9.8} = \dfrac{3.6}{x}$

**44.** $\dfrac{x}{3.6} = \dfrac{4.5}{6}$

**45.** $\dfrac{250}{24.8} = \dfrac{x}{1.75}$

**46.** $\dfrac{4.75}{17} = \dfrac{43}{x}$

## Chapter 6 Ratio, Proportion, and Line/Angle/Triangle Relationships

*Find the unknown number in each proportion. Write your answers as whole or mixed numbers when possible. See Example 5.*

47. $\dfrac{15}{1\frac{2}{3}} = \dfrac{9}{x}$

48. $\dfrac{x}{\frac{3}{10}} = \dfrac{2\frac{2}{9}}{1}$

49. $\dfrac{2\frac{1}{3}}{1\frac{1}{2}} = \dfrac{x}{2\frac{1}{4}}$

50. $\dfrac{1\frac{5}{6}}{x} = \dfrac{\frac{3}{14}}{\frac{6}{7}}$

*Solve each proportion two different ways. First change all the numbers to decimal form and solve. Then change all the numbers to fraction form and solve; write your answers in lowest terms.*

51. $\dfrac{\frac{1}{2}}{x} = \dfrac{2}{0.8}$

52. $\dfrac{\frac{3}{20}}{0.1} = \dfrac{0.03}{x}$

53. $\dfrac{x}{\frac{3}{50}} = \dfrac{0.15}{1\frac{4}{5}}$

54. $\dfrac{8\frac{4}{5}}{1\frac{1}{10}} = \dfrac{x}{0.4}$

---

**RELATING CONCEPTS (Exercises 55–56)**   **FOR INDIVIDUAL OR GROUP WORK**

**Work Exercises 55–56 in order.** *First prove that the proportions are **not** true. Then create four true proportions for each exercise by changing only one number at a time.*

55. $\dfrac{10}{4} = \dfrac{5}{3}$

56. $\dfrac{6}{8} = \dfrac{24}{30}$

## 6.4 PROBLEM SOLVING WITH PROPORTIONS

**OBJECTIVE**
1. Use proportions to solve application problems.

**1** **Use proportions to solve application problems.** Proportions can be used to solve a wide variety of problems. Watch for problems in which you are given a ratio or rate and then asked to find part of a corresponding ratio or rate. Remember that a ratio or rate compares two quantities and often includes one of these indicator words.

in    for    on    per    from    to

Use the six problem-solving steps from **Section 3.3**. When setting up the proportion, use a variable to represent the unknown number. We have used the letter $x$, but you may use any letter you like.

### Example 1  Solving a Proportion Application

Mike's car can travel 163 **miles** on 6.4 **gallons** of gas. How far can it travel on a full tank of 14 **gallons** of gas? Round to the nearest mile.

**Step 1** **Read** the problem. It is about how far a car can travel on a certain amount of gas.

    Unknown: miles traveled on 14 gallons of gas
    Known: 163 miles traveled on 6.4 gallons of gas

**Step 2** **Assign a variable.** There is only one unknown, so let $x$ be the number of miles traveled on 14 gallons.

**Step 3** **Write an equation.** The equation is in the form of a proportion. Decide what is being compared. This problem compares **miles** to **gallons**. Write the two rates described in the problem. Be sure that *both* rates compare miles to gallons in the same order. In other words, miles is in both numerators and gallons is in both denominators.

This rate compares miles to gallons. $\quad \dfrac{163 \text{ miles}}{6.4 \text{ gallons}} = \dfrac{x \text{ miles}}{14 \text{ gallons}} \quad$ This rate compares miles to gallons.

**Step 4** **Solve** the equation. Ignore the units while finding the cross products.

$(6.4)(x) = (163)(14)$    Show that cross products are equivalent.

$(6.4)(x) = 2282$

$\dfrac{(6.4)(x)}{6.4} = \dfrac{2282}{6.4}$    Divide both sides by 6.4.

$x = 356.5625$    Round to 357.

**Step 5** **State the answer.** Mike's car can travel 357 miles, rounded to the nearest mile.

*Continued on Next Page*

**1** Set up and solve a proportion for each problem using the six problem-solving steps.

(a) If 2 pounds of fertilizer will cover 50 square feet of garden, how many pounds are needed for 225 square feet?

(b) A U.S. map has a scale of 1 inch to 75 miles. Lake Superior is 4.75 inches long on the map. What is the lake's actual length in miles?

(c) Cough syrup is to be given at the rate of 30 milliliters for each 100 pounds of body weight. How much should be given to a 34-pound child? Round to the nearest whole milliliter.

*Step 6* **Check** the solution by putting it back into the original problem. The car traveled 163 miles on 6.4 gallons of gas; 14 gallons is a little more than *twice as much* gas, so the car should travel a little more than *twice as far*.

$$(2)(163 \text{ miles}) = 326 \text{ miles} \quad \leftarrow \text{Estimate}$$

The solution, 357 miles, is just a little more than the estimate of 326 miles, so it is reasonable.

> **CAUTION**
>
> When setting up the proportion, do *not* mix up the units in the rates.
>
> $\left.\begin{array}{r}\text{compares } \textbf{miles}\\ \text{to } \textbf{gallons}\end{array}\right\}\ \dfrac{163 \textbf{ miles}}{6.4 \textbf{ gallons}} = \dfrac{14 \textbf{ gallons}}{x \textbf{ miles}}\ \left\{\begin{array}{l}\text{compares } \textbf{gallons}\\ \text{to } \textbf{miles}\end{array}\right.$
>
> These rates do *not* compare things in the same order and *cannot* be set up as a proportion.

Work Problem **1** at the Side.

### Example 2  Solving a Proportion Application

A newspaper report says that 7 out of 10 people surveyed watch the news on TV. At that rate, how many of the 3200 people in town would you expect to watch the news?

*Step 1* **Read** the problem. It is about people watching the news on TV.

Unknown: how many people in town are expected to watch the news on TV

Known: 7 out of 10 people surveyed watched the news on TV.

*Step 2* **Assign a variable.** There is only one unknown, so let $x$ be the number of people in town who watch news on TV.

*Step 3* **Write an equation.** Set up the two rates as a proportion. You are comparing people who watch the news to people surveyed. Write the two rates described in the example. Be sure that both rates make the same comparison. "People who watch the news" is mentioned first, so it should be in the numerator of *both* rates.

$\begin{array}{r}\text{People who watch news} \rightarrow\\ \text{Total group} \rightarrow\\ \text{(people surveyed)}\end{array} \dfrac{7}{10} = \dfrac{x}{3200} \begin{array}{l}\leftarrow \text{People who watch news}\\ \leftarrow \text{Total group}\\ \text{(people in town)}\end{array}$

*Step 4* **Solve** the equation.

$(10)(x) = (7)(3200)$  Show that cross products are equivalent.

$(10)(x) = 22{,}400$

$\dfrac{\cancel{(10)}^{1}(x)}{\cancel{10}_{1}} = \dfrac{22{,}400}{10}$  Divide both sides by 10.

$x = 2240$

*Continued on Next Page*

**ANSWERS**

1. (a) $\dfrac{2 \text{ pounds}}{50 \text{ sq. feet}} = \dfrac{x \text{ pounds}}{225 \text{ sq. feet}}$
$x = 9$ pounds

(b) $\dfrac{1 \text{ inch}}{75 \text{ miles}} = \dfrac{4.75 \text{ inches}}{x \text{ miles}}$
$x = 356.25$ miles or $x \approx 356$ miles

(c) $\dfrac{30 \text{ milliliters}}{100 \text{ pounds}} = \dfrac{x \text{ milliliters}}{34 \text{ pounds}}$
$x \approx 10$ milliliters

*Step 5* **State the answer.** You would expect 2240 people in town to watch the news on TV.

*Step 6* **Check** the solution by putting it back into the original problem. Notice that 7 out of 10 people is more than half the people, but less than all the people. Half of the 3200 people in town is $3200 \div 2 = 1600$, so between 1600 and 3200 people would be expected to watch the news on TV. The solution, 2240 people, is between 1600 and 3200, so it is reasonable.

**CAUTION**

Always check that your answer is reasonable. If it isn't, look at the way your proportion is set up. Be sure you have matching units in the numerators and matching units in the denominators.

For example, suppose you set up the last proportion *incorrectly*, as shown below.

$$\frac{7}{10} = \frac{3200}{x} \quad \leftarrow \text{Incorrect setup}$$

$$(7)(x) = (10)(3200)$$

$$\frac{(\cancel{7})(x)}{\cancel{7}} = \frac{32{,}000}{7}$$

$$x \approx 4571 \text{ people} \quad \leftarrow \text{Unreasonable answer}$$

This answer is *unreasonable* because there are only 3200 people in the town; it is **not** possible for 4571 people to watch the news.

═══════════ Work Problem ❷ at the Side.

❷ Solve each problem to find a reasonable answer. Then flip one side of your proportion to see what answer you get with an incorrect setup. Explain why the second answer is unreasonable.

(a) A survey showed that 2 out of 3 people would like to lose weight. At this rate, how many people in a group of 150 want to lose weight?

(b) In one state, 3 out of 5 college students receive financial aid. At this rate, how many of the 4500 students at Central Community College receive financial aid?

(c) An advertisement says that 9 out of 10 dentists recommend sugarless gum. If the ad is true, how many of the 60 dentists in our city would recommend sugarless gum?

**ANSWERS**

2. (a) 100 people (reasonable); incorrect setup gives 225 people (only 150 people in the group).
   (b) 2700 students (reasonable); incorrect setup gives 7500 students (only 4500 students at the college).
   (c) 54 dentists (reasonable); incorrect setup gives ≈67 dentists (only 60 dentists in the city).

# Focus on Real-Data Applications

## Feeding Hummingbirds

A recipe can be used to make as much of a mixture as you might need as long as the ingredients are kept proportional. Use the recipe for a homemade mixture of sugar water for hummingbird feeders to answer these problems.

1. What is the ratio of sugar to water in the recipe?

   What is the ratio of water to sugar in the recipe?

2. Complete each table.

   | Sugar | Water |
   |---|---|
   | 1 cup | 4 cups |
   |  | 5 cups |
   |  | 6 cups |
   |  | 7 cups |
   | 2 cups | 8 cups |

   | Sugar | Water |
   |---|---|
   | 1 cup | 4 cups |
   |  | 3 cups |
   |  | 2 cups |
   |  | 1 cup |

3. How much water would you need

   (a) if you wanted to use 3 cups of sugar?

   (b) if you wanted to use 4 cups of sugar?

   (c) if you wanted to use $\frac{1}{3}$ cup of sugar?

4. One cup of sugar weighs about 200 grams and one cup of water weighs about 235 grams. If you mix 1 cup of sugar and 4 cups of water, what is the approximate weight of the resulting mixture?

5. The article says that the nectar from wildflowers visited by hummingbirds has an average sugar concentration of 21 percent. That represents a ratio of 21:100. If you wanted to mix a sugar solution in the same proportion as the wildflower nectar, how much water should you use for 1 cup of sugar? What proportion did you set up to solve this problem?

### Feeding Hummingbirds

After getting a hummingbird feeder, the next step is to fill it! You have two choices at this point: you can either buy one of the commercial mixtures or you can make your own solution:

> **Recipe for Homemade Mixture:**
> 1 part sugar (not honey)
> 4 parts water
> Boil for 1 to 2 minutes. Cool.
> Store extra in refrigerator.

The concentration of the sugar is important. A 1 to 4 ratio of sugar to water is recommended because it approximates the ratio of sugar to water found in the nectar of many hummingbird flowers. A recent study of native California wildflowers visited by hummingbirds showed that their nectar had an average sugar concentration of 21 percent. This is sweet enough to attract the hummers without being too sweet. If you increase the concentration of sugar, it may be harder for the birds to digest; if you decrease the concentration, they may lose interest.

Boiling the solution helps retard fermentation. Sugar-and-water solutions are subject to rapid spoiling, especially in hot weather.

*Source: The Hummingbird Book.*

6. As you change the amounts of water and sugar, should you change the length of time that you boil the mixture? Explain your answer.

7. Will the length of time it takes to get the water hot enough to start boiling change? Explain your answer.

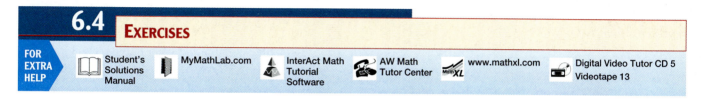

*Set up and solve a proportion for each application problem. See Example 1.*

1. Caroline can sketch four cartoon strips in five hours. How long will it take her to sketch 18 strips?

2. The Cosmic Toads recorded eight songs on their first CD in 26 hours. How long will it take them to record 14 songs for their second CD?

3. Sixty newspapers cost $27. Find the cost of 16 newspapers.

4. Twenty-two guitar lessons cost $396. Find the cost of 12 lessons.

5. If three pounds of fescue grass seed cover about 350 square feet of ground, how many pounds are needed for 4900 square feet?

6. Anna earns $1242.08 in 14 days. How much does she earn in 260 days?

7. Tom makes $455.75 in five days. How much does he make in three days?

8. If 5 ounces of a medicine must be mixed with 8 ounces of water, how many ounces of medicine would be mixed with 20 ounces of water?

9. The bag of rice noodles shown below makes 7 servings. (*Source:* Everfresh Foods.) At that rate, how many ounces of noodles do you need for 12 servings, to the nearest ounce?

10. This can of sweet potatoes is enough for four servings. (*Source:* Moody Dunbar, Inc.) How many ounces are needed for 9 servings, to the nearest ounce?

11. Three quarts of a latex enamel paint will cover about 270 square feet of wall surface. (*Source:* Thompson and Formby, Inc.) How many quarts will you need to cover 350 square feet of wall surface in your kitchen and 100 square feet of wall surface in your bathroom?

12. One gallon of clear gloss wood finish covers about 550 square feet of surface. (*Source:* The Flecto Company.) If you need to apply three coats of finish to 400 square feet of surface, how many gallons do you need, to the nearest tenth?

*Use the floor plan shown to answer Exercises 13–16. On the plan, one inch represents four feet.*

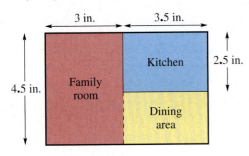

**13.** What is the actual length and width of the kitchen?

**14.** What is the actual length and width of the family room?

**15.** What is the actual length and width of the dining area?

**16.** What is the actual length and width of the entire floor plan?

**17.** Vince Carter scored 18 points during the 28 minutes that he played in the Olympic basketball game between the United States and Lithuania. (*Source:* Associated Press.) At that rate, how many points would he make if he played the entire game (40 minutes), to the nearest whole point?

**18.** In the U.S.–France gold medal Olympic basketball game, Alonzo Mourning played 26 minutes and made 7 rebounds. (*Source:* Associated Press.) How many rebounds would you expect him to make in 40 minutes (a complete Olympic game), to the nearest whole number?

*Set up a proportion to solve each problem. Check to see whether your answer is reasonable. Then flip one side of your proportion to see what answer you get with an incorrect setup. Explain why the second answer is unreasonable. See Example 2.*

**19.** About 7 out of 10 people entering our community college need to take a refresher math course. If we have 2950 entering students, how many will probably need refresher math? (*Source:* Minneapolis Community and Technical College.)

**20.** In a survey, only 3 out of 100 people like their eggs poached. At that rate, how many of the 60 customers who ordered eggs at Soon-Won's restaurant this morning asked to have them poached? Round to the nearest whole person.

**21.** Nearly 4 out of 5 people choose vanilla as their favorite ice cream flavor. (*Source:* Baskin-Robbins.) If 238 people attend an ice cream social, how many would you expect to choose vanilla? Round to the nearest whole person.

**22.** In a test of 200 sewing machines, only one had a defect. At that rate, how many of the 5600 machines shipped from the factory have defects?

**23.** About 9 out of 10 U.S. households have TV remote controls, according to a survey. There were 102,500,000 U.S. households in 1998. If the survey is accurate, how many U.S. households had TV remote controls? (*Source:* Magnavox, U.S. Bureau of the Census.)

**24.** In a survey, 3 out of 100 dog owners washed their pets by having the dogs go into the shower with them. If the survey is accurate, how many of the 31,200,000 dog owners in the United States use this method? (*Source:* Teledyne Water Pik, American Veterinary Medical Association.)

*Set up and solve a proportion for each problem.*

**25.** The stock market report says that five stocks went up for every six stocks that went down. If 750 stocks went down yesterday, how many went up?

**26.** The human body contains 90 pounds of water for every 100 pounds of body weight. How many pounds of water are in a child who weighs 80 pounds?

**27.** The ratio of the length of an airplane wing to its width is 8 to 1. If the length of a wing is 32.5 meters, how wide must it be? Round to the nearest hundredth.

**28.** The Rosebud School District wants a student-to-teacher ratio of 19 to 1. How many teachers are needed for 1850 students? Round to the nearest whole number.

**29.** The number of calories you burn depends on your weight. A 150-pound person burns 222 calories during 30 minutes of tennis. How many calories would a 210-pound person burn, to the nearest whole number? (*Source: Wellness Encyclopedia.*)

**30.** (Complete Exercise 29 first.) A 150-pound person burns 189 calories during 45 minutes of grocery shopping. How many calories would a 115-pound person burn, to the nearest whole number? (*Source: Wellness Encyclopedia.*)

**31.** At 3 P.M., Coretta's shadow is 1.05 meters long. Her height is 1.68 meters. At the same time, a tree's shadow is 6.58 meters long. How tall is the tree? Round to the nearest hundredth.

**32.** (Complete Exercise 31 first.) Later in the day, Coretta's shadow was 2.95 meters long. How long a shadow did the tree have at that time? Round to the nearest hundredth.

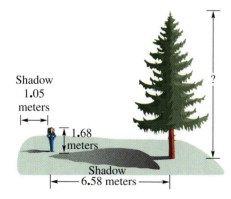

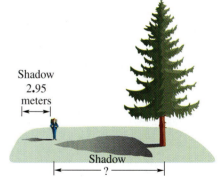

33. Can you set up a proportion to solve this problem? Explain why or why not. Jim is 25 years old and weighs 180 pounds. How much will he weigh when he is 50 years old?

34. Write your own application problem that can be solved by setting up a proportion. Also show the proportion and the steps needed to solve your problem.

35. A survey of college students shows that 4 out of 5 drink coffee. Of the students who drink coffee, 1 out of 8 adds cream to it. How many of the 46,000 students at the University of Minnesota would be expected to use cream in their coffee?

36. About 9 out of 10 adults think it is a good idea to exercise regularly. But of the ones who think it is a good idea, only 1 in 6 actually exercise at least three times a week. At this rate, how many of the 300 employees in our company exercise regularly?

37. The nutrition information on a bran cereal box says that a $\frac{1}{3}$ cup serving provides 80 calories and 8 grams of dietary fiber. (*Source:* Kraft Foods, Inc.) At that rate, how many calories and grams of fiber are in a $\frac{1}{2}$ cup serving?

38. A $\frac{2}{3}$ cup serving of penne pasta has 210 calories and 2 grams of dietary fiber. (*Source:* Borden Foods.) How many calories and grams of fiber would be in a 1 cup serving?

### RELATING CONCEPTS (Exercises 39–42) FOR INDIVIDUAL OR GROUP WORK

*A box of instant mashed potatoes has the list of ingredients shown in the table. Use this information to* **work Exercises 39–42 in order.**

| Ingredient | For 12 Servings |
| --- | --- |
| Water | $3\frac{1}{2}$ cups |
| Margarine | 6 tablespoons |
| Milk | $1\frac{1}{2}$ cups |
| Potato flakes | 4 cups |

*Source:* General Mills.

39. Find the amount of each ingredient for six servings. Show *two* different methods for finding the amounts. One method should use proportions.

40. Find the amount of each ingredient for 18 servings. Show *two* different methods for finding the amounts, one using proportions and one using your answers from Exercise 39.

41. Find the amount of each ingredient for three servings, using your answers from either Exercise 39 or Exercise 40.

42. Find the amount of each ingredient for nine servings, using your answers from either Exercise 40 or Exercise 41.

# 6.5 Geometry: Lines and Angles

Geometry starts with the idea of a point. A **point** can be described as a location in space. It has no length or width. A point is represented by a dot and is named by writing a capital letter next to the dot.

Point P

**1** **Identify lines, line segments, and rays.** A **line** is a straight row of points that goes on forever in both directions. A line is drawn by using arrowheads to show that it never ends. The line is named by using the letters of any two points on the line.

Line AB, written $\overleftrightarrow{AB}$

A piece of a line that has two endpoints is called a **line segment**. A line segment is named for its endpoints. The segment with endpoints P and Q is shown below. It can be named $\overline{PQ}$ or $\overline{QP}$.

Line segment PQ, written $\overline{PQ}$

A **ray** is a part of a line that has only one endpoint and goes on forever in one direction. A ray is named by using the endpoint and some other point on the ray. The endpoint is always mentioned first.

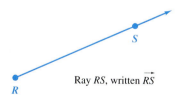
Ray RS, written $\overrightarrow{RS}$

### OBJECTIVES

1. Identify lines, line segments, and rays.
2. Identify parallel and intersecting lines.
3. Identify and name angles.
4. Classify angles as right, acute, straight, or obtuse.
5. Identify perpendicular lines.
6. Identify complementary and supplementary angles and find the measure of a complement or supplement of a given angle.
7. Identify congruent and vertical angles and use this knowledge to find the measures of angles.
8. Identify corresponding and alternate interior angles and use this knowledge to find the measures of angles.

**Example 1** Identifying Lines, Rays, and Line Segments

Identify each figure as a line, line segment, or ray.

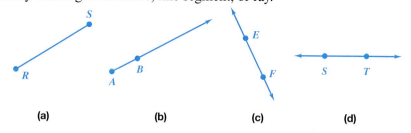

(a)  (b)  (c)  (d)

Figure **(a)** has two endpoints, so it is a line segment.

Figure **(b)** starts at point A and goes on forever in one direction, so it is a ray.

Figures **(c)** and **(d)** go on forever in both directions, so they are lines.

**Work Problem 1 at the Side.**

**1** Identify each figure as a line, line segment, or ray.

(a)

(b)

(c)
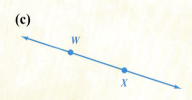

(d)
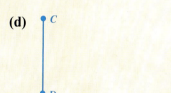

**ANSWERS**

1. **(a)** line segment **(b)** ray **(c)** line
   **(d)** line segment

**458** Chapter 6 Ratio, Proportion, and Line/Angle/Triangle Relationships

**❷** Label each pair of lines that appear to be parallel, and each pair that are intersecting.

(a)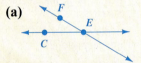

**2** **Identify parallel and intersecting lines.** A *plane* is an infinitely large flat surface. A floor or a wall is part of a plane. Lines that are in the *same plane*, but that never intersect (never cross), are called **parallel lines**, while lines that cross are called **intersecting lines**. (Think of an intersection, where two streets cross each other.)

**Example 2** Identifying Parallel and Intersecting Lines

Label each pair of lines that appear to be parallel, and each pair that are intersecting.

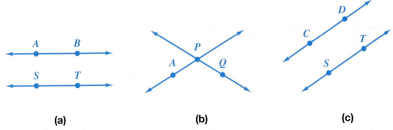

The lines in Figures **(a)** and **(c)** do not intersect; they appear to be parallel lines. The lines in Figure **(b)** cross at *P*, so they are intersecting lines.

(b)

**CAUTION**

Appearances may be deceiving! Do not assume lines are parallel unless it is stated that they are parallel.

**Work Problem ❷ at the Side.**

**3** **Identify and name angles.** An **angle** is made up of two rays that start at a common endpoint. This common endpoint is called the *vertex*.

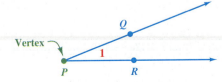

The rays *PQ* and *PR* are called *sides*. The angle can be named in four different ways, as shown below.

(c)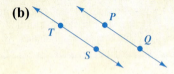

**Naming an Angle**

To name an angle, write the vertex alone or write the vertex in the middle of two other points, one from each side. If two or more angles have the *same vertex*, as in Example 3 on the next page, do *not* use the vertex alone to name an angle.

**ANSWERS**
**2.** (a) intersecting  (b) appear to be parallel
 (c) appear to be parallel

Section 6.5 Geometry: Lines and Angles   459

**Example 3** Identifying and Naming an Angle

Name the highlighted angle.

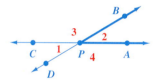

The angle can be named ∠BPA, ∠APB, or ∠2. It cannot be named ∠P, using the vertex alone, because four different angles have P as their vertex.

━━━━━━━━━━━━━━━ Work Problem ❸ at the Side.

**4** Classify angles as right, acute, straight, or obtuse. Angles can be measured in **degrees**. The symbol for degrees is a small, raised circle °. Think of the minute hand on a clock as a ray of an angle. Suppose it is at 12:00. During one hour of time, the minute hand moves around in a complete circle. It moves 360 **degrees**, or 360°. In half an hour, at 12:30, the minute hand has moved half way around the circle, or 180°. An angle of 180° is called a **straight angle.** When two rays go in opposite directions, the rays form a straight angle.

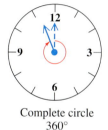
Complete circle
360°

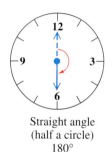
Straight angle
(half a circle)
180°

In a quarter of an hour, at 12:15, the minute hand has moved $\frac{1}{4}$ of the way around the circle, or 90°. An angle of 90° is called a **right angle.** The rays of a right angle form one corner of a square. So, to show that an angle is a **right angle**, we draw a **small square** at the vertex.

Right angle
($\frac{1}{4}$ of a circle)
90°

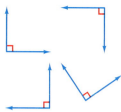
A small square at the
vertex identifies right angles.

In one minute, the *minute hand* moves 6°. From this you can tell that an angle of 1° is very small.

1° angle

❸ **(a)** Name the highlighted angle in three different ways.

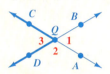

**(b)** Darken the rays that make up ∠ZTW.

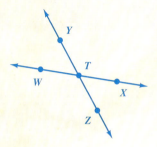

**(c)** Name this angle in four different ways.

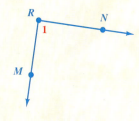

**ANSWERS**
3. **(a)** ∠3, ∠CQD, ∠DQC
   **(b)**

   **(c)** ∠1, ∠R, ∠MRN, ∠NRM

**4** Label each angle as acute, right, obtuse, or straight. State the number of degrees in the right angle and in the straight angle.

(a)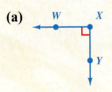

(b)

(c)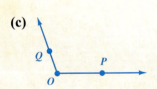

(d)

Some other terms used to describe angles are shown below.

**Acute angles** measure less than 90°.

Examples of acute angles

**Obtuse angles** measure more than 90° but less than 180°.

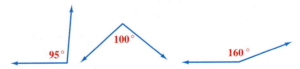

Examples of obtuse angles

**Section 9.2** shows you how to use a tool called a *protractor* to measure the number of degrees in an angle.

### Classifying Angles

**Acute angles** measure less than 90°.
**Right angles** measure exactly 90°.
**Obtuse angles** measure more than 90° but less than 180°.
**Straight angles** measure exactly 180°.

**NOTE**
Angles can also be measured in radians, which you will learn about in a later math course.

**Example 4**  Classifying an Angle

Label each angle as acute, right, obtuse, or straight.

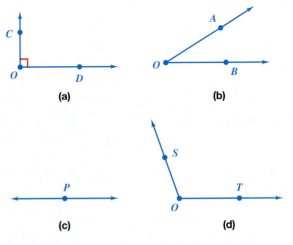

Figure **(a)** shows a *right angle* (exactly 90° and identified by a small square at the vertex).
Figure **(b)** shows an *acute angle* (less than 90°).
Figure **(c)** shows a *straight angle* (exactly 180°).
Figure **(d)** shows an *obtuse angle* (more than 90° but less than 180°).

**Work Problem 4 at the Side.**

---

ANSWERS
**4.** (a) right; 90°  (b) straight; 180°
(c) obtuse  (d) acute

Section 6.5 Geometry: Lines and Angles   461

**5** **Identify perpendicular lines.** Two lines are called **perpendicular lines** if they intersect to form a right angle.

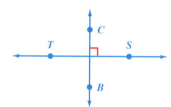

Lines *CB* and *ST* are **perpendicular** because they intersect at right angles. This can be written in the following way: $\overleftrightarrow{CB} \perp \overleftrightarrow{ST}$.

**Example 5** Identifying Perpendicular Lines

Which pairs of lines are perpendicular?

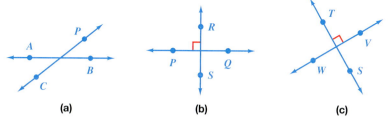

(a)   (b)   (c)

The lines in Figures **(b)** and **(c)** are perpendicular to each other, because they intersect at right angles. The lines in Figure **(a)** are intersecting lines, but they are not perpendicular because they do not form a right angle.

— **Work Problem 5 at the Side.**

**6** **Identify complementary and supplementary angles.** Two angles are called **complementary angles** if their sum is 90°. If two angles are complementary, each angle is the *complement* of the other.

**Example 6** Identifying Complementary Angles

Identify each pair of complementary angles.

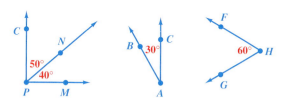

∠*MPN* (40°) and ∠*NPC* (50°) are complementary angles because

$$40° + 50° = 90°.$$

∠*CAB* (30°) and ∠*FHG* (60°) are complementary angles because

$$30° + 60° = 90°.$$

— **Work Problem 6 at the Side.**

**5** Which pair of lines is perpendicular? How can you describe the other pair of lines?

(a)
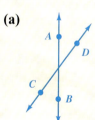

(b)
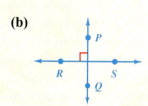

**6** Identify each pair of complementary angles.

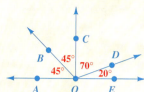

**ANSWERS**
5. Figure **(b)** shows perpendicular lines; Figure **(a)** shows intersecting lines.
6. ∠*AOB* and ∠*BOC*; ∠*COD* and ∠*DOE*

**462** Chapter 6 Ratio, Proportion, and Line/Angle/Triangle Relationships

**7** Find the complement of each angle.

(a) 35°

(b) 80°

**Example 7** Finding the Complement of Angles

Find the complement of each angle.

(a) 30°
The complement of 30° is 60°, because **90°** − 30° = 60°.

(b) 40°
The complement of 40° is 50°, because **90°** − 40° = 50°.

Work Problem **7** at the Side.

Two angles are called **supplementary angles** if their sum is 180°. If two angles are supplementary, each angle is the *supplement* of the other.

**Example 8** Identifying Supplementary Angles

Identify each pair of supplementary angles.

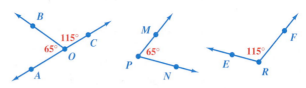

∠BOA and ∠BOC, because 65° + 115° = **180°**.
∠BOA and ∠ERF, because 65° + 115° = **180°**.
∠BOC and ∠MPN, because 115° + 65° = **180°**.
∠MPN and ∠ERF, because 65° + 115° = **180°**.

Work Problem **8** at the Side.

**8** Identify each pair of supplementary angles.

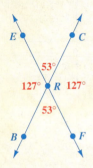

**9** Find the supplement of each angle.

(a) 175°

(b) 30°

**Example 9** Finding the Supplement of Angles

Find the supplement of each angle.

(a) 70°
The supplement of 70° is 110°, because **180°** − 70° = 110°.

(b) 140°
The supplement of 140° is 40°, because **180°** − 140° = 40°.

Work Problem **9** at the Side.

**7** Identify congruent and vertical angles and use this knowledge to find the measures of angles. Two angles are called **congruent angles** if they measure the same number of degrees. If two angles are congruent, this is written as ∠A ≅ ∠B and read as, "angle A **is congruent to** angle B." Here is an example.

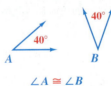

∠A ≅ ∠B

Example of congruent angles

**ANSWERS**
7. (a) 55°  (b) 10°
8. ∠CRF and ∠BRF; ∠CRE and ∠ERB; ∠BRF and ∠BRE; ∠CRE and ∠CRF
9. (a) 5°  (b) 150°

Section 6.5 Geometry: Lines and Angles   463

### Example 10 — Identifying Congruent Angles

Identify the angles that are congruent.

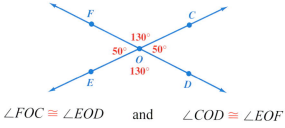

∠FOC ≅ ∠EOD   and   ∠COD ≅ ∠EOF

**(10)** Identify the angles that are congruent.

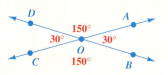

───── Work Problem **10** at the Side. ─────

Angles that share a common side are called *adjacent* angles, such as ∠FOC and ∠COD in Example 10 above. Angles that do *not* share a common side are called *nonadjacent* angles. Two nonadjacent angles formed by intersecting lines are called **vertical angles**.

### Example 11 — Identifying Vertical Angles

Identify the vertical angles in this figure.

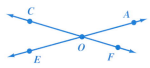

∠AOF and ∠COE are vertical angles because they do *not* share a common side and they are formed by two intersecting lines ($\overleftrightarrow{CF}$ and $\overleftrightarrow{EA}$).

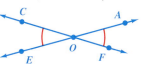

∠COA and ∠EOF are also vertical angles.

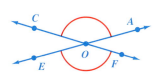

**(11)** Identify the vertical angles.

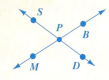

───── Work Problem **11** at the Side. ─────

Look back at Example 10 above. Notice that the two congruent angles that measure 130° are also vertical angles. Also, the two congruent angles that measure 50° are vertical angles. This illustrates the following property.

### Congruent Angles

If two angles are vertical angles, they are congruent; that is, they measure the same number of degrees.

**ANSWERS**
10. ∠BOC ≅ ∠AOD; ∠AOB ≅ ∠DOC
11. ∠SPB and ∠MPD; ∠BPD and ∠SPM

**12** In the figure below, find the measure of each unlabeled angle.

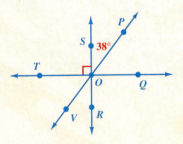

(a) ∠VOR

(b) ∠POQ

(c) ∠QOR

(d) ∠TOV

**ANSWERS**
12. (a) 38°   (b) 52°   (c) 90°   (d) 52°

**Example 12** Finding the Measures of Vertical Angles

In the figure below, find the measure of each unlabeled angle.

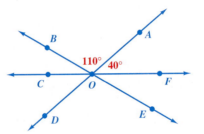

(a) ∠COD
∠COD and ∠AOF are vertical angles, so they are congruent. This means they measure the same number of degrees.

The measure of ∠AOF is 40°   so   the measure of ∠COD is 40° also.

(b) ∠DOE
∠DOE and ∠BOA are vertical angles, so they are congruent.

The measure of ∠BOA is 110°   so   the measure of ∠DOE is 110° also.

(c) ∠COB
Look at ∠COB, ∠BOA, and ∠AOF. Notice that $\overrightarrow{OC}$ and $\overrightarrow{OF}$ go in opposite directions. Therefore, ∠COF is a straight angle and measures 180°. To find the measure of ∠COB, subtract the sum of the other two angles from 180°.

$$180° - (110° + 40°) = 180° - (150°) = 30°$$

The measure of ∠COB is 30°.

(d) ∠EOF
∠EOF and ∠COB are vertical angles, so they are congruent. We know from part (c) above that

the measure of ∠COB is 30°   so   the measure of ∠EOF is 30° also.

**Work Problem** 12 **at the Side.**

**8** Identify corresponding and alternate interior angles and use this knowledge to find the measures of angles. We can also find congruent angles (angles with the same measure) when two *parallel lines* are crossed by a third line, called a *transversal*. When a transversal crosses two *parallel* lines, eight angles are formed, as shown below. There are special names for certain pairs of angles.

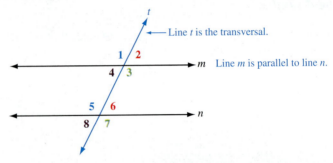

∠1 and ∠5 are called **corresponding angles.** Notice that they are both on the same side of the transversal and in the same relative position. *Corresponding angles are congruent*, so ∠1 and ∠5 measure the same number of degrees. There are four pairs of corresponding angles.

∠1 and ∠5 are corresponding angles, so ∠1 ≅ ∠5.
∠2 and ∠6 are corresponding angles, so ∠2 ≅ ∠6.
∠3 and ∠7 are corresponding angles, so ∠3 ≅ ∠7.
∠4 and ∠8 are corresponding angles, so ∠4 ≅ ∠8.

**13** In the figure below, line $m$ is parallel to line $n$. Identify all pairs of corresponding angles and all pairs of alternate interior angles.

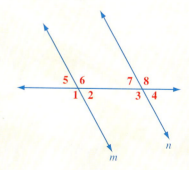

When a transversal crosses two parallel lines, angles 3, 4, 5, and 6 are called *interior angles*. You can see that they are "inside" the *parallel* lines.

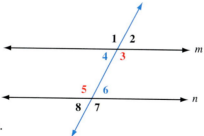

∠3 and ∠5 are alternate interior angles.
∠4 and ∠6 are alternate interior angles.

When two lines are *parallel,* then **alternate interior angles** *are congruent* (they have the same measure). Notice that alternate interior angles are on opposite sides of the transversal.

∠3 ≅ ∠5   and   ∠4 ≅ ∠6

### Angles Formed by Parallel Lines and a Transversal

When two parallel lines are crossed by a transversal:
1. Corresponding angles are congruent, and
2. Alternate interior angles are congruent.

**Example 13** Identifying Corresponding Angles and Alternate Interior Angles

In each figure, line $m$ is parallel to line $n$. Identify all pairs of corresponding angles and all pairs of alternate interior angles.

(a)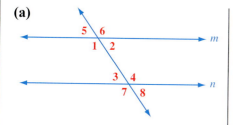

(b)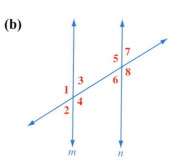

There are four pairs of corresponding angles:
∠5 and ∠3      ∠6 and ∠4
∠1 and ∠7      ∠2 and ∠8

Alternate interior angles:
∠1 and ∠4      ∠2 and ∠3

Corresponding angles:
∠1 and ∠5      ∠3 and ∠7
∠2 and ∠6      ∠4 and ∠8

Alternate interior angles:
∠3 and ∠6      ∠4 and ∠5

— Work Problem **13** at the Side.

**ANSWERS**
**13.** corresponding angles: ∠5 and ∠7; ∠6 and ∠8; ∠1 and ∠3; ∠2 and ∠4
alternate interior angles: ∠6 and ∠3; ∠2 and ∠7

**466** Chapter 6 Ratio, Proportion, and Line/Angle/Triangle Relationships

**14** In each figure below, line *m* is parallel to line *n*.

(a) The measure of ∠6 is 150°. Find the measures of the other angles.

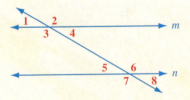

Recall that two angles are supplementary angles if the sum of their measures is 180°. Also remember that the two sides of a 180° angle form a straight line. Now you can combine your knowledge about supplementary angles with the information on parallel lines.

### Example 14 Working with Parallel Lines

In the figure at the right, line *m* is parallel to line *n* and the measure of ∠4 is 70°. Find the measures of the other angles.

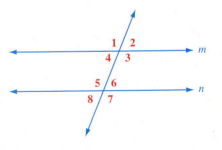

As you find the measure of each angle, write it on the figure.

∠4 ≅ ∠8 (corresponding angles), so the measure of ∠8 is also 70°.
∠4 ≅ ∠6 (alternate interior angles), so the measure of ∠6 is also 70°.
∠6 ≅ ∠2 (corresponding angles), so the measure of ∠2 is also 70°.

Notice that the exterior sides of ∠4 and ∠3 form a straight line, that is, a straight angle of 180°. Therefore, ∠4 and ∠3 are supplementary angles and their sum is 180°. If ∠4 is 70° then ∠3 must be 110° because 180° − 70° = 110°. So the measure of ∠3 is 110°.

∠3 ≅ ∠7 (corresponding angles), so the measure of ∠7 is also 110°.
∠3 ≅ ∠5 (alternate interior angles), so the measure of ∠5 is also 110°.
∠5 ≅ ∠1 (corresponding angles), so the measure of ∠1 is also 110°.

(b) The measure of ∠1 is 45°. Find the measures of the other angles.

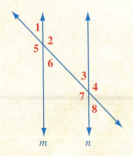

With the measures of all the angles labeled, you can double check that each pair of angles that forms a straight angle also adds up to 180°.

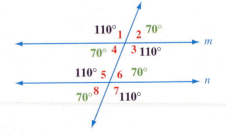

**Work Problem 14 at the Side.**

**ANSWERS**
14. (a) ∠1, ∠4, ∠5, and ∠8 all measure 30°.
∠2, ∠3, ∠6, and ∠7 all measure 150°.
(b) ∠1, ∠3, ∠6, and ∠8 all measure 45°.
∠5, ∠7, ∠2, and ∠4 all measure 135°.

## 6.5 EXERCISES

*Name each line, line segment, or ray using the appropriate symbol. See Example 1.*

1.

2.

3.

4.

5.

6.

*Label each pair of lines as* parallel, perpendicular, *or* intersecting. *See Examples 2 and 5.*

7.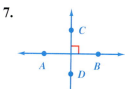

8.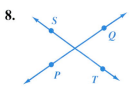

9.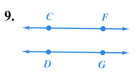

10.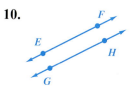

11.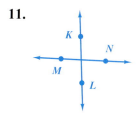

12.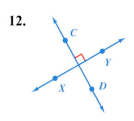

*Name each highlighted angle by using the three-letter form of identification. See Example 3.*

13.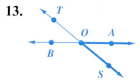

14.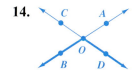

15.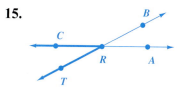

16.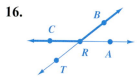

17.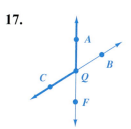

18.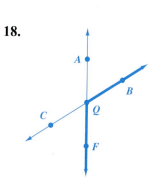

**468** Chapter 6 Ratio, Proportion, and Line/Angle/Triangle Relationships

*Label each angle as* acute, right, obtuse, *or* straight. *For right angles and straight angles, indicate the number of degrees in the angle. See Example 4.*

19.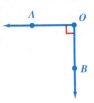

20.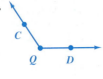

21.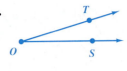

22.

23.

24.

*Identify each pair of complementary angles. See Example 6.*

25.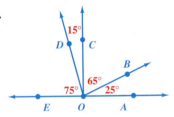

26.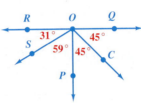

*Identify each pair of supplementary angles. See Example 8.*

27.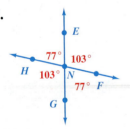

28.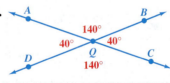

*Find the complement of each angle. See Example 7.*

**29.** 40°   **30.** 35°   **31.** 86°   **32.** 59°

*Find the supplement of each angle. See Example 9.*

**33.** 130°   **34.** 75°   **35.** 90°   **36.** 5°

*In each figure, identify the angles that are congruent. See Examples 10–12.*

**37.**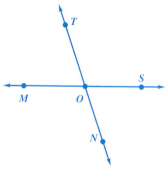

**38.**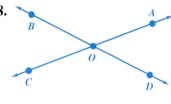

**39.** In the figure below, ∠AOH measures 37° and ∠COE measures 63°. Find the measure of each of the other angles.

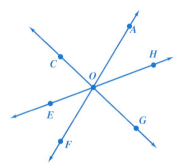

**40.** In the figure below, ∠POU measures 105° and ∠UOT measures 40°. Find the measure of each of the other angles.

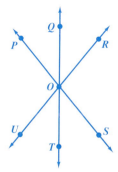

**RELATING CONCEPTS (Exercises 41–46)** **FOR INDIVIDUAL OR GROUP WORK**

*Use the figure to* **work Exercises 41–46 in order.** *Decide whether each statement is* **true** *or* **false.** *If it is true, explain why. If it is false, rewrite it to make a true statement.*

**41.** ∠UST is 90°.

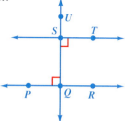

**42.** $\overleftrightarrow{SQ}$ and $\overleftrightarrow{PQ}$ are perpendicular.

**43.** The measure of ∠USQ is less than the measure of ∠PQR.

**44.** $\overleftrightarrow{ST}$ and $\overleftrightarrow{PR}$ are intersecting.

**45.** $\overleftrightarrow{QU}$ and $\overleftrightarrow{TS}$ are parallel.

**46.** ∠UST and ∠UQR measure the same number of degrees.

**470** Chapter 6 Ratio, Proportion, and Line/Angle/Triangle Relationships

*In each figure, line m is parallel to line n. Identify all pairs of corresponding angles and all pairs of alternate interior angles. See Example 13.*

**47.**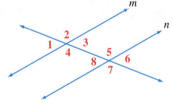

**48.**

*In each figure, line m is parallel to line n. Find the measure of each angle. See Example 14.*

**49.** ∠8 measures 130°.

**50.** ∠2 measures 80°.

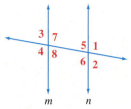

**51.** ∠6 measures 47°.

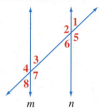

**52.** ∠2 measures 108°.

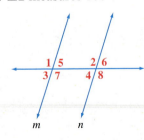

**53.** ∠6 measures 114°.

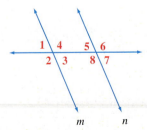

**54.** ∠3 measures 59°.

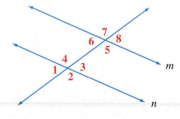

*In each figure, $\overrightarrow{AB}$ is parallel to $\overrightarrow{CD}$. Find the measure of each numbered angle. See Example 14.*

**55.**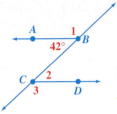

**56.**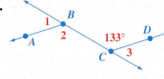

# 6.6 GEOMETRY APPLICATIONS: CONGRUENT AND SIMILAR TRIANGLES

Two useful concepts in geometry are *congruence* and *similarity*. If two figures are *identical*, both in *shape* and in *size*, we say the figures are **congruent**. In other words, the figures are perfect duplicates of each other, like getting two prints from the same negative on a roll of film. If two figures have the *same shape* but are *different sizes*, we say the figures are **similar**, like getting a print from a roll of film and then an enlargement of the same print. We'll explore the ideas of congruence and similarity using triangles.

**OBJECTIVES**

1. Identify corresponding parts of congruent triangles.
2. Prove that triangles are congruent using SAS, SSS, and ASA.
3. Identify corresponding parts of similar triangles.
4. Find the unknown lengths of sides in similar triangles.
5. Solve problems involving similar triangles.

**1 Identify corresponding parts of congruent triangles.** The two triangles shown below are *congruent* because they are the *same shape* and the *same size*.

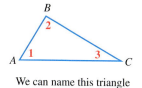
We can name this triangle △ABC.

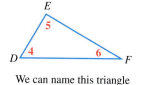
We can name this triangle △DEF.

Suppose you picked up △ABC and slid it over on top of △DEF. You would see that the two triangles are a perfect match. ∠1 would be on top of ∠4, so they are called *corresponding angles*. Similarly, ∠2 and ∠5 are corresponding angles, and ∠3 and ∠6 are corresponding angles. You would see that corresponding angles have the same measure, as indicated below.

$$m\angle 1 = m\angle 4 \qquad m\angle 2 = m\angle 5 \qquad m\angle 3 = m\angle 6$$

The abbreviation for measure is *m*, so $m\angle 1$ is read, "the measure of angle 1."

When you put △ABC on top of △DEF, you would also see that side AB is on top of side DE. We say that $\overline{AB}$ and $\overline{DE}$ are *corresponding sides*. Similarly, $\overline{BC}$ and $\overline{EF}$ are corresponding sides, and $\overline{AC}$ and $\overline{DF}$ are corresponding sides. You would see that corresponding sides have the same length.

$$AB = DE \qquad BC = EF \qquad AC = DF$$

Because corresponding angles have the same measure, and corresponding sides have the same length, we know that △ABC **is congruent to** △DEF. We can write this as △ABC ≅ △DEF.

### Congruent Triangles

If two triangles are congruent, then
1. Corresponding angles have the same measure, and
2. Corresponding sides have the same length.

### Example 1 Identifying Corresponding Parts in Congruent Triangles

Each pair of triangles is congruent. List the corresponding angles and the corresponding sides.

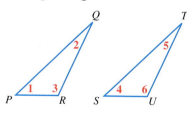

(a) If you picked up △PQR and slid it over on top of △STU, the two triangles would match.
The corresponding parts are congruent, so:

$m\angle 1 = m\angle 4 \qquad PQ = ST$

$m\angle 2 = m\angle 5 \qquad PR = SU$

$m\angle 3 = m\angle 6 \qquad QR = TU$

*Continued on Next Page*

**472** Chapter 6 Ratio, Proportion, and Line/Angle/Triangle Relationships

❶ Each pair of triangles is congruent. List the corresponding angles and the corresponding sides.

(a)
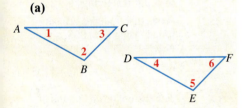

(b) G J L
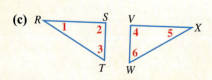

(*Hint:* Rotate △FGH, then slide it on top of △JLK.)

(c)

(*Hint:* Flip △RST over, then slide it on top of △VWX.)

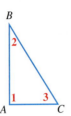

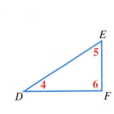

**(b)** If you picked up △ABC and slid it over on top of △DEF, it wouldn't match. But if you *rotate* △ABC before sliding it on top of △DEF, it *will* match.

The corresponding parts are congruent, so:

$m\angle 1 = m\angle 6$   $BC = DE$
$m\angle 2 = m\angle 4$   $BA = DF$
$m\angle 3 = m\angle 5$   $CA = EF$

**Work Problem ❶ at the Side.**

❷ **Prove that triangles are congruent using SAS, SSS, and ASA.** One way to prove that two triangles are congruent would be to measure all the angles and all the sides. If the measures of the corresponding angles and sides are equal, then the triangles are congruent. But here are three quicker methods to prove that two triangles are congruent.

### Proving That Two Triangles Are Congruent

**1. Angle–Side–Angle (ASA) Method**
If two angles and the side that connects them on one triangle measure the same as the corresponding parts on another triangle, the triangles are congruent.

If $m\angle 1 = m\angle 3$ and $m\angle 2 = m\angle 4$ and $a = x$ then the two triangles are congruent.

**2. Side–Side–Side (SSS) Method**
If three sides of one triangle measure the same as the corresponding sides of another triangle, the triangles are congruent.

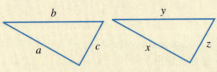

If $a = x$ and $b = y$ and $c = z$ then the two triangles are congruent.

**3. Side–Angle–Side (SAS) Method**
If two sides and the angle between them on one triangle measure the same as the corresponding parts on another triangle, the triangles are congruent.

If $a = x$ and $b = y$ and $m\angle 1 = m\angle 2$ then the two triangles are congruent.

**ANSWERS**
1. (a) $m\angle 1 = m\angle 4$, $m\angle 2 = m\angle 5$, $m\angle 3 = m\angle 6$;
   $AC = DF$, $AB = DE$, $BC = EF$,
   (b) $m\angle 1 = m\angle 6$, $m\angle 2 = m\angle 5$, $m\angle 3 = m\angle 4$;
   $FG = KL$, $FH = JL$, $GH = JK$,
   (c) $m\angle 1 = m\angle 5$, $m\angle 2 = m\angle 4$, $m\angle 3 = m\angle 6$;
   $RS = VX$, $RT = WX$, $ST = VW$

## Example 2 Proving That Two Triangles Are Congruent

Explain which method can be used to prove that each pair of triangles is congruent. Choose from ASA, SSS, and SAS.

(a)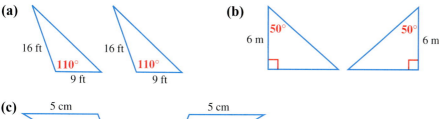

(b)

(c)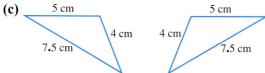

(a) On both triangles, two corresponding sides and the angle between them measure the same, so the Side–Angle–Side (SAS) method can be used to prove that the triangles are congruent.

(b) On both triangles, two corresponding angles and the side that connects them measure the same, so the Angle–Side–Angle (ASA) method can be used to prove that the triangles are congruent.

(c) Each pair of corresponding sides has the same length, so the Side–Side–Side (SSS) method can be used to prove that the triangles are congruent.

**Work Problem ❷ at the Side.**

**3** Identify corresponding parts of similar triangles. Now that you've worked with *congruent* triangles, let's look at *similar* triangles. Remember that congruent triangles match exactly, both in shape and in size. Similar triangles, on the other hand, have the same shape but are *different sizes*. Three pairs of similar triangles are shown here.

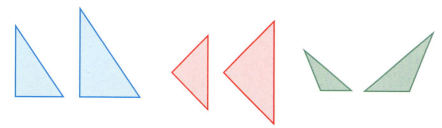

Each pair of triangles has the same shape because the corresponding angles have the same measure. But the corresponding sides are *not* the same length, so the triangles are of *different sizes*.

Two similar triangles are shown to the right. Notice that corresponding angles have the same measure, but corresponding sides have different lengths.

$\overline{CB}$ corresponds to $\overline{RQ}$. Similarly, $\overline{CA}$ corresponds to $\overline{RP}$, and $\overline{BA}$ corresponds to $\overline{QP}$. Notice that each side in the larger triangle is *twice* the length of the corresponding side in the smaller triangle.

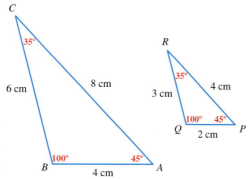

**Work Problem ❸ at the Side.**

---

❷ Determine which method can be used to prove that each pair of triangles is congruent.

(a)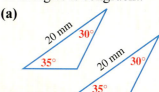

(b)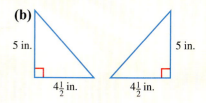

(c)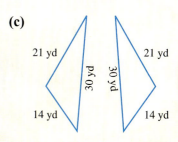

❸ Identify corresponding angles and sides in these similar triangles.

(a)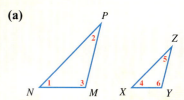

Angles:          Sides:
1 and _____     $\overline{PN}$ and _____
2 and _____     $\overline{PM}$ and _____
3 and _____     $\overline{NM}$ and _____

(b)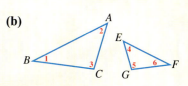

Angles:          Sides:
1 and _____     $\overline{AB}$ and _____
2 and _____     $\overline{BC}$ and _____
3 and _____     $\overline{AC}$ and _____

**ANSWERS**
2. (a) ASA  (b) SAS  (c) SSS
3. (a) 4; 5; 6; $\overline{ZX}$; $\overline{ZY}$; $\overline{XY}$
   (b) 6; 4; 5; $\overline{EF}$; $\overline{FG}$; $\overline{EG}$

④ Find the length of $\overline{EF}$ in Example 3 by setting up and solving a proportion. Let $x$ represent the unknown length.

**4** **Find the unknown lengths of sides in similar triangles.** Similar triangles are useful because of the following definition.

### Similar Triangles

If two triangles are similar, then
1. Corresponding angles have the same measure, and
2. The ratios of the lengths of corresponding sides are equal.

**Example 3**  Finding the Unknown Lengths of Sides in Similar Triangles

Find the length of $y$ in the smaller triangle. Assume the triangles are similar.

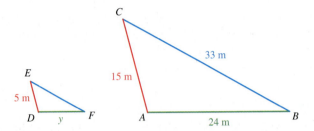

$\overline{DF}$, the length you want to find, corresponds to $\overline{AB}$ in the larger triangle. Then, notice that $\overline{ED}$ and $\overline{CA}$ are corresponding sides and you know their lengths. Since the ratios of the lengths of corresponding sides are equal, you can set up a proportion. (Recall that a proportion states that two ratios are equal.)

$$\text{Corresponding sides} \begin{cases} DF \to \\ AB \to \end{cases} \frac{y}{24} = \frac{5}{15} \begin{cases} \leftarrow ED \\ \leftarrow CA \end{cases} \text{Corresponding sides}$$

$$\frac{y}{24} = \frac{1}{3} \quad \text{Write } \tfrac{5}{15} \text{ in lowest terms.}$$

Find the cross products.

$$\frac{y}{24} \times \frac{1}{3} \qquad \begin{array}{l} 24 \cdot 1 = 24 \\ y \cdot 3 \end{array}$$

Show that the cross products are equivalent.

$$y \cdot 3 = 24$$

Divide both sides by 3.

$$\frac{y \cdot \cancel{3}}{\cancel{3}} = \frac{24}{3}$$

$$y = 8$$

$\overline{DF}$ has a length of 8 m.

**Work Problem** ④ **at the Side.**

ANSWERS

**4.** $\dfrac{x}{33} = \dfrac{1}{3}$; $x = 11$ m

### Example 4 — Finding an Unknown Length and the Perimeter

Find the perimeter of the smaller triangle. Assume the triangles are similar.

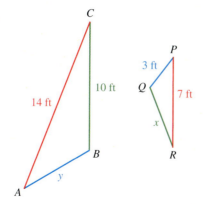

First find $x$, the length of $\overline{RQ}$, then add the lengths of all three sides to find the perimeter.

The smaller triangle is turned "upside down" compared to the larger triangle, so be careful when identifying corresponding sides. $\overline{AC}$ is the longest side in the larger triangle, and $\overline{PR}$ is the longest side in the smaller triangle. So $\overline{PR}$ and $\overline{AC}$ are corresponding sides and you know their lengths. $\overline{QR}$, the length you want to find, corresponds to $\overline{BC}$. The ratios of the lengths of corresponding sides are equal, so you can set up a proportion.

$$\begin{array}{c} QR \rightarrow \\ BC \rightarrow \end{array} \dfrac{x}{10} = \dfrac{7}{14} \begin{array}{c} \leftarrow PR \\ \leftarrow AC \end{array}$$

$$\dfrac{x}{10} = \dfrac{1}{2} \quad \text{Write } \tfrac{7}{14} \text{ in lowest terms.}$$

Find the cross products.

$$\dfrac{x}{10} \bowtie \dfrac{1}{2} \quad \begin{array}{c} 10 \cdot 1 = 10 \\ \\ x \cdot 2 \end{array}$$

Show that the cross products are equivalent.

$$x \cdot 2 = 10$$

Divide both sides by 2.

$$\dfrac{x \cdot \cancel{2}}{\cancel{2}} = \dfrac{10}{2}$$

$$x = 5$$

$\overline{RQ}$ has a length of 5 ft.

Now add the lengths of all three sides to find the perimeter of the smaller triangle.

$$\text{Perimeter} = 5 \text{ ft} + 3 \text{ ft} + 7 \text{ ft} = 15 \text{ ft}$$

*Work Problem* **5** *at the Side.*

---

**5** (a) Find the perimeter of triangle $ABC$ in Example 4.

(b) Find the perimeter of each triangle. Assume the triangles are similar.

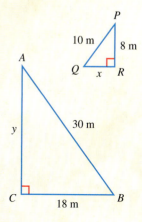

**ANSWERS**
5. (a) $\overline{AB}$ is 6 ft; perimeter = 14 ft + 10 ft + 6 ft = 30 ft
(b) $x = 6$ m, perimeter = 24 m; $y = 24$ m, perimeter = 72 m

**6** Find the height of each flagpole.

(a)
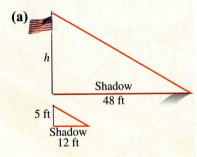

**5** Solve problems involving similar triangles. The next example shows an application of similar triangles.

### Example 5   Using Similar Triangles in an Application

A flagpole casts a shadow 99 m long at the same time that a pole 10 m tall casts a shadow 18 m long. Find the height of the flagpole.

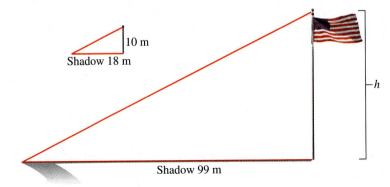

The triangles shown are similar, so write a proportion to find $h$.

Height in larger triangle → $\dfrac{h}{10} = \dfrac{99}{18}$ ← Shadow in larger triangle
Height in smaller triangle →                   ← Shadow in smaller triangle

Find the cross products and show that they are equivalent.

$$h \cdot 18 = 10 \cdot 99$$
$$h \cdot 18 = 990$$

Divide both sides by 18.

$$\dfrac{h \cdot \cancel{18}}{\cancel{18}} = \dfrac{990}{18}$$

$$h = 55$$

The flagpole is 55 m high.

(b)
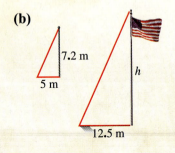

> **NOTE**
>
> There are several other correct ways to set up the proportion in Example 5. One way is to simply flip the ratios on *both* sides of the equal sign.
>
> $$\dfrac{10}{h} = \dfrac{18}{99}$$
>
> But there is another option, shown below.
>
> Height in larger triangle → $\dfrac{h}{99} = \dfrac{10}{18}$ ← Height in smaller triangle
> Shadow in larger triangle →                  ← Shadow in smaller triangle
>
> Notice that both ratios compare *height* to *shadow* in the same order. The ratio on the left describes the larger triangle, and the ratio on the right describes the smaller triangle.

Work Problem **6** at the Side.

**ANSWERS**
6. (a) $h = 20$ ft   (b) $h = 18$ m

## 6.6 EXERCISES

Label each pair of triangles as *congruent, similar, or neither*.

1.
2.
3.

4.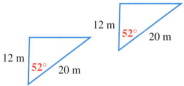
5.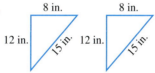
6.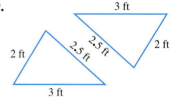

Determine which of these methods can be used to prove that each pair of triangles is congruent: Angle–Side–Angle (ASA), Side–Side–Side (SSS), or Side–Angle–Side (SAS). See Examples 1 and 2.

7.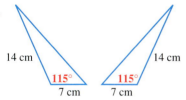
8.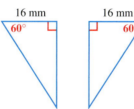
9.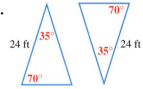

10.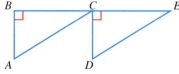
11.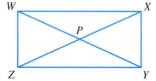
12. 

### RELATING CONCEPTS (Exercises 13–16) FOR INDIVIDUAL OR GROUP WORK

**Work Exercises 13–16 in order.** Given the information in each exercise, explain how you can prove that the indicated triangles are congruent. Note: A midpoint divides a segment into two congruent parts.

13.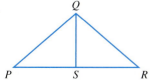

C is the midpoint of $\overline{BE}$ and $CD = BA$. Prove that $\triangle ABC \cong \triangle DCE$.

14. P is the midpoint of both $\overline{WY}$ and $\overline{XZ}$; $WZ = XY$. Prove that $\triangle WPZ \cong \triangle YPX$.

15.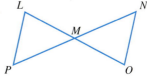

$\overline{QS} \perp \overline{PR}$ and S is the midpoint of $\overline{PR}$. Prove that $\triangle PQS \cong \triangle RQS$.

16. M is the midpoint of both $\overline{LO}$ and $\overline{PN}$. Prove that $\triangle PLM \cong \triangle NOM$.

*Write the ratio for each pair of corresponding sides in the similar triangles shown below. Write the ratios as fractions in lowest terms. See Example 3.*

17. $\dfrac{AB}{PQ}$; $\dfrac{AC}{PR}$; $\dfrac{BC}{QR}$

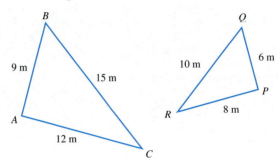

18. $\dfrac{AB}{PQ}$; $\dfrac{AC}{PR}$; $\dfrac{BC}{QR}$

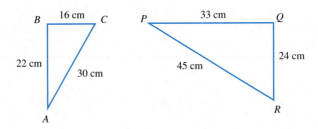

*Find the unknown lengths in each pair of similar triangles. See Example 3.*

19.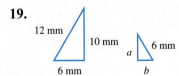

20.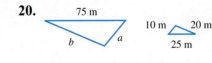

21.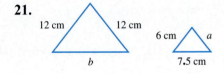

22.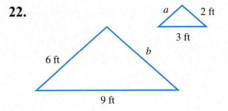

*Find the perimeter of each triangle. Assume the triangles are similar. See Example 4.*

23.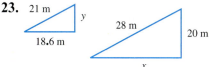

24.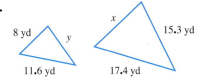

25. Triangles *CDE* and *FGH* are similar. Find the perimeter and area of triangle *FGH*. Note: The heights of similar triangles have the same ratio as corresponding sides. Round to the nearest tenth when necessary.

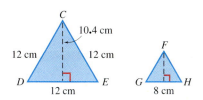

26. Triangles *JKL* and *MNO* are similar. Find the perimeter and area of triangle *MNO*.

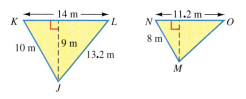

*Solve each application problem. See Example 5.*

27. The height of the house shown here can be found by comparing its shadow to the shadow cast by a 3-foot stick. Find the height of the house by writing a proportion and solving it.

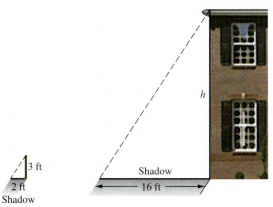

28. A fire lookout tower provides an excellent view of the surrounding countryside. The height of the tower can be found by lining up the top of the tower with the top of a 2-meter stick. Use similar triangles to find the height of the tower.

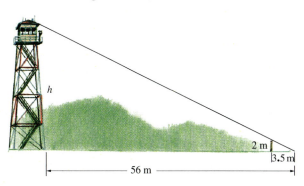

**29.** Look up the word *similar* in a dictionary. What is the nonmathematical definition of this word? Describe two examples of similar objects at home, school, or work.

**30.** Look up the word *congruent* in a dictionary. What is the nonmathematical definition of this word? Describe two examples of congruent objects at home, school, or work.

*Find the unknown length in Exercises 31–34. Round your answers to the nearest tenth. Note: When a line is drawn parallel to one side of a triangle, the smaller triangle that is formed will be similar to the original triangle. In Exercises 31–32, the red segments are parallel.*

**31.**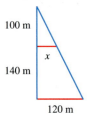

**32.**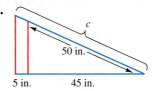

 **33.** Use similar triangles and a proportion to find the length of the lake shown here. (*Hint:* The side 100 m long in the smaller triangle corresponds to a side of 100 m + 120 m = 220 m in the larger triangle.)

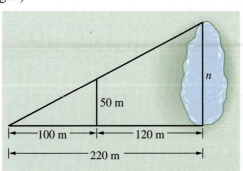

 **34.** To find the height of the tree, find $y$ and then add $5\frac{1}{2}$ ft for the distance from the ground to the eye level of the person.

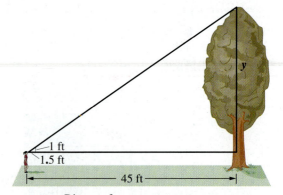

Distance from person to tree

## Chapter 6

### SUMMARY

**KEY TERMS**

| | | |
|---|---|---|
| **6.1** | **ratio** | A ratio compares two quantities that have the same type of units. For example, the ratio of 6 apples to 11 apples is written in fraction form as $\frac{6}{11}$. The common units (apples) divide out. |
| **6.2** | **rate** | A rate compares two measurements with different types of units. Examples are 96 dollars for 8 hours, or 450 miles on 18 gallons. |
| | **unit rate** | A unit rate has 1 in the denominator. |
| | **cost per unit** | Cost per unit is a rate that tells how much you pay for one item or one unit. The lowest cost per unit is the best buy. |
| **6.3** | **proportion** | A proportion states that two ratios or rates are equivalent. |
| | **cross products** | Multiply along one diagonal and then along the other diagonal to find the cross products of a proportion. If the cross products are equal, the proportion is true. |
| **6.5** | **point** | A point is a location in space. *Example:* Point P at the right. |
| | **line** | A line is a straight row of points that goes on forever in both directions. *Example:* Line AB, written $\overleftrightarrow{AB}$, at the right. |
| | **line segment** | A line segment is a piece of a line with two endpoints. *Example:* Line segment PQ, written $\overline{PQ}$, at the right. |
| | **ray** | A ray is a part of a line that has one endpoint and extends forever in one direction. *Example:* Ray RS, written $\overrightarrow{RS}$, at the right. |
| | **parallel lines** | Parallel lines are two lines in the same plane that never intersect (never cross). *Example:* $\overleftrightarrow{AB}$ is parallel to $\overleftrightarrow{ST}$ at the right. |
| | **intersecting lines** | Intersecting lines cross. *Example:* $\overleftrightarrow{RQ}$ intersects $\overleftrightarrow{AB}$ at point P at the right. |
| | **angle** | An angle is made up of two rays that have a common endpoint called the vertex. *Example:* Angle 1 at the right. |
| | **degrees** | A system used to measure angles in which a complete circle is 360 degrees, written 360°. |
| | **straight angle** | A straight angle is an angle that measures exactly 180°; its sides form a straight line. *Example:* Angle G at the right. |
| | **right angle** | A right angle is an angle that measures exactly 90°. *Example:* Angle AOB at the right. |
| | **acute angle** | An acute angle is an angle that measures less than 90°. *Example:* Angle E at the right. |
| | **obtuse angle** | An obtuse angle is an angle that measures more than 90° but less than 180°. *Example:* Angle F at the right. |
| | **perpendicular lines** | Perpendicular lines are two lines that intersect to form a right angle. *Example:* $\overleftrightarrow{PQ}$ is perpendicular to $\overleftrightarrow{RS}$ at the right. |

*(continued)*

## Key Terms

| | |
|---|---|
| **complementary angles** | Complementary angles are two angles whose measures add up to 90°. |
| **supplementary angles** | Supplementary angles are two angles whose measures add up to 180°. |
| **congruent angles** | Congruent angles are angles that measure the same number of degrees. |
| **vertical angles** | Vertical angles are two nonadjacent congruent angles formed by intersecting lines. *Example:* ∠COA and ∠EOF are vertical angles at the right. |
| **corresponding angles** | Corresponding angles are formed when two parallel lines are crossed by a transversal; corresponding angles are congruent and are on the same side of the transversal and in the same relative position. *Example:* In the figure at the right, line *m* is parallel to line *n*. The pairs of corresponding angles are ∠1 and ∠5, ∠2 and ∠6, ∠3 and ∠7, ∠4 and ∠8. |
| **alternate interior angles** | When two parallel lines are crossed by a transversal, there are two pairs of alternate interior angles and each pair is congruent. They are on opposite sides of the transversal. *Example:* In the figure at the right, line *m* is parallel to line *n*. The pairs of alternate interior angles are ∠3 and ∠5, ∠4 and ∠6. |
| 6.6 **congruent figures** | Congruent figures are identical both in shape and in size. |
| **similar figures** | Similar figures have the same shape but are different sizes. |
| **congruent triangles** | Congruent triangles are triangles with the same shape and the same size; corresponding angles measure the same number of degrees and corresponding sides have the same length. |
| **similar triangles** | Similar triangles are triangles with the same shape but not necessarily the same size; corresponding angles measure the same number of degrees and the *ratios* of the lengths of corresponding sides are equal. |

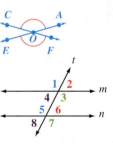
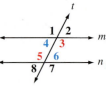

## Test Your Word Power

*See how well you have learned the vocabulary in this chapter. Answers follow the Quick Review.*

1. A **ratio**
   (a) can be written only as a fraction
   (b) compares two quantities that have the same type of units
   (c) compares two quantities that have different types of units
   (d) is the reciprocal of a rate.

2. A **rate**
   (a) can be written only as a decimal
   (b) compares two quantities that have the same type of units
   (c) compares two quantities that have different types of units
   (d) is the reciprocal of a ratio.

3. **Cost per unit** is
   (a) the best buy
   (b) a ratio written in lowest terms
   (c) found by comparing cross products
   (d) the price of one item or one unit.

4. A **proportion**
   (a) shows that two ratios or rates are equivalent
   (b) contains only whole numbers or decimals
   (c) always has one unknown number
   (d) states that two improper fractions are equivalent.

5. An **obtuse angle**
   (a) is formed by perpendicular lines
   (b) is congruent to a right angle
   (c) measures more than 90° but less than 180°
   (d) measures less than 90°.

6. Two angles that are **complementary**
   (a) have measures that add up to 180°
   (b) are always congruent

   (c) form a straight angle
   (d) have measures that add up to 90°.

7. **Perpendicular lines**
   (a) intersect to form a right angle
   (b) intersect to form an acute angle
   (c) never intersect
   (d) have a common endpoint called the vertex.

8. In a pair of **similar triangles**
   (a) corresponding sides have the same length
   (b) all the angles have the same measure
   (c) the perimeters are equal
   (d) the ratios of the lengths of corresponding sides are equal.

## Quick Review

*Concepts*

### 6.1 Writing a Ratio
A ratio compares two quantities that have the same type of units. A ratio is usually written as a fraction with the number that is mentioned first in the numerator. The common units divide out and are not written in the answer. Check that the fraction is in lowest terms.

### 6.1 Using Mixed Numbers in a Ratio
If a ratio has mixed numbers, change the mixed numbers to improper fractions. Rewrite the problem in horizontal form, using the "÷" symbol for division. Finally, multiply by the reciprocal of the divisor.

### 6.1 Using Measurements in Ratios
When a ratio compares measurements, both measurements must be in the *same* units. It is usually easier to compare the measurements using the smaller unit, for example, inches instead of feet.

### 6.2 Writing Rates
A rate compares two measurements with different types of units. The units do *not* divide out, so you must write them as part of the rate.

*Examples*

Write this ratio as a fraction in lowest terms.

60 ounces of medicine **to** 160 ounces of medicine

$$\frac{60 \text{ ounces}}{160 \text{ ounces}} = \frac{60 \div 20}{160 \div 20} = \frac{3}{8} \leftarrow \text{Lowest terms}$$

↑ Divide out common units.

Write as a ratio of whole numbers in lowest terms.

$$2\frac{1}{2} \quad \text{to} \quad 3\frac{3}{4}$$

$$\frac{2\frac{1}{2}}{3\frac{3}{4}} \quad \text{Ratio with mixed numbers}$$

$$= \frac{\frac{5}{2}}{\frac{15}{4}} \quad \text{Ratio with improper fractions}$$

$$= \frac{5}{2} \div \frac{15}{4} = \frac{5}{2} \cdot \frac{4}{15} = \frac{5 \cdot 2 \cdot 2}{2 \cdot 3 \cdot 5} = \frac{2}{3} \quad \text{Lowest terms}$$

Reciprocals

Write 8 inches to 6 feet as a ratio in lowest terms.

Compare using the smaller unit, inches. Because 1 foot has 12 inches, 6 feet is

$$6 \cdot 12 \text{ inches} = 72 \text{ inches.}$$

The ratio is shown below.

$$\frac{8 \text{ inches}}{72 \text{ inches}} = \frac{8 \div 8}{72 \div 8} = \frac{1}{9}$$

Write the rate as a fraction in lowest terms.

475 miles in 10 hours

$$\frac{475 \text{ miles} \div 5}{10 \text{ hours} \div 5} = \frac{95 \text{ miles}}{2 \text{ hours}} \quad \text{Must write units: miles and hours}$$

| Concepts | Examples |
|---|---|
| **6.2 Finding a Unit Rate**<br>A unit rate has 1 in the denominator. To find the unit rate, divide the numerator by the denominator. Write unit rates using the word **per** or a / mark. | Write as a unit rate: $1278 in 9 days.<br><br>$$\frac{\$1278}{9 \text{ days}} \leftarrow \text{Fraction bar indicates division.}$$<br><br>$9\overline{)1278}\; = 142$ so $\dfrac{\$1278 \div 9}{9 \text{ days} \div 9} = \dfrac{\$142}{1 \text{ day}}$<br><br>Write the answer as $142 **per** day or $142/day. |
| **6.2 Finding the Best Buy**<br>The best buy is the item with the lowest cost per unit. Divide the price by the number of units. Round to thousandths when necessary. Then compare to find the lowest cost per unit. | Find the best buy on cheese.<br><br>2 pounds for $2.25<br>3 pounds for $3.40<br><br>Find cost per unit (cost per pound).<br><br>$$\frac{\$2.25}{2} = \$1.125 \text{ per pound}$$<br><br>$$\frac{\$3.40}{3} \approx \$1.133 \text{ per pound}$$<br><br>The lower cost per pound is $1.125, so 2 pounds of cheese is the best buy. |
| **6.3 Writing Proportions**<br>A proportion states that two ratios or rates are equivalent. The proportion "5 is to 6 as 25 is to 30" is written as shown below.<br><br>$$\frac{5}{6} = \frac{25}{30}$$<br><br>To see whether a proportion is true or false, multiply along one diagonal, then multiply along the other diagonal. If the two cross products are equal, the proportion is true. If the two cross products are unequal, the proportion is false. | Write as a proportion: 8 is to 40 as 32 is to 160.<br><br>$$\frac{8}{40} = \frac{32}{160}$$<br><br>Is this proportion true or false?<br><br>$$\frac{6}{8\frac{1}{2}} = \frac{24}{34}$$<br><br>Find the cross products.<br><br>$8\frac{1}{2} \cdot 24 = \frac{17}{2} \cdot \frac{\cancel{2} \cdot 12}{1} = \mathbf{204}$<br><br>$6 \cdot 34 = \mathbf{204}$ ← Equal<br><br>The cross products are equal, so the proportion is true. |

| Concepts | Examples |
|---|---|
| **6.3 Solving Proportions**  Solve for an unknown number in a proportion by using these steps. | Find the unknown number.  $\dfrac{12}{x} = \dfrac{6}{8}$  } Write $\dfrac{6}{8}$ in lowest terms.  $\dfrac{12}{x} = \dfrac{3}{4}$ |
| *Step 1* Find the cross products. (If desired, you can rewrite the ratios in lowest terms before finding the cross products.) | *Step 1* $\dfrac{12}{x} = \dfrac{3}{4}$ — Find the cross products. ($x \cdot 3$ and $12 \cdot 4$) |
| *Step 2* Show that the cross products are equivalent. | *Step 2* $x \cdot 3 = 12 \cdot 4$ — Show that cross products are equivalent.  $x \cdot 3 = 48$ |
| *Step 3* Divide both sides of the equation by the coefficient of the variable term. | *Step 3* $\dfrac{x \cdot 3}{3} = \dfrac{48}{3}$  Divide both sides by 3.  $x = 16$ |
| *Step 4* Check by writing the solution in the proportion and finding the cross products. | *Step 4*  $x$ is 16. → $\dfrac{12}{16} = \dfrac{6}{8}$  $16 \cdot 6 = 96$  $12 \cdot 8 = 96$  } Equal  The cross products are equal, so 16 is the correct solution. |
| **6.4 Applications of Proportions**  Use the six problem-solving steps from Section 3.3. | If 3 pounds of grass seed cover 450 square feet of lawn, how much seed is needed for 1500 square feet of lawn? |
| *Step 1* **Read** the problem. | *Step 1* The problem is about the amount of grass seed needed for a lawn.  Unknown: pounds of seed needed for 1500 square feet of lawn  Known: 3 pounds cover 450 square feet of lawn |
| *Step 2* **Assign a variable.** | *Step 2* There is only one unknown, so let $x$ be the pounds of seed needed for 1500 square feet of lawn. |
| *Step 3* **Write an equation.** | *Step 3* The equation is in the form of a proportion. Be sure that both rates in the proportion compare pounds to square feet in the same order.  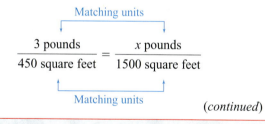 |

(continued)

| Concepts | Examples |
|---|---|
| **6.4 Applications of Proportions (continued)** | |
| Step 4  **Solve** the equation. | Step 4  Ignore the units while finding the cross products and solving for $x$. |

$$450 \cdot x = 3 \cdot 1500 \quad \text{Show that cross products are equivalent.}$$
$$450 \cdot x = 4500$$
$$\frac{\overset{1}{\cancel{450}} \cdot x}{\cancel{450}} = \frac{4500}{450} \quad \text{Divide both sides by 450.}$$
$$x = 10$$

Step 5  **State the answer.**

Step 5  10 pounds of seed are needed for 1500 square feet of lawn.

Step 6  **Check** the solution by putting it back into the original problem.

Step 6  450 square feet of lawn needs 3 pounds of seed; 1500 square feet is about *three times as much* lawn, so about *three times as much* seed is needed.

$$(3)(3 \text{ pounds}) = 9 \text{ pounds} \leftarrow \text{Estimate}$$

The solution, 10 pounds, is close to the estimate of 9 pounds, so it is reasonable.

**6.5 Lines**

A *line* is straight row of points that goes on forever in both directions. If part of a line has one endpoint, it is a *ray*. If it has two endpoints, it is a *line segment*.

Identify each figure as a line, line segment, or ray.

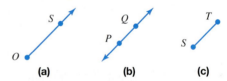

Figure **(a)** shows a ray, **(b)** shows a line, and **(c)** shows a line segment.

If two lines intersect at right angles, they are *perpendicular*. If two lines in the same plane never intersect, they are *parallel*.

Label each pair of lines as parallel or perpendicular.

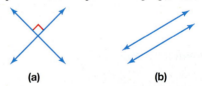

Figure **(a)** shows two perpendicular lines (they intersect at 90°), and figure **(b)** shows two parallel lines (they never intersect).

Chapter 6 Summary **487**

| Concepts | Examples |
|---|---|
| **6.5 Angles**<br>If the sum of the measures of two angles is 90°, they are *complementary*.<br>If the sum of the measures of two angles is 180°, they are *supplementary*.<br><br>If two angles measure the same number of degrees, the angles are *congruent*. The symbol for congruent is ≅.<br><br>Two nonadjacent angles formed by intersecting lines are called *vertical angles*. Vertical angles are congruent. | Find the complement and supplement of a 35° angle.<br>$$90° - 35° = 55°\text{ (the complement)}$$<br>$$180° - 35° = 145°\text{ (the supplement)}$$<br><br>Identify the vertical angles in this figure. Which angles are congruent?<br>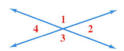<br>∠1 and ∠3 are vertical angles.<br>∠2 and ∠4 are vertical angles.<br>Vertical angles are congruent, so ∠1 ≅ ∠3 and ∠2 ≅ ∠4. |
| **6.5 Parallel Lines**<br>When two parallel lines are crossed by a transversal, corresponding angles are congruent, and alternate interior angles are congruent. Use this information to find the measures of the other angles.<br><br>Read $m\angle 1$ as "the measure of angle 1." | Line *m* is parallel to line *n* and the measure of ∠4 is 125°. Find the measures of the other angles.<br>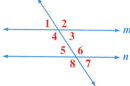<br>∠4 ≅ ∠8 (corresponding angles), so $m\angle 8 = 125°$.<br>∠4 ≅ ∠6 (alternate interior angles), so $m\angle 6 = 125°$.<br>∠6 ≅ ∠2 (corresponding angles), so $m\angle 2 = 125°$.<br>∠4 and ∠3 are supplements, so $m\angle 3 = 180° - 125° = 55°$.<br>∠3 ≅ ∠7 (corresponding angles), so $m\angle 7 = 55°$.<br>∠3 ≅ ∠5 (alternate interior angles), so $m\angle 5 = 55°$.<br>∠5 ≅ ∠1 (corresponding angles), so $m\angle 1 = 55°$. |
| **6.6 Proving That Two Triangles Are Congruent**<br>Congruent triangles are identical both in shape and in size. This means that corresponding angles have the same measure and corresponding sides have the same length.<br><br>Here are three ways to prove that two triangles are congruent.<br>1. Angle–Side–Angle (ASA) method: If two angles and the side that connects them on one triangle measure the same as the corresponding parts on another triangle, the triangles are congruent.<br>2. Side–Side–Side (SSS) method: If three sides of one triangle measure the same as the corresponding sides of another triangle, the triangles are congruent.<br>3. Side–Angle–Side (SAS) method: If two sides and the angle between them on one triangle measure the same as the corresponding parts on another triangle, the triangles are congruent. | Determine which method can be used to prove that each pair of triangles is congruent.<br>**(a)**   **(b)**<br>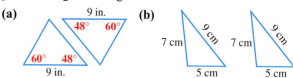<br>**(c)**<br><br><br>**(a)** On both triangles, two corresponding angles and the side that connects them measure the same, so use ASA.<br>**(b)** Each pair of corresponding sides has the same length, so use SSS.<br>**(c)** On both triangles, two corresponding sides and the angle between them measure the same, so use SAS. |

| Concepts | Examples |
|---|---|
| **6.6 Finding the Unknown Lengths in Similar Triangles** Use the fact that in similar triangles, the ratios of the lengths of corresponding sides are equal. Write a proportion. Then find the cross products and show that they are equivalent. Finish solving for the unknown length. | Find $x$ and $y$ if the triangles are similar. $$\frac{x}{8} = \frac{5}{10}$$ $$x \cdot 10 = 8 \cdot 5$$ $$\frac{x \cdot \cancel{10}}{\cancel{10}} = \frac{40}{10}$$ $$x = 4 \text{ m}$$ $$\frac{y}{12} = \frac{5}{10}$$ $$y \cdot 10 = 12 \cdot 5$$ $$\frac{y \cdot \cancel{10}}{\cancel{10}} = \frac{60}{10}$$ $$y = 6 \text{ m}$$ 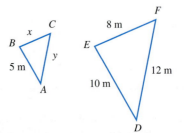 |

**ANSWERS TO TEST YOUR WORD POWER**

1. **(b)** *Example:* The ratio of 3 miles to 4 miles is $\frac{3}{4}$; the common units (miles) divide out.
2. **(c)** *Example:* $4.50 for 3 pounds is a rate comparing dollars to pounds.
3. **(d)** *Example:* $1.65 per gallon tells the price of one gallon (one unit).
4. **(a)** *Example:* $\frac{5}{6} = \frac{25}{30}$ is a proportion because $\frac{5}{6}$ is equivalent to $\frac{25}{30}$.
5. **(c)** *Examples:* Angles that measure 91°, 120°, and 175° are all obtuse angles.
6. **(d)** *Example:* If $\angle 1$ measures 35° and $\angle 2$ measures 55°, the angles are complementary because 35° + 55° = 90°.
7. **(a)** *Example:* $\overleftrightarrow{EF}$ is perpendicular to $\overleftrightarrow{GH}$.

8. **(d)** *Example:* Triangle $ABC$ is similar to triangle $DEF$, so the ratios of corresponding sides are equal.

$$\frac{AB}{DE} = \frac{3 \text{ m}}{6 \text{ m}} = \frac{1}{2} \qquad \frac{BC}{EF} = \frac{2 \text{ m}}{4 \text{ m}} = \frac{1}{2} \qquad \frac{AC}{DF} = \frac{3.5 \text{ m}}{7 \text{ m}} = \frac{1}{2}$$

# Chapter 6 Review Exercises

**[6.1]** *Write each ratio as a fraction in lowest terms. Change to the same units when necessary, using the table of measurement comparisons in* **Section 6.1.** *Use the information in the graph to answer Exercises 1–3.*

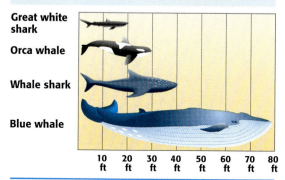

**Average Length of Sharks and Whales**

Source: *Grolier Multimedia Encyclopedia.*

1. Orca whale's length to whale shark's length

2. Blue whale's length to great white shark's length

3. Great white shark's length to whale shark's length

4. $2.50 to $1.25

5. $0.30 to $0.45

6. $1\frac{2}{3}$ cups to $\frac{2}{3}$ cup

7. $2\frac{3}{4}$ miles to $16\frac{1}{2}$ miles

8. 5 hours to 100 minutes

9. 9 in. to 2 ft

10. 1 ton to 1500 pounds

11. 8 hours to 3 days

12. Jake sold $350 worth of his kachina figures. Ramona sold $500 worth of her pottery. What is the ratio of her sales to his?

13. Ms. Wei's new car gets 35 miles per gallon. Her old car got 25 miles per gallon. Find the ratio of the new car's mileage to the old car's mileage.

14. This fall, 60 students are taking math and 72 students are taking English. Find the ratio of math students to English students.

**[6.2]** *Write each rate as a fraction in lowest terms.*

**15.** $88 for 8 dozen

**16.** 96 children in 40 families

**17.** In his keyboarding class, Patrick can type four pages in 20 minutes. Give his rate in pages per minute and minutes per page.

**18.** Elena made $24 in three hours. Give her earnings in dollars per hour and hours per dollar.

*Find the best buy.*

**19.** Minced onion

    13 ounces for $2.29

    8 ounces for $1.45

    3 ounces for $0.95

**20.** Dog food; you have a coupon for $1 off on 25 pounds or more.

    50 pounds for $19.95

    25 pounds for $10.40

    8 pounds for $3.40

**[6.3]** *Use either the method of writing in lowest terms or of finding cross products to decide whether each proportion is* **true** *or* **false**.

**21.** $\dfrac{6}{10} = \dfrac{9}{15}$

**22.** $\dfrac{6}{48} = \dfrac{9}{36}$

**23.** $\dfrac{47}{10} = \dfrac{98}{20}$

**24.** $\dfrac{1.5}{2.4} = \dfrac{2}{3.2}$

**25.** $\dfrac{3\frac{1}{2}}{2\frac{1}{3}} = \dfrac{6}{4}$

*Find the unknown number in each proportion. Round answers to the nearest hundredth when necessary.*

**26.** $\dfrac{4}{42} = \dfrac{150}{x}$

**27.** $\dfrac{16}{x} = \dfrac{12}{15}$

**28.** $\dfrac{100}{14} = \dfrac{x}{56}$

**29.** $\dfrac{5}{8} = \dfrac{x}{20}$

**30.** $\dfrac{x}{24} = \dfrac{11}{18}$

**31.** $\dfrac{7}{x} = \dfrac{18}{21}$

**32.** $\dfrac{x}{3.6} = \dfrac{9.8}{0.7}$

**33.** $\dfrac{13.5}{1.7} = \dfrac{4.5}{x}$

**34.** $\dfrac{0.82}{1.89} = \dfrac{x}{5.7}$

## [6.4] Set up and solve a proportion for each application problem.

**35.** The ratio of cats to dogs at the animal shelter is 3 to 5. If there are 45 dogs, how many cats are there?

**36.** Danielle had 8 hits in 28 times at bat during last week's games. If she continues to hit at the same rate, how many hits will she get in 161 times at bat?

**37.** If 3.5 pounds of ground beef cost $9.77, what will 5.6 pounds cost? Round to the nearest cent.

**38.** About 4 out of 10 students are expected to vote in campus elections. There are 8247 students. How many are expected to vote? Round to the nearest whole number.

**39.** The scale on Brian's model railroad is 1 inch to 16 feet. One of the scale model boxcars is 4.25 inches long. What is the length of a real boxcar in feet?

**40.** In the hospital pharmacy, Michiko sees that a certain medicine is to be given at the rate of 3.5 milligrams for every 50 pounds of body weight. How much medicine should be given to a patient who weighs 210 pounds?

**41.** A 180-pound person burns 284 calories playing basketball for 25 minutes. How many calories would the person burn in 45 minutes, to the nearest whole number? (*Source: Wellness Encyclopedia.*)

**42.** Marvette makes necklaces to sell at a local gift shop. She made 2 dozen necklaces in $16\frac{1}{2}$ hours. How long will it take her to make 40 necklaces?

## [6.5] Name each line, line segment, or ray.

**43.**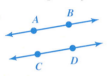

**44.**

**45.**

*Label each pair of lines as* parallel, perpendicular, *or* intersecting.

**46.**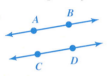

**47.**

**48.**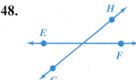

*Label each angle as an* acute, right, obtuse, *or* straight angle. *For right and straight angles, indicate the number of degrees in the angle.*

49.
50.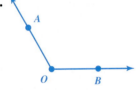
51.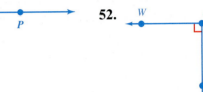
52.

*Find the complement or supplement of each angle.*

53. Find each complement.
    (a) 80°
    (b) 45°
    (c) 7°

54. Find each supplement.
    (a) 155°
    (b) 90°
    (c) 33°

55. In the figure below, ∠2 measures 60°. Find the measure of each of the other angles.

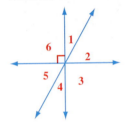

56. Line *m* is parallel to line *n*. ∠8 measures 160°. Find the measures of the other angles.

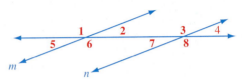

**[6.6]** *Determine which method can be used to prove that each pair of triangles is congruent.*

57.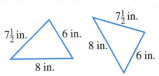
58.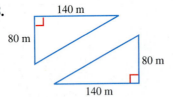
59.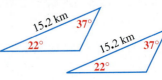

*Find the unknown lengths in each pair of similar triangles. Then find the perimeter of the larger triangle in each pair.*

60.
61.
62.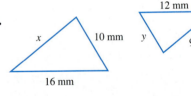

## MIXED REVIEW EXERCISES

*Find the unknown number in each proportion. Round to hundredths when necessary.*

**63.** $\dfrac{x}{45} = \dfrac{70}{30}$

**64.** $\dfrac{x}{52} = \dfrac{0}{20}$

**65.** $\dfrac{64}{10} = \dfrac{x}{20}$

**66.** $\dfrac{15}{x} = \dfrac{65}{100}$

**67.** $\dfrac{7.8}{3.9} = \dfrac{13}{x}$

**68.** $\dfrac{34.1}{x} = \dfrac{0.77}{2.65}$

*Write each ratio as a fraction in lowest terms. Change to the same units when necessary.*

**69.** 4 dollars to 10 quarters

**70.** $4\dfrac{1}{8}$ inches to 10 inches

**71.** 10 yards to 8 feet

**72.** $3.60 to $0.90

**73.** 12 eggs to 15 eggs

**74.** 37 meters to 7 meters

**75.** 3 pints to 4 quarts

**76.** 15 minutes to 3 hours

**77.** $4\dfrac{1}{2}$ miles to $1\dfrac{3}{10}$ miles

*Set up and solve a proportion for each application problem.*

**78.** Nearly 7 out of 8 fans buy something to drink at rock concerts. How many of the 28,500 fans at today's concert would be expected to buy a beverage? Round to the nearest hundred fans.

**79.** Emily spent $150 on car repairs and $400 on car insurance. What is the ratio of the amount spent on insurance to the amount spent on repairs?

**80.** Antonio is choosing among three packages of plastic wrap. Is the best buy 25 feet for $0.78; 75 feet for $1.99; or 100 feet for $2.59? He has a coupon for 50¢ off that is good for either of the larger two packages.

**81.** On this scale drawing of a backyard patio, 0.5 inch represents 6 feet. If the patio measures 1.25 inches long on the drawing, what will be the actual length of the patio when it is built?

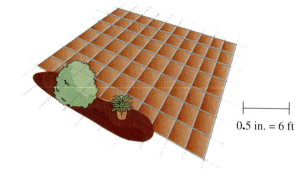

0.5 in. = 6 ft

**82.** A vitamin supplement for cats is to be given at the rate of 1000 milligrams for a 5-pound cat. (*Source:* St. Jon Pet Care Products.)

(a) How much should be given to a 7-pound cat?

(b) How much should be given to an 8-ounce kitten?

**83.** Charles made 251 points during 169 minutes of playing time last year. If he plays 14 minutes in tonight's game, how many points would you expect him to make? Round to the nearest whole number.

**84.** An antibiotic is to be given at the rate of $1\dfrac{1}{2}$ teaspoons for every 24 pounds of body weight. How much should be given to an infant who weighs 8 pounds?

*Label each figure. Choose from these labels:* line segment, ray, parallel lines, perpendicular lines, intersecting lines, acute angle, right angle, straight angle, obtuse angle. *Indicate the number of degrees in the right angle and the straight angle.*

**85.**

**86.**

**87.**

**88.**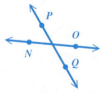

**89.**

**90.**

**91.**

**92.**

**93.**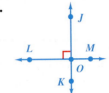

**94.** Explain what is happening in each sentence.
  **(a)** The road was so slippery that my car did a 360.

  **(b)** After the election, the governor's view on new taxes took a 180° turn.

**95. (a)** Can two obtuse angles be supplementary? Explain why or why not.

  **(b)** Can two acute angles be complementary? Explain why or why not.

**96.** In the figure below, ∠2 measures 45° and ∠7 measures 55°. Find the measure of each of the other angles.

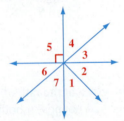

**97.** In the figure below, $\overline{AB}$ is parallel to $\overline{DC}$ and $E$ is the midpoint of $\overline{AC}$. Explain how you can prove that $\triangle ABE \cong \triangle CDE$.

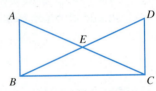

# Chapter 6 TEST

*Write each rate or ratio as a fraction in lowest terms. Change to the same units when necessary.*

1. $15 for 75 minutes

2. 3 hours to 40 minutes

3. The little theater has 320 seats. The auditorium has 1200 seats. Find the ratio of auditorium seats to theater seats.

4. Find the best buy on spaghetti sauce. You have a coupon for 75¢ off Brand X and a coupon for 25¢ off Brand Y.
   28 ounces of Brand X for $3.89
   18 ounces of Brand Y for $1.89
   13 ounces of Brand Z for $1.29

*Find the unknown number in each proportion. Round answers to the nearest hundredth when necessary.*

5. $\dfrac{5}{9} = \dfrac{x}{45}$

6. $\dfrac{3}{1} = \dfrac{8}{x}$

7. $\dfrac{x}{20} = \dfrac{6.5}{0.4}$

8. $\dfrac{2\frac{1}{3}}{x} = \dfrac{\frac{8}{9}}{4}$

*Set up and solve a proportion for each application problem.*

9. Pedro types 240 words in five minutes. How many words can he type in 12 minutes?

10. About 2 out of every 15 people are left-handed. How many of the 650 students in our school would you expect to be left-handed? Round to the nearest whole number.

11. A medication is given at the rate of 8.2 grams for every 50 pounds of body weight. How much should be given to a 145-pound person? Round to the nearest tenth.

12. On a scale model, 1 inch represents 8 feet. If a building in the model is 7.5 inches tall, what is the actual height of the building in feet?

1. _____

2. _____

3. _____

4. _____

5. _____

6. _____

7. _____

8. _____

9. _____

10. _____

11. _____

12. _____

13. ______________

14. ______________

15. ______________

16. ______________

17. ______________

18. ______________

19. ______________

20. ______________

21. ______________

22. ______________

23. ______________

24. ______________

25. ______________

*Choose the figure that matches each label. For right and straight angles, indicate the number of degrees in the angle.*

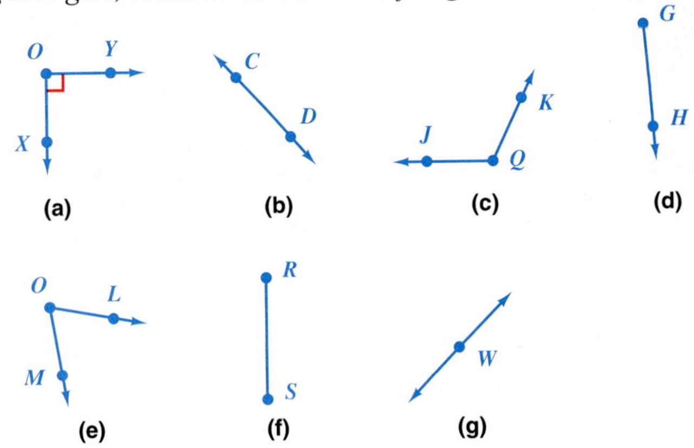
O
Y
X
(a)
C
D
(b)
K
J
Q
(c)
G
H
(d)
O
L
M
(e)
R
S
(f)
W
(g)

**13.** Acute angle is figure ______

**14.** Right angle is figure ______ and its measure is ______

**15.** Ray is figure ______

**16.** Straight angle is figure ______ and its measure is ______

**17.** Write a definition of parallel lines and a definition of perpendicular lines. Make a sketch to illustrate each definition.

**18.** Find the complement of an 81° angle.

**19.** Find the supplement of a 20° angle.

**20.** In the figure below, ∠4 measures 50° and ∠6 measures 95°. Find the measures of the other angles.

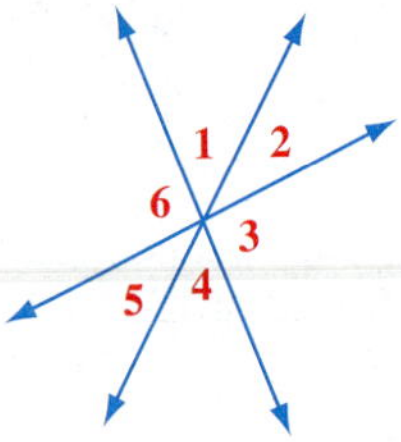
1
2
3
4
5
6

**21.** In the figure below, line $m$ is parallel to line $n$ and ∠3 measures 65°. Find the measures of the other angles.

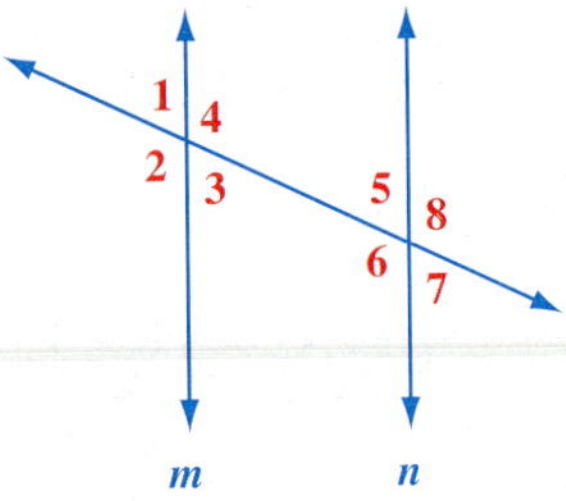
1
4
2
3
5
8
6
7
m
n

*Determine which method can be used to prove that each pair of triangles is congruent.*

**22.**

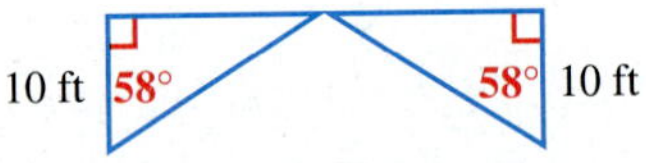
10 ft
58°
58°
10 ft

**23.**

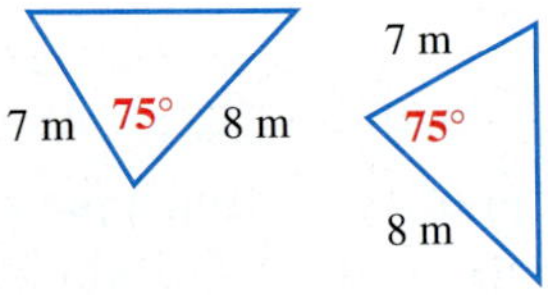
7 m
75°
8 m
7 m
75°
8 m

**24.** Find the unknown lengths in these similar triangles.

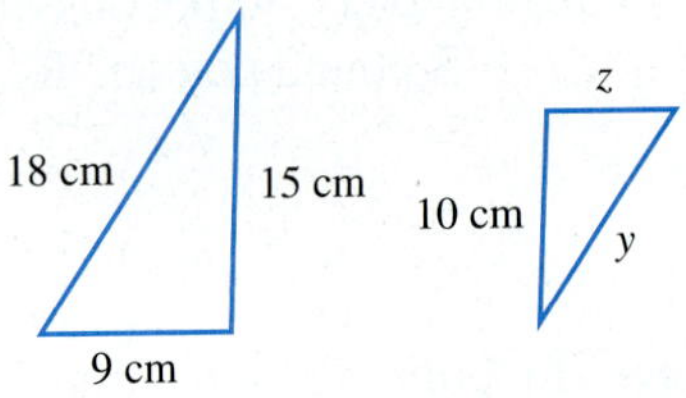
18 cm
15 cm
9 cm
z
10 cm
y

**25.** Find the perimeter of each of these similar triangles.

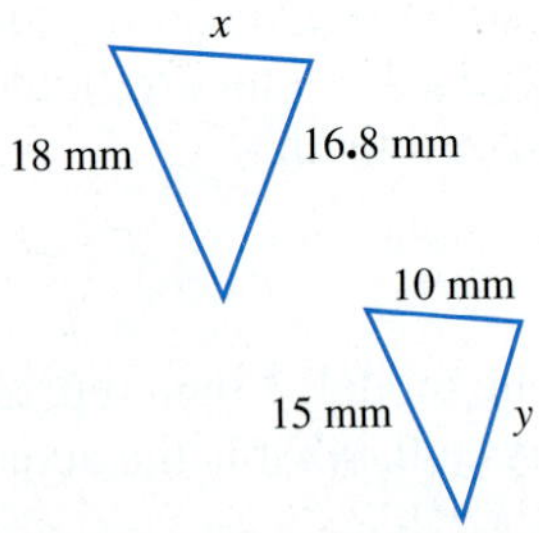
x
18 mm
16.8 mm
10 mm
15 mm
y

# Cumulative Review Exercises  Chapters 1–6

1. Write these numbers in words.
   (a) 77,001,000,805

   (b) 0.02

2. Write these numbers using digits.
   (a) Three and forty thousandths

   (b) Five hundred million, thirty-seven thousand

*Simplify.*

3. $\dfrac{0}{-16}$

4. $|0| + |-6 - 8|$

5. $4\dfrac{3}{4} - 1\dfrac{5}{6}$

6. $\dfrac{h}{5} - \dfrac{3}{10}$

7. $100 - 0.0095$

8. $\dfrac{-4 + 7}{9 - 3^2}$

9. $-6 + 3(0 - 4)$

10. $\dfrac{-8}{\frac{4}{7}}$

11. $\dfrac{5n}{6m^3} \div \dfrac{10}{3m^2}$

12. $(0.06)(-0.007)$

13. $\dfrac{x}{14y} \cdot \dfrac{7}{xy}$

14. $\dfrac{9}{n} + \dfrac{2}{3}$

15. $-40 + 8(-5) + 2^4$

16. $5.8 - (-0.6)^2 \div 0.9$

17. $\left(-\dfrac{1}{2}\right)^3 \left(\dfrac{2}{3}\right)^2$

*Solve each equation. Show your work.*

18. $7y + 5 = -3 - y$

19. $-2 + \dfrac{3}{5}x = 7$

20. $\dfrac{4}{x} = \dfrac{14}{35}$

*Solve each application problem using the six problem-solving steps from Chapter 3.*

21. Tuyen heard on the radio that the temperature had risen 15 degrees by noon, then dropped 23 degrees due to a storm, then risen 5 degrees and was now at 71 degrees. What was the starting temperature?

22. The width of a rectangular swimming pool is 14 ft less than the length. The perimeter of the pool is 100 ft. Find the length and the width.

**498** Chapter 6 Ratio, Proportion, and Line/Angle/Triangle Relationships

*Find the unknown length, perimeter, area, or volume. When necessary, use 3.14 as the approximate value of π and round answers to the nearest tenth.*

23. Find the perimeter and the area.

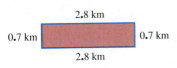

24. Find the diameter, circumference, and area.

25. Find the volume and surface area.

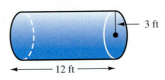

*Use the circle graph of one college's enrollment to complete Exercises 26–29. Where indicated, use front end rounding to estimate the answer. Then find the exact answer.*

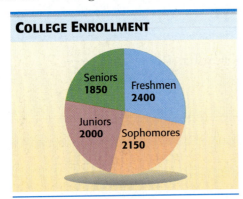

26. Find the total enrollment at the college.

    *Estimate:*

    *Exact:*

27. The college has budgeted $186,400 for freshman orientation. How much is spent on each freshman, to the nearest dollar?

    *Estimate:*

    *Exact:*

28. (a) Write the ratio of freshmen to the total enrollment as a fraction in lowest terms.

    (b) Write the ratio of freshmen and sophomores to juniors and seniors as a fraction in lowest terms.

29. The college collects a $3.75 technology fee from each student to support the computer lab. What total amount is collected?

    *Estimate:*

    *Exact:*

30. The distance around Dunning Pond is $1\frac{1}{10}$ miles. Norma ran around the pond four times in the morning and $2\frac{1}{2}$ times in the afternoon. How far did she run in all?

    *Estimate:*

    *Exact:*

31. Rodney bought 49.8 gallons of gas for his truck while driving 896.5 miles on a vacation. How many miles per gallon did he get, rounded to the nearest tenth?

    *Estimate:*

    *Exact:*

32. The honor society has a goal of collecting 1500 pounds of food to fill Thanksgiving baskets. So far they've collected $\frac{5}{6}$ of their goal. How many more pounds do they need?

33. The directions on a can of plant food call for $\frac{1}{2}$ teaspoon in two quarts of water. How much plant food is needed for five quarts?

# Percent

There are more than 844,000 restaurants in the United States. On a typical day, 4 out of 10 adults eat a meal in a restaurant. (*Source: National Restaurant Association.*) If you're the one paying the bill, how can you quickly estimate a 15% or 20% tip for the server, without pulling out a calculator? To find out, see Section 7.5, Example 3. Then get some practice in Section 7.5, Exercises 11–16. Next time you eat out, you'll know you're leaving the right amount for the tip.

**7.1** The Basics of Percent
**7.2** The Percent Proportion
**7.3** The Percent Equation
**7.4** Problem Solving with Percent
**7.5** Consumer Applications: Sales Tax, Tips, Discounts, and Simple Interest

Summary Exercises on Percent

# 7.1 THE BASICS OF PERCENT

**OBJECTIVES**

1. Learn the meaning of percent.
2. Write percents as decimals.
3. Write decimals as percents.
4. Write percents as fractions.
5. Write fractions as percents.
6. Use 100% and 50%.

**1** Learn the meaning of percent. You have probably seen percents frequently in daily life. The symbol for percent is %. For example, during one day you may leave a 15% tip for the waitress at dinner, pay 7% sales tax on a CD player, and buy shoes at 25% off the regular price. The next day your score on a math test may be 89% correct.

### The Meaning of Percent

A **percent** is a ratio with a denominator of 100. So percent means "per 100" or "how many out of 100." The symbol for percent is %. Read 15% as "fifteen percent."

**Example 1** Understanding Percent

Write a percent to describe each situation.

(a) If you left a $15 tip when the restaurant bill was $100, then you left $15 per $100 or $\frac{15}{100}$ or 15%.

(b) If you pay $7 in tax on a $100 CD player, then the tax rate is $7 per $100 or $\frac{7}{100}$ or 7%.

(c) If you earn 89 points on a 100-point math test, then your score is 89 out of 100 or $\frac{89}{100}$ or 89%.

**Work Problem** ❶ **at the Side.**

---

**1** Write a percent to describe each situation.

(a) You leave a $20 tip for a restaurant bill of $100. What percent tip did you leave?

(b) The tax on a $100 graphing calculator is $5. What is the tax rate?

(c) You earn 94 points on a 100 point test. What percent of the points did you earn?

---

**2** Write percents as decimals. In order to work with percents, you will need to write them as decimals or as fractions. We'll start by writing percents as equivalent decimal numbers. Twenty-five percent, or 25%, means 25 parts out of 100 parts, or $\frac{25}{100}$. Remember that the fraction bar indicates division. So we can write $\frac{25}{100}$ as $25 \div 100$. When you do the division, $25 \div 100$ is $0.25$.

Indicates division → $\frac{25}{100}$ can be written as $25 \div 100 = 0.25$
↑
Indicates division

Another way to remember how to write a percent as a decimal is to use the meaning of the word *percent*. The first part of the word, *per*, is an indicator word for *division*. The last part of the word, *cent*, comes from the Latin word for *hundred*. (Recall that there are 100 *cent*s in a dollar, and 100 years in a *cent*ury.)

25% is 25 **percent**
↓ ↓ ↓
25 ÷ 100 = 0.25

### Writing a Percent as an Equivalent Decimal

To write a percent as a decimal, drop the % symbol and divide by 100.

---

**ANSWERS**

**1.** (a) 20% (b) 5% (c) 94%

Section 7.1 The Basics of Percent **501**

> **Example 2** Writing Percents as Decimals
>
> Write each percent as a decimal.
>
> (a) 47%     47% = 47 ÷ 100 = 0.47    Decimal form
> (b) 3%      3% = 3 ÷ 100 = 0.03      Decimal form
> (c) 28.2%   28.2% = 28.2 ÷ 100 = 0.282   Decimal form
> (d) 100%    100% = 100 ÷ 100 = 1.00  Decimal form
> (e) 135%    135% = 135 ÷ 100 = 1.35  Decimal form

**CAUTION**

In Example 2(d) notice that 100% is 1.00, or 1, which is a whole number. Whenever you have a percent that is *100% or greater*, the equivalent decimal number will be *1 or greater*. Notice in Example 2(e) that 135% is 1.35 (greater than 1).

**❷** Write each percent as a decimal.

(a) 68%

(b) 5%

(c) 40.6%

(d) 200%

(e) 350%

Work Problem ❷ at the Side.

In the exercise set for **Section 5.5**, you discovered a shortcut for *dividing* by 100: Move the decimal point *two* places to the *left*. You can use this shortcut when writing percents as decimals.

> **Example 3** Changing Percents to Decimals by Moving the Decimal Point
>
> Write each percent as a decimal by moving the decimal point two places to the left.
>
> (a) 17%
>
> 17% = 17.%      ← Decimal point starts at far right side.
>
> .17            ← Percent symbol is dropped.
>
> ← Decimal point is moved two places to the left. This is a quick way to divide by 100.
>
> 17% = 0.17
>
> (b) 160%
>
> 160% = 160.% = 1.60 or 1.6     Decimal point starts at far right side. 1.60 is equivalent to 1.6.
>
> (c) 4.9%
>
> 04.9%     0 is attached so the decimal point can be moved two places to the left.
>
> 4.9% = 0.049
>
> (d) 0.6%
>
> 00.6% = 0.006    0 is attached so the decimal point can be moved two places to the left.

**Answers**

**2.** (a) 0.68  (b) 0.05  (c) 0.406
       (d) 2.00 or 2  (e) 3.50 or 3.5

**3** Write each percent as a decimal.

(a) 90%

(b) 9%

(c) 900%

(d) 9.9%

(e) 0.9%

**NOTE**

In Example 3(d) notice that 0.6% is less than 1%. Because 1% is equivalent to 0.01 or $\frac{1}{100}$, any fraction of a percent smaller than 1% is less than 0.01. The decimal equivalent of 0.6% is 0.006, which is less than 0.01.

**Work Problem 3 at the Side.**

**3** **Write decimals as percents.** You can write a decimal as a percent. For example, the decimal 0.25 is the same as the fraction $\frac{25}{100}$.

This fraction means 25 out of 100 parts, or 25%. Notice that multiplying 0.25 by 100 gives the same result.

$$(0.25)(100) = 25\%$$

This result makes sense because we are doing the opposite of what we did to change a percent to a decimal.

To change a percent to a decimal, we *divide* by 100. So, to *reverse* the process and change a decimal to a percent, we *multiply* by 100 instead of divide.

**Writing a Decimal as a Percent**

To write a decimal as a percent, multiply by 100 and attach a % symbol.

**NOTE**

A quick way to *multiply* a number by 100 is to move the decimal point *two* places to the *right*. Notice that this is the opposite of the shortcut for *dividing* by 100.

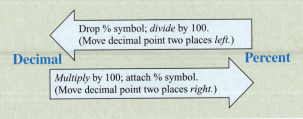

**Example 4** Changing Decimals to Percents by Moving the Decimal Point

Write each decimal as a percent.

(a) 0.21

0.21 ⟶ Decimal point is moved two places to the right.

0.21 = 21% ← Percent symbol is attached after decimal point is moved.
⟶ Decimal point is not written with whole number percents.

(b) 0.529 = 52.9% ← Percent symbol is attached after decimal point is moved.

(c) 1.92 = 192% ← Percent symbol is attached after decimal point is moved.

*Continued on Next Page*

**ANSWERS**

**3.** (a) 0.90 or 0.9  (b) 0.09  (c) 9.00 or 9
(d) 0.099  (e) 0.009

**(d)** 2.5

$$2.\underset{\smile}{50}$$    0 is attached so the decimal point can be moved two places to the right.

$$2.5 = 250\%$$

**(e)** 3

$$3. = 3.\underset{\smile}{00} \quad \text{so} \quad 3 = 300\%$$

### CAUTION

In Examples 4(c), 4(d), and 4(e), notice that 1.92, 2.5, and 3 are greater than 1. Because the number 1 is equivalent to 100%, all *numbers greater than 1* will be equivalent to *percents greater than 100%*.

Work Problem ❹ at the Side.

**4**   **Write percents as fractions.**   Percents can also be written as fractions. Recall that a percent is a ratio with a denominator of 100. For example, 89% is $\frac{89}{100}$. Because the fraction bar indicates division, we are dividing the percent by 100, just as we did when writing a percent as a decimal.

### Writing a Percent as a Fraction

To write a percent as a fraction, drop the % symbol and write the number over 100. Then write the fraction in lowest terms.

**Example 5**   **Writing Percents as Fractions**

Write each percent as a fraction or mixed number in lowest terms.

**(a)** 25%    Drop the % symbol and write 25 over 100.

$$25\% = \frac{25}{100} \leftarrow 25 \text{ per } 100$$

$$= \frac{25 \div 25}{100 \div 25} = \frac{1}{4} \leftarrow \text{Lowest terms}$$

As a check, write 25% as a decimal.

$$25\% = 25 \div 100 = 0.25 \leftarrow \text{Percent sign dropped}$$

Because 0.25 means 25 hundredths,

$$0.25 = \frac{25 \div 25}{100 \div 25} = \frac{1}{4} \leftarrow \text{Same result as above}$$

**(b)** 76%    Drop the % symbol and write 76 over 100.

$$76\% = \frac{76}{100}$$

↑ The percent becomes the numerator.
← The *denominator* is always 100 because percent means parts per 100.

Write $\frac{76}{100}$ in lowest terms.    $\frac{76 \div 4}{100 \div 4} = \frac{19}{25} \leftarrow$ Lowest terms

*Continued on Next Page*

---

**❹** Write each number as a percent.

**(a)** 0.95

**(b)** 0.18

**(c)** 0.09

**(d)** 0.617

**(e)** 0.834

**(f)** 5.34

**(g)** 2.8

**(h)** 4

**ANSWERS**
**4.** **(a)** 95%   **(b)** 18%   **(c)** 9%   **(d)** 61.7%
     **(e)** 83.4%   **(f)** 534%   **(g)** 280%
     **(h)** 400%

**5** Write each percent as a fraction or mixed number in lowest terms.

(a) 50%

(b) 19%

(c) 80%

(d) 6%

(e) 125%

(f) 300%

(c) 150%

$$150\% = \frac{150}{100} = \frac{150 \div 50}{100 \div 50} = \frac{3}{2} = 1\frac{1}{2}  \leftarrow \text{Mixed number}$$

(d) $100\% = \frac{100}{100} = 1 \leftarrow \text{Whole number}$

**NOTE**

Remember that percent means *per 100*.

Work Problem **5** at the Side.

Example 6 shows how to write decimal percents and fraction percents as fractions.

### Example 6 Writing Decimal Percents or Fraction Percents as Fractions

Write each percent as a fraction in lowest terms.

(a) 15.5%

Drop the % symbol and write 15.5 over 100.

$$15.5\% = \frac{15.5}{100}$$

To get a whole number in the numerator, multiply the numerator and denominator by 10. (Multiplying by $\frac{10}{10}$ is the same as multiplying by 1.)

$$\frac{15.5}{100} = \frac{(15.5)(10)}{(100)(10)} = \frac{155}{1000}$$

Write the fraction in lowest terms.

$$\frac{155 \div 5}{1000 \div 5} = \frac{31}{200}$$

(b) $33\frac{1}{3}\%$

Drop the % symbol and write $33\frac{1}{3}$ over 100.

$$33\frac{1}{3}\% = \frac{33\frac{1}{3}}{100}$$

When there is a mixed number in the numerator, write it as an improper fraction. So $33\frac{1}{3}$ is $\frac{100}{3}$.

$$\frac{33\frac{1}{3}}{100} = \frac{\frac{100}{3}}{100}$$

Continued on Next Page

**ANSWERS**

5. (a) $\frac{1}{2}$ (b) $\frac{19}{100}$ (c) $\frac{4}{5}$ (d) $\frac{3}{50}$

  (e) $1\frac{1}{4}$ (f) 3

Now you have a complex fraction (see **Section 4.6**). Rewrite the complex fraction using the ÷ symbol for division. Then follow the steps for dividing fractions.

$$\frac{\frac{100}{3}}{100} = \frac{100}{3} \div 100 = \frac{100}{3} \div \frac{100}{1} = \frac{100}{3} \cdot \frac{1}{100} = \frac{\overset{1}{100} \cdot 1}{3 \cdot \underset{1}{100}} = \frac{1}{3}$$

Reciprocals

> **NOTE**
>
> In Example 6(a) we could have changed 15.5% to $15\frac{1}{2}$% and then written it as the improper fraction $\frac{31}{2}$ over 100. But it is usually easier to work with decimal percents as they are.

Work Problem ❻ at the Side.

**5** **Write fractions as percents.** Recall that to write a percent as a fraction, you *drop* the percent symbol and *divide* by 100, as in Examples 5 and 6. So, to reverse the process and change a fraction to a percent, you *multiply* by 100 and *attach* a percent symbol.

### Writing a Fraction as a Percent

To write a fraction as a percent, multiply by 100 and attach a % symbol. This is the same as multiplying by 100%.

> **NOTE**
>
> Look back at Example 5(d) to see that 100% = 1. Recall that multiplying a number by 1 does *not* change the value of the number. So multiplying by 100% does not change the value of a number; it just gives us an *equivalent percent*.

**Example 7** Writing Fractions as Percents

Write each fraction as a percent. Round to the nearest tenth of a percent, if necessary.

**(a)** $\frac{2}{5}$   Multiply $\frac{2}{5}$ by 100%.

$$\frac{2}{5} = \frac{2}{5} \cdot 100\% = \frac{2}{5} \cdot \frac{100}{1}\% = \frac{2}{5} \cdot \frac{5 \cdot 20}{1}\% = \frac{2 \cdot \overset{1}{\cancel{5}} \cdot 20}{\underset{1}{\cancel{5}} \cdot 1}\%$$

$$= \frac{40}{1}\% = 40\%$$

To check the result, write 40% as $\frac{40}{100}$ and reduce to lowest terms.

$$40\% = \frac{40}{100} = \frac{40 \div 20}{100 \div 20} = \frac{2}{5} \quad \leftarrow \text{Original fraction}$$

*Continued on Next Page*

❻ Write each percent as a fraction in lowest terms.

(a) 18.5%

(b) 87.5%

(c) 6.5%

(d) $66\frac{2}{3}$%

(e) $12\frac{1}{3}$%

(f) $62\frac{1}{2}$%

**Answers**

6. (a) $\frac{37}{200}$   (b) $\frac{7}{8}$   (c) $\frac{13}{200}$   (d) $\frac{2}{3}$
   (e) $\frac{37}{300}$   (f) $\frac{5}{8}$

**7** Write each fraction as a percent. If you're using a calculator, first work each one by hand. Then use your calculator and round to the nearest tenth of a percent if necessary.

(a) $\dfrac{1}{2}$

(b) $\dfrac{3}{4}$

(c) $\dfrac{1}{10}$

(d) $\dfrac{7}{8}$

(e) $\dfrac{5}{6}$

(f) $\dfrac{2}{3}$

**ANSWERS**

7. (a) 50%  (b) 75%  (c) 10%
(d) $87\dfrac{1}{2}$% or 87.5% (Both are exact answers.)

(e) exactly $83\dfrac{1}{3}$%, or 83.3% (rounded)

(f) exactly $66\dfrac{2}{3}$%, or 66.7% (rounded)

---

(b) $\dfrac{5}{8}$    Multiply $\dfrac{5}{8}$ by 100%.

$$\dfrac{5}{8} = \dfrac{5}{8} \cdot 100\% = \dfrac{5}{8} \cdot \dfrac{100}{1}\% = \dfrac{5}{2\cdot 4} \cdot \dfrac{4\cdot 25}{1}\% = \dfrac{5\cdot \cancel{4}\cdot 25}{2\cdot \cancel{4}\cdot 1}\%$$

$$= \dfrac{125}{2}\% = 62\dfrac{1}{2}\%$$

You can also do the last step of simplifying $\dfrac{125}{2}$ on your calculator. Enter $\dfrac{125}{2}$ as 125 ÷ 2 =. The result is 62.5, so $\dfrac{5}{8} = 62\dfrac{1}{2}\%$ or $\dfrac{5}{8} = 62.5\%$.

(c) $\dfrac{1}{6}$    Multiply $\dfrac{1}{6}$ by 100%.

$$\dfrac{1}{6} = \dfrac{1}{6}\cdot 100\% = \dfrac{1}{6}\cdot\dfrac{100}{1}\% = \dfrac{1}{2\cdot 3}\cdot\dfrac{2\cdot 50}{1}\% = \dfrac{1\cdot \cancel{2}\cdot 50}{\cancel{2}\cdot 3\cdot 1}\%$$

$$= \dfrac{50}{3}\% = 16\dfrac{2}{3}\%$$

To simplify $\dfrac{50}{3}$ on your calculator, enter 50 ÷ 3 =. The result is 16.66666666, with the 6 continuing to repeat. The directions say to round to the nearest *tenth* of a percent.

16.66666666%   rounds to   **16.7%**
   ↑
Tenths place

So $\dfrac{1}{6} = 16\dfrac{2}{3}\%$ or $\dfrac{1}{6} \approx 16.7\%$.

---

**Calculator Tip**   In Example 7(a) on the previous page you can use your calculator to write $\dfrac{2}{5}$ as a percent.

*Step 1*   Enter $\dfrac{2}{5}$ as 2 ÷ 5 =. Your calculator shows   **0.4**.
                                                                    ↑
                                                    Decimal equivalent of $\dfrac{2}{5}$

*Step 2*   Change the decimal number 0.4 to a percent by moving the decimal point two places to the right (multiplying by 100).

0.40 = **40%** ← Attach a % symbol.

Try this technique on Examples 7(b) and 7(c) on this page.

For $\dfrac{5}{8}$, enter 5 ÷ 8 =. Your calculator shows   **0.625**.

Move the decimal point in 0.625 two places to the right.

0.625 = **62.5%** ← Attach a % symbol.

For $\frac{1}{6}$, enter 1 ÷ 6 =. Your calculator shows **0.166666666**.

Some calculators show 7 in last place.

Move the decimal point two places to the right. Then round to the nearest tenth.

$$0.166666666 = 16.6666666\% \approx 16.7\%$$ ← Attach a % symbol.

With a fraction such as $\frac{1}{6}$, your calculator gives only an *approximate* answer. You can't get the exact answer of $16\frac{2}{3}\%$ by using your calculator.

**Work Problem ❼ on the Previous Page at the Side.**

**6** **Use 100% and 50%.** When working with percents, it is helpful to have several reference points. 100% and 50% are two helpful reference points.

100% means 100 parts out of 100 parts. That's *all* of the parts. If you pay 100% of a $45 dentist bill, you pay $45 (*all* of it).

**Example 8** **Finding 100% of a Number**

Fill in the blanks.

(a) 100% of $34 is _____.
100% is *all* of the money.
So 100% of $34 is $34.

(b) 100% of 4 cats is _____.
100% is *all* of the cats.
So 100% of 4 cats is 4 cats.

**Work Problem ❽ at the Side.**

50% means 50 parts out of 100 parts, which is *half* of the parts ($\frac{50}{100} = \frac{1}{2}$). 50% of $12 is $6 (*half* of the money).

**Example 9** **Finding 50% of a Number**

Fill in the blanks.

(a) 50% of $20 is _____.
50% is *half* of the money.
So 50% of $20 is $10.

(b) 50% of 280 miles is _____.
50% is *half* of the miles.
So 50% of 280 miles is 140 miles.

**Work Problem ❾ at the Side.**

❽ Fill in the blanks.

(a) 100% of $3.95 is _____.

(b) 100% of 3000 students is _____.

(c) 100% of 7 pages is _____.

(d) 100% of 305 miles is _____.

(e) 100% of $10\frac{1}{2}$ hours is _____.

❾ Fill in the blanks.

(a) 50% of $10 is _____.

(b) 50% of 36 cookies is _____.

(c) 50% of 6000 women is _____.

(d) 50% of 8 hours is _____.

(e) 50% of $2.50 is _____.

**Answers**
8. (a) $3.95  (b) 3000 students  (c) 7 pages
   (d) 305 miles  (e) $10\frac{1}{2}$ hours
9. (a) $5  (b) 18 cookies  (c) 3000 women
   (d) 4 hours  (e) $1.25

# Focus on Real-Data Applications

## Decimals, Percents, and Quilt Patterns

*Please complete the Quilt Pattern activity on page 286 before starting this activity.*

Four of the quilt patterns you worked with on page 286 are repeated here. For each pattern, complete the table by copying the fractions you found and then changing the fractions to decimals and to percents. Use your calculator and round decimals to the nearest thousandth and percents to the nearest tenth when necessary. In the last row of the table, add the fractions. Then add the decimals, and finally, add the percents.

1.

Barbara Fritchie Star

| Color | Fraction | Decimal | Percent |
|---|---|---|---|
| Blue | | | |
| White | | | |
| Total | | | |

2.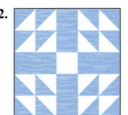

Handy Andy

| Color | Fraction | Decimal | Percent |
|---|---|---|---|
| Blue | | | |
| White | | | |
| Total | | | |

3.

Peace and Plenty

| Color | Fraction | Decimal | Percent |
|---|---|---|---|
| Aqua | | | |
| Yellow | | | |
| Whitish | | | |
| Total | | | |

4.

Cobweb's

| Color | Fraction | Decimal | Percent |
|---|---|---|---|
| Mauve | | | |
| Purple | | | |
| Yellow | | | |
| Total | | | |

5. What pattern do you notice in the Total row in each table?

6. Explain why some of the totals in the Cobweb's table don't quite fit the pattern.

# 7.1 Exercises

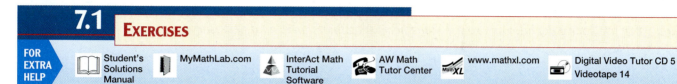

*Write each percent as a decimal. See Examples 2 and 3.*

1. 25%   2. 35%   3. 30%   4. 20%

5. 6%   6. 3%   7. 140%   8. 250%

9. 7.8%   10. 6.7%   11. 100%   12. 600%

13. 0.5%   14. 0.2%   15. 0.35%   16. 0.076%

*Write each decimal as a percent. See Example 4.*

17. 0.5   18. 0.6   19. 0.62   20. 0.18

21. 0.03   22. 0.09   23. 0.125   24. 0.875

25. 0.629   26. 0.494   27. 2   28. 5

29. 2.6   30. 1.8   31. 0.0312   32. 0.0625

*Write each percent as a fraction or mixed number in lowest terms. See Examples 5 and 6.*

33. 20%   34. 40%   35. 50%   36. 75%

37. 55%   38. 35%   39. 37.5%   40. 87.5%

41. 6.25%   42. 43.75%   43. $16\frac{2}{3}$%   44. $83\frac{1}{3}$%

45. 130%   46. 175%   47. 250%   48. 325%

*Write each fraction as a percent. If you're using a calculator, first work each one by hand. Then use your calculator and round to the nearest tenth of a percent if necessary. See Example 7.*

49. $\dfrac{1}{4}$   50. $\dfrac{1}{5}$   51. $\dfrac{3}{10}$   52. $\dfrac{9}{10}$

53. $\dfrac{3}{5}$   54. $\dfrac{3}{4}$   55. $\dfrac{37}{100}$   56. $\dfrac{63}{100}$

57. $\dfrac{3}{8}$   58. $\dfrac{1}{8}$   59. $\dfrac{1}{20}$   60. $\dfrac{1}{50}$

61. $\dfrac{5}{9}$   62. $\dfrac{7}{9}$   63. $\dfrac{1}{7}$   64. $\dfrac{5}{7}$

*In each statement, write percents as decimals and decimals as percents. See Examples 2–4.*

65. In 1900, only 8% of U.S. homes had a telephone. (*Source: Harper's Index.*)

66. In 1900, only 14% of homes in the United States had a bathtub. (*Source: Harper's Index.*)

67. Tornadoes can occur on any day of the year, but 42% of them appear in May and June. (*Source:* National Severe Storms Laboratory.)

68. Only 2.1% of tornadoes occur in December, making it the least likely month for twisters. (*Source:* National Severe Storms Laboratory.)

69. The property tax rate in Alpine County is 0.035.

70. A church building fund has 0.49 of the money needed.

71. The number of people taking CPR training this session is 2 times that of the last session.

72. Attendance at this year's company picnic is 3 times last year's attendance.

Section 7.1   **511**

*Write a fraction and a percent for the shaded part of each figure. Then write a fraction and a percent for the unshaded part of each figure.*

**73.**    **74.**    **75.**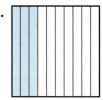

**76.**    **77.** 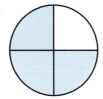   **78.**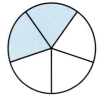

*Complete this table. Write fractions and mixed numbers in lowest terms. See Examples 2–7.*

| | Fraction | Decimal | Percent |
|---|---|---|---|
| 79. | $\frac{1}{100}$ | _____ | _____ |
| 80. | $\frac{1}{10}$ | _____ | _____ |
| 81. | _____ | 0.2 | _____ |
| 82. | _____ | 0.25 | _____ |
| 83. | _____ | _____ | 30% |
| 84. | _____ | _____ | 40% |
| 85. | $\frac{1}{2}$ | _____ | _____ |
| 86. | $\frac{3}{4}$ | _____ | _____ |
| 87. | _____ | _____ | 90% |
| 88. | _____ | _____ | 100% |
| 89. | _____ | 1.5 | _____ |
| 90. | _____ | 2.25 | _____ |

*College students were asked to rank the most important issues they would like presidential candidates to address. The bar graph shows the ranking of these issues and the percent of students selecting each one. Use this graph to answer Exercises 91–94. If the answer is a percent, also write it as a fraction in lowest terms.*

**91.** (a) What was the fifth most important issue?

(b) What portion of the students selected that issue?

**92.** (a) What was the top issue for college students?

(b) What portion of the students selected that issue?

**93.** (a) Which issue was selected by $\frac{1}{2}$ of the students?

(b) Which other issue(s) were selected by *about* $\frac{1}{2}$ of the students?

**EDUCATION IS TOPS**

Education ranks as the most important issue that college students would like presidential candidates to address.

- Education 76%
- Health care 51%
- Gun control 50%
- Environmental issues 49%
- Human rights 44%
- Political reform 39%

*Source:* Greenfield Online, YouthStream: Pulsefinder On-Campus Market Study.

**94.** (a) If education had been selected by 75% instead of 76% of students, what fraction would that be?

(b) If political reform had been selected by 40% instead of 39% of students, what fraction would that be?

*The diagram shows the different types of teeth in an adult's mouth. Use the diagram to answer Exercises 95–100. Write each answer in Exercises 95–98 as a fraction in lowest terms, as a decimal, and as a percent.*

**95.** What portion of an adult's teeth are incisors, designed to bite and cut?

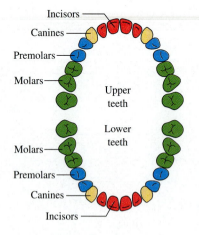

*Source:* Time-Life *Human Body.*

**96.** The pointy canine teeth tear and rip food. They are what portion of an adult's teeth?

**97.** The molars, which grind up food, are what portion of an adult's teeth?

**98.** What portion of an adult's teeth are premolars?

**99.** Some people have four fewer molars than shown in the tooth diagram on the previous page. For these people, the canines are what portion of their teeth? Write your answer as a fraction in lowest terms and as a percent.

**100.** For the adults who have four fewer molars than shown, the incisors are what portion of their teeth? Write your answer as a fraction in lowest terms and as a percent.

**101.** Explain and correct the errors made by students when they used their calculators on these problems.

(a) Write $\frac{7}{20}$ as a percent.

Student entered 7 ÷ 20 =, and the result was 0.35. So $\frac{7}{20} = 0.35\%$

(b) Write $\frac{16}{25}$ as a percent.

Student entered 25 ÷ 16 =, and the result was 1.5625. So $\frac{16}{25} = 156.25\%$

**102.** Explain and correct the errors made by students who moved decimal points to solve these problems.

(a) Write 3.2 as a percent.

03.2 = 0.032   so   3.2 = 0.032%

(b) Write 60% as a decimal.

00.60 = 0.0060   so   60% = 0.0060

*Fill in the blanks. Remember that 100% is* all *of something and 50% is* half *of it. See Examples 8 and 9.*

**103.** (a) 100% of $78 is _____.

(b) 50% of $78 is _____.

**104.** (a) 100% of 5 hours is _____.

(b) 50% of 5 hours is _____.

**105.** (a) 100% of 15 inches is _____.

(b) 50% of 15 inches is _____.

**106.** (a) 100% of $6000 is _____.

(b) 50% of $6000 is _____.

**107.** There are 20 children in the preschool class. 100% of the children are served breakfast and lunch. How many children are served both meals? _____.

**108.** The company owns 345 vans. 100% of the vans are painted white with blue lettering. How many vans are painted white with blue lettering? _____.

**109.** (a) 100% of 2.8 miles is _____.

(b) 50% of 2.8 miles is _____.

**110.** (a) 100% of $2.50 is _____.

(b) 50% of $2.50 is _____.

**111.** John owes $285 for tuition. Financial aid will pay 50% of the cost. Financial aid will pay _____.

**112.** The Animal Humane Society took in 20,000 animals last year. About 50% of them were dogs. The number of dogs taken in was about _____.

**113.** About 50% of the 8200 college students work more than 20 hours per week. How many students is this? _____

**114.** Dylan's test score was 100%. How many of the 35 problems did he work correctly?
_____

**115.** Describe a shortcut way to find 100% of a number. Include two examples to illustrate the shortcut.

**116.** Describe a shortcut way to find 50% of a number. Include two examples to illustrate the shortcut.

# 7.2 THE PERCENT PROPORTION

We will show you two ways to solve percent problems. One is the proportion method, which we discuss in this section. The other is the percent equation method, which we explain in **Section 7.3.**

**OBJECTIVES**

1. Identify the percent, whole, and part.
2. Solve percent problems using the percent proportion.

**1 Identify the percent, whole, and part.** You have learned that a statement of two equivalent ratios is called a proportion (see **Section 6.3**). For example, the fraction $\frac{3}{5}$ is the same as the ratio 3 to 5, and 60% is the ratio 60 to 100. As the figure below shows, these two ratios are equivalent and make a proportion.

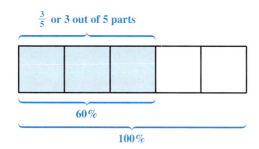

The **percent proportion** can be used to solve percent problems.

### The Percent Proportion

*Percent* is to *100*  **as**  *part* is to *whole*.

$$\underset{\substack{\text{Always 100} \\ \text{because percent} \\ \text{means "per 100."}}}{\frac{\text{percent}}{100}} = \frac{\text{part}}{\text{whole}}$$

In some textbooks the percent proportion is written with the terms *amount* and *base*.

$$\frac{\text{percent}}{100} = \frac{\text{amount}}{\text{base}}$$

Here is the proportion for the figure at the top of the page.

$$\left.\begin{array}{c}60\% \text{ means 60 parts} \\ \text{out of 100 parts.}\end{array}\right\} \quad \frac{60}{100} = \frac{3}{5} \begin{array}{l}\leftarrow \text{Shaded (3 parts)} \\ \leftarrow \text{Whole (5 parts)}\end{array}$$

If we write $\frac{60}{100}$ in lowest terms, it is equal to $\frac{3}{5}$, so the proportion is true.

$$\frac{60}{100} = \frac{60 \div 20}{100 \div 20} = \frac{3}{5} \quad \begin{array}{l}\leftarrow \text{Matches ratio on} \\ \text{right side of proportion.}\end{array}$$

In order to use the percent proportion to solve problems, you must be able to pick out the *percent,* the *whole,* and the *part*. Look for the percent first, because it is the easiest to identify.

### Identifying the Percent

The **percent** is a ratio of a part to a whole, with 100 as the denominator. In a problem, the percent appears with the word *percent* or with the symbol **%** after it.

**1** Identify the percent.

(a) Of the $2000, 15% will be spent on a washing machine.

(b) 60 employees is what percent of 750 employees?

(c) The state sales tax is $6\frac{1}{2}$ percent of the $590 price.

(d) $30 is 150% of what amount of money?

(e) 75 of the 110 rental cars were rented today. What percent were rented?

**2** Identify the whole.

(a) Of the $2000, 15% will be spent on a washing machine.

(b) 60 employees is what percent of 750 employees?

(c) The state sales tax is $6\frac{1}{2}$ percent of the $590 price.

(d) $30 is 150% of what amount of money?

(e) 75 of the 110 rental cars were rented today. What percent were rented?

ANSWERS

1. (a) 15  (b) unknown  (c) $6\frac{1}{2}$
   (d) 150  (e) unknown
2. (a) $2000  (b) 750  (c) $590
   (d) unknown  (e) 110

### Example 1  Identifying the Percent in Percent Problems

Identify the percent in each problem.

(a) 32% of the 900 women were retired. How many were retired?
   ↓
   Percent

The percent is 32. The number 32 appears with the symbol %.

(b) $150 is 25 percent of what number?
   ↓
   Percent

The percent is 25 because 25 appears with the word *percent*.

(c) If 7 students failed, what percent of the 350 students failed?
   ↓
   Percent (unkown)

The word *percent* has no number with it, so the percent is the unknown part of the problem.

**Work Problem 1 at the Side.**

The second thing to look for in a percent problem is the *whole* (sometimes called the *base*).

### Identifying the Whole

The **whole** is the entire quantity. In a problem the *whole* often appears after the word **of**.

### Example 2  Identifying the Whole in Percent Problems

These problems are the same as those in Example 1. Now identify the *whole*.

(a) 32% **of** the 900 women were retired. How many were retired?
   ↓        ↓
   Percent  Whole

The whole is 900. The number 900 appears after the word *of*.

(b) $150 is 25 percent **of** what number?
   ↓              ↓
   Percent        Whole (unknown; follows *of*)

(c) If 7 students failed, what percent **of** the 350 students failed?
   ↓                          ↓
   Percent (unknown)          Whole (follows *of*)

**Work Problem 2 at the Side.**

The third and final thing to identify in a percent problem is the *part* (sometimes called the *amount*).

### Identifying the Part

The **part** is the number being compared to the whole.

**NOTE**

If you have trouble identifying the *part*, find the *whole* and the *percent* first. The remaining number is the *part*.

**Example 3** Identifying the Part in Percent Problems

These problems are the same as those in Examples 1 and 2. Identify the *part*.

(a) 32% **of** the 900 women were retired. How many were retired?
   ↓        ↓                              ↓
  Percent  Whole                      Part (unknown)

The part of the women who were retired is unknown. In other words, some part of 900 women were retired.

(b) $150 is 25 percent **of** what number?
    ↓     ↓              ↓
   Part  Percent    Whole (unknown)

150 is the remaining number, so 150 is the part.

(c) If 7 students failed, what percent **of** 350 students failed?
    ↓                    ↓                  ↓
   Part            Percent (unknown)    Whole

The part of the students who failed is 7 students.

════ Work Problem ❸ at the Side.

**2** Solve percent problems using the percent proportion.

**Example 4** Using the Percent Proportion to Find the Part

Use the percent proportion to answer this question.

15% of $165 is how much money?

Recall that the percent proportion is $\frac{\text{percent}}{100} = \frac{\text{part}}{\text{whole}}$. First identify the percent by looking for the % symbol or the word *percent*. Then look for the *whole* (usually follows the word *of*). Finally, identify the *part*.

15% **of** $165 is how much money?
 ↓      ↓              ↓
Percent Whole      Part (unknown)
      (follows *of*)

Set up the percent proportion. Here we use $n$ as the variable representing the unknown part. You may use any letter you like.

Percent → $\dfrac{15}{100} = \dfrac{n}{165}$ ← Part (unknown)
Always 100 →              ← Whole

Recall from **Section 6.3** that the first step in solving a proportion is to find the cross products.

Step 1  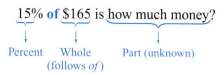   Find the cross products.

$$\frac{15}{100} = \frac{n}{165} \quad \begin{array}{l} 100 \cdot n \\ 15 \cdot 165 \end{array}$$

*Continued on Next Page*

❸ Identify the part.

(a) Of the $2000, 15% will be spent on a washing machine.

(b) 60 employees is what percent of 750 employees?

(c) The state sales tax is $6\frac{1}{2}$ percent of the $590 price.

(d) $30 is 150% of what amount of money?

(e) 75 of the 110 rental cars were rented today. What percent were rented?

**ANSWERS**
3. (a) unknown  (b) 60  (c) unknown
   (d) $30  (e) 75

**4** Use the percent proportion to answer these questions.

(a) 9% of 3250 miles is how many miles?

(b) What is 20% of 180 calories?

(c) 78% of $5.50 is how much?

(d) What is $12\frac{1}{2}\%$ of 400 homes? (*Hint:* Write $12\frac{1}{2}\%$ as 12.5%.)

**5** Use the percent proportion to answer these questions.

(a) 1200 books is what percent of 5000 books?

(b) What percent of $6.50 is $0.52?

(c) 20 athletes is what percent of 32 athletes?

**ANSWERS**

4. (a) $\frac{9}{100} = \frac{n}{3250}$  The part is 292.5 miles.
   (b) $\frac{20}{100} = \frac{n}{180}$  The part is 36 calories.
   (c) $\frac{78}{100} = \frac{n}{5.50}$  The part is $4.29.
   (d) $\frac{12.5}{100} = \frac{n}{400}$  The part is 50 homes.

5. (a) $\frac{p}{100} = \frac{1200}{5000}$; 24%
   (b) $\frac{p}{100} = \frac{0.52}{6.50}$; 8%
   (c) $\frac{p}{100} = \frac{20}{32}$; 62.5% or $62\frac{1}{2}\%$

---

Step 2  $100 \cdot n = 15 \cdot 165$   Show that the cross products are equivalent.
$100 \cdot n = 2475$

Step 3  $\dfrac{100 \cdot n}{100} = \dfrac{2475}{100}$   Divide both sides by 100, the coefficient of the variable term. On the left side, divide out the common factor of 100.

$n = 24.75$   On the right side, $2475 \div 100$ is 24.75.

The part is **$24.75**, so 15% of $165 is $24.75.

### CAUTION
When you use the percent proportion, do *not* move the decimal point in the percent or in the answer.

**Work Problem 4 at the Side.**

### Example 5  Using the Percent Proportion to Find the Percent

Use the percent proportion to answer this question.

8 pounds is what percent of 160 pounds?
↓              ↓                   ↓
Part     Percent (unknown)    Whole
                              (follows *of*)

Set up the percent proportion using $p$ as the variable representing the unknown percent. Then find the cross products.

Percent (unknown) → $\dfrac{p}{100} = \dfrac{8}{160}$ ← Part
Always 100 →                              ← Whole

$\dfrac{p}{100} \times \dfrac{8}{160}$   $100 \cdot 8 = 800$   Cross products
                                          $p \cdot 160$

$p \cdot 160 = 800$   Show that the cross products are equivalent.

$\dfrac{p \cdot 160}{160} = \dfrac{800}{160}$   Divide both sides by 160.

$p = 5$

The percent is **5%**. So 8 pounds is 5% of 160 pounds.

### CAUTION
When you're finding an unknown percent, be careful to label your answer with the % symbol. Do *not* add a decimal point or move the decimal point in your answer.

**Work Problem 5 at the Side.**

Section 7.2 The Percent Proportion   519

### Example 6  Using the Percent Proportion to Find the Whole

Use the percent proportion to answer this question.

$$\underbrace{162}_{\text{Part}} \text{ credits is } \underbrace{90\%}_{\text{Percent}} \underbrace{\text{of how many credits?}}_{\text{Whole (unknown); (follows } of\text{)}}$$

Percent → $\dfrac{90}{100} = \dfrac{162}{n}$ ← Part
Always 100 → ← Whole (unknown)

$\dfrac{90}{100} = \dfrac{162}{n}$   $100 \cdot 162 = 16{,}200$
$90 \cdot n$   } Cross products

$90 \cdot n = 16{,}200$   Show that the cross products are equivalent.

$\dfrac{\overset{1}{\cancel{90}} \cdot n}{\underset{1}{\cancel{90}}} = \dfrac{16{,}200}{90}$   Divide both sides by 90.

$n = 180$

The whole is **180 credits**. So 162 credits is 90% of 180 credits.

━━━━━━━━━━━━━ Work Problem ❻ at the Side.

So far in all the examples, the part has been *smaller* than the whole. This is because all the percents have been less than 100%. Recall that 100% of something is *all* of it. When the percent is *less* than 100%, you have *less* than all of it.

Now let's look at percents *greater* than 100%. For example,

100% of $20 is all of the money, or $20.

150% of $20 is *more* than $20.

| 100% | + | 50% | = | 150% |
| of the money is | | of the money is | | of the money is |
| $20 | + | $10 | = | $30 |

So 150% of $20 is $30.

When the percent is *greater* than 100%, the part is *larger* than the whole, as you'll see in Example 7.

---

❻ Use the percent proportion to answer these questions.

(a) 37 cars is 74% of how many cars?

(b) 45% of how much money is $139.59?

(c) 1.2 tons is $2\tfrac{1}{2}\%$ of how many tons?

---

ANSWERS

6. (a) $\dfrac{74}{100} = \dfrac{37}{n}$; 50 cars

(b) $\dfrac{45}{100} = \dfrac{139.59}{n}$; $310.20

(c) $\dfrac{2.5}{100} = \dfrac{1.2}{n}$; 48 tons

**7** Use the percent proportion to answer each question.

(a) 350% of $6 is how much?

(b) 23 hours is what percent of 20 hours?

(c) What percent of $47.32 is $106.47?

---

**Example 7** Working with Percents Greater Than 100%

Use the percent proportion to answer each question.

(a) How many students is 210% of 40 students?

↓ ↓ ↓
Part (unknown)  Percent  Whole (follows *of*)

Percent → $\dfrac{210}{100} = \dfrac{n}{40}$ ← Part (unknown)
Always 100 →                                 ← Whole

$\dfrac{210}{100} \bowtie \dfrac{n}{40}$   $\begin{array}{l}100 \cdot n \\ 210 \cdot 40 = 8400\end{array}$  Cross products

$100 \cdot n = 8400$   Show that the cross products are equivalent.

$\dfrac{\cancel{100} \cdot n}{\cancel{100}} = \dfrac{8400}{100}$   Divide both sides by 100.

$n = 84$

The part is **84 students**, which is *more* than the whole of 40 students. This result makes sense because the percent is 210%. If it was exactly 200%, we would have *2 times the whole,* and 2 times 40 students is 80 students. So 210% should be even a little more than 80 students. Our answer of 84 students is reasonable.

(b) $68 is what percent of $50?

↓ ↓ ↓
Part   Percent   Whole
       (unknown) (follows *of*)

Percent (unknown) → $\dfrac{p}{100} = \dfrac{68}{50}$ ← Part
Always 100 →                                         ← Whole

$\dfrac{p}{100} \bowtie \dfrac{68}{50}$   $\begin{array}{l}100 \cdot 68 = 6800 \\ p \cdot 50\end{array}$  Cross products

$p \cdot 50 = 6800$   Show that the cross products are equivalent.

$\dfrac{p \cdot \cancel{50}}{\cancel{50}} = \dfrac{6800}{50}$   Divide both sides by 50.

$p = 136$

The percent is **136%**. This result makes sense because $68 is *more* than $50, so $68 has to be *more than 100%* of $50.

**Work Problem 7 at the Side.**

---

**ANSWERS**

**7.** (a) $\dfrac{350}{100} = \dfrac{n}{6}$; $21   (b) $\dfrac{p}{100} = \dfrac{23}{20}$; 115%

(c) $\dfrac{p}{100} = \dfrac{106.47}{47.32}$; 225%

## 7.2 Exercises

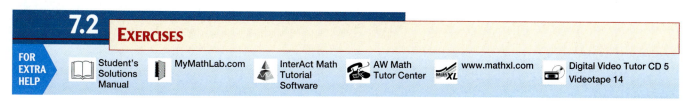

*Use the percent proportion to answer these questions. If necessary, round money answers to the nearest cent and percent answers to the nearest tenth of a percent. See Examples 1–7.*

1. What is 10% of 3000 runners?

2. What is 35% of 2340 volunteers?

3. 4% of 120 feet is how many feet?

4. 9% of $150 is how much money?

5. 16 pepperoni pizzas is what percent of 32 pizzas?

6. 35 hours is what percent of 140 hours?

7. What percent of 200 calories is 16 calories?

8. What percent of 350 parking spaces is 7 handicapped parking spaces?

9. 495 successful students is 90% of what number of students?

10. 84 letters is 28% of what number of letters?

11. $12\frac{1}{2}$% of what amount is $3.50?

12. $5\frac{1}{2}$% of what amount is $17.60?

13. 250% of 7 hours is how long?

14. What is 130% of 60 trees?

15. What percent of $172 is $32?

16. $14 is what percent of $398?

**17.** 748 books is 110% of what number of books?

**18.** 145% of what number of inches is 11.6 inches?

**19.** What is 14.7% of $274?

**20.** 8.3% of $43 is how much?

**21.** 105 employees is what percent of 54 employees?

**22.** What percent of 46 animals is 100 animals?

**23.** $0.33 is 4% of what amount?

**24.** 6% of what amount is $0.03?

**25.** A student turned in the following answers on a test. You can see that *two* of the answers are incorrect *without working the problems*. Find the incorrect answers and explain how you identified them (without actually solving the problems).

$$50\% \text{ of } \$84 \text{ is } \underline{\$42}$$
$$150\% \text{ of } \$30 \text{ is } \underline{\$20}$$
$$25\% \text{ of } \$16 \text{ is } \underline{\$32}$$
$$100\% \text{ of } \$217 \text{ is } \underline{\$217}$$

**26.** Name the three parts in a percent problem. For each of these three parts, write a sentence telling how you will identify it.

**27.** Explain and correct the *two* errors that a student made when solving this problem: $14 is what percent of $8?

$$\frac{p}{100} = \frac{8}{14} \qquad p \cdot 14 = 100 \cdot 8$$

$$\frac{p \cdot \cancel{14}^1}{\cancel{14}_1} = \frac{800}{14}$$

$$p \approx 57.1$$

The answer is 57.1 (rounded).

**28.** Explain and correct the *two* errors that a student made when solving this problem: 9 children is 30% of what number of children?

$$\frac{30}{100} = \frac{n}{9} \qquad 100 \cdot n = 30 \cdot 9$$

$$\frac{\cancel{100}^1 \cdot n}{\cancel{100}_1} = \frac{270}{100}$$

$$n = 2.7$$

The answer is 2.7%.

## 7.3 THE PERCENT EQUATION

**OBJECTIVES**

1. Estimate answers to percent problems involving 25%, 10%, and 1%.
2. Solve basic percent problems by using the percent equation.

**1** Estimate answers to percent problems involving 25%, 10%, and 1%. Before showing you the percent equation, we need to do some more estimation. As you have learned when working with integers, fractions, and decimals, it is always a good idea to estimate the answer. Doing so helps you catch mistakes. Also, when you're out shopping or eating in a restaurant, you will be able to estimate the sales tax, discount, or tip.

In **Section 7.1** we used shortcuts for 100% of a number (all of the number) and 50% of a number (divide the number by 2). Now let's look at quick ways to work with 25%, 10%, and 1%.

25% means 25 parts out of 100 parts, or $\frac{25}{100}$, which is the same as $\frac{1}{4}$.

25% of $40 would be $\frac{1}{4}$ of $40, or $10.

A quick way to find $\frac{1}{4}$ of a number is to *divide it by 4*. Recall that the denominator, 4, tells you that the whole is divided into 4 equal parts.

### Example 1  Estimating 25% of a Number

Estimate the answer to each question.

**(a)** What is 25% of $817?
Use front end rounding to round $817 to $800. Then divide $800 by 4. The estimate is **$200**.

**(b)** Find 25% of 19.7 miles.
Use front end rounding to round 19.7 miles to 20 miles. Then divide 20 miles by 4. The estimate is **5 miles**.

**(c)** 25% of 49 days is how long?
You could round 49 days to 50 days, using front end rounding. Then divide 50 by 4 to get an estimate of **12.5 days**.

However, the division step is simpler if you notice that 48 is a multiple of 4. You can round 49 days to 48 days and divide by 4 to get an estimate of **12 days**. Either way gives you a fairly good idea of the correct answer.

**1** Estimate the answer to each question.

**(a)** What is 25% of $110.38?

**(b)** Find 25% of 7.6 hours.

**(c)** 25% of 34 pounds is how many pounds?

== Work Problem **1** at the Side. ==

Ten percent, or 10%, means 10 parts out of 100 parts or $\frac{10}{100}$, which is the same as $\frac{1}{10}$. A quick way to find $\frac{1}{10}$ of a number is to *divide it by 10*. The denominator, 10, tells you that the whole is divided into 10 equal parts. The shortcut for dividing by 10 is to move the decimal point *one* place to the *left*.

### Example 2  Finding 10% of a Number by Moving the Decimal Point

Find the *exact* answer to each question by moving the decimal point. Then use front end rounding to get an *estimated* answer.

**(a)** What is 10% of $817?
To find 10% of $817, divide $817 by 10. Do the division by moving the decimal point *one* place to the *left*. The decimal point starts at the far right side of $817.

$$10\% \text{ of } \$817. = \$81.70 \leftarrow \text{Exact}$$

↑ Write this 0 because it's money.

The *exact* answer is **$81.70**.

*Continued on Next Page*

**ANSWERS**

1. **(a)** $100 ÷ 4 gives an estimate of $25.
   **(b)** 8 hours ÷ 4 gives an estimate of 2 hours.
   **(c)** 30 pounds ÷ 4 gives an estimate of 7.5 pounds, or 32 pounds ÷ 4 gives an estimate of 8 pounds.

❷ First, find the *exact* answer to each question by moving the decimal point. Then, round to get an *estimated* answer.

(a) What is 10% of $110.38?

(b) Find 10% of 7.6 hours.

(c) 10% of 34 pounds is how many pounds?

❸ First find the *exact* answer to each question by moving the decimal point. Then round to get an *estimated* answer.

(a) What is 1% of $110.38?

(b) Find 1% of 7.6 hours.

(c) 1% of 34 pounds is how many pounds?

ANSWERS
2. (a) $110.38 = $11.038 (exact); estimate is $10.
   (b) 7.6 hours = 0.76 hour (exact); estimate is 0.8 hour.
   (c) 34 pounds = 3.4 pounds (exact); estimate is 3 pounds.
3. (a) $110.38 = $1.1038 (exact); estimate is $1.
   (b) 07.6 hours = 0.076 hour (exact); estimate is 0.08 hour.
   (c) 34 pounds = 0.34 pound (exact); estimate is 0.3 pound.

---

For an *estimate*, you could round $817 to **$800** and then move the decimal point.

$$10\% \text{ of } \$800. = \$80 \leftarrow \text{Estimate}$$

(b) Find 10% of 19.7 miles.
   Move the decimal point *one* place to the *left*.

$$10\% \text{ of } 19.7 \text{ miles} = 1.97 \text{ miles} \leftarrow \text{Exact}$$

The *exact* answer is **1.97 miles**.

For an *estimate*, you could round 19.7 miles to **20 miles** and then move the decimal point.

$$10\% \text{ of } 20. \text{ miles} = 2.0 \quad \text{or} \quad 2 \text{ miles} \leftarrow \text{Estimate}$$

**Work Problem ❷ at the Side.**

One percent, or 1%, is 1 part out of 100 parts or $\frac{1}{100}$. This time the denominator of 100 tells you that the whole is divided into 100 parts. Recall that a quick way to divide by 100 is to move the decimal point *two* places to the *left*.

**Example 3** Finding 1% of a Number by Moving the Decimal Point

Find the *exact* answer to each question by moving the decimal point. Then use front end rounding to get an *estimated* answer.

(a) What is 1% of $817?
   To find 1% of $817, divide $817 by 100. Do the division by moving the decimal point *two* places to the *left*.

$$1\% \text{ of } \$817. = \$8.17 \leftarrow \text{Exact}$$

The *exact* answer is **$8.17**.

For an *estimate*, you could round $817 to **$800** and then move the decimal point.

$$1\% \text{ of } \$800. = \$8.00 \quad \text{or} \quad \$8 \leftarrow \text{Estimate}$$

(b) Find 1% of 19.7 miles.
   Move the decimal point *two* places to the *left*.

$$1\% \text{ of } 19.7 \text{ miles} = 0.197 \text{ mile} \leftarrow \text{Exact}$$

The *exact* answer is **0.197 mile**.

For an *estimate*, you could round 19.7 mile to **20 miles** and then move the decimal point.

$$1\% \text{ of } 20. \text{ miles} = 0.20 \quad \text{or} \quad 0.2 \text{ miles} \leftarrow \text{Estimate}$$

**Work Problem ❸ at the Side.**

Here is a summary of the shortcuts you can use with percents.

### Percent Shortcuts

**200% of a number** is 2 times the number; **300% of a number** is 3 times the number, and so on.
**100% of a number** is the entire number.
To find **50% of a number**, divide the number by 2.
To find **25% of a number**, divide the number by 4.
To find **10% of a number**, divide the number by 10. To do the division, move the decimal point in the number *one* place to the *left*.
To find **1% of a number**, divide the number by 100. To do the division, move the decimal point in the number *two* places to the *left*.

**2** **Solve basic percent problems by using the percent equation.** In **Section 7.2** you used a proportion to solve percent problems. Now you will learn how to solve these problems by using the percent equation.

> **Percent Equation**
>
> $$\text{percent } \textbf{of} \text{ whole} = \text{part}$$
>
> The word **of** indicates multiplication, so the **percent equation** becomes
>
> $$\text{percent} \cdot \text{whole} = \text{part}$$
>
> Be sure to write the percent as a decimal or fraction before using the equation.

The percent equation is just a rearrangement of the percent proportion. Recall that in the proportion you wrote the percent over 100. Because there is no 100 in the equation, you have to change the percent to a decimal or fraction by dividing by 100 *before* using the equation.

> **NOTE**
>
> Once you have set up a percent equation, we encourage you to use your calculator to do the multiplying or dividing needed to solve the equation. For this reason, we will always write the percent as a decimal. If you're doing the problems by hand, changing the percent to a fraction may be easier at times. Either method will work.

Examples 4, 5, and 6 ask the same percent questions that were asked in the examples in **Section 7.2.** There we used a proportion to answer each question. Now we will use an equation to answer them. You can then compare the equation method with the proportion method.

### Example 4 — Using the Percent Equation to Find the Part

Write and solve a percent equation to answer each question.

**(a)** 15% of $165 is how much money?

Translate the sentence into an equation. Recall that *of* indicates multiplication and *is* translates to the equal sign. The percent must be written in decimal form. Use any letter you like to represent the unknown quantity. (We will use $n$ for an unknown number and $p$ for an unknown percent.)

15.% of $165 is how much money?

Write the percent as a decimal.

$$0.15 \cdot 165 = n$$

To solve the equation, simplify the left side, multiplying 0.15 by 165.

$$0.15(165) = n$$
$$24.75 = n$$

So 15% of $165 is **$24.75**, which matches the answer obtained by using a proportion (see Example 4 in **Section 7.2**).

*Continued on Next Page*

**4** Write and solve an equation to answer each question.

(a) 9% of 3250 miles is how many miles?

(b) 78% of $5.50 is how much?

(c) What is $12\frac{1}{2}\%$ of 400 homes? (*Hint:* Write $12\frac{1}{2}\%$ as 12.5%. Then move the decimal point two places to the left.)

(d) How much is 350% of $6?

**Answers**
4. (a) $0.09\ (3250) = n$; 292.5 miles
(b) $0.78\ (5.50) = n$; $4.29
(c) $n = 0.125\ (400)$; 50 homes
(d) $n = 3.5\ (6)$; $21

**Check** Use estimation to check that the solution is reasonable. First find 10% of $165 by moving the decimal point.

10% of 165. is $16.50 and

5% of $165 would be half as much, that is, half of $16.50 or about $8.

So the *estimate* for 15% of $165 is $16.50 + $8 = $24.50. The exact answer of $24.75 is very close to this estimate, so it is reasonable.

(b) How many students is 210% of 40 students?
   Translate the sentence into an equation. Write the percent in decimal form.

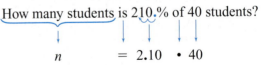

$$n = 2.10 \cdot 40$$

This time the two sides of the percent equation are reversed, so

$$\text{part} = \text{percent} \cdot \text{whole}.$$

Recall that the variable may be on either side of the equal sign. To solve the equation, simplify the right side, multiplying 2.10 by 40.

$$n = 2.10\ (40)$$
$$n = 84$$

So **84 students** is 210% of 40 students. This matches the answer obtained by using a proportion (see Example 7(a) in **Section 7.2**).

**Check** Use estimation to check that the solution is reasonable. 210% is close to 200%.

200% of 40 students is 2 times 40 students = 80 students ← Estimate

The exact answer of 84 students is reasonable.

**Work Problem 4 at the Side.**

**Example 5** Using the Percent Equation to Find the Percent

Write and solve a percent equation to answer each question.

(a) 8 pounds is what percent of 160 pounds?
   Translate the sentence into an equation. This time the percent is unknown. Do *not* move the decimal point in the other numbers.

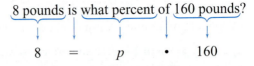

$$8 = p \cdot 160$$

To solve the equation, divide both sides by 160.

On the left side, divide 8 by 160.   $\dfrac{8}{160} = \dfrac{p \cdot \cancel{160}}{\cancel{160}}$   On the right side, divide out the common factor of 160.

Solution in *decimal* form   $0.05 = p$

Now multiply the solution by 100 to change it from a *decimal* to a *percent*.

$$0.05 = 5\%$$

So 8 pounds is **5%** of 160 pounds. This matches the answer obtained by using a proportion (see Example 5 in **Section 7.2**).

*Continued on Next Page*

**Check** The solution makes sense because 10% of 160 pounds would be 16 pounds.

$$10\% \text{ of } 160 \text{ pounds is } 16 \text{ pounds} \quad \text{so}$$

5% of 160 pounds is half as much, that is, half of 16 pounds, or 8 pounds.

8 pounds matches the number given in the original problem, so 5% is the correct solution.

**(b)** What percent of $50 is $68?
Translate the sentence into an equation and solve it.

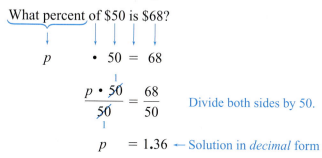

$$\frac{p \cdot 50}{50} = \frac{68}{50} \quad \text{Divide both sides by 50.}$$

$$p = 1.36 \leftarrow \text{Solution in } decimal \text{ form}$$

Now multiply the solution by 100 to change it from a *decimal* to a *percent*.

$$1.36 = 136\%$$

So **136%** of $50 is $68.

**Check** The solution makes sense because 100% of $50 would be $50 (all of it), and 200% of $50 would be 2 times $50, or $100. So $68 has to be between 100% and 200%.

136% is between 100% and 200%. → 100% of $50 = $50   ← $68 is between $50 and $100.
200% of $50 = $100

The solution of 136% fits the conditions.

---

**CAUTION**

When you use an equation to solve for an unknown percent, *the solution will be in decimal form.* Remember to **multiply the solution by 100** to change it from decimal form to a percent. The shortcut is to move the decimal point in the solution *two* places to the *right* and attach a % symbol.

Work Problem ❺ at the Side.

---

❺ Write and solve an equation to answer each question.

**(a)** 1200 books is what percent of 5000 books?

**(b)** 23 hours is what percent of 20 hours?

**(c)** What percent of $6.50 is $0.52?

---

**ANSWERS**

**5. (a)** $1200 = p \cdot 5000$; $0.24 = 24\%$

   **(b)** $23 = p \cdot 20$; $1.15 = 115\%$

   **(c)** $p \cdot 6.50 = 0.52$; $0.08 = 8\%$

**6** Write and solve an equation to answer each question.

(a) 74% of how many cars is 37 cars?

(b) 1.2 tons is $2\frac{1}{2}$% of how many tons?

(c) 216 calculators is 160% of how many calculators?

**Example 6  Using the Percent Equation to Find the Whole**

Write and solve a percent equation to answer each question.

(a) 162 credits is 90% of how many credits?
Translate the sentence into an equation. Write the percent in decimal form.

162 credits is 90.% of how many credits?

$$162 = 0.90 \cdot n$$

Recall that 0.90 is equivalent to 0.9, so use 0.9 in the equation.

$$\frac{162}{0.9} = \frac{0.9 \cdot n}{0.9} \quad \text{Divide both sides by 0.9.}$$

$$180 = n$$

So 162 credits is 90% of **180 credits**.

**Check** The solution makes sense because 90% of 180 credits should be 10% less than 100% of the credits, and 10% of 180. credits is 18 credits.

| 100% of 180 credits | − | 10% of 180 credits | = | 90% of 180 credits |
|---|---|---|---|---|
| ↓ | | ↓ | | ↓ |
| **180 credits** | − | **18 credits** | = | **162 credits** ← Matches the number given in the original problem |

(b) 250% of what amount is $75?
Translate the sentence into an equation. Write the percent in decimal form.

250.% of what amount is $75.

$$2.5 \cdot n = 75$$

$$\frac{2.5 \cdot n}{2.5} = \frac{75}{2.5} \quad \text{Divide both sides by 2.5.}$$

$$n = 30$$

So 250% of **$30** is $75.

**Check** The solution makes sense because 200% of $30 is 2 times $30 = $60, and 50% of $30 is $30 ÷ 2 = $15.

| 200% of $30 | + | 50% of $30 | = | 250% of $30 |
|---|---|---|---|---|
| ↓ | | ↓ | | ↓ |
| **$60** | + | **$15** | = | **$75** ← Matches the number given in the original problem |

**Work Problem** 6 **at the Side.**

---

**ANSWERS**
6. (a) $0.74 \cdot n = 37$;  50 cars
   (b) $1.2 = 0.025 \cdot n$;  48 tons
   (c) $216 = 1.6 \cdot n$;  135 calculators

Section 7.3   **529**

## 7.3 EXERCISES

FOR EXTRA HELP

  Student's Solutions Manual   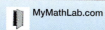 MyMathLab.com    InterAct Math Tutorial Software    AW Math Tutor Center    www.mathxl.com    Digital Video Tutor CD 5 Videotape 14

*Use your estimation skills and the percent shortcuts to select the most reasonable answers. Circle your choices. Do **not** write an equation and solve it. See Examples 1–3.*

1. Find 50% of 3000 patients.

   150 patients   1500 patients   300 patients

2. What is 50% of 192 pages?

   48 pages   384 pages   96 pages

3. 25% of $60 is how much?

   $15   $6   $30

4. Find 25% of $2840.

   $28.40   $710   $284

5. What is 10% of 45 pounds?

   0.45 pounds   22.5 pounds   4.5 pounds

6. 10% of 7 feet is how many feet?

   0.7 feet   3.5 feet   14 feet

7. Find 200% of $3.50.

   $0.35   $1.75   $7.00

8. What is 300% of $12?

   $4   $36   $1.20

9. 1% of 5200 students is how many students?

   520 students   52 students   2600 students

10. Find 1% of 460 miles.

    0.46 mile   46 miles   4.6 miles

11. Find 10% of 8700 cell phones.

    8700 phones   4350 phones   870 phones

12. 25% of 128 CDs is how many CDs?

    64 CDs   32 CDs   1280 CDs

13. What is 25% of 19 hours?

    4.75 hours   1.9 hours   2.5 hours

14. What is 1% of $37?

    $370   $3.70   $0.37

15. **(a)** Describe a shortcut for finding 10% of a number and explain *why* your shortcut works.

    **(b)** Once you know 10% of a certain number, explain how you could use that information to find 20% and 30% of the same number.

16. **(a)** Describe a shortcut for finding 1% of a number and explain *why* it works.

    **(b)** Once you know 1% of a certain number, explain how you could use that information to find 2% and 3% of the same number.

Chapter 7 Percent

*Write and solve an equation to answer each question. See Examples 4–6.*

**17.** 35% of 660 programs is how many programs?

**18.** 55% of 740 canisters is how many canisters?

**19.** 70 truckloads is what percent of 140 truckloads?

**20.** 30 crew members is what percent of 75 crew members?

**21.** 476 circuits is 70% of what number of circuits?

**22.** 621 tons is 45% of what number of tons?

**23.** $12\frac{1}{2}$% of what number of people is 135 people?

**24.** $6\frac{1}{2}$% of what number of bottles is 130 bottles?

**25.** What is 65% of 1300 species?

**26.** What is 75% of 360 dosages?

**27.** 4% of $520 is how much?

**28.** 7% of $480 is how much?

**29.** 38 styles is what percent of 50 styles?

**30.** 75 offices is what percent of 125 offices?

**31.** What percent of $264 is $330?

**32.** What percent of $480 is $696?

33. 141 employees is 3% of what number of employees?

34. 16 books is 8% of what number of books?

35. 32% of 260 quarts is how many quarts?

36. 44% of 430 liters is how many liters?

37. $1.48 is what percent of $74?

38. $0.51 is what percent of $8.50?

39. How many tablets is 140% of 500 tablets?

40. How many patients is 175% of 540 patients?

41. 40% of what number of salads is 130 salads?

42. 75% of what number of wrenches is 675 wrenches?

43. What percent of 160 liters is 2.4 liters?

44. What percent of 600 miles is 7.5 miles?

45. 225% of what number of gallons is 11.25 gallons?

46. 180% of what number of ounces is 6.3 ounces?

47. What is 12.4% of 8300 meters?

48. What is 13.2% of 9400 acres?

**49.** Explain and correct the error in each of these solutions.

(a) 3 hours is what percent of 15 hours?

$$3 = p \cdot 15$$

$$\frac{3}{15} = \frac{p \cdot \cancel{15}}{\cancel{15}}$$

$$0.2 = p$$

The answer is 0.2%.

(b) $50 is what percent of $20?

$$50 \cdot p = 20$$

$$\frac{\cancel{50} \cdot p}{\cancel{50}} = \frac{20}{50}$$

$$p = 0.40 = 40\%$$

The answer is 40%.

**50.** Explain and correct the error in each of these solutions.

(a) 12 inches is 5% of what number of inches?

$$12 \cdot 0.05 = n$$

$$0.6 = n$$

The answer is 0.6 inch.

(b) What is 4% of 30 pounds?

$$n = 4 \cdot 30$$

$$n = 120$$

The answer is 120 pounds.

---

**RELATING CONCEPTS (Exercises 51–52)** **FOR INDIVIDUAL OR GROUP WORK**

*Use your knowledge of fractions to work Exercises 51 and 52 in order.*

**51.** Suppose that you have this problem: $33\frac{1}{3}\%$ of $162 is how much?

(a) First, change $33\frac{1}{3}\%$ to a fraction. (See **Section 7.1** for help.) Then, write an equation and solve it.

(b) Now solve the problem by changing $33\frac{1}{3}\%$ to a decimal. (*Hint:* Look at part (a) to see what fraction is equivalent to $33\frac{1}{3}\%$. Change the fraction to a decimal, using your calculator. Keep *all* the decimal places shown in the calculator's display window. Now write the equation and solve it, using your calculator.)

(c) Compare your answers from part (a) and part (b). How different are they?

**52.** Now suppose that you have this problem: 22 cans is $66\frac{2}{3}\%$ of what number of cans?

(a) First, change $66\frac{2}{3}\%$ to a fraction. (See **Section 7.1** for help.) Then, write an equation and solve it. (See **Section 4.7** for help.)

(b) Now solve the problem by changing $66\frac{2}{3}\%$ to a decimal. Use your calculator and keep all the decimal places shown. Now write the equation and solve it.

(c) Compare your answers from part (a) and part (b). How different are they?

# 7.4 PROBLEM SOLVING WITH PERCENT

**OBJECTIVES**
1. Solve percent application problems.
2. Solve problems involving percent of increase or decrease.

**1** Solve percent application problems. Solving percent problems involves identifying three items: the *percent*, the *whole*, and the *part*. Then you can write a percent equation and solve it to answer the question in the problem. Use the six problem-solving steps from **Section 3.3**.

### Example 1  Finding the Part

A new low-income housing project charges 30% of a family's income as rent. The Smiths' family income is $1260 per month. How much will the Smiths pay for rent?

*Step 1* **Read the problem.** It is about a family paying part of its income for rent.

> Unknown: amount of rent
> Known: 30% of income paid for rent; $1260 monthly income

*Step 2* **Assign a variable.** There is only one unknown, so let *n* be the amount paid for rent.

*Step 3* **Write an equation.** Use the percent equation, percent • whole = part. Recall that the *whole* often follows the word *of*.

The percent is given in the problem: 30%. The key word **of** appears *right after* 30%, which means that you can use the phrase "30% **of** a family's income" to help you write one side of the equation. Write the percent as a decimal.

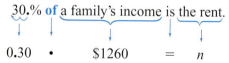

$$0.30 \cdot \$1260 = n$$

*Step 4* **Solve the equation.** Simplify the left side, multiplying 0.30 by 1260.

$$0.30(1260) = n$$
$$378 = n$$

*Step 5* **State the answer.** The Smith family will pay **$378** for rent.

*Step 6* **Check** the solution. The solution of $378 makes sense because 10% of $1260 is $126, so 30% would be 3 times $126, or $378.

                              Work Problem **1** at the Side.

● About 65% of the students at City Center College receive some form of financial aid. How many of the 9280 students enrolled this year are receiving aid? Use the six problem-solving steps.

### Example 2  Finding the Part

When Britta received her first $180 paycheck as a math tutor, $12\frac{1}{2}\%$ was withheld for federal income tax. How much was withheld?

*Step 1* **Read** the problem. It is about part of Britta's pay being withheld for taxes.

> Unknown: amount withheld for taxes
> Known: $12\frac{1}{2}\%$ of earnings withheld; $180 in pay

*Step 2* **Assign a variable.** There is only one unknown, so let *n* be the amount withheld for taxes.

*Continued on Next Page*

**ANSWERS**
1. Let *n* be number of students receiving aid.
   $0.65(9280) = n$
   6032 students receive aid.
   *Check:* 10% of 9280 students is 928, which rounds to 900 students. So 60% would be 6 times 900 students = 5400 students, and 70% would be 7 • 900 = 6300. The solution falls between 5400 and 6300 students, so it is reasonable.

❷ There were 50 points on the first math test. Hue's score was 83% correct. How many points did Hue earn? Use the six problem-solving steps.

*Step 3* **Write an equation.** Use the percent equation. The *percent* is given: $12\frac{1}{2}\%$. Write $12\frac{1}{2}\%$ as $12.5\%$ and then move the decimal point two places to the left.

The key word *of* doesn't appear after $12\frac{1}{2}\%$. Instead, think about whether you know the *whole* or the *part*. You know Britta's *whole* paycheck is $180, but you do *not* know what *part* of it was withheld.

$$\text{percent} \cdot \text{whole} = \text{part}$$
$$12.5\% \cdot \$180 = n$$

*Step 4* **Solve.**
$$0.125 \, (180) = n$$
$$22.5 = n$$

*Step 5* **State the answer.** **$22.50** was withheld from Britta's paycheck.

*Step 6* **Check the solution.** The solution of $22.50 makes sense because 10% of $180 is $18, so a little more than $18 should be withheld.

Work Problem ❷ at the Side.

### Example 3 — Finding the Percent

On a 15-point quiz, Zenitia earned 13 points. What percent correct is this, to the nearest whole percent?

*Step 1* **Read** the problem. It is about points earned on a quiz.

Unknown: percent correct
Known: earned 13 out of 15 points

*Step 2* **Assign a variable.** Let *p* be the unknown percent.

*Step 3* **Write an equation.** Use the percent equation. There is no number with a % symbol in the problem. The question, "What percent is this?" tells you that the *percent* is unknown. The *whole* is all the points on the quiz (15 points), and 13 points is the *part* of the quiz that Zenitia did correctly.

$$\text{percent} \cdot \text{whole} = \text{part}$$
$$p \cdot 15 \text{ points} = 13 \text{ points}$$

❸ The Los Angeles Lakers made 47 of 80 field goal attempts in one game. What percent is this, to the nearest whole percent? Use the six problem-solving steps.

*Step 4* **Solve** the equation.

$$\frac{p \cdot \cancel{15}}{\cancel{15}} = \frac{13}{15} \quad \text{Divide both sides by 15.}$$

$$p = 0.8\overline{6} \quad \leftarrow \text{The solution is a repeating decimal.}$$

Multiply the solution by 100 to change it from a *decimal* to a *percent*.

$$0.866666667 \approx 86.6666667\% \approx 87\% \leftarrow \text{Rounded}$$

*Step 5* **State the answer.** Zenitia had **87%** correct, rounded to the nearest whole percent.

*Step 6* **Check the solution.** The solution of 87% makes sense because she earned most of the possible points, so the percent should be fairly close to 100%.

Work Problem ❸ at the Side.

---

**ANSWERS**

**2.** Let *n* be the points earned.
$0.83(50) = n$
Hue earned 41.5 points.
*Check:* 10% of 50 points is 5 points, so 80% would be 8 times 5 points = 40 points. Hue earned a little more than 80%, so 41.5 points is reasonable.

**3.** Let *p* be the unknown percent.
$p \cdot 80 = 47$
$p = 0.5875$
$0.5875 = 58.75\% \approx 59\%$
The Lakers made 59% of their field goals, to the nearest whole percent.

*Check:* The Lakers made a little more than half of their field goals. $\frac{1}{2} = 50\%$, so the solution of 59% is reasonable.

### Example 4 Finding the Percent

The rainfall in the Red River Valley was 33 inches this year. The average rainfall is 30 inches. This year's rainfall is what percent of the average rainfall?

*Step 1* **Read the problem.** It is about comparing this year's rainfall to the average rainfall.

Unknown: This year's rain is what percent of the average?
Known: 33 inches this year; 30 inches is average

*Step 2* **Assign a variable.** Let $p$ be the unknown percent.

*Step 3* **Write an equation.** The percent is unknown. The key word **of** appears *right after* the word *percent,* so you can use that sentence to help you write the equation.

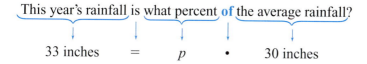

$$33 \text{ inches} = p \cdot 30 \text{ inches}$$

*Step 4* **Solve the equation.**

$$33 = p \cdot 30$$

$$\frac{33}{30} = \frac{p \cdot \cancel{30}}{\cancel{30}} \qquad \text{Divide both sides by 30.}$$

Solution in *decimal* form → $1.1 = p$

Multiply the solution by 100 to change it from a *decimal* to a *percent.*

$1.10 = 110\%$

*Step 5* **State the answer.** This year's rainfall is **110%** of the average rainfall.

*Step 6* **Check the solution.** The solution of 110% makes sense because 33 inches is *more* than 30 inches (more than 100% of the average rainfall), so 33 inches must be *more* than 100% of 30 inches.

Work Problem ❹ at the Side.

---

❹ Valley College predicted that 1200 new students would enroll in the fall. It actually had 1620 new students enroll. The actual enrollment is what percent of the predicted number? Use the six problem-solving steps.

**ANSWERS**

**4.** Let $p$ be the unknown percent.
$p \cdot 1200 = 1620$
$p = 1.35$
$1.35 = 135\%$

Enrollment is 135% of the predicted number.
*Check:* More than 1200 students enrolled (more than 100%), so 1620 students must be more than 100% of the predicted number. The solution of 135% is reasonable.

**5** Use the six problem-solving steps to answer each question.

(a) Ezra did 15 problems correctly on a test, giving him a score of $62\frac{1}{2}\%$. How many problems were on the test?

(b) A frozen dinner advertises that only 18% of its calories are from fat. If the dinner contains 55 calories from fat, what is the total number of calories in the dinner? Round to the nearest whole number.

---

**Example 5** Finding the Whole

A newspaper article stated that 648 pints of blood were donated at the blood bank last month, which was only 72% of the number of pints needed. How many pints of blood were needed?

*Step 1* **Read the problem.** It is about blood donations.

    Unknown: number of pints needed
    Known: 648 pints were donated;
          648 pints is 72% of the number needed.

*Step 2* **Assign a variable.** Let $n$ be the number of pints needed.

*Step 3* **Write an equation.** The percent is given in the problem: 72%. The key word *of* appears *right after* 72%, so you can use the phrase "72% *of* the number of pints needed" to help you write one side of the equation.

72% *of* the number of pints needed is 648 pints.

$$0.72 \cdot n = 648$$

*Step 4* **Solve.**

$$\frac{0.72 \cdot n}{0.72} = \frac{648}{0.72} \quad \text{Divide both sides by 0.72.}$$

$$n = 900$$

*Step 5* **State the answer. 900 pints** of blood were needed.

*Step 6* **Check the solution.** The solution of 900 pints makes sense because 10% of 900 pints is 90 pints, so 70% would be 7 times 90 pints, or 630 pints, which is close to the number given in the problem (648 pints).

**Work Problem 5 at the Side.**

---

**2** Solve problems involving percent of increase or decrease. We are often interested in looking at increases or decreases in prices, earnings, population, and many other numbers. This type of problem involves finding the percent of change. Use the following steps to find the **percent of increase.**

**Finding the Percent of Increase**

*Step 1* Use subtraction to find the *amount* of increase.

*Step 2* Use a form of the percent equation to find the *percent* of increase.

percent *of* whole = part

percent *of* original value = amount of increase

---

**ANSWERS**

**5.** (a) Let $n$ be number of problems on the test.
$0.625 \cdot n = 15$
There were 24 problems on the test.
*Check:* 50% of 24 problems is $24 \div 2 = 12$ problems correct, so it is reasonable that $62\frac{1}{2}\%$ would be 15 problems correct.

(b) Let $n$ be total number of calories.
$0.18 \cdot n = 55$
There were 306 calories (rounded) in the dinner. *Check:* 10% of 306 calories is about 30 calories, so 20% would be $2 \cdot 30 = 60$ calories, which is close to the number given (55 calories).

### Example 6  Finding the Percent of Increase

Brad's hourly wage as assistant manager of a fast-food restaurant was raised from $9.40 to $9.87. What was the percent of increase?

*Step 1*  **Read the problem.** It is about an increase in wages.

> Unknown: percent of increase
> Known: Original hourly wage was $9.40; new hourly wage is $9.87.

*Step 2*  **Assign a variable.** Let $p$ be the percent of increase.

*Step 3*  **Write an equation.** First subtract $9.87 − $9.40 to find how much Brad's wage went up. That is the *amount* of increase. Then write an equation to find the unknown *percent* of increase. Be sure to use his *original* wage ($9.40) in the equation. Do *not* use the new wage of $9.87 in the equation.

$$\$9.87 - 9.40 = \$0.47 \leftarrow \text{Amount of increase}$$

percent **of** original wage = amount of increase

$$p \cdot \$9.40 = \$0.47$$

*Step 4*  **Solve.**

$$\frac{p \cdot 9.40}{9.40} = \frac{0.47}{9.40} \quad \text{Divide both sides by 9.40.}$$

$$p = 0.05 \leftarrow \text{Solution in } decimal \text{ form}$$

Multiply the solution by 100 to change it from a *decimal* to a *percent*.

$$0.05 = 5\%$$

*Step 5*  **State the answer.** Brad's hourly wage increased **5%**.

*Step 6*  **Check the solution.** 10% of $9.40 would be a $0.94 raise. A raise of $0.47 is half as much, and half of 10% is 5%, so the solution checks.

━━━━━━━━━━━━━━ Work Problem **6** at the Side.

Use a similar procedure to find the **percent of decrease**.

### Finding the Percent of Decrease

*Step 1*  Use subtraction to find the *amount* of decrease.

*Step 2*  Use a form of the percent equation to find the *percent* of decrease.

percent **of** whole = part

percent **of** original value = amount of decrease

---

**6** Use the six problem-solving steps to answer each question.

**(a)** Over the last two years, Duyen's rent has increased from $650 per month to $767. What is the percent increase?

**(b)** A shopping mall increased the number of handicapped parking spaces from 8 to 20. What is the percent increase?

**ANSWERS**

6. **(a)** Let $p$ be percent increase.
$767 − $650 = $117 increase
$p \cdot 650 = 117$
$p = 0.18 = 18\%$
Duyen's rent increased 18%.
*Check:* 10% increase would be $650. = $65; 20% increase would be 2 • $65 = $130; so an 18% increase is reasonable.

**(b)** Let $p$ be percent increase.
20 − 8 = 12 space increase
$p \cdot 8 = 12$
$p = 1.50 = 150\%$
The number of parking spaces increased 150%.
*Check:* 150% is $1\frac{1}{2}$, and

$1\frac{1}{2} \cdot 8 = 12$ space increase,

so it checks.

**7** Use the six problem-solving steps to answer each question. Round answers to the nearest whole percent.

(a) During a severe winter storm, average daily attendance at an elementary school fell from 425 students to 200 students. What was the percent decrease?

(b) The makers of a brand of spaghetti sauce claim that the number of calories from fat in each serving has been reduced by 20%. Is the claim correct if the number of calories from fat dropped from 70 calories to 60 calories per serving? Explain your answer.

### Example 7 Finding the Percent of Decrease

Rozenia has been training for six months to run in a marathon race. Her weight has dropped from 137 pounds to 122 pounds. What is the percent of decrease? Round to the nearest whole percent.

Step 1  **Read the problem.** It is about a decrease in weight.

   Unknown: percent of decrease
   Known: Original weight was 137 pounds; new weight is 122 pounds.

Step 2  **Assign a variable.** Let $p$ be the percent of decrease.

Step 3  **Write an equation.** First subtract 137 pounds − 122 pounds to find how much Rozenia's weight went down. That is the *amount* of decrease. Then write an equation to find the unknown *percent* of decrease. Be sure to use her *original* weight (137 pounds) in the equation. Do *not* use the new weight of 122 pounds in the equation.

   137 pounds − 122 pounds = 15 pounds  ← Amount of decrease

   percent **of** original weight = amount of decrease

   $p \cdot 137 \text{ pounds} = 15 \text{ pounds}$

Step 4  **Solve.**

   $$\frac{p \cdot 137}{137} = \frac{15}{137} \quad \text{Divide both sides by 137.}$$

   $p = 0.109489051$ ← Solution in *decimal* form

   Multiply the solution by 100 to change it from a *decimal* to a *percent*.

   $0.109489051 = 10.9489051\% \approx 11\%$ (Rounded to nearest whole percent)

Step 5  **State the answer.** Rozenia's weight decreased approximately **11%**.

Step 6  **Check the solution.** A 10% decrease would be 13.7 = 13.7 pounds, so an 11% decrease is a reasonable solution.

Work Problem **7** at the Side.

---

**ANSWERS**

7. (a) Let $p$ be percent of decrease.
   425 − 200 = 225 student decrease
   $p \cdot 425 = 225$
   $p \approx 0.529 \approx 53\%$ (rounded)

   Daily attendance decreased 53%.
   *Check:* A 50% decrease would be 425 ÷ 2 ≈ 212, so a 53% decrease is reasonable.

(b) Let $p$ be percent of decrease.
   70 − 60 = 10 calorie decrease
   $p \cdot 70 = 10$
   $p \approx 0.143 \approx 14\%$ (rounded)

   The claim of a 20% decrease is not true; the decrease in calories is about 14%.
   *Check:* A 10% decrease would be 70. = 7 calories; a 20% decrease would be 2 · 7 = 14 calories, so a 14% decrease is reasonable.

## 7.4 Exercises

*Use the six problem-solving steps to answer each question. Round percent answers to the nearest tenth of a percent. See Examples 1–5.*

1. Robert Garrett, who works part-time, earns $110 per week and has 18% of this amount withheld for taxes, Social Security, and Medicare. Find the amount withheld.

2. Most shampoos contain 75% to 90% water. (*Source: Consumer Reports.*) If a 16-ounce bottle of shampoo contains 78% water, find the number of ounces of water in the bottle. Round to the nearest tenth of an ounce.

3. Sharon needs 64 credits to graduate from her community college. So far she has earned 40 credits. What percent of the required credits does she have?

4. There are about 55,000 words in *Webster's Dictionary*, but most educated people can identify only 20,000 of these words. What percent of the words in the dictionary can these people identify?

5. A survey at an intersection found that, of 2200 drivers, 38% were wearing seat belts. How many drivers in the survey were wearing seat belts?

6. For a tour of the eastern United States, a travel agent promised a trip of 3300 miles, with 35% of the trip by air. How many miles would be traveled by air?

7. The guided-missile destroyer USS *Sullivans* has a 335-person crew of which 44 are female. What percent of the crew is female? What percent of the crew is male? (*Source:* U.S. Navy.)

8. In a test by *Consumer Reports,* 6 of the 123 cans of tuna analyzed contained more than the 30-microgram intake limit of mercury. What percent of the cans contained an excessive level of mercury? What percent of the cans contained less than or equal to 30 micrograms of mercury?

9. The U.S. Bureau of the Census reported that Americans who are 65 years of age or older make up 12.7% of the total population in 2000. It said that there are 35.7 million Americans in this group. Find the total U.S. population. (Round to the nearest million.)

10. Julie Ward has 8.5% of her earnings deposited into the credit union. If this amounts to $263.50 per month, find her monthly and annual earnings.

11. The campus honor society hoped to raise $50,000 in donations from businesses for scholarships. It actually raised $69,000. This amount was what percent of the goal?

12. Doug had budgeted $120 for textbooks but ended up spending $172.80. The amount he spent was what percent of his budget?

13. Alfonso earned a score of 87.5% on his test. He did 35 problems correctly. How many problems were on the test?

14. In the 2000–2001 basketball season, Kevin Garnett made 76.4% of his free throws. He made 357 free throws. How many did he attempt, to the nearest whole number? (*Source:* NBA.)

15. In the United States, 15 of the 50 states limit the blood alcohol level for drivers to 0.08%. The remaining states limit the level to 0.10%. (*Source: Wall Street Journal.*) What percent of the states have a blood alcohol limit of 0.08%? What percent have a limit of 0.10%?

16. Among the 50 companies receiving the greatest number of U.S. patents in 2000, 18 were Japanese companies. (*Source: Wall Street Journal.*) What percent of the companies were Japanese? What percent of the companies were not Japanese?

17. An ad for steel-belted radial tires promises 15% better mileage. If Sheera's SUV has gotten 20.6 miles per gallon in the past, what mileage can she expect after the new tires are installed? (Round to the nearest tenth of a mile.)

18. Spam and Spam Lite have yearly sales of $92 million, which is about 62% of all canned lunch meat sales. (*Source:* Hormel Foods.) Find the total yearly sales of canned lunch meat, to the nearest million.

*The graph (pictograph) shows the percent of chicken noodle soup sold during the cold and flu season. Use this information to answer Exercises 19–22.*

**SOUP'S ON**
350 million cans of chicken noodle soup are sold each year. More than half are bought during cold-and-flu season, with January being the number one month. The percent sold during each flu-season month is shown.

10%  9%  8%  15%  11%  7%
October  November  December  January  February  March
*Source: USA Today.*

**19.** Which of the flu season months had the lowest sales of chicken noodle soup? How many cans were bought that month?

**20.** What percent of the chicken noodle soup sales take place during the flu months of October through March? What percent of sales take place in the *non-flu* season months?

**21.** Find the number of cans of soup sold in the highest sales month and in the second highest sales month.

**22.** How many more cans of soup were sold in October than in November? How many more were sold in November than December?

*Use the six problem-solving steps to find the percent increase or decrease. Round your answers to the nearest tenth of a percent. See Examples 6 and 7.*

**23.** The price per share of Toys "R" Us stock fell from $35.50 to close at $33.50. Find the percent of decrease in price.

**24.** In the past five years, the cost of generating electricity from the sun has been brought down from 24 cents per kilowatt hour to 8 cents (less than the newest nuclear power plants). Find the percent of decrease.

**25.** Students at Lane College were charged $1449 as tuition this semester. If the tuition was $1228 last semester, find the percent of increase.

**26.** Americans are eating more fish. This year the average American will eat 15.5 pounds compared to only 12.5 pounds per year a decade ago. Find the percent of increase. (*Source: Consumer Reports.*)

**27.** Jordan's part-time work schedule has been reduced to 18 hours per week. He had been working 30 hours per week. What is the percent decrease?

**28.** Janis works as a hair stylist. During January, she cut her price on haircuts from $28 to $25.50 to try to get more customers. By what percent did she decrease the price?

**29.** Harley-Davidson, the only major U.S.-based motorcycle manufacturer, says that it expects to build 145,000 motorcycles this year, up from 131,000 last year. Find the percent of increase in production. (*Source:* Harley-Davidson, Inc.)

**30.** The world population was estimated at 6,080,000,000 people in 2000. It is projected to reach 7,841,000,000 people by 2025. By what percent will the world's population increase in those 25 years? (*Source:* United Nations.)

**31.** You can have an *increase* of 150% in the price of something. Could there be a 150% *decrease* in its price? Explain why or why not.

**32.** Show how to use a shortcut to find 25% of $80. Then explain how to use the result to find 75% of $80 and 125% of $80 *without* solving a proportion or equation.

### RELATING CONCEPTS (Exercises 33–36) FOR INDIVIDUAL OR GROUP WORK

*As you* **work Exercises 33–36 in order,** *explain why each solution does* **not** *make sense. Then find and correct the error.*

**33.** The recommended maximum daily amount of dietary fat is 65 grams. George ate 78 grams of fat today. He ate what percent of the recommended amount?

$$p \cdot 78 = 65$$

$$\frac{p \cdot 78}{78} = \frac{65}{78}$$

$$p = 0.833 = 83.3\%$$

**34.** The Goblers soccer team won 18 of its 25 games this season. What percent did the team win?

$$p \cdot 25 = 18$$

$$\frac{p \cdot 25}{25} = \frac{18}{25}$$

$$p = 0.72\%$$

**35.** The human brain is $2\frac{1}{2}\%$ of total body weight. How much would the brain of a 150-pound person weigh?

$$2\frac{1}{2}\% \text{ of } 150 = n$$
$$2.5 \cdot 150 = n$$
$$375 = n$$

**36.** Yesterday, because of an ice storm, 80% of the students were absent. How many of the 800 students made it to class?

$$80\% \text{ of } 800 = n$$
$$0.80 \cdot 800 = n$$
$$640 = n$$

The answer is 640 students.

## 7.5 CONSUMER APPLICATIONS: SALES TAX, TIPS, DISCOUNTS, AND SIMPLE INTEREST

Four of the more common uses of percent in daily life are sales taxes, tips, discounts, and simple interest on loans.

**OBJECTIVES**
1. Find sales tax and total cost.
2. Calculate restaurant tips.
3. Find the discount and sale price.
4. Calculate simple interest and the total amount due on a loan.

**1 Find sales tax and total cost.** Most states collect **sales taxes** on the purchases you make in stores. Your county or city may also add on a small amount of sales tax. For example, your state may charge $6\frac{1}{2}\%$ on purchases and your city may add on another $\frac{1}{2}\%$ for a total of 7%. The exact percent varies from place to place but is usually from 4% to 8%. The stores collect the tax and send it to the city or state government where it is used to pay for things like road repair, public schools, parks, police and fire protection, and so on.

You can use a form of the percent equation to calculate sales tax. The **tax rate** is the *percent*. The cost of the item(s) you are buying is the *whole*. The amount of tax you pay is the *part*.

### Finding Sales Tax and Total Cost

Use a form of the percent equation to find sales tax.

$$\text{percent} \cdot \text{whole} = \text{part}$$
$$\text{tax rate} \cdot \text{cost of item} = \text{amount you pay in sales tax}$$

Then add to find how much you will pay in all.

$$\text{cost of item} + \text{sales tax} = \text{total cost paid by you}$$

### Example 1 Finding Sales Tax and Total Cost

Suppose that you buy a CD player for $289 from A-1 Electronics. The sales tax rate in your state is $6\frac{1}{2}\%$. How much is the tax? What is the total cost of the CD player?

*Step 1* **Read the problem.** It asks for the sales tax on a CD player and the total cost.

*Step 2* **Assign a variable.** Let $n$ be the amount of tax.

*Step 3* **Write an equation.** Use the sales tax equation. Write $6\frac{1}{2}\%$ as 6.5% and then move the decimal point two places to the left.

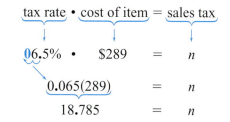

*Step 4* **Solve.**
$$0.065(289) = n$$
$$18.785 = n$$

The store will round the tax to the nearest cent, so $18.785 rounds to **$18.79**.

Now add the sales tax to the cost of the CD player to find your total cost.

$$\text{cost of item} + \text{sales tax} = \text{total cost}$$
$$\$289 + \$18.79 = \mathbf{\$307.79}$$

*Continued on Next Page*

**1** Find the sales tax and the total cost. Round the sales tax to the nearest cent if necessary. Check your answer by estimating the sales tax.

(a) $495 camcorder; $5\frac{1}{2}$% sales tax

(b) $29.98 watch; 7% sales tax

(c) $1.19 candy bar; 4% sales tax

**2** Find the sales tax rate on each purchase. Then use estimation to check your solution.

(a) The tax on a $57 textbook is $3.42.

(b) The tax on a $4 notebook is $0.18.

(c) The tax on a $998 sofa is $49.90.

**ANSWERS**

1. (a) sales tax = $27.23 (rounded)
 total cost = $522.23
 Check: 1% of $500 is $5;
 
 6% is 6 times $5 = $30
 (b) sales tax = $2.10 (rounded)
 total cost = $32.08
 Check: 1% of $30 is $0.30;
 
 7% is 7 times $0.30 = $2.10
 (c) Sales tax = $0.05 (rounded)
 total cost = $1.24
 Check: 1% of $1 is $0.01;
 
 4% is 4 times $0.01 = $0.04
2. (a) 6%  Check: 1% of $60 is $0.60;
 
 6% would be 6 times $0.60, or $3.60 (which is close to $3.42 given in the problem).
 (b) 4.5%  Check: 1% of $4 is $0.04; round 4.5% to 5%; 5% of $4 is 5 times $0.04, or $0.20 (which is close to $0.18 given in the problem).
 (c) 5%  Check: 10% of $1000 is $100;
 
 5% is half of that, or $50 (which is close to $49.90 given in the problem).

*Step 5* **State the answer.** The tax is **$18.79** and the total cost of the CD player, including tax, is **$307.79**.

*Step 6* **Check:** Use estimation to check that the amount of sales tax is reasonable. Round $289 to $300. Then 1% of $300 is $3. Round $6\frac{1}{2}$% to 7%. Then 7% would be 7 times $3 or $21 for sales tax. This is close to our solution of $18.79.

**Work Problem** ❶ **at the Side.**

### Example 2 — Finding the Sales Tax Rate

Ms. Ortiz bought a $21,950 pickup truck. She paid an additional $1646.25 in sales tax. What was the sales tax rate?

*Step 1* **Read** the problem. It asks for the sales tax rate on a truck purchase.

*Step 2* **Assign a variable.** Let $p$ be the tax rate (the percent).

*Step 3* **Write an equation.** Use the sales tax equation.

$$\text{tax rate} \cdot \text{cost of item} = \text{sales tax}$$
$$p \cdot \$21{,}950 = \$1646.25$$

*Step 4* **Solve.**  $\dfrac{p \cdot 21{,}950}{21{,}950} = \dfrac{1646.25}{21{,}950}$  Divide both sides by 21,950.

$p = 0.075$ ← Solution in *decimal* form

Multiply the solution by 100 to change it from a decimal to a percent: $0.075 = 7.5\%$.

*Step 5* **State the answer.** The sales tax rate is **7.5%** (or $7\frac{1}{2}$%).

*Step 6* **Check.** Use estimation to check that the solution is reasonable. If the tax rate was 1%, then 1% of $21950 = $219.50, or about $200.

Round 7.5% to 8%. Then 8% would be 8 times $200, or $1600.

The tax amount given in the original problem, $1646.25, is close to the estimate, so our solution of 7.5% is reasonable.

**Work Problem** ❷ **at the Side.**

**2** **Calculate restaurant tips.** Waiters and waitresses rely on tips as a major part of their income. The general rule of thumb is to leave 15% of your bill for food and beverages as a tip for the server. If you receive exceptional service or are eating in an upscale restaurant, consider leaving a 20% tip.

### Example 3 — Estimating 15% and 20% Tips

First estimate each tip. Then calculate the exact amount.

(a) Kirby took his wife to dinner at a nice restaurant to celebrate her promotion at work. The bill came to $77.85. How much should he leave for a 20% tip?

*Estimate:* Round $77.85 to $80. Then 10% of $80 is $8.

20% would be 2 times $8, or **$16**. ← Estimate

**Continued on Next Page**

*Exact:* Use the percent equation. Write 20% as a decimal. The bill for food and beverages is the *whole* and the tip is the *part*.

$$\text{percent} \cdot \text{whole} = \text{part}$$
$$20.\% \cdot \$77.85 = n$$
$$0.20(77.85) = n$$
$$15.57 = n$$

20% of $77.85 is **$15.57**, which is close to the estimate of $16.

A tip is usually rounded off to a convenient amount, such as the nearest quarter or nearest dollar, so Kirby left $16.

**(b)** Linda, Peggy, and Mary ordered similarly priced lunches and agreed to split the bill plus a 15% tip. How much should each woman pay if the bill is $21.63?

*Estimate:* Round $21.63 to $20. Then 10% of $20 is $2.

5% of $20 would be half as much, that is, half of $2, or $1.

So an estimate of the 15% tip is $2 + $1 = $3. ←Estimate

An estimate of the amount each woman should pay is ($20 + $3) ÷ 3 ≈ $8.

*Exact:* Use the percent equation to calculate the 15% tip. Add the tip to the bill. Then divide the total by 3 to find the amount each woman should pay.

$$\text{percent} \cdot \text{whole} = \text{part}$$
$$15.\% \cdot \$21.63 = n$$
$$0.15(21.63) = n$$
$$3.2445 = n$$

Round $3.2445 to **$3.24** (nearest cent), which is close to the estimate of $3 for the tip.

Add: $21.63 + $3.24 = $24.87 ← Total cost of lunch and tip

Divide: $24.87 ÷ 3 = **$8.29** ← Amount paid by each woman

━━━━━━━━━━━━━━━━━━ **Work Problem ❸ at the Side.**

**❸ Find the discount and sale price.** Most people prefer buying things when they are on sale. A store will reduce prices, or **discount,** to attract additional customers. You can use a form of the percent equation to calculate the discount. The rate of discount is the *percent*. The original price is the *whole*. The amount that will be discounted (subtracted from the original price) is the *part*.

### Finding the Discount and Sale Price

Use a form of the percent equation to find the discount.

$$\text{percent} \cdot \text{whole} = \text{part}$$
$$\underbrace{\text{rate of discount}} \cdot \underbrace{\text{original price}} = \underbrace{\text{amount of discount}}$$

Then subtract to find the sale price.

$$\text{original price} - \text{amount of discount} = \text{sale price}$$

---

❸ First estimate each tip. Then calculate the exact tip.

**(a)** 20% tip on a bill of $58.37

**(b)** 15% tip on a bill of $11.93

**(c)** A bill of $89.02 plus a 15% tip shared equally by four friends. How much will each friend pay?

**ANSWERS**
3. **(a)** *Estimate:* 10% of $60 is $6; 20% is 2 times $6, or $12.
   *Exact:* $11.67 (rounded to nearest cent)
   **(b)** *Estimate:* 10% of $12 is $1.20; 5% is half of $1.20, or $0.60; $1.20 + 0.60 = $1.80.
   *Exact:* $1.79 (rounded to nearest cent)
   **(c)** *Estimate* of tip: 10% of $90 is $9; 5% is half of $9, or $4.50; $9 + $4.50 = $13.50.
   *Exact* tip: $13.35 (rounded to nearest cent). Each friend pays: ($89.02 + $13.35) ÷ 4 = $25.59 (rounded).

4 Find the amount of the discount and the sale price for each item.

(a) An Easy-Boy leather recliner originally priced at $950 is offered at a 35% discount. What is the sale price?

(b) Eastside Department Store has women's swimsuits on sale at 40% off. One swimsuit was originally priced at $34. Another suit was originally priced at $72. What is the sale price of each suit?

### Example 4  Finding the Discount and Sale Price

The Oak Mill Furniture Store has an oak entertainment center with an original price of $840 on sale at 15% off. Find the sale price of the entertainment center.

*Step 1*  **Read** the problem. It asks for the sale price on an entertainment center.

*Step 2*  **Assign a variable.** Let $n$ be the amount of discount.

*Step 3*  **Write an equation.** Use a form of the percent equation to find the discount. Write 15% as 0.15.

rate of discount • original price = amount of discount

0.15 • 840 = $n$

*Step 4*  **Solve.**    126 = $n$

The amount of discount is **$126**. Find the sale price of the entertainment center by subtracting the amount of the discount ($126) from the original price.

original price − amount of discount = sale price

$840 − $126 = $714

*Step 5*  **State the answer.** During the sale, you can buy the entertainment center for **$714**.

*Step 6*  **Check.** Round $840 to $800. Then 10% of $800 is $80 and 5% is half as much, or $40. So 15% of $800 is $80 + $40 = $120. An estimate of the sale price is $800 − $120 = $680, so the exact answer of $714 is reasonable.

**Work Problem 4 at the Side.**

**Calculator Tip**  In Example 4, you can use a *scientific* calculator to find the amount of discount and subtract the discount from the original price.

840 ⊖ .15 ⊗ 840 ⊜ 714
↑          ↑          ↑
Original   Amount of   Sale
price      discount    price

Your *scientific* calculator observes the order of operations, so it will automatically do the multiplication before the subtraction. (Recall that simple, 4-function calculators *may not* follow the order of operations; they would give an incorrect result.)

**4** Calculate simple interest and the total amount due on a loan.  **Interest** is a fee paid, or a charge made, for lending or borrowing money. The amount of money borrowed is called the **principal.** The charge for interest is usually given as a percent, called the **interest rate.** The interest rate is assumed to be *per year* (for *one* year) unless stated otherwise.

In most cases, interest is calculated on the original principal and is called **simple interest.** Use the following **interest formula** to find simple interest.

**ANSWERS**
4. (a) Discount is $332.50; sale price is $617.50.
   (b) Discount is $13.60, sale price is $20.40; discount is $28.80; sale price is $43.20.

## Formula for Simple Interest

Interest = principal • rate • time

The formula is usually written using variables.

$$I = p \cdot r \cdot t \quad \text{or} \quad I = prt$$

**NOTE**

Simple interest calculations are used for most short-term business loans, most real estate loans, and many automobile and consumer loans.

### Example 5 — Finding Simple Interest for 1 Year

Find the simple interest on a $2000 loan at 6% for 1 year.

The amount borrowed, or principal ($p$), is $2000. The interest rate ($r$) is 6%, which is 0.06 as a decimal, and the time of the loan ($t$) is 1 year. Use the interest formula.

$$I = p \cdot r \cdot t$$
$$I = (2000)(0.06)(1)$$
$$I = 120$$

The interest is **$120**.

*Work Problem ❺ at the Side.*

### Example 6 — Finding Simple Interest for More Than 1 Year

Find the simple interest on a $4200 loan at $8\frac{1}{2}$% for $3\frac{1}{2}$ years. The principal ($p$) is $4200.

The rate ($r$) is $8\frac{1}{2}$%, which is the same as 8.5%. Move the decimal point two places to the left to change 8.5% to a decimal.

$$8\tfrac{1}{2}\% = 8.5\% = 08.5 = 0.085$$

The time ($t$) is $3\frac{1}{2}$ or 3.5 years. Use the formula.

$$I = p \cdot r \cdot t$$
$$I = (4200)(0.085)(3.5)$$
$$I = 1249.50$$

The interest is **$1249.50**.

**CAUTION**

Be careful when changing a mixed number percent, like $8\frac{1}{2}$%, to a decimal. Writing $8\frac{1}{2}$% as 8.5% is only the first step. There is a decimal point in 8.5% but there is still a % sign. You must divide by 100 before dropping the % sign. *Remember to move the decimal point two places to the left*, as in Example 6 above.

*Work Problem ❻ at the Side.*

---

❺ Find the simple interest.

**(a)** $500 at 4% for 1 year

**(b)** $1850 at $9\frac{1}{2}$% for 1 year
(*Hint:* Write $9\frac{1}{2}$% as 9.5%. Then *move the decimal point* two places to the left to change 9.5% to a decimal.)

❻ Find the simple interest.

**(a)** $340 at 5% for $3\frac{1}{2}$ years

**(b)** $2450 at 8% for $3\frac{1}{4}$ years
(*Hint:* Write $3\frac{1}{4}$ years as 3.25 years.)

**(c)** $14,200 at $7\frac{1}{2}$% for $2\frac{3}{4}$ years

**ANSWERS**
5. **(a)** $20  **(b)** $175.75
6. **(a)** $59.50  **(b)** $637  **(c)** $2928.75

**7** Find the simple interest.

(a) $1600 at 7% for 4 months

(b) $25,000 at $10\frac{1}{2}$% for 3 months

**8** Find the total amount due on each loan.

(a) $2500 at $7\frac{1}{2}$% for 6 months

(b) $10,800 at 6% for 4 years

(c) $4350 at $10\frac{1}{4}$% for $2\frac{1}{2}$ years

**ANSWERS**
7. (a) $37.33 (rounded)   (b) $656.25
8. (a) $2593.75   (b) $13,392
   (c) $5464.69 (rounded)

Interest rates are given *per year*. For loan periods of less than one year, be careful to express the time as a fraction of a year.

If the time is given in months, use a denominator of 12, because there are 12 months in a year. A loan of 9 months would be for $\frac{9}{12}$ of a year, a loan of 7 months would be for $\frac{7}{12}$ of a year, and so on.

### Example 7  Finding Simple Interest for Less Than 1 Year

Find the simple interest on $840 at $9\frac{3}{4}$% for 7 months.
The principal is $840. The rate is $9\frac{3}{4}$% or 0.0975.

$$9\frac{3}{4}\% = 9.75\% = 09.75 = 0.0975$$

The time is $\frac{7}{12}$ of a year. Use the formula $I = prt$.

$$I = (840)(0.0975)\left(\frac{7}{12}\right) \quad \text{7 months} = \tfrac{7}{12} \text{ of a year}$$

$$= (81.9)\left(\frac{7}{12}\right)$$

$$= \frac{81.9}{1} \cdot \frac{7}{12} \quad \text{Multiply numerators. Multiply denominators.}$$

$$= \frac{573.3}{12} = 47.775 \quad \text{Divide 573.3 by 12.}$$

The interest is **$47.78**, rounded to the nearest cent.

**Calculator Tip**  The calculator solution to Example 7 uses chain calculations.

840 ×  .0975 × 7 ÷ 12 = 47.775  ← Round to $47.78.

**Work Problem 7 at the Side.**

When you repay a loan, the interest is added to the original principal to find the total amount due.

### Finding the Total Amount Due

amount due = principal + interest

### Example 8  Calculating the Total Amount Due

Charlesetta borrowed $1080 at 8% for three months to pay for tuition and books. Find the total amount due on her loan.

First find the interest. Use $I = prt$. Write 8% as 0.08.

$$I = (1080)(0.08)\left(\frac{3}{12}\right) \quad \text{3 months} = \tfrac{3}{12} \text{ of a year}$$

$$I = 21.60$$

Now add the principal and the interest to find the total amount due.

$$\textbf{amount due} = \text{principal} + \textbf{interest}$$
$$= \$1080 + \$21.60$$
$$= \$1101.60$$

The total amount due is **$1101.60**.

**Work Problem 8 at the Side.**

Section 7.5  **549**

## 7.5 Exercises

Find the amount of the sales tax or the tax rate and the total cost. Round money answers to the nearest cent. See Examples 1 and 2.

| Cost of Item | Tax Rate | Amount of Tax | Total Cost |
|---|---|---|---|
| 1. $100 | 6% | | |
| 2. $200 | 4% | | |
| 3. $68 | | $2.04 | |
| 4. $185 | | $9.25 | |
| 5. $365.98 | 6% | | |
| 6. $28.49 | 7% | | |
| 7. $2.10 | $5\frac{1}{2}\%$ | | |
| 8. $7.00 | $7\frac{1}{2}\%$ | | |
| 9. $12,600 | | $567 | |
| 10. $21,800 | | $1417 | |

*For each restaurant bill, estimate a 15% tip and a 20% tip. Then find the exact amounts for a 15% tip and a 20% tip. Round exact amounts to the nearest cent if necessary. See Example 3.*

| Bill | Estimate of 15% Tip | Exact 15% Tip | Estimate of 20% Tip | Exact 20% Tip |
|---|---|---|---|---|
| 11. $32.17 | | | | |
| 12. $21.94 | | | | |
| 13. $78.33 | | | | |
| 14. $67.85 | | | | |
| 15. $9.55 | | | | |
| 16. $52.61 | | | | |

*Find the amount or rate of discount and the sale price. Round money answers to the nearest cent if necessary. See Example 4.*

| Original Price | Rate of Discount | Amount of Discount | Sale Price |
|---|---|---|---|
| 17. $100 | 15% | | |
| 18. $200 | 20% | | |
| 19. $180 | | $54 | |
| 20. $38 | | $9.50 | |
| 21. $17.50 | 25% | | |
| 22. $76 | 60% | | |
| 23. $37.88 | 10% | | |
| 24. $59.99 | 40% | | |

*Find the simple interest and total amount due on each loan. See Examples 5–8.*

| Principal | Rate | Time | Interest | Total Amount Due |
|---|---|---|---|---|
| 25. $300 | 14% | 1 year | _____ | _____ |
| 26. $600 | 11% | 6 months | _____ | _____ |
| 27. $740 | 6% | 9 months | _____ | _____ |
| 28. $1180 | 9% | 2 years | _____ | _____ |
| 29. $1500 | 10% | 18 months | _____ | _____ |
| 30. $3000 | 15% | 5 months | _____ | _____ |
| 31. $17,800 | $7\frac{1}{2}$% | 9 months | _____ | _____ |
| 32. $20,500 | $5\frac{1}{2}$% | 6 months | _____ | _____ |

*Solve each application problem. Round money answers to the nearest cent if necessary.*

**33.** Diamonds at Discounts sells diamond engagement rings at 40% off the regular price. Find the sale price of a $\frac{1}{2}$-carat diamond ring normally priced at $1950.

**34.** A Palm IIIx$^e$ Connected Organizer originally priced at $332 is marked down 25%. Find the price of the organizer after the markdown.

**35.** Evelina Jones lends $7500 to her son Rick, the owner of Rick's Limousine Service. He will repay the loan at the end of 9 months at $8\frac{1}{2}$% simple interest. What is the total amount that Rick will owe his mother?

**36.** The owners of Delta Trucking purchased four diesel-powered tractors for cross-country hauling at a cost of $87,500 per tractor. If they borrowed the purchase price for $1\frac{1}{2}$ years at 11% simple interest, find the total amount due.

**37.** A Uniden digital cordless phone with caller ID and call waiting is priced at $99.99. The sales tax rate is $6\frac{1}{2}$%. Find the total cost of the phone. (*Source:* www.target.com)

**38.** A "golf/breakfast special" includes breakfast, 18 holes of golf, and use of a cart for $25.95 plus tax per person. If the sales tax rate is $7\frac{1}{2}$%, find the total cost per person. How much will three friends pay to play golf? (*Source:* Eagle Trace Golfers Club.)

**39.** An Anderson wood frame French door is priced at $1980 with a sales tax of $99. Find the sales tax rate.

**40.** Textbooks for two classes cost $135 plus sales tax of $8.10. Find the sales tax rate.

**41.** A "super 45% off sale" begins today. What is the sale price of a ski parka normally priced at $135?

**42.** What is the sale price of a $549 Maytag dishwasher with a discount of 35%?

**43.** Ricia and Seitu split a $43.70 dinner bill plus 15% tip. How much did each person pay?

**44.** Marvette took her brother out to dinner for his birthday. The bill for food was $58.36 and for wine was $15.44. How much was her 20% tip, rounded to the nearest dollar?

**45.** An 8-millimeter camcorder normally priced at $590 is on sale for 18% off. Find the discount and the sale price.

**46.** This week minivans are offered at 15% off manufacturers' suggested price. Find the discount and the sale price of a minivan originally priced at $23,500.

**47.** Ms. Henderson owes $1900 in taxes. She is charged a penalty of $12\frac{1}{4}$% annual interest and pays the taxes and penalty after 6 months. Find the total amount she must pay.

**48.** Norell Di Loreto, owner of Sunset Realtors, borrows $27,000 to update her office computer system. If the loan is for 24 months at $7\frac{3}{4}$%, find the total amount due on the loan.

**49.** Vincente and Samuel ordered a large deep dish pizza for $17.98. How much did they give the delivery person to pay for the pizza and a 15% tip, rounded to the nearest dollar?

**50.** Cher, Maya, and Adara shared a $25.50 bill for a buffet lunch. Because the server only brought their beverages, they left a 10% tip instead of the usual 15%. How much did each person pay?

*Use the information in the store ad to answer Exercises 51–54. Round sale prices and sales tax to the nearest cent if necessary.*

**51.** Danika bought a computer modem originally priced at $129 and a $60 pair of earrings. What was her bill for the two items?

**52.** Find David's total bill for a $189 jacket and a $75 graphing calculator.

**53.** Sergei purchased a television originally priced at $287.95, two pairs of $48 jeans, and a $95 ring. Find his total bill.

**54.** Richard picked out three pairs of $15 running shorts, two $28 shirts, and a "point and shoot" camera originally priced at $99.99. How much did he pay in all?

### RELATING CONCEPTS (Exercises 55–56) FOR INDIVIDUAL OR GROUP WORK

*Use your knowledge of percent to **work Exercises 55 and 56 in order**.*

**55. (a)** College students are offered a 6% discount on a dictionary that sells for $18.50. If the sales tax is 6%, find the cost of the dictionary, including the sales tax, to the nearest cent.

**(b)** In part (a) the rate of discount and the sales tax rate are the same percent. Explain why the answer did *not* end up back at $18.50.

**56. (a)** A FAX machine priced at $398 is marked down 7% to promote the new model. If the sales tax is also 7%, find the cost of the FAX machine, including sales tax, to the nearest cent.

**(b)** What rate of sales tax would have made the answer in part (a) end up back at $398? Round your answer to the nearest hundredth of a percent.

# Focus on Real-Data Applications

## Make Your Investments Grow—Compound Interest

*Simple interest* is paid only on the original principal. But savings accounts and most investments earn *compound interest*. In that case, interest is paid on the principal *and* the interest earned. Calculating compound interest can be quite tedious. For this reason compound interest tables have been developed.

Suppose you deposit $1 in a savings account today that earns 4% compounded annually and you allow the deposit to remain for three years. The diagram below shows the compound amount in your account at the end of each of the 3 years.

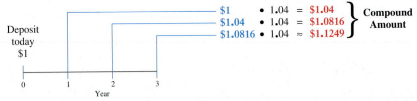

Using the compound amounts for $1, a table can be formed. Look at the table below and find the column headed 4%. The first three numbers for years 1, 2, and 3 are the same as those we have calculated for $1 at 4% for 3 years.

### Compound Interest Table

| Years | 3.00% | 3.50% | 4.00% | 4.50% | 5.00% | 5.50% | 6.00% | 8.00% |
|---|---|---|---|---|---|---|---|---|
| 1 | 1.0300 | 1.0350 | 1.0400 | 1.0450 | 1.0500 | 1.0550 | 1.0600 | 1.0800 |
| 2 | 1.0609 | 1.0712 | 1.0816 | 1.0920 | 1.1025 | 1.1130 | 1.1236 | 1.1664 |
| 3 | 1.0927 | 1.1087 | 1.1249 | 1.1412 | 1.1576 | 1.1742 | 1.1910 | 1.2597 |
| 4 | 1.1255 | 1.1475 | 1.1699 | 1.1925 | 1.2155 | 1.2388 | 1.2625 | 1.3605 |
| 5 | 1.1593 | 1.1877 | 1.2167 | 1.2462 | 1.2763 | 1.3070 | 1.3382 | 1.4693 |
| 6 | 1.1941 | 1.2293 | 1.2653 | 1.3023 | 1.3401 | 1.3788 | 1.4185 | 1.5869 |
| 7 | 1.2299 | 1.2723 | 1.3159 | 1.3609 | 1.4071 | 1.4547 | 1.5036 | 1.7138 |
| 8 | 1.2668 | 1.3168 | 1.3686 | 1.4221 | 1.4775 | 1.5347 | 1.5938 | 1.8509 |
| 9 | 1.3048 | 1.3629 | 1.4233 | 1.4861 | 1.5513 | 1.6191 | 1.6895 | 1.9990 |
| 10 | 1.3439 | 1.4106 | 1.4802 | 1.5530 | 1.6289 | 1.7081 | 1.7908 | 2.1589 |

To find the compound amount for $1000 deposited at 4% interest for 3 years, multiply the number from the table (1.1249) times the principal ($1000).

($1000)(**1.1249**) = $1124.90 ← Amount in account after 3 years

1. (a) Use the table to find the compound amount for $1000 deposited at 6% for 3 years, and the compound amount at 8% for 3 years.

    (b) How much more do you earn in 3 years on $1000 if the interest rate is 6% instead of 4%? How much more if the interest rate is 8% instead of 4%?

2. (a) Use the table to find the compound amount on $15,000 invested for 10 years at 4%, at 6%, and at 8%.

    (b) How much more do you earn at 6% than 4%? How much more at 8% than 4%?

3. Compound interest makes a tremendous difference when you invest money for retirement. Suppose a person plans to retire at age 65. Use the part of the compound interest table shown at the right. Find the compound amount on $5000 invested at 8% when the person is 45 years old, and left in the account until age 65. Then find the compound amount if the investment had been made at age 35, if the investment had been made at age 25, and if the investment had been made at age 15.

| Years | 8.00% |
|---|---|
| 20 | 4.6610 |
| 30 | 10.0627 |
| 40 | 21.7245 |
| 50 | 46.9016 |

## Summary Exercises on PERCENT

1. Complete this table. Write fractions in lowest terms and as mixed numbers when possible.

|     | Fraction | Decimal | Percent |
|-----|----------|---------|---------|
| (a) | $\frac{3}{100}$ |  |  |
| (b) |  |  | 30% |
| (c) |  | 0.375 |  |
| (d) |  |  | 160% |
| (e) | $\frac{1}{16}$ |  |  |
| (f) |  |  | 5% |

2. Use percent shortcuts to answer these questions.

   (a) 10% of 35 ft is _____

   (b) 100% of 19 miles is _____

   (c) 50% of 210 cows is _____

   (d) 1% of $8 is _____

   (e) 25% of 2000 women is _____

   (f) 300% of $15 is _____

*Use the percent proportion or percent equation to answer each question. If necessary, round money amounts to the nearest cent and percent answers to the nearest tenth of a percent.*

3. 9 Web sites is what percent of 72 Web sites?

4. 30 DVDs is 40% of what number of DVDs?

5. 6% of $8.79 is how much?

6. 945 students is what percent of 540 students?

7. $3\frac{1}{2}$% of 168 pounds is how much, to the nearest tenth of a pound?

8. 1.25% of what number of hours is 7.5 hours?

9. What percent of 80,000 deer is 40,000 deer?

10. 465 camp sites is 93% of what number of camp sites?

11. What number of golf balls is 280% of 35 golf balls?

12. What percent of $66 is $1.80?

13. 9% of what number of apartments is 207 apartments?

14. What weight is 84% of 0.75 ounce?

*The circle graph shows the average costs for various wedding expenses. Use the graph to answer Exercises 15–18. Round money answers to the nearest cent if necessary.*

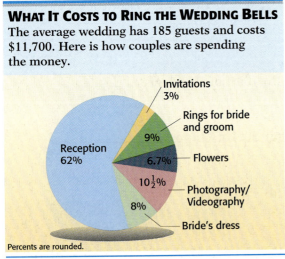

**WHAT IT COSTS TO RING THE WEDDING BELLS**
The average wedding has 185 guests and costs $11,700. Here is how couples are spending the money.

*Source: Bride's magazine.*

**15.** Which item is least expensive, and how much is spent on it?

**16.** What amount is paid for flowers?

**17.** Find the cost of photography and videography.

**18.** How much, on average, is spent for each guest at the reception?

*Solve each application problem. Round money answers to the nearest cent and percent answers to the nearest tenth of a percent if necessary.*

**19.** Inline skates regularly priced at $109.99 are on special at 55% off. What is the sale price of the skates? (*Source:* Sportmart.)

**20.** From 1990 to 2000, the number of people age 100 or older went from 37,000 to 70,000. What was the percent increase during that period? (*Source:* U.S. Bureau of the Census.)

**21.** Steven and Heather's lunch bill at a Fisherman's Wharf restaurant in San Francisco was $28.19. Show how to estimate a 15% tip and then find the exact tip. If they decided to leave a $5 bill as the tip, what percent did they leave?

**22.** Vineeta has two choices for a $4500 loan to pay her spring semester tuition. She can get a 9-month loan at 11% simple interest, or a 1-year loan at $8\frac{1}{2}$% simple interest. Which loan will cost her less in interest, and by how much?

**23.** LaTroy can buy an 11-piece set of Wilson golf clubs for $329.99 plus 7% sales tax in his home state. Or he can buy the same set while on vacation in a state with $5\frac{1}{2}$% sales tax. What is the total cost at home and on vacation? How much could he save?

**24.** In a survey on how people learn to parent, 360 parents out of 800 said they were most influenced by relatives, friends, and spouses. What percent learned to parent this way? (*Source:* Bama Research.)

# Chapter 7

## SUMMARY

### KEY TERMS

| | | |
|---|---|---|
| 7.1 | **percent** | Percent means "per one hundred." A percent is a ratio with a denominator of 100. |
| 7.2 | **percent proportion** | The proportion used to solve percent problems is $\dfrac{\text{percent}}{100} = \dfrac{\text{part}}{\text{whole}}$. |
| | **whole** | The *whole* in a percent problem is the entire quantity or the total. It is sometimes called the *base*. |
| | **part** | The *part* in a percent problem is the number being compared to the *whole*. |
| 7.3 | **percent equation** | The percent equation is percent • whole = part. It can be used instead of the percent proportion to solve percent problems. |
| 7.4 | **percent of increase or decrease** | Percent of increase or decrease is the amount of change (increase or decrease) expressed as a percent of the original value. |
| 7.5 | **sales tax** | Sales tax is a percent of the total sales charged as a tax. |
| | **tax rate** | The tax rate is the percent used when calculating the amount of tax. |
| | **discount** | Discount is often expressed as a percent of the original price; it is then deducted from the original price, resulting in the sale price. |
| | **interest** | Interest is a fee paid or a charge made for lending or borrowing money. |
| | **principal** | Principal is the amount of money on which interest is earned. |
| | **interest rate** | Often referred to as *rate*, it is the charge for interest and is given as a percent. |
| | **simple interest** | When interest is calculated on the original principal, it is called simple interest. |
| | **interest formula** | The interest formula is used to calculate simple interest. It is Interest = principal • rate • time or $I = prt$. |

### NEW FORMULAS

Finding simple interest: $I = prt$

### TEST YOUR WORD POWER

See how well you have learned the vocabulary in this chapter. Answers follow the Quick Review.

1. **Percent** means
   (a) per one thousand
   (b) part divided by whole
   (c) per one hundred
   (d) part times whole.

2. The **whole** in a percent problem is
   (a) the entire quantity or total
   (b) a ratio with a denominator of 100
   (c) the amount of change
   (d) always 100.

3. When calculating sales tax, the **rate** is
   (a) the part
   (b) the whole
   (c) the total cost
   (d) the percent.

4. The **interest formula** is
   (a) tax rate • cost of item = sales tax
   (b) $I = prt$
   (c) percent • whole = part
   (d) $\dfrac{\text{percent}}{100} = \dfrac{\text{part}}{\text{whole}}$.

5. When calculating interest on a loan, the **principal** is the
   (a) amount of money borrowed
   (b) total amount due
   (c) rate charged for borrowing money
   (d) amount of simple interest.

6. The **percent of increase or decrease** compares the amount of change to
   (a) 100
   (b) the percent
   (c) the original value before the change
   (d) the new value after the change.

7. The **percent equation** is
   (a) tax rate • cost of item = sales tax
   (b) $I = prt$
   (c) percent • whole = part
   (d) $\dfrac{\text{percent}}{100} = \dfrac{\text{part}}{\text{whole}}$.

8. A **discount** is
   (a) added to the original price
   (b) divided by 100
   (c) multiplied by the percent
   (d) subtracted from the original price.

## Quick Review

**Concepts**

### 7.1 Basics of Percent

**Writing a Percent as a Decimal**
To write a percent as a decimal, move the decimal point *two* places to the *left* and drop the % sign.

**Writing a Decimal as a Percent**
To write a decimal as a percent, move the decimal point *two* places to the *right* and attach a % sign.

**Writing a Percent as a Fraction**
To write a percent as a fraction, drop the % symbol and write the number over 100. Then write the fraction in lowest terms.

**Writing a Fraction as a Percent**
To write a fraction as a percent, multiply by 100 and attach a % symbol. This is the same as multiplying by 100%.

**Examples**

$50\% = 50.\% = 0.50$ or $0.5$

$3\% = 03.\% = 0.03$

$0.75 = 0.75 = 75\%$

$3.6 = 3.60 = 360\%$

$35\% = \frac{35}{100} = \frac{35 \div 5}{100 \div 5} = \frac{7}{20}$ ← Lowest terms

$125\% = \frac{125}{100} = \frac{125 \div 25}{100 \div 25} = \frac{5}{4} = 1\frac{1}{4}$

$\frac{3}{5} = \frac{3}{5} \cdot \frac{100}{1}\% = \frac{3 \cdot 5 \cdot 20}{5 \cdot 1}\% = \frac{60}{1}\% = 60\%$

$\frac{7}{8} = \frac{7}{8} \cdot \frac{100}{1}\% = \frac{7 \cdot 4 \cdot 25}{2 \cdot 4 \cdot 1}\% = \frac{175}{2}\% = 87\frac{1}{2}$ or $87.5\%$

### 7.2 Using the Percent Proportion

Percent is to 100 as part is to whole.

$$\underset{\text{Always 100} \longrightarrow}{\text{percent}} \frac{\phantom{1}}{100} = \frac{\text{part}}{\text{whole}}$$

Identify the percent first. It appears with the word *percent* or the % symbol.

The *whole* is the entire quantity or total. It often appears after the word *of*.

The *part* is the number being compared to the whole.

30 children is what percent **of** 75 children?

Percent (unknown) → $\frac{p}{100} = \frac{30}{75}$ ← Part
Always 100 →            ← Whole (follows *of*)

To solve the proportion, find the cross products.

$\frac{p}{100} = \frac{30}{75}$    $100 \cdot 30 = 3000$    Cross products
           $p \cdot 75$

$p \cdot 75 = 3000$    Show that the cross products are equivalent.

$\frac{p \cdot 75}{75} = \frac{3000}{75}$    Divide both sides by 75.

$p = 40$

30 children is **40%** of 75 children.

### 7.3 Percent Shortcuts

200% of a number is 2 times the number.

100% of a number is the entire number.

To find 50% of a number, divide the number by 2.

To find 25% of a number, divide the number by 4.

To find 10% of a number, move the decimal point *one* place to the *left*.

To find 1% of a number, move the decimal point *two* places to the *left*.

200% of $35 is 2 times $35, or $70.

100% of 600 women is *all* the women (600 women).

50% of $8000 is $8000 ÷ 2 = $4000.

25% of 40 pens is 40 ÷ 4 = 10 pens.

10% of $92.40 is $9.24.

1% of 62. miles is 0.62 mile.

| Concepts | Examples |
|---|---|
| **7.3–7.4 Using the Percent Equation**<br>Use the six problem-solving steps from Chapter 3. | Todd's regular pay is $540 per week but $43.20 is taken out of each paycheck for medical insurance. What percent is that? |
| Step 1  **Read** the problem. | Step 1  The problem asks for the percent of Todd's pay that is taken out for insurance.<br>Unknown: percent taken out<br>Known: Whole paycheck is $540; $43.20 taken out |
| Step 2  **Assign a variable.** | Step 2  Let $p$ be the unknown percent. |
| Step 3  **Write an equation.** | Step 3  Use the percent equation. The *whole* is Todd's entire pay of $540 and the *part* is $43.20 (the part taken out of his paycheck).<br><br>$$\text{percent} \cdot \text{whole} = \text{part}$$<br>$$p \cdot 540 = 43.20$$ |
| Step 4  **Solve the equation.** | Step 4  $$\frac{p \cdot 540}{540} = \frac{43.20}{540} \quad \text{Divide both sides by 540.}$$<br>$$p = 0.08 \quad \leftarrow \text{Decimal form}$$<br>Multiply 0.08 by 100 so that $0.08 = 8\%$. |
| Step 5  **State the answer.** | Step 5  8% of Todd's pay is taken out for medical insurance. |
| Step 6  **Check** the solution. | Step 6  Use estimation. If 10% of Todd's pay were withheld, then 10% of $540 = $54 which is a little more than the $43.20 actually withheld. So 8% is a reasonable solution. |
| **7.4 Finding Percent of Increase or Decrease**<br>Use subtraction to find the *amount* of increase or decrease. When writing the equation, be careful to use the *original* value as the base. | Enrollment rose from 3820 students to 5157 students. Find the percent of increase.<br><br>5157 students − 3820 students = 1337 students (Amount of increase)<br><br>percent **of** original value = amount of increase<br>$$p \cdot 3820 = 1337$$<br>$$p \cdot 3820 = 1337$$<br>$$\frac{p \cdot 3820}{3820} = \frac{1337}{3820} \quad \text{Divide both sides by 3820.}$$<br>$$p = 0.35 \quad \leftarrow \text{Decimal form}$$<br>$$0.35 = 35\% \quad \leftarrow \text{Percent increase}$$<br>The enrollment increased 35%. |

| Concepts | Examples |
|---|---|
| **7.5 Consumer Applications** | |

**Finding Sales Tax**

To find sales tax, use this equation.

$$\text{tax rate} \cdot \text{cost of item} = \text{sales tax}$$

Write the percent as a decimal.

Find the sales tax on an $89 pair of binoculars if the sales tax rate is 6%.

$$\text{tax rate} \cdot \text{cost of item} = \text{sales tax}$$
$$06.\% \cdot \$89 = n$$
$$0.06(89) = n$$
$$5.34 = n$$

The sales tax is **$5.34**.

**Estimating a Restaurant Tip**

To estimate a 15% tip, first find 10% of the food bill by moving the decimal point *one* place to the *left*. Then add half of that amount for the other 5%.

To estimate a 20% tip, first find 10% by moving the decimal point *one* place to the *left*. Then double the amount.

To find the exact tip, write the percent as a decimal. The bill for food and beverages is the *whole* and the tip is the *part*.

Estimate a 15% tip on a restaurant bill of $38.72. Then find the exact tip.

*Estimate:* Round $38.72 to $40.

10% of $40. is $4.

Half of $4 is $2.

So an estimate of the 15% tip is $4 + $2 = $6.

*Exact:*

$$\text{percent} \cdot \text{whole} = \text{part}$$
$$15.\% \cdot 38.72 = n$$
$$0.15 \cdot 38.72 = n$$
$$5.808 = n$$

The exact tip is **$5.81** (rounded to nearest cent).

**Finding a Discount**

To find a discount, use this formula.

$$\text{rate of discount} \cdot \text{original price} = \text{amount of discount}$$

Then subtract to find the sale price.

$$\text{original price} - \text{amount of discount} = \text{sale price}$$

All calculators are on sale at 20% off. Find the sale price of a calculator originally marked $35.

$$20\% \cdot \$35 = n$$
$$0.20 \cdot 35 = n$$
$$7 = n$$

The amount of discount is $7.
Then $35 - $7 = **$28**. ← Sale price

**Finding Simple Interest**

To find the simple interest on a loan, use the formula $I = prt$.

$$\text{Interest} = \text{principal} \cdot \text{rate} \cdot \text{time}$$

Time ($t$) is in years. When the time is given in months, use a fraction with 12 in the denominator because there are 12 months in a year.

Write the rate (the percent) as a decimal.

$2800 is borrowed at 8% for 5 months. Find the amount of interest.

$$I = p \cdot r \cdot t$$
$$= (2800)(0.08)\left(\frac{5}{12}\right)$$
$$= (224)\left(\frac{5}{12}\right) = \frac{(224)(5)}{12} \approx \$93.33$$

## Answers to Test Your Word Power

1. **(c)** *Example:* 7% means 7 per 100, or, 7 out of 100.

2. **(a)** *Example:* In the problem "15 computers is what percent of 75 computers," the whole is 75 computers, which is the total group of computers.

3. **(d)** *Example:* Houston has a tax rate of 8.25%; Minneapolis has a tax rate of 7%.

4. **(b)** *Example:* The interest formula is Interest = principal • rate • time. If you borrow $4000 for 2 years at a rate of 9%, then $I = (4000)(0.09)(2) = \$720$.

5. **(a)** *Example:* If you borrow $4000 for 2 years at a rate of 9%, the principal is $4000.

6. **(c)** *Example:* If your rent increased from $800 to $850, the *percent of increase* compares the amount of change ($50) to the *original* rent ($800).

7. **(c)** *Example:* To answer the question, "What percent of 40 Web pages is 12 Web pages," let $p$ be the unknown percent and write the equation as: $p \cdot 40 = 12$.

8. **(d)** *Example:* If sunglasses are on sale at 10% off, then a pair of sunglasses regularly priced at $25 will have a discount of $2.50 subtracted from the price; you will pay $22.50. To find the amount of discount, multiply 0.10($25) to get $2.50.

# Focus on Real-Data Applications

## Educational Tax Incentives

The government sponsors tax incentive programs to make education more affordable. To qualify for the programs, you have to have an adjusted gross income below a certain level (typically $40,000). You can find specific information at the Internal Revenue Service Web site: www.irs.ustreas.gov.

- The Hope Scholarship offers 100% of the first $1000 spent for certain expenses, such as tuition and books, during the first year of college, plus 50% of the next $1000 incurred during the second year of college. The scholarship money is payable as a tax refund. The student cannot have completed the first two years of post-secondary education and must meet certain educational goals and workload criteria.

- Lifetime Learning Credits are based on qualified expenses, including tuition and books, and equal 20% of the first $5000 in expenses. It is not based on a student's workload and is not limited to only two years.

- Only one of the credits can be claimed for each student.

Suppose you are paying your own educational costs, and your adjusted gross income meets the guidelines to qualify for the Hope Scholarship or Lifetime Learning Credits. Your goals are to earn an Associate of Arts degree from a community college and then transfer to a state university to complete a Bachelor's degree. Tuition costs for resident students at North Harris Montgomery Community College District (NHMCCD) in Texas are used as an example of educational expenses. (Expenses vary among schools, and you can easily find that information in the college's catalog or Internet site.)

Residents of NHMCCD pay $12 registration fee for each semester enrolled plus $30 per semester hour tuition and fees. Assume that you must study a total of 15 semester hours in developmental work in mathematics, reading, and writing, and to complete an Associate of Arts degree you must study 60 additional semester hours. You decide to limit your course load to 15 credit hours each semester. Assume that one course is 3 semester hours, and you will have to purchase books at an approximate cost of $75 per course.

1. How many semesters and how many courses will it take you to finish the requirements for an Associate of Arts degree?

2. What is the total cost to complete the Associate of Arts degree for **(a)** books and **(b)** tuition and fees?

3. Calculate the total costs for tuition, fees, and books during the first two years (four semesters). What is the maximum tax incentive payable under the Hope Scholarship during **(a)** the first year and **(b)** the second year?

4. Assume that you are returning to school and do not qualify for the Hope Scholarship. What is the maximum tax incentive payable under the Lifetime Learning Credits during **(a)** the first year and **(b)** the second year?

5. How much additional tax incentive would be payable under the Lifetime Learning Credits for the remaining coursework to complete the Associate of Arts degree? How much additional tax incentive would be payable under the Lifetime Learning Credits to complete a Bachelor's degree?

# Chapter 7 Review Exercises

**[7.1]** *Write each percent as a decimal and each decimal as a percent.*

1. 25%
2. 180%
3. 12.5%
4. 0.085%

5. 2.65
6. 0.02
7. 0.875
8. 0.002

*Write each percent as a fraction or mixed number in lowest terms. Write each fraction as percent.*

9. 12%
10. 37.5%
11. 250%
12. 5%

13. $\frac{3}{4}$
14. $\frac{5}{8}$
15. $3\frac{1}{4}$
16. $\frac{3}{50}$

*Complete this table.*

| Fraction | Decimal | Percent |
|---|---|---|
| $\frac{1}{8}$ | 17. _____ | 18. _____ |
| 19. _____ | 0.15 | 20. _____ |
| 21. _____ | 22. _____ | 180% |

*Use percent shortcuts to fill in the blanks.*

23. 100% of $46 is _____.

24. 50% of $46 is _____.

25. 100% of 9 hours is _____.

26. 50% of 9 hours is _____.

**[7.2]** *Use a percent proportion to answer each question. If necessary, round percent answers to the nearest tenth of a percent.*

27. 338.8 meters is 140% of what number of meters?

28. 2.5% of what number of cases is 425 cases?

29. What is 6% of 450 cellular phones?

30. 60% of 1450 reference books is how many books?

31. What percent of 380 pairs is 36 pairs?

32. 1440 cans is what percent of 640 cans?

**564** Chapter 7 Percent

**[7.3]** *Use the percent equation to answer each question. Round money answers to the nearest cent if necessary.*

**33.** 11% of $23.60 is how much?

**34.** What is 125% of 64 days?

**35.** 1.28 ounces is what percent of 32 ounces?

**36.** $46 is 8% of what number of dollars?

**37.** 8 people is 40% of what number of people?

**38.** What percent of 174 ft is 304.5 ft?

**[7.4]** *Use the six problem-solving steps to answer each question.*

**39.** A medical clinic found that 16.8% of the patients were late for their appointments last month. The number of patients who were late was 504. Find the total number of patients.

**40.** Coreen budgeted $280 for food on her vacation. She actually spent 130% of that amount. How much did she spend on food?

**41.** In a tree-planting project, 640 of the 800 trees planted were still living one year later. What percent of the trees planted were still living?

**42.** Scientists tell us that there are 9600 species of birds and that 1000 of these species are in danger of extinction. What percent of the bird species are in danger of extinction, to the nearest tenth of a percent?

The ivory-billed wood pecker is near extinction or may already be extinct.

**[7.5]** *Find the amount of sales tax or the tax rate and the total cost. Round to the nearest cent if necessary.*

| Amount of Sale | Tax Rate | Amount of Tax | Total Cost |
|---|---|---|---|
| **43.** $2.79 | 4% | _____ | _____ |
| **44.** $780 | _____ | $58.50 | _____ |

*For each restaurant bill, estimate a 15% tip and a 20% tip. Then find the exact amount for a 15% tip and a 20% tip. Round to the nearest cent if necessary.*

| Bill | Estimated 15% | Exact 15% | Estimated 20% | Exact 20% |
|---|---|---|---|---|
| **45.** $42.73 | _____ | _____ | _____ | _____ |
| **46.** $8.05 | _____ | _____ | _____ | _____ |

*Find the amount or rate of discount and the sale price.*

| Original Price | Rate of Discount | Amount of Discount | Sale Price |
|---|---|---|---|
| **47.** $37.50 | 10% | _____ | _____ |
| **48.** $252 | _____ | $63 | _____ |

*Find the simple interest and total amount due on each loan.*

| Principal | Rate | Time | Interest | Total Amount Due |
|---|---|---|---|---|
| **49.** $350 | $6\frac{1}{2}\%$ | 3 years | _____ | _____ |
| **50.** $1530 | 16% | 9 months | _____ | _____ |

**MIXED REVIEW EXERCISES**

*The bar graph shows the types of electronic/computer games that adults like to play. Use the information in the graph to answer Exercises 51–53.*

**CARE FOR A GAME OF CARDS?**
One in three adults say they play electronic or computer games. Types of games they play:

- Cards 68%
- Sports/driving 46%
- Action/combat 43%
- Board 39%
- Sci-fi/simulation 37%

*Source:* Cable & Telecommunications Association for Marketing.

**51.** Write the portion of adults who like to play electronic/computer games as a fraction and as a percent.

**52.** What type of game is most popular? Write the portion of players who picked this type as a percent, a decimal, and a fraction in lowest terms.

**53. (a)** If 830 adults were surveyed, how many of them play electronic/computer games, to the nearest whole number?

**(b)** Using your answer from part (a), how many of the playing adults chose the most popular and least popular type of game?

*The table shows the number of animals received during the first nine months of the year by the local Humane Society and the number placed in new homes. Use the table to answer Exercises 54–56. Round percent answers to the nearest tenth of a percent.*

| Animals Received | Placed in New Homes |
|---|---|
| 5371 dogs | 2599 dogs |
| 6447 cats | 2346 cats |
| 2223 other | 406 other |

**54. (a)** What percent of the dogs were placed in new homes?

**(b)** The number of dogs received so far this year is 75% of the total number expected. How many dogs are expected, to the nearest whole number?

**55. (a)** What percent of the cats were placed in new homes?

**(b)** During the first nine months of last year, 2300 cats were received. What is the percent increase in cats received from last year to this year?

**56. (a)** What percent of all the animals received were placed in new homes?

**(b)** The Society's goal was to place 40% of all animals received. So far they have missed their goal by how many animals? Round to the nearest whole number.

# Chapter 7 TEST

 *Study Skills Workbook*
Activity 12

*Write each percent as a decimal and each decimal as a percent.*

1. 75%
2. 0.6
3. 1.8
4. 0.075
5. 300%
6. 2%

*Write each percent as a fraction or mixed number in lowest terms.*

7. 62.5%
8. 240%

*Write each fraction or mixed number as a percent.*

9. $\frac{1}{20}$
10. $\frac{7}{8}$
11. $1\frac{3}{4}$

*Write and solve a proportion or an equation to answer each question. Show your work.*

12. 16 laptops is 5% of what number of laptops?

13. $192 is what percent of $48?

14. Erica Green has saved 75% of the amount needed for a down payment on a condominium. If she has saved $14,625, find the total down payment needed.

15. The price of a used car is $5680 plus sales tax of $6\frac{1}{2}$%. Find the total cost of the car including sales tax.

16. Enrollment in mathematics courses increased from 1440 students last semester to 1925 students this semester. Find the percent of increase, to the nearest whole percent.

1. _____
2. _____
3. _____
4. _____
5. _____
6. _____
7. _____
8. _____
9. _____
10. _____
11. _____
12. _____
13. _____
14. _____
15. _____
16. _____

17. Explain a shortcut for finding 50% of a number and a shortcut for finding 25% of a number. Show an example of how to use each shortcut.

17. _____

18. Explain how you would *estimate* a 15% tip on a restaurant bill of $31.94. Then explain how you would *estimate* a 20% tip on the same bill.

18. _____

19. Find the exact 15% tip, to the nearest cent, for the restaurant bill in Problem 18. If you and two friends are sharing the bill and the exact tip, how much will each person pay?

19. _____

*Find the amount of discount and the sale price of each item. Round answers to the nearest cent if necessary. Show your work.*

20. Jeremy plans to use his 8% employee discount to buy a $48 clock radio so he can get to work on time.

20. _____

21. A Sony 4-head VCR regularly priced at $229.95 is on sale at 35% off.

21. _____

22. Jamal found a $1089 computer on sale at 30% off because it was a "discontinued" model. The store will let him pay for it over 6 months with no interest charge. How much will each monthly payment need to be to cover the discounted price plus 7% sales tax?

22. _____

23. What is the simple interest on a four-year loan of $5000 at $8\frac{1}{4}$%?

23. _____

24. Kendra borrowed $860 to pay medical expenses. The loan is for 6 months at 12% simple interest. Find the total amount due on the loan.

24. _____

# Cumulative Review Exercises — Chapters 1–7

1. Write these numbers in words.
   (a) 90.105
   (b) 125,000,670

2. Write these numbers in digits.
   (a) Thirty billion, five million
   (b) Seventy-eight ten-thousandths

3. Round each number as indicated.
   (a) 49,617 to the nearest thousand
   (b) 0.7039 to the nearest hundredth
   (c) 8945 to the nearest hundred

4. Name the property illustrated by each example.
   (a) $-7(0.8) = 0.8(-7)$
   (b) $\left(\dfrac{2}{3} + \dfrac{3}{4}\right) + \dfrac{1}{2} = \dfrac{2}{3} + \left(\dfrac{3}{4} + \dfrac{1}{2}\right)$

5. Write > or < between each pair of numbers to make true statements.
   $-18$ ____ $-8$      $0$ ____ $-5$

6. Arrange in order from smallest to largest.
   $0.705 \quad 0.755 \quad \dfrac{3}{4} \quad 0.7005$

7. Find the mean and median for this set of data on rent prices: $710, $780, $650, $785, $1125, $695, $740, $685.

8. Find the best buy on cereal.
   17 ounces of Brand A for $2.89
   21 ounces of Brand B for $3.59
   15 ounces of Brand C for $2.79

*Simplify.*

9. $50 - 1.099$

10. $(-3)^2 + 2^3$

11. $\dfrac{3b}{10a} \cdot \dfrac{15ab}{4}$

12. $3\dfrac{3}{10} - 2\dfrac{4}{5}$

13. $0 + 2(-6 + 1)$

14. $\dfrac{4}{5} + \dfrac{x}{4}$

15. $-20 - 20$

16. $(-0.5)(0.002)$

17. $\dfrac{-\dfrac{10}{11}}{-\dfrac{5}{6}}$

18. $\dfrac{3}{8}$ of 328

19. $\dfrac{-16 + 2^4}{-3 - 2}$

20. $\dfrac{7}{8} - \dfrac{2}{m}$

**21.** $\dfrac{4.8}{-0.16}$

**22.** $\dfrac{8}{9} \div 2n$

**23.** $1\dfrac{5}{6} + 1\dfrac{2}{3}$

**24.** $5 - 1\dfrac{7}{9}$

**25.** $|10 - 30| + (-4)^3$

**26.** $0.6 \div 12(3.6 - 4)$

**27.** $\dfrac{3}{10} - \left(\dfrac{1}{4} - \dfrac{3}{4}\right)^2 + \left(\dfrac{1}{2}\right)^2$

*Evaluate each expression when w is −4, x is −2, and y is 3.*

**28.** $-6w - 5$

**29.** $5x + 3y$

**30.** $x^3 y$

**31.** $-5w^2 x$

*Simplify each expression.*

**32.** $-2x^2 + 5x - 7x^2$

**33.** $ab - ab$

**34.** $-10(4w^3)$

**35.** $3(h - 4) + 2$

*Solve each equation. Show your work.*

**36.** $2n - 3n = 0 - 5$

**37.** $12 - h = -3h$

**38.** $5a - 0.6 = 10.4$

**39.** $-\dfrac{7}{8} = \dfrac{3}{16} y$

**40.** $3 + \dfrac{1}{10} b = 5$

**41.** $\dfrac{0.2}{3.25} = \dfrac{10}{x}$

**42.** $32 - 3h = 5h + 8$

**43.** $3 + 2(x + 4) = -4x + 7 + 2x$

*Translate each sentence into an equation and solve it.*

**44.** If 5 is subtracted from four times a number, the result is −17. What is the number?

**45.** The sum of a number and 31 is three times the number plus 1. Find the number.

*Solve each application problem using the six problem-solving steps from Chapter 3.*

**46.** Lawrence brought home three packages of disposable diapers. He used 17 diapers the first day and 19 the next day. There were 12 diapers left. How many diapers were originally in each package?

**47.** Susanna and Neoka are splitting $1620 for painting a house. Neoka worked twice as many hours, so she should earn twice as much. How much should each woman receive?

*Find the unknown length, perimeter, circumference, area, or volume. When necessary, use 3.14 as the approximate value of π and round answers to the nearest tenth.*

**48.** Find the perimeter and the area.

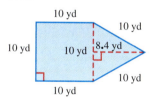

**49.** Find the circumference and the area.

**50.** Find the unknown length in these similar triangles.

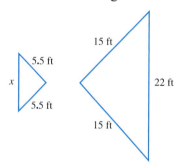

**51.** Find the volume and surface area.

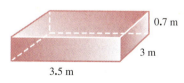

**52.** Find the volume.

**53.** Find the unknown length.

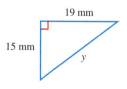

*Solve each application problem.*

**54.** A survey of the 5600 students on our campus found that $\frac{3}{8}$ of the students work 20 hours or more per week. How many students is that?

**55.** The Americans With Disabilities Act provides the single parking space design below. Find the perimeter and area of this parking space, including the accessible aisle.

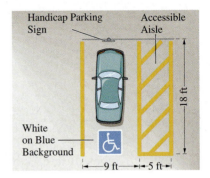

**56.** The Jackson family is making three kinds of holiday cookies that require brown sugar. The recipes call for $2\frac{1}{4}$ cups, $1\frac{1}{2}$ cups, and $\frac{3}{4}$ cup, respectively. They bought two packages of brown sugar, each holding $2\frac{1}{3}$ cups. The amount bought is how much more or less than the amount needed?

**57.** Leather jackets are on sale at 30% off the regular price. Tracy likes a jacket with a regular price of $189. Find the amount of discount and the sale price she will pay.

**58.** On the Illinois map, one centimeter represents 12 kilometers. The center of Springfield is 7.8 cm from the center of Bloomington on the map. What is the actual distance in kilometers?

**59.** On a 35-problem math test, Juana solved 31 problems correctly. What percent of the problems were correct? Round to the nearest tenth of a percent.

**60.** Daniel had a roll of 35 mm film developed. He received 24 prints for $10.25. What is the cost per print, to the nearest cent?

*This table shows information about the five U.S. colleges that accepted the lowest percent of student applications for the 2000–2001 academic year. Use the table to answer Exercises 61–64. Round percent answers to the nearest tenth of a percent.*

**THE CHOOSIEST COLLEGES IN THE UNITED STATES**

| Rank | College | Number Applied | Percent Accepted |
|---|---|---|---|
| 1 | U.S. Coast Guard Academy | 5458 | 9.6 |
| 2 | Julliard School of Music | 1176 | 10.9 |
| 3 | Harvard College | 18,161 | 11.4 |
| 4 | Princeton University | 14,875 | 11.4 |
| 5 | Cooper Union | 2216 | 13.1 |

Source: Kaplan/Newsweek *College Catalog 2001*.

**61.** (a) How many applications were received by all the schools combined?

(b) Harvard's applications were what percent of all the applications?

**62.** How many more applications were received by the third ranking college than by the fourth?

**63.** How many students were accepted at the top ranking college? How many were rejected? Round to the nearest whole number.

**64.** The number of applicants at Cooper Union is what percent of the applicants at Julliard?

# Measurement 8

8.1 Problem Solving with English Measurement

8.2 The Metric System—Length

8.3 The Metric System—Capacity and Weight (Mass)

8.4 Problem Solving with Metric Measurement

8.5 Metric–English Conversions and Temperature

At the 2000 Sydney Olympics, Michael Johnson won a pair of gold medals in track events to go with his gold shoes. The shoes, made especially for Johnson by 3M Company, contained 5 grams of pure gold. (*Source:* 3M Company.) But did the gold weigh Johnson down, slowing his speed? To find out, read about grams in Section 8.3. Then find out how to convert grams to ounces in Section 8.5. (See Section 8.5, Exercise 13.)

# 8.1 Problem Solving with English Measurement

**OBJECTIVES**

1. Learn the basic measurement units in the English system.
2. Convert among measurement units using multiplication or division.
3. Convert among measurement units using unit fractions.
4. Solve application problems using English measurement.

We measure things all the time: the distance traveled on vacation, the floor area we want to cover with carpet, the amount of milk in a recipe, the weight of the bananas we buy at the store, the number of hours we work, and many more.

In the United States we still use the **English system** of measurement for many everyday activities. Examples of English units are inches, feet, quarts, ounces, and pounds. However, the fields of science, medicine, sports, and manufacturing use the **metric system** (meters, liters, and grams). And, because the rest of the world uses only the metric system, U.S. businesses are beginning to change to the metric system in order to compete internationally.

**1** Learn the basic measurement units in the English system. Until the switch to the metric system is complete, we still need to know how to use the English system of measurement. The table below lists the relationships you should memorize. The time relationships are used in both the English and metric systems.

### English Measurement Relationships

**Length**
1 foot (ft) = 12 inches (in.)
1 yard (yd) = 3 feet (ft)
1 mile (mi) = 5280 feet (ft)

**Weight**
1 pound (lb) = 16 ounces (oz)
1 ton (T) = 2000 pounds (lb)

**Capacity**
1 cup (c) = 8 fluid ounces (fl oz)
1 pint (pt) = 2 cups (c)
1 quart (qt) = 2 pints (pt)
1 gallon (gal) = 4 quarts (qt)

**Time**
1 minute (min) = 60 seconds (sec)
1 hour (hr) = 60 minutes (min)
1 day = 24 hours (hr)
1 week (wk) = 7 days

As you can see, there is no simple or "natural" way to convert among these various measures. The units evolved over hundreds of years and were based on a variety of "standards." For example, one yard was the distance from the tip of a king's nose to his thumb when his arm was outstretched. An inch was three dried barleycorns laid end to end.

**Example 1** Knowing English Measurement Units

Memorize the English measurement conversions. Then answer these questions.

(a) _____ day = 24 hr    Answer: 1 day
(b) 1 yd = _____ ft    Answer: 3 ft

**Work Problem 1 at the Side.**

**1** After memorizing the measurement conversions, answer these questions.

(a) 1 c = _____ fl oz

(b) _____ qt = 1 gal

(c) 1 wk = _____ days

(d) _____ ft = 1 yd

(e) 1 ft = _____ in.

(f) _____ oz = 1 lb

(g) 1 T = _____ lb

(h) _____ min = 1 hr

(i) 1 pt = _____ c

(j) _____ hr = 1 day

(k) 1 min = _____ sec

(l) 1 qt = _____ pt

(m) _____ ft = 1 mi

**2** Convert among measurement units using multiplication or division. You often need to convert from one unit of measure to another. Two methods of converting measurements are shown here. Study each way and use the method you prefer. The first method involves deciding whether to multiply or divide.

### Converting among Measurement Units

1. *Multiply* when converting from a larger unit to a smaller unit.
2. *Divide* when converting from a smaller unit to a larger unit.

**ANSWERS**
1. (a) 8  (b) 4  (c) 7  (d) 3  (e) 12
   (f) 16  (g) 2000  (h) 60  (i) 2  (j) 24
   (k) 60  (l) 2  (m) 5280

Section 8.1 Problem Solving with English Measurement  **575**

**Example 2** Converting from One Unit of Measure to Another

Convert each measurement.

**(a)** 7 ft to inches
You are converting from a *larger* unit to a *smaller* unit (feet to inches), so multiply.
Because *1 ft = **12** in.*, multiply by 12.
$$7 \text{ ft} = 7 \cdot 12 = 84 \text{ in.}$$

**(b)** $3\frac{1}{2}$ lb to ounces
You are converting from a *larger* unit to a *smaller* unit (pounds to ounces), so multiply.
Because *1 lb = **16** oz*, multiply by 16.
$$3\frac{1}{2} \text{ lb} = 3\frac{1}{2} \cdot 16 = \frac{7}{2} \cdot \frac{\cancel{16}^{8}}{1} = \frac{56}{1} = 56 \text{ oz}$$

**(c)** 20 qt to gallons
You are converting from a *smaller* unit to a *larger* unit (quarts to gallons), so divide.
Because *4 qt = 1 gal*, divide by 4.
$$20 \text{ qt} = \frac{20}{4} = 5 \text{ gal}$$
Divide by 4.

**(d)** 45 min to hours
You are converting from a *smaller* unit to a *larger* unit (minutes to hours), so divide.
Because *60 min = 1 hr*, divide by 60 and write the fraction in lowest terms.
$$45 \text{ min} = \frac{45}{60} = \frac{45 \div 15}{60 \div 15} = \frac{3}{4} \text{ hr} \quad \leftarrow \text{Lowest terms}$$
Divide by 60.

**Work Problem ❷ at the Side.**

**3** **Convert among measurement units using unit fractions.** If you have trouble deciding whether to multiply or divide when converting measurements, use *unit fractions* to solve the problem. You'll also find this method useful in science classes. A **unit fraction** is equivalent to 1. Here is an example.

$$\frac{12 \text{ in.}}{12 \text{ in.}} = \frac{\cancel{12}^{1} \text{ in.}}{\cancel{12}_{1} \text{ in.}} = 1$$

Use the table of measurement relationships on the first page of this section to find that 12 in. is the same as 1 ft. So in the numerator you can substitute 1 ft for 12 in., or in the denominator you can substitute 1 ft for 12 in. This makes two useful unit fractions.

$$\frac{\mathbf{1 \text{ ft}}}{12 \text{ in.}} = 1 \quad \text{or} \quad \frac{12 \text{ in.}}{\mathbf{1 \text{ ft}}} = 1$$

To convert from one measurement unit to another, just multiply by the appropriate unit fraction. Remember, a unit fraction is equivalent to 1. Multiplying something by 1 does *not* change its value.

❷ Convert each measurement using multiplication or division.

**(a)** $5\frac{1}{2}$ ft to inches

**(b)** 64 oz to pounds

**(c)** 6 yd to feet

**(d)** 2 T to pounds

**(e)** 35 pt to quarts

**(f)** 20 min to hours

**(g)** 4 wk to days

**ANSWERS**
**2. (a)** 66 in. **(b)** 4 lb **(c)** 18 ft
**(d)** 4000 lb **(e)** $17\frac{1}{2}$ qt **(f)** $\frac{1}{3}$ hr
**(g)** 28 days

**3** First write the unit fraction needed to make each conversion. Then complete the conversion.

(a) 36 in. to feet

unit fraction } $\dfrac{1 \text{ ft}}{12 \text{ in.}}$

(b) 14 ft to inches

unit fraction } $\dfrac{\text{in.}}{\text{ft}}$

(c) 60 in. to feet

unit fraction } _____

(d) 4 yd to feet

unit fraction } _____

(e) 39 ft to yards

unit fraction } _____

(f) 2 mi to feet

unit fraction } _____

---

Use these guidelines to choose the correct unit fraction.

### Choosing a Unit Fraction

The *numerator* should use the measurement unit you want in the *answer*.

The *denominator* should use the measurement unit you want to *change*.

**Example 3**  Using Unit Fractions with Length Measurements

(a) Convert 60 in. to feet.

Use a unit fraction with feet (the unit for your answer) in the numerator, and inches (the unit being changed) in the denominator. Because *1 ft = 12 in.*, the necessary unit fraction is

$$\dfrac{1 \text{ ft}}{12 \text{ in.}} \begin{array}{l} \leftarrow \text{Unit for your answer is feet.} \\ \leftarrow \text{Unit being changed is inches.} \end{array}$$

Next, multiply 60 in. times this unit fraction. Write 60 in. as the fraction $\dfrac{60 \text{ in.}}{1}$. Then divide out common units and factors wherever possible.

$$60 \text{ in.} \cdot \dfrac{1 \text{ ft}}{12 \text{ in.}} = \dfrac{\overset{5}{\cancel{60} \text{ in.}}}{1} \cdot \dfrac{1 \text{ ft}}{\underset{1}{\cancel{12} \text{ in.}}} = \dfrac{5 \cdot 1 \text{ ft}}{1} = 5 \text{ ft}$$

These units should match.

— Divide out inches.
— Divide 60 and 12 by 12.

(b) Convert 9 ft to inches.

Select the correct unit fraction to change 9 ft to inches.

$$\dfrac{12 \text{ in.}}{1 \text{ ft}} \begin{array}{l} \leftarrow \text{Unit for your answer is inches.} \\ \leftarrow \text{Unit being changed is feet.} \end{array}$$

Multiply 9 ft times the unit fraction.

$$9 \text{ ft} \cdot \dfrac{12 \text{ in.}}{1 \text{ ft}} = \dfrac{9 \cancel{\text{ ft}}}{1} \cdot \dfrac{12 \text{ in.}}{1 \cancel{\text{ ft}}} = \dfrac{9 \cdot 12 \text{ in.}}{1} = 108 \text{ in.}$$

These units should match.

— Divide out feet.

**CAUTION**

If no units will divide out, you made a mistake in choosing the unit fraction.

Work Problem **3** at the Side.

**Example 4**  Using Unit Fractions with Capacity and Weight Measurements

(a) Convert 9 pt to quarts.
First select the correct unit fraction.

$$\dfrac{1 \text{ qt}}{2 \text{ pt}} \begin{array}{l} \leftarrow \text{Unit for your answer is quarts.} \\ \leftarrow \text{Unit being changed is pints.} \end{array}$$

Continued on Next Page

---

ANSWERS

3. (a) 3 ft  (b) $\dfrac{12 \text{ in.}}{1 \text{ ft}}$; 168 in.
(c) $\dfrac{1 \text{ ft}}{12 \text{ in.}}$; 5 ft  (d) $\dfrac{3 \text{ ft}}{1 \text{ yd}}$; 12 ft
(e) $\dfrac{1 \text{ yd}}{3 \text{ ft}}$; 13 yd  (f) $\dfrac{5280 \text{ ft}}{1 \text{ mi}}$; 10,560 ft

Now multiply.

$$9 \text{ pt} \cdot \frac{1 \text{ qt}}{2 \text{ pt}} = \frac{9 \text{ pt}}{1} \cdot \frac{1 \text{ qt}}{2 \text{ pt}} = \frac{9}{2} \text{ qt} = 4\frac{1}{2} \text{ qt}$$

These units should match. Divide out pints. Write as mixed number.

**(b)** Convert $7\frac{1}{2}$ gal to quarts.

Write as an improper fraction.

$$\frac{7\frac{1}{2} \text{ gal}}{1} \cdot \frac{4 \text{ qt}}{1 \text{ gal}} = \frac{15}{2} \cdot \frac{4}{1} \text{ qt}$$

$$= \frac{15}{2} \cdot \frac{4}{1} \text{ qt}$$

$$= 30 \text{ qt}$$

**(c)** Convert 36 oz to pounds.

$$\frac{36 \text{ oz}}{1} \cdot \frac{1 \text{ lb}}{16 \text{ oz}} = \frac{9}{4} \text{ lb} = 2\frac{1}{4} \text{ lb}$$

**NOTE**

In Example 4(c) you get $\frac{9}{4}$ lb. Recall that $\frac{9}{4}$ means $9 \div 4$. If you do $9 \div 4$ on your calculator, you get **2.25 lb**. English measurements usually use fractions or mixed numbers, like $2\frac{1}{4}$ lb. However, 2.25 lb is also correct and is the way grocery stores often show weights of produce, meat, and cheese.

Work Problem ④ at the Side.

**Example 5** Using Several Unit Fractions

Sometimes you may need to use two or three unit fractions to complete a conversion.

**(a)** Convert 63 in. to yards.

Use the unit fraction $\frac{1 \text{ ft}}{12 \text{ in.}}$ to change inches to feet and the unit fraction $\frac{1 \text{ yd}}{3 \text{ ft}}$ to change feet to yards. Notice how all the units divide out except yards, which is the unit you want in the answer.

$$\frac{63 \text{ in.}}{1} \cdot \frac{1 \text{ ft}}{12 \text{ in.}} \cdot \frac{1 \text{ yd}}{3 \text{ ft}} = \frac{63}{36} \text{ yd} = 1\frac{3}{4} \text{ yd}$$

*Continued on Next Page*

④ Convert using unit fractions.

**(a)** 16 qt to gallons

**(b)** 3 c to pints

**(c)** $3\frac{1}{2}$ T to pounds

**(d)** $1\frac{3}{4}$ lb to ounces

**(e)** 4 oz to pounds

**ANSWERS**

**4. (a)** 4 gal  **(b)** $1\frac{1}{2}$ pt or 1.5 pt  **(c)** 7000 lb

**(d)** 28 oz  **(e)** $\frac{1}{4}$ lb or 0.25 lb

**5** Convert using two or three unit fractions.

(a) 4 T to ounces

(b) 3 mi to inches

(c) 36 pt to gallons

(d) 2 wk to minutes

You can also divide out common factors in the numbers.

$$\frac{\overset{7}{\cancel{\overset{21}{\cancel{63}}}}}{1} \cdot \frac{1}{\underset{4}{\cancel{12}}} \cdot \frac{1}{\underset{1}{\cancel{3}}} = \frac{7}{4} = 1\frac{3}{4} \text{ yd}$$

Instead of changing $\frac{7}{4}$ to $1\frac{3}{4}$, you can enter $7 \div 4$ on your calculator to get 1.75 yd. Both answers are correct because 1.75 is equivalent to $1\frac{3}{4}$.

**(b)** Convert 2 days to seconds.

Use three unit fractions. The first one changes days to hours, the second one changes hours to minutes, and the third one changes minutes to seconds. All the units divide out except seconds, which is what you want in your answer.

$$\frac{2 \text{ days}}{1} \cdot \frac{24 \text{ hr}}{1 \text{ day}} \cdot \frac{60 \text{ min}}{1 \text{ hr}} \cdot \frac{60 \text{ seconds}}{1 \text{ min}} = 172{,}800 \text{ seconds}$$

Divide out **days**.
Divide out **hr**.
Divide out **min**.

**Work Problem 5 at the Side.**

**4** **Solve application problems using English measurement.** To solve measurement application problems, we will use the steps summarized here.

*Step 1* **Read** the problem.
*Step 2* **Work out a plan.**
*Step 3* **Estimate** a reasonable answer.
*Step 4* **Solve** the problem.
*Step 5* **State the answer.**
*Step 6* **Check** your work.

Because measurement applications often involve conversions, writing an equation may not be the most helpful way to solve the problem. Therefore, Steps 2 and 3 are different from the ones you learned in Chapter 3; the other steps are the same.

**Example 6** Solving English Measurement Applications

(a) A 36 oz can of coffee is on sale at Cub Foods for $4.98. What is the cost per pound, to the nearest cent? (*Source:* Cub Foods.)

*Step 1* **Read** the problem. The problem asks for the cost per *pound* of coffee.

*Step 2* **Work out a plan.** The given amount of coffee is in *ounces* but the answer must be cost per *pound*. Convert ounces to pounds. The word *per* indicates division. You need to divide the cost by the number of pounds.

*Step 3* **Estimate** a reasonable answer. To estimate, round $4.98 to $5. Then, there are 16 oz in a pound, so 36 oz is a little more than 2 lb. So, $5 ÷ 2 = $2.50 per pound as our estimate.

**Continued on Next Page**

**ANSWERS**
**5.** (a) 128,000 oz  (b) 190,080 in.
  (c) $4\frac{1}{2}$ gal or 4.5 gal  (d) 20,160 min

*Step 4* **Solve** the problem. Use a unit fraction to convert 36 oz to pounds.

$$\frac{\overset{9}{\cancel{36}} \text{ oz}}{1} \cdot \frac{1 \text{ lb}}{\underset{4}{\cancel{16}} \text{ oz}} = \frac{9}{4} \text{ lb} = 2.25 \text{ lb}$$

Then divide.

$$\frac{\$4.98}{2.25 \text{ lb}} = 2.21\overline{3} \approx 2.21 \quad \text{Rounded}$$

*Step 5* **State the answer.** The coffee costs $2.21 per pound (to the nearest cent).

*Step 6* **Check** your work. The answer, $2.21, is close to our estimate of $2.50.

**(b)** Bilal's favorite dessert recipe uses $1\frac{2}{3}$ cups of milk. If he makes six desserts for a bake sale at his son's school, how many quarts of milk will he need?

*Step 1* **Read** the problem. The problem asks for the number of *quarts* of milk needed for six desserts.

*Step 2* **Work out a plan.** Multiply to find the number of *cups* of milk for six desserts. Then convert *cups* to *quarts* (the unit required in the answer).

*Step 3* **Estimate** a reasonable answer. To estimate, round $1\frac{2}{3}$ cups to 2 cups. Then, 2 cups times 6 = 12 cups. There are 4 cups in a quart, so 12 cups ÷ 4 = 3 qt as our estimate.

*Step 4* **Solve** the problem. First multiply. Then use unit fractions to convert.

$$1\frac{2}{3} \cdot 6 = \frac{5}{\cancel{3}} \cdot \frac{\overset{2}{\cancel{6}}}{1} = \frac{10}{1} = 10 \text{ cups}$$

$$\frac{\overset{5}{\cancel{10 \text{ cups}}}}{1} \cdot \frac{1 \text{ pt}}{\underset{1}{\cancel{2 \text{ cups}}}} \cdot \frac{1 \text{ qt}}{2 \text{ pt}} = \frac{5}{2} \text{ qt} = 2\frac{1}{2} \text{ qt}$$

*Step 5* **State the answer.** Bilal needs $2\frac{1}{2}$ qt (or 2.5 qt) of milk.

*Step 6* **Check** your work. The answer, $2\frac{1}{2}$ qt, is close to our estimate of 3 qt.

**NOTE**

In *Step 2* above, we *first multiplied* $1\frac{2}{3}$ cups times 6 to find the number of cups needed, then *converted* 10 cups to $2\frac{1}{2}$ quarts. It would also work to *first convert* $1\frac{2}{3}$ cups to $\frac{5}{12}$ qt, then multiply $\frac{5}{12}$ qt times 6 to get $2\frac{1}{2}$ qt.

Work Problem ❻ at the Side.

---

❻ Solve each application problem using the six problem-solving steps.

**(a)** Kristin paid $3.29 for 12 oz of extra sharp cheddar cheese. What is the price per pound, to the nearest cent?

**(b)** A moving company estimates 11,000 lb of furnishings for an average 3-bedroom house. (*Source:* North American Van Lines.) If the company made five such moves last week, how many tons of furnishings did they move?

**ANSWERS**

**6. (a)** $4.39 per pound (rounded)

**(b)** 27.5 or $27\frac{1}{2}$ T

# Focus on Real-Data Applications

## Growing Sunflowers

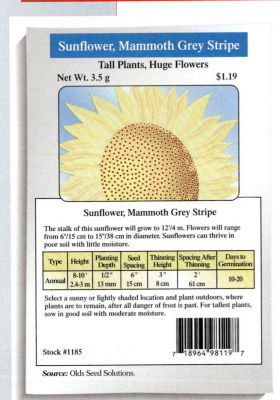

The front and back of a seed packet for sunflowers are shown. Look at the front of the packet first.

1. There were 42 seeds in the packet. If 40 of the seeds sprouted, what was the cost per sprout, to the nearest cent?

2. If vegetable and flower seeds were on sale at 30% off, what was the cost per sprout, to the nearest cent?

3. What percent of the seeds sprouted, to the nearest whole percent?

4. How many seeds would weigh 1 gram?

5. The table on the back of the packet uses the symbol (') for feet, and the symbol (") for inches.

    (a) How tall will the plants grow, in feet?

    (b) How tall will they grow in inches?

    (c) How tall will they grow in yards?

6. If you plant all 42 seeds in one long row, using the spacing given on the package, how long will your row be in feet?

7. How many inches tall should the plants be when you thin them (remove less vigorous plants to give others room to grow)? How tall is that in feet?

8. What is the range in the diameter of the flowers, in inches, and in feet? Diameter is the distance across the circular flower.

9. (a) Using the information in the article on the right, how many gum wrappers are needed to make 1 foot of chain?

    (b) To make 1 inch of chain?

    (c) How many inches of chain, to the nearest hundredth, are made from one wrapper?

10. Is the article correct in saying that 125 miles is 34 million gum wrappers?

### Wrapped Up in Work?

Sometimes a person's life's work makes it into a museum. So it figures that Michael Knutson's 128-foot chain of gum wrappers now resides in the Yellow Medicine County Museum in Granite Falls, Minnesota.

According to the *Redwood Gazette*, the wrapper chain started in 1974 when Knutson, now a 37-year-old woodworker, became bored during study hall. "I've never been much of a gum chewer, but I found most of the wrappers on the streets of the city," he said. (FYI: It takes 6602 wrappers to make a 128-foot chain.)

Granite Falls is about 125 miles—or 34 million gum wrappers—west of the Twin Cities.

*Source: Minneapolis Star Tribune.*

## 8.1 Exercises

*Fill in the blanks with the measurement relationships you have memorized. See Example 1.*

1. 1 yd = _____ ft
2. 1 ft = _____ in.
3. _____ fl oz = 1 c
4. _____ qt = 1 gal
5. 1 mi = _____ ft
6. 1 wk = _____ days
7. _____ lb = 1 T
8. _____ oz = 1 lb
9. 1 min = _____ sec
10. 1 day = _____ hr

*Convert each measurement using unit fractions. See Examples 3 and 4.*

11. 120 sec = _____ min
12. 180 min = _____ hr
13. 8 qt = _____ gal
14. 6 gal = _____ qt

15. An adult sperm whale may weigh 38 to 40 tons. How many pounds could it weigh? (*Source: Grolier Multimedia Encyclopedia.*)

16. Recent fossil finds in Argentina indicate that the largest meat-eating dinosaur may have been 45 ft long. How many yards long was this dinosaur? (*Source: Washington Post.*)

17. 9 yd = _____ ft
18. 20,000 lb = _____ T
19. 7 lb = _____ oz
20. 96 oz = _____ lb

**21.** 5 qt = _____ pt

**22.** 26 pt = _____ qt

**23.** 90 min = _____ hr

**24.** 45 sec = _____ min

**25.** 3 in. = _____ ft

**26.** 30 in. = _____ ft

**27.** 24 oz = _____ lb

**28.** 36 oz = _____ lb

**29.** 5 c = _____ pt

**30.** 15 qt = _____ gal

**31.** Mr. Kashpaws and his son worked for 12 hours doing traditional harvesting of wild rice. What part of a day did they work?

**32.** The starflower blooms in spring in woods throughout the northeastern United States. Its genus name, *Trientalis*, comes from a Latin word meaning "one-third of a foot." The name refers to the height of the plant. How tall is the starflower in inches? (*Source: Peterson Field Guides: Wildflowers.*)

**33.** $2\frac{1}{2}$ T = _____ lb

**34.** $4\frac{1}{2}$ pt = _____ c

**35.** $4\frac{1}{4}$ gal = _____ qt

**36.** $2\frac{1}{4}$ hr = _____ min

**37.** Our premature baby weighed $2\frac{3}{4}$ lb at birth. How many ounces did our baby weigh?

**38.** Michelle prepares 4 oz hamburgers at a fast-food restaurant. Each hamburger is what part of a pound?

*Use two or three unit fractions to make each conversion. See Example 5.*

**39.** 6 yd = _____ in.

**40.** 2 T = _____ oz

**41.** 112 c = _____ qt

**42.** 336 hr = _____ wk

**43.** 6 days = _____ sec

**44.** 5 gal = _____ c

**45.** $1\frac{1}{2}$ T = _____ oz

**46.** $3\frac{1}{3}$ yd = _____ in.

**47.** The statement 8 = 2 is *not* true. But with appropriate measurement units, it *is* true.

$$8 \text{ quarts} = 2 \text{ gallons}$$

Attach measurement units to these numbers to make the statements true.

(a) 1 _____ = 16 _____
(b) 10 _____ = 20 _____
(c) 120 _____ = 2 _____
(d) 2 _____ = 24 _____
(e) 6000 _____ = 3 _____
(f) 35 _____ = 5 _____

**48.** Explain in your own words why you can add 2 feet + 12 inches to get 3 feet, but you cannot add 2 feet + 12 pounds.

*Convert the following. See Example 5.*

**49.** $2\frac{3}{4}$ mi = _____ in.

**50.** $5\frac{3}{4}$ tons = _____ oz

**51.** $6\frac{1}{4}$ gal = _____ fl oz

**52.** $3\frac{1}{2}$ days = _____ sec

**53.** 24,000 oz = _____ T

**54.** 57,024 in. = _____ mi

*Solve each application problem. Show your work. See Example 6.*

**55.** Geralyn bought a 20 oz box of strawberries for $2.29. What was the price per pound for the strawberries, to the nearest cent?

**56.** Zach paid $0.79 for a 1.6 oz candy bar. (*Source:* Byerly's Foods.) What was the cost per pound?

**57.** Dan orders supplies for the science labs. Each of the 24 stations in the chemistry lab needs 2 ft of rubber tubing. If rubber tubing sells for $8.75 per yard, how much will it cost to equip all the stations?

**58.** In 1998, Marquette, Michigan, had 136 in. of snowfall, while Detroit, Michigan, had 14 in. (*Source: World Almanac,* 2000.) What was the difference in snowfall between the two cities, in feet? Round to the nearest tenth.

**59.** At the day care center, each of the 15 toddlers drinks about $\frac{2}{3}$ cup of milk with lunch. The center is open 5 days a week.
  **(a)** How many quarts of milk will the center need for one week of lunches?
  **(b)** If the center buys milk in gallon jugs, how many jugs should be ordered for one week?

**60.** A snail moves at an average speed of 2 feet every 3 minutes. (*Source: Beakman and Jax.*) At that rate, how long would it take the snail to travel one mile? Give your answer
  **(a)** in hours

  **(b)** in days.

**61.** Bob's Candies in Albany, Georgia, makes 135,000 pounds of candy canes each day. (*Source:* Bob's Candies, Inc.)
  **(a)** How many tons of candy canes are produced during a 5-day workweek?
  **(b)** The plant operates 24 hours per day. How many tons of candy canes are produced each hour, to the nearest tenth?

**62.** An Olympic gold medal weighs a total of 7 oz and is made of silver covered with about 0.2 oz of gold (*Source:* Gold and Silver Institute.) During the 2000 Olympics in Sydney, Australia, 301 gold medals were awarded.
  **(a)** How many pounds of gold were used to make all the medals, to the nearest tenth?
  **(b)** How many pounds of silver were used to make all the medals, to the nearest tenth?

### RELATING CONCEPTS (Exercises 63–66) For Individual or Group Work

*People often complain about the price of a gallon of gasoline. In response,* AutoWeek *magazine asked readers to look at the price per gallon of other liquids. See what they discovered as you* **work Exercises 63–66 in order.**

**63.** If 16 fl oz of Ocean Spray cranberry juice costs $1.25, how could you find the cost per gallon? Here is one way.
  *Step 1* Use unit fractions to find the number of fluid ounces in one gallon.
  *Step 2* Set up and solve a proportion. (*Hint:* One side of the proportion should compare 16 fl oz to $1.25.)

**64.** Find the cost per gallon for Evian water if an 11.2 fl oz bottle sells for $1.49. Use the same method as in Exercise 63.

**65.** If you pay $3.85 for a bottle of Pepto Bismol containing 4 fl oz, what is the cost per gallon? Use a different method to find the answer than you used in Exercises 63 and 64.

**66.** Does it make sense to compare the price per gallon of gasoline to the price per gallon of cranberry juice or Pepto Bismol? Explain why or why not.

## 8.2 THE METRIC SYSTEM—LENGTH

Around 1790, a group of French scientists developed the metric system of measurement. It is an organized system based on multiples of 10, like our number system and our money. After you are familiar with metric units, you will see that they are easier to use than the hodgepodge of English measurement relationships you used in **Section 8.1**.

**OBJECTIVES**
1. Learn the basic metric units of length.
2. Use unit fractions to convert among units.
3. Move the decimal point to convert among units.

**1** Learn the basic metric units of length. The basic unit of length in the metric system is the **meter** (also spelled *metre*). Use the symbol **m** for meter; do not put a period after it. If you put five of the pages from this textbook side by side, they would measure about 1 meter. Or, look at a yardstick—a meter is just a little longer. A yard is 36 inches long; a meter is about 39 inches long.

| book page | book page | book page | book page | book page |

|← 1 meter (about 39 in.) →|
|← 1 yard (36 in.) →|

In the metric system you use meters for things like buying fabric for sewing projects, measuring the length of your living room, talking about heights of buildings, or describing track and field athletic events.

Buy 2 m of fabric (about 2 yd)     6 m (about 20 ft)     15 m (about 49 ft)

**Work Problem ❶ at the Side.**

To make longer or shorter length units in the metric system, **prefixes** are written in front of the word *meter*. For example, the prefix *kilo* means 1000, so a *kilo*meter is 1000 meters. The table below shows how to use the prefixes for length measurements. It is helpful to memorize the prefixes because they are also used with weight and capacity measurements. The blue boxes are the units you will use most often in daily life.

| Prefix | kilo- meter | hecto- meter | deka- meter | meter | deci- meter | centi- meter | milli- meter |
|---|---|---|---|---|---|---|---|
| Meaning | 1000 meters | 100 meters | 10 meters | 1 meter | $\frac{1}{10}$ of a meter | $\frac{1}{100}$ of a meter | $\frac{1}{1000}$ of a meter |
| Symbol | **k**m | **h**m | **da**m | m | **d**m | **c**m | **m**m |

Units that are used most often

Here are some comparisons to help you get acquainted with the commonly used length units: km, m, cm, mm.

*Kilo*meters are used instead of miles. A kilometer is 1000 meters. It is about 0.6 mile (a little more than half a mile) or about 5 to 6 city blocks. If you participate in a 10 km run, you'll run about 6 miles.

❶ Circle the items that measure about 1 meter.

Length of a pencil

Length of a baseball bat

Height of doorknob from the floor

Height of a house

Basketball player's arm length

Length of a paper clip

**ANSWERS**
1. baseball bat, height of doorknob, basketball player's arm length

## Chapter 8 Measurement

**2** Write the most reasonable metric unit in each blank. Choose from km, m, cm, and mm.

(a) The woman's height is 168 _____.

(b) The man's waist is 90 _____ around.

(c) Louise ran the 100 _____ dash in the track meet.

(d) A postage stamp is 22 _____ wide.

(e) Michael paddled his canoe 2 _____ down the river.

(f) The pencil lead is 1 _____ thick.

(g) A stick of gum is 7 _____ long.

(h) The highway speed limit is 90 _____ per hour.

(i) The classroom was 12 _____ long.

(j) A penny is about 18 _____ across.

A meter is divided into 100 smaller pieces called *centi*meters. Each centimeter is $\frac{1}{100}$ of a meter. Centimeters are used instead of inches. A centimeter is a little shorter than $\frac{1}{2}$ inch. The cover of this textbook is about 21 cm wide. A nickel is about 2 cm across. Measure the width and length of your little finger on this centimeter ruler. The width of your little finger is probably about 1 cm.

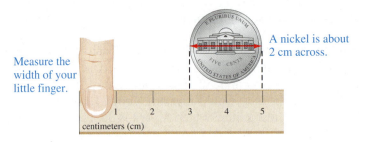

Measure the width of your little finger.

A nickel is about 2 cm across.

A meter is divided into 1000 smaller pieces called *milli*meters. Each millimeter is $\frac{1}{1000}$ of a meter. It takes 10 mm to equal 1 cm, so it is a very small length. The thickness of a dime is about 1 mm. Measure the width of your pen or pencil and the width of your little finger on this millimeter ruler.

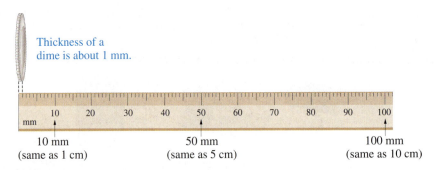

Thickness of a dime is about 1 mm.

10 mm (same as 1 cm)   50 mm (same as 5 cm)   100 mm (same as 10 cm)

### Example 1  Using Metric Length Units

Write the most reasonable metric unit in each blank. Choose from km, m, cm, and mm.

(a) The distance from home to work is 20 _____.

20 <u>km</u> because kilometers are used instead of miles.

20 km is about 12 miles.

(b) My wedding ring is 4 _____ wide.

4 <u>mm</u> because the width of a ring is very small.

(c) The newborn baby is 50 _____ long.

50 <u>cm</u>, which is half of a meter; a meter is about 39 inches so half a meter is around 20 inches.

**Work Problem** ❷ **at the Side.**

---

**ANSWERS**

2. (a) cm  (b) cm  (c) m  (d) mm  (e) km
   (f) mm  (g) cm  (h) km  (i) m  (j) mm

### Section 8.2 The Metric System—Length

**2** **Use unit fractions to convert among units.** You can convert among metric length units using unit fractions. Keep these relationships in mind when setting up the unit fractions.

**Metric Length Relationships**

| 1 km = 1000 m so the unit fractions are: | 1 m = 1000 mm so the unit fractions are: |
|---|---|
| $\dfrac{1 \text{ km}}{1000 \text{ m}}$ or $\dfrac{1000 \text{ m}}{1 \text{ km}}$ | $\dfrac{1 \text{ m}}{1000 \text{ mm}}$ or $\dfrac{1000 \text{ mm}}{1 \text{ m}}$ |
| 1 m = 100 cm so the unit fractions are: | 1 cm = 10 mm so the unit fractions are: |
| $\dfrac{1 \text{ m}}{100 \text{ cm}}$ or $\dfrac{100 \text{ cm}}{1 \text{ m}}$ | $\dfrac{1 \text{ cm}}{10 \text{ mm}}$ or $\dfrac{10 \text{ mm}}{1 \text{ cm}}$ |

**Example 2   Using Unit Fractions to Convert Length Measurements**

Convert the following.

**(a)** 5 km to m

Put the unit for the answer (meters) in the numerator of the unit fraction; put the unit you want to change (km) in the denominator.

Unit fraction equivalent to 1 $\left\{ \dfrac{1000 \text{ m}}{1 \text{ km}} \right.$ ← Unit for answer
← Unit being changed

Multiply. Divide out common units where possible.

$$5 \text{ km} \cdot \dfrac{1000 \text{ m}}{1 \text{ km}} = \dfrac{5 \text{ km}}{1} \cdot \dfrac{1000 \text{ m}}{1 \text{ km}} = \dfrac{5 \cdot 1000 \text{ m}}{1} = 5000 \text{ m}$$

These units should match.

The answer makes sense because a kilometer is much longer than a meter, so 5 km will contain many meters.

**(b)** 18.6 cm to m

Multiply by a unit fraction that allows you to divide out centimeters.

$$\dfrac{18.6 \text{ cm}}{1} \cdot \overbrace{\dfrac{1 \text{ m}}{100 \text{ cm}}}^{\text{Unit fraction}} = \dfrac{18.6}{100} \text{ m} = 0.186 \text{ m}$$

There are 100 cm in a meter, so 18.6 cm will be a small part of a meter. The answer makes sense.

Work Problem ❸ at the Side.

**❸** First write the unit fraction needed to make each conversion. Then complete the conversion.

**(a)** 3.67 m to cm

unit fraction $\left\{ \dfrac{100 \text{ cm}}{1 \text{ m}} \right.$

**(b)** 92 cm to m

unit fraction $\left\{ \dfrac{\text{m}}{\text{cm}} \right.$

**(c)** 432.7 cm to m

unit fraction $\left\{ \rule{2cm}{0.4pt} \right.$

**(d)** 65 mm to cm

unit fraction $\left\{ \rule{2cm}{0.4pt} \right.$

**(e)** 0.9 m to mm

unit fraction $\left\{ \rule{2cm}{0.4pt} \right.$

**(f)** 2.5 cm to mm

unit fraction $\left\{ \rule{2cm}{0.4pt} \right.$

**ANSWERS**

**3.** (a) 367 cm  (b) $\dfrac{1 \text{ m}}{100 \text{ cm}}$; 0.92 m

(c) $\dfrac{1 \text{ m}}{100 \text{ cm}}$; 4.327 m

(d) $\dfrac{1 \text{ cm}}{10 \text{ mm}}$; 6.5 cm

(e) $\dfrac{1000 \text{ mm}}{1 \text{ m}}$; 900 mm

(f) $\dfrac{10 \text{ mm}}{1 \text{ cm}}$; 25 mm

## 588 Chapter 8 Measurement

**4** Do each multiplication or division by hand or on a calculator. Compare your answer to the one you get by moving the decimal point.

(a) $(43.5)(10) =$ _____

43.5 gives 435.

(b) $43.5 \div 10 =$ _____

43.5 gives _____.

(c) $(28)(100) =$ _____

28.00 gives _____.

(d) $28 \div 100 =$ _____

28. gives _____.

(e) $(0.7)(1000) =$ _____

0.700 gives _____.

(f) $0.7 \div 1000 =$ _____

000.7 gives _____.

**Answers**
4. (a) 435  (b) 4.35; 4.35  (c) 2800; 2800
(d) 0.28; 0.28  (e) 700; 700
(f) 0.0007; 0.0007

**3** **Move the decimal point to convert among units.** By now you have probably noticed that conversions among metric units are made by multiplying or dividing by 10, by 100, or by 1000. A quick way to *multiply* by 10 is to move the decimal point one place to the *right*. Move it two places to the right to multiply by 100, three places to multiply by 1000. *Dividing* is done by moving the decimal point to the *left* in the same manner.

**Work Problem 4 at the Side.**

An alternate conversion method to unit fractions is moving the decimal point using this **metric conversion line.**

Here are the steps for using the conversion line.

### Using the Metric Conversion Line

*Step 1* Find the unit you are given on the metric conversion line.

*Step 2* Count the number of places to get from the unit you are given to the unit you want in the answer.

*Step 3* Move the decimal point the **same number of places** and in the **same direction** as you did on the conversion line.

**Example 3** Using the Metric Conversion Line

Use the metric conversion line to make the following conversions.

(a) 5.702 km to m

Find **km** on the metric conversion line. To get to **m**, you move *three places* to the *right*. So move the decimal point in 5.702 *three places* to the *right*.

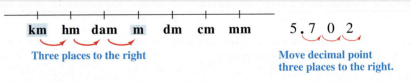

5.702 km = 5702 m

(b) 69.5 cm to m

Find **cm** on the conversion line. To get to **m**, move *two places* to the *left*.

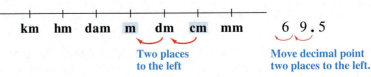

69.5 cm = 0.695 m

*Continued on Next Page*

**(c)** 8.1 cm to mm

From **cm** to **mm** is *one place* to the *right*.

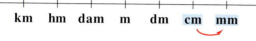

8.1 cm = 81 mm

*Work Problem 5 at the Side.*

### Example 4 — Practicing Length Conversions

Convert using the metric conversion line.

**(a)** 1.28 m to mm

Moving from **m** to **mm** is going *three places to the right*. In order to move the decimal point in 1.28 three places to the right, you must write a 0 as a placeholder.

    1.28**0**  Zero is written in as a placeholder.

 Move decimal point
 three places to the right.

    1.28 m = 1280 mm

**(b)** 60 cm to m

From **cm** to **m** is two places to the left. The decimal point in 60 starts at the *far right side* because 60 is a whole number. Then move it two places to the left.

  60.     60.

Decimal point starts here.  Move decimal point two places to the left.

60 cm = 0.60 m, which is equivalent to 0.6 m.

**(c)** 8 m to km

From **m** to **km** is three places to the left. The decimal point in 8 starts at the far right side. In order to move it three places to the left, you must write two zeros as placeholders.

  8.     **00**8.  Two zeros are written in as placeholders.

Decimal point starts here.  Move decimal point three places to the left.

8 m = 0.008 km.

*Work Problem 6 at the Side.*

---

**5** Convert using the metric conversion line.

**(a)** 12.008 km to m

**(b)** 561.4 m to km

**(c)** 20.7 cm to m

**(d)** 20.7 cm to mm

**(e)** 4.66 m to cm

**(f)** 85.6 mm to cm

**6** Convert using the metric conversion line.

**(a)** 9 m to mm

**(b)** 3 cm to m

**(c)** 14.6 km to m

**(d)** 5 mm to cm

**(e)** 70 m to km

**(f)** 0.8 m to cm

**ANSWERS**
**5. (a)** 12,008 m  **(b)** 0.5614 km
 **(c)** 0.207 m  **(d)** 207 mm
 **(e)** 466 cm  **(f)** 8.56 cm
**6. (a)** 9000 mm  **(b)** 0.03 m
 **(c)** 14,600 m  **(d)** 0.5 cm
 **(e)** 0.07 km  **(f)** 80 cm

# Focus on Real-Data Applications

## Measuring Up

1. How much do nails grow in one week? one month? one year?

2. How much does scalp hair grow in one week? one month? one year? (Use metric units.)

3. When you have finished Section 8.5, come back to this article. Is the statement about hair growing 6 inches a year accurate? Explain your answer.

### Hair and Nail Growth

*Q* How fast do hair and nails grow? Do they grow faster in the summer?

*A* Fingernails grow, on average, about one-tenth of a millimeter per day, although there is considerable variation among individuals. Fingernails grow faster than toenails, and nails on the longest fingers appear to grow the fastest.

Fingernails, as well as hair and skin, grow faster in the summer, presumably under the influence of sunlight, which expands blood vessels, bringing more oxygen and nutrients to the area and allowing for faster growth.

The rate the scalp hair grows is 0.3 to 0.4 millimeter per day, or about 6 inches a year.

*Source: Minneapolis Star Tribune.*

### New Device Measures Distances within Billionths of an Inch

U.S. officials have unveiled "the ultimate ruler," a measuring device that can gauge distances to within billionths of an inch—the length of five individual atoms—and may help revolutionize high-tech manufacturing.

Developed at a cost of $8 million by the National Institute of Standards and Technology with other agencies, the Molecular Measuring Machine can measure distances to within 40 nanometers, or billionths of a meter. After further refinement it is expected to measure within 1 nanometer.

By way of comparison, the period at the end of this sentence is about 300,000 nanometers wide.

The machine is expected to prove a boon to U.S. manufacturers of computer chips and other tiny high-tech items that must meet exacting specifications. It can also help manufacturers better calibrate their own super-accurate equipment so that, for example, even more devices could be put onto silicon chips to increase their computing power.

"This is a key project where the United States is a long way in front of any other country," said Trevor Howe, a University of Connecticut professor of metallurgy and director of its Precision Manufacturing Center.

*Source: Boston Globe.*

4. The article on the left states, "the Molecular Measuring Machine can measure distances to within 40 nanometers, or billionths of a meter." This suggests that the prefix *nano* means what?

5. Write the two unit fractions you would use to convert between meters and nanometers. Use your unit fractions to change 40 nanometers to meters, and to change 300,000 nanometers to meters.

6. Why do you suppose the headline and first paragraph talk about "billionths of an inch" when the rest of the article specifies nanometers, which are billionths of meter? Would one-billionth of an inch be the same as one-billionth of a meter? How different would they be?

7. Use the information in the article to find the length of one atom. Will you use inches or meters?

## 8.2 Exercises

*Use your knowledge of the meaning of metric prefixes to fill in the blanks.*

1. *kilo* means _____ so

   1 km = _____ m

2. *deka* means _____ so

   1 dam = _____ m

3. *milli* means _____ so

   1 mm = _____ m

4. *deci* means _____ so

   1 dm = _____ m

5. *centi* means _____ so

   1 cm = _____ m

6. *hecto* means _____ so

   1 hm = _____ m

*Use this ruler to measure your thumb and hand.*

7. The width of your hand in centimeters

8. The width of your hand in millimeters

9. The width of your thumb in millimeters

10. The width of your thumb in centimeters

*Write the most reasonable metric length unit in each blank. Choose from km, m, cm, and mm. See Example 1.*

11. The child was 91 _____ tall.

12. The cardboard was 3 _____ thick.

13. Ming-Na swam in the 200 _____ backstroke race.

14. The bookcase is 75 _____ wide.

15. Adriana drove 400 _____ on her vacation.

16. The door is 2 _____ high.

17. An aspirin tablet is 10 _____ across.

18. Lamard jogs 4 _____ every morning.

19. A paper clip is about 3 _____ long.

20. My pen is 145 _____ long.

21. Dave's truck is 5 _____ long.

22. Wheelchairs need doorways that are at least 80 _____ wide.

23. Describe at least three examples of metric length units that you have come across in your daily life.

24. Explain one reason why the metric system would be easier for a child to learn than the English system.

Chapter 8 Measurement

*Convert each measurement. Use unit fractions or the metric conversion line. See Examples 2–4.*

**25.** 7 m to cm

**26.** 18 m to cm

**27.** 40 mm to m

**28.** 6 mm to m

**29.** 9.4 km to m

**30.** 0.7 km to m

**31.** 509 cm to m

**32.** 30 cm to m

**33.** 400 mm to cm

**34.** 25 mm to cm

**35.** 0.91 m to mm

**36.** 4 m to mm

**37.** Is 82 cm greater than or less than 1 m? What is the difference in the lengths?

**38.** Is 1022 m greater than or less than 1 km? What is the difference in the lengths?

**39.** Many cameras use film that is 35 mm wide. Film for movie theaters may be 70 mm wide. Using the ruler on the previous page, draw a line that is 35 mm long and a line 70 mm long. Then convert each measurement to centimeters.

**40.** Gold wedding bands may be very narrow or quite wide. Common widths are 3 mm, 5 mm, and 10 mm. Using the ruler on the previous page, draw lines that are 3 mm, 5 mm, and 10 mm long. Then convert each measurement to centimeters.

**41.** The Roe River near Great Falls, Montana, is the shortest river in the world, with a north fork that is just under 18 m long. (*Source: Guinness Book of Amazing Nature.*) How many kilometers long is the north fork of the river?

**42.** There are 60,000 km of blood vessels in the human body. (*Source: Big Book of Knowledge.*) How many meters of blood vessels are in the body?

**43.** Use two unit fractions to convert 5.6 mm to km.

**44.** Use two unit fractions to convert 16.5 km to mm.

# 8.3 The Metric System—Capacity and Weight (Mass)

We use capacity units to measure liquids, such as the amount of milk in a recipe, the gasoline in our car tank, and the water in an aquarium. (The English capacity units used in the United States are cups, pints, quarts, and gallons.) The basic metric unit for capacity is the **liter** (also spelled *litre*). The capital letter **L** is the symbol for liter, to avoid confusion with the numeral 1.

**OBJECTIVES**
1. Learn the basic metric units of capacity.
2. Convert among metric capacity units.
3. Learn the basic metric units of weight (mass).
4. Convert among metric weight (mass) units.
5. Distinguish among basic metric units of length, capacity, and weight (mass).

**1** **Learn the basic metric units of capacity.** The liter is related to metric length in this way: A box that measures 10 cm on every side holds exactly one liter. (The volume of the box is 1000 cubic centimeters. Volume is discussed in **Section 4.8**.) A liter is just a little more than 1 quart.

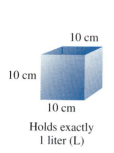
Holds exactly 1 liter (L)

About 1 liter of milk

About 1 liter of oil for your car

A liter is a little more than one quart (just $\frac{1}{4}$ cup more).

In the metric system you use liters for things like buying milk at the store, filling a pail with water, and describing the size of your home aquarium.

Buy a 4 L jug of milk.  Use a 12 L pail to wash floors.  Watch the fish in your 40 L aquarium.

**1** Which things would you measure in liters?

Amount of water in the bathtub

Length of the bathtub

Width of your car

Amount of gasoline you buy for your car

Weight of your car

Height of a pail

Amount of water in a pail

Work Problem **1** at the Side.

To make larger or smaller capacity units, we use the same **prefixes** as we did with length units. For example, *kilo* means 1000 so a *kilo*meter is 1000 meters. In the same way, a *kilo*liter is 1000 liters.

| Prefix | kilo- liter | hecto- liter | deka- liter | liter | deci- liter | centi- liter | milli- liter |
|---|---|---|---|---|---|---|---|
| Meaning | 1000 liters | 100 liters | 10 liters | 1 liter | $\frac{1}{10}$ of a liter | $\frac{1}{100}$ of a liter | $\frac{1}{1000}$ of a liter |
| Symbol | kL | hL | daL | L | dL | cL | mL |

Used most often (liter through mL)

**ANSWERS**
1. water in bathtub, gasoline, water in a pail

**594** Chapter 8 Measurement

**2** Write the most reasonable metric unit in each blank. Choose from L and mL.

(a) I bought 8 _____ of milk at the store.

(b) The nurse gave me 10 _____ of cough syrup.

(c) This is a 100 _____ garbage can.

(d) It took 10 _____ of paint to cover the bedroom walls.

(e) My car's gas tank holds 50 _____.

(f) I added 15 _____ of oil to the pancake mix.

(g) The can of orange soda holds 350 _____.

(h) My friend gave me a 30 _____ bottle of expensive perfume.

**ANSWERS**
2. (a) L  (b) mL  (c) L
   (d) L  (e) L  (f) mL
   (g) mL (h) mL

The capacity units you will use most often in daily life are liters (L) and *milli*liters (mL). A tiny box that measures 1 cm on every side holds exactly one milliliter. (In medicine, this small amount is also called 1 cubic centimeter, or 1 cc for short.) It takes 1000 mL to make 1 L. Here are some useful comparisons.

Holds exactly 1 milliliter (mL)   Teaspoon holds 5 mL   One cup holds about 250 mL

**Example 1** Using Metric Capacity Units

Write the most reasonable metric unit in each blank. Choose from L and mL.

(a) The bottle of shampoo held 500 _____.
500 mL because 500 L would be about 500 quarts, which is too much.

(b) I bought a 2 _____ carton of orange juice.
2 L because 2 mL would be less than a teaspoon.

**Work Problem 2 at the Side.**

**2** Convert among metric capacity units. Just as with length units, you can convert between milliliters and liters using unit fractions.

**Metric Capacity Relationships**

1 L = 1000 mL, so the unit fractions are:

$$\frac{1\text{ L}}{1000\text{ mL}} \quad \text{or} \quad \frac{1000\text{ mL}}{1\text{ L}}$$

Or you can use a metric conversion line to decide how to move the decimal point.

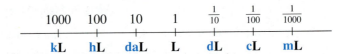

**Example 2** Converting among Metric Capacity Units

Convert using the metric conversion line or unit fractions.

(a) 2.5 L to mL
Using the metric conversion line:
From **L** to **mL** is *three places* to the *right*.

2.500   Write two zeros as placeholders.

2.5 L = 2500 mL

Using unit fractions:

Multiply by a unit fraction that allows you to divide out liters.

$$\frac{2.5 \cancel{L}}{1} \cdot \frac{1000\text{ mL}}{1 \cancel{L}} = 2500\text{ mL}$$

**Continued on Next Page**

**(b)** 80 mL to L

Using the metric conversion line:

From **mL** to **L** is *three places* to the *left*.

80.　　　　080.
↑　　　　　～～～
Decimal point　Move three
starts here.　　places left.

80 mL = 0.080 L or 0.08 L

Using unit fractions:

Multiply by a unit fraction that allows you to divide out mL.

$$\frac{80 \text{ mL}}{1} \cdot \frac{1 \text{ L}}{1000 \text{ mL}}$$

$$= \frac{80}{1000} \text{ L} = 0.08 \text{ L}$$

➤ Work Problem ❸ at the Side.

### 3 Learn the basic metric units of weight (mass).

The **gram** is the basic metric unit for *mass*. Although we often call it "weight," there is a difference. Weight is a measure of the pull of gravity; the farther you are from the center of the earth, the less you weigh. In outer space you become weightless, but your mass, the amount of matter in your body, stays the same regardless of where you are. In science courses, it will be important to distinguish between the weight of an object and its mass. But for everyday purposes, we will use the word *weight*.

The gram is related to metric length in this way: The weight of the water in a box measuring 1 cm on every side is 1 gram. This is a very tiny amount of water (1 mL) and a very small weight. One gram is also the weight of a dollar bill or a single raisin. A nickel weighs 5 grams. A regular hamburger weighs from 175 to 200 grams.

The 1 mL of water in this box weighs 1 gram.

A nickel weighs 5 grams.

(Now look back at the information about Michael Johnson's golden shoes at the start of this chapter.)

A dollar bill weighs 1 gram.

A hamburger weighs 175 to 200 grams.

➤ Work Problem ❹ at the Side.

❸ Convert.

(a) 9 L to mL

(b) 0.75 L to mL

(c) 500 mL to L

(d) 5 mL to L

(e) 2.07 L to mL

(f) 3275 mL to L

❹ Which things would weigh about 1 gram?

A small paperclip

A pair of scissors

One playing card from a deck of cards

A calculator

An average-sized apple

The check you wrote at the grocery store

**ANSWERS**
3. (a) 9000 mL　(b) 750 mL　(c) 0.5 L
　(d) 0.005 L　(e) 2070 mL　(f) 3.275 L
4. paperclip, playing card, check

**5** Write the most reasonable metric unit in each blank. Choose from kg, g, and mg.

(a) A thumbtack weighs 800 _____.

(b) A teenager weighs 50 _____.

(c) This large cast-iron frying pan weighs 1 _____.

(d) Jerry's basketball weighed 600 _____.

(e) Tamlyn takes a 500 _____ calcium tablet every morning.

(f) On his diet, Greg can eat 90 _____ of meat for lunch.

(g) One strand of hair weighs 2 _____.

(h) One banana might weigh 150 _____.

**ANSWERS**
5. (a) mg  (b) kg  (c) kg
   (d) g   (e) mg  (f) g
   (g) mg  (h) g

To make larger or smaller weight units, we use the same **prefixes** as we did with length and capacity units. For example, *kilo* means 1000 so a *kilo*meter is 1000 meters, a *kilo*liter is 1000 liters, and a *kilo*gram is 1000 grams.

| Prefix | kilo-gram | hecto-gram | deka-gram | gram | deci-gram | centi-gram | milli-gram |
|---|---|---|---|---|---|---|---|
| Meaning | 1000 grams | 100 grams | 10 grams | 1 gram | $\frac{1}{10}$ of a gram | $\frac{1}{100}$ of a gram | $\frac{1}{1000}$ of a gram |
| Symbol | kg | hg | dag | g | dg | cg | mg |

Units that are used most often

The units you will use most often in daily life are kilograms (kg), grams (g), and milligrams (mg). *Kilo*grams are used instead of pounds. A kilogram is 1000 grams. It is about 2.2 pounds. Two packages of butter plus one stick of butter weigh about 1 kg. An average newborn baby weighs 3 to 4 kg; a college football player might weigh 100 to 130 kg.

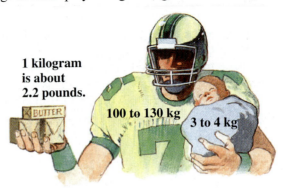

1 kilogram is about 2.2 pounds.
100 to 130 kg
3 to 4 kg

Extremely small weights are measured in *milli*grams. It takes 1000 mg to make 1 g. Recall that a dollar bill weighs about 1 g. Imagine cutting it into 1000 pieces; the weight of one tiny piece would be 1 mg. Dosages of medicine and vitamins are given in milligrams. You will also use milligrams in science classes.

Cut a dollar bill into 1000 pieces. One tiny piece weighs 1 milligram.

**Example 3** Using Metric Weight Units

Write the most reasonable metric unit in each blank. Choose from kg, g, and mg.

(a) Ramon's suitcase weighed 20 _____.
20 kg because kilograms are used instead of pounds. 20 kg is about 44 pounds.

(b) LeTia took a 350 _____ aspirin tablet.
350 mg because 350 g would be more than the weight of a hamburger, which is too much.

(c) Jenny mailed a letter that weighed 30 _____.
30 g because 30 kg would be much too heavy and 30 mg is less than the weight of a dollar bill.

**Work Problem 5 at the Side.**

Section 8.3 The Metric System—Capacity and Weight (Mass)

**4** **Convert among metric weight (mass) units.** As with length and capacity, you can convert among metric weight units by using unit fractions. The unit fractions you need are shown here.

**Metric Weight (Mass) Relationships**

| 1 kg = 1000 g so the unit fractions are: | 1 g = 1000 mg so the unit fractions are: |
|---|---|
| $\frac{1 \text{ kg}}{1000 \text{ g}}$ or $\frac{1000 \text{ g}}{1 \text{ kg}}$ | $\frac{1 \text{ g}}{1000 \text{ mg}}$ or $\frac{1000 \text{ mg}}{1 \text{ g}}$ |

Or you can use a metric conversion line to decide how to move the decimal point.

### Example 4  Converting among Metric Weight Units

Convert using the metric conversion line or unit fractions.

**(a)** 7 mg to g

Using the metric conversion line:
From **mg** to **g** is *three places* to the *left*.

7.          007.
↑            ~~~
Decimal point    Move three
starts here.     places left.

7 mg = 0.007 g

Using unit fractions:

Multiply by a unit fraction that allows you to divide out mg.

$$\frac{7 \text{ mg}}{1} \cdot \frac{1 \text{ g}}{1000 \text{ mg}} = \frac{7}{1000} \text{ g}$$

$$= 0.007 \text{ g}$$

**(b)** 13.72 kg to g

Using the metric conversion line:
From **kg** to **g** is *three* places to the *right*.

13.720     Decimal point moves
           three places to
           the right.

13.72 kg = 13,720 g
                ↑
             A comma
        (not a decimal point)

Using unit fractions:

Multiply by a unit fraction that allows you to divide out kg.

$$\frac{13.72 \text{ kg}}{1} \cdot \frac{1000 \text{ g}}{1 \text{ kg}} = 13{,}720 \text{ g}$$
                                              ↑
                                           A comma
                                    (not a decimal point)

Work Problem **6** at the Side.

**6** Convert.

(a) 10 kg to g

(b) 45 mg to g

(c) 6.3 kg to g

(d) 0.077 g to mg

(e) 5630 g to kg

(f) 90 g to kg

**ANSWERS**
**6.** (a) 10,000 g  (b) 0.045 g  (c) 6300 g
     (d) 77 mg  (e) 5.63 kg  (f) 0.09 kg

**7** First decide which type of units are needed: length, capacity, or weight. Then write the most appropriate unit in the blank. Choose from km, m, cm, mm, L, mL, kg, g, and mg.

(a) Gail bought a 4 _____ can of paint.

Use _____ units.

(b) The bag of chips weighed 450 _____.

Use _____ units.

(c) Give the child 5 _____ of cough syrup.

Use _____ units.

(d) The width of the window is 55 _____.

Use _____ units.

(e) Akbar drives 18 _____ to work.

Use _____ units.

(f) Each computer weighs 5 _____.

Use _____ units.

(g) A credit card is 55 _____ wide.

Use _____ units.

**ANSWERS**

7. (a) L; capacity  (b) g; weight
   (c) mL; capacity  (d) cm; length
   (e) km; length  (f) kg; weight
   (g) mm; length

---

**5** **Distinguish among basic metric units of length, capacity, and weight (mass).** As you encounter things to be measured at home, on the job, or in your classes at school, be careful to use the correct type of measurement unit.

Use *length units* (kilometers, meters, centimeters, millimeters) to measure:

| how long | how high | how far away |
| how wide | how tall | how far around (perimeter) |
| how deep | distance | |

Use *capacity units* (liters, milliliters) to measure liquids (things that can be poured) such as:

| water | shampoo | gasoline |
| milk | perfume | oil |
| soft drinks | cough syrup | paint |

Also use liters and milliliters to describe how much liquid something can hold, such as an eyedropper, measuring cup, pail, or bathtub.

Use *weight units* (kilograms, grams, milligrams) to measure:

the weight of something    how heavy something is

In Chapters 3 and 4 you used square units (such as $cm^2$ and $m^2$) to measure area, and cubic units (such as $cm^3$ and $m^3$) to measure volume.

**Example 5** Using a Variety of Metric Units

First decide which type of units are needed: length, capacity, or weight. Then write the most appropriate metric unit in the blank. Choose from km, m, cm, mm, L, mL, kg, g, and mg.

(a) The letter needs another stamp because it weighs 40 _____.

Use _____ units.

The letter weighs 40 **grams** because 40 mg is less than the weight of a dollar bill and 40 kg would be about 88 pounds.

Use **weight** units because of the word "weighs."

(b) The swimming pool is 3 _____ deep at the deep end.

Use _____ units.

The pool is 3 **meters** deep because 3 cm is only about an inch and 3 km is more than a mile.

Use **length** units because of the word "deep."

(c) This is a 340 _____ can of juice.

Use _____ units.

It is a 340 **milliliter** can because 340 liters would be more than 340 quarts.

Use **capacity** units because juice is a liquid.

**Work Problem 7 at the Side.**

## 8.3 Exercises

*Write the most reasonable metric unit in each blank. Choose from L, mL, kg, g, and mg.*
*See Examples 1 and 3.*

1. The glass held 250 _____ of water.

2. Hiromi used 20 _____ of water to wash the kitchen floor.

3. Dolores can make 10 _____ of soup in that pot.

4. Jay gave 2 _____ of vitamin drops to the baby.

5. Our yellow Labrador dog grew up to weigh 40 _____.

6. A small safety pin weighs 750 _____.

7. Lori caught a small sunfish weighing 150 _____.

8. One dime weighs 2 _____.

9. Andre donated 500 _____ of blood today.

10. Barbara bought the 2 _____ bottle of cola.

11. The patient received a 250 _____ tablet of medication each hour.

12. The 8 people on the elevator weighed a total of 500 _____.

13. The gas can for the lawn mower holds 4 _____.

14. Kevin poured 10 _____ of vanilla into the mixing bowl.

15. Pam's backpack weighs 5 _____ when it is full of books.

16. One grain of salt weighs 2 _____.

*Today, medical measurements are usually given in the metric system. Since we convert among metric units of measure by moving the decimal point, it is possible that mistakes can be made. Examine the following dosages and indicate whether they are reasonable or unreasonable. If a dose is unreasonable, indicate whether it is too much or too little.*

17. Drink 4.1 L of Kaopectate after each meal.

18. Drop 1 mL of solution into the eye twice a day.

19. Soak your feet in 5 kg of Epsom salts per liter of water.

20. Inject 0.5 L of insulin each morning.

21. Take 15 mL of cough syrup every four hours.

22. Take 200 mg of vitamin C each day.

23. Take 350 mg of aspirin three times a day.

24. Buy a tube of ointment weighing 0.002 g.

25. Describe at least two examples of metric capacity units and two examples of metric weight units that you have come across in your daily life.

26. Explain in your own words how the meter, liter, and gram are related.

27. Describe how you decide which unit fraction to use when converting 6.5 kg to grams.

28. Write an explanation of each step you would use to convert 20 mg to grams using the metric conversion line.

*Convert each measurement. Use unit fractions or the metric conversion line. See Examples 2 and 4.*

29. 15 L to mL

30. 6 L to mL

31. 3000 mL to L

32. 18,000 mL to L

33. 925 mL to L

34. 200 mL to L

35. 8 mL to L

36. 25 mL to L

37. 4.15 L to mL

38. 11.7 L to mL

39. 8000 g to kg

40. 25,000 g to kg

41. 5.2 kg to g

42. 12.42 kg to g

43. 0.85 g to mg

44. 0.2 g to mg

45. 30,000 mg to g

46. 7500 mg to g

47. 598 mg to g

48. 900 mg to g

49. 60 mL to L

50. 6.007 kg to g

51. 3 g to kg

52. 12 mg to g

53. 0.99 L to mL

54. 13,700 mL to L

*Write the most appropriate metric unit in each blank. Choose from km, m, cm, mm, L, mL, kg, g, and mg. See Example 5.*

55. The masking tape is 19 _____ wide.

56. The roll has 55 _____ of tape on it.

57. Buy a 60 _____ jar of acrylic paint for art class.

58. One onion weighs 200 _____.

59. My waist measurement is 65 _____.

60. Add 2 _____ of windshield washer fluid to your car.

61. A single postage stamp weighs 90 _____.

62. The hallway is 10 _____ long.

*Solve each application problem. Show your work. (Source for Exercises 63–68: Top 10 of Everything, 2000.)*

63. Human skin has about 3 million sweat glands, which release an average of 300 mL of sweat per day. How many liters of sweat are released each day?

64. In hot climates, the sweat glands in a person's skin may release up to 3.5 L of sweat in one day. How many milliliters is that?

65. The average weight of a human brain is 1.34 kg. How many grams is that?

66. In the Victorian era, people believed that heavier brains meant greater intelligence. They were impressed that Otto von Bismarck's brain weighed 1907 g, which is how many kilograms?

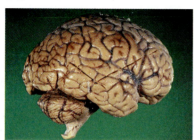

67. A healthy human heart pumps about 70 mL of blood per beat. How many liters of blood does it pump per beat?

68. On average, we breathe in and out roughly 900 mL of air every 10 seconds. How many liters of air is that?

**69.** A small adult cat weighs from 3000 g to 4000 g. How many kilograms it that? (*Source:* Lyndale Animal Hospital.)

**70.** If the letter you are mailing weighs 29 g, you must put additional postage on it. (*Source:* U.S. Postal Service.) How many kilograms does the letter weigh?

**71.** Is 1005 mg greater than or less than 1 g? What is the difference in the weights?

**72.** Is 990 mL greater than or less than 1 L? What is the difference in the amounts?

**73.** One nickel weighs 5 g. How many nickels are in 1 kg of nickels?

**74.** Seawater contains about 3.5 g of salt per 1000 mL of water. How many grams of salt would be in 1 L of seawater?

### RELATING CONCEPTS (Exercises 75–78) FOR INDIVIDUAL OR GROUP WORK

*Recall that the prefix **kilo** means 1000, so a **kilo**meter is 1000 meters. You'll learn about other prefixes for numbers greater than 1000 as you **work Exercises 75–78 in order**.*

**75.** **(a)** The prefix *mega* (abbreviated with a capital M) means one million. So a *mega*meter (Mm) is how many meters?

1 Mm = _____ m

**(b)** Figure out a unit fraction that you can use to convert megameters to meters. Then use it to convert 3.5 Mm to meters.

**76.** **(a)** The prefix *giga* (abbreviated with a capital G) means one billion. So a *giga*meter (Gm) is how many meters?

1 Gm = _____ m

**(b)** Figure out a unit fraction you can use to convert meters to gigameters. Then use it to convert 2500 m to gigameters.

**77.** **(a)** The prefix *tera* (abbreviated with a capital T) means one trillion.

So 1 Tm = _____ m.

**(b)** Think carefully before you fill in the blanks.

1 Tm = _____ Gm

1 Tm = _____ Mm

**78.** A computer's memory is measured in *bytes*. A byte can represent a single letter, a digit, or a punctuation mark. The memory for a desktop computer may be measured in megabytes (abbreviated MB) or gigabytes (abbreviated GB). Using the meanings of *mega* and *giga,* it would seem that

1 MB = _____ bytes   and

1 GB = _____ bytes.

However, because computers use a base 2 or binary system, 1 MB is actually $2^{20}$ and 1 GB is $2^{30}$. Use your calculator to find the actual values.

$2^{20}$ = _____    $2^{30}$ = _____

## 8.4 Problem Solving with Metric Measurement

**OBJECTIVE**
1. Solve application problems involving metric measurements.

**1** Solve application problems involving metric measurements. One advantage of the metric system is the ease of comparing measurements in application situations. Just be sure that you are comparing similar units: mg to mg, km to km, and so on.

We will use the same problem-solving steps you used for English measurement applications in **Section 8.1**.

### Example 1  Solving a Metric Application

Cheddar cheese is on sale at $8.99 per kilogram. Jake bought 350 g of the cheese. How much did he pay, to the nearest cent?

*Step 1*  **Read** the problem. The problem asks for the cost of 350 g of cheese.

*Step 2*  **Work out a plan.** The price is $8.99 per *kilogram,* but the amount Jake bought is given in *grams*. Convert grams to kilograms (the unit in the price). Then multiply the weight times the cost per kilogram.

*Step 3*  **Estimate** a reasonable answer. Round the cost of 1 kg from $8.99 to $9. There are 1000 g in a kilogram, so 350 g is about $\frac{1}{3}$ of a kilogram. Jake is buying about $\frac{1}{3}$ of a kilogram, so $\frac{1}{3}$ of $9 = $3 as our estimate.

*Step 4*  **Solve** the problem. Use a unit fraction to convert 350 g to kilograms.

$$\frac{350 \text{ g}}{1} \cdot \frac{1 \text{ kg}}{1000 \text{ g}} = \frac{350}{1000} \text{ kg} = 0.35 \text{ kg}$$

Now multiply 0.35 kg times the cost per kilogram.

$$\frac{\$8.99}{1 \text{ kg}} \cdot \frac{0.35 \text{ kg}}{1} = \$3.1465 \approx \$3.15 \quad \text{Rounded}$$

*Step 5*  **State the answer.** Jake paid $3.15, rounded to the nearest cent.

*Step 6*  **Check** your work. The answer, $3.15, is close to our estimate of $3.

 Satin ribbon is on sale at $0.89 per meter. How much will 75 cm cost, to the nearest cent?

Work Problem **1** at the Side.

### Example 2  Solving a Metric Application

Olivia has 2.6 m of lace. How many centimeters of lace can she use to trim each of six hair ornaments? Round to the nearest tenth of a centimeter.

*Step 1*  **Read** the problem. The problem asks for the number of centimeters of lace for each of six hair ornaments.

*Step 2*  **Work out a plan.** The given amount of lace is in *meters,* but the answer must be in *centimeters*. Convert meters to centimeters, then divide by 6 (the number of hair ornaments).

Continued on Next Page

**ANSWER**
1. $0.67 (rounded)

**2** Lucinda's doctor wants her to take 1.2 g of medication each day in three equal doses. How many milligrams should be in each dose?

*Step 3* **Estimate** a reasonable answer. To estimate, round 2.6 m of lace to 3 m. Then, 3 m = 300 cm, and 300 cm ÷ 6 = 50 cm as our estimate.

*Step 4* **Solve** the problem. On the metric conversion line, moving from **m** to **cm** is two places to the right, so move the decimal point in 2.6 m two places to the right. Then divide by 6.

$$2.60 \text{ m} = 260 \text{ cm} \qquad \frac{260 \text{ cm}}{6 \text{ ornaments}} \approx 43.3 \text{ cm per ornament}$$

*Step 5* **State the answer.** Olivia can use about 43.3 cm of lace on each ornament.

*Step 6* **Check** your work. The answer, 43.3 cm, is close to our estimate of 50 cm.

**Work Problem** ❷ **at the Side.**

### NOTE

In Example 1 we used a unit fraction to convert the measurement, and in Example 2 we moved the decimal point. Use whichever method you prefer. Also, there is more than one way to solve an application problem. Another way to solve Example 2 is to divide 2.6 m by 6 to get 0.4333 m of lace for each ornament. Then convert 0.4333 m to 43.3 cm (rounded to the nearest tenth).

**3** Andrea has two pieces of fabric. One measures 2 m 35 cm and the other measures 1 m 85 cm. How many meters of fabric does she have in all?

### Example 3  Solving a Metric Application

Rubin measured a board and found that the length was 3 m plus an additional 5 cm. He cut off a piece measuring 1 m 40 cm for a shelf. Find the length in meters of the remaining piece of board.

*Step 1* **Read** the problem. Part of a board is cut off. The problem asks what length of board, in meters, is left over.

*Step 2* **Work out a plan.** The lengths involve two units, m and cm. Rewrite both lengths in meters (the unit called for in the answer), and then subtract.

*Step 3* **Estimate** a reasonable answer. To estimate, 3 m 5 cm can be rounded to 3 m, because 5 cm is less than half of a meter (less than 50 cm). Round 1 m 40 cm down to 1 m. Then, 3 m − 1 m = 2 m as our estimate.

*Step 4* **Solve** the problem. Rewrite the lengths in meters. Then subtract.

$$\begin{array}{r} 3\text{m} \rightarrow \quad 3.00 \text{ m} \\ \text{plus 5 cm} \rightarrow +\,0.05 \text{ m} \\ \hline 3.05 \text{ m} \end{array} \qquad \begin{array}{r} 1\text{ m} \rightarrow \quad 1.0 \text{ m} \\ \text{plus 40 cm} \rightarrow +\,0.4 \text{ m} \\ \hline 1.4 \text{ m} \end{array}$$

$$\begin{array}{r} \text{Subtract to find} \quad 3.05 \text{ m} \leftarrow \text{Board} \\ \text{leftover length.} \quad -\,1.40 \text{ m} \leftarrow \text{Shelf} \\ \hline 1.65 \text{ m} \leftarrow \text{Leftover piece} \end{array}$$

*Step 5* **State the answer.** The length of the remaining piece of board is 1.65 m.

*Step 6* **Check** your work. The answer, 1.65 m, is close to our estimate of 2 m.

**Work Problem** ❸ **at the Side.**

**ANSWERS**
2. 400 mg per dose
3. 4.2 m

# 8.4 Exercises

*Solve each application problem. Show your work. Round money answers to the nearest cent. See Examples 1–3.*

1. Bulk rice is on special at $0.65 per kilogram. Pam scooped some rice into a bag and put it on the scale. How much will she pay for 2 kg 50 g of rice?

2. Lanh is buying a piece of plastic tubing measuring 3 m 15 cm for the science lab. The price is $4.75 per meter. How much will Lanh pay?

3. A miniature Yorkshire terrier, one of the smallest dogs, may weigh only 500 g. But a St. Bernard, the heaviest dog, could easily weigh 90 kg. What is the difference in the weights of the two dogs, in kilograms? (*Source: Big Book of Knowledge.*)

4. The world's longest insect is the giant stick insect of Indonesia, measuring 33 cm. The fairy fly, the smallest insect, is just 0.2 mm long. How much longer is the giant stick insect, in millimeters? (*Source: Big Book of Knowledge.*)

Giant stick insect

5. An adult human body contains about 5 L of blood. If each beat of the heart pumps 70 mL of blood, how many times must the heart beat to pass all the blood through the heart? Round to the nearest whole number of beats. (*Source: Harper's Index.*)

6. A floor tile measures 30 cm by 30 cm and weighs 185 g. How many kilograms would a stack of 24 tiles weigh? How much would five stacks of tile weigh? (*Source: The Tile Shop.*)

7. Each piece of lead for a mechanical pencil has a thickness of 0.5 mm and is 60 mm long. Find the total length in centimeters of the lead in a package with 30 pieces. (*Source:* Pentel.) If the price of the package is $3.29, find the cost per centimeter for the lead.

8. The apartment building caretaker puts 750 mL of chlorine into the swimming pool every day. How many liters should he order to have a one-month (30-day) supply on hand? If chlorine is sold in containers that hold 4 L, how many containers should be ordered for one month? How much chlorine will be left at the end of the month?

9. Rosa is building a bookcase. She has one board that is 2 m 8 cm long and another that is 2 m 95 cm long. How long are the two boards together, in meters?

10. Janet has 10 m 30 cm of fabric. She wants to make curtains for three windows that are all the same size. How much fabric is available for each window, to the nearest tenth of a meter?

11. In a chemistry lab, each of the 45 students needs 85 mL of acid. How many one-liter bottles of acid need to be ordered?

12. James needs 3 m 80 cm of wood molding to frame a picture. The price is $5.89 per meter plus a 7% sales tax. How much will James pay?

13. As a fund raiser, the PTA bought 40 kg of nuts for $113.50. They sold the nuts in 250 g bags for $2.95 each. Find the amount of profit.

14. Which case of shampoo is the better buy: a $16 case that holds twelve 1 L bottles or an $18 case that holds thirty-six 400 mL bottles?

**RELATING CONCEPTS (Exercises 15–19)** FOR INDIVIDUAL OR GROUP WORK

*It is difficult to weigh very light objects, such as a single sheet of paper or a single staple (unless you have a very expensive scientific scale). One way around this problem is to weigh a large number of the items and then divide to find the weight of one item. Of course, before dividing, you must subtract the weight of the box or wrapper that the items are packaged in to find the net weight. To complete the table,* **work Exercises 15–18 in order.** *Then answer Exercise 19.*

|     | Item | Total Weight | Weight of Packaging | Net Weight | Weight of One Item in Grams | Weight of One Item in Milligrams |
|-----|------|--------------|---------------------|------------|-----------------------------|----------------------------------|
| 15. | Box of 50 envelopes | 255 g | 40 g | _____ | _____ | _____ |
| 16. | Box of 1000 staples | 350 g | 20 g | _____ | _____ | _____ |
| 17. | Ream of paper (500 sheets) | _____ | 50 g | _____ | _____ | 3000 mg |
| 18. | Box of 100 small paper clips | _____ | 5 g | _____ | _____ | 500 mg |

19. Bonus question: One million seeds from an aspen tree weigh about 0.45 kg. (*Source:* Scenic State Park.) How much does one seed weigh in milligrams?

## 8.5 Metric–English Conversions and Temperature

**1** **Use unit fractions to convert between metric and English units.** Until the United States has switched completely from the English system to the metric system, it will be necessary to make conversions from one system to the other. *Approximate* conversions can be made with the help of the following table, in which the values have been rounded to the nearest hundredth or thousandth. (The only value that is exact, not rounded, is 1 inch = 2.54 cm.)

| Metric to English | English to Metric |
|---|---|
| 1 kilometer ≈ 0.62 mile | 1 mile ≈ 1.61 kilometers |
| 1 meter ≈ 1.09 yards | 1 yard ≈ 0.91 meter |
| 1 meter ≈ 3.28 feet | 1 foot ≈ 0.30 meter |
| 1 centimeter ≈ 0.39 inch | 1 inch = 2.54 centimeters |
| 1 liter ≈ 0.26 gallon | 1 gallon ≈ 3.78 liters |
| 1 liter ≈ 1.06 quarts | 1 quart ≈ 0.95 liter |
| 1 kilogram ≈ 2.20 pounds | 1 pound ≈ 0.45 kilogram |
| 1 gram ≈ 0.035 ounce | 1 ounce ≈ 28.35 grams |

### OBJECTIVES

**1** Use unit fractions to convert between metric and English units.

**2** Learn common temperatures on the Celsius scale.

**3** Convert temperatures by using formulas and following the order of operations.

**❶** Convert using unit fractions. Round your answers to the nearest tenth.

**(a)** 23 m to yards

### Example 1 Converting between Metric and English Length Units

Convert 10 m to yards using unit fractions. Round your answer to the nearest tenth if necessary.

We're changing from a *metric* unit to an *English* unit. In the "Metric to English" side of the table, you see that 1 meter ≈ 1.09 yards. Two unit fractions can be written using that information.

$$\frac{1 \text{ m}}{1.09 \text{ yd}} \quad \text{or} \quad \frac{1.09 \text{ yd}}{1 \text{ m}}$$

Multiply by the unit fraction that allows you to divide out meters (that is, meters is in the denominator).

$$10 \text{ m} \cdot \frac{1.09 \text{ yd}}{1 \text{ m}} = \frac{10 \text{ m}}{1} \cdot \frac{1.09 \text{ yd}}{1 \text{ m}} = \frac{(10)(1.09 \text{ yd})}{1} = 10.9 \text{ yd}$$

These units should match.

10 m ≈ 10.9 yd

**(b)** 40 cm to inches

**(c)** 5 mi to kilometers (Look at the "English to Metric" side of the table.)

### NOTE

You could also use the other numbers from the table involving meters and yards: 1 yard ≈ 0.91 meter.

$$\frac{10 \text{ m}}{1} \cdot \frac{1 \text{ yd}}{0.91 \text{ m}} = \frac{10}{0.91} \text{ yd} \approx 10.99 \text{ yd}$$

The answer is slightly different because the values in the table are rounded. Also, you have to divide instead of multiply, which is usually more difficult to do without a calculator. We will use the first method in this chapter.

**(d)** 12 in. to centimeters

Work Problem ❶ at the Side.

**ANSWERS**
**1.** **(a)** 23 m ≈ 25.1 yd   **(b)** 40 cm ≈ 15.6 in.
  **(c)** 5 mi ≈ 8.1 km   **(d)** 12 in. ≈ 30.5 cm

**2** Convert. Use the values from the table on the previous page to make unit fractions. Round answers to the nearest tenth.

(a) 17 kg to pounds

(b) 5 L to quarts

(c) 90 g to ounces

(d) 3.5 gal to liters

(e) 145 lb to kilograms

(f) 8 oz to grams

**Example 2**    **Converting between Metric and English Weight and Capacity Units**

Convert using unit fractions. Round your answers to the nearest tenth.

**(a)** 3.5 kg to pounds

Look in the "Metric to English" side of the table on the previous page to see that 1 kilogram ≈ 2.20 pounds. Use this information to write a unit fraction that allows you to divide out kilograms.

$$\frac{3.5 \text{ kg}}{1} \cdot \frac{2.20 \text{ lb}}{1 \text{ kg}} = \frac{(3.5)(2.20 \text{ lb})}{1} = 7.7 \text{ lb}$$

3.5 kg ≈ 7.7 lb

**(b)** 18 gal to liters

Look in the "English to Metric" side of the table to see that 1 gallon ≈ 3.78 liters. Write a unit fraction that will allow you to divide out gallons.

$$\frac{18 \text{ gal}}{1} \cdot \frac{3.78 \text{ L}}{1 \text{ gal}} = \frac{(18)(3.78 \text{ L})}{1} = 68.04 \text{ L}$$

68.04 rounded to the nearest tenth is 68.0.
18 gal ≈ 68.0 L

**(c)** 300 g to ounces

In the "Metric to English" side of the table, 1 gram ≈ 0.035 ounce.

$$\frac{300 \text{ g}}{1} \cdot \frac{0.035 \text{ oz}}{1 \text{ g}} = 10.5 \text{ oz}$$

300 g ≈ 10.5 oz

**CAUTION**

Because the metric and English systems were developed independently, almost all comparisons are approximate. Your answers should be written with the "≈" symbol to show they are approximate.

Work Problem **2** at the Side.

**2**    **Learn common temperatures on the Celsius scale.** In the metric system, temperature is measured on the **Celsius scale.** On the Celsius scale, water freezes at 0 °C and boils at 100 °C. The small raised circle stands for "degrees" and capital **C** is for Celsius. Read the temperatures like this:

Water freezes at 0 degrees Celsius (0 °C).

Water boils at 100 degrees Celsius (100 °C).

The English temperature system, used only in the United States, is measured on the **Fahrenheit scale.** On this scale:

Water freezes at 32 degrees Fahrenheit (32 °F).

Water boils at 212 degrees Fahrenheit (212 °F).

**ANSWERS**
2. (a) 17 kg ≈ 37.4 lb   (b) 5 L ≈ 5.3 qt
   (c) 90 g ≈ 3.2 oz   (d) 3.5 gal ≈ 13.2 L
   (e) 145 lb ≈ 65.3 kg   (f) 8 oz ≈ 226.8 g

Section 8.5 Metric–English Conversions and Temperature   609

The thermometer below shows some typical temperatures in both Celsius and Fahrenheit. For example, comfortable room temperature is about 20 °C or 68 °F, and normal body temperature is about 37 °C or 98.6 °F.

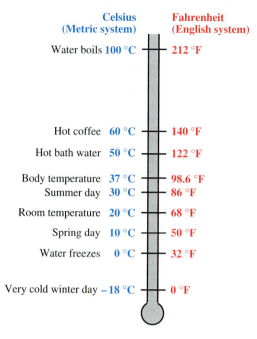

**NOTE**

The freezing and boiling temperatures are exact. The other temperatures are approximate. Even normal body temperature varies slightly from person to person.

**Example 3**  Using Celsius Temperatures

Circle the Celsius temperature that is most reasonable for each situation.

**(a)** Warm summer day   29 °C   64 °C   90 °C

29 °C is reasonable. 64 °C and 90 °C are too hot; they're both above the temperature of hot bath water (above 122 °F).

**(b)** Inside a freezer   −10 °C   3 °C   25 °C

−10 °C is the reasonable temperature because it is the only one below the freezing point of water (0 °C). Your frozen foods would start thawing at 3 °C or 25 °C.

━━━━━━━━━━━━━━━ Work Problem ❸ at the Side.

**3** **Convert temperatures by using formulas and following the order of operations.** You can use these formulas to convert between Celsius and Fahrenheit temperatures.

**Celsius–Fahrenheit Conversion Formulas**

Converting from Fahrenheit (F) to Celsius (C)

$$C = \frac{5(F - 32)}{9}$$

Converting from Celsius (C) to Fahrenheit (F)

$$F = \frac{9C}{5} + 32$$

❸ Circle the Celsius temperature that is *most* reasonable for each situation.

(a) Set the living room thermostat at:
11 °C   21 °C   71 °C

(b) The baby has a fever of:
29 °C   39 °C   49 °C

(c) Wear a sweater outside because it's:
15 °C   25 °C   50 °C

(d) My iced tea is:
−5 °C   5 °C   30 °C

(e) Time to go swimming! It's:
95 °C   65 °C   35 °C

(f) Inside a refrigerator (not the freezer) it's:
−15 °C   0 °C   3 °C

(g) There's a blizzard outside. It's:
10 °C   0 °C   −20 °C

(h) I need hot water to get these clothes clean. It should be:
55 °C   105 °C   200 °C

**ANSWERS**
3. (a) 21 °C   (b) 39 °C   (c) 15 °C
(d) 5 °C   (e) 35 °C   (f) 3 °C
(g) −20 °C   (h) 55 °C

## Chapter 8 Measurement

**④ Convert to Celsius. Round your answers to the nearest degree if necessary.**

(a) 72 °F

(b) 20 °F

(c) 212 °F

(d) 98.6 °F

**⑤ Convert to Fahrenheit. Round your answers to the nearest degree if necessary.**

(a) 100 °C

(b) −25 °C

(c) 32 °C

(d) 5 °C

---

As you use these formulas, be sure to follow the order of operations from **Section 1.8.**

### Order of Operations

*Step 1* Work inside **parentheses** or **other grouping symbols.**

*Step 2* Simplify any expressions with **exponents.**

*Step 3* Do the remaining **multiplications and divisions** as they occur from left to right.

*Step 4* Do the remaining **additions and subtractions** as they occur from left to right.

**Example 4**  Converting Fahrenheit to Celsius

Convert 10 °F to Celsius. Round your answer to the nearest degree.

Use the formula and the order of operations.

$$C = \frac{5(\mathbf{F} - 32)}{9} \quad \text{Replace F with 10.}$$

$$= \frac{5(\mathbf{10} - 32)}{9} \quad \text{Work inside parentheses first.}\\ 10 - 32 \text{ becomes } 10 + (-32).$$

$$= \frac{5(-22)}{9} \quad \text{Multiply; positive times negative gives a negative product.}$$

$$= \frac{-110}{9} \quad \text{Divide; negative divided by positive gives a negative quotient.}$$

$$= -12.\overline{2} \quad \text{Round to } -12.$$

Thus, 10 °C ≈ −12 °F.

**Work Problem ④ at the Side.**

**Example 5**  Converting Celsius to Fahrenheit

Convert 15 °C to Fahrenheit.

Use the formula and follow the order of operations.

$$F = \frac{9\mathbf{C}}{5} + 32 \quad \text{Replace C with 15.}$$

$$= \frac{9 \cdot \mathbf{15}}{5} + 32$$

$$= \frac{9 \cdot 3 \cdot \cancel{5}}{\cancel{5}} + 32 \quad \text{Divide out the common factor.}\\ \text{Multiply in the numerator.}$$

$$= 27 + 32 \quad \text{Add.}$$

$$= 59$$

Thus, 15 °C = 59 °F.

**Work Problem ⑤ at the Side.**

---

**Answers**

4. (a) 22 °C (rounded)  (b) −7 °C (rounded)
   (c) 100 °C  (d) 37 °C
5. (a) 212 °F  (b) −13 °F
   (c) 90 °F (rounded)  (d) 41 °F

# 8.5 Exercises

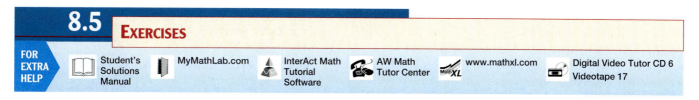

*Use the table on the first page of this section and unit fractions to make approximate conversions from metric to English or English to metric. Round your answers to the nearest tenth. See Examples 1 and 2.*

1. 20 m to yards
2. 8 km to miles
3. 80 m to feet

4. 85 cm to inches
5. 16 ft to meters
6. 3.2 yd to meters

7. 150 g to ounces
8. 2.5 oz to grams
9. 248 lb to kilograms

10. 7.68 kg to pounds
11. 28.6 L to quarts
12. 15.75 L to gallons

13. On the first page of this chapter, we said that the 3M Company used 5 g of pure gold to coat Michael Johnson's Olympic track shoes.
    (a) How many ounces of gold were used, to the nearest tenth?
    (b) Was this enough extra weight to slow him down?

14. The fastest nerve signals in the human body travel 120 m per second. (*Source: Big Book of Knowledge.*) How many feet per second do the signals travel?

15. The heavy duty wash cycle in a dishwasher uses 8.4 gal of water. How many liters does it use, to the nearest tenth? (*Source:* Frigidaire.)

16. The rinse-and-hold cycle in a dishwasher uses only 4.5 L of water. How many gallons does it use, to the nearest tenth? (*Source:* Frigidaire.)

**17.** The smallest pet fish are dwarf gobies, which are half an inch long. (*Source:* Big Book of Knowledge.) How many centimeters long is a dwarf gobie, to the nearest tenth? (*Hint:* Write half an inch in decimal form.)

**18.** A Toshiba Protégé laptop computer weighs 4.4 lb. How many kilograms does it weigh, to the nearest tenth? (*Source:* Toshiba.)

**19.** On Northwest Airlines flights, a piece of carry-on luggage cannot exceed 40 lb in weight or measure more than 22 in. by 14 in. by 8 in. What measurements would be given to Northwest passengers from countries outside the United States? Round to the nearest tenth of a kilogram and nearest centimeter if necessary.

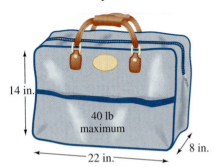

**20.** Northwest Airlines has a limit of 70 lb for each checked piece of luggage. The maximum size is 62 in., which is length + width + height. Convert these measurements to metric for Northwest flights to countries outside the United States. Round to the nearest tenth of a kilogram and nearest centimeter if necessary.

**21.** Part of a tea bag packet imported from Scotland is shown below. (*Source:* Heath and Heather.) Are the two weights equivalent? Explain your answer.

**Ingredients**
Cinnamon, Liquorice Root, Camomile, Hibiscus, Rosehips, Apples, Flavourings, Chicory Root, Citric Acid, Banana Flakes.
Net Wt. 0.09 oz  2.5 g
Produce of more than one country.
Heath & Heather, 9 Kinnaird Park, Edinburgh, Scotland, EH15 3RF.

**22.** Part of the wrapper from a ball of twine is shown below. (*Source:* Lehigh Group.) The manufacturer has designed the label so the twine can be sold worldwide. Are the two lengths equivalent? Explain your answer.

*The BabyBjörn is a popular baby carrier imported from Sweden. Use the information from the instruction sheet that comes with the carrier to answer Exercises 23–24. (Source: BabyBjörn, Sweden.)*

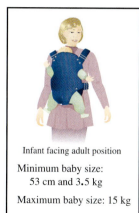

Infant facing adult position
Minimum baby size: 53 cm and 3.5 kg
Maximum baby size: 15 kg

**23.** Can the carrier be safely used for a newborn infant who weighs 8 lb and is 19.5 in. long? Explain your answer.

**24.** Can the carrier be safely used for a baby who weighs 30 lb?

*Circle the more reasonable temperature for each situation. See Example 3.*

**25.** A snowy day
    28 °C    28 °F

**26.** Brewing coffee
    80 °C    80 °F

**27.** A high fever
    40 °C    40 °F

**28.** Swimming pool water
    78 °C    78° F

**29.** Oven temperature
    150 °C    150 °F

**30.** Light jacket weather
    10 °C    10 °F

*Use the conversion formulas from this section and the order of operations to convert Fahrenheit temperatures to Celsius and Celsius temperatures to Fahrenheit. Round your answers to the nearest degree if necessary. See Examples 4 and 5.*

**31.** 60 °F

**32.** 80 °F

**33.** −4 °F

**34.** 15 °F

**35.** 8 °C

**36.** 18 °C

**37.** −5 °C

**38.** 0 °C

*Solve each application problem. Round your answers to the nearest degree if necessary.*

**39.** The highest temperature ever recorded on Earth was 136 °F at Aziza, Libya, in 1922. The lowest was −129 °F in Antarctica. Convert these temperatures to Celsius. (*Source: World Almanac, 2001.*)

**40.** Hummingbirds have a normal body temperature of 107 °F. But on cold nights they go into a state of torpor where their body temperature drops to 39 °F. What are these temperatures in Celsius? (*Source: Wildbird.*)

**41.** The directions for a self-stick clothes hook with adhesive on the back are as follows: "Apply to surfaces above 50 °F. Adhesive could soften and lose adhesion above 105 °F." What are these temperatures in the metric system? (*Source: 3M Company.*)

**42.** A box of imported Belgian chocolates carries a warning to keep the box dry and at <18 °C. Translate this warning into the English system. (*Source: Chocolaterie Guylian N.V.*)

**43.** Would a drop in temperature of 20 Celsius degrees be more or less than a drop of 20 Fahrenheit degrees? Explain your answer.

**44.** Describe one advantage of switching from the Fahrenheit temperature scale to the Celsius scale. Describe one disadvantage.

**45. (a)** Here is the tag on a pair of hiking boots. In what kind of weather would you wear these boots?

Comfort range
24 °C to 4 °C

*Source:* Sorrel.

**(b)** For what Fahrenheit temperatures are the boots designed?

**(c)** What range of metric temperatures would you have in January where you live?

**46. (a)** What are the picture directions on this tea bag package telling you to do?

*Source:* Pickwick Teas.

**(b)** What Fahrenheit temperature would give the same result?

---

### Relating Concepts (Exercises 47–50) — For Individual or Group Work

*After years of discussion, the National Collegiate Athletic Association (NCAA) decided to "go metric" in 2000 by changing the lengths of swimming events from yards to meters. This will help American swimmers prepare for international competitions, such as the Olympics, where all events are in meters. (Source: NCAA Swimming and Diving Committee.)* **Work Exercises 47–50 in order,** *rounding answers to the nearest tenth.*

**47.** All 25 yd swimming races will now be 25 m. Is the new distance longer or shorter than the old distance? By how many yards? (Use the table on the first page of this section. Round your answer to the nearest tenth.)

**48.** All 50 yd races will now be 50 m and all 100 yd races will be 100 m. Use your rounded answer from Exercise 47 (not the table) to find how much longer or shorter each new race will be, in yards.

**49.** Now use the table to find how much longer or shorter the 50 m and 100 m races will be. Why are your answers different from the ones you got in Exercise 48?

**50.** The 500 yd race will now be 400 m. Is the new distance longer or shorter than the old distance? By how many yards?

# Chapter 8

## SUMMARY

### KEY TERMS

**8.1** **English system** — The English system of measurement (United States system of units) is used for many daily activities only in the United States. Common units in this system include quarts, pounds, feet, miles, and degrees Fahrenheit.

**metric system** — The metric system of measurement is an international system used in manufacturing, science, medicine, sports, and other fields. Common units in this system include meters, liters, grams, and degrees Celsius.

**unit fraction** — A unit fraction involves measurement units and is equivalent to 1. Unit fractions are used to convert among different measurements.

**8.2** **meter** — The meter is the basic unit of length in the metric system. The symbol **m** is used for meter. One meter is a little longer than a yard.

**prefixes** — Attaching a prefix to meter, liter, or gram produces larger or smaller units. For example, the prefix *kilo* means 1000, so a *kilo*meter is 1000 meters.

**metric conversion line** — The metric conversion line is a line showing the various metric measurement prefixes and their size relationship to each other.

**8.3** **liter** — The liter is the basic unit of capacity in the metric system. The symbol **L** is used for liter. One liter is a little more than one quart.

**gram** — The gram is the basic unit of weight (mass) in the metric system. The symbol **g** is used for gram. One gram is the weight of 1 milliliter of water or one dollar bill.

**8.5** **Celsius** — The Celsius scale is used to measure temperature in the metric system. Water boils at 100 °C and freezes at 0 °C.

**Fahrenheit** — The Fahrenheit scale is used to measure temperature in the English system. Water boils at 212 °F and freezes at 32 °F.

### NEW FORMULAS

Converting from Celsius to Fahrenheit: $F = \dfrac{9C}{5} + 32$

Converting from Fahrenheit to Celsius: $C = \dfrac{5(F - 32)}{9}$

### TEST YOUR WORD POWER

*See how well you have learned the vocabulary in this chapter. Answers follow the Quick Review.*

1. The **metric system**
   (a) uses meters, liters, and degrees Fahrenheit
   (b) is based on multiples of 10
   (c) is used only in the United States
   (d) has evolved over centuries.

2. The **English system** of measurement
   (a) is used throughout the world
   (b) is based on multiples of 12
   (c) uses feet, inches, quarts, and pounds
   (d) was developed by a group of scientists in 1790.

3. A **unit fraction**
   (a) has the unit you want to change in the numerator
   (b) has a denominator of 1
   (c) must be written in lowest terms
   (d) is equivalent to 1.

4. A **gram** is
   (a) the weight of 1 mL of water
   (b) abbreviated gm
   (c) equivalent to 1000 kg
   (d) approximately equal to 2.2 pounds.

5. A **meter** is
   (a) equivalent to 1000 cm
   (b) approximately equal to $\frac{1}{2}$ inch
   (c) abbreviated m with no period after it
   (d) the basic unit of capacity in the metric system.

6. The **Celsius** temperature scale
   (a) shows water freezing at 32°
   (b) is used in the English system of measurement
   (c) shows water boiling at 100°
   (d) cannot be converted to the Fahrenheit temperature scale.

## QUICK REVIEW

*Concepts*

### 8.1 The English System of Measurement
Memorize the basic measurement relationships. Then, to convert units, multiply when changing from a larger unit to a smaller unit; divide when changing from a smaller unit to a larger unit.

*Examples*

Convert each measurement.

(a) 5 ft to inches

$$5 \text{ ft} = 5 \cdot 12 = 60 \text{ in.}$$

(b) 3 lb to ounces

$$3 \text{ lb} = 3 \cdot 16 = 48 \text{ oz}$$

(c) 15 qt to gallons

$$15 \text{ qt} = \frac{15}{4} = 3\frac{3}{4} \text{ gal}$$

### 8.1 Using Unit Fractions
Another, more useful, conversion method is multiplying by a unit fraction. The unit you want in the answer should be in the numerator. The unit you want to change should be in the denominator.

Convert 32 oz to pounds.

$$\frac{32 \text{ oz}}{1} \cdot \frac{1 \text{ lb}}{16 \text{ oz}} \quad \} \text{ Unit fraction}$$

$$= \frac{\overset{2}{\cancel{32} \cancel{\text{oz}}}}{1} \cdot \frac{1 \text{ lb}}{\underset{1}{\cancel{16} \cancel{\text{oz}}}} \quad \begin{array}{l} \text{Divide out ounces.} \\ \text{Divide out common factors.} \end{array}$$

$$= 2 \text{ lb}$$

### 8.1 Solving English Measurement Application Problems
To solve measurement application problems, use these problem-solving steps.

Use the six steps to solve this problem.

Mr. Green has 10 yd of rope. He is cutting it into eight pieces so his sailing class can practice knot tying. How many feet of rope will each of his eight students get?

*Step 1* **Read** the problem.

*Step 1* The problem asks how many feet of rope can be given to each of eight students.

*Step 2* **Work out a plan.**

*Step 2* Convert 10 yd to feet (the unit required in the answer). Then divide by eight students.

*Step 3* **Estimate** a reasonable answer.

*Step 3* There are 3 ft in one yard, so there are 30 ft in 10 yd. Then 30 ft ÷ 8 ≈ 4 ft as our estimate.

*Step 4* **Solve** the problem.

*Step 4* Use a unit fraction to convert 10 yd to feet, then divide.

$$\frac{10 \text{ yd}}{1} \cdot \frac{3 \text{ ft}}{1 \text{ yd}} = 30 \text{ ft}$$

$$\frac{30 \text{ ft}}{8 \text{ students}} = 3\frac{3}{4} \text{ ft or } 3.75 \text{ ft per student}$$

*Step 5* **State the answer.**

*Step 5* Each student gets $3\frac{3}{4}$ ft or 3.75 ft.

*Step 6* **Check** your work.

*Step 6* The answer, 3.75 ft, is close to our estimate of 4 ft.

# Chapter 8 Summary

| Concepts | Examples |
|---|---|
| **8.2 Basic Metric Length Units** <br> Use approximate comparisons to judge which units are appropriate: <br><br> 1 mm is the thickness of a dime. <br> 1 cm is about $\frac{1}{2}$ inch. <br> 1 m is a little more than 1 yard. <br> 1 km is about 0.6 mile. | Write the most reasonable metric unit in each blank. Choose from km, m, cm, and mm. <br><br> The room is 6 __m__ long. <br> A paper clip is 30 __mm__ long. <br> He drove 20 __km__ to work. |

**8.2 and 8.3 Converting within the Metric System**

**Using Unit Fractions**
One conversion method is to multiply by a unit fraction. Use a fraction with the unit you want in the answer in the numerator and the unit you want to change in the denominator.

Convert 9 g to kg.

$$\frac{9 \cancel{g}}{1} \cdot \frac{1 \text{ kg}}{1000 \cancel{g}} = \frac{9}{1000} \text{ kg} = 0.009 \text{ kg}$$

Convert 3.6 m to cm.

$$\frac{3.6 \cancel{m}}{1} \cdot \frac{100 \text{ cm}}{1 \cancel{m}} = 360 \text{ cm}$$

**Using the Metric Conversion Line**
Another conversion method is to find the unit you are given on the metric conversion line. Count the number of places to get from the unit you are given to the unit you want. Move the decimal point the same number of places and in the same direction.

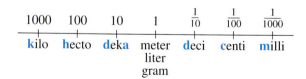

Convert.

(a) 68.2 kg to g
From **kg** to **g** is three places to the right.

6 8.2 0 0   Decimal point is moved three places to the right.

68.2 kg = 68,200 g

(b) 300 mL to L
From **mL** to **L** is three places to the left.

3 0 0.   Decimal point is moved three places to the left.

300 mL = 0.3 L

(c) 825 cm to m
From **cm** to **m** is two places to the left.

8 2 5.   Decimal point is moved two places to the left.

825 cm = 8.25 m

| Concepts | Examples |

### 8.3 Basic Metric Capacity Units

Use approximate comparisons to judge which units are appropriate:

    1 L is a little more than 1 quart.

    1 mL is the amount of water in a cube 1 cm on each side.

    5 mL is about one teaspoon.

    250 mL is about one cup.

Write the most appropriate metric unit in each blank. Choose from L and mL.

The pail holds 12 __L__ .

The milk carton from the vending machine holds 250 __mL__ .

### 8.3 Basic Metric Weight (Mass) Units

Use approximate comparisons to judge which units are appropriate:

    1 kg is about 2.2 pounds.

    1 g is the weight of 1 mL of water or one dollar bill.

    1 mg is $\frac{1}{1000}$ of a gram; very tiny!

Write the most appropriate metric unit in each blank. Choose from kg, g, and mg.

The wrestler weighed 95 __kg__ .

She took a 500 __mg__ aspirin tablet.

One banana weighs 150 __g__ .

### 8.4 Solving Metric Application Problems

Convert units so you are comparing kg to kg, cm to cm, and so on. When a measurement involves two units, such as 6 m 20 cm, write it in terms of the unit called for in the answer (6.2 m or 620 cm).

Use these problem-solving steps.

*Step 1*   **Read** the problem.

*Step 2*   **Work out a plan.**

*Step 3*   **Estimate** a reasonable answer.

*Step 4*   **Solve** the problem.

*Step 5*   **State the answer.**

*Step 6*   **Check** your work.

Use the six steps to solve this problem.

George cut 1 m 35 cm off of a 3 m board. How long was the leftover piece, in meters?

*Step 1*   The problem asks for the length of the leftover piece in meters.

*Step 2*   Convert the cut-off length to meters (the unit required in the answer), and then subtract to find the "leftover."

*Step 3*   To estimate, round 1 m 35 cm to 1 m, because 35 cm is less than half a meter. Then, 3 m − 1 m = 2 m as our estimate.

*Step 4*   Convert the cut-off measurement to meters, then subtract.

    1 m →   1.00 m       3.00 m ← Board
plus 35 cm → +0.35 m     − 1.35 m ← Cut off
              1.35 m       1.65 m ← Left

*Step 5*   The leftover piece is 1.65 m long.

*Step 6*   The answer, 1.65 m, is close to our estimate of 2 m.

| Concepts | Examples |
|---|---|
| **8.5 Converting between Metric and English Units**<br>Use the values in the table of conversion factors (on the first page of **Section 8.5**) to write a unit fraction. Because the values in the table are rounded, your answers will be approximate. | Convert. Round answers to the nearest tenth.<br><br>**(a)** 23 m to yards<br>From the table, 1 meter ≈ 1.09 yards.<br>$$\frac{23 \text{ m}}{1} \cdot \frac{1.09 \text{ yd}}{1 \text{ m}} = 25.07 \text{ yd}$$<br>25.07 rounds to 25.1, so 23 m ≈ 25.1 yd.<br><br>**(b)** 4 oz to grams<br>From the table, 1 ounce ≈ 28.35 grams.<br>$$\frac{4 \text{ oz}}{1} \cdot \frac{28.35 \text{ g}}{1 \text{ oz}} = 113.4 \text{ g}$$<br>So 4 oz ≈ 113.4 g. |
| **8.5 Common Celsius Temperatures**<br>Use approximate and exact comparisons to judge which temperatures are appropriate.<br><br>Exact comparisons:<br>0 °C is the freezing point of water (32 °F).<br>100 °C is the boiling point of water (212 °F).<br><br>Approximate comparisons:<br>10 °C for a spring day (50 °F)<br>20 °C for room temperature (68 °F)<br>30 °C for summer day (86 °F)<br>37 °C for body temperature (98.6 °F) | Circle the Celsius temperature that is most reasonable.<br><br>**(a)** Hot summer day:<br>**(35 °C)**    90 °C    110 °C<br><br>**(b)** The first snowy day in winter:<br>−20 °C    **(0 °C)**    15 °C |
| **8.5 Converting between Fahrenheit and Celsius Temperatures**<br>Use this formula to convert from Fahrenheit (F) to Celsius (C).<br>$$C = \frac{5(F - 32)}{9}$$ | Convert 100 °F to Celsius. Round your answer to the nearest degree if necessary.<br><br>$C = \frac{5(\mathbf{100} - 32)}{9}$    Replace F with 100.<br><br>$= \frac{5(68)}{9}$<br><br>$= \frac{340}{9}$<br><br>$= 37.\overline{7}$    Round to 38.<br><br>100 °F ≈ 38 °C |

*(continued)*

| Concepts | Examples |
|---|---|
| **8.5 Converting between Fahrenheit and Celsius Temperatures** (*continued*)<br>Use this formula to convert from Celsius (C) to Fahrenheit (F).<br>$$F = \frac{9C}{5} + 32$$ | Convert $-8$ °C to Fahrenheit. Round your answer to the nearest degree if necessary.<br><br>$F = \dfrac{9(-8)}{5} + 32$  Replace C with $-8$.<br>$\phantom{F} = \dfrac{-72}{5} + 32$<br>$\phantom{F} = -14.4 + 32$<br>$\phantom{F} = 17.6$  Round to 18.<br><br>$-8$ °C $\approx 18$ °F |

### ANSWERS TO TEST YOUR WORD POWER

1. **(b)** *Examples:* 10 meters = 1 dekameter; 100 meters = 1 hectometer; 1000 meters = 1 kilometer.
2. **(c)** *Examples:* feet and inches are used to measure length, quarts to measure capacity, pounds to measure weight.
3. **(d)** *Example:* Because 12 in. = 1 ft, the unit fraction $\dfrac{12 \text{ in.}}{1 \text{ ft}}$ is equivalent to $\dfrac{12 \text{ in.}}{12 \text{ in.}} = 1$.
4. **(a)** *Example:* A small box measuring 1 cm on every edge holds exactly 1 mL of water, and the water weighs 1 g.
5. **(c)** *Example:* A measurement of 16 meters is written 16 m (without a period).
6. **(c)** *Example:* In the metric system, water freezes at 0 °C and boils at 100 °C. The English system uses the Fahrenheit temperature scale where water freezes at 32 °F and boils at 212 °F.

# Chapter 8 REVIEW EXERCISES

**[8.1]** *Fill in the blanks with the measurement relationships you have memorized.*

1. 1 lb = _____ oz
2. _____ ft = 1 yd
3. 1 T = _____ lb
4. _____ qt = 1 gal
5. 1 hr = _____ min
6. 1 c = _____ fl oz
7. _____ sec = 1 min
8. _____ ft = 1 mi
9. _____ in. = 1 ft

*Convert using unit fractions.*

10. 4 ft = _____ in.
11. 6000 lb = _____ T
12. 64 oz = _____ lb
13. 18 hr = _____ day
14. 150 min = _____ hr
15. $1\frac{3}{4}$ lb = _____ oz
16. $6\frac{1}{2}$ ft = _____ in.
17. 7 gal = _____ c
18. 4 days = _____ sec

19. The average depth of the world's oceans is 12,460 ft. (*Source: Handy Ocean Answer Book.*)
    (a) What is the average depth in yards?
    (b) What is the average depth in miles, to the nearest tenth?

20. During the first year of a program to recycle office paper, a company recycled 123,260 pounds of paper. The company received $40 per ton for the paper. (*Source: I. C. System.*) How much money did the company make?

**[8.2]** *Write the most reasonable metric length unit in each blank. Choose from km, m, cm, and mm.*

21. My thumb is 20 _____ wide.
22. Her waist measurement is 66 _____.
23. The two towns are 40 _____ apart.
24. A basketball court is 30 _____ long.
25. The height of the picnic bench is 45 _____.
26. The eraser on the end of my pencil is 5 _____ long.

*Convert using unit fractions or the metric conversion line.*

27. 5 m to cm
28. 8.5 km to m
29. 85 mm to cm

**30.** 370 cm to m      **31.** 70 m to km      **32.** 0.93 m to mm

**[8.3]** *Write the most reasonable metric unit in each blank. Choose from L, mL, kg, g, and mg.*

**33.** The eyedropper holds 1 _____.

**34.** I can heat 3 _____ of water in this pan.

**35.** Loretta's hammer weighed 650 _____.

**36.** Yongshu's suitcase weighed 20 _____ when it was packed.

**37.** My fish tank holds 80 _____ of water.

**38.** I'll buy the 500 _____ bottle of mouthwash.

**39.** Mara took a 200 _____ antibiotic pill.

**40.** This piece of chicken weighs 100 _____.

*Convert using unit fractions or the metric conversion line.*

**41.** 5000 mL to L      **42.** 8 L to mL      **43.** 4.58 g to mg

**44.** 0.7 kg to g      **45.** 6 mg to g      **46.** 35 mL to L

**[8.4]** *Solve each application problem. Show your work.*

**47.** Each serving of punch at the wedding reception will be 180 mL. How many liters of punch are needed for 175 servings?

**48.** Jason is serving a 10 kg turkey to 28 people. How many grams of meat is he allowing for each person? Round to the nearest whole gram.

**49.** Yerald weighed 92 kg. Then he lost 4 kg 750 g. What is his weight now, in kilograms?

**50.** Young-Mi bought 2 kg 20 g of onions. The price was $1.49 per kilogram. How much did she pay, to the nearest cent?

**[8.5]** *Use the table on the first page of **Section 8.5** and unit fractions to make approximate conversions. Round your answers to the nearest tenth if necessary.*

**51.** 6 m to yards      **52.** 30 cm to inches

**53.** 108 km to miles

**54.** 800 mi to kilometers

**55.** 23 qt to liters

**56.** 41.5 L to quarts

*Write the appropriate **metric** temperature in each blank.*

**57.** Water freezes at _____.

**58.** Water boils at _____.

**59.** Normal body temperature is about _____.

**60.** Comfortable room temperature is about _____.

*Use the conversion formulas in **Section 8.5** to convert each temperature to Fahrenheit or to Celsius. Round to the nearest degree if necessary.*

**61.** 77 °F

**62.** 92 °F

**63.** −2 °C

**64.** Water coming into a dishwasher should be at least 49 °C to clean the dishes properly. What Fahrenheit temperature is that? (*Source:* Frigidaire.)

## MIXED REVIEW EXERCISES

*Write the most reasonable metric unit in each blank. Choose from km, m, cm, mm, L, mL, kg, g, and mg.*

**65.** I added 1 _____ of oil to my car.

**66.** The box of books weighed 15 _____.

**67.** Larry's shoe is 30 _____ long.

**68.** Jan used 15 _____ of shampoo on her hair.

**69.** My fingernail is 10 _____ wide.

**70.** I walked 2 _____ to school.

**71.** The tiny bird weighed 15 _____.

**72.** The new library building is 18 _____ wide.

**73.** The cookie recipe uses 250 _____ of milk.

**74.** Renee's pet mouse weighs 30 _____.

**75.** One postage stamp weighs 90 _____.

**76.** I bought 30 _____ of gas for my car.

*Convert the following using unit fractions, the metric conversion line, or the temperature conversion formulas.*

**77.** 10.5 cm to millimeters

**78.** 45 min to hours

**79.** 90 in. to feet

**80.** 1.3 m to centimeters

**81.** 25 °C to Fahrenheit

**82.** $3\frac{1}{2}$ gal to quarts

**83.** 700 mg to grams

**84.** 0.81 L to milliliters

**85.** 5 lb to ounces

**86.** 60 kg to grams

**87.** 1.8 L to milliliters

**88.** 30 °F to Celsius

**89.** 0.36 m to centimeters

**90.** 55 mL to liters

*Solve each application problem. Show your work.*

**91.** Peggy had a board measuring 2 m 4 cm. She cut off 78 cm. How long is the board now, in meters?

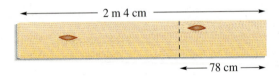

**92.** During the 12-day Minnesota State Fair, one of the biggest in the United States, Sweet Martha's booth sells an average of 3000 pounds of cookies per day. How many tons of cookies are sold in all? (*Source: Minneapolis Star Tribune.*)

**93.** Olivia is sending a recipe to her mother in Mexico. Among other things, the recipe calls for 4 oz of rice and a baking temperature of 350 °F. Convert these measurements to metric, rounding to the nearest gram and nearest degree.

**94.** While on vacation in Canada, Jalo became ill and went to a health clinic. They said he weighed 80.9 kg and was 1.83 m tall. Find his weight in pounds and height in feet. Round to the nearest tenth.

*Fill in the blank spaces in this table. Then use the information in the table to answer Exercises 99–101.*

| | Cell Phone | Weight ounces (nearest tenth) | Weight grams (nearest whole) | Battery Life Talking | Battery Life Standby |
|---|---|---|---|---|---|
| **95.** | Audiovox CDM 9000 | 4.8 oz | | 180 min | 170 hr |
| **96.** | Motorola V2282 | | 150 g | 3.5 hr | 135 hr |
| **97.** | NeoPoint 1000 | | 180 g | 2.5 hr | $1\frac{2}{3}$ days |
| **98.** | Samsung SCH-3500 | 5.5 oz | | 168 min | $6\frac{1}{4}$ days |

*Source: Access.*

**99.** Suppose you want a lightweight cell phone. List the phones in order from the lightest to the heaviest.

**100. (a)** List the phones in order from the one with the longest battery life for talking to the one with the shortest.

**(b)** What is the difference in length of talking battery life between the longest and shortest?

**101.** Which cell phone would you choose, based on the table? Explain your answer.

# Chapter 8 TEST

*Convert each measurement.*

1. 9 gal = _____ qt
2. 45 ft = _____ yd
3. 135 min = _____ hr
4. 9 in. = _____ ft
5. $3\frac{1}{2}$ lb = _____ oz
6. 5 days = _____ min

*Write the most reasonable metric unit in each blank. Choose from km, m, cm, mm, L, mL, kg, g, and mg.*

7. My husband weighs 75 _____.
8. I hiked 5 _____ this morning.
9. She bought 125 _____ of cough syrup.
10. This apple weighs 180 _____.
11. This page is about 21 _____ wide.
12. My watch band is 10 _____ wide.
13. I bought 10 _____ of soda for the picnic.
14. The bracelet is 16 _____ long.

*Convert the following measurements. Show your work.*

15. 250 cm to meters
16. 4.6 km to meters
17. 5 mm to centimeters
18. 325 mg to grams
19. 16 L to milliliters
20. 0.4 kg to grams
21. 10.55 m to centimeters
22. 95 mL to liters

1. _____
2. _____
3. _____
4. _____
5. _____
6. _____
7. _____
8. _____
9. _____
10. _____
11. _____
12. _____
13. _____
14. _____
15. _____
16. _____
17. _____
18. _____
19. _____
20. _____
21. _____
22. _____

**23.** The rainiest place in the world is Mount Waialeale in Hawaii, which receives 460 in. of rain each year. (*Source:* National Geographic Society.) What is the average rainfall per month, in feet, to the nearest tenth?

**24.** A 6-inch Subway "Vegie Delite" sandwich has 590 mg of sodium. A 6-inch "Super Subway Melt" sandwich has 2.9 g of sodium. (*Source:* Subway.)

(a) How much more sodium is in the "Super Melt" sandwich, in milligrams, than in the "Vegie Delite"?

(b) The recommended amount of sodium is less than 2400 mg daily. How much more or less sodium does the "Super Melt" have than the recommended daily amount?

*Pick the metric temperature that is most appropriate in each situation.*

**25.** The water is almost boiling.
210 °C    155 °C    95 °C

**26.** The tomato plants may freeze tonight.
30 °C    20 °C    0 °C

*Use the table from* **Section 8.5** *and unit fractions to convert each measurement. Round your answers to the nearest tenth if necessary.*

**27.** 6 ft to meters

**28.** 125 lb to kilograms

**29.** 50 L to gallons

**30.** 8.1 km to miles

*Use the conversion formulas to convert each temperature. Round your answers to the nearest degree if necessary.*

**31.** 74 °F to Celsius

**32.** −12 °C to Fahrenheit

*Solve this application problem. Show your work.*

**33.** Denise is making five matching pillows. She needs 1 m 20 cm of braid to trim each pillow. If the braid costs $3.98 per yard, how much will she spend to trim the pillows, to the nearest cent? (First find the number of meters of braid Denise needs.)

**34.** Describe two benefits the United States would achieve by switching entirely to the metric system.

# Cumulative Review Exercises    Chapters 1–8

1. Write these numbers in words.
   (a) 603,005,040,000
   (b) 9.040

2. Write these numbers in digits.
   (a) Eighty and eight hundredths
   (b) Two hundred million, sixty-five thousand, four

3. Round each number as indicated.
   (a) 0.9802 to the nearest tenth
   (b) 495 to the nearest ten
   (c) 306,472,000 to the nearest million

4. Find the mean and median, to the nearest tenth, for this set of data on hours worked: 22, 18, 40, 18, 20, 21, 45, 25.

*Simplify.*

5. $(-2)^3 - 3^2$

6. $2\dfrac{2}{5} - \dfrac{3}{4}$

7. $\dfrac{-4}{0.16}$

8. $\dfrac{7}{6x^2} \cdot \dfrac{9x}{14y}$

9. $(-0.003)(-0.05)$

10. $0.083 - 42$

11. $\dfrac{3}{c} - \dfrac{5}{6}$

12. $-15 - 15$

13. $10 - 3(6 - 7)$

14. $\dfrac{-3(-4)}{27 - 3^3}$

15. $\dfrac{3}{5}$ of $(-400)$

16. $3\dfrac{7}{12} - 4$

17. $\dfrac{9y^2}{8x} \div \dfrac{y}{6x^2}$

18. $\dfrac{2}{3} + \dfrac{n}{m}$

19. $1\dfrac{1}{6} + 1\dfrac{2}{3}$

20. $\dfrac{\frac{14}{15}}{-6}$

21. $\dfrac{12 \div (2 - 5) + 12(-1)}{2^3 - (-4)^2}$

22. $(-0.8)^2 \div (0.8 - 1)$

23. $\left(-\dfrac{1}{3}\right)^2 - \dfrac{1}{4}\left(\dfrac{4}{9}\right)$

*Evaluate each expression when $k = -6$, $m = -1$, and $n = 2$.*

**24.** $-3k + 4$

**25.** $4k - 5n$

**26.** $3m^3n$

*Simplify each expression.*

**27.** $3p - 3p^2 - 4p$

**28.** $-5(x + 2) - 4$

**29.** $7(-2r^3)$

*Solve each equation. Show your work.*

**30.** $2b + 2 = -5 + 5$

**31.** $12.92 - a = 4.87$

**32.** $7(t + 6) = 42$

**33.** $-5n = n - 12$

**34.** $2 = \frac{1}{4}w - 3$

**35.** $\frac{1.5}{45} = \frac{x}{12}$

**36.** $4y - 3 = 7y + 12$

**37.** $3(k - 6) - 4 = -2(k + 1)$

*Translate each sentence into an equation and solve it.*

**38.** If eight times a number is subtracted from eleven times the number, the result is $-9$. Find the number.

**39.** When twice a number is decreased by 8, the result is the number increased by 7. Find the number.

*Solve each application problem using the six problem-solving steps from Chapter 3.*

**40.** The perimeter of a rectangle is 124 cm. The width is 25 cm. Find the length.

**41.** A 90 ft pipe is cut into two pieces so that one piece is 6 ft shorter than the other. Find the length of each piece.

*Find the unknown length, perimeter, circumference, area, or volume. When necessary, use 3.14 as the approximate value of $\pi$ and round answers to the nearest tenth.*

**42.** Find the perimeter and area.

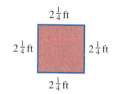

**43.** Find the circumference and the area.

**44.** Find the perimeter and the area.

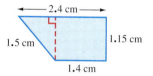

**45.** Find the unknown length.

**46.** Find the volume and the surface area.

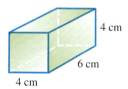

**47.** Find the unknown length in these similar triangles.

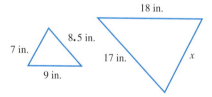

*Convert the following using unit fractions, the metric conversion line, or the temperature conversion formulas.*

**48.** $4\frac{1}{2}$ ft to inches

**49.** 72 hours to days

**50.** 3.7 kg to grams

**51.** 60 cm to meters

**52.** 7 mL to liters

**53.** $-20\ °C$ to Fahrenheit

*Solve each application problem. Show your work.*

**54.** Mei Ling must earn 60 credits to receive an associate of arts degree. She has 35 credits. What percent of the necessary credits does she have? Round to the nearest whole percent.

**55.** Which bag of chips is the best buy: Brand T is $15\frac{1}{2}$ ounces for $2.99, Brand F is 14 ounces for $2.49, and Brand H is 18 ounces for $3.89. You have a coupon for 40¢ off Brand H and another for 30¢ off Brand T.

**56.** A coffee can has a diameter of 13 cm and a height of 17 cm. Find the volume of the can. Use 3.14 for $\pi$ and round your answer to the nearest tenth.

**57.** Bags of slivered almonds weigh 115 g each. They are packed in a carton that weighs 450 g. How many kilograms would a carton containing 48 bags weigh?

**58.** Steven bought $4\frac{1}{2}$ yards of canvas material to repair the tents used by the scout troop. He used $1\frac{2}{3}$ yards on one tent and $1\frac{3}{4}$ yards on another. How much material is left?

**59.** Mark bought 650 grams of maple sugar candy on his vacation in Montreal. The candy is priced at $14.98 per kilogram. How much did Mark pay, to the nearest cent?

**60.** Calbert works at a Wal-Mart store and used his employee discount to buy a $189.94 DVD player at 10% off. Find the amount of his discount, to the nearest cent, and the sale price. (*Source:* Wal-Mart.)

**61.** Akuba is knitting a scarf. Six rows of knitting result in 5 cm of scarf. At that rate, how many rows will she knit to make a 100 cm scarf?

**62.** A loan of $3500 will be paid back with $7\frac{1}{2}$% simple interest at the end of six months. Find the total amount due.

**63.** A shelter for homeless people received a $400 donation to buy blankets. If the blankets cost $17 each, how many blankets can be purchased? How much money will be left over?

# Graphs 9

**9.1** Problem Solving with Tables and Pictographs

**9.2** Reading and Constructing Circle Graphs

**9.3** Bar Graphs and Line Graphs

**9.4** The Rectangular Coordinate System

**9.5** Introduction to Graphing Linear Equations

The Internet can be a great place to do research. Whether you're writing term papers and speeches for college courses, or plan on a career as a writer, you'll need to find information at various Web sites, interpret the data, and communicate it to other people. The data will often be presented in the form of a table (see Section 9.1, Examples 1 and 2) or a graph (see Sections 9.2 and 9.3). Interpreting the data accurately is critical to the success of your report, article, or speech. (See Section 9.3, Exercises 31–34 and 41–42.)

## 9.1 Problem Solving with Tables and Pictographs

**Objectives**

1. Read and interpret data presented in a table.
2. Read and interpret data from a pictograph.

Throughout this book you have used numbers, expressions, formulas, and equations to communicate rules or information. In this chapter you'll see how *tables, pictographs, circle graphs, bar graphs,* and *line graphs* are also used to communicate information.

**1 Read and interpret data presented in a table.** A **table** presents data organized into rows and columns. The advantage of a table is that you can find very specific, exact values. The disadvantages are that you may have to spend some time searching through the table to find what you want, and it may not be easy to see trends or patterns.

The table below shows information about the performance of the seven largest U.S. airlines during the second quarter of 2000 (April–June).

**PERFORMANCE DATA FOR THE LARGEST U.S. AIRLINES**
**APRIL–JUNE 2000**

| Airline | On-Time Performance | Luggage Handling* |
|---|---|---|
| American | 72% | 4.3 |
| Continental | 77% | 3.0 |
| Delta | 78% | 3.1 |
| Northwest | 78% | 3.7 |
| Trans World | 74% | 4.0 |
| United | 57% | 5.3 |
| US Airways | 71% | 4.1 |

*Luggage problems per 1000 passengers.
*Source:* Department of Transportation Air Travel Consumer Report.

For example, by reading from left to right along the row marked Delta, you first see that 78% of Delta's flights were on time during April–June 2000. The next number is 3.1 and the heading at the top of that column is *Luggage Handling\**. The little star (asterisk) tells you to look below the table for more information. Next to the asterisk below the table it says "Luggage problems per 1000 passengers." So, just over 3 passengers out of every 1000 passengers on Delta flights had some sort of problem with their luggage.

**Example 1** Reading and Interpreting Data from a Table

Use the table on airline performance to answer these questions.

**(a)** What percent of Continental's flights were on time?
Look across the row labeled Continental to see that 77% of its flights were on time.

**(b)** Which airline had the worst luggage handling record?
Look down the column headed Luggage Handling. To find the *worst* record, look for the *highest* number of luggage problems. The highest number is 5.3. Then look to the left to find the airline, which is United.

*Continued on Next Page*

(c) What was the average percent of on-time flights for the four airlines with the best performance?

Look down the column headed On-Time Performance to find the four highest numbers: 78%, 78%, 77%, and 74%. To find the average, add the values and divide by 4.

$$\frac{78 + 78 + 77 + 74}{4} = \frac{307}{4} \approx 77 \text{ (rounded to nearest whole number)}$$

The average on-time performance for the four best airlines was about 77%.

============ Work Problem ❶ at the Side.

### Example 2  Interpreting Data from a Table

This table shows the maximum cab fares in five different cities in March 2000. The "flag drop" charge is made when the driver starts the meter. "Wait time" is the charge for having to wait in the middle of a ride.

MAXIMUM TAXICAB FARES ALLOWED IN SELECTED CITIES IN MARCH 2000

| City | Flag Drop | Price per Mile | Wait Time per Hour |
|---|---|---|---|
| Chicago | $1.60 | $1.40 | $16 |
| Denver | $1.40–$1.60 | $1.60 | $16 to $18 |
| Detroit | $1.40 | $1.40 | $13.80 |
| New York | $2 | $1.50 | $12 |
| St. Paul | $2 | $1.60 | $21 |

*Source:* City of St. Paul, Minnesota.

Use the table to answer these questions.

**(a)** What is the range of fares you could pay for a 3-mile cab ride (with no wait time) in Denver?

The table shows that the price per mile in Denver is $1.60, so the cost for 3 miles is 3($1.60) = $4.80.

The flag drop charge ranges from $1.40 (lowest) to $1.60 (highest), and this must be added to $4.80.

Lowest → $1.40 + $4.80 = $6.20  
Highest → $1.60 + $4.80 = $6.40  
The range is $6.20 to $6.40.

**(b)** What is the maximum fare for a 9-mile ride in New York that includes having the cab wait 15 minutes while you pick up a package at a store?

The price per mile in New York is $1.50, so the cost for 9 miles is 9($1.50) = $13.50. Then, add the flag drop charge of $2. Finally, figure out the cost of the wait time. One way to do that is to set up a proportion. Recall that 1 hour is 60 minutes, so each side of the proportion compares the cost to the number of minutes of wait time.

Cost → $12 / 60 min = $x / 15 min ← Cost  
Wait time →                        ← Wait time

$60 \cdot x = 12 \cdot 15$   Find the cross products.

$\dfrac{60x}{60} = \dfrac{180}{60}$   Divide both sides by 60.

$x = \$3$  ← Charge for waiting 15 minutes

Total fare = $2 + $13.50 + $3 = $18.50

============ Work Problem ❷ at the Side.

---

❶ Use the table of airline performance to answer these questions.

(a) What percent of American flights were on time?

(b) Which airline had the best on-time performance?

(c) Which airline had the best record for luggage handling?

(d) Which airline(s) had fewer than 4 luggage handling problems per 1000 passengers?

(e) What was the average number of luggage problems for all seven airlines, to the nearest tenth?

❷ Use the table of cab fares to answer these questions.

(a) What is the difference in the maximum fare for a cab ride of 6.5 miles in Chicago compared to New York? Assume there is no wait time.

(b) What is the maximum fare for a 12-mile cab ride in St. Paul that includes 10 minutes of wait time?

(c) Find the range of fares for a cab ride of 4.5 miles in Denver that includes 30 minutes of wait time.

**ANSWERS**
1. (a) 72%  (b) Delta and Northwest tied  (c) Continental  (d) Continental, Delta, Northwest  (e) 27.5 ÷ 7 ≈ 3.9 (rounded)
2. (a) Chicago $10.70; New York $11.75; New York fare is $1.05 higher.  (b) $24.70  (c) $16.60 (lowest) to $17.80 (highest)

**2** **Read and interpret data from pictographs.** Tables show numbers in rows and columns. Graphs, on the other hand, are a *visual* way to communicate data; that is, they show a *picture* of the information rather than a list of numbers. In this section you'll work with one type of graph, *pictographs*, and in the next two sections you'll learn about circle graphs, bar graphs, and line graphs.

The advantage of a graph is that you can easily make comparisons or see trends just by looking. The disadvantage is that the graph may not give you the specific, more exact numbers you need in some situations.

### Example 3  Reading and Interpreting a Pictograph

A **pictograph** uses symbols or pictures to represent various amounts. The pictograph below shows the population of five U.S. metropolitan areas (cities with their surrounding suburbs) in 2000. The *key* at the bottom of the graph tells you that each symbol of a person represents 2 million people. A fractional part of a symbol represents a fractional part of 2 million people.

*Source:* U.S. Bureau of the Census, U.S. Department of Commerce.

Use the pictograph to answer these questions.

**(a)** What is the approximate population of Chicago?

The population of Chicago is represented by 4 whole symbols (4 • 2 million = 8 million) plus half of a symbol ($\frac{1}{2}$ of 2 million is 1 million) for a total of 9 million.

Recall that 2 million can be written as 2,000,000, so another way to get the answer is to multiply 4.5 (the number of symbols) times 2,000,000 for a total of 9,000,000. So the approximate population of Chicago is 9 million people (or 9,000,000).

**(b)** How much greater is the population of Los Angeles than Philadelphia?

Los Angeles shows 5 more symbols than Philadelphia, and 5 • 2 million = 10 million. So Los Angeles has 10 million (or 10,000,000) more people than Philadelphia.

**Continued on Next Page**

**NOTE**

One disadvantage of a pictograph is that it is difficult to draw and interpret fractional parts of a symbol. The data on metropolitan area populations were rounded a great deal in order to use only whole symbols and half symbols.

For example, in a *table* of metropolitan area populations in 2000, the figure for Los Angeles is 16,402,079 people. We could start by rounding this to 16,400,000 or 16.4 million. We would then draw 8 whole symbols and $\frac{1}{5}$ of a symbol. But a reader might easily think the fractional symbol represented $\frac{1}{10}$ or $\frac{1}{4}$. So we rounded the population even further, to the *nearest million*. The population of Los Angeles is closer to 16 million than to 17 million, so there are 8 whole symbols in the pictograph. That means our pictograph is off by about 400,000 people. Therefore, if you need fairly accurate numbers for a particular situation, it's better to use a table than a pictograph.

→ **Work Problem ❸ at the Side.**

❸ Use the pictograph in Example 3 to answer these questions.

(a) What is the approximate population of Philadelphia?

(b) What is the approximate population of New York?

(c) Approximately how much greater is the population of New York than Chicago?

(d) The population of Dallas is approximately how much less than Philadelphia?

**Answers**

3. (a) 6 million people (or 6,000,000)
   (b) 21 million people (or 21,000,000)
   (c) 12 million people (or 12,000,000)
   (d) 1 million people (or 1,000,000)

# Focus on Real-Data Applications

## Currency Exchange

When you travel between countries, you will exchange U.S. dollars for the local currency. The exchange rate between currencies changes daily, and you can easily find the updated rates using the Internet or any major newspaper. The table shown below has been extracted from the Bloomberg Currency Calculator Web page.

NORTH AMERICA/CARIBBEAN CURRENCY RATES

| Currency | Symbol | Value | Currency per 1 unit of USD | |
|---|---|---|---|---|
| | | | Net Chg | Pct Chg |
| Canadian Dollar | CAD | 1.4967 | +0.0024 | +0.1606 |
| Cayman Islands | KYD | 0.8282 | — | — |
| Jamaica Dollar | JMD | 45.1 | +0.1 | +0.2222 |
| Mexican Peso | MXN | 9.761 | −0.014 | −0.1432 |
| United States Dollar | USD | 1.00 | | |

On February 4, 2001, the currency exchange rate from U.S. dollars to Mexican pesos was given as follows:

$1.00 U.S. was equivalent to 9.761 Mexican pesos

You can set up a proportion to convert dollars to pesos. For example, suppose you want to determine the number of pesos that is equivalent to $50.00.

$$\frac{\$1}{9.761 \text{ pesos}} = \frac{\$50}{x \text{ pesos}} \quad \text{or} \quad \frac{1}{9.761} = \frac{50}{x}$$

$$(1)(x) = (9.761)(50)$$

$$x = 488.05 \text{ pesos}$$

So $50 buys 488 pesos and 5 centavos.

1. Based on the currency exchange rates for February 4, 2001, find the amount of each local currency that is equivalent to $50 U.S. and find the number of U.S. dollars that is equivalent to 200 units of each local currency. Round your answers to the nearest hundredth.

    (a) $50 = _____ Canadian dollars, and 200 Canadian dollars = _____ U.S. dollars.

    (b) $50 = _____ Cayman Island dollars, and
        200 Cayman Island dollars = _____ U.S. dollars.

    (c) $50 = _____ Jamaican dollars, and 200 Jamaican dollars = _____ U.S. dollars.

2. Set up a proportion to find the number of U.S. dollars that was equivalent to 1 Mexican Peso. 1 Mexican peso was equivalent to $ _____ (U.S.).

3. From Problem 2, you should recognize the conversion rate based on 1 Mexican peso as the expression $\frac{1}{9.761}$. What is the mathematical word that describes the relationship between the conversion rates 9.761 and $\frac{1}{9.761}$?

# 9.1 Exercises

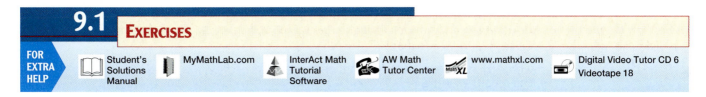

*This table lists the basketball players in the NBA with the highest scoring average at the start of the 2000–2001 season. (Players must have a minimum of 10,000 points or 400 games.) Use the table to answer Exercises 1–10. See Examples 1 and 2.*

**ALL-TIME NBA STATISTICAL LEADERS—SCORING AVERAGE**
**(AT THE START OF THE 2000–2001 SEASON)**

| Player | Games | Points | Average Points Per Game |
|---|---|---|---|
| Michael Jordan | 930 | 29,277 | 31.5 |
| Wilt Chamberlain | 1045 | 31,419 | 30.1 |
| Shaquille O'Neal* | 534 | 14,687 | 27.5 |
| Elgin Baylor | 846 | 23,419 | 27.4 |
| Jerry West | 932 | 25,192 | |
| Bob Pettit | 792 | 20,880 | |

*Player active in 2000–2001 season.
*Source:* National Basketball Association.

**1.** (a) How many points did Michael Jordan score during his NBA career?

(b) Which player(s) scored more points than Jordan?

**2.** (a) How many games did Elgin Baylor play in during his career in the NBA?

(b) Which player is closest to Baylor in number of games?

**3.** (a) Which player has been in the greatest number of games?

(b) Which player has been in the second greatest number of games?

**4.** Which players have scored more than 25,000 points? List them in order, starting with the player with the greatest number of points.

**5.** What is the difference in points scored between the player with the greatest number of points and the player with the least number of points?

**6.** How many fewer games has Shaquille O'Neal played in than Wilt Chamberlain?

**7.** Complete the table by finding the average number of points scored per game by Jerry West and by Bob Pettit. Look at the other averages in the table to decide how to round your answer.

**8.** Find the overall scoring average (points per game) for all six players listed in the table. Use the numbers in the *Games* column and the *Points* column to calculate your answer.

**9.** Which player(s) will have a different number of games and points by the end of the 2000–2001 season? Explain why.

**10.** Explain why Michael Jordan is listed first in the table even though Wilt Chamberlain scored more points.

This table shows the number of calories burned during 30 minutes of various types of exercise. The table also shows how the number of calories burned varies according to the weight of the person doing the exercise. Use the table to answer Exercises 11–20. See Examples 1 and 2.

**CALORIES BURNED DURING 30 MINUTES OF EXERCISE BY PEOPLE OF DIFFERENT WEIGHTS**

| Activity | Calories Burned in 30 Minutes | | |
|---|---|---|---|
| | 110 Pounds | 140 Pounds | 170 Pounds |
| Moderate jogging | 322 | 410 | 495 |
| Moderate walking | 110 | 140 | 170 |
| Moderate bicycling | 140 | 180 | 220 |
| Aerobic dance | 200 | 255 | 310 |
| Racquetball | 210 | 268 | 325 |
| Tennis | 160 | 205 | 250 |

*Source:* Fairview Health Services.

**11.** A person weighing 140 pounds is looking at the table.
   **(a)** How many calories will be burned during 30 minutes of aerobic dance?
   **(b)** Which activity burns the most calories?

**12.** A person weighing 170 pounds is looking at the table.
   **(a)** How many calories are burned during 30 minutes of tennis?
   **(b)** Which activity burns the fewest calories?

**13. (a)** Which activities can a 110-pound person do to burn at least 200 calories in 30 minutes?
   **(b)** Which activities can a 170-pound person do to burn at least 200 calories in 30 minutes?

**14. (a)** Which activities would burn less than 200 calories in 30 minutes for a 140-pound person?
   **(b)** Which activities would burn less than 300 calories in 30 minutes for a 170-pound person?

**15.** How many total calories will a 140-pound person burn during 15 minutes of bicycling and 60 minutes of walking?

**16.** How many total calories will a 110-pound person burn during 90 minutes of tennis and 15 minutes of aerobic dance?

*Set up and solve proportions to answer Exercises 17–20.*

**17.** How many calories would you expect a 125-pound person to burn
   **(a)** during 30 minutes of moderate jogging?
   **(b)** during 30 minutes of racquetball? Round to the nearest whole number.

**18.** How many calories would you expect a 185-pound person to burn
   **(a)** during 30 minutes of walking?
   **(b)** during 30 minutes of bicycling? Round to the nearest ten.

**19.** How many more calories would a 158-pound person burn during 15 minutes of aerobic dance than during 20 minutes of walking? Round to the nearest whole number.

**20.** How many fewer calories would a 196-pound person burn during 25 minutes of walking than during 20 minutes of tennis? Round to the nearest whole number.

*This pictograph shows the approximate number of passenger arrivals and departures at selected U.S. airports in 1999. Use the pictograph to answer Exercises 21–28. See Example 3.*

*Source:* Airports Council International—North America.

21. Approximately how many passenger arrivals and departures took place at the
    (a) St. Louis airport?
    (b) Atlanta airport?

22. Approximately how many passenger arrivals and departures took place at the
    (a) Dallas airport?
    (b) Miami airport?

23. What is the approximate total number of arrivals and departures at the two busiest airports?

24. What is the difference in the number of arrivals and departures at the busiest airport and the least busy airport?

25. How many fewer arrivals and departures did Miami's airport have compared to San Francisco's airport?

26. Find the approximate total number of arrivals and departures for the three least busy airports.

27. What is the approximate total number of arrivals and departures for all five airports?

28. Find the average number of arrivals and departures for the five airports.

### RELATING CONCEPTS (Exercises 29–34)   FOR INDIVIDUAL OR GROUP WORK

*Look back at the first table in this section, Performance Data for the Largest U.S. Airlines in April–June 2000. Use the table as you* **work Exercises 29–34 in order.**

**29.** Suppose you are planning a business trip where you will fly to a new city each day on a tight schedule. You'll travel light, carrying one small bag on the plane rather than checking it. If you can choose any one of the airlines in the table, which one would you pick? Explain why.

**30.** Suppose you are planning the business trip described in Exercise 29 and the only airline that goes to the cities you want is United. What could you do to minimize the problems caused by the possibility of late flights?

**31.** Now you are planning a two-week vacation trip to a beachfront resort. You'll be checking several bags and your expensive golf clubs. If you can choose any one of the airlines in the table, which one would you pick? Explain why.

**32.** Suppose you are planning the vacation trip described in Exercise 31 and American is the only airline that goes to the city you want. What could you do to minimize possible luggage handling problems?

**33.** Think of four possible reasons why an airline might have a lower percentage of on-time flights during a particular three-month period than they usually do. What, if anything, could the airline do to resolve each of the problems you listed?

**34.** Describe three other factors you might consider when selecting an airline, other than on-time performance and luggage handling problems.

# 9.2 Reading and Constructing Circle Graphs

A *circle graph* is another way to show a *picture* of a set of data. This picture can often be understood faster and more easily than a formula or a list of numbers.

**OBJECTIVES**
1. Read a circle graph.
2. Use a circle graph.
3. Use a protractor to draw a circle graph.

**1 Read a circle graph.** A **circle graph** is used to show how a total amount is divided into parts. The following circle graph shows you how 24 hours in the life of a college student are divided among different activities.

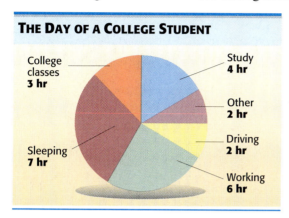

**THE DAY OF A COLLEGE STUDENT**
- College classes 3 hr
- Study 4 hr
- Other 2 hr
- Driving 2 hr
- Working 6 hr
- Sleeping 7 hr

Work Problem ❶ at the Side.

**2 Use a circle graph.** The circle graph above uses pie-shaped pieces called *sectors* to show the amount of time spent on each activity (the total must be 24 hours); a circle graph can therefore be used to compare the time spent on one activity to the total number of hours in the day.

### Example 1 Using a Circle Graph

Find the ratio of time spent in college classes to the total number of hours in a day. Write the ratio as a fraction in lowest terms. (See **Section 6.1**.)

The circle graph shows that 3 of the 24 hours in a day are spent in class. The ratio of class time to the hours in a day is shown below.

$$\frac{3 \text{ hours (college classes)}}{24 \text{ hours (whole day)}} = \frac{3 \text{ hours}}{24 \text{ hours}} = \frac{\cancel{3}}{\cancel{3} \cdot 8} = \frac{1}{8} \leftarrow \text{Lowest terms}$$

Work Problem ❷ at the Side.

The circle graph above can also be used to find the ratio of the time spent on one activity to the time spent on any other activity. See the next example.

### Example 2 Finding a Ratio from a Circle Graph

Find the ratio of working time to class time.
The circle graph shows 6 hours spent working and 3 hours spent in class. The ratio of working time to class time is shown below.

$$\frac{6 \text{ hours (working)}}{3 \text{ hours (class)}} = \frac{6 \text{ hours}}{3 \text{ hours}} = \frac{\cancel{3} \cdot 2}{\cancel{3}} = \frac{2}{1} \leftarrow \text{Ratio in lowest terms}$$

Work Problem ❸ at the Side.

---

❶ Use the circle graph to answer each question.

(a) The greatest number of hours is spent in which activity?

(b) How many more hours are spent working than studying?

(c) Find the total number of hours spent studying, working, and attending classes.

❷ Use the circle graph to find each ratio. Write the ratios as fractions in lowest terms.

(a) Hours spent driving to whole day

(b) Hours spent studying to whole day

(c) Hours spent sleeping and doing other to whole day

❸ Use the circle graph to find each ratio. Write the ratios as fractions in lowest terms.

(a) Hours spent studying to hours spent working

(b) Hours spent working to hours spent sleeping

(c) Hours spent studying to hours spent driving

**ANSWERS**
1. (a) sleeping (b) 2 hours (c) 13 hours
2. (a) $\frac{1}{12}$ (b) $\frac{1}{6}$ (c) $\frac{3}{8}$
3. (a) $\frac{2}{3}$ (b) $\frac{6}{7}$ (c) $\frac{2}{1}$

4 Use the circle graph on frozen pizza sales to find the amount of sales for each company.

(a) Kraft

(b) Van De Kamps

(c) Tony's Pizza Service

(d) Pillsbury Corp.

A circle graph often shows data as percents. For example, total U.S. sales of frozen pizza are $2 billion each year. The circle graph below shows how sales are divided among various companies that make frozen pizza. The entire circle represents $2 billion in sales. Each sector represents the sales of one company as a percent of the total sales. The total in a circle graph must be 100%, although it may be slightly more or less due to rounding the percent for each sector.

**HOT SALES OF FROZEN PIZZA**
Americans eat $2 billion worth of frozen pizzas each year. The percent of total sales for each company is rounded to the nearest whole percent.

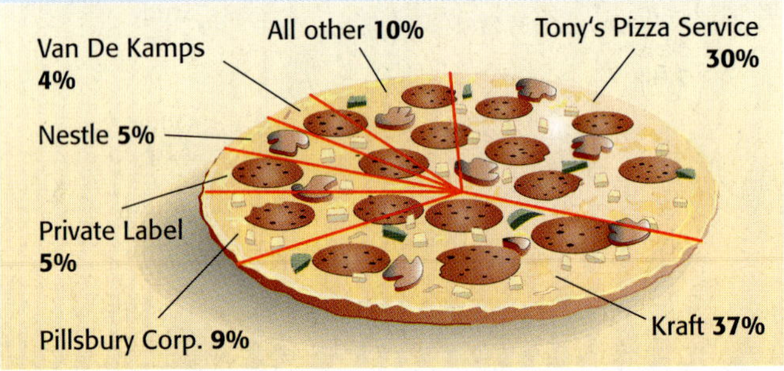

Van De Kamps 4%
All other 10%
Tony's Pizza Service 30%
Nestle 5%
Private Label 5%
Pillsbury Corp. 9%
Kraft 37%

*Source:* Information Resources, Inc. (1/16/2000).

### Example 3   Calculating an Amount by Using a Circle Graph

Use the circle graph on frozen pizza sales to find the amount of sales for Nestle.

Recall the percent equation.

$$\text{percent} \cdot \text{whole} = \text{part}$$

The percent for Nestle is 5%. Rewrite 5% as a decimal. The *whole* is the total sales of $2 billion (the entire circle).

$$\text{percent} \cdot \text{whole} = \text{part}$$

$$05.\% \cdot \$2 \text{ billion} = n \quad \text{Write \$2 billion as \$2,000,000,000.}$$

$$(0.05)(\$2,000,000,000) = n$$

$$\$100,000,000 = n$$

The sales for Nestle are $100,000,000.

**Work Problem 4 at the Side.**

**3** Use a protractor to draw a circle graph. The coordinator of the Fair Oaks Youth Soccer League organizes teams in five age groups. She counts the number of registered players in each age group as shown in the table below. Then she calculates what percent of the total each group represents. For example, there are 59 players in the "Under 8" group, out of 298 total players.

$$\text{percent} \cdot \text{whole} = \text{part}$$

$$p \cdot 298 = 59$$

$$\frac{p \cdot 298}{298} = \frac{59}{298}$$

$$p \approx 0.197 \approx 20\%$$

**ANSWERS**
4. (a) $740,000,000  (b) $80,000,000
   (c) $600,000,000  (d) $180,000,000

| Age Group | Number of Players | Percent of Total (rounded to nearest whole percent) |
|---|---|---|
| Under 8 years | 59 | 20%  ← 59 players ≈ 20% of 298 players |
| Ages 8–9 | 46 | 15% |
| Ages 10–11 | 75 | 25% |
| Ages 12–13 | 74 | 25% |
| Ages 14–15 | 44 | 15% |
| Total | 298 | 100% |

You can show these percents by using a circle graph. Recall that a circle has 360 degrees (written 360°). The 360° represents the entire league, or 100% of the soccer players.

### Example 4  Drawing a Circle Graph

Using the data on *age groups,* find the number of degrees in the sector that would represent the "Under 8" group, and begin constructing a circle graph.

A complete circle has 360°. Because the "Under 8" group makes up 20% of the total number of players, the number of degrees needed for the "Under 8" sector of the circle graph is 20% of 360°.

$$20.\% \text{ of } 360° = n$$
$$0.20 \cdot 360° = n$$
$$72° = n$$

Use a tool called a **protractor** to make a circle graph. First, using a ruler or straight edge, draw a line from the center of the circle to the left edge. Place the hole in the protractor over the center of the circle, making sure that 0 on the protractor lines up with the line that was drawn. Find 72° and make a mark as shown in the illustration. Then remove the protractor and use the straight edge to draw a line from the center of the circle to the 72° mark at the edge of the circle.

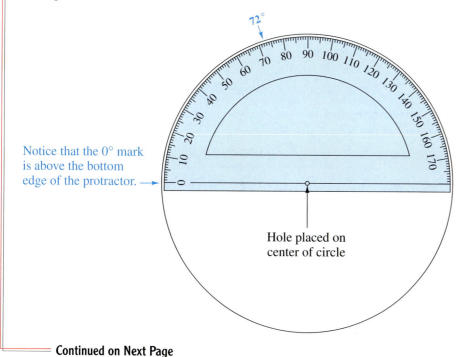

Notice that the 0° mark is above the bottom edge of the protractor.

Hole placed on center of circle

**Continued on Next Page**

**5** Using the information on the soccer age groups in the table, find the number of degrees needed for each sector. Then complete the circle graph at the bottom right on this page.

**(a)** "Ages 10–11" sector

**(b)** "Ages 12–13" sector

**(c)** "Ages 14–15" sector

To draw the "Ages 8–9" sector, begin by finding the number of degrees in the sector, which is 15% of the total circle.

$$15\% \text{ of } 360° = n$$
$$0.15 \cdot 360° = n$$
$$54° = n$$

Again, place the hole of the protractor at the center of the circle, but this time align 0 on the second line that was drawn. Make a mark at 54° and draw a line as before. This sector is 54° and represents the "Ages 8–9" group.

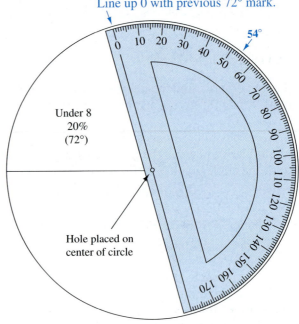

### CAUTION
You must be certain that the hole in the protractor is placed on the exact center of the circle each time you measure the size of a sector.

**Work Problem 5 at the Side.**

Use this circle for Problem 5 at the Side.

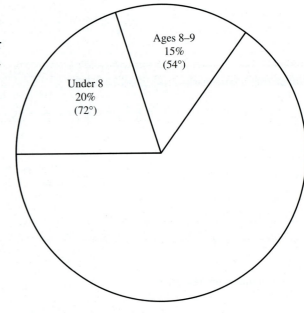

**ANSWERS**
**5.** (a) 90°  (b) 90°  (c) 54°

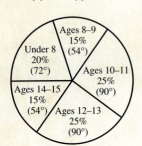

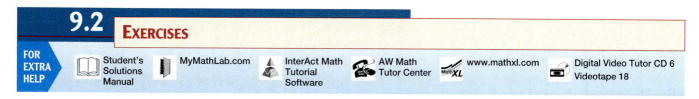

*This circle graph shows the cost of adding an art studio to an existing building. Use this circle graph to answer Exercises 1–6. Write ratios as fractions in lowest terms. See Examples 1 and 2.*

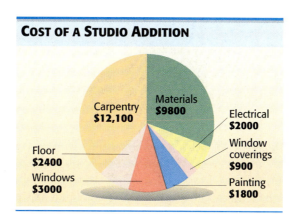

1. **(a)** Find the total cost of adding the art studio.

   **(b)** What is the largest single expense?

2. **(a)** What is the second largest expense in adding the studio?

   **(b)** What is the smallest expense?

3. **(a)** Find the ratio of the cost of materials to the total remodeling cost.

   **(b)** Find the ratio of the cost of windows to the cost of electrical.

4. **(a)** Find the ratio of the cost of painting to the total remodeling cost.

   **(b)** Find the ratio of the cost of windows to the cost of window coverings.

5. Find the ratio of the cost of carpentry, windows, and window coverings to the total remodeling cost.

6. Find the ratio of the cost of windows and electrical to the cost of the floor and painting.

This circle graph, adapted from *USA Today*, shows the number of people in a survey who gave various reasons for eating dinner at restaurants. Use this circle graph to answer Exercises 7–14. See Examples 1 and 2.

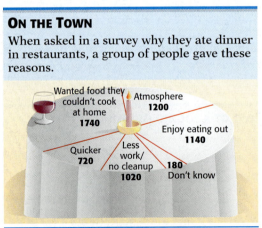

*Source:* Market Facts for Tyson Foods.

7. **(a)** Which reason was given by the least number of people?

   **(b)** Which reason was given by the second fewest number of people?

8. **(a)** Which reason was given by the highest number of people?

   **(b)** Which reason was given by the second highest number of people?

*Find each ratio in Exercises 9–14. Write the ratios as fractions in lowest terms.*

9. Those who said dining out is "Quicker" to total people in the survey

10. Those who said "Enjoy eating out" to the total people in the survey

11. Those who said "Less work/no clean up" to those who said "Atmosphere"

12. Those who said "Don't know" to those who said "Quicker"

13. Those who said "Wanted food they couldn't cook at home" to those who said "Don't know"

14. Those who said "Atmosphere" to those who said "Enjoy eating out"

*This circle graph shows the costs necessary to comply with the Americans with Disabilities Act (ADA) at the Dos Pueblos College. Each cost item is expressed as a percent of the total cost of* $1,740,000. *Use the graph to find the dollar amount spent for each item in Exercises 15–20. See Example 3.*

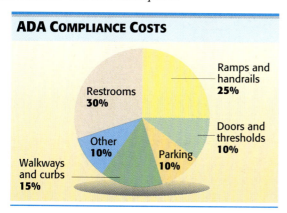

**15.** Restrooms

**16.** Ramps and handrails

**17.** Doors and thresholds

**18.** Parking

**19.** Walkways and curbs

**20.** Other

*This circle graph, adapted from* USA Today, *shows how consumers in a survey said they prefer to pay and keep track of their bills. If* 10,860 *people were surveyed, use the graph to find the number who made each of the choices listed in Exercises 21–26. Round to the nearest whole number. See Example 3.*

**21.** Checks and paper records

**22.** Checks and personal accounting software

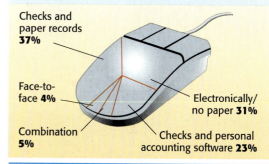

**23.** Combination

**24.** Electronically/no paper

**25.** Face-to-face

**26.** Face-to-face and Checks and personal accounting software combined

**648** Chapter 9 Graphs

**27.** Describe the procedure for determining how large each sector must be to represent each of the items in a circle graph.

**28.** A protractor is the tool used to draw a circle graph. Give a brief explanation of what the protractor does and how you would use it to measure and draw each sector in the circle graph.

*During one semester Kara Diano spent $5460 for school expenses as shown in this table. Find all numbers missing from the table.*

| Item | Dollar Amount | Percent of Total | Degrees of a Circle |
|---|---|---|---|
| **29.** Rent | $1365 | 25% | _____ |
| **30.** Food | $1092 | _____ | 72° |
| **31.** Clothing | $ 546 | _____ | _____ |
| **32.** Books | $ 546 | 10% | _____ |
| **33.** Entertainment | $ 819 | 15% | _____ |
| **34.** Savings | $ 273 | _____ | _____ |
| **35.** Other | $ 819 | _____ | 54° |

**36.** Draw a circle graph by using the above information. See Example 4.

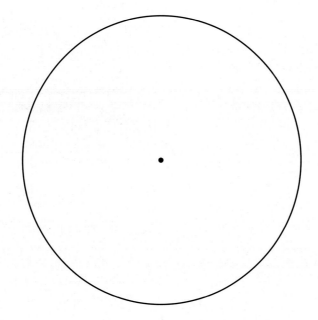

**37.** White Water Rafting Company divides its annual sales into five categories as follows.

| Category | Annual Sales |
|---|---|
| Adventure classes | $12,500 |
| Grocery and provision sales | $40,000 |
| Equipment rentals | $60,000 |
| Rafting tours | $50,000 |
| Equipment sales | $37,500 |

(a) Find the total sales for the year.

(b) Find the number of degrees in a circle graph for each item.

(c) Make a circle graph showing this information.

**38.** A book publisher had 25% of total sales in mysteries, 10% in biographies, 15% in cookbooks, 15% in romance novels, 20% in science, and the rest in business books.

(a) Find the number of degrees in a circle graph for each type of book.

(b) Draw a circle graph, using the information given.

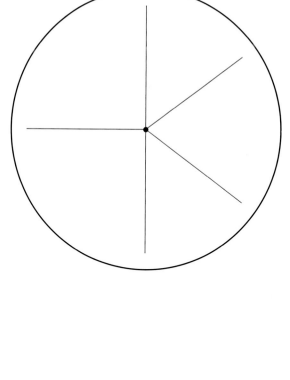

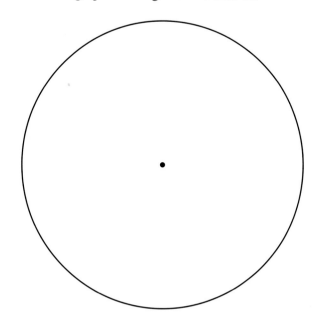

**39.** The Pathfinder Research Group asked 4488 Americans how they fall asleep. The results are shown in the visual on the right.

Use this information to complete the table and draw a circle graph. Round to the nearest whole percent and to the nearest degree.

| | Position | Number | Percent of Total | Number of Degrees |
|---|---|---|---|---|
| (a) | Side | _____ | _____ | _____ |
| (b) | Back | _____ | _____ | _____ |
| (c) | Stomach | _____ | _____ | _____ |
| (d) | Varies | _____ | _____ | _____ |
| (e) | Not sure | _____ | _____ | _____ |

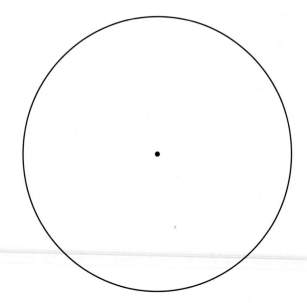

*Source:* Pathfinder Research Group.

**(f)** Add up the percents. Is the total 100%? Explain why or why not.

**(g)** Add up the degrees. Is the total 360°? Explain why or why not.

Section 9.3 Bar Graphs and Line Graphs **651**

## 9.3 BAR GRAPHS AND LINE GRAPHS

**OBJECTIVES**

Read and understand
1. a bar graph;
2. a double-bar graph;
3. a line graph;
4. a comparison line graph.

**1** Read and understand a bar graph. A **bar graph** is useful for showing comparisons. For example, the bar graph below compares the number of college graduates who continued taking advanced courses in their major field during each of five years.

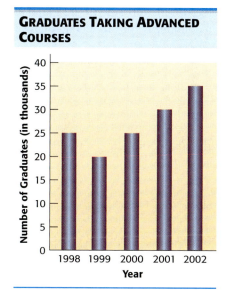

**GRADUATES TAKING ADVANCED COURSES**

**① Use the bar graph in the text to find the number of college graduates who took advanced classes in their major field in each of these years.**

(a) 1998

### Example 1 Using a Bar Graph

How many college graduates took advanced classes in their major field in 2000?

The bar for 2000 rises to 25. Notice the label along the left side of the graph that says "Number of Graduates (in thousands)." The phrase *in thousands* means you have to multiply 25 by 1000 to get 25,000. So, 25,000 (not 25) graduates took advanced classes in their major field in 2000.

➤ Work Problem ① at the Side.

(b) 1999

**2** Read and understand a double-bar graph. A **double-bar graph** can be used to compare two sets of data. The following graph shows the number of DSL (digital subscriber line) installations each quarter for two different years.

(c) 2001

**DIGITAL SUBSCRIBER LINE INSTALLATIONS**

(d) 2002

**ANSWERS**

1. (a) 25,000 graduates  (b) 20,000 graduates
   (c) 30,000 graduates  (d) 35,000 graduates

**652** Chapter 9 Graphs

❷ Use the double-bar graph to find the number of DSL installations in 2001 and 2002 for each quarter.

(a) 1st quarter

(b) 3rd quarter

(c) 4th quarter

(d) Find the greatest number of installations. Identify the quarter and the year in which they occurred.

❸ Use the line graph in the text to find the number of trout stocked in each month.

(a) June

(b) May

(c) April

(d) July

**ANSWERS**
2. (a) 4000; 3000 (b) 7000; 8000
   (c) 5000; 4000
   (d) 8000; 3rd quarter of 2002
3. (a) 55,000 trout (b) 30,000 trout
   (c) 40,000 trout (d) 60,000 trout

### Example 2 Reading a Double-Bar Graph

Use the double-bar graph on the previous page to find the following.

(a) The number of DSL installations in the second quarter of 2001

There are two bars for the second quarter. The color code in the upper right-hand corner of the graph tells you that the **red bars** represent 2001. So the **red bar** on the *left* is for the 2nd quarter of 2001. It rises to 6. Multiply 6 by 1000 because the label on the left side of the graph says *in thousands*. So there were 6000 DSL installations for the second quarter in 2001.

(b) The number of DSL installations in the second quarter of 2002

The **green bar** for the second quarter rises to 5 and 5 times 1000 is 5000. So, in the second quarter of 2002, there were 5000 DSL installations.

**CAUTION**

Use a ruler or straight edge to line up the top of the bar with the number on the left side of the graph.

Work Problem ❷ at the Side.

**3** Read and understand a line graph. A **line graph** is often useful for showing a trend. The line graph below shows the number of trout stocked along the Feather River over a five month period. Each dot indicates the number of trout stocked during the month directly below that dot.

**TROUT STOCKED IN FEATHER RIVER**

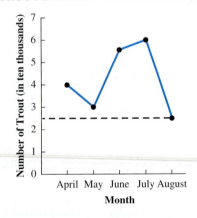

### Example 3 Understanding a Line Graph

Use the line graph above to answer each question.

(a) In which month were the least number of trout stocked?
The lowest point on the graph is the dot directly over August, so the least number of trout were stocked in August.

(b) How many trout were stocked in August?
Use a ruler or straight edge to line up the August dot with the numbers along the left edge of the graph. The August dot is halfway between the 2 and 3. Notice the label on the left side says *in ten thousands*. So August is halfway between (2 • 10,000) and (3 • 10,000). It is halfway between 20,000 and 30,000. That means 25,000 trout were stocked in August.

Work Problem ❸ at the Side.

### 4  Read and understand a comparison line graph.
Two sets of data can also be compared by drawing two line graphs together as a **comparison line graph**. For example, the following line graph compares the number of analog cell phones and the number of digital cell phones sold during each of five years.

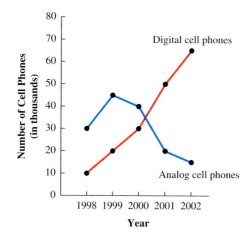

SALES OF ANALOG AND DIGITAL CELL PHONES

### Example 4  Interpreting a Comparison Line Graph

Use the comparison line graph above to find the following.

**(a)** The number of analog cell phones sold in 1999

Find the dot on the **blue line** above 1999. Use a ruler or straight edge to line up the dot with the numbers along the left edge. The dot is halfway between 40 and 50, which is 45. Then, 45 times 1000 is 45,000 analog cell phones sold in 1999.

**(b)** The number of digital cell phones sold in 2002

The **red line** on the graph shows that 65,000 digital cell phones were sold in 2002.

#### NOTE
Both the double-bar graph and the comparison line graph are used to compare two or more sets of data.

Work Problem ❹ at the Side.

❹ Use the comparison line graph in the text to find the following.

(a) The number of analog cell phones sold in 1998, 2000, 2001, and 2002

(b) The number of digital cell phones sold in 1998, 1999, 2000, and 2001

(c) The first full year in which the number of digital cell phones sold was greater than the number of analog cell phones sold

**ANSWERS**
4. (a) 30,000; 40,000; 20,000; 15,000
   (b) 10,000; 20,000; 30,000; 50,000
   (c) 2001

# Focus on Real-Data Applications

## Grocery Shopping

**GENDER BUYS**
On average, each person spends $32 weekly on groceries. Men average $35 and women $32. Saving strategies include:

Stock up on bargains — 26% / 28%
Stick to list — 15% / 23%
Check ads for specials — 21% / 33%
Mail/newspaper coupons — 21% / 30%
Buy store brands — 21% / 21%
In-store coupons — 20% / 25%

Men ▬  Women ▬

*Source:* Food Marketing Institute.

1. The graph at the right is a double-bar graph. What is it about? Write a brief paragraph describing the general purpose of the graph.

2. Which two strategies do women use most often?

3. Which two strategies do men use most often?

4. Which two strategies show the greatest difference in use between the men and women surveyed?

5. Which two strategies show the least difference?

6. The sum of the percents for the women's responses and the sum for the men's responses do not equal 100%. Why?

7. Is it possible to decide how many of the people in the survey use *none* of the strategies listed? Why?

8. Sometimes people "jump to conclusions" without enough evidence. Which of these conclusions are reasonable, *based on the information in the graph*?
   (a) Men and women use a variety of savings strategies when grocery shopping.
   (b) Men spend more for groceries than women.
   (c) Men eat more groceries than women.
   (d) Women are better grocery shoppers than men.
   (e) There are some differences in the grocery shopping strategies used by men and women.

9. Conduct a survey of your class members. Find out how many of them regularly use each of the saving strategies shown in the graph. Complete the table below.

10. Make a double-bar graph showing your survey data. How is your data similar to the graph shown above? How is it different?

| Strategy | Number of Women Using Strategy | Percent of Women Using Strategy | Number of Men Using Strategy | Percent of Men Using Strategy |
|---|---|---|---|---|
| Stock up on bargains | | | | |
| Stick to list | | | | |
| Check ads for specials | | | | |
| Mail/newspaper coupons | | | | |
| Buy store brands | | | | |
| In-store coupons | | | | |
| Total | | | | |

## 9.3 Exercises

*This bar graph shows the top seven reasons why people say they shop on-line. Use the graph to answer Exercises 1–6. See Example 1.*

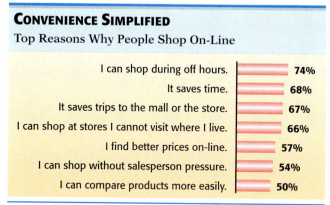

Source: EMARKETER, 2000.

1. What is the top reason why people shop on-line? What percent gave this reason?

2. What is the second most popular reason for shopping on-line? What percent gave this reason?

3. What percent of the people say they find better prices on-line? If 600 people were surveyed, how many gave this answer?

4. What percent say it saves trips to the mall or store? If 600 people were surveyed, how many gave this answer?

5. Which reason(s) were given by $\frac{1}{2}$ of the people? Which reason(s) were given by nearly $\frac{3}{4}$ of the people?

6. Which reason(s) were given by about $\frac{2}{3}$ of the people?

*This double-bar graph shows the number of workers who were unemployed in a city during the first six months of 2001 and 2002. Use this graph to answer Exercises 7–12. See Example 2.*

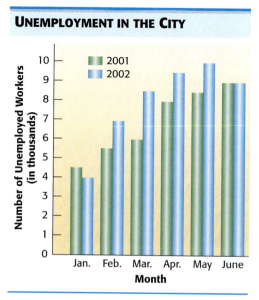

7. Which month in 2002 had the greatest number of unemployed workers? What was the total number unemployed in that month?

8. How many workers were unemployed in January of 2001?

9. How many more workers were unemployed in February of 2002 than in February of 2001?

10. How many fewer workers were unemployed in March of 2001 than in March of 2002?

11. Find the increase in the number of unemployed workers from February 2001 to April 2001.

12. Find the increase in the number of unemployed workers from January 2002 to June 2002.

*This double-bar graph shows sales of super unleaded and supreme unleaded gasoline at a service station for each of five years. Use this graph to answer Exercises 13–18. See Example 2.*

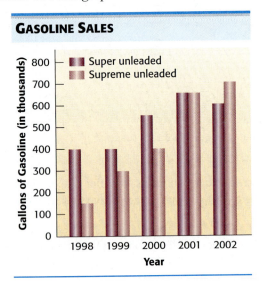

13. How many gallons of supreme unleaded gasoline were sold in 1998?

14. How many gallons of super unleaded gasoline were sold in 2001?

15. In which year did the greatest difference in sales between super unleaded and supreme unleaded gasoline occur? Find the difference.

16. In which year did the sales of supreme unleaded gasoline surpass the sales of super unleaded gasoline?

17. Find the increase in supreme unleaded gasoline sales from 1998 to 2002.

18. Find the increase in super unleaded gasoline sales from 1998 to 2002.

*This line graph shows the cost of a 3-minute phone call (in 1990 dollars) from New York to London over the last 70 years. Use the graph to answer Exercises 19–24. See Example 3.*

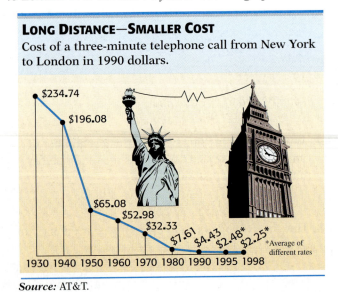

19. Find the cost of a 3-minute phone call from New York to London in 1940.

20. What was the cost of a 3-minute phone call from New York to London in 1970?

21. Find the amount of decrease in the cost of a 3-minute phone call from New York to London from 1970 to 1990. What was the percent of decrease, to the nearest whole percent?

22. How much less was the average cost of a 3-minute phone call from New York to London in 1998 than in 1980? What was the percent of decrease, to the nearest whole percent?

23. Give two possible explanations for the decrease in long-distance phone rates.

24. Give two possible conditions that could result in higher long-distance phone rates in the future.

*This comparison line graph shows the number of compact discs (CDs) sold by two different chain stores during each of five years. Use this graph to find the annual number of CDs sold in each year listed in Exercises 25–30. See Example 4.*

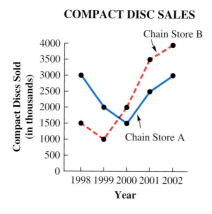

**25.** Chain Store A in 2002

**26.** Chain Store A in 2001

**27.** Chain Store A in 2000

**28.** Chain Store B in 2002

**29.** Chain Store B in 2001

**30.** Chain Store B in 2000

### RELATING CONCEPTS (Exercises 31–34) FOR INDIVIDUAL OR GROUP WORK

*Use the comparison line graph above on Compact Disc Sales as you work Exercises 31–34 in order.*

**31.** Describe the pattern(s) or trend(s) you see in the graph.

**32.** Store B used to have lower sales than Store A. What might have happened to cause this change? Give four possible explanations.

**33.** Which store would you like to own? Explain why.

**34.** *Based on the graph,* what amount of sales would you predict for each store in 2003? in 2004?

*This comparison line graph shows the sales and profits of Tacos-To-Go for each of four years. Use the graph to answer Exercises 35–42. See Example 4.*

**35.** Total sales in 2002

**36.** Total sales in 2001

**37.** Total sales in 2000

**38.** Profit in 2002

**39.** Profit in 2001

**40.** Profit in 2000

**41.** Give two possible explanations for the decrease in sales from 1999 to 2000 and two possible explanations for the increase in sales from 2000 to 2002.

**42.** *Based on the graph,* what conclusion can you make about the relationship between sales and profits for Tacos-To-Go?

# 9.4 THE RECTANGULAR COORDINATE SYSTEM

**1** Plot a point, given the coordinates, and find the coordinates, given a point. A bar graph or line graph shows the relationship between two things. The line graph below is from Example 3 in **Section 9.3.** It shows the relationship between the month of the year and the number of trout stocked in the Feather River.

### OBJECTIVES

**1** Plot a point, given the coordinates, and find the coordinates, given a point.

**2** Identify the four quadrants and determine which points lie within each one.

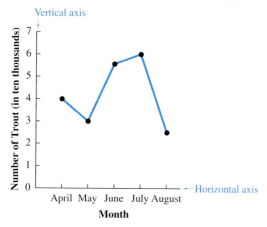

Each black dot on the graph represents a particular month paired with a particular number of trout. This is an example of **paired data.** We write each pair inside parentheses, with a comma separating the two items. To be consistent, we will always list the item on the *horizontal axis* first. In this case, the months are shown on the **horizontal axis** (the line that goes "left and right"), and the number of trout is shown along the **vertical axis** (the line that goes "up and down").

**Paired Data from Line Graph on Trout Stocked in Feather River**
(Apr, 40,000)   (May, 30,000)   (June, 55,000)   (July, 60,000)   (Aug, 25,000)

Each data pair gives you the location of a particular spot on the graph, and that spot is marked with a dot. This idea of paired data can be used to locate particular places on any flat surface.

Think of a small town laid out in a grid of square blocks, as shown below. To tell a taxi driver where to go, you could say, "the corner of 4th Avenue and 2nd Street" or just "4th and 2nd." As an *ordered pair,* it would be (4, 2). Of course, both you and the taxi driver need to know that the avenue is mentioned first (the number on the horizontal axis) and that the street is mentioned second (the number on the vertical axis). If the driver goes to (2, 4) instead, you'll be at the wrong corner.

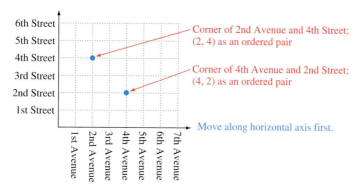

**660** Chapter 9 Graphs

① Plot each point on the grid. Write the ordered pair next to each point.

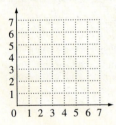

(a) (1, 4)

(b) (5, 2)

(c) (4, 1)

(d) (3, 3)

### Example 1  Plotting Points on a Grid

Use the grid at the right to plot each point.

**(a)** (3, 5)

Start at 0. Move *to the right* along the horizontal axis until you reach 3. Then move *up* 5 units so that you are aligned with 5 on the vertical axis. Make a dot. This is the plot, or graph, of the point (3, 5).

**(b)** (5, 3)

Start at 0. Move *to the right* along the horizontal axis until you reach 5. Then move *up* 3 units so that you are aligned with 3 on the vertical axis. Make a dot. This is the plot, or graph, of the point (5, 3).

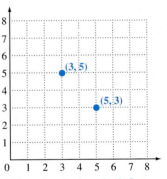

Move along horizontal axis first.

**CAUTION**

In Example 1, the points (3, 5) and (5, 3) are *not* the same. The "address" of a point is called an *ordered pair* because the *order* within the pair is important. Always move along the horizontal axis first.

Work Problem ① at the Side.

You have been using both positive and negative numbers throughout this book. We can extend our grid system to include negative numbers, as shown below. The horizontal axis is now a number line with 0 at the center, positive numbers extending to the right and negative numbers to the left. This horizontal number line is called the **x-axis**.

The vertical axis is also a number line, with positive numbers extending upward and negative numbers extending downward from 0. The vertical axis is called the **y-axis**. Together, the *x*-axis and the *y*-axis form a rectangular **coordinate system**. The center point (0, 0) is where the *x*-axis crosses the *y*-axis; it is called the **origin**.

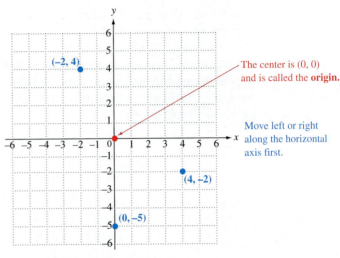

**Rectangular Coordinate System**

ANSWERS
1.

Section 9.4 The Rectangular Coordinate System  661

## Example 2  Plotting Points on a Rectangular Coordinate System

Plot each point on the rectangular coordinate system shown at the bottom of the previous page.

**(a)** $(4, -2)$

Start at 0. Then move left or right along the horizontal *x*-axis first. Because 4 is *positive,* move *to the right* until you reach 4. Now, because the 2 is *negative,* move *down* 2 units so that you are aligned with $-2$ on the *y*-axis. Make a dot and label it $(4, -2)$.

**(b)** $(-2, 4)$

Starting at 0, move left or right along the horizontal *x*-axis first. In this case, move *to the left* until you reach $-2$. Then move *up* 4 units. Make a dot and label it $(-2, 4)$. Notice that $(-2, 4)$ is **not** the same as $(4, -2)$.

**(c)** $(0, -5)$

Move left or right along the horizontal *x*-axis first. However, because the first number is 0, stay right at the center of the coordinate system. Then move *down* 5 units. Make a dot and label it $(0, -5)$.

**NOTE**

When the *first* number in an ordered pair is 0, the point is on the *y*-axis, as in Example 2(c) above. When the *second* number in an ordered pair is 0, the point is on the *x*-axis—for example $(4, 0)$.

Work Problem ❷ at the Side.

Once we have drawn a coordinate system, we can show the location of any point with an ordered pair. The numbers in the ordered pair are called the **coordinates** of the point.

## Example 3  Finding the Coordinates of Points

Find the coordinates of points *A*, *B*, *C,* and *D*.

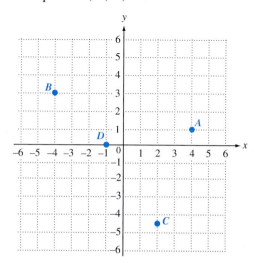

To reach point *A* from the origin, move 4 units *to the right;* then move *up* 1 unit. The coordinates are $(4, 1)$.

To reach point *B* from the origin, move 4 units *to the left;* then move *up* 3 units. The coordinates are $(-4, 3)$.

— Continued on Next Page

❷ Plot each point on the coordinate system shown. Write the ordered pair next to each point.

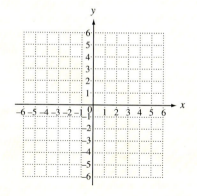

**(a)** $(5, -3)$

**(b)** $(-5, 3)$

**(c)** $(0, 3)$

**(d)** $(-4, -4)$

**(e)** $(-2, 0)$

**ANSWERS**
1.
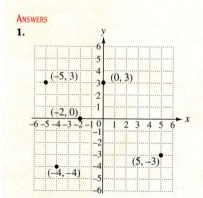

❸ Find the coordinates of points A, B, C, D, and E.

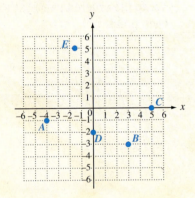

❹ (a) All points in the fourth quadrant are similar in what way? Give two examples of points in the fourth quadrant.

(b) In which quadrant is each point located: $(-2, -6)$; $(0, 5)$; $(-3, 1)$; $(4, -1)$?

To reach point C from the origin, move 2 units *to the right;* then move *down* approximately $4\frac{1}{2}$ units. The approximate coordinates are $(2, -4\frac{1}{2})$.

To reach point D from the origin, move 1 unit *to the left;* then do *not* move either up or down. The coordinates are $(-1, 0)$.

**NOTE**

If a point is between the lines on the coordinate system, you can use fractions to give the approximate coordinates. For example, the approximate coordinates of point C above are $(2, -4\frac{1}{2})$.

Work Problem ❸ at the Side.

**2** Identify the four quadrants and determine which points lie within each one. The *x*-axis and *y*-axis divide the coordinate system into four regions, called **quadrants.** These quadrants are numbered with Roman numerals, as shown below. Points on the axes themselves are not in any quadrant.

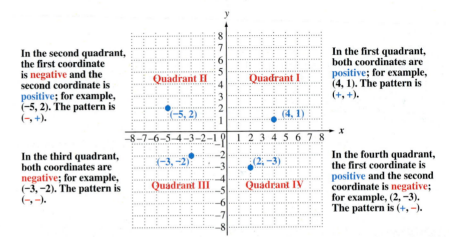

In the second quadrant, the first coordinate is **negative** and the second coordinate is **positive**; for example, $(-5, 2)$. The pattern is $(-, +)$.

In the third quadrant, both coordinates are **negative**; for example, $(-3, -2)$. The pattern is $(-, -)$.

In the first quadrant, both coordinates are **positive**; for example, $(4, 1)$. The pattern is $(+, +)$.

In the fourth quadrant, the first coordinate is **positive** and the second coordinate is **negative**; for example, $(2, -3)$. The pattern is $(+, -)$.

**Example 4** Working with Quadrants

(a) All points in the third quadrant are similar in what way? Give two examples of points in the third quadrant.

For all points in quadrant III, both coordinates are negative. The pattern is $(-, -)$. There are many possible examples, such as $(-2, -5)$ and $(-4, -4)$. Just be sure that both numbers are negative.

(b) In which quadrant is each point located: $(3, 5)$; $(1, -6)$; $(-4, 0)$?

For $(3, 5)$ the pattern is $(+, +)$, so the point is in **quadrant I.**

For $(1, -6)$ the pattern is $(+, -)$, so the point is in **quadrant IV.**

The point corresponding to $(-4, 0)$ is on the *x*-axis, so it isn't in any quadrant.

Work Problem ❹ at the Side.

**ANSWERS**

3. A is $(-4, -1)$; B is $(3, -3)$; C is $(5, 0)$; D is $(0, -2)$; E is approximately $(-1\frac{1}{2}, 5)$.

4. (a) The pattern for all points in quadrant IV is $(+, -)$. Examples will vary; just be sure that they fit the $(+, -)$ pattern.
   (b) III; no quadrant; II; IV

## 9.4 EXERCISES

*Plot each point on the rectangular coordinate system. Label each point with its coordinates. See Examples 1 and 2.*

**1.** (3, 7)   (−2, 2)   (−3, −7)   (2, −2)   (0, 6)
(6, 0)   (0, −4)   (−4, 0)

**2.** (5, 2)   (−3, −3)   (4, −1)   (−4, 1)   (−1, 0)
(0, 3)   (2, 0)   (0, −5)

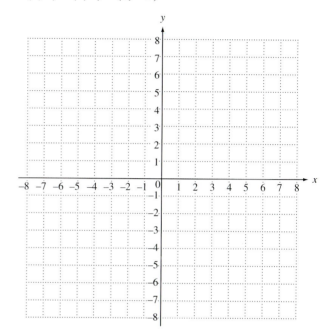

**3.** (−5, 3)   (4, 4)   (−2½, 0)   (3, −5)   (0, 0)
(2, ½)   (−7, −5)   (−1, −6)

**4.** (1, 7)   (0, 3½)   (−5, −1)   (6, −2)   (−2, 6)
(0, 0)   (−3, 3)   (−½, −2)

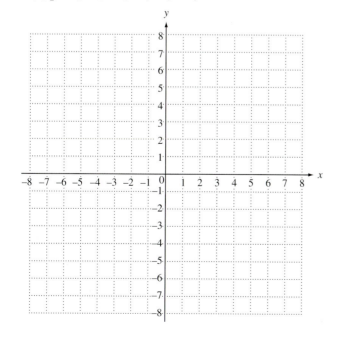

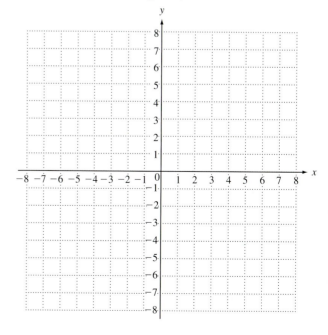

*Give the coordinates of each point. See Example 3.*

5.

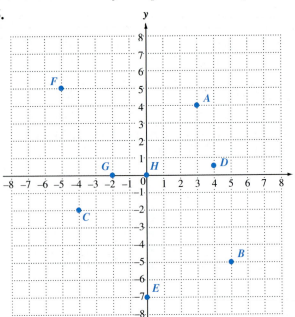

6.

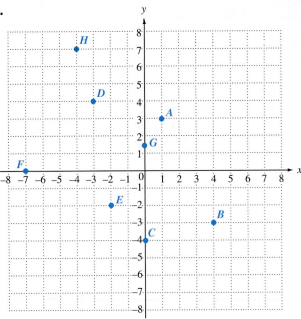

*Identify the quadrant in which each point is located. See Example 4.*

7. In which quadrant is each point located?
   $(-3, -7)$ $(0, 4)$ $(10, -16)$ $(-9, 5)$

8. In which quadrant is each point located?
   $(1, 12)$ $(20, -8)$ $(-5, 0)$ $(-14, 14)$

*Complete each ordered pair with a number that will make the point fall in the specified quadrant.*

9. (a) Quadrant II   $(-4, \underline{\phantom{xx}})$

   (b) Quadrant IV   $(7, \underline{\phantom{xx}})$

   (c) No quadrant   $(\underline{\phantom{xx}}, -2)$

   (d) Quadrant III  $(\underline{\phantom{xx}}, -1\frac{1}{2})$

   (e) Quadrant I    $(3\frac{1}{4}, \underline{\phantom{xx}})$

10. (a) Quadrant III  $(-5, \underline{\phantom{xx}})$

    (b) Quadrant I    $(\underline{\phantom{xx}}, 3)$

    (c) Quadrant IV   $(\underline{\phantom{xx}}, -\frac{1}{2})$

    (d) No quadrant   $(6, \underline{\phantom{xx}})$

    (e) Quadrant II   $(\underline{\phantom{xx}}, 1\frac{3}{4})$

11. Explain how to graph the ordered pair $(a, b)$, where $a$ and $b$ are positive or negative integers.

12. Explain how to graph the ordered pair $(a, b)$ where $a$ is 0 and $b$ is an integer. Explain how to graph $(a, b)$ when $a$ is an integer and $b$ is 0.

## 9.5 INTRODUCTION TO GRAPHING LINEAR EQUATIONS

In Chapters 2–7 you solved equations that had only one variable, such as $2n - 3 = 7$ or $\frac{1}{3}x = 10$. Each of these equations had exactly one solution; $n$ is 5 in the first equation, and $x$ is 30 in the second equation. In other words, there was only *one* number that could replace the variable and make the equation balance. As you take more algebra courses, you will work with equations that have two variables and many different numbers that will make the equation balance. This section will get you started.

**OBJECTIVES**

1. Graph linear equations in two variables.
2. Identify the slope of a line as positive or negative.

**1. Graph linear equations in two variables.** Suppose that you have 6 hours of study time planned during a weekend. You plan to study math and psychology. For example, you could spend 4 hours on math and then 2 hours on psychology, for a total of 6 hours. Or you could spend $1\frac{1}{2}$ hours on math and then $4\frac{1}{2}$ hours on psychology, for a total of 6 hours. Here is a list of *some* of the possible combinations.

| Hours on Math | + | Hours on Psychology | = | Total Hours Studying |
|---|---|---|---|---|
| 0 | + | 6 | = | 6 |
| 1 | + | 5 | = | 6 |
| $1\frac{1}{2}$ | + | $4\frac{1}{2}$ | = | 6 |
| 3 | + | 3 | = | 6 |
| 4 | + | 2 | = | 6 |
| $5\frac{1}{2}$ | + | $\frac{1}{2}$ | = | 6 |
| 6 | + | 0 | = | 6 |

We can write an equation to represent this situation.

$$\text{hours studying math} + \text{hours studying psychology} = \text{total of 6 hours}$$

$$m + p = 6$$

This equation has *two* variables. The hours spent on math ($m$) can vary, and the hours spent on psychology ($p$) can vary.

As you can see, there is more than one solution for this equation. We can list possible solutions as *ordered pairs*. The first number in the pair is the value of $m$, and the second number in the pair is the corresponding value of $p$.

$(m, p)$   $(m, p)$   $(m, p)$   $(m, p)$   $(m, p)$   $(m, p)$   $(m, p)$
↓ ↓    ↓ ↓    ↓ ↓    ↓ ↓    ↓ ↓    ↓ ↓    ↓ ↓
$(0, 6)$   $(1, 5)$   $\left(1\frac{1}{2}, 4\frac{1}{2}\right)$   $(3, 3)$   $(4, 2)$   $\left(5\frac{1}{2}, \frac{1}{2}\right)$   $(6, 0)$

Another way to show the solutions is to plot the ordered pairs, as you learned to do in **Section 9.4.** This method will give us a "picture" of the solutions that we listed on the previous page.

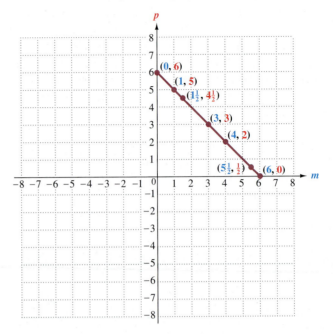

Notice that all the solutions (all the ordered pairs) lie on a straight line. When you draw a line connecting the ordered pairs, you have graphed the solutions. **Every point on the line is a solution.** You can use the line to find additional solutions besides the ones that we listed. For example, the point (5, 1) is on the line. This point tells you that another solution is 5 hours on math and 1 hour on psychology. The fact that the line is a *straight* line tells you that $m + p = 6$ is a *linear equation.* (The word *line* is part of the word *line*ar.) Later on in algebra you will work with equations whose solutions form a curve rather than a straight line when you graph them.

To draw the line for $m + p = 6$, we really needed only two solutions (two ordered pairs). But it's a good idea to use a third ordered pair as a check. If the three ordered pairs are *not* in a straight line, there is an error in your work.

### Graphing a Linear Equation

To **graph a linear equation,** find at least three ordered pairs that satisfy the equation. Then plot the ordered pairs on a coordinate system and connect them with a straight line. *Every* point on the line is a solution of the equation.

### Example 1  Graphing a Linear Equation

Graph $x + y = 3$ by finding three solutions and plotting the ordered pairs. Then use the graph to find a fourth solution of the equation.

There are many possible solutions. Start by picking three different values for *x*. You can choose any numbers you like, but 0 and small numbers usually are easy to use. Then find the value of *y* that will make the sum equal to 3. Set up a table to organize the information.

*Continued on Next Page*

| x | y | Check that x + y = 3 | Ordered Pair (x, y) |
|---|---|---|---|
| 0 | 3 | 0 + 3 = 3 | (0, 3) |
| 1 | 2 | 1 + 2 = 3 | (1, 2) |
| 2 | 1 | 2 + 1 = 3 | (2, 1) |

Plot the ordered pairs and draw a line through the points, extending it in both directions as shown below.

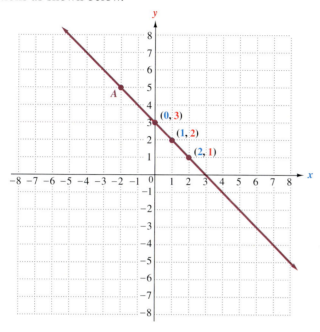

❶ Graph $x + y = 5$ by finding three solutions and plotting the ordered pairs. Then use the graph to find *two* other solutions of the equation.

| x | y | Check that x + y = 5 | Ordered Pair (x, y) |
|---|---|---|---|
| 0 | | | |
| 1 | | | |
| 2 | | | |

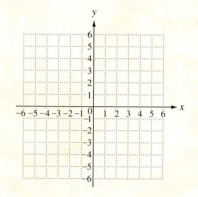

Two other solutions are (__, __) and (__, __).

Now you can use the graph to find more solutions of $x + y = 3$. *Every* point on the line is a solution. Suppose that you pick **point A**. The coordinates are $(-2, 5)$.

To check that $(-2, 5)$ is a solution, substitute $-2$ for $x$ and $5$ for $y$ in the original equation.

$$x + y = 3$$
$$-2 + 5 = 3$$
$$3 = 3 \quad \text{Balances}$$

The equation balances, so $(-2, 5)$ is another solution of $x + y = 3$.

### NOTE

The line in Example 1 above was extended in both directions because *every* point on the line is a solution of $x + y = 3$. However, when we graphed the line for the hours spent studying, $m + p = 6$, we did *not* extend the line. That is because the variables $m$ and $p$ represented hours, and hours can only be positive numbers; all the solutions had to be in the first quadrant.

Work Problem ❶ at the Side.

**ANSWERS**
1. Plot (0, 5), (1, 4), and (2, 3).

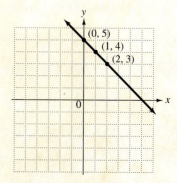

There are many other solutions. Some possibilities are $(-1, 6); (3, 2); (4, 1);$ $(5, 0); (6, -1)$.

**2** Graph $y = 2x$ by finding three solutions and plotting the ordered pairs. Then use the graph to find *two* other solutions of the equation.

| x | y (must be 2 • x) | Ordered Pair (x, y) |
|---|---|---|
| 0 | | |
| 1 | | |
| 2 | | |

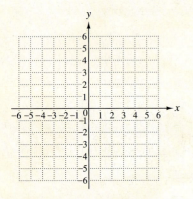

Two other solutions are (___, ___) and (___, ___).

### Example 2  Graphing a Linear Equation

Graph $y = -3x$ by finding three solutions and plotting the ordered pairs. Then use the graph to find a fourth solution of the equation.

You can choose any three values for *x*, but small numbers such as 0, 1, and 2 are easy to use. Then $y = -3x$ tells you that *y* is $-3$ times the value of *x*.

$y = -3x$   $-3x$ means $-3$ times *x*.

*y* is $-3$ times *x*

First set up a table.

| x | y (must be $-3$ times x) | Ordered Pair (x, y) |
|---|---|---|
| 0 | $-3 \cdot 0$ is $0$ | (0, 0) |
| 1 | $-3 \cdot 1$ is $-3$ | (1, $-3$) |
| 2 | $-3 \cdot 2$ is $-6$ | (2, $-6$) |

Plot the ordered pairs and draw a line through the points. Be sure to draw arrows on both ends of the line to show that it continues in both directions.

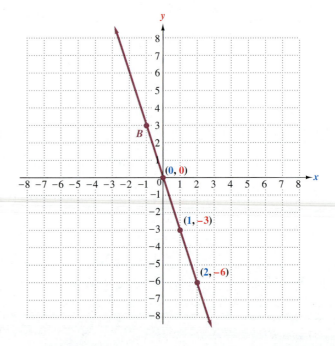

Now use the graph to find more solutions. *Every* point on the line is a solution.
  Suppose that you pick **point B**. The coordinates are $(-1, 3)$. To check that $(-1, 3)$ is a solution, substitute $-1$ for *x* and 3 for *y* in the original equation.

$y = -3x$
$3 = -3(-1)$
$3 = 3$    Balances

The equation balances, so $(-1, 3)$ is another solution of $y = -3x$.

**Work Problem** ➁ **at the Side.**

---

**ANSWERS**

**2.** Plot (0, 0), (1, 2), and (2, 4).

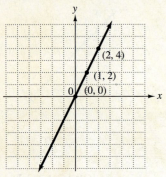

There are many other solutions. Some possibilities are (3, 6); $(-1, -2)$; $(-2, -4)$; $(-3, -6)$.

## Example 3  Graphing a Linear Equation

Graph $y = \frac{1}{2}x$.

Complete the table. The coefficient of $x$ is $\frac{1}{2}$, so chose even numbers like 2, 4, and 6 as values for $x$ because they are easy to divide in half. The equation $y = \frac{1}{2}x$ tells you that $y$ is $\frac{1}{2}$ times the value of $x$, or $\frac{1}{2}$ of $x$.

| $x$ | $y$ (must be $\frac{1}{2}$ of $x$) | Ordered Pair $(x, y)$ |
|---|---|---|
| 2 | $\frac{1}{2}$ of 2 is 1 | (2, 1) |
| 4 | $\frac{1}{2}$ of 4 is 2 | (4, 2) |
| 6 | $\frac{1}{2}$ of 6 is 3 | (6, 3) |

Plot the ordered pairs and draw a line through the points.

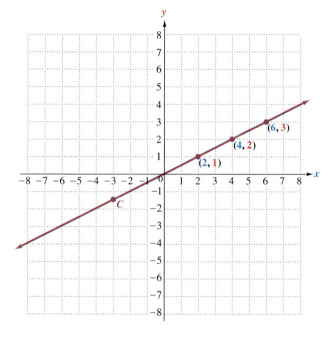

Now use the graph to find more solutions. *Every* point on the line is a solution.

Suppose that you pick **point C**. The coordinates are $(-3, -1\frac{1}{2})$. Check that $(-3, -1\frac{1}{2})$ is a solution by substituting $-3$ for $x$ and $-1\frac{1}{2}$ for $y$.

$$y = \frac{1}{2}x$$
$$-1\frac{1}{2} = \frac{1}{2}(-3)$$
$$-1\frac{1}{2} = -\frac{3}{2} \quad \text{Balances; } -1\frac{1}{2} \text{ is equivalent to } -\frac{3}{2}.$$

The equation balances, so $(-3, -1\frac{1}{2})$ is another solution of $y = \frac{1}{2}x$.

**Work Problem ③ at the Side.**

③ Graph $y = -\frac{1}{2}x$ by finding three solutions and plotting the ordered pairs. Then use the graph to find *two* more solutions.

| $x$ | $y$ (must be $-\frac{1}{2} \cdot x$) | Ordered Pair $(x, y)$ |
|---|---|---|
| 2 | | |
| 4 | | |
| 6 | | |

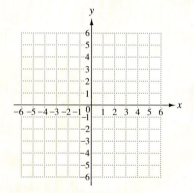

Two other solutions are (__, __) and (__, __).

**ANSWERS**

**3.** Plot $(2, -1)$, $(4, -2)$, and $(6, -3)$.

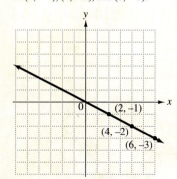

There are many other solutions. Some possibilities are $(0, 0)$; $(-2, 1)$; $(-4, 2)$; $(-6, 3)$; $(1, -\frac{1}{2})$; $(3, -1\frac{1}{2})$; $(5, -2\frac{1}{2})$.

**4** Graph the equation $y = x - 5$ by finding three solutions and plotting the ordered pairs. Then use the graph to find *two* more solutions.

| x | y (must be 5 less than x) | Ordered Pair (x, y) |
|---|---|---|
| 1 | | |
| 2 | | |
| 3 | | |

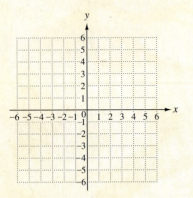

Two other solutions are (__, __) and (__, __).

**ANSWERS**

**4.** Plot $(1, -4)$, $(2, -3)$, and $(3, -2)$.

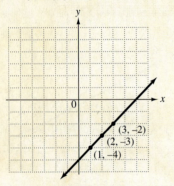

There are many other solutions. Some possibilities are $(-1, -6)$; $(0, -5)$; $(4, -1)$; $(5, 0)$; $(6, 1)$.

### Example 4 — Graphing a Linear Equation

Graph the equation $y = x + 4$ by finding three solutions and plotting the ordered pairs. Then use the graph to find two more solutions of the equation.

First set up a table. The equation $y = x + 4$ tells you that $y$ must be 4 more than the value of $x$.

| x | y (must be 4 more than x) | Ordered Pair (x, y) |
|---|---|---|
| 0 | 0 + 4 is **4** | (0, **4**) |
| 1 | 1 + 4 is **5** | (1, **5**) |
| 2 | 2 + 4 is **6** | (2, **6**) |

Plot the ordered pairs and draw a line through the points.

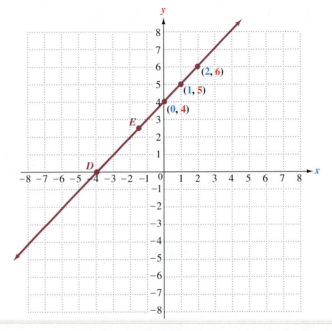

Now use the graph to find two more solutions. *Every* point on the line is a solution. Suppose that you pick **point D** at $(-4, 0)$ and **point E** at $(-1\tfrac{1}{2}, 2\tfrac{1}{2})$. Check that both ordered pairs are solutions.

**Check** $(-4, 0)$

$y = x + 4$
$0 = -4 + 4$
$0 = 0$  Balances

**Check** $\left(-1\tfrac{1}{2}, 2\tfrac{1}{2}\right)$

$y = x + 4$
$2\tfrac{1}{2} = -1\tfrac{1}{2} + 4$
$\tfrac{5}{2} = -\tfrac{3}{2} + \tfrac{8}{2}$
$\tfrac{5}{2} = \tfrac{5}{2}$  Balances

Both equations balance, so $(-4, 0)$ and $(-1\tfrac{1}{2}, 2\tfrac{1}{2})$ are also solutions of $y = x + 4$.

**Work Problem 4 at the Side.**

**2** **Identify the slope of a line as positive or negative.** Let's look again at some of the lines that we graphed for various equations. All are straight lines, but some are almost flat and some tilt steeply upward or downward.

Graph from Example 1:   $x + y = 3$

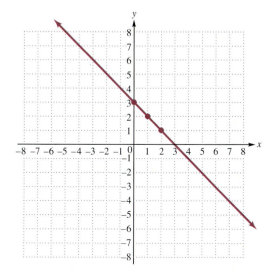

As you move from *left to right*, the line slopes downward, as if you were walking down a hill. When a line tilts downward, we say that it has a *negative slope*.

Now look at the table of solutions we used to draw the line.

| x | y |
|---|---|
| 0 | 3 |
| 1 | 2 |
| 2 | 1 |

The value of *x* is *increasing* from 0 to 1 to 2.

The value of *y* is *decreasing* from 3 to 2 to 1.

As the value of *x* increases, the value of *y* does the *opposite*—it *decreases*. Whenever one variable increases while the other variable decreases, the line will have a negative slope.

Graph from Example 3:   $y = \dfrac{1}{2}x$

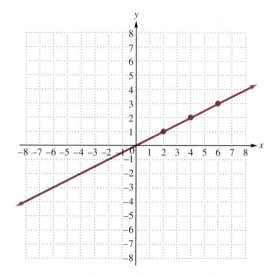

As you move from *left to right*, this line slopes upward, as if you were walking up a hill. When a line tilts upward, we say that it has a *positive slope*.

**5** Look back at the graphs in Margin Problems 2 and 3. Then complete these sentences.

(a) The graph of $y = 2x$ has a _____ slope. As the value of $x$ increases, the value of $y$ _____.

(b) The graph of $y = -\frac{1}{2}x$ has a _____ slope. As the value of $x$ increases, the value of $y$ _____.

Now look at the table of solutions we used to draw the line.

| x | y |
|---|---|
| 2 | 1 |
| 4 | 2 |
| 6 | 3 |

The value of $x$ is *increasing* from 2 to 4 to 6.

The value of $y$ is *increasing* from 1 to 2 to 3.

As the value of *x* increases, the value of *y* does the *same* thing—it also *increases*. Whenever both variables do the same thing (both increase or both decrease) the line will have a positive slope.

### Positive and Negative Slopes

As you move from left to right, a line with a *positive* slope tilts *upward* or rises.

As you move from left to right, a line with a *negative* slope tilts *downward* or falls.

### Example 5   Identifying Positive or Negative Slope in a Line

Look back at the graph of $y = -3x$ in Example 2. Then complete these sentences.

The graph of $y = -3x$ has a _____ slope. As the value of *x* increases, the value of *y* _____.

The graph of $y = -3x$ has a <u>negative</u> slope because it tilts downward. As the value of *x* increases, the value of *y* <u>decreases</u> (does the opposite).

**Work Problem 5 at the Side.**

---

**ANSWERS**
**5.** (a) positive; increases
(b) negative; decreases

## 9.5 Exercises

*Graph each equation by completing the table to find three solutions and plotting the ordered pairs. Then use the graph to find* two *other solutions. See Example 1.*

**1.** $x + y = 4$

| x | y | Ordered Pair (x, y) |
|---|---|---|
| 0 | | |
| 1 | | |
| 2 | | |

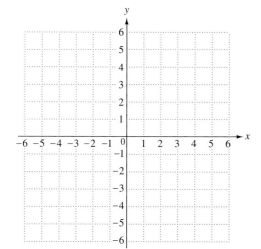

Two other solutions are (__, __) and (__, __).

**2.** $x + y = -4$

| x | y | Ordered Pair (x, y) |
|---|---|---|
| 0 | | |
| 1 | | |
| 2 | | |

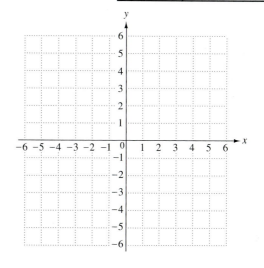

Two other solutions are (__, __) and (__, __).

**3.** $x + y = -1$

| x | y | Ordered Pair (x, y) |
|---|---|---|
| 0 | | |
| 1 | | |
| 2 | | |

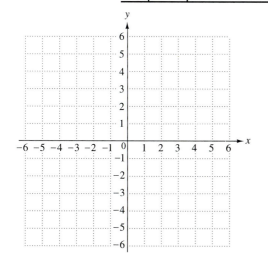

Two other solutions are (__, __) and (__, __).

**4.** $x + y = 1$

| x | y | Ordered Pair (x, y) |
|---|---|---|
| 0 | | |
| 1 | | |
| 2 | | |

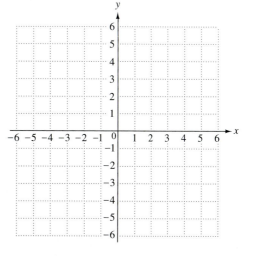

Two other solutions are (__, __) and (__, __).

**5.** The line in Exercise 1 crosses the *y*-axis at what point? _____. The line in Exercise 3 crosses the *y*-axis at what point? _____. Based on these examples, where would the graph of $x + y = -6$ cross the *y*-axis? Where would the graph of $x + y = 99$ cross the *y*-axis?

**6.** Look at where the line crosses the *x*-axis and where it crosses the *y*-axis in Exercises 2 and 4. What pattern do you see?

*Graph each equation. Make your own table using the listed values of x. See Examples 2 and 4.*

**7.** $y = x - 2$

Use 1, 2, and 3 as the values of *x*.

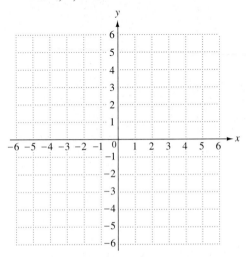

**8.** $y = x + 1$

Use 1, 2, and 3 as the values of *x*.

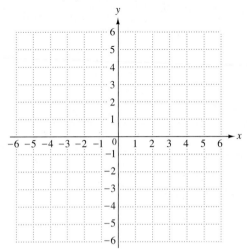

**9.** $y = x + 2$

Use 0, $-1$, and $-2$ as the values of *x*.

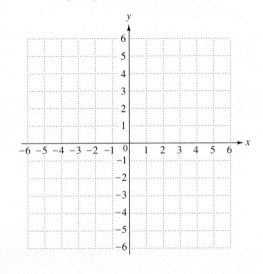

**10.** $y = x - 1$

Use 0, $-1$, and $-2$ as the values of *x*.

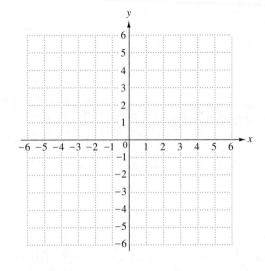

**11.** $y = -3x$

Use 0, 1, and 2 as the values of $x$.

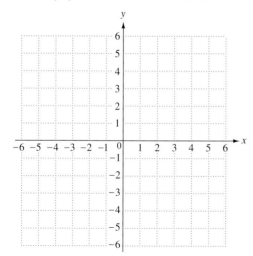

**12.** $y = -2x$

Use 0, 1, and 2 as the values of $x$.

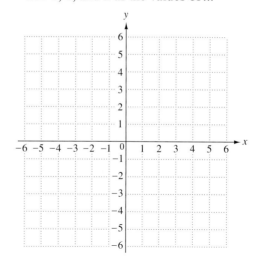

**13.** Look back at the graphs in Exercises 1, 3, 7, 9, and 11. Which lines have a positive slope? Which lines have a negative slope?

**14.** Look back at the graphs in Exercises 2, 4, 8, 10, and 12. Which lines have a positive slope? Which lines have a negative slope?

*Graph each equation. Make your own table using the listed values of x. See Examples 1–4.*

**15.** $y = \dfrac{1}{3}x$

Use 0, 3, and 6 as the values of $x$.

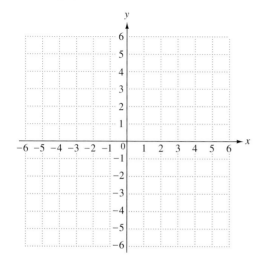

**16.** $y = \dfrac{1}{2}x$

Use 0, 2, and 4 as the values of $x$.

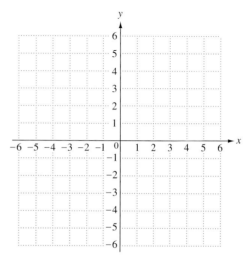

**17.** $y = x$

Use $-1, -2,$ and $-3$ as the values of $x$.

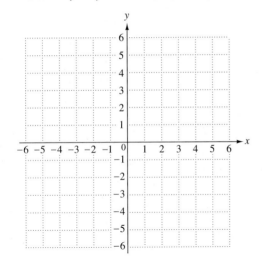

**18.** $x + y = 0$

Use $1, 2,$ and $3$ as the values of $x$.

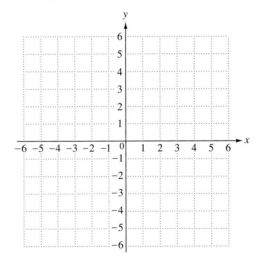

**19.** $y = -2x + 3$

Use $0, 1,$ and $2$ as the values of $x$.

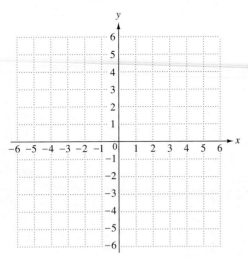

**20.** $y = 3x - 4$

Use $0, 1,$ and $2$ as the values of $x$.

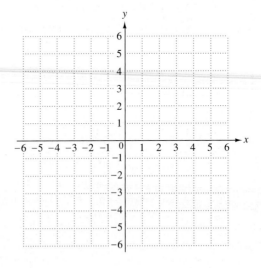

*Graph each equation. Choose three values for x. Make a table showing your x values and the corresponding y values. After graphing the equation, state whether the line has a positive or negative slope. See Examples 1–5.*

**21.** $x + y = -3$

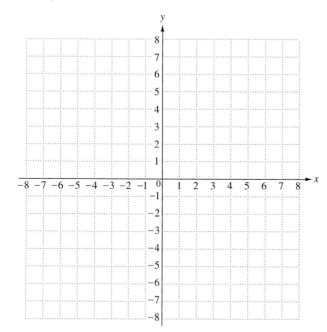

**22.** $x + y = 2$

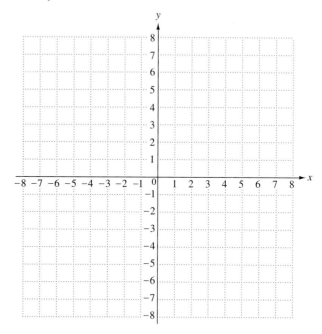

**23.** $y = \dfrac{1}{4}x$ (*Hint:* Try using multiples of 4 as the values of *x*.)

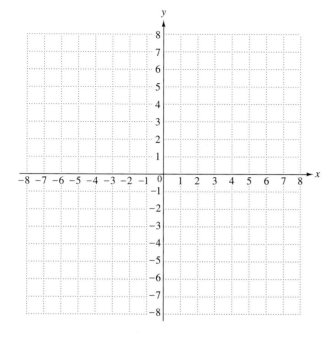

**24.** $y = -\dfrac{1}{3}x$ (*Hint:* Try using multiples of 3 as the values of *x*.)

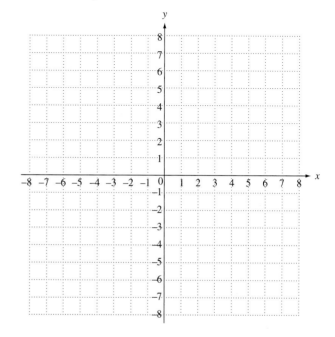

**25.** $y = x - 5$

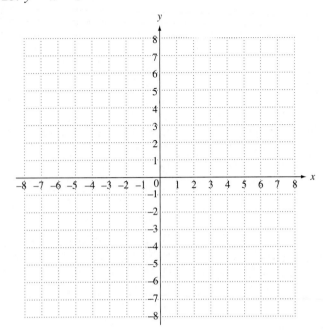

**26.** $y = x + 4$

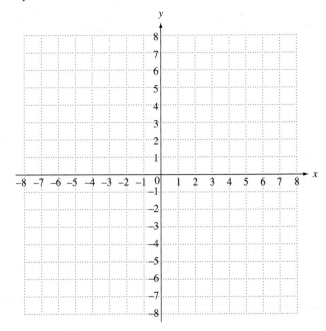

**27.** $y = -3x + 1$

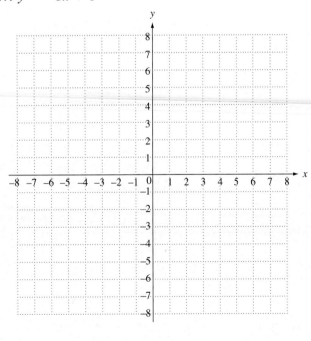

**28.** $y = 2x - 2$

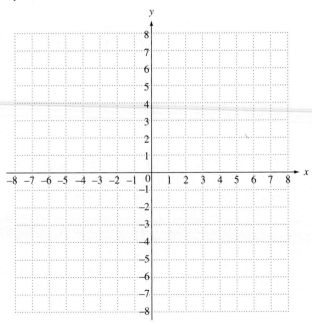

# Chapter 9

## SUMMARY

### KEY TERMS

**9.1** **table** — A table presents data organized into rows and columns.

**pictograph** — A pictograph uses symbols or pictures to represent various amounts.

**9.2** **circle graph** — A circle graph shows how a total amount is divided into parts or sectors. It is based on percents of 360°.

**protractor** — A protractor is a tool (usually in the shape of a half-circle) used to measure the number of degrees in an angle or parts of a circle.

**9.3** **bar graph** — A bar graph uses bars of various heights to show quantity or frequency.

**double-bar graph** — A double-bar graph compares two sets of data by showing two sets of bars.

**line graph** — A line graph uses dots connected by lines to show trends.

**comparison line graph** — A comparison line graph shows how two or more sets of data relate to each other by showing a line graph for each set of data.

**9.4** **paired data** — When each number in a set of data is matched with another number by some rule of association, we call it paired data.

**horizontal axis** — The horizontal axis is the number line in a coordinate system that goes "left and right."

**vertical axis** — The vertical axis is the number line in a coordinate system that goes "up and down."

**$x$-axis** — The horizontal axis is called the $x$-axis.

**$y$-axis** — The vertical axis is called the $y$-axis.

**coordinate system** — Together, the $x$-axis and the $y$-axis form a rectangular coordinate system. *Example:* See figure at the right.

**origin** — The center point of a rectangular coordinate system is $(0, 0)$ and is called the origin. *Example:* See red dot marking $(0, 0)$ in figure at the right.

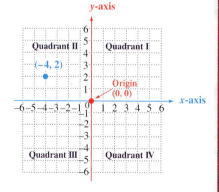

**coordinates** — Coordinates are the numbers in the ordered pair that specify the location of a point on a rectangular coordinate system. *Example:* See the point $(-4, 2)$ at the right.

**quadrants** — The $x$-axis and the $y$-axis divide the coordinate system into four regions called quadrants; they are designated with Roman numerals. *Example:* The point $(-4, 2)$ at the right is in quadrant II.

**9.5** **graph a linear equation** — All the solutions of a linear equation (all the ordered pairs that satisfy the equation) lie along a straight line. When you draw the line, you have graphed the equation.

## Test Your Word Power

*See how well you have learned the vocabulary in this chapter. Answers follow the Quick Review.*

1. A **circle graph**
   (a) uses symbols to represent various amounts
   (b) is useful for showing trends
   (c) shows how a total amount is divided into parts
   (d) compares two sets of data.

2. A **line graph**
   (a) uses symbols to represent various amounts
   (b) is useful for showing trends
   (c) shows how a total amount is divided into parts
   (d) compares two sets of data.

3. A **protractor** is used to
   (a) measure degrees in an angle
   (b) draw circles of various sizes
   (c) measure lengths
   (d) calculate circumference.

4. Two sets of data can be compared using a
   (a) circle graph
   (b) pictograph
   (c) line graph
   (d) double-bar graph.

5. The **x-axis** in a rectangular coordinate system is the
   (a) number line that goes "up and down"
   (b) number line that goes "left and right"
   (c) center point of the grid
   (d) vertical axis.

6. **Quadrants** are the
   (a) numbers in an ordered pair
   (b) solutions of a linear equation
   (c) four regions in a coordinate system
   (d) paired data shown on a graph.

7. **Coordinates** are
   (a) points in a straight line on a coordinate system
   (b) designated with Roman numerals
   (c) the solutions of a linear equation
   (d) numbers used to locate a point in a coordinate system.

8. A **rectangular coordinate system**
   (a) is formed by the $x$-axis and $y$-axis
   (b) is divided into eight quadrants
   (c) is the solution of a linear equation
   (d) has only positive numbers.

## Quick Review

| Concepts | Examples |
|---|---|
| **9.1 Reading a Table**<br>The data in a table is organized into rows and columns. As you read from left to right along each row, check the heading at the top of each column. | Use the table below to answer the question.<br><br>**PER CAPITA CONSUMPTION OF SELECTED BEVERAGES IN GALLONS**<br><br>|  | 1980 | 1985 | 1990 | 1995 |<br>|---|---|---|---|---|<br>| Milk | 27.6 | 26.7 | 25.7 | 24.3 |<br>| Coffee | 26.7 | 27.4 | 26.9 | 20.5 |<br>| Bottled water | 2.4 | 4.5 | 8.0 | 11.6 |<br>| Soft drinks | 35.1 | 35.7 | 46.3 | 51.6 |<br><br>*Source:* U.S. Dept. of Agriculture Economic Research Service.<br><br>In which years was the consumption of coffee greater than the consumption of milk?<br><br>Read across the rows labeled "milk" and "coffee" from left to right, comparing the numbers in each column. In 1985 and 1990, the figure for coffee is greater than for milk. |

| Concepts | Examples |
|---|---|
| **9.1 Reading a Pictograph**<br>A pictograph uses symbols or pictures to represent various amounts. The *key* tells you how much each symbol represents. A fractional part of a symbol represents a fractional part of the symbol's value. | Use the pictograph to answer these questions.<br>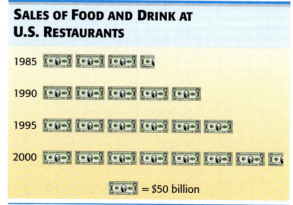<br>**Source:** National Restaurant Association.<br><br>**(a)** How much was spent at U.S. restaurants in 1985?<br><br>Sales for 1985 are represented by 3 whole symbols (3 • $50 billion = $150 billion) plus half of a symbol ($\frac{1}{2}$ of $50 billion = $25 billion) for a total of $175 billion.<br><br>**(b)** How much did restaurant expenditures increase from 1985 to 2000?<br><br>2000 shows four more symbols than 1985, so the increase is 4 • $50 billion = $200 billion. |

## Quick Review

### Concepts

**9.2 Constructing a Circle Graph**

*Step 1* Determine the percent of the total for each item.

*Step 2* Find the number of degrees out of 360° that each percent represents.

*Step 3* Use a protractor to measure the number of degrees for each item in the circle.

### Examples

Construct a circle graph for the following table, which lists expenses for a business trip.

| Item | Amount |
|---|---|
| Transportation | $200 |
| Lodging | $300 |
| Food | $250 |
| Entertainment | $150 |
| Other | $100 |
| Total | **$1000** |

| Item | Amount | Percent of Total | Sector Size |
|---|---|---|---|
| Transportation | $200 | $\frac{\$200}{\$1000} = \frac{1}{5} =$ **20%** so  20% • 360° = (0.20)(360) | = 72° |
| Lodging | $300 | $\frac{\$300}{\$1000} = \frac{3}{10} =$ **30%** so  30% • 360° = (0.30)(360) | = 108° |
| Food | $250 | $\frac{\$250}{\$1000} = \frac{1}{4} =$ **25%** so  25% • 360° = (0.25)(360) | = 90° |
| Entertainment | $150 | $\frac{\$150}{\$1000} = \frac{3}{20} =$ **15%** so  15% • 360° = (0.15)(360) | = 54° |
| Other | $100 | $\frac{\$100}{\$1000} = \frac{1}{10} =$ **10%** so  10% • 360° = (0.10)(360) | = 36° |

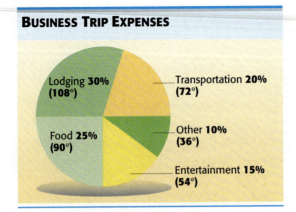

**BUSINESS TRIP EXPENSES**

Lodging 30% (108°)
Transportation 20% (72°)
Food 25% (90°)
Other 10% (36°)
Entertainment 15% (54°)

| Concepts | Examples |
|---|---|
| **9.3 Reading a Bar Graph**<br>The height of the bar is used to show the quantity or frequency (number) in a specific category. Use a ruler or straight edge to line up the top of each bar with the numbers on the left side of the graph. | Use the bar graph below to determine the number of students who earned each letter grade.<br>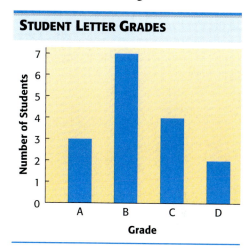<br><br>\| Grade \| Number of Students \|<br>\|---\|---\|<br>\| A \| 3 \|<br>\| B \| 7 \|<br>\| C \| 4 \|<br>\| D \| 2 \| |
| **9.3 Reading a Line Graph**<br>A dot is used to show the number or quantity in a specific class. The dots are connected with lines. This kind of graph is used to show a trend. | The line graph below shows the annual sales for the Fabric Supply Center for each of four years.<br>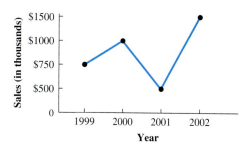<br>Find the sales in each year.<br><br>\| Year \| Total Sales \|<br>\|---\|---\|<br>\| 1999 \| $750 • 1000 = $ 750,000 \|<br>\| 2000 \| $1000 • 1000 = $1,000,000 \|<br>\| 2001 \| $500 • 1000 = $ 500,000 \|<br>\| 2002 \| $1500 • 1000 = $1,500,000 \| |

| Concepts | Examples |
|---|---|
| **9.4 Plotting Points**<br>Start at the center of the coordinate system (the origin). The first number in an ordered pair tells you how far to move *left* or *right* along the horizontal axis; *positive* numbers are to the *right*, *negative* numbers to the *left*. The second number in an ordered pair tells you how far to move *up* or *down*; *positive* numbers are *up*, *negative* numbers are *down*. | To plot $(3, -2)$, move to the *right* 3 units and then move *down* 2 units.<br><br>To plot $(-2, 3)$, move to the *left* 2 units and then move *up* 3 units.<br>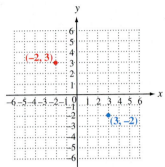 |
| **9.4 Identifying Quadrants**<br>The *x*-axis and *y*-axis divide the coordinate system into four regions called quadrants. The quadrants are designated with Roman numerals.<br><br>Points in the first quadrant fit the pattern $(+, +)$.<br><br>Points in the second quadrant fit the pattern $(-, +)$.<br><br>Points in the third quadrant fit the pattern $(-, -)$.<br><br>Points in the fourth quadrant fit the pattern $(+, -)$. | 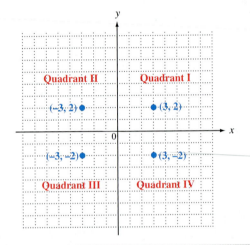 |
| **9.5 Graphing Linear Equations**<br>Choose any three values for *x*. Then find the corresponding values of *y*. Plot the three ordered pairs on a coordinate system. Draw a line through the points, extending it in both directions. If the three points do *not* lie on a straight line, there is an error in your work. Every point on the line is a solution of the given equation. | Graph $y = -2x$.<br><br>| x | y (must be −2 times x) | Ordered Pair (x, y) |<br>|---|---|---|<br>| 1 | $-2 \cdot 1$ is $-2$ | $(1, -2)$ |<br>| 2 | $-2 \cdot 2$ is $-4$ | $(2, -4)$ |<br>| 3 | $-2 \cdot 3$ is $-6$ | $(3, -6)$ |<br><br>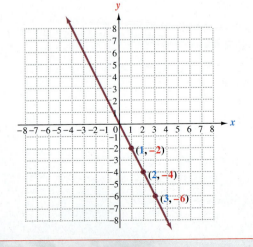 |

**ANSWERS TO TEST YOUR WORD POWER**

**1. (c)** *Example:* A circle graph can show how a 24-hour day is divided among various activities.
**2. (b)** *Example:* A line graph can show changes in the amount of sales over several months or several years.
**3. (a)** *Example:* A protractor is a tool, usually in the shape of a half circle, used to measure or draw angles of a certain number of degrees. A tool called a compass is used to draw circles of various sizes.
**4. (d)** *Example:* Two sets of bars in different colors can compare monthly unemployment figures for two different years.
**5. (b)** *Example:* The $x$-axis is the horizontal number line with 0 at the center, negative numbers extending to the left, and positive numbers extending to the right.
**6. (c)** *Example:* The $x$-axis and the $y$-axis divide the coordinate system into four regions; each region is designated by a Roman numeral.
**7. (d)** *Example:* To locate the point $(2, -3)$, start at the origin and move 2 units to the right along the $x$-axis, then move down 3 units.
**8. (a)** *Example:* Together, a horizontal number line ($x$-axis) and vertical number line ($y$-axis) form a rectangular coordinate system.

# Focus on Real-Data Applications

## Surfing the Net

1. Look at the "Source" information at the bottom of the graph. How were the numbers in the graph obtained?

2. A researcher seeks information about members of a *population*. The individuals who are polled must be representative of the *population*. Describe the population that was targeted by this survey.

3. How many people in the poll said they cut back on television viewing to find time to use the Internet?

4. Find the number of people in the poll who cut back on each of the other activities listed in the graph.

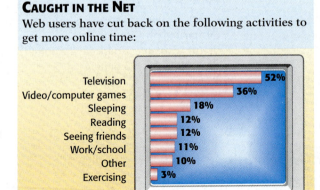

**CAUGHT IN THE NET**
Web users have cut back on the following activities to get more online time:

- Television 52%
- Video/computer games 36%
- Sleeping 18%
- Reading 12%
- Seeing friends 12%
- Work/school 11%
- Other 10%
- Exercising 3%

*Note:* Respondents could choose more than one activity.

**Source:** NUKE InterNETWORK poll of 500 regular users.

5. Add up all the responses to the poll from Problem 4. Why is the total more than the 500 people that were in the poll?

6. Suppose you took a similar poll of 100 students at your school. Would you expect the results to be similar to those shown in the graph? Why or why not?

7. Suppose you first asked students if they regularly used the Internet, and then took a similar poll of 100 of those students. Would you expect the results to be similar to those shown in the graph? Why or why not?

8. Conduct a survey of your class members. First find out if they regularly use the Internet. Ask those who regularly use the Internet which of the activities they cut back on to have more surfing time. Each person polled can select more than one activity.

   (a) How many students are in your class poll?

   (b) Complete the table using the responses from those who regularly use the Internet.

| Activity | Number Who Cut Back on the Activity | Percent Who Cut Back on the Activity |
|---|---|---|
| Television | | |
| Video/computer games | | |
| Sleeping | | |
| Reading | | |
| Seeing friends | | |
| Work/school | | |
| Other | | |
| Exercising | | |

9. Make a bar graph showing your survey data. How is your data similar to the graph shown above? How is it different?

# Chapter 9 Review Exercises

**[9.1]** *Use the table at the right to answer Exercises 1–4.*

1. Which sport had the
   - (a) fewest men's teams?
   - (b) second greatest number of women's teams?

2. Which sport had
   - (a) about 10,000 female athletes?
   - (b) about 7000 male athletes?

### PARTICIPATION IN SELECTED NCAA SPORTS

| Sport | Males | | | Females | | |
| --- | --- | --- | --- | --- | --- | --- |
| | Teams | Athletes | Average Squad | Teams | Athletes | Average Squad |
| Basketball | 950 | 15,141 | 15.9 | 966 | 13,392 | 13.9 |
| Cross Country | 792 | 10,271 | 13.0 | 838 | 10,141 | 12.1 |
| Golf | 678 | 7197 | 10.6 | 282 | 2323 | 8.2 |
| Gymnastics | 28 | 413 | 14.8 | 91 | 1311 | 14.4 |
| Volleyball | 74 | 1052 | | 923 | 12,284 | |

*Source:* The National Collegiate Athletic Association (NCAA).

3. (a) How many more men participated in basketball than cross country?
   (b) How many fewer women participated in gymnastics than basketball?

4. Find the average squad size for men's and women's volleyball teams and complete the table. Round to the nearest tenth.

*Use the pictograph to answer Exercises 5–8.*

5. What is the average yearly snowfall in
   - (a) Juneau?
   - (b) Washington, D.C.?

6. What is the average yearly snowfall in
   - (a) Minneapolis?
   - (b) Cleveland?

7. Find the difference in the average yearly snowfall between
   - (a) Buffalo and Cleveland.
   - (b) Memphis and Minneapolis.

8. Find the difference in the average yearly snowfall between the city with the greatest amount and the city with the least amount.

**AVERAGE YEARLY SNOWFALL IN SELECTED U.S. CITIES**

*Source:* National Climatic Data Center.

**[9.2]** *Use this circle graph for Exercises 9–13.*

9. (a) What was the largest single expense of the vacation? How much was that item?

   (b) What was the second most expensive item? How much was that item?

   (c) What was the total cost of the vacation?

*Find each ratio. Write the ratio as a fraction in lowest terms.*

10. Cost of food to the total cost of the vacation

11. Cost of gasoline to the total cost of the vacation

12. Cost of sightseeing to the total cost of the vacation

13. Cost of gasoline to the cost of the *other* category

**[9.3]** This bar graph shows recent trends in employee benefits. It includes the most frequently offered "work perks" and the percent of the responding companies offering them. The survey was conducted on-line and included 4800 companies ranging in size from 2 to 5000 employees. Use this graph to find the number of companies offering each work perk listed in Exercises 14–17 and to answer Exercises 18 and 19. (Source: Work Perks Survey, Ceridian Employer Services, www.ces.ceridian.com)

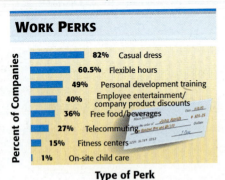

14. Casual dress

15. Free food/beverages

16. Fitness centers

17. Flexible hours

18. Which two work perks do companies offer least often? Give one possible explanation why these work perks are not offered.

19. Which two work perks do companies offer most often? Give one possible explanation why these work perks are so popular.

*This double-bar graph shows the number of acre-feet of water in Lake Natoma for each of the first six months of 2001 and 2002. Use this graph to answer Exercises 20–25.*

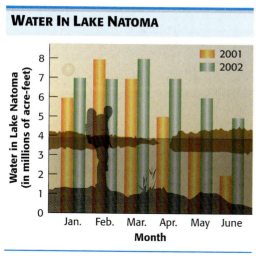

**20.** During which month in 2002 was the greatest amount of water in the lake? How much was there?

**21.** During which month in 2001 was the least amount of water in the lake? How much was there?

**22.** How many acre-feet of water were in the lake in June of 2002?

**23.** How many acre-feet of water were in the lake in January of 2001?

**24.** Find the decrease in the amount of water in the lake from March 2001 to June 2001.

**25.** Find the decrease in the amount of water in the lake from April 2002 to June 2002.

*This comparison line graph shows the annual floor-covering sales of two different home improvement centers during each of five years. Use this graph to find the amount of sales in each year listed in Exercises 26–29 and to answer Exercises 30 and 31.*

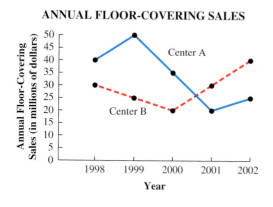

**26.** Center A in 1999

**27.** Center A in 2001

**28.** Center B in 2000

**29.** Center B in 2002

**30.** What trend do you see in Center A's sales from 1999 to 2002? Why might this have happened?

**31.** What trend do you see in Center B's sales starting in 2000? Why might this have happened?

**[9.2]** *The Broadway Hair Salon spent a total of* $22,400 *to open a new shop. The breakdown of expenditures for various items is shown. Find all the missing numbers in Exercises 32–36.*

| Item | Dollar Amount | Percent of Total | Degrees of Circle |
|---|---|---|---|
| 32. Plumbing and electrical changes | $2240 | 10% | _____ |
| 33. Work stations | $7840 | _____ | _____ |
| 34. Small appliances | $4480 | _____ | _____ |
| 35. Interior decoration | $5600 | _____ | _____ |
| 36. Supplies | _____ | _____ | 36° |

**37.** Draw a circle graph by using the information in Exercises 32–36.

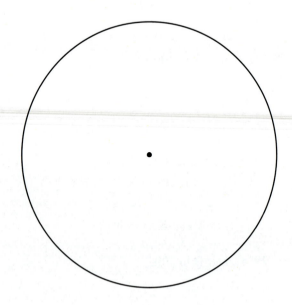

**[9.4]** *In Exercise 38, plot each point on the rectangular coordinate system and label it with its coordinates. In Exercise 39, give the coordinates of each point.*

**38.** $(1, 7)$  $\left(-1\frac{1}{2}, 0\right)$  $(-4, -2)$  $(0, 3)$  $\left(2, -\frac{1}{2}\right)$
$(-7, 1)$  $(5, -5)$  $(0, -4)$

**39.**

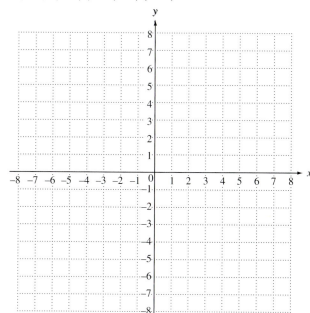
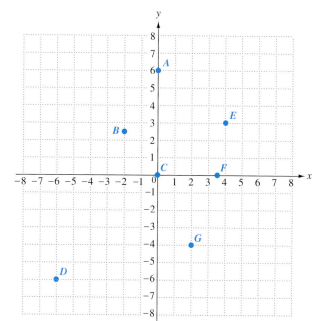

**[9.5]** *Graph each equation. Complete the table using the listed values of x. State whether each line has a positive or negative slope. Use the graph to find two other solutions of the equation.*

**40.** $x + y = -2$

Use 0, 1, and 2 as the values of $x$.

| x | y | (x, y) |
|---|---|--------|
|   |   |        |
|   |   |        |
|   |   |        |

The graph of $x + y = -2$ has a _____ slope.

Two other solutions of $x + y = -2$ are (____, ____) and (____, ____).

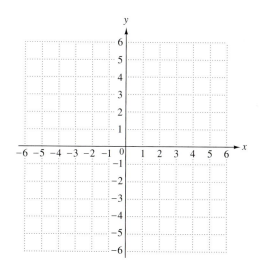

**41.** $y = x + 3$

Use 0, 1, and 2 as the values of $x$.

| x | y | (x, y) |
|---|---|--------|
|   |   |        |
|   |   |        |
|   |   |        |

The graph of $y = x + 3$ has a _____ slope.

Two other solutions for $y = x + 3$ are
(____, ____) and (____, ____).

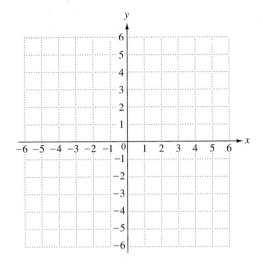

**42.** $y = -4x$

Use $-1$, 0, and 1 as the values of $x$.

| x | y | (x, y) |
|---|---|--------|
|   |   |        |
|   |   |        |
|   |   |        |

The graph of $y = -4x$ has a _____ slope.

Two other solutions of $y = -4x$ are
(____, ____) and (____, ____).

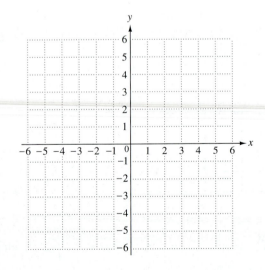

# Chapter 9 TEST

*Study Skills Workbook* 
Activity 12

*Use this table to answer Problems 1–3.*

**AMOUNT OF CALCIUM IN SELECTED FOODS**

| Food | Measure | Calories | Calcium (mg) |
|---|---|---|---|
| Swiss cheese | 1 oz | 95 | 219 |
| Cream cheese | 1 oz | 100 | 23 |
| Yogurt, fruit flavor | 8 oz | 230 | 345 |
| Skim milk | 1 cup | 85 | 302 |
| Sardines | 3 oz | 175 | 371 |

*Source:* U.S. Department of Agriculture.

1. Which food has **(a)** the greatest amount of calcium? **(b)** the least amount of calcium?

    1. (a) _____
    (b) _____

2. A packet of Swiss cheese contains 1.75 oz. How many calories are in that amount of Swiss cheese? Round to the nearest whole number.

    2. _____

3. Find the amount of calcium in a 6 oz container of fruit-flavored yogurt. Round to the nearest whole number.

    3. _____

*Use this pictograph to answer Problems 4–6.*

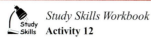

**NUMBER OF U.S. ENDANGERED WILDLIFE SPECIES**

Mammals: 🐺🐺🐺🐺🐺🐺
Birds: 🐺🐺🐺🐺🐺🐺🐺🐺
Reptiles: 🐺🐺
Fishes: 🐺🐺🐺🐺🐺🐺🐺🐺
Insects: 🐺🐺🐺

🐺 = 10 Species

*Source:* U.S. Fish and Wildlife Service.

4. **(a)** How many species of insects are endangered?
   **(b)** How many species of birds are endangered?

    4. (a) _____
    (b) _____

5. How many more species of mammals are endangered than reptiles?

    5. _____

6. What is the total number of endangered species shown in the pictograph?

    6. _____

*This circle graph shows the advertising budget for Lakeland Amusement Park. Find the dollar amount budgeted for each category listed in Exercises 7–10. The total advertising budget is $2,800,000.*

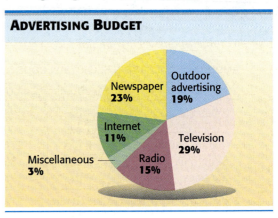

7. Which form of advertising has the largest budget? What amount is this?

8. Which form of advertising has the smallest budget? How much is this?

9. How much is budgeted for Internet advertising?

10. How much is budgeted for newspaper ads?

*This graph shows one student's income and expenses for four years.*

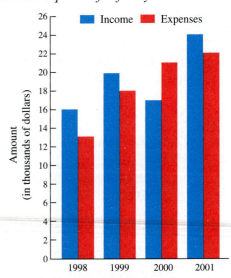

11. In what year did the student's expenses exceed income? By how much?

12. How much did the student's expenses increase from 1998 to 2001?

13. In what year did the student's income decline? Give two possible explanations for the decline.

*This graph shows enrollment at two community colleges.*

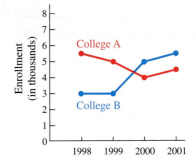

14. What was College A's enrollment in 1998? What was College B's enrollment in 1999?

15. Which college had higher enrollment in 2001? How much higher?

16. Give two possible explanations for the fact that College B's enrollment passed College A's.

During a one-year period, Oak Mill Furniture Sales had a total of $480,000 in expenses. Find all the numbers missing from the table.

| Item | Dollar Amount | Percent of Total | Degrees of a Circle | |
|---|---|---|---|---|
| 17. Salaries | $144,000 | _____ | _____ | 17. _____ |
| 18. Delivery expense | $48,000 | _____ | _____ | 18. _____ |
| 19. Advertising | $96,000 | _____ | _____ | 19. _____ |
| 20. Rent | $144,000 | _____ | _____ | 20. _____ |
| 21. Other | _____ | _____ | 36° | 21. _____ |

22. Draw a circle graph using a protractor and the information in Problems 17–21.

22.

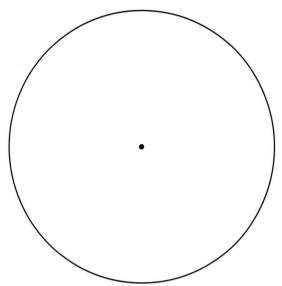

*Plot each point on the coordinate system below. Label each point with its coordinates.*

23. $(-5, 3)$

24. $(1, -4)$

25. $(0, 6)$

26. $(2, 0)$

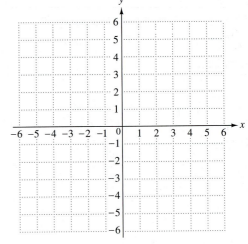

*Give the coordinates of each lettered point shown below, and state which quadrant the point is in.*

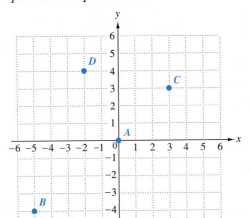

27. Point A

28. Point B

29. Point C

30. Point D

31. Graph $y = x - 4$. Make your own table using 0, 1, and 2 as the values of $x$.

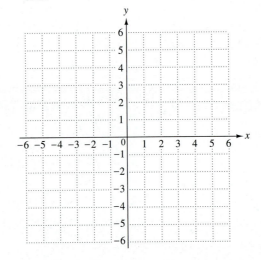

32. Use the graph in Problem 31 to find *two* other solutions of $y = x - 4$.

# Cumulative Review Exercises  Chapters 1–9

1. Write these numbers in words.
   (a) 0.0602
   (b) 300,000,560

2. Write these numbers in digits.
   (a) Seventy billion, five million, forty-three
   (b) Eighteen and nine hundredths

3. Round each number as indicated.
   (a) 3.049 to the nearest tenth
   (b) 0.7982 to the nearest hundredth
   (c) 68,592,000 to the nearest million

4. Find the mean, median, and mode for this set of data on student ages in years: 23, 29, 18, 23, 36, 62, 23, 19, 27, 30.

*Simplify.*

5. $\dfrac{4}{5} + 2\dfrac{1}{3}$

6. $\dfrac{0.8}{-3.2}$

7. $5^2 + (-4)^3$

8. $(0.002)(-0.05)$

9. $\dfrac{4a}{9} \cdot \dfrac{6b}{2a^3}$

10. $1\dfrac{1}{4} - 3\dfrac{5}{6}$

11. $-13 + 2.993$

12. $\dfrac{2}{7} - \dfrac{8}{x}$

13. $-3 - 33$

14. $\dfrac{10m}{3n^2} \div \dfrac{2m^2}{5n}$

15. $\dfrac{-3}{-\dfrac{9}{10}}$

16. $\dfrac{3(-7)}{2^4 - 16}$

17. $10 - 2\dfrac{5}{8}$

18. $\dfrac{3}{w} + \dfrac{x}{6}$

19. $8 + 4(2 - 5)$

20. $\dfrac{7}{8}$ of 960

21. $\dfrac{(-4)^2 + 8(0 - 2)}{8 \div 2(-3 + 5) - 10}$

22. $0.5 - 0.25(3.2)^2$

23. $6\left(-\dfrac{1}{2}\right)^3 + \dfrac{2}{3}\left(\dfrac{3}{5}\right)$

*Evaluate each expression when $a = 2$, $b = 3$, and $c = -5$.*

**24.** $20 + 4c$

**25.** $7b - 4c$

**26.** $-2ac^2$

*Simplify each expression.*

**27.** $-4x + x^2 - x$

**28.** $3(y - 4) + 2y$

**29.** $-5(8h^2)$

*Solve each equation. Show your work.*

**30.** $2x - 3 = -20 + 3$

**31.** $-12 = 3(y + 2)$

**32.** $6x = 14 - x$

**33.** $-8 = \dfrac{2}{3}m + 2$

**34.** $\dfrac{2}{13.5} = \dfrac{2.4}{n}$

**35.** $-20 = -w$

**36.** $3.4x - 6 = 8 + 1.4x$

**37.** $2(h - 1) = -3(h + 12) - 11$

*Translate each sentence into an equation and solve it.*

**38.** If five times a number is subtracted from 12, the result is the number. Find the number.

**39.** When $-8$ is added to twice a number, the result is $-28$. What is the number?

*Solve each application problem using the six problem-solving steps from Chapter 3.*

**40.** An $1800 lottery prize is to be split between two people so that one person gets $500 more than the other. How much will each person receive?

**41.** The length of a rectangular movie theater is three times the width. The perimeter is 280 ft. Find the width and the length.

*Find the unknown length, perimeter, circumference, area, or volume. When necessary, use 3.14 as the approximate value of $\pi$, and round answers to the nearest tenth.*

**42.** Find the perimeter and area.

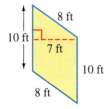

**43.** Find the perimeter and area.

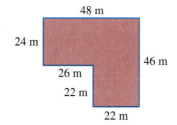

**44.** Find the unknown length, the perimeter, and the area.

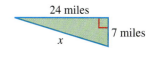

**45.** A circular hot tub has a diameter of 6 ft. Find its circumference and the area of the tub's floor.

**46.** A cylindrical oil tank is 12 m tall and has a diameter of 20 m. Find its volume.

**47.** Find the volume of a storage shed that is 4 yd long, 3 yd wide, and $2\frac{1}{2}$ yd high.

*Convert each measurement using unit fractions, the metric conversion line, or the temperature conversion formulas.*

**48.** $2\frac{1}{4}$ hours to minutes

**49.** 54 in. to feet

**50.** 1.85 L to milliliters

**51.** 35 mm to centimeters

**52.** 10 g to kilograms

**53.** 25 °F to Celsius; round to the nearest whole degree.

*Solve each application problem.*

**54.** A survey found that 19 out of 25 adults are nonsmokers. If Mathtronic has 732 employees, how many would be expected to be nonsmokers? Round to the nearest whole number.

**55.** Esther bought a cellular phone at a 15% discount. The regular price was $129. She also paid $6\frac{1}{2}$% sales tax. What was her total cost for the phone, to the nearest cent?

**56.** Abiola earned 167 points out of 180 on the prealgebra final exam. What percent of the points did she earn, to the nearest tenth?

**57.** Century College received a $30,000 technology grant.

(a) How many $957 computers were they able to buy with the grant money? How much was left over?

(b) How many $19 calculators could they buy with the money left over from the computer purchase? How much was left then?

**58.** Wayne bought $2\frac{1}{4}$ pounds of lunch meat. He used $\frac{1}{6}$ pound of meat on each of five sandwiches. How much meat is left?

**59.** Plot the following points and label each one with its coordinates. Which of the points are in the second quadrant? Which are in the third quadrant?

$(3, -5)$  $(0, -4)$  $(-2, 3\frac{1}{2})$  $(1, 0)$  $(4, 2)$

**60.** Graph $y = x + 6$. Use 0, $-1$, and $-2$ as the values of $x$. After graphing the equation, state whether the line has a positive or negative slope, and find two other solutions of $y = x + 6$.

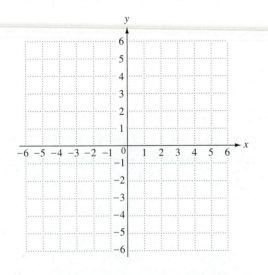

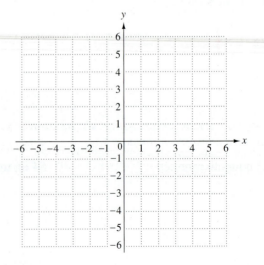

# Exponents and Polynomials

The expression $100t - 13t^2$ gives the distance in feet that a car going approximately 68 mph will skid in $t$ seconds. This expression in $t$ is an example of a *polynomial*. Accident investigators use polynomials like this to determine the length of a skid or the elapsed time during a skid. (See Section 10.4, Exercises 65–68, where we use this polynomial to approximate skidding distance.)

**10.1** The Product Rule and Power Rules for Exponents

**10.2** Integer Exponents and the Quotient Rule

**10.3** An Application of Exponents: Scientific Notation

**10.4** Adding and Subtracting Polynomials

**10.5** Multiplying Polynomials: An Introduction

## 10.1 THE PRODUCT RULE AND POWER RULES FOR EXPONENTS

**OBJECTIVES**

1. Review the use of exponents.
2. Use the product rule for exponents.
3. Use the rule $(a^m)^n = a^{mn}$.
4. Use the rule $(ab)^m = a^m b^m$.
5. Use the rule $\left(\dfrac{a}{b}\right)^m = \dfrac{a^m}{b^m}$.

① Write $2 \cdot 2 \cdot 2 \cdot 2$ in exponential form, and find its value.

② Evaluate each exponential expression. Name the base and the exponent.

(a) $(-2)^5$  (b) $-2^5$

(c) $-4^2$  (d) $(-4)^2$

**ANSWERS**
1. $2^4 = 16$
2. (a) $-32; -2; 5$  (b) $-32; 2; 5$
   (c) $-16; 4; 2$  (d) $16; -4; 2$

---

**1** **Review the use of exponents.** In **Section 1.8** we used exponents to write repeated products. Recall that in the expression $5^2$, the number 5 is called the *base* and 2 is called the *exponent* or *power*. The expression $5^2$ is written in *exponential form*. Usually we do not write a quantity with an exponent of 1, but sometimes it is convenient to do so. In general, for any quantity $a$, we can write $a = a^1$.

### Example 1  Review of Using Exponents

Write $3 \cdot 3 \cdot 3 \cdot 3 \cdot 3$ in exponential form, and find the value of the exponential expression.

Since 3 occurs as a factor five times, the base is 3 and the exponent is 5. The exponential expression is $3^5$, read "3 to the fifth power" or simply "3 to the fifth." To find the value of $3^5$, actually do the multiplication.

$$3^5 = 3 \cdot 3 \cdot 3 \cdot 3 \cdot 3 = 243$$

**Work Problem** ① **at the Side.**

### Example 2  Evaluating Exponential Expressions

Evaluate each exponential expression. Name the base and the exponent.

|  | Base | Exponent |
|---|---|---|
| (a) $5^4 = 5 \cdot 5 \cdot 5 \cdot 5 = 625$ | 5 | 4 |
| (b) $-5^4 = -(5 \cdot 5 \cdot 5 \cdot 5) = -625$ | 5 | 4 |
| (c) $(-5)^4 = (-5)(-5)(-5)(-5) = 625$ | $-5$ | 4 |

**CAUTION**

It is important to understand the difference between parts (b) and (c) of Example 2. In $-5^4$ the lack of parentheses shows that the exponent 4 applies only to the base 5, and not $-5$. In $(-5)^4$ the parentheses show that the exponent 4 applies to the base $-5$. In summary, $-a^n$ and $(-a)^n$ are *not* always the same. The exponent applies only to what is immediately to the left of it.

| Expression | Base | Exponent | Example |
|---|---|---|---|
| $-a^n$ | $a$ | $n$ | $-3^2 = -(3 \cdot 3) = -9$ |
| $(-a)^n$ | $-a$ | $n$ | $(-3)^2 = (-3)(-3) = 9$ |

**Work Problem** ② **at the Side.**

**2** **Use the product rule for exponents.** To develop the product rule, we use the definition of an exponential expression.

$$2^4 \cdot 2^3 = \overbrace{(2 \cdot 2 \cdot 2 \cdot 2)}^{4 \text{ factors}} \overbrace{(2 \cdot 2 \cdot 2)}^{3 \text{ factors}}$$
$$= \underbrace{2 \cdot 2 \cdot 2 \cdot 2 \cdot 2 \cdot 2 \cdot 2}_{4 \text{ factors} + 3 \text{ factors} = 7 \text{ factors}}$$
$$= 2^7$$

Here is another example.

$$6^2 \cdot 6^3 = (6 \cdot 6)(6 \cdot 6 \cdot 6)$$
$$= 6 \cdot 6 \cdot 6 \cdot 6 \cdot 6$$
$$= 6^5 \quad \leftarrow \text{2 factors + 3 factors = 5 factors}$$

Generalizing from these examples, $2^4 \cdot 2^3 = 2^{4+3} = 2^7$ and $6^2 \cdot 6^3 = 6^{2+3} = 6^5$. In each case, adding the exponents gives the exponent of the product, suggesting the **product rule for exponents.**

### Product Rule for Exponents

If $m$ and $n$ are positive integers, then $a^m \cdot a^n = a^{m+n}$.
(Keep the same base and add the exponents.)

*Example:* $6^2 \cdot 6^5 = 6^{2+5} = 6^7$

### CAUTION

Avoid the common error of multiplying the *bases* when using the product rule.

$$6^2 \cdot 6^5 \neq 36^7$$

Keep the *same* base and add the exponents.

**Example 3** Using the Product Rule

Use the product rule for exponents to find each product, if possible.

(a) $6^3 \cdot 6^5 = 6^{3+5} = 6^8$  by the product rule.
(b) $(-4)^7(-4)^2 = (-4)^{7+2} = (-4)^9$  by the product rule.
(c) $x^2 \cdot x = x^2 \cdot x^1 = x^{2+1} = x^3$   Recall that $x$ can be written as $x^1$.
(d) $m^4 \cdot m^3 = m^{4+3} = m^7$
(e) $2^3 \cdot 3^2$
   The product rule does *not* apply to the product $2^3 \cdot 3^2$ because the *bases are different.*

$$2^3 \cdot 3^2 = (2 \cdot 2 \cdot 2)(3 \cdot 3) = (8)(9) = 72$$

(f) $2^3 + 2^4$
   The product rule does *not* apply to $2^3 + 2^4$ because it is a *sum*, not a *product*.

$$2^3 + 2^4 = (2 \cdot 2 \cdot 2) + (2 \cdot 2 \cdot 2 \cdot 2) = (8) + (16) = 24$$

### CAUTION

The bases must be the same before we can apply the product rule for exponents.

Work Problem ❸ at the Side.

❸ Find each product by the product rule, if possible. Write answers in exponential form.

(a) $8^2 \cdot 8^5$

(b) $(-7)^5 \cdot (-7)^3$

(c) $y^3 \cdot y$

(d) $4^2 \cdot 3^5$

(e) $6^4 + 6^2$

ANSWERS
3. (a) $8^7$  (b) $(-7)^8$  (c) $y^4$
   (d) Cannot use the product rule because the bases are different.
   (e) Cannot use the product rule because it is a sum, not a product.

**704** Chapter 10 Exponents and Polynomials

**4** Multiply.

(a) $5m^2 \cdot 2m^6$

(b) $3p^5 \cdot 9p^4$

(c) $-7p^5 \cdot (3p^8)$

**5** Simplify each expression. Write answers in exponential form.

(a) $(5^3)^4$

(b) $(6^2)^5$

(c) $(3^2)^4$

(d) $(a^6)^5$

---

**Example 4** Using the Product Rule

Multiply $2x^3$ and $3x^7$.

Recall that $2x^3$ means $2 \cdot x^3$ and $3x^7$ means $3 \cdot x^7$. Now use the associative and commutative properties and the product rule.

$$2x^3 \cdot 3x^7 = 2 \cdot 3 \cdot x^3 \cdot x^7 = 6x^{10}$$

**CAUTION**

Be sure you understand the difference between *adding* and *multiplying* exponential expressions. Here is a comparison.

Adding $\quad 8x^3 + 5x^3 = (8 + 5)x^3 = 13x^3$

Multiplying $\quad (8x^3)(5x^3) = (8 \cdot 5)x^{3+3} = 40x^6$

Work Problem **4** at the Side.

**3** Use the rule $(a^m)^n = a^{mn}$. We can simplify an expression such as $(8^3)^2$ with the product rule for exponents, as follows.

$$(8^3)^2 = (8^3)(8^3) = 8^{3+3} = 8^6$$

Notice that the *product* of the original exponents, $3 \cdot 2$, gives 6, the exponent in $8^6$. Here is another example.

$$(5^2)^4 = 5^2 \cdot 5^2 \cdot 5^2 \cdot 5^2 \quad \text{Definition of exponent}$$
$$= 5^{2+2+2+2} \quad \text{Product rule}$$
$$= 5^8$$

Notice that the *product* of the original exponents gives the exponent of the result: $2 \cdot 4 = 8$. These examples suggest **power rule (a) for exponents.**

**Power Rule (a) for Exponents**

If $m$ and $n$ are positive integers, then $(a^m)^n = a^{mn}$.
(Raise a power to a power by multiplying exponents.)

*Example:* $(3^2)^4 = 3^{2 \cdot 4} = 3^8$

**Example 5** Using Power Rule (a)

Use power rule (a) for exponents to simplify each expression. Write answers in exponential form.

(a) $(2^5)^3 = 2^{5 \cdot 3} = 2^{15}$ 

(b) $(5^7)^2 = 5^{7 \cdot 2} = 5^{14}$

(c) $(x^2)^5 = x^{2 \cdot 5} = x^{10}$ 

(d) $(n^3)^2 = n^{3 \cdot 2} = n^6$

Work Problem **5** at the Side.

---

**ANSWERS**

**4.** (a) $10m^8$  (b) $27p^9$  (c) $-21p^{13}$

**5.** (a) $5^{12}$  (b) $6^{10}$  (c) $3^8$  (d) $a^{30}$

**4** Use the rule $(ab)^m = a^m b^m$. We can use the definition of an exponential expression and the commutative and associative properties to develop two more rules for exponents. Here is an example.

$(4x)^3 = (4x)(4x)(4x)$     Definition of exponent
$\phantom{(4x)^3} = 4 \cdot 4 \cdot 4 \cdot x \cdot x \cdot x$     Commutative and associative properties
$\phantom{(4x)^3} = 4^3 x^3$     Definition of exponent

This example suggests **power rule (b) for exponents.**

### Power Rule (b) for Exponents

If $m$ is a positive integer, then $(ab)^m = a^m b^m$.
(Raise a product to a power by raising each factor to the power.)
*Example:* $(2p)^5 = 2^5 p^5$

### Example 6   Using Power Rule (b)

Use power rule (b) to simplify each expression.

(a) $(3xy)^2 = 3^2 x^2 y^2$     Power rule (b)
$\phantom{(3xy)^2} = 9x^2 y^2$     $3^2 = 3 \cdot 3 = 9$

(b) $9(pq)^2 = 9(p^2 q^2)$     Power rule (b); notice that the 9 is *not* inside the parentheses.
$\phantom{9(pq)^2} = 9p^2 q^2$

(c) $5(2m^2 p^3)^4 = 5[2^4 (m^2)^4 (p^3)^4]$     Power rule (b)
$\phantom{5(2m^2 p^3)^4} = 5(2^4 m^8 p^{12})$     Power rule (a)
$\phantom{5(2m^2 p^3)^4} = 5 \cdot 2^4 m^8 p^{12}$     $5 \cdot 2^4 = 5 \cdot 16 = 80$
$\phantom{5(2m^2 p^3)^4} = 80 m^8 p^{12}$

### CAUTION

Power rule (b) *does not* apply to a *sum*.
$(x + 4)^2 \neq x^2 + 4^2$

**6** Simplify.

(a) $5(mn)^3$

(b) $(4ab)^2$

(c) $(3a^2 b^4)^5$

(d) $2(3xy^3)^4$

**Answers**
6. (a) $5m^3 n^3$   (b) $4^2 a^2 b^2 = 16 a^2 b^2$
(c) $3^5 a^{10} b^{20} = 243 a^{10} b^{20}$
(d) $2 \cdot 3^4 x^4 y^{12} = 162 x^4 y^{12}$

Chapter 10 Exponents and Polynomials

**7** Simplify. Assume all variables represent nonzero real numbers.

(a) $\left(\dfrac{5}{2}\right)^4$

(b) $\left(\dfrac{p}{q}\right)^2$

(c) $\left(\dfrac{r}{t}\right)^3$

**5** Use the rule $\left(\dfrac{a}{b}\right)^m = \dfrac{a^m}{b^m}$. Since the quotient $\dfrac{a}{b}$ can be written as $a \cdot \dfrac{1}{b}$, we can use power rule (b), together with some of the properties of real numbers, to get **power rule (c) for exponents.**

### Power Rule (c) for Exponents

If $m$ is a positive integer, then $\left(\dfrac{a}{b}\right)^m = \dfrac{a^m}{b^m}$ $(b \neq 0)$.

(Raise a quotient to a power by raising both the numerator and the denominator to the power.)

*Example:* $\left(\dfrac{5}{3}\right)^2 = \dfrac{5^2}{3^2}$

### Example 7 Using Power Rule (c)

Simplify each expression.

(a) $\left(\dfrac{2}{3}\right)^5 = \dfrac{2^5}{3^5} = \dfrac{32}{243}$

(b) $\left(\dfrac{m}{n}\right)^4 = \dfrac{m^4}{n^4}$, $n \neq 0$    Recall that division by 0 is undefined, so the denominator cannot be equal to 0.

Work Problem **7** at the Side.

Here is a list of the rules for exponents discussed in this section. These rules are basic to the study of algebra and should be *memorized*.

### Rules for Exponents

If $m$ and $n$ are positive integers, then:                     *Examples*

**Product rule**     $a^m \cdot a^n = a^{m+n}$                $6^2 \cdot 6^5 = 6^{2+5} = 6^7$

**Power rule (a)**   $(a^m)^n = a^{mn}$                       $(3^2)^4 = 3^{2 \cdot 4} = 3^8$

**Power rule (b)**   $(ab)^m = a^m b^m$                       $(2p)^5 = 2^5 p^5$

**Power rule (c)**   $\left(\dfrac{a}{b}\right)^m = \dfrac{a^m}{b^m}$ $(b \neq 0)$   $\left(\dfrac{5}{3}\right)^2 = \dfrac{5^2}{3^2}$

ANSWERS

7. (a) $\dfrac{5^4}{2^4} = \dfrac{625}{16}$   (b) $\dfrac{p^2}{q^2}$   (c) $\dfrac{r^3}{t^3}$

## 10.1 Exercises

1. What is the understood exponent for $x$ in the expression $xy^2$?

2. How are the expressions $3^2$, $5^3$, and $7^4$ read?

*Decide whether each statement is true or false. If a statement is false, correct it.*

3. $3^3 = 9$

4. $(-2)^4 = 2^4$

5. $(a^2)^3 = a^5$

6. $\left(\dfrac{1}{4}\right)^2 = \dfrac{1}{4^2}$

*Write each expression using exponents. See Example 1.*

7. $(-2)(-2)(-2)(-2)(-2)$

8. $w \cdot w \cdot w \cdot w \cdot w \cdot w$

9. $\left(\dfrac{1}{2}\right)\left(\dfrac{1}{2}\right)\left(\dfrac{1}{2}\right)\left(\dfrac{1}{2}\right)\left(\dfrac{1}{2}\right)\left(\dfrac{1}{2}\right)$

10. $\left(-\dfrac{1}{4}\right)\left(-\dfrac{1}{4}\right)\left(-\dfrac{1}{4}\right)\left(-\dfrac{1}{4}\right)\left(-\dfrac{1}{4}\right)$

11. $(-8p)(-8p)$

12. $(-7x)(-7x)(-7x)(-7x)$

13. Explain how the expressions $(-3)^4$ and $-3^4$ are different.

14. Explain how the expressions $(5x)^3$ and $5x^3$ are different.

*Identify the base and the exponent for each exponential expression. In Exercises 15–18, also evaluate the expression. See Example 2.*

15. $3^5$

16. $2^7$

17. $(-3)^5$

18. $(-2)^7$

19. $(-6x)^4$

20. $(-8x)^4$

21. $-6x^4$

22. $-8x^4$

23. Explain why the product rule does *not* apply to the expression $5^2 + 5^3$. Then evaluate the expression by finding the individual powers and adding the results.

24. Explain why the product rule does *not* apply to the expression $3^2 \cdot 4^3$. Then evaluate the expression by finding the individual powers and multiplying the results.

*Use the product rule to simplify each expression. Write each answer in exponential form. See Examples 3 and 4.*

25. $5^2 \cdot 5^6$
26. $3^6 \cdot 3^7$
27. $4^2 \cdot 4^7 \cdot 4^3$
28. $5^3 \cdot 5^8 \cdot 5^2$

29. $(-7)^3(-7)^6$
30. $(-9)^8(-9)^5$
31. $t^3 \cdot t^8 \cdot t^{13}$
32. $n^5 \cdot n^6 \cdot n^9$

33. $(-8r^4)(7r^3)$
34. $(10a^7)(-4a^3)$
35. $(-6p^5)(-7p^5)$
36. $(-5w^8)(-9w^8)$

*For each group of terms, first add the given terms. Then start over and multiply them.*

37. $5x^4, 9x^4$
38. $8t^5, 3t^5$
39. $-7a^2, 2a^2, 10a^2$
40. $6x^3, 9x^3, -2x^3$

*Use the power rules for exponents to simplify each expression. Write each answer in exponential form. See Examples 5–7.*

41. $(4^3)^2$
42. $(8^3)^6$
43. $(t^4)^5$
44. $(y^6)^5$

45. $(7r)^3$
46. $(11x)^4$
47. $(5xy)^5$
48. $(9pq)^6$

49. $8(qr)^3$
50. $4(vw)^5$
51. $\left(\dfrac{1}{2}\right)^3$
52. $\left(\dfrac{1}{3}\right)^5$

53. $\left(\dfrac{a}{b}\right)^3 \ (b \neq 0)$
54. $\left(\dfrac{r}{t}\right)^4 \ (t \neq 0)$
55. $\left(\dfrac{9}{5}\right)^8$
56. $\left(\dfrac{12}{7}\right)^3$

57. $(-2x^2y)^3$
58. $(-5m^4p^2)^3$
59. $(3a^3b^2)^2$
60. $(4x^3y^5)^4$

*Find the area of each figure. For help, refer to the formulas found on the inside covers.*

61.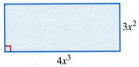

62.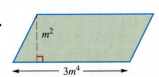

## 10.2 INTEGER EXPONENTS AND THE QUOTIENT RULE

In the last section we studied the product rule for exponents. In all our earlier work, exponents were positive integers. Now we want to develop meaning for exponents that are not positive integers.

Consider the following list of exponential expressions.

$$2^4 = 16$$
$$2^3 = 8$$
$$2^2 = 4$$

Do you see the pattern in the values? Each time we reduce the exponent by 1, the value is divided by 2 (the base). Using this pattern, we can continue the list to smaller and smaller integer exponents.

$$2^1 = 2$$
$$2^0 = 1$$
$$2^{-1} = \frac{1}{2}$$

**Work Problem ❶ at the Side.**

From the list above and the answers to Problem 1 at the side, it appears that we should define $2^0$ as 1 and negative exponents as reciprocals.

**1** **Use 0 as an exponent.** We want the definitions of 0 and negative exponents to satisfy the rules for exponents from **Section 10.1.** For example, if $6^0 = 1$,

$$6^0 \cdot 6^2 = 1 \cdot 6^2 = 6^2 \quad \text{and} \quad 6^0 \cdot 6^2 = 6^{0+2} = 6^2,$$

so the product rule is satisfied. Check that the power rules are also valid for a 0 exponent. Thus, we define a 0 exponent as follows.

### Zero Exponent

If $a$ is any *nonzero* real number, then $a^0 = 1$.

*Example:* $17^0 = 1$

### Example 1 Using Zero Exponents

Evaluate each exponential expression.

(a) $60^0 = 1$
(b) $(-60)^0 = 1$
(c) $-60^0 = -(1) = -1$
(d) $y^0 = 1$, if $y \neq 0$
(e) $6y^0 = 6(1) = 6$, if $y \neq 0$
(f) $(6y)^0 = 1$, if $y \neq 0$

### CAUTION

Notice the difference between parts (b) and (c) of Example 1. In Example 1(b) the base is $-60$ and the exponent is 0. Any nonzero base raised to the exponent zero is 1. But in Example 1(c), the base is 60. Then $60^0 = 1$, and $-60^0 = -1$.

### OBJECTIVES

1. Use 0 as an exponent.
2. Use negative numbers as exponents.
3. Use the quotient rule for exponents.
4. Use the product rule with negative exponents.

❶ Continue the list of exponentials using $-2, -3,$ and $-4$ as exponents.

$2^{-2} =$ _____

$2^{-3} =$ _____

$2^{-4} =$ _____

**ANSWERS**

**1.** $2^{-2} = \frac{1}{4}$; $2^{-3} = \frac{1}{8}$; $2^{-4} = \frac{1}{16}$

**2** Evaluate.

(a) $28^0$

(b) $(-16)^0$

(c) $-7^0$

(d) $m^0, m \neq 0$

**3** Write with positive exponents.

(a) $4^{-3}$

(b) $6^{-2}$

(c) $2^{-1} + 5^{-1}$

(d) $m^{-5}, m \neq 0$

**ANSWERS**
2. (a) 1  (b) 1  (c) $-1$  (d) 1
3. (a) $\frac{1}{4^3}$  (b) $\frac{1}{6^2}$  (c) $\frac{1}{2} + \frac{1}{5} = \frac{7}{10}$  (d) $\frac{1}{m^5}$

Work Problem **2** at the Side.

## **2** Use negative numbers as exponents.

In Margin Problem 1 on the previous page, we saw that $2^{-2} = \frac{1}{4}$, $2^{-3} = \frac{1}{8}$ and $2^{-4} = \frac{1}{16}$. If we write the denominators as powers of 2, we see an interesting pattern in the exponents.

$$2^{-2} = \frac{1}{4} = \frac{1}{2^2} \qquad 2^{-3} = \frac{1}{8} = \frac{1}{2^3} \qquad 2^{-4} = \frac{1}{16} = \frac{1}{2^4}$$

So it seems that $2^{-n}$ should equal $\frac{1}{2^n}$. Is the product rule valid in such cases? For example, if we use the product rule to multiply $6^{-2}$ by $6^2$, we get the following.

$$6^{-2} \cdot 6^2 = 6^{-2+2} = 6^0 = 1$$

The expression $6^{-2}$ behaves as if it were the reciprocal of $6^2$, because their product is 1. The reciprocal of $6^2$ may be written $\frac{1}{6^2}$, leading us to define $6^{-2}$ as $\frac{1}{6^2}$. This example illustrates the definition of negative exponents.

### Negative Exponents

If $a$ is any *nonzero* real number and $n$ is any integer, then $\quad a^{-n} = \frac{1}{a^n}$.

*Example:* $\quad 3^{-2} = \frac{1}{3^2}$

### Example 2  Using Negative Exponents

Simplify by writing each expression with positive exponents.

(a) $4^{-2} = \frac{1}{4^2} = \frac{1}{16}$

(b) $5^{-3} = \frac{1}{5^3} = \frac{1}{125}$

(c) $4^{-1} - 2^{-1} = \frac{1}{4} - \frac{1}{2}$    Apply the exponents first.

$\qquad\qquad\quad = \frac{1}{4} - \frac{2}{4}$    Then subtract.

$\qquad\qquad\quad = -\frac{1}{4}$

(d) $p^{-2} = \frac{1}{p^2}, p \neq 0$

### CAUTION

A negative exponent does **not** indicate a negative number; negative exponents lead to reciprocals.

| Expression | Example |
|---|---|
| $a^{-n}$ | $3^{-2} = \frac{1}{3^2} = \frac{1}{9}$    Not negative |

Work Problem **3** at the Side.

**3** Use the quotient rule for exponents. What about the quotient of two exponential expressions with the same base? We know that

$$\frac{6^5}{6^3} = \frac{\cancel{6}\cdot\cancel{6}\cdot\cancel{6}\cdot 6 \cdot 6}{\cancel{6}\cdot\cancel{6}\cdot\cancel{6}} = \frac{6^2}{1} = 6^2$$

Notice that the difference between the exponents, $5 - 3 = 2$, is the exponent in the quotient. Here is another example.

$$\frac{6^2}{6^4} = \frac{\cancel{6}\cdot\cancel{6}}{\cancel{6}\cdot\cancel{6}\cdot 6 \cdot 6} = \frac{1}{6^2} = 6^{-2}$$

In this case, $2 - 4 = -2$. These examples suggest the quotient rule for exponents.

### Quotient Rule for Exponents

If $a$ is any nonzero real number, and $m$ and $n$ are integers, then

$$\frac{a^m}{a^n} = a^{m-n}.$$

(Keep the base and subtract the exponents.)

*Example:* $\dfrac{5^8}{5^4} = 5^{8-4} = 5^4$

### CAUTION

A common *error* is to write $\frac{5^8}{5^4} = 1^{8-4} = 1^4$. Notice that by the quotient rule, the quotient should have the *same base*, 5.

$$\frac{5^8}{5^4} = 5^{8-4} = 5^4$$

If you are not sure, use the definition of an exponent to write out the factors.

$$\frac{5^8}{5^4} = \frac{\cancel{5}\cdot\cancel{5}\cdot\cancel{5}\cdot\cancel{5}\cdot 5 \cdot 5 \cdot 5 \cdot 5}{\cancel{5}\cdot\cancel{5}\cdot\cancel{5}\cdot\cancel{5}} = \frac{5^4}{1} = 5^4$$

### Example 3  Using the Quotient Rule for Exponents

Simplify, using the quotient rule for exponents. Write answers with positive exponents.

(a) $\dfrac{5^8}{5^6} = 5^{8-6} = 5^2$

(b) $\dfrac{4^2}{4^9} = 4^{2-9} = 4^{-7} = \dfrac{1}{4^7}$

(c) $\dfrac{5^{-3}}{5^{-7}} = 5^{-3-(-7)} = 5^4$

(d) $\dfrac{q^5}{q^{-3}} = q^{5-(-3)} = q^8,\ q \neq 0$

──── Work Problem **4** at the Side.

**4** Simplify. Write answers with positive exponents.

(a) $\dfrac{5^{11}}{5^8}$

(b) $\dfrac{4^7}{4^{10}}$

(c) $\dfrac{6^{-5}}{6^{-3}}$

(d) $\dfrac{a^4}{a^{-2}},\ a \neq 0$

**ANSWERS**

4. (a) $5^3$  (b) $4^{-3} = \dfrac{1}{4^3}$  (c) $6^{-2} = \dfrac{1}{6^2}$

(d) $a^6$

## Chapter 10 Exponents and Polynomials

**⑤** Simplify. Assume all variables represent nonzero real numbers. Write answers with positive exponents.

(a) $10^{-7}(10^9)$

(b) $(10^{-3})(10^{-3})$

(c) $x^{-3} \cdot x^{-2} \cdot x^4$

(d) $(a^{-2})(a^{12})(a^{-4})$

**4** **Use the product rule with negative exponents.** As shown in the next example, we may use the product rule to simplify expressions with negative exponents.

### Example 4  Using the Product Rule with Negative Exponents

Simplify each expression. Assume all variables represent nonzero real numbers. Write answers with positive exponents.

(a) $10^5(10^{-4}) = 10^{5+(-4)}$    Using the product rule, add the exponents.
$= 10^1$ or $10$

(b) $(10^{-5})(10^{-8}) = 10^{-5+(-8)}$    Product rule
$= 10^{-13} = \dfrac{1}{10^{13}}$    Definition of negative exponent

(c) $y^{-2} \cdot y^5 \cdot y^{-8} = y^{-2+5} \cdot y^{-8}$    Product rule
$= y^3 \cdot y^{-8}$
$= y^{3+(-8)}$    Product rule
$= y^{-5}$
$= \dfrac{1}{y^5}$    Definition of negative exponent

**Work Problem ⑤ at the Side.**

The definitions and rules for exponents given in this section and **Section 10.1** are summarized below.

### Definitions and Rules for Exponents

| If $m$ and $n$ are any integers, | | Examples |
|---|---|---|
| **Product rule** | $a^m \cdot a^n = a^{m+n}$ | $7^4 \cdot 7^5 = 7^9$ |
| **Zero exponent** | $a^0 = 1$ $(a \neq 0)$ | $(-3)^0 = 1$ |
| **Negative exponent** | $a^{-n} = \dfrac{1}{a^n}$ $(a \neq 0)$ | $5^{-3} = \dfrac{1}{5^3}$ |
| **Quotient rule** | $\dfrac{a^m}{a^n} = a^{m-n}$ $(a \neq 0)$ | $\dfrac{2^2}{2^5} = 2^{2-5} = 2^{-3} = \dfrac{1}{2^3}$ |
| **Power rule (a)** | $(a^m)^n = a^{mn}$ | $(4^2)^3 = 4^6$ |
| **Power rule (b)** | $(ab)^m = a^m b^m$ | $(3k)^4 = 3^4 k^4$ |
| **Power rule (c)** | $\left(\dfrac{a}{b}\right)^m = \dfrac{a^m}{b^m}$ $(b \neq 0)$ | $\left(\dfrac{2}{3}\right)^2 = \dfrac{2^2}{3^2}$ |

**ANSWERS**

**5.** (a) $10^2$   (b) $\dfrac{1}{10^6}$   (c) $\dfrac{1}{x^1}$ or $\dfrac{1}{x}$   (d) $a^6$

# 10.2 Exercises

*Decide whether each expression is positive, negative, or 0.*

1. $(-2)^{-3}$
2. $(-3)^{-2}$
3. $-2^4$
4. $-3^6$
5. $(-2)^6$
6. $(-2)^3$
7. $1 - 5^0$
8. $1 - 7^0$

*Each expression is equal to either 0, 1, or $-1$. Decide which is correct. See Example 1.*

9. $(-4)^0$
10. $(-10)^0$
11. $-9^0$
12. $-5^0$
13. $(-2)^0 - 2^0$
14. $(-8)^0 - 8^0$
15. $\dfrac{0^{10}}{10^0}$
16. $\dfrac{0^5}{5^0}$

*Evaluate each expression. See Examples 1 and 2.*

17. $7^0 + 9^0$
18. $8^0 + 6^0$
19. $b^0, b \neq 0$
20. $x^0, x \neq 0$
21. $15x^0$
22. $12g^0$
23. $4^{-3}$
24. $5^{-4}$
25. $5^{-1} + 3^{-1}$
26. $6^{-1} + 2^{-1}$
27. $x^{-4}$
28. $b^{-3}$

## Relating Concepts (Exercises 29–32) For Individual or Group Work

In Objective 1, we used the product rule to motivate the definition of a 0 exponent. We can also use the quotient rule. To see this, **work Exercises 29–32 in order.**

29. Consider the expression $\frac{25}{25}$. What is its simplest form?

30. Rewrite $\frac{25}{25}$ using the fact that $25 = 5^2$.

31. Apply the quotient rule for exponents to your answer for Exercise 30. Give the answer as a power of 5.

32. Because your answers for Exercises 29 and 31 both represent $\frac{25}{25}$, they must be equal. Write this equality. What definition does it support?

*Use the quotient rule to simplify each expression. Write each expression with positive exponents. Assume that all variables represent nonzero real numbers. See Example 3.*

33. $\dfrac{6^7}{6^2}$
34. $\dfrac{8^5}{8^3}$
35. $\dfrac{10^4}{10}$
36. $\dfrac{7^8}{7}$

37. $\dfrac{y^2}{y^6}$
38. $\dfrac{d^9}{d^5}$
39. $\dfrac{c^6}{c^5}$
40. $\dfrac{x^4}{x^3}$

41. $\dfrac{5^3}{5^7}$
42. $\dfrac{9^2}{9^8}$
43. $\dfrac{m^7}{m^8}$
44. $\dfrac{n^2}{n^3}$

45. $\dfrac{3^{-4}}{3^{-8}}$
46. $\dfrac{5^{-3}}{5^{-10}}$
47. $\dfrac{a^{-2}}{a^{-5}}$
48. $\dfrac{b^{-4}}{b^{-6}}$

49. $\dfrac{2^{-10}}{2^{-2}}$
50. $\dfrac{4^{-7}}{4^{-3}}$
51. $\dfrac{r^{-12}}{r^{-8}}$
52. $\dfrac{v^{-10}}{v^{-5}}$

53. $\dfrac{10^6}{10^{-4}}$
54. $\dfrac{10^4}{10^{-2}}$
55. $\dfrac{10^{-2}}{10^3}$
56. $\dfrac{10^{-1}}{10^7}$

57. $\dfrac{10^3}{10^{-8}}$
58. $\dfrac{10^5}{10^{-7}}$
59. $\dfrac{10^{-4}}{10^4}$
60. $\dfrac{10^{-6}}{10^6}$

*Simplify each expression. Write answers with only positive exponents. Assume that all variables represent nonzero real numbers. See Example 4.*

61. $10^6(10^{-2})$
62. $10^{-7}(10^4)$
63. $a^{-4}(a^3)$
64. $w^9(w^{-8})$

65. $(2^{-4})(2^{-4})$
66. $(2^{-1})(2^{-5})$
67. $(x^{-10})(x^{-1})$
68. $(h^{-5})(h^{-5})$

69. $10^8 \cdot 10^{-2} \cdot 10^{-4}$
70. $10^{-2} \cdot 10^7 \cdot 10^{-2}$
71. $(y^{-3})(y^5)(y^{-4})$
72. $(n^{-8})(n^{-2})(n^5)$

73. $m^{-6} \cdot m^3 \cdot m^{12}$
74. $b^{-3} \cdot b^3 \cdot b^8$
75. $(10^{-1})(10^{-2})(10^{-3})$
76. $(10^{-4})(10^{-5})(10^{-6})$

## 10.3 An Application of Exponents: Scientific Notation

**OBJECTIVES**
1. Express numbers in scientific notation.
2. Convert numbers in scientific notation to numbers without exponents.
3. Use scientific notation in calculations.
4. Solve application problems using scientific notation.

**1** **Express numbers in scientific notation.** An important example of the use of exponents comes from science. The numbers occurring in science are often extremely large (such as the distance from Earth to the sun, 93,000,000 miles) or extremely small (the wavelength of yellow-green light, approximately 0.0000006 m). Because of the difficulty of working with many zeros, scientists often express such numbers with exponents. Each number is written as $a \times 10^n$, where $1 \leq |a| < 10$ and $n$ is an integer. This form is called **scientific notation.** There is always one nonzero digit before the decimal point. Here are some examples.

$$35 \quad \text{is written} \quad 3.5 \times 10^1 \quad \text{or} \quad 3.5 \times 10$$

$$56{,}200 \quad \text{is written} \quad 5.62 \times 10^4$$
$$\text{because } 56{,}200 = 5.62 \times \mathbf{10{,}000} = 5.62 \times \mathbf{10^4}$$

$$0.09 \quad \text{is written} \quad 9 \times 10^{-2}$$
$$\text{because } 0.09 = \frac{9}{100} = 9 \times \frac{\mathbf{1}}{\mathbf{100}} = 9 \times \frac{1}{10^2} = 9 \times \mathbf{10^{-2}}$$

The steps involved in writing a number in scientific notation are given below. For negative numbers, follow these steps using the absolute value of the number; then make the result negative.

### Writing a Number in Scientific Notation

To write a number in scientific notation (as $a \times 10^n$), move the decimal point to the right of the first nonzero digit. If the decimal point is moved $n$ places to the left, and this makes the number smaller, $n$ is positive; otherwise, $n$ is negative. If the decimal point is not moved, $n$ is 0.

**Example 1** Writing Numbers in Scientific Notation

Write each number in scientific notation.

**(a)** 93,000,000

Move the decimal point to follow the first nonzero digit. Count the number of places the decimal point was moved.

$$9.3\,000\,000 \quad \text{7 places}$$

We will write the number as $9.3 \times 10^n$. Notice that we moved the decimal point **7** places to the *left*, and that we made the number *smaller* (9.3 is smaller than 93,000,000). So the exponent $n$ will be **positive 7**.

$$93{,}000{,}000 = 9.3 \times 10^7$$

**(b)** $302{,}100 = 3.021 \times 10^5$    Move the decimal point 5 places to the left; makes the number smaller.

*Continued on Next Page*

**1** Write each number in scientific notation.

(a) 63,000

(b) 5,870,000

(c) 0.0571

(d) 0.000062

**2** Write without exponents.

(a) $4.2 \times 10^3$

(b) $8.7 \times 10^5$

(c) $6.42 \times 10^{-3}$

**3** Simplify, and write without exponents.

(a) $(2.6 \times 10^4)(2 \times 10^{-6})$

(b) $\dfrac{4.8 \times 10^2}{2.4 \times 10^{-3}}$

**ANSWERS**
1. (a) $6.3 \times 10^4$  (b) $5.87 \times 10^6$
   (c) $5.71 \times 10^{-2}$  (d) $6.2 \times 10^{-5}$
2. (a) 4200  (b) 870,000  (c) 0.00642
3. (a) 0.052  (b) 200,000

---

(c) 0.00462

Move the decimal point to follow the first nonzero digit and count the number of places the decimal point was moved.

0 004.62    3 places

Notice that we moved the decimal point 3 places to the *right*, and that we made the number *larger* (4.62 is larger than 0.00462). So the exponent will be **negative 3**.

$$0.00462 = 4.62 \times 10^{-3}$$

(d) $0.0000762 = 7.62 \times 10^{-5}$   Move the decimal point 5 places to the right; makes the number larger.

Work Problem **1** at the Side.

**2** **Convert numbers in scientific notation to numbers without exponents.** To convert a number written in scientific notation to a number without exponents, work in reverse. Multiplying a number by a positive power of 10 will make the number larger; multiplying by a negative power of 10 will make the number smaller.

**Example 2** Writing Numbers without Exponents

Write each number without exponents.

(a) $6.2 \times 10^3$

Since the exponent is positive, make 6.2 larger by moving the decimal point 3 places to the *right*.

$$6.2 \times 10^3 = 6.200 = 6200$$

(b) $4.283 \times 10^5 = 4.28300 = 428,300$   Move 5 places to the *right*.

(c) $9.73 \times 10^{-2} = 09.73 = 0.0973$   Move 2 places to the *left*.

As these examples show, the exponent tells the number of places and the direction that the decimal point is moved.

Work Problem **2** at the Side.

**3** **Use scientific notation in calculations.** The next example shows how scientific notation can be used with products and quotients.

**Example 3** Multiplying and Dividing with Scientific Notation

Write each product or quotient without exponents.

(a) $(6 \times 10^3)(5 \times 10^{-4})$
$= (6 \times 5)(10^3 \times 10^{-4})$   Commutative and associative properties
$= 30 \times 10^{-1}$   Product rule for exponents
$= 30. = 3.0$   or   3   Write without exponents.

(b) $\dfrac{6 \times 10^{-5}}{2 \times 10^3} = \dfrac{6}{2} \times \dfrac{10^{-5}}{10^3} = 3 \times 10^{-8} = 0.00000003$

Work Problem **3** at the Side.

**Section 10.3** An Application of Exponents: Scientific Notation

**4** Solve application problems using scientific notation. Sometimes application problems may involve one or more numbers written in scientific notation and other numbers written without exponents. By writing all the numbers in scientific notation, we can use the product and quotient rules for exponents to help solve the problem.

### Example 4  Solving an Application Problem

There are approximately $3.03 \times 10^{23}$ molecules in one gram of hydrogen gas. How many molecules are in 2000 g of hydrogen gas? Write your answer both in scientific notation and as a number without exponents.

First write 2000 in scientific notation.

$$2000 = 2.000 \quad \text{Move 3 places to the left; makes the number smaller.}$$
$$= 2.0 \times 10^3$$

Now multiply the number of molecules in one gram times the number of grams.

$$(3.03 \times 10^{23})(2.0 \times 10^3)$$
$$= (3.03 \times 2.0)(10^{23} \times 10^3) \quad \text{Use the product rule for exponents.}$$
$$= 6.06 \times 10^{26}$$

Written in scientific notation, there are $6.06 \times 10^{26}$ molecules in 2000 g of hydrogen gas. Written without an exponent, the number is 606,000,000,000,000,000,000,000,000 molecules.

**4** There are $4.356 \times 10^4$ square feet in one acre. (*Source: World Almanac and Book of Facts 2001.*) How many square feet are in 1500 acres? Write your answer both in scientific notation and as a number without exponents.

— Work Problem **4** at the Side.

**Calculator Tip**  Most scientific calculators have a key that allows you to enter numbers in scientific notation. Usually this key is labeled (EE) or (EXP). Then you could solve Example 4, above, by pressing:

3.03 (EXP) 23 (×) 2000 (=)    6.06²⁶

The exponent in the answer is 26. You have to write the "× 10" part of the answer to get $6.06 \times 10^{26}$.

If you need to enter a number in scientific notation that has a negative exponent, use the *change of sign* key. For example, to enter $8.45 \times 10^{-6}$, press:

8.45 (EXP) 6 (+/−)    8.45⁻⁰⁶

### Example 5  Solving an Application Problem

At a printing shop, a stack of $5 \times 10^3$ sheets of paper measures 0.38 m high. What is the thickness of one sheet of paper? Write your answer both in scientific notation and as a number without exponents.

First write 0.38 in scientific notation.

$$0.38 = 0\,3.8 \quad \text{Move 1 place to the right; makes the number larger.}$$
$$= 3.8 \times 10^{-1}$$

Then divide the total height of the stack by the number of sheets in the stack.

$$\frac{3.8 \times 10^{-1}}{5 \times 10^3} = \frac{3.8}{5} \times \frac{10^{-1}}{10^3} = 0.76 \times 10^{-4}$$

— Continued on Next Page

**ANSWERS**
4. $6.534 \times 10^7$ square feet, or 65,340,000 square feet

**5** In a laboratory experiment, a test group of 950 common house spiders weighed $2.09 \times 10^{-1}$ pound. Find the weight of each spider. Write your answer both in scientific notation and as a number without exponents.

The answer $0.76 \times 10^{-4}$ is *not* in proper scientific notation yet, because $0.76$ is not between 1 and 10. The decimal point in $0.76$ must be moved *1 place* to the *right* to get $7.6$. Then, because we made the number larger, add *negative* 1 to the exponent.

$$0.76 \times 10^{-4} = 7.6 \times 10^{(-4)+(-1)} = 7.6 \times 10^{-5}$$

Move 1 place to the right.

The thickness of one sheet of paper is $7.6 \times 10^{-5}$ m. Written without an exponent, the thickness is $0.000076$ m.

**Work Problem 5 at the Side.**

---

**ANSWERS**
5. $2.2 \times 10^{-4}$ pound, or $0.00022$ pound

## 10.3 EXERCISES

*Write the numbers (other than dates) mentioned in the following statements in scientific notation.*

1. NASA has budgeted $13,750,000,000 in each of the years 2003 and 2004 for the international space station. (*Source:* U.S. National Aeronautics and Space Administration.)

2. The mass of Pluto, the smallest planet, is 0.0021 times that of Earth; the mass of Jupiter, the largest planet, is 317.83 times that of Earth. (*Source: World Almanac and Book of Facts,* 2000.)

3. In 1998, the federal government spent $66,600,000,000 on research and development. Industry spent $143,700,000,000 in that same year. (*Source:* U.S. National Science Foundation.)

4. The risk to industrial workers at the Hansom Landfill at the Kennedy Space Center depends on the reference doses of materials dumped there. For thallium, the reference dose is 700,000 mg/kg per day, and the reference dose for beryllium is 5000 mg/kg per day. (*Source: NASA-AMATYC-NSF Math Explorations I,* Capital Community College, 1999.)

*Determine whether or not the given number is written in scientific notation as defined in Objective 1. If it is not, write it as such.*

5. $4.56 \times 10^3$  
6. $7.34 \times 10^5$  
7. 5,600,000  
8. 34,000

9. 0.004  
10. 0.0007  
11. $0.8 \times 10^2$  
12. $0.9 \times 10^3$

13. Explain in your own words how to write a very large number in scientific notation, and how to write a very small number in scientific notation.

14. Explain how to multiply a number by a positive power of ten. Then explain how to multiply a number by a negative power of ten.

*Write each number in scientific notation. See Example 1.*

15. 5,876,000,000  
16. 9,994,000,000  
17. 82,350  
18. 78,330

19. 0.000007  
20. 0.0000004  
21. 0.00203  
22. 0.0000578

*Write each number without exponents. See Example 2.*

23. $7.5 \times 10^5$  
24. $8.8 \times 10^6$  
25. $5.677 \times 10^{12}$  
26. $8.766 \times 10^9$

**27.** $6.21 \times 10^0$  **28.** $8.56 \times 10^0$  **29.** $7.8 \times 10^{-4}$

**30.** $8.9 \times 10^{-5}$  **31.** $5.134 \times 10^{-9}$  **32.** $7.123 \times 10^{-10}$

*Perform the indicated operations, and write the answers in scientific notation and then without exponents. See Example 3.*

**33.** $(2 \times 10^8)(3 \times 10^3)$  **34.** $(4 \times 10^7)(3 \times 10^3)$  **35.** $(5 \times 10^4)(3 \times 10^2)$

**36.** $(8 \times 10^5)(2 \times 10^3)$  **37.** $(3.15 \times 10^{-4})(2.04 \times 10^8)$  **38.** $(4.92 \times 10^{-3})(2.25 \times 10^7)$

*Perform the indicated operations, and write the answers in scientific notation. See Example 3.*

**39.** $\dfrac{9 \times 10^{-5}}{3 \times 10^{-1}}$  **40.** $\dfrac{12 \times 10^{-4}}{4 \times 10^{-3}}$  **41.** $\dfrac{8 \times 10^3}{2 \times 10^2}$

**42.** $\dfrac{5 \times 10^4}{1 \times 10^3}$  **43.** $\dfrac{(2.6 \times 10^{-3})(7.0 \times 10^{-1})}{(2 \times 10^2)(3.5 \times 10^{-3})}$  **44.** $\dfrac{(9.5 \times 10^{-1})(2.4 \times 10^4)}{(5 \times 10^3)(1.2 \times 10^{-2})}$

*Work each problem. Give answers both in scientific notation and without exponents. See Examples 4 and 5.*

**45.** The population of China in 2000 was about $1.26 \times 10^9$ people. The country's land area is 3,700,000 square miles. (*Source:* World Almanac and Book of Facts 2001.) What is the average number of people per square mile in China?

**46.** The mass of Earth is approximately $5.98 \times 10^{24}$ kg, and the mass of the moon is approximately $7.36 \times 10^{22}$ kg. (*Source:* Time Life *The Universe.*) Earth's mass is how many times greater than the moon's mass?

**47.** The average diameter of a hydrogen atom is $1 \times 10^{-10}$ m. (*Source: FactFinder.*)

(a) If one million hydrogen atoms are lined up in a row, how long would the row be?

(b) Find the length of a row of one billion hydrogen atoms.

**48.** The speed of light is about $1.86 \times 10^5$ miles per second. (*Source:* Time Life *The Universe.*)

(a) How far does light travel in one minute?

(b) How far does light travel in one hour?

**49.** There were $6.3 \times 10^{10}$ dollars spent to attend motion pictures in a recent year. Approximately $1.3 \times 10^8$ adults attended a motion picture theatre at least once. (*Source:* U.S. National Endowment for the Arts.) What was the average amount spent per person that year?

**50.** The body of a 150-pound person contains about $2.3 \times 10^{-4}$ pound of copper. How much copper is contained in the bodies of 1200 such people?

# 10.4 Adding and Subtracting Polynomials

Recall from **Section 2.2** that in an expression such as

$$4x^3 + 6x^2 + 5x + 8,$$

the quantities that are added, $4x^3$, $6x^2$, $5x$, and $8$ are called *terms*. In the term $4x^3$, the number 4 is called the *numerical coefficient*, or simply the *coefficient*, of $x^3$. In the same way, 6 is the coefficient of $x^2$ in the term $6x^2$, 5 is the coefficient of $x$ in the term $5x$, and 8 is the constant term.

**① Review combining like terms.** In **Section 2.2**, we saw that *like terms* are terms with exactly the same variables, with the same exponents on the variables. Only the coefficients may differ. Like terms are combined, or *added*, by adding their coefficients using the distributive property.

### Example 1  Review of Adding Like Terms

Simplify each expression by adding like terms.

(a) $-4x^3 + 6x^3 = (-4 + 6)x^3$    Distributive property
    $= 2x^3$

(b) $9x^6 - 14x^6 + x^6 = (9 - 14 + 1)x^6 = -4x^6$

(c) $12m^2 + 5m + 4m^2 = (12 + 4)m^2 + 5m$
    $= 16m^2 + 5m$

(d) $3x^2y + 4x^2y - x^2y = (3 + 4 - 1)x^2y = 6x^2y$

In Example 1(c), we cannot add $16m^2$ and $5m$. These two terms are *unlike* because the exponents on the variables are different. *Unlike terms* have different variables or different exponents on the same variables.

**Work Problem ① at the Side.**

**② Know the vocabulary for polynomials.** A **polynomial in $x$** is a term, or the sum of several terms, of the form $ax^n$, where $a$ is any real number and $n$ is any whole number. For example,

$$16x^8 - 7x^6 + 5x^4 - 3x^2 + 4$$

is a polynomial in $x$. This polynomial is written in **descending powers**, because the exponents on $x$ decrease from left to right. On the other hand,

$$2x^3 - x^2 + \frac{4}{x}$$

is not a polynomial, since a variable appears in a denominator. Of course, we could define a *polynomial* using any variable, not just $x$, as in Example 1(c) above. In fact, polynomials may have terms with more than one variable, as in Example 1(d) above.

**Work Problem ② at the Side.**

---

**OBJECTIVES**

1. Review combining like terms.
2. Know the vocabulary for polynomials.
3. Evaluate polynomials.
4. Add polynomials.
5. Subtract polynomials.

**①** Add like terms.

(a) $5x^4 + 7x^4$

(b) $9pq + 3pq - 2pq$

(c) $r^2 + 3r + 5r^2$

(d) $8t + 6w$

**②** Choose all descriptions that apply for each of the expressions in parts (a)–(d).
A. Polynomial
B. Polynomial written in descending powers
C. Not a polynomial

(a) $3m^5 + 5m^2 - 2m + 1$

(b) $2p^4 + p^6$

(c) $\frac{1}{x} + 2x^2 + 3$

(d) $x - 3$

**ANSWERS**
1. (a) $12x^4$  (b) $10pq$  (c) $6r^2 + 3r$
   (d) cannot be added—unlike terms
2. (a) A and B  (b) A  (c) C  (d) A and B

**3** For each polynomial, first simplify if possible. Then give the degree and tell whether the polynomial is a monomial, binomial, trinomial, or none of these.

(a) $3x^2 + 2x - 4$

(b) $x^3 + 4x^3$

(c) $x^8 - x^7 + 2x^8$

The **degree of a term** is the sum of the exponents on the variables. A constant term has degree 0. For example, $3x^4$ has degree 4, while $6x^{17}$ has degree 17. The term $5x$ has degree 1, $-7$ has degree 0, and $2x^2y$ has degree $2 + 1 = 3$ (because $y$ has an exponent of 1). The **degree of a polynomial** is the highest degree of any nonzero term of the polynomial. For example, $3x^4 - 5x^2 + 6$ is a polynomial of degree 4, the polynomial $5x + 7$ is of degree 1, 3 is of degree 0, and $x^2y + xy - 5xy^2$ is of degree 3.

Three types of polynomials are very common and are given special names. A polynomial with only one term is called a **monomial.** (*Mon(o)-* means "one," as in *mono*rail.) Some examples are shown below.

$$9m \quad -6y^5 \quad a^2 \quad 6 \quad \text{Monomials}$$

A polynomial with exactly two terms is called a **binomial.** (*Bi-* means "two," as in *bi*cycle.) Examples are shown below.

$$-9x^4 + 9x^3 \quad 8m^2 + 6m \quad 3m^5 - 9m^2 \quad \text{Binomials}$$

A polynomial with exactly three terms is called a **trinomial.** (*Tri-* means "three," as in *tri*angle.) Examples are shown below.

$$9m^3 - 4m^2 + 6 \quad -3m^5 - 9m^2 + 2 \quad \text{Trinomials}$$

**Example 2** Classifying Polynomials

For each polynomial, first simplify if possible by combining like terms. Then give the degree and tell whether the polynomial is a monomial, a binomial, a trinomial, or none of these.

(a) $2x^3 + 5$
The polynomial cannot be simplified. The degree is 3. The polynomial is a binomial.

(b) $4x - 5x + 2x$
Add like terms to simplify: $4x - 5x + 2x = x$. The degree is 1 (since $x = x^1$). The simplified polynomial is a monomial.

**Work Problem 3 at the Side.**

**3** Evaluate polynomials. A polynomial usually represents different numbers for different values of the variable, as shown in the next example.

**Example 3** Evaluating a Polynomial

Find the value of $5x^3 - 4x - 4$ when $x = -2$ and when $x = 3$.

First, substitute $-2$ for $x$.
$$5x^3 - 4x - 4 = 5(-2)^3 - 4(-2) - 4$$
$$= 5(-8) - 4(-2) - 4 \quad \text{Apply exponents.}$$
$$= -40 + 8 - 4 \quad \text{Multiply.}$$
$$= -36 \quad \text{Add and subtract.}$$

Next, replace $x$ with 3.
$$5x^3 - 4x - 4 = 5(3)^3 - 4(3) - 4$$
$$= 5 \cdot 27 - 4(3) - 4$$
$$= 135 - 12 - 4$$
$$= 119$$

**Answers**
3. (a) cannot be simplified; degree 2; trinomial
   (b) simplify to $5x^3$; degree 3; monomial
   (c) simplify to $3x^8 - x^7$; degree 8; binomial

Section 10.4 Adding and Subtracting Polynomials   **723**

**CAUTION**

Notice the use of parentheses around the numbers that are substituted for the variable in Example 3. This is particularly important when substituting a negative number for a variable that is raised to a power, so the sign of the product is correct.

**Work Problem ❹ at the Side.**

**❹** Add polynomials. Polynomials may be added, subtracted, multiplied, and divided.

### Adding Polynomials

To add two polynomials, add like terms.

**Example 4** Adding Polynomials Vertically

(a) Add $6x^3 - 4x^2 + 3$ and $-2x^3 + 7x^2 - 5$.

Write like terms in columns. Then add column by column.

$$\begin{array}{r} 6x^3 - 4x^2 + 3 \\ -2x^3 + 7x^2 - 5 \\ \hline 4x^3 + 3x^2 - 2 \end{array}$$

(b) Add $2x^2 - 4x + 3$ and $x^3 + 5x$.

Write like terms in columns and add column by column.

$$\begin{array}{r} 2x^2 - 4x + 3 \\ x^3 \phantom{{}+2x^2} + 5x \phantom{{}+3} \\ \hline x^3 + 2x^2 + x + 3 \end{array}$$ Leave spaces for missing terms.

**Work Problem ❺ at the Side.**

The polynomials in Example 4 also could be added horizontally by combining like terms, as shown in the next example.

**Example 5** Adding Polynomials Horizontally

(a) Add $6x^3 - 4x^2 + 3$ and $-2x^3 + 7x^2 - 5$. Combine like terms.

$$(6x^3 - 4x^2 + 3) + (-2x^3 + 7x^2 - 5)$$

The sum is the same as in Example 4(a) above: $4x^3 + 3x^2 - 2$

(b) Add $2x^2 - 4x + 3$ and $x^3 + 5x$.

$$(2x^2 - 4x + 3) + (x^3 + 5x) = 2x^2 - 4x + 3 + x^3 + 5x$$
$$= x^3 + 2x^2 + x + 3 \quad \text{Combine like terms.}$$

*Continued on Next Page*

**❹** Find the value of $2y^3 + 8y - 6$ in each case.

(a) When $y = -1$

(b) When $y = 4$

**❺** Add each pair of polynomials.

(a) $\begin{array}{r} 4x^3 - 3x^2 + 2x \\ 6x^3 + 2x^2 - 3x \\ \hline \end{array}$

(b) $\begin{array}{r} x^2 - 2x + 5 \\ 4x^2 \phantom{{}-2x} - 2 \\ \hline \end{array}$

**ANSWERS**
**4.** (a) $-16$  (b) $154$
**5.** (a) $10x^3 - x^2 - x$  (b) $5x^2 - 2x + 3$

**6** Find each sum.

(a) $(2x^4 - 6x^2 + 7)$
   $+ (-3x^4 + 5x^2 + 2)$

(b) $(3x^2 + 4x + 2)$
   $+ (6x^3 - 5x - 7)$

**7** Subtract, and check your answers by addition.

(a) $(14y^3 - 6y^2 + 2y - 5)$
   $- (2y^3 - 7y^2 - 4y + 6)$

(b) $(7y^2 - 11y + 8)$
   $- (2y^2 - 6y + 6)$

---

**NOTE**

Notice the order of the terms in the final answer to Example 5(b) on the previous page. Mathematicians generally write polynomials in descending order of power. The answers in this textbook will also be shown that way.

Work Problem **6** at the Side.

**5** **Subtract polynomials.** Recall from **Section 1.4** that we changed subtraction problems to addition problems. To subtract two numbers, we *added* the first number to the *opposite* of the second number. Here are two examples.

$$7 - 2 \qquad\qquad -8 - (-5)$$
$$7 + (-2) = 5 \qquad -8 + (+5) = -3$$

A similar method is used to subtract polynomials.

### Subtracting Polynomials

To subtract two polynomials, change *all* the signs of the second polynomial and add the result to the first polynomial.

**Example 6** Subtracting Polynomials

(a) Perform the subtraction $(5x - 2) - (3x - 8)$.

$(5x - 2) - (3x - 8)$   Change subtraction to addition.
                                 Change every sign in the second polynomial to its opposite.
$= (5x - 2) + (-3x + 8)$
$= 2x + 6$        Combine like terms.

(b) Subtract $6x^3 - 4x^2 + 2$ from $11x^3 + 2x^2 - 8$.

Write the problem with the polynomials in the correct order.

$$(11x^3 + 2x^2 - 8) - (6x^3 - 4x^2 + 2)$$

Change every sign in the *second* polynomial and *add* the two polynomials.

$$(11x^3 + 2x^2 - 8) + (-6x^3 + 4x^2 - 2)$$
$$= 5x^3 + 6x^2 - 10$$

To check a subtraction problem, use the fact that if $a - b = c$, then $a = b + c$. For example, $6 - 2 = 4$, so check by writing $6 = 2 + 4$, which is correct.

Check the polynomial subtraction above by adding $6x^3 - 4x^2 + 2$ and $5x^3 + 6x^2 - 10$. Since the sum is $11x^3 + 2x^2 - 8$, the subtraction was done correctly.

Work Problem **7** at the Side.

---

**Answers**
6. (a) $-x^4 - x^2 + 9$
   (b) $6x^3 + 3x^2 - x - 5$
7. (a) $12y^3 + y^2 + 6y - 11$
   (b) $5y^2 - 5y + 2$

## 10.4 Exercises

*Fill in each blank with the correct response.*

1. In the term $7x^5$, the coefficient is _____ and the exponent is _____.

2. The expression $5x^3 - 4x^2$ has _____ term(s).
   (How many?)

3. The degree of the term $-4x^8$ is _____.

4. The polynomial $4x^2 - y^2$ _____ an example of a trinomial.
   (is/is not)

5. When $x^2 + 10$ is evaluated for $x = 4$, the result is _____.

6. _____ is an example of a monomial with coefficient 5, in the variable $x$, having degree 9.

*For each polynomial, determine the number of terms, and identify the coefficient of each term.*

7. $6x^4$

8. $-9y^5$

9. $t^4$

10. $s^7$

11. $-19r^2 - r$

12. $2y^3 - y$

13. $x + 8x^2$

14. $v - 2v^3$

*In each polynomial, combine like terms whenever possible. Write the result with descending powers. See Example 1.*

15. $-3m^5 + 5m^5$

16. $-4y^3 + 3y^3$

17. $2r^5 + (-3r^5)$

18. $-19y^2 + 9y^2$

19. $0.2m^5 - 0.5m^2$

20. $-0.9y + 0.9y^2$

21. $-3x^5 + 2x^5 - 4x^5$

22. $6x^3 - 8x^3 + 9x^3$

23. $-4p^7 + 8p^7 + 5p^9$

24. $-3a^8 + 4a^8 - 3a^2$

25. $-4y^2 + 3y^2 - 2y^2 + y^2$

26. $3r^5 - 8r^5 + r^5 + 2r^5$

*For each polynomial, first simplify, if possible, and write it with descending powers. Then give the degree of the simplified polynomial, and tell whether it is a* monomial, *a* binomial, *a* trinomial, *or* none of these. *See Example 2.*

27. $6x^4 - 9x$

28. $7t^3 - 3t$

29. $5m^4 - 3m^2 + 6m^5 - 7m^3$

30. $6p^5 + 4p^3 - 8p^4 + 10p^2$

31. $\dfrac{5}{3}x^4 - \dfrac{2}{3}x^4 + \dfrac{1}{3}x^2 - 4$

32. $\dfrac{4}{5}r^6 + \dfrac{1}{5}r^6 - r^4 + \dfrac{2}{5}r$

33. $0.8x^4 - 0.3x^4 - 0.5x^4 + 7$

34. $1.2t^3 - 0.9t^3 - 0.3t^3 + 9$

*Find the value of each polynomial* **(a)** *when* $x = 2$ *and* **(b)** *when* $x = -1$. *See Example 3.*

35. $-2x + 3$

36. $5x - 4$

37. $2x^2 + 5x + 1$

38. $-3x^2 + 14x - 2$

39. $2x^5 - 4x^4 + 5x^3 - x^2$

40. $x^4 - 6x^3 + x^2 + 1$

41. $-4x^5 + x^2$

42. $2x^6 - 4x$

*Add. See Example 4.*

43. $3m^2 + 5m$
    $\underline{2m^2 - 2m}$

44. $4a^3 - 4a^2$
    $\underline{6a^3 + 5a^2}$

45. $-6x^2 - 4x + 1$
    $\underline{-4x^2 + 5x + 5}$

46. $\phantom{-}7n^2 + 2n - 4$
    $\underline{-6n^2 - 3n - 8}$

47. $3w^3 - 2w^2 + 8w$
    $\underline{3w^3 + 2w^2\phantom{ + 8w}}$

48. $-5r^3 - 4r^2 - 3r$
    $\underline{-5r^3 + \phantom{4}r^2 + 3r}$

*Perform the indicated operations. See Examples 5 and 6.*

**49.** $(12x^4 - x^2) - (8x^4 + 3x^2)$

**50.** $(13y^5 - y^3) - (7y^5 + 5y^3)$

**51.** $(2r^2 + 3r - 12) + (6r^2 + 2r)$

**52.** $(3r^2 + 5r - 6) + (2r - 5r^2)$

**53.** $(8m^2 - 7m) - (3m^2 + 7m - 6)$

**54.** $(x^2 + x) - (3x^2 + 2x - 1)$

**55.** $(16x^3 - x^2 + 3x) + (-12x^3 + 3x^2 + 2x)$

**56.** $(-2b^6 + 3b^4 - b^2) + (b^6 + 2b^4 + 2b^2)$

**57.** $(12m^3 - 8m^2 + 6m + 7) - (5m^2 - 4)$

**58.** $(-3a^3 + 2a^2 - a + 6) - (a^2 + a - 1)$

**59.** Subtract $9x^2 - 3x + 7$ from $-2x^2 - 6x + 4$.

**60.** Subtract $-5w^3 + 5w^2 - 7$ from $6w^3 + 8w + 5$.

*Find the perimeter of each rectangle or triangle.*

**61.** 

Rectangle: top side $4x^2 + 3x + 1$, right side $x + 2$

**62.**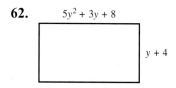

Rectangle: top side $5y^2 + 3y + 8$, right side $y + 4$

**63.**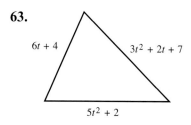

Triangle with sides $6t + 4$, $3t^2 + 2t + 7$, $5t^2 + 2$

**64.**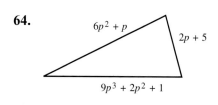

Triangle with sides $6p^2 + p$, $2p + 5$, $9p^3 + 2p^2 + 1$

**728** Chapter 10 Exponents and Polynomials

**RELATING CONCEPTS (Exercises 65–68)** FOR INDIVIDUAL OR GROUP WORK

*At the start of this chapter, we gave a polynomial that models the distance in feet that a car going approximately 68 mph will skid in t seconds. If we let D represent the skidding distance, then we have this formula.*

$$D = 100t - 13t^2$$

*Use this formula as you* **work Exercises 65–68 in order.**

**65.** To find the distance that the car will skid in 1 second, evaluate the polynomial when $t = 1$. Finish the work that is started below.

$D = 100t - 13t^2$     Replace $t$ with 1.
$D = 100(1) - 13(1)^2$
$D =$

**66.** From Exercise 65, you should have found that when $t$ is 1 second, then $D$ is 87 ft. Now evaluate the polynomial for the other values of $t$ listed in the table at the right. Write the ordered pairs in the table. Round distance ($D$) to the nearest whole number.

| t (sec) | D (ft) | (t, D) |
|---|---|---|
| 0 | | |
| 0.5 | | |
| 1 | 87 | (1, 87) |
| 1.5 | | |
| 2 | | |
| 2.5 | | |
| 3 | | |
| 3.5 | | |

**67.** Graph each of the ordered pairs by making a dot at the appropriate point on the grid.

**68.** Connect the points on the grid with a smooth curve. Describe how this graph is different from the graphs of linear equations in **Section 9.5.**

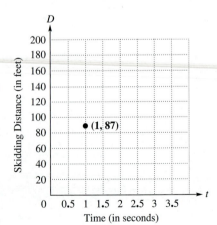

## 10.5 MULTIPLYING POLYNOMIALS: AN INTRODUCTION

**OBJECTIVES**
1. Multiply a monomial and a polynomial.
2. Multiply two polynomials.

### 1 Multiply a monomial and a polynomial.
As shown in **Section 10.1**, we find the product of two monomials by using the rules for exponents and the commutative and associative properties. Here is an example.

$$(-8m^6)(-9n^6) = (-8)(-9)(m^6)(n^6) = 72m^6n^6$$

**CAUTION**

Do not confuse *addition* of terms with *multiplication* of terms. For example,

$$7q^5 + 2q^5 = (7+2)q^5 = 9q^5, \quad \text{but} \quad (7q^5)(2q^5) = (7 \cdot 2)q^{5+5} = 14q^{10}.$$

To find the product of a monomial and a polynomial with more than one term, we use the distributive property and multiplication of monomials.

### Example 1 Multiplying a Monomial and a Polynomial

Use the distributive property to find each product.

**(a)** $4x^2(3x + 5)$

$$4x^2(3x + 5) = 4x^2(3x) + 4x^2(5) \quad \text{Distributive property}$$
$$= 12x^3 + 20x^2 \quad \text{Multiply monomials.}$$

**(b)** $-8m^3(4m^3 + 3m^2 + 2m)$

$$= -8m^3(4m^3) + (-8m^3)(3m^2) + (-8m^3)(2m) \quad \text{Distributive property}$$
$$= -32m^6 + (-24m^5) + (-16m^4) \quad \text{Multiply monomials.}$$
$$= -32m^6 - 24m^5 - 16m^4$$

**Work Problem ❶ at the Side.**

### 2 Multiply two polynomials.
We can use the distributive property repeatedly to find the product of any two polynomials. For example, to find the product of the polynomials $x^2 + 3x + 5$ and $x - 4$, think of $x - 4$ as a single quantity and use the distributive property as follows.

$$(x^2 + 3x + 5)(x - 4) = x^2(x - 4) + 3x(x - 4) + 5(x - 4)$$

Now use the distributive property three times to find $x^2(x - 4)$, $3x(x - 4)$, and $5(x - 4)$.

$$x^2(x - 4) + 3x(x - 4) + 5(x - 4)$$
$$= x^2(x) + x^2(-4) + 3x(x) + 3x(-4) + 5(x) + 5(-4)$$
$$= (x^3) + (-4x^2) + (3x^2) + (-12x) + (5x) + (-20)$$
$$= x^3 - x^2 - 7x - 20 \quad \text{Combine like terms.}$$

This example suggests the following rule.

---

**❶** Find each product.

**(a)** $5m^3(2m + 7)$

**(b)** $2x^4(3x^2 + 2x - 5)$

**(c)** $-4y^2(3y^3 + 2y^2 - 4y + 8)$

**ANSWERS**
1. (a) $10m^4 + 35m^3$ (b) $6x^6 + 4x^5 - 10x^4$
   (c) $-12y^5 - 8y^4 + 16y^3 - 32y^2$

## 2 Multiply.

(a) $(k + 6)(k + 7)$

(b) $(n - 2)(4n + 3)$

(c) $(2m + 1)(2m^2 + 4m - 3)$

(d) $(6p^2 - 2p - 4)(3p^2 + 5)$

## 3 Find the product.

$3x^2 + 4x - 5$
$x + 4$

**ANSWERS**

2. (a) $k^2 + 13k + 42$  (b) $4n^2 - 5n - 6$
   (c) $4m^3 + 10m^2 - 2m - 3$
   (d) $18p^4 - 6p^3 + 18p^2 - 10p - 20$
3. $3x^3 + 16x^2 + 11x - 20$

---

### Multiplying Polynomials

To multiply two polynomials, multiply each term of the second polynomial by each term of the first polynomial and add the products.

#### Example 2 Multiplying Two Polynomials

(a) Multiply $(x + 1)(x - 4)$.

Multiply each term of the second polynomial by each term of the first. Then combine like terms.

$$(x + 1)(x - 4) = x(x) + x(-4) + 1(x) + 1(-4)$$
$$= x^2 + (-4x) + x + (-4)$$
$$= x^2 - 3x - 4 \quad \text{Combine like terms.}$$

(b) Multiply $(m^2 + 5)(4m^3 - 2m^2 + 4m)$.

Multiply each term of the second polynomial by each term of the first.

$$(m^2 + 5)(4m^3 - 2m^2 + 4m)$$
$$= m^2(4m^3) + m^2(-2m^2) + m^2(4m) + 5(4m^3) + 5(-2m^2) + 5(4m)$$
$$= (4m^5) + (-2m^4) + (4m^3) + (20m^3) + (-10m^2) + (20m)$$

Now combine like terms.
$$= 4m^5 - 2m^4 + 24m^3 - 10m^2 + 20m$$

**Work Problem 2 at the Side.**

When at least one of the factors in a product of polynomials has three or more terms, the multiplication can be simplified by writing one polynomial above the other vertically.

#### Example 3 Multiplying Polynomials Vertically

Multiply $(x^3 + 2x^2 + 4x + 1)(3x + 5)$ using the vertical method.
Write the polynomials as follows.

$$x^3 + 2x^2 + 4x + 1$$
$$3x + 5$$

It is not necessary to line up terms in columns, because any terms may be multiplied (not just like terms). Here are the steps we will take.

*Step 1*  Multiply each term in the top row by 5.

*Step 2*  Multiply each term in the top row by $3x$. Be careful to place like terms in columns.

*Step 3*  Add like terms.

$$
\begin{array}{r}
x^3 + 2x^2 + 4x + 1 \\
3x + 5 \\
\hline
5x^3 + 10x^2 + 20x + 5 \\
3x^4 + 6x^3 + 12x^2 + 3x \\
\hline
3x^4 + 11x^3 + 22x^2 + 23x + 5
\end{array}
$$

← Step 1  $5(x^3 + 2x^2 + 4x + 1)$
← Step 2  $3x(x^3 + 2x^2 + 4x + 1)$
← Step 3  Add like terms.

The product is $3x^4 + 11x^3 + 22x^2 + 23x + 5$.

**Work Problem 3 at the Side.**

## 10.5 Exercises

*Find each product. See Example 1.*

1. $5y(8y^2 - 3)$
2. $3a(4a^2 + 6a)$
3. $-2m(3m + 2)$
4. $-6p(3p - 1)$
5. $4x^2(6x^2 - 3x + 2)$
6. $7n^2(5n^2 + n - 4)$
7. $-3k^3(2k^3 - 3k^2 - k + 1)$
8. $-2b^4(-3b^3 - b^2 + 6b - 8)$

*Find each product. See Examples 2 and 3.*

9. $(n - 2)(n + 3)$
10. $(r - 6)(r + 8)$
11. $(4r + 1)(2r - 3)$
12. $(5x + 2)(2x - 7)$
13. $(6x + 1)(2x^2 + 4x + 1)$
14. $(9y - 2)(8y^2 - 6y + 1)$
15. $(4m + 3)(5m^3 - 4m^2 + m - 5)$
16. $(y + 4)(3y^3 - 2y^2 + y + 3)$
17. $(5x^2 + 2x + 1)(x^2 - 3x + 5)$
18. $(2m^2 + m - 3)(m^2 - 4m + 5)$

**19.** $(x + 3)(x + 4)$

**20.** $(x + 5)(x + 2)$

**21.** $(2x + 1)(x^2 + 3x)$

**22.** $(x + 4)(3x^2 + 2x)$

*Match each product with the correct polynomial.*

**23.** 
| Product | | Polynomial |
|---|---|---|
| $(a + 4)(a + 5)$ | ___ | **A.** $a^2 - a - 20$ |
| $(a - 4)(a + 5)$ | ___ | **B.** $a^2 + a - 20$ |
| $(a + 4)(a - 5)$ | ___ | **C.** $a^2 - 9a + 20$ |
| $(a - 4)(a - 5)$ | ___ | **D.** $a^2 + 9a + 20$ |

**24.**
| Product | | Polynomial |
|---|---|---|
| $(x - 5)(x + 3)$ | ___ | **A.** $x^2 + 8x + 15$ |
| $(x + 5)(x + 3)$ | ___ | **B.** $x^2 - 8x + 15$ |
| $(x - 5)(x - 3)$ | ___ | **C.** $x^2 - 2x - 15$ |
| $(x + 5)(x - 3)$ | ___ | **D.** $x^2 + 2x - 15$ |

### RELATING CONCEPTS (Exercises 25–30)  FOR INDIVIDUAL OR GROUP WORK

**Work Exercises 25–30 in order.** *(All units are in yards.) Refer to the figure as necessary.*

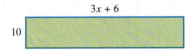

**25.** Find a polynomial that represents the area of the rectangle.

**26.** Suppose you know that the area of the rectangle is 600 yd². Use this information and the polynomial from Exercise 25 to write an equation in $x$, and solve it.

**27.** What are the dimensions of the rectangle?

**28.** Suppose the rectangle represents a lawn and it costs $3.50 per square yard to lay sod on the lawn. How much will it cost to sod the entire lawn?

**29.** Use the result of Exercise 27 to find the perimeter of the lawn.

**30.** Again, suppose the rectangle represents a lawn and it costs $9.00 per yard to fence the lawn. How much will it cost to fence the lawn?

**31.** Perform the following multiplications: $(x + 4)(x - 4)$; $(y + 2)(y - 2)$; $(r + 7)(r - 7)$. Observe your answers, and explain the pattern that can be found in the answers.

# Chapter 10

## SUMMARY

### KEY TERMS

**10.3** **scientific notation**    A number written as $a \times 10^n$, where $1 \leq |a| < 10$ and $n$ is an integer, is in scientific notation.

**10.4** **polynomial**    A polynomial is a term or the sum of terms with whole number exponents.

**descending powers**    A polynomial in $x$ is written in descending powers if the exponents on $x$ in its terms are in decreasing order.

**degree of a term**    The degree of a term is the sum of the exponents on the variables in that term.

**degree of a polynomial**    The degree of a polynomial is the highest degree of any term of the polynomial.

**monomial**    A monomial is a polynomial with one term.

**binomial**    A binomial is a polynomial with two terms.

**trinomial**    A trinomial is a polynomial with three terms.

### NEW SYMBOLS

$a^{-n}$    $a$ to the negative $n$ power

$a^0$    $a$ to the zero power

### TEST YOUR WORD POWER

*See how well you have learned the vocabulary in this chapter. Answers follow the Quick Review.*

1. A **polynomial** is an algebraic expression made up of
   (a) a term or the product of several terms with whole number exponents
   (b) a term or the sum of several terms with whole number exponents
   (c) the product of two or more terms with negative exponents
   (d) the sum of two or more terms with negative exponents.

2. The **degree of a term** is
   (a) the number of variables in the term
   (b) the product of the exponents on the variables
   (c) the smallest exponent on the variables
   (d) the sum of the exponents on the variables.

3. A **trinomial** is a polynomial with
   (a) only one term
   (b) exactly two terms
   (c) exactly three terms
   (d) more than three terms.

4. A **binomial** is a polynomial with
   (a) only one term
   (b) exactly two terms
   (c) exactly three terms
   (d) more than three terms.

5. A **monomial** is a polynomial with
   (a) only one term
   (b) exactly two terms
   (c) exactly three terms
   (d) more than three terms.

6. An example of a polynomial written in **descending powers** is
   (a) $7x + 6x^2 - 4x^3 - 2x^4$
   (b) $x^6 - 2x^5 + 3x^2 - 6x$
   (c) $16 + x^5 - x^3 + x^2$
   (d) $3x^4 - 3x^2 - 3 - x$.

7. An example of a number written in **scientific notation** is
   (a) $0.25 \times 10^5$
   (b) $57.6 \times 10^{-3}$
   (c) $4.42 \times 100^6$
   (d) $3.61 \times 10^{-4}$.

## QUICK REVIEW

| Concepts | Examples |
|---|---|

**10.1 The Product Rule and Power Rules for Exponents**

For any integers $m$ and $n$:

**Product rule**   $a^m \cdot a^n = a^{m+n}$

**Power rule (a)**   $(a^m)^n = a^{mn}$

**Power rule (b)**   $(ab)^m = a^m b^m$

**Power rule (c)**   $\left(\dfrac{a}{b}\right)^m = \dfrac{a^m}{b^m}$   $(b \neq 0)$

$2^4 \cdot 2^5 = 2^{4+5} = 2^9$

$(3^4)^2 = 3^{4 \cdot 2} = 3^8$

$(6a)^5 = 6^5 a^5$

$\left(\dfrac{2}{3}\right)^4 = \dfrac{2^4}{3^4}$

---

**10.2 Integer Exponents and the Quotient Rule**

If $a \neq 0$, for integers $m$ and $n$:

**Zero exponent**   $a^0 = 1$

**Negative exponent**   $a^{-n} = \dfrac{1}{a^n}$

**Quotient rule**   $\dfrac{a^m}{a^n} = a^{m-n}$

$15^0 = 1$

$5^{-2} = \dfrac{1}{5^2} = \dfrac{1}{25}$

$\dfrac{4^8}{4^3} = 4^{8-3} = 4^5$

---

**10.3 An Application of Exponents: Scientific Notation**

To write a number in scientific notation (as $a \times 10^n$), move the decimal point to the right of the first nonzero digit. If the decimal point is moved $n$ places to the left, and this makes the number smaller, $n$ is positive; otherwise, $n$ is negative. If the decimal point is not moved, $n$ is 0.

$247 = 2.47 \times 10^2$

$0.0051 = 5.1 \times 10^{-3}$

$4.8 = 4.8 \times 10^0$

$3.25 \times 10^5 = 325{,}000$

$8.44 \times 10^{-6} = 0.00000844$

---

**10.4 Adding and Subtracting Polynomials**

**Addition:** Add like terms.

Add: $\begin{array}{r} 2x^2 + 5x - 3 \\ 5x^2 - 2x + 7 \\ \hline 7x^2 + 3x + 4 \end{array}$

**Subtraction:** Change the signs of all the terms in the second polynomial and add to the first polynomial.

$(2x^2 + 5x - 3) - (5x^2 - 2x + 7)$
$= (2x^2 + 5x - 3) + (-5x^2 + 2x - 7)$
$= -3x^2 + 7x - 10$

---

**10.5 Multiplying Polynomials**

Multiply each term of the first polynomial by each term of the second polynomial. Then add like terms.

Multiply:
$\begin{array}{r} 3x^3 - 4x^2 + 2x - 7 \\ 4x + 3 \\ \hline 9x^3 - 12x^2 + 6x - 21 \\ 12x^4 - 16x^3 + 8x^2 - 28x \\ \hline 12x^4 - 7x^3 - 4x^2 - 22x - 21 \end{array}$

---

**ANSWERS TO TEST YOUR WORD POWER**

**1. (b)** *Example:* $5x^3 + 2x^2 - 7$   **2. (d)** *Examples:* The term 6 has degree 0, the term $3x$ has degree 1, the term $-2x^8$ has degree 8, and the term $5x^2y^4$ has degree 6.   **3. (c)** *Example:* $2a^2 - 3ab + b^2$
**4. (b)** *Example:* $3t^3 + 5t$   **5. (a)** *Examples:* $-5$ and $4xy^5$   **6. (b)** *Example:* $3n^4 + n^3 - 2n + 8$
**7. (d)** *Example:* $2.5 \times 10^4$ is written in scientific notation.

# Chapter 10 REVIEW EXERCISES

**[10.1]** *Use the product rule or power rules to simplify each expression. Write the answer in exponential form.*

1. $4^3 \cdot 4^8$
2. $(-5)^6(-5)^5$
3. $(-8x^4)(9x^3)$
4. $(2x^2)(5x^3)(x^9)$

5. $(19x)^5$
6. $(-4y)^7$
7. $5(pt)^4$
8. $\left(\dfrac{7}{5}\right)^6$

9. $(3x^2y^3)^3$
10. Explain why the product rule for exponents does not apply to the expression $7^2 + 7^4$.

**[10.2]** *Evaluate each expression.*

11. $5^0 + 8^0$
12. $2^{-5}$
13. $10w^0$
14. $4^{-2} + 4^{-1}$

*Simplify. Write each answer in exponential form, using only positive exponents. Assume all variables are nonzero.*

15. $\dfrac{6^{-3}}{6^{-5}}$
16. $\dfrac{x^{-7}}{x^{-9}}$
17. $\dfrac{p^{-8}}{p^4}$
18. $\dfrac{r^{-2}}{r^{-6}}$

19. $\dfrac{5^4}{5^5}$
20. $10^3(10^{-10})$
21. $n^{-2} \cdot n^3 \cdot n^{-4}$
22. $(10^4)(10^{-1})$

23. $(2^{-4})(2^{-4})$
24. $h^{10} \cdot h^{-4} \cdot h^{-3}$
25. $\dfrac{10^{-4}}{10^9}$
26. $\dfrac{x^2}{x^{-4}}$

**[10.3]** *Write each number in scientific notation.*

27. 48,000,000
28. 28,988,000,000
29. 0.000065
30. 0.0000000824

*Write each number without exponents.*

31. $2.4 \times 10^4$
32. $7.83 \times 10^7$
33. $8.97 \times 10^{-7}$
34. $9.95 \times 10^{-12}$

**Chapter 10** Exponents and Polynomials

*Perform the indicated operations and write the answers in scientific notation and then without exponents.*

**35.** $(2 \times 10^{-3})(4 \times 10^5)$

**36.** $\dfrac{8 \times 10^4}{2 \times 10^{-2}}$

**37.** $\dfrac{12 \times 10^{-8}}{4 \times 10^{-3}}$

**38.** $\dfrac{(2.5 \times 10^5)(4.8 \times 10^{-4})}{(7.5 \times 10^8)(1.6 \times 10^{-5})}$

*Solve each application problem. Write your answers both in scientific notation and as numbers without exponents.*

**39.** The population of the United States in 2000 was about $2.81 \times 10^8$ people spread over $3.72 \times 10^6$ square miles of land. What is the average number of people per square mile? (*Source:* U.S. Bureau of the Census.)

**40.** It takes about 36 seconds at a speed of $3.0 \times 10^5$ km per second for light from the sun to reach Venus. (*Source:* World Almanac and Book of Facts, 2001.) How far is Venus from the sun?

**[10.4]** *Combine terms where possible in each polynomial. Write the answer in descending powers of the variable. Give the degree of the answer. Identify the polynomial as a* monomial, binomial, trinomial, *or* none of these.

**41.** $9m^2 + 11m^2 + 2m^2$

**42.** $-4p + p^3 - p^2 + 8p + 2$

**43.** $12a^5 - 9a^4 + 8a^3 + 2a^2 - a + 3$

**44.** $-7y^5 - 8y^4 - y^5 + y^4 + 9y$

*Add or subtract as indicated.*

**45.** Add.
$-2a^3 + 5a^2$
$\underline{-3a^3 - \phantom{5}a^2}$

**46.** Add.
$4r^3 - 8r^2 + 6r$
$\underline{-2r^3 + 5r^2 + 3r}$

**47.** Subtract $-5y^2 + 2y - 7$ from $6y^2 - 8y + 2$.

**48.** $(2m^3 - 8m^2 + 4) + (8m^3 + 2m^2 - 7)$

**49.** $(-5y^2 + 3y + 11) + (4y^2 - 7y + 15)$

**50.** $(6p^2 - p - 8) - (-4p^2 + 2p + 3)$

**51.** $(12r^4 - 7r^3 + 2r^2) - (5r^4 - 3r^3 + 2r^2 + 1)$

**[10.5]** *Find each product.*

**52.** $5x(2x + 14)$

**53.** $-3p^3(2p^2 - 5p)$

**54.** $(3r - 2)(2r^2 + 4r - 3)$

**55.** $(2y + 3)(4y^2 - 6y + 9)$

**56.** $(5p^2 + 3p)(p^3 - p^2 + 5)$

**57.** $(3k - 6)(2k + 1)$

**58.** $(6p - 3q)(2p - 7q)$

**59.** $(m^2 + m - 9)(2m^2 + 3m - 1)$

**MIXED REVIEW EXERCISES**

*Perform the indicated operations. Write answers with positive exponents. Assume that no denominators are equal to 0.*

**60.** $19^0 - 3^0$

**61.** $(3p)^4$

**62.** $7^{-2}$

**63.** $-m^5(8m^2 + 10m + 6)$

**64.** $2^{-1} + 4^{-1}$

**65.** $(a + 2)(a^2 - 4a + 1)$

**66.** $(5y^3 - 8y^2 + 7) - (-3y^3 + y^2 + 2)$

**67.** Write a polynomial that represents the perimeter and one that represents the area of this rectangle.

$2x - 3$

$x + 2$

## Focus on Real-Data Applications

### Earthquake Intensities Measured by the Richter Scale

Charles F. Richter devised a scale in 1935 to compare the intensities, or relative power, of earthquakes. The **intensity** of an earthquake is measured relative to the intensity of a standard **zero-level** earthquake of intensity $I_0$. The relationship is equivalent to $I = I_0 \times 10^R$, where R is the **Richter scale** measure. For example, if an earthquake has magnitude 5.0 on the Richter scale, then its intensity is calculated as $I = I_0 \times 10^{5.0} = I_0 \times 100{,}000$, which is 100,000 times as intense as a zero-level earthquake. The following diagram illustrates the intensities of earthquakes and their Richter scale magnitudes.

| Intensity | $I_0$ | $I_0 \times 10^1$ | $I_0 \times 10^2$ | $I_0 \times 10^3$ | $I_0 \times 10^4$ | $I_0 \times 10^5$ | $I_0 \times 10^6$ | $I_0 \times 10^7$ | $I_0 \times 10^8$ |
|---|---|---|---|---|---|---|---|---|---|
| Richter Scale | 0 | 1 | 2 | 3 | 4 | 5 | 6 | 7 | 8 |

To compare two earthquakes to each other, a ratio of the intensities is calculated. For example, to compare an earthquake that measures 8.0 on the Richter scale to one that measures 5.0, simply find the ratio of the intensities.

$$\frac{\text{intensity } 8.0}{\text{intensity } 5.0} = \frac{I_0 \times 10^{8.0}}{I_0 \times 10^{5.0}} = \frac{10^8}{10^5} = 10^{8-5} = 10^3 = 1000$$

Therefore an earthquake that measures 8.0 on the Richter scale is 1000 times as intense as one that measures 5.0.

### For Group Discussion

The table gives Richter scale measurements for several earthquakes.

| | Earthquake | Richter Scale Measurement |
|---|---|---|
| 1960 | Concepción, Chile | 9.5 |
| 1906 | San Francisco, California | 8.3 |
| 1939 | Erzincan, Turkey | 8.0 |
| 1998 | Sumatra, Indonesia | 7.0 |
| 1998 | Adana, Turkey | 6.3 |

*Source: World Almanac and Book of Facts, 2001.*

1. Compare the intensity of the 1939 Erzincan earthquake to the 1998 Sumatra earthquake.

2. Compare the intensity of the 1998 Adana earthquake to the 1906 San Francisco earthquake.

3. Compare the intensity of the 1939 Erzincan earthquake to the 1998 Adana earthquake.

4. Suppose an earthquake measures 7.2 on the Richter scale. How would the intensity of a second earthquake compare if its Richter scale measure differed by $+3.0$? by $-1.0$?

# Chapter 10 Test

*Study Skills Workbook*
**Activity 12**

*Evaluate each expression.*

1. $5^{-4}$

2. $(-3)^0 + 4^0$

3. $4^{-1} + 3^{-1}$

*Simplify, and write each answer using only positive exponents. Assume that variables represent nonzero numbers.*

4. $6^{-3} \cdot 6^4$

5. $12(xy)^3$

6. $\dfrac{10^5}{10^9}$

7. $r^{-4} \cdot r^{-4} \cdot r^3$

8. $(7x^2)(-3x^5)$

9. $\dfrac{m^{-5}}{m^{-8}}$

10. $(3a^4b)^2$

11. $3^4 + 3^2$

*Write each number in scientific notation.*

12. 344,000,000,000

13. 0.00000557

*Write each number without exponents.*

14. $2.96 \times 10^7$

15. $6.07 \times 10^{-8}$

1. _____
2. _____
3. _____
4. _____
5. _____
6. _____
7. _____
8. _____
9. _____
10. _____
11. _____
12. _____
13. _____
14. _____
15. _____

**16.** The mass of the sun is about 330,000 times the mass of Earth. If Earth's mass is about $5.98 \times 10^{24}$ kg, find the approximate mass of the sun. Write your answer in scientific notation.

*For each polynomial, combine like terms when possible, and write the polynomial in descending powers of the variable. Give the degree of the simplified polynomial. Decide whether the simplified polynomial is a monomial, binomial, trinomial, or none of these.*

**17.** $5x^2 + 8x - 12x^2$

**18.** $13n^3 - n^2 + n^4 + 3n^4 - 9n^2$

*Perform the indicated operations.*

**19.** $(5t^4 - 3t^2 + 3) - (t^4 - t^2 + 3)$

**20.** $(2y^2 - 8y + 8) + (-3y^2 + 2y + 3)$

**21.** Subtract $9t^3 + 8t^2 - 6$ from $9t^3 - 4t^2 + 2$.

**22.** $-4x^3(3x^2 - 5x)$

**23.** $(y + 2)(3y - 1)$

**24.** $(2r - 3)(r^2 + 2r - 5)$

**25.** Write a polynomial that represents the perimeter of this square, and a polynomial that represents the area.

$3x + 9$

# Cumulative Review Exercises  Chapters 1–10

1. Write these numbers in words.
   (a) 10.035
   (b) 410,000,351,109

2. Write these numbers in digits.
   (a) Three hundred million, six thousand, eighty
   (b) Fifty-five ten-thousandths

3. Round each number as indicated.
   (a) 0.8029 to the nearest hundredth
   (b) 340,519,000 to the nearest million
   (c) 14.973 to the nearest tenth

4. Find the mean, median, and mode for this set of ticket prices: $41, $65, $37, $90, $41, $65, $48, $41, $59, $40.

*Simplify.*

5. $-12 + 7.829$

6. $7 + 5(3 - 8)$

7. $1\dfrac{2}{3} + \dfrac{3}{5}$

8. $\dfrac{(-6)^2 + 9(0 - 4)}{10 \div 5(-7 + 5) - 10}$

9. $\dfrac{8x^2}{9} \cdot \dfrac{12w}{2x^3}$

10. $\dfrac{-0.7}{5.6}$

11. $\dfrac{10^5}{10^6}$

12. $2\dfrac{1}{4} - 2\dfrac{5}{6}$

13. $(-6)^2 + (-2)^3$

14. $(-4y^3)(3y^4)$

15. $\dfrac{5b}{2a^2} \div \dfrac{3b^2}{10a}$

16. $\dfrac{r}{8} + \dfrac{6}{t}$

17. $\dfrac{-4(6)}{3^3 - 27}$

18. $\dfrac{5}{6}$ of 900

19. $(-0.003)(0.04)$

20. $(2a^3b^2)^4$

21. $\dfrac{4}{9} - \dfrac{6}{m}$

22. $t^5 \cdot t^{-2} \cdot t^{-4}$

23. $\dfrac{-\dfrac{15}{16}}{-6}$

24. $\left(-\dfrac{1}{2}\right)^4 + 6\left(\dfrac{2}{3}\right)$

25. $\dfrac{n^{-3}}{n^{-4}}$

26. $-8 - 88$

27. $8^0 + 2^{-1}$

28. $9 - 7\dfrac{4}{5}$

*Evaluate each expression when $w = -2$, $x = 4$, and $y = -1$.*

**29.** $-3xy^3$

**30.** $15y - 6w$

**31.** $x^2 - 2xy + 6$

*Solve each equation. Show your work.*

**32.** $4 + h = 3h - 6$

**33.** $-1.65 = 0.5x + 2.3$

**34.** $3(a - 5) = -3 + a$

**35.** Translate into an equation and solve. If three times a number is subtracted from 20, the result is twice the number. Find the number.

**36.** Use the six problem-solving steps from Chapter 3. A mouse takes 16 times as many breaths as an elephant. (*Source: Dinosaurs, Spitfires, and Dragons.*) If the two animals together take 170 breaths per minute, how many breaths does each take?

**37.** The Cardinals' pitcher gave up 78 runs in 234 innings. At that rate, how many runs will he give up in a 9-inning game?

**38.** This month Dallas received 2.61 inches of rain. This is what percent of the normal rainfall of 1.8 inches?

**39.** Calbert bought an automatic focus camera for $64.95. He paid $7\frac{1}{2}\%$ sales tax. Find the amount of tax, to the nearest cent, and the total cost of the camera.

**40.** The distance that light will travel in one year is $5.87 \times 10^{12}$ miles. How far will light travel in one day? (Assume there are 365 days in a year.) Write your answer both in scientific notation and without exponents.

**41.** Use the table on toddler shoe sizes.

**TODDLER SHOES**
Trace bare foot, then measure from heel to longest toe.

| Toddler | |
|---|---|
| Foot Length (in.) | Shoe Size |
| $4\frac{13}{16}$ | 5 |
| $5\frac{1}{8}$ | 6 |
| $5\frac{1}{2}$ | 7 |
| $5\frac{13}{16}$ | 8 |

*Source:* Lands' End.

(a) What foot length corresponds to size 6 shoes?

(b) If Kamesha measured her child's foot at $5\frac{7}{16}$ in., what shoe size should she order?

(c) What is the difference in length between a size 5 and a size 6 shoe?

**42.** Find the perimeter and area of this shape.

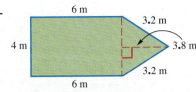

**43.** Find the circumference and area of this circle. Use 3.14 for $\pi$ and round answers to the nearest tenth.

# Whole Numbers Computation: Pretest

This test will check your skills in doing whole numbers computation, using paper and pencil. Each part of the test is keyed to a section in the Review Chapter, which follows this test. Based on your test results, work the appropriate section(s) in the Review Chapter *before* you start Chapter 1.

**Adding Whole Numbers**   *(Do not use a calculator.)*

1. 368
   + 22

2. 7093
   + 6073

3. 85
   + 2968

4. 57,208
   915
   + 59,387

5. 714 + 3728 + 9 + 683,775

1. _____
2. _____
3. _____
4. _____
5. _____

**Subtracting Whole Numbers**   *(Do not use a calculator.)*

1. 426
   − 76

2. 3358
   − 2729

3. 30,602
   − 5708

4. 4006 − 97

5. 679,420 − 88,033

1. _____
2. _____
3. _____
4. _____
5. _____

**Multiplying Whole Numbers**   *(Do not use a calculator.)*

1. 3 × 3 × 0 × 6

2. 3841
   × 7

3. (520)(3000)

1. _____
2. _____
3. _____

743

*Do not use a calculator and show your work.*

**4.** 71
  × 26

**5.** Multiply 359 and 48.

**6.** 853 × 609

**Dividing Whole Numbers** *(Do not use a calculator and show your work.)*

**1.** $3\overline{)69}$

**2.** 12 ÷ 0

**3.** $\dfrac{25{,}036}{4}$

**4.** $7\overline{)5655}$

**5.** $52\overline{)1768}$

**6.** 45,000 ÷ 900

**7.** $38\overline{)2300}$

**8.** $83\overline{)44{,}799}$

Now check your answers on page A–29 in the Answers section at the back of the book. Record the number of problems you worked correctly in each part of the test.

Adding Whole Numbers: _____ correct out of 5.
  If you got 0, 1, or 2 correct, work **Section R.1** in the Review Chapter.
Subtracting Whole Numbers: _____ correct out of 5.
  If you got 0, 1, or 2 correct, work **Section R.2** in the Review Chapter.
Multiplying Whole Numbers: _____ correct out of 6.
  If you got 0, 1, 2, or 3 correct, work **Section R.3** in the Review Chapter.
Dividing Whole Numbers: _____ correct out of 8.
  If you got 0, 1, or 2 correct, work **Sections R.4 and R.5** in the Review Chapter.
  If you got 3 or 4 correct, work **Section R.5** in the Review Chapter.

# Whole Numbers Review

**NOTE**

Use the Whole Numbers Computation Pretest on pages 743–44 to determine which sections you need to work in this chapter.

- R.1 Adding Whole Numbers
- R.2 Subtracting Whole Numbers
- R.3 Multiplying Whole Numbers
- R.4 Dividing Whole Numbers
- R.5 Long Division

## R.1 ADDING WHOLE NUMBERS

There are 4 triangles at the left and 2 at the right. In all, there are 6 triangles.

**OBJECTIVES**

1. Add two single-digit numbers.
2. Add more than two numbers.
3. Add when carrying is not required.
4. Add with carrying.
5. Use addition to solve application problems.
6. Check the sum in addition.

The process of finding the total is called *addition*. Here 4 and 2 were added to get 6. Addition is written with a + sign, so that

$$4 + 2 = 6.$$

**1** Add two single-digit numbers. In addition, the numbers being added are called **addends**, and the resulting answer is called the **sum** or **total**.

$$\begin{array}{r} 4 \leftarrow \text{Addend} \\ +2 \leftarrow \text{Addend} \\ \hline 6 \leftarrow \text{Sum (answer)} \end{array}$$

Addition problems can also be written horizontally as follows.

$$\underset{\text{Addend}}{4} \; + \; \underset{\text{Addend}}{2} \; = \; \underset{\text{Sum}}{6}$$

### Commutative Property of Addition

The **commutative property of addition** states that changing the *order* of the addends in an addition problem does not change the sum.

For example, the sum of $4 + 2$ is the same as the sum of $2 + 4$. This allows the addition of the same numbers in a different order.

745

# Chapter R  Whole Numbers Review

**1** Add. Then use the commutative property to write another addition problem and find the sum.

(a) 3 + 4

(b) 9 + 9

(c) 7 + 8

(d) 6 + 9

**2** Add the following columns of numbers.

(a) 5
4
6
9
+2

(b) 7
5
1
2
+6

(c) 9
2
1
3
+4

(d) 3
8
6
4
+8

**ANSWERS**

1. (a) 7: 4 + 3 = 7   (b) 18: no change
   (c) 15: 8 + 7 = 15   (d) 15: 9 + 6 = 15
2. (a) 26  (b) 21  (c) 19  (d) 29

### Example 1  Adding Two Single-Digit Numbers

Add. Then use the commutative property to write another addition problem and find the sum.

(a) 6 + 2 = 8   and   2 + 6 = 8

(b) 5 + 9 = 14   and   9 + 5 = 14

(c) 8 + 3 = 11   and   3 + 8 = 11

(d) 8 + 8 = 16 (No change occurs when the commutative property is used.)

**Work Problem 1 at the Side.**

### Associative Property of Addition

By the **associative property of addition,** changing the *grouping* of addends does not change the sum.

For example, the sum of 3 + 5 + 6 may be found in several ways.

(3 + 5) + 6 = 8 + 6 = 14    Parentheses tell you to add 3 + 5 first.

3 + (5 + 6) = 3 + 11 = 14    Parentheses tell you to add 5 + 6 first.

Either method gives a sum of 14.

**2** Add more than two numbers.   To add several numbers, first write them in a column. Add the first number to the second. Add this sum to the third number; continue until all the numbers are used.

### Example 2  Adding More Than Two Numbers

Add 2, 5, 6, 1, and 4.

2   ⎤ 2 + 5 = 7
5   ⎦
⑥         7 + 6 = 13
①              13 + 1 = 14
+ ④                  14 + 4 = 18
18

**Work Problem 2 at the Side.**

**NOTE**

By the commutative and associative properties of addition, you may also add numbers by starting at the bottom of a column. Adding down from the top or adding up from the bottom will give the same answer.

**3** Add when carrying is not required.   If numbers have two or more digits, first you must arrange the numbers in columns so that the ones digits are in the same column, tens are in the same column, hundreds are in the same column, and so on. Next, you add column by column, starting at the right.

## Example 3 Adding without Carrying

Add 511 + 23 + 154 + 10.

First line up the numbers in columns, with the ones column at the right.

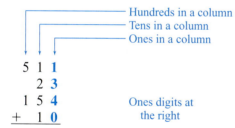

Now start at the right and add the ones digits. Add the tens digits next, and finally, the hundreds digits.

```
  5 1 1
    2 3
  1 5 4
+   1 0
  6 9 8
```
— Sum of ones
— Sum of tens
— Sum of hundreds

The sum of the four numbers is 698.

**Work Problem 3 at the Side.**

**4** **Add with carrying.** If the sum of the digits in a column is more than 9, use **carrying**.

## Example 4 Adding with Carrying

Add 47 and 29.

Add the digits in the ones column.

$$\begin{array}{r} 47 \\ +29 \end{array}$$

— Sum of ones is 16.

Because 16 is 1 ten plus 6 ones, write 6 ones in the ones column and *carry* 1 ten to the tens column.

Carry 1 ten.
```
  1
  47     7 + 9 = 16
+ 29
   6
```
Write 6 ones in ones column.

Add the digits in the tens column, including the carried 1.

```
  1
  47
+ 29
  76
```
— Sum of digits in tens column

**Work Problem 4 at the Side.**

---

**3** Add.

(a)  25
    +73

(b)  364
    +532

(c)  42,305
    +11,563

**4** Add by using carrying.

(a)  69
    +26

(b)  76
    +18

(c)  56
    +37

(d)  34
    +49

**ANSWERS**
3. (a) 98  (b) 896  (c) 53,868
4. (a) 95  (b) 94  (c) 93  (d) 83

**748** Chapter R Whole Numbers Review

**5** Add, carrying when necessary.

(a) 481
79
38
+395

(b) 4271
372
8976
+ 162

(c) 57
4
392
804
51
+ 27

(d) 7821
435
72
305
+1693

**6** Add with mental carrying.

(a) 278
825
14
3
7
+9275

(b) 3305
650
708
29
40
6
+ 3

(c) 15,829
765
78
15
9
7
+13,179

**Answers**
5. (a) 993  (b) 13,781  (c) 1335
   (d) 10,326
6. (a) 10,402  (b) 4741  (c) 29,882

## Example 5  Adding with Carrying

Add 324 + 7855 + 23 + 7 + 86.

*Step 1*  Add the digits in the ones column.

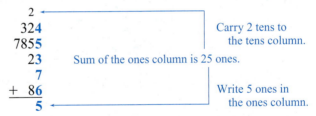

In 25, the 5 represents 5 ones and is written in the ones column, while the 2 represents 2 tens and is carried to the tens column.

*Step 2*  Now add the digits in the tens column, including the carried 2.

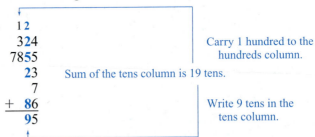

*Step 3*  Add the hundreds column, including the carried 1.

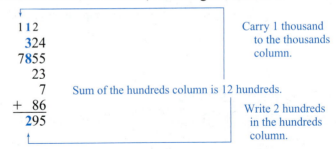

*Step 4*  Add the thousands column, including the carried 1.

```
  1 1 2
   324
  7855
    23
     7
  + 86
  8295
```
Sum of the thousands column is 8 thousands.

Thus, 324 + 7855 + 23 + 7 + 86 = 8295.

Work Problem **5** at the Side.

**NOTE**
For additional speed, try to carry mentally. Do not write the carried numbers, but just remember them. Try this method in Problem 6 at the side. If it works for you, use it.

Work Problem **6** at the Side.

Section R.1 Adding Whole Numbers    749

**5** Use addition to solve application problems.

**Example 6** Applying Addition Skills

On this map, the distance in miles from one location to another is written alongside the road. Find the shortest distance from Altamonte Springs to Clear Lake.

*Approach* Add the mileage along various routes to determine the distances from Altamonte Springs to Clear Lake. Then select the shortest route.

*Solution* One way from Altamonte Springs to Clear Lake is through Orlando. Add the mileage numbers along this route.

```
  8    Altamonte Springs to Pine Hills
  5    Pine Hills to Orlando
+ 8    Orlando to Clear Lake
 21 →  miles from Altamonte Springs to Clear Lake,
       going through Orlando
```

Another way is through Bertha and Winter Park. Add the mileage numbers along this route.

```
  5    Altamonte Springs to Castleberry
  6    Castleberry to Bertha
  7    Bertha to Winter Park
+ 7    Winter Park to Clear Lake
 25 →  miles from Altamonte Springs to Clear Lake
       through Bertha and Winter Park
```

The shortest way from Altamonte Springs to Clear Lake is 21 miles through Orlando.

**Work Problem 7 at the Side.**

**Example 7** Applying Addition Skills

Using the map in Example 6, find the total distance from Shadow Hills to Castleberry to Orlando and back to Shadow Hills through Clear Lake.

*Approach* Add the mileage from Shadow Hills to Castleberry to Orlando and back to Shadow Hills to find the total distance.

*Solution* Use the numbers from the map.

```
     9    Shadow Hills to Bertha
     6    Bertha to Castleberry
     5    Castleberry to Altamonte Springs
     8    Altamonte Springs to Pine Hills
     5    Pine Hills to Orlando
     8    Orlando to Clear Lake
  + 11    Clear Lake to Shadow Hills
    52 →  miles from Shadow Hills to Castleberry to
          Orlando and back to Shadow Hills
```

**Work Problem 8 at the Side.**

**7** Use the map to find the shortest distance from Conway to Pine Hills.

**8** The road is closed between Orlando and Clear Lake, so this route cannot be used. Find the next shortest distance from Orlando to Clear Lake.

**ANSWERS**
7. 29 miles through Belle Isle
8. Orlando to Pine Hills          5
   Pine Hills to Altamonte Springs   8
   Altamonte Springs to Castleberry  5
   Castleberry to Bertha          6
   Bertha to Winter Park          7
   Winter Park to Clear Lake    + 7
                                 38 miles

## Chapter R  Whole Numbers Review

**9** Check each addition. If the sum is incorrect, find the correct sum.

(a)  32
     8
     5
   +14
   ———
    59

(b)  872
     539
      46
   +152
   ———
   1609

(c)   79
     218
       7
   +639
   ———
    953

(d)  21,892
     11,746
   +43,925
   ———————
     79,563

**6** **Check the sum in addition.** Checking the answer is an important part of problem solving. A common method for checking addition is to re-add from the bottom to top. This is an application of the commutative and associative properties of addition.

### Example 8  Checking Addition

Check this sum.

Add down.
```
  1428
   738
    63
   125
    17
 + 485
 ―――――
  1428
```
To check, add up.

Adding down and adding up should give the same sum. In this case, the answers agree, so the sum is probably correct.

### Example 9  Checking Addition

Check each sum.

(a)
```
   785      1033
    63       785
 + 185        63
 ―――――     + 185
  1033     ―――――
             1033
```
Correct, because both answers are the same.
To check, add up.

(b)
```
   635      2454
    73       635
   831        73
 + 915       831
 ―――――     + 915
  2444     ―――――
             2444
```
Error, because answers are different.
To check, add up.

Re-add to find that the correct sum is 2454.

**Work Problem 9 at the Side.**

---

**ANSWERS**

9. (a) correct  (b) correct
   (c) incorrect, should be 943
   (d) incorrect, should be 77,563

# R.1 Exercises

Section R.1   751

**FOR EXTRA HELP**  Student's Solutions Manual  MyMathLab.com  InterAct Math Tutorial Software  AW Math Tutor Center  www.mathxl.com  Digital Video Tutor CD 8 Videotape 21

*Add. Then use the commutative property to write another addition problem and find the sum. See Example 1.*

1. $3213 + 5715$

2. $6344 + 1655$

3. $38{,}204 + 21{,}020$

4. $63{,}251 + 36{,}305$

*Add, carrying as necessary. See Examples 2–5.*

5. $\begin{array}{r} 67 \\ +83 \\ \hline \end{array}$

6. $\begin{array}{r} 78 \\ +36 \\ \hline \end{array}$

7. $\begin{array}{r} 746 \\ +905 \\ \hline \end{array}$

8. $\begin{array}{r} 621 \\ +359 \\ \hline \end{array}$

9. $\begin{array}{r} 798 \\ +206 \\ \hline \end{array}$

10. $\begin{array}{r} 172 \\ +156 \\ \hline \end{array}$

11. $\begin{array}{r} 7968 \\ +1285 \\ \hline \end{array}$

12. $\begin{array}{r} 1768 \\ +8275 \\ \hline \end{array}$

13. $\begin{array}{r} 7896 \\ +3728 \\ \hline \end{array}$

14. $\begin{array}{r} 9382 \\ +7586 \\ \hline \end{array}$

15. $\begin{array}{r} 3705 \\ 3916 \\ +9037 \\ \hline \end{array}$

16. $\begin{array}{r} 6629 \\ 6076 \\ +8218 \\ \hline \end{array}$

17. $\begin{array}{r} 32 \\ +4977 \\ \hline \end{array}$

18. $\begin{array}{r} 402 \\ +9938 \\ \hline \end{array}$

19. $\begin{array}{r} 3077 \\ 8 \\ +421 \\ \hline \end{array}$

20. $\begin{array}{r} 56 \\ 7721 \\ +172 \\ \hline \end{array}$

21. $\begin{array}{r} 9056 \\ 78 \\ 6089 \\ +731 \\ \hline \end{array}$

22. $\begin{array}{r} 4022 \\ 709 \\ 8621 \\ +37 \\ \hline \end{array}$

23. $\begin{array}{r} 18 \\ 708 \\ 9286 \\ +636 \\ \hline \end{array}$

24. $\begin{array}{r} 1708 \\ 321 \\ 61 \\ +8926 \\ \hline \end{array}$

*Check each sum by adding from bottom to top. If an answer is incorrect, find the correct sum. See Examples 8 and 9.*

25. $\begin{array}{r} 179 \\ 214 \\ +376 \\ \hline 759 \end{array}$

26. $\begin{array}{r} 17 \\ 296 \\ 713 \\ +94 \\ \hline 1220 \end{array}$

27. $\begin{array}{r} 4713 \\ 28 \\ 615 \\ +64 \\ \hline 5420 \end{array}$

28. $\begin{array}{r} 6\ 215 \\ 744 \\ 36 \\ +4\ 284 \\ \hline 11{,}279 \end{array}$

*Using the map below, find the shortest distance between the following cities. See Examples 6 and 7.*

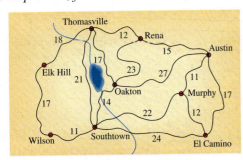

**29.** Southtown and Rena

**30.** Elk Hill and Oakton

**31.** Thomasville and Murphy

**32.** Murphy and Thomasville

*Solve each application problem.*

**33.** A basic auto tune-up costs $79, tire rotation is $24, and an oil change is $19. Find the total cost for all the services.

**34.** Jane Lim bought an 11-piece set of golf clubs for $120, a dozen golf balls for $9, golf shoes for $45, and a golf glove for $12. (*Source:* Sportmart.) How much did she spend in all?

**35.** There are 413 women and 286 men on the sales staff. How many people are on the sales staff?

**36.** One department in an office building has 283 employees while another department has 218 employees. How many employees are in the two departments?

**37.** According to the 2000 census, the two states with the highest population are California with 33,871,648 people and Texas with 20,851,820 people. How many people live in those two states? (*Source:* U.S. Bureau of the Census.)

**38.** The two states with the smallest population in 2000 were Wyoming with 493,782 people and Vermont with 608,827 people. What is the total population of the two states? (*Source:* U.S. Bureau of the Census.)

*Find the perimeter of (total distance around) each figure.*

**39.**

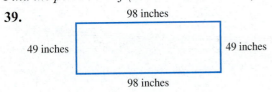

**40.**

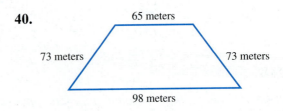

**41.**

**42.**

# R.2 SUBTRACTING WHOLE NUMBERS

Suppose you have $18, and you spend $15 for gasoline. You then have $3 left. There are two different ways of looking at these numbers.

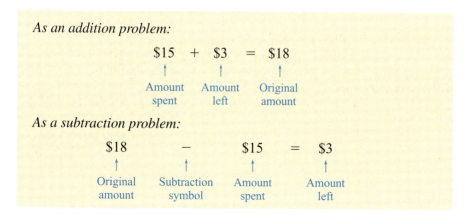

**OBJECTIVES**

1. Change addition problems to subtraction problems or the reverse.
2. Identify the minuend, subtrahend, and difference.
3. Subtract when no borrowing is needed.
4. Use addition to check subtraction answers.
5. Subtract with borrowing.
6. Use subtraction to solve application problems.

**1** **Change addition problems to subtraction problems or the reverse.** As this example shows, an addition problem can be changed to a subtraction problem and a subtraction problem can be changed to an addition problem.

### Example 1 Changing Addition Problems to Subtraction

Change each addition problem to a subtraction problem.

**(a)** $4 + 1 = 5$

Two subtraction problems are possible, as shown below.
$$5 - 1 = 4 \quad \text{or} \quad 5 - 4 = 1$$

These figures show each subtraction problem.

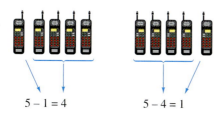

$5 - 1 = 4$     $5 - 4 = 1$

**(b)** $10 + 17 = 27$
$$27 - 17 = 10 \quad \text{or} \quad 27 - 10 = 17$$

**Work Problem 1 at the Side.**

**1** Write two subtraction problems for each addition problem.

(a) $4 + 3 = 7$

(b) $6 + 5 = 11$

(c) $150 + 220 = 370$

(d) $623 + 55 = 678$

### Example 2 Changing Subtraction Problems to Addition

Change each subtraction problem to an addition problem.

**(a)** $8 - 3 = 5$
$8 = 3 + 5$

It is also correct to write $8 = 5 + 3$ (using the commutative property).

*Continued on Next Page*

**ANSWERS**
1. (a) $7 - 3 = 4$ or $7 - 4 = 3$
   (b) $11 - 5 = 6$ or $11 - 6 = 5$
   (c) $370 - 220 = 150$ or $370 - 150 = 220$
   (d) $678 - 55 = 623$ or $678 - 623 = 55$

**2** Write an addition problem for each subtraction problem.

(a) $5 - 3 = 2$

(b) $8 - 3 = 5$

(c) $21 - 15 = 6$

(d) $58 - 42 = 16$

**3** Subtract.

(a) $\phantom{-}56$
$\underline{-31}$

(b) $\phantom{-}38$
$\underline{-14}$

(c) $\phantom{-}378$
$\underline{-235}$

(d) $\phantom{-}3927$
$\underline{-2614}$

(e) $\phantom{-}5464$
$\underline{-324}$

**ANSWERS**

2. (a) $5 = 3 + 2$ (b) $8 = 3 + 5$
   (c) $21 = 15 + 6$ (d) $58 = 42 + 16$
3. (a) $25$ (b) $24$ (c) $143$ (d) $1313$
   (e) $5140$

(b) $19 - 14 = 5$
$19 = 14 + 5$

(c) $290 - 130 = 160$
$290 = 130 + 160$

**Work Problem 2 at the Side.**

**2** **Identify the minuend, subtrahend, and difference.** In subtraction, as in addition, the numbers in a problem have names. For example, in the problem $8 - 5 = 3$, the number 8 is the **minuend,** 5 is the **subtrahend,** and 3 is the **difference** or answer.

$$8 - 5 = 3 \leftarrow \text{Difference (answer)}$$
$$\uparrow \phantom{-} \uparrow$$
$$\text{Minuend} \phantom{-} \text{Subtrahend}$$

$$\phantom{-}8 \leftarrow \text{Minuend}$$
$$\underline{-5} \leftarrow \text{Subtrahend}$$
$$\phantom{-}3 \leftarrow \text{Difference}$$

**3** **Subtract when no borrowing is needed.** Subtract two numbers by lining up the numbers in columns so that the digits in the ones place are in the same column. Next subtract by columns, starting at the right with the ones column.

**Example 3** Subtracting Two Numbers without Borrowing

Subtract.

(a) Ones digits are lined up in the same column.
$$\phantom{-}53$$
$$\underline{-21}$$
$$\phantom{-}32$$

$3 - 1 = 2$
$5 \text{ tens} - 2 \text{ tens} = 3 \text{ tens}$

(b) Ones digits are lined up.
$$\phantom{-}385$$
$$\underline{-161}$$
$$\phantom{-}224 \leftarrow 5 - 1 = 4$$

$8 \text{ tens} - 6 \text{ tens} = 2 \text{ tens}$
$3 \text{ hundreds} - 1 \text{ hundred} = 2 \text{ hundreds}$

(c) $$\phantom{-}9431$$
$$\underline{-\phantom{0}210}$$
$$\phantom{-}9221 \leftarrow 1 - 0 = 1$$

$3 \text{ tens} - 1 \text{ ten} = 2 \text{ tens}$
$4 \text{ hundreds} - 2 \text{ hundreds} = 2 \text{ hundreds}$
$9 \text{ thousands} - 0 \text{ thousands} = 9 \text{ thousands}$

**Work Problem 3 at the Side.**

**4** **Use addition to check subtraction answers.** You can check $8 - 3 = 5$ by *adding* 3 and 5.

$3 + 5 = 8$, so $8 - 3 = 5$ is correct.

## Example 4  Checking Subtraction

Check each answer.

(a)  89
    −47
    ───
     42

Rewrite as an addition problem, as shown in Example 2.

Subtraction problem $\begin{cases} 89 \\ -47 \\ \overline{42} \end{cases}$  Addition problem  $\begin{array}{r} 47 \\ +42 \\ \hline 89 \end{array}$

Because $47 + 42 = 89$, the subtraction was done correctly.

(b) $72 − 41 = 21$

Rewrite as an addition problem.

$$72 = 41 + 21$$

But, $41 + 21 = 62$, *not* 72, so the subtraction was done *incorrectly*. Rework the original subtraction to get the correct answer of 31. Then, $41 + 31 = 72$.

(c)   374 ←── Match ──┐
    − 141            │
    ─────  } $141 + 233 = 374$
     233  ←──────────┘

The answer checks.

━━━━━━━━━━━━━━ Work Problem ❹ at the Side.

**5** **Subtract with borrowing.** If a digit in the minuend is *less* than the one directly below it, we cannot subtract, so **borrowing** is necessary.

## Example 5  Subtracting with Borrowing

Subtract 19 from 57.

Write the problem in vertical format.

$$\begin{array}{r} 5\;7 \\ -\;1\;9 \\ \hline \end{array}$$

In the ones column, 7 is less than 9, so, in order to subtract, you must borrow 1 ten from the 5 tens.

5 tens − 1 ten = 4 tens ────→ 4  17 ←── 1 ten = 10 ones and 10 + 7 = 17
                              5̸  7̸
                            − 1  9
                            ──────

Now subtract $17 − 9$ in the ones column and then subtract 4 tens − 1 ten in the tens column.

         4  17
         5̸  7̸
       − 1  9  ←── Difference
       ──────
         3  8

Thus, $57 − 19 = 38$. Check by adding 19 and 38. You should get 57.

━━━━━━━━━━━━━━ Work Problem ❺ at the Side.

---

❹ Decide whether each answer is correct. If incorrect, find the correct answer.

(a)   65
    − 23
    ────
     42

(b)   46
    − 32
    ────
     24

(c)   374
    − 251
    ─────
     113

(d)  7531
   − 4301
   ──────
    3230

❺ Subtract.

(a)   67
    − 38

(b)   97
    − 29

(c)   31
    − 17

(d)  863
    − 47

(e)  762
   − 157

**Answers**
4. (a) correct  (b) incorrect, should be 14
   (c) incorrect, should be 123  (d) correct
5. (a) 29  (b) 68  (c) 14  (d) 816  (e) 605

**6** Subtract.

(a) 354
   − 82

(b) 457
   − 68

(c) 874
   − 486

(d) 1437
   − 988

(e) 8739
   − 3892

**ANSWERS**
6. (a) 272  (b) 389  (c) 388  (d) 449
   (e) 4847

### Example 6 — Subtracting with Borrowing

Subtract, borrowing when necessary.

(a)  7856
    − 137

Borrow 1 ten. → 1 ten = 10 ones and 10 + 6 = 16

$$\begin{array}{r} 4\ 16 \\ 7\ 8\ \cancel{5}\ \cancel{6} \\ -\ 1\ 3\ 7 \\ \hline 7\ 7\ 1\ 9 \end{array}$$ ← Difference

(b)  635
    − 546

Borrow 1 ten. → 1 ten = 10 ones; 10 + 5 = 15

$$\begin{array}{r} 2\ 15 \\ 6\ \cancel{3}\ \cancel{5} \\ -\ 5\ 4\ 6 \\ \hline 9 \end{array}$$ Need to borrow farther because 2 is less than 4 in tens column.

Borrow 1 hundred. 1 hundred = 10 tens; 10 tens + 2 tens = 12 tens

$$\begin{array}{r} 5\ 12\ 15 \\ \cancel{6}\ \cancel{3}\ \cancel{5} \\ -\ 5\ 4\ 6 \\ \hline 8\ 9 \end{array}$$ ← Difference

(c)  412
    − 225

$$\begin{array}{r} 0\ 12 \\ 4\ \cancel{1}\ \cancel{0} \\ -\ 2\ 2\ 5 \\ \hline 7 \end{array}$$ Need to borrow farther because 0 is less than 2.

$$\begin{array}{r} 3\ 10\ 12 \\ \cancel{4}\ \cancel{1}\ \cancel{2} \\ -\ 2\ 2\ 5 \\ \hline 1\ 8\ 7 \end{array}$$ ← Difference

**Work Problem 6 at the Side.**

Sometimes a minuend has zeros in some of the positions. In such cases, borrowing may be a little more complicated than what we have shown so far.

### Example 7 — Borrowing with Zeros

Subtract.

   4607
 − 3168

It is not possible to borrow from the tens position. Instead, you must first borrow from the hundreds position.

Borrow 1 hundred. → 1 hundred is 10 tens.

$$\begin{array}{r} 5\ 10 \\ 4\ \cancel{6}\ \cancel{0}\ 7 \\ -\ 3\ 1\ 6\ 8 \end{array}$$

Now borrow from the tens position.

$$\begin{array}{r} 9 \\ 5\ \cancel{10}\ 17 \\ 4\ \cancel{6}\ \cancel{0}\ \cancel{7} \\ -\ 3\ 1\ 6\ 8 \\ \hline 9 \end{array}$$

← 10 tens − 1 ten = 9 tens
← 1 ten = 10 ones; 10 + 7 = 17

*Continued on Next Page*

Complete the problem.

$$\begin{array}{r} \overset{9}{\underset{5\ \cancel{10}\ 17}{4\ \cancel{6}\ \cancel{0}\ \cancel{7}}} \\ -\ 3\ 1\ 6\ 8 \\ \hline 1\ 4\ 3\ 9 \end{array}$$ ← Difference

Check by adding 1439 and 3168; you should get 4607.

**Work Problem 7 at the Side.**

### Example 8  Borrowing with Zeros

Subtract.

(a) $\quad 708$
$\quad -\ 149$

1 hundred is 10 tens. — Borrow 1 ten.
Borrow 1 hundred. → 1 ten is 10 ones; $10 + 8 = 18$.

$$\begin{array}{r} 6\ \cancel{10}\ 18 \\ \cancel{7}\ \cancel{0}\ \cancel{8} \\ -\ 1\ 4\ 9 \\ \hline 5\ 5\ 9 \end{array}$$

(b) $\quad 380$
$\quad -\ 276$

Borrow 1 ten. — 1 ten is 10 ones.

$$\begin{array}{r} 7\ 10 \\ 3\ \cancel{8}\ \cancel{0} \\ -\ 2\ 7\ 6 \\ \hline 1\ 0\ 4 \end{array}$$

(c) $\quad 9000$
$\quad -\ 6999$

$$\begin{array}{r} 9\ 9 \\ 8\ \cancel{10}\ \cancel{10}\ 10 \\ \cancel{9}\ \cancel{0}\ \cancel{0}\ \cancel{0} \\ -\ 6\ 9\ 9\ 9 \\ \hline 2\ 0\ 0\ 1 \end{array}$$

**Work Problem 8 at the Side.**

Recall that an answer to a subtraction problem can be checked by adding.

### Example 9  Checking Subtraction

Check each answer.

(a) $\quad\begin{array}{r}613 \\ -\ 275 \\ \hline 338\end{array}$  Check  $\begin{array}{r}275 \\ +\ 338 \\ \hline 613\end{array}$  Matches  Correct

**Continued on Next Page**

---

**7** Subtract.

(a) $\quad 308$
$\quad -\ 285$

(b) $\quad 206$
$\quad -\ 148$

(c) $\quad 5073$
$\quad -\ 1632$

**8** Subtract.

(a) $\quad 405$
$\quad -\ 267$

(b) $\quad 370$
$\quad -\ 163$

(c) $\quad 1570$
$\quad -\ 983$

(d) $\quad 7001$
$\quad -\ 5193$

(e) $\quad 4000$
$\quad -\ 1782$

**ANSWERS**
7. (a) 23  (b) 58  (c) 3441
8. (a) 138  (b) 207  (c) 587  (d) 1808
   (e) 2218

**9** Check each answer. If an answer is incorrect, find the correct answer.

(a)   425
    − 368
    ─────
       57

(b)   670
    − 439
    ─────
      241

(c)  14,726
    −  8 839
    ───────
       5 887

(b)   1915          Check
    − 1635          1635
    ─────   Matches + 280
      280           ─────
                    1915   Correct

(c)  15,803          Check
    −  7 325         7 325
    ───────  Does not + 8 578
      8 578  match   ──────
                    15,903   Error

Rework the original problem to get the correct answer, 8478.

**Work Problem 9 at the Side.**

**6** Use subtraction to solve application problems.

**Example 10** Applying Subtraction Skills

Diana Lopez drives a United Parcel Service delivery truck. Using the table below, decide how many more deliveries were made by Lopez on Monday than on Thursday.

**PACKAGE DELIVERY (LOPEZ)**

| Day | Number of Deliveries |
|---|---|
| Monday | 137 |
| Tuesday | 126 |
| Wednesday | 119 |
| Thursday | 89 |
| Friday | 147 |
| Saturday | 0 |

Lopez made 137 deliveries on Monday, but had only 89 deliveries on Thursday. Find how many more deliveries were made on Monday than on Thursday by subtracting 89 from 137.

    137   ← Deliveries on Monday
  −  89   ← Deliveries on Thursday
  ─────
     48   ← More deliveries on Monday

Lopez made 48 more deliveries on Monday than she made on Thursday.

**Work Problem 10 at the Side.**

**10** Using the table from Example 10, how many more deliveries did Lopez make

(a) on Friday than on Tuesday?

(b) on Tuesday than on Wednesday?

**ANSWERS**
9. (a) correct  (b) incorrect, should be 231
   (c) correct
10. (a) 21 more deliveries
    (b) 7 more deliveries

## R.2 Exercises

**FOR EXTRA HELP**   Student's Solutions Manual   MyMathLab.com   InterAct Math Tutorial Software   AW Math Tutor Center   www.mathxl.com  Digital Video Tutor CD 8 Videotape 21

*Use addition to check each subtraction. If an answer is incorrect, find the correct answer. See Examples 3 and 4.*

1.  89
    − 27
    ─────
    63

2.  47
    − 35
    ─────
    13

3.  382
    − 261
    ─────
    131

4.  838
    − 516
    ─────
    322

*Subtract, borrowing when necessary. See Examples 5–8.*

5.  36
    − 28

6.  97
    − 39

7.  83
    − 58

8.  65
    − 28

9.  45
    − 29

10. 93
    − 37

11. 719
    − 658

12. 916
    − 618

13. 771
    − 252

14. 973
    − 788

15. 9861
    −  684

16. 6171
    − 1182

17. 9988
    − 2399

18. 3576
    − 1658

19. 38,335
    − 29,476

20. 61,278
    −  3,559

21. 40
    − 37

22. 80
    − 73

23. 60
    − 37

24. 70
    − 27

25. 6020
    − 4078

26. 7050
    − 6045

27. 8503
    − 2816

28. 16,004
    −  5 087

29. 80,705
    − 61,667

30. 72,000
    − 44,234

31. 66,000
    −    444

32. 77,000
    −    308

33. 20,080
    −     96

34. 80,056
    −     69

*Use addition to check each subtraction. If an answer is incorrect, find the correct answer. See Example 9.*

35. 3070
    − 576
    ─────
    2596

36. 1439
    − 1169
    ─────
     270

37. 27,600
    −    807
    ──────
    26,793

38. 34,021
    − 33,708
    ──────
       727

*Solve each application problem. See Example 10.*

**39.** A man burns 103 calories during 30 minutes of bowling while a woman burns 88 calories. How many fewer calories does a woman burn than a man?

**40.** Lynn Couch had $553 in her checking account. She wrote a check for $308 for school fees. How much is left in her account?

**41.** An airplane was carrying 254 passengers. When it landed in Atlanta, 133 passengers get off. How many passengers were left on the plane?

**42.** On Tuesday, 5822 people went to a soccer game, and on Friday, 7994 people went to a soccer game. How many more people went to the game on Friday?

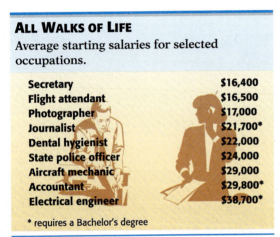

**Source:** U.S. Department of Labor/U.S. Bureau of the Census.

*This table shows the average yearly starting salaries for various occupations. Use the table to answer Exercises 43 and 44.*

**43.** How much less does a flight attendant earn than a state police officer?

**44.** What is the difference in the starting salaries of an electrical engineer and an accountant?

**45.** Downtown Toronto's skyline is dominated by the CN Tower which rises 1815 feet. The Sears Tower in Chicago is 1450 feet high. Find the difference in height between the two structures. (*Source: World Almanac*, 2001.)

**46.** The fastest animal in the world, the peregrine falcon, dives at 185 miles per hour. (*Source: Top 10 of Everything*, 2000.) A Boeing 747 cruises at 580 miles per hour. How much faster does the plane fly than the falcon?

# R.3 Multiplying Whole Numbers

Suppose we want to know the total number of computers in a computer lab. The computers are arranged in four rows with three computers in each row. Adding the number 3 a total of 4 times gives 12.

$$3 + 3 + 3 + 3 = 12$$

This result can also be shown with a figure.

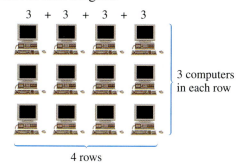

4 rows; 3 computers in each row

**1** **Identify the parts of a multiplication problem.** Multiplication is a shortcut for repeated addition. The numbers being multiplied are called **factors.** The answer is called the **product.** For example, the product of 3 and 4 can be written with the symbol ×, a raised dot, parentheses, or, in computer work, an asterisk.

$$\begin{array}{r} 3 \\ \times\ 4 \\ \hline 12 \end{array}$$ ← Factor (also called *multiplicand*)
← Factor (also called *multiplier*)
← Product (answer)

$3 \times 4 = 12 \quad 3 \cdot 4 = 12 \quad (3)(4) = 12 \quad 3 * 4 = 12$
(In computer work)

**Work Problem 1 at the Side.**

### Commutative Property of Multiplication

By the **commutative property of multiplication,** changing the *order* of two factors does not change the product.

For example: $3 \times 5 = 15$ and $5 \times 3 = 15$.
Both products are 15.

**CAUTION**

Recall that addition also has a commutative property. Remember that $4 + 2$ is the same as $2 + 4$. Subtraction, however, is *not* commutative.

### Example 1 Multiplying Two Numbers

Multiply. (Remember that a raised dot means to multiply.) Do the work mentally.

(a) $3 \times 4 = 12$   By the commutative property, $4 \times 3 = 12$ also.
(b) $6 \cdot 0 = 0$   The product of any number and 0 is 0; if you give no money to each of 6 relatives, you give no money.
(c) $(4)(8) = 32$   By the commutative property, $(8)(4) = 32$ also.

**Work Problem 2 at the Side.**

## Objectives

1. Identify the parts of a multiplication problem.
2. Do chain multiplications.
3. Multiply by single-digit numbers.
4. Multiply quickly by numbers ending in zeros.
5. Multiply by numbers having more than one digit.
6. Use multiplication to solve application problems.

**1** Identify the factors and the product in each multiplication problem.

(a) $3 \times 6 = 18$

(b) $8 \times 4 = 32$

(c) $5 \cdot 7 = 35$

(d) $(3)(9) = 27$

**2** Multiply. Do the work mentally. Then use the commutative property to write another multiplication problem and find the product.

(a) $4 \times 7$

(b) $0 \times 9$

(c) $8 \cdot 6$

(d) $5 \cdot 5$

(e) $(3)(8)$

**Answers**
1. (a) factors: 3, 6; product: 18
   (b) factors: 8, 4; product: 32
   (c) factors: 5, 7; product: 35
   (d) factors: 3, 9; product: 27
2. (a) 28; $7 \times 4 = 28$  (b) 0; $9 \times 0 = 0$
   (c) 48; $6 \cdot 8 = 48$  (d) 25; no change
   (e) 24; $(8)(3) = 24$

**3** Multiply.

(a) $2 \times 3 \times 4$

(b) $6 \cdot 1 \cdot 5$

(c) $(8)(3)(0)$

(d) $3 \times 3 \times 7$

(e) $4 \cdot 2 \cdot 8$

(f) $(2)(2)(9)$

**ANSWERS**
3. (a) 24  (b) 30  (c) 0  (d) 63
   (e) 64  (f) 36

**2** **Do chain multiplications.** Some multiplications involve more than two factors.

### Associative Property of Multiplication

By the **associative property of multiplication,** changing the *grouping* of factors does not change the product.

**Example 2** Multiplying Three Numbers

Multiply $2 \times 3 \times 5$.

$(2 \times 3) \times 5$    Parentheses tell what to do first.
$\quad 6 \quad \times 5 = 30$

Also,

$2 \times (3 \times 5)$
$2 \times \quad 15 \quad = 30$

By the associative property, either grouping results in the same product.

**Calculator Tip** The calculator approach to Example 2 uses chain calculations. Notice that you can enter *all* the factors before pressing the $=$ key.

$2 \; \otimes \; 3 \; \otimes \; 5 \; \ominus \; 30$

Work Problem **3** at the Side.

**3** **Multiply by single-digit numbers.** Carrying may be needed in multiplication problems with larger factors.

**Example 3** Carrying with Multiplication

Multiply.

(a)     53
   $\times \quad 4$

Start by multiplying in the ones column. Multiply 4 times 3 ones.

$\quad\;\; 1$
$\quad\; 53$
$\times \;\; 4 \quad\quad 4 \times 3 = 12$
$\overline{\quad\;\; 2}$

Carry 1 ten to the tens column.

Write 2 ones in the ones column.

Next, multiply 4 times 5 tens.

$\quad\;\; 1$
$\quad\; 53$
$\times \;\; 4 \quad\quad 4 \times 5 \text{ tens} = 20 \text{ tens}$
$\overline{\quad\;\; 2}$

Add the 1 that was carried to the tens column.

$\quad\;\; 1$
$\quad\; 53$
$\times \;\; 4$
$\overline{\; 212} \quad\quad 20 \text{ tens} + 1 \text{ ten} = 21 \text{ tens}$

*Continued on Next Page*

**(b)** 724
    × 5

Work as shown below.

```
    1 2
    724
  ×   5
   3620
```
← 5 × 4 = 20 ones; write 0 ones and carry 2 tens.

5 × 2 tens = 10 tens; add the 2 carried tens to get 12 tens; write 2 tens and carry 1 hundred.

5 × 7 hundreds = 35 hundreds; add the 1 carried hundred to get 36 hundreds.

========= Work Problem ❹ at the Side.

**4** **Multiply quickly by numbers ending in zeros.** The product of two whole-number factors is also called a **multiple** of either factor. For example, since 4 • 2 = 8, the number 8 is a multiple of 4 and 8 is also a multiple of 2. Multiples of 10 are very useful when multiplying. A *multiple of 10* is a whole number that ends in 0, such as 10, 20, or 30; 100, 200, or 300; 1000, 2000, or 3000. There is a short way to multiply by multiples of 10. Look at the following examples.

$$26 \times 1 = 26$$
$$26 \times 10 = 260$$
$$26 \times 100 = 2600$$
$$26 \times 1000 = 26{,}000$$

Do you see a pattern in the multiplications? These examples suggest the following rule.

### Multiplying by Multiples of 10

To multiply a whole number by 10, by 100, or by 1000, attach one, two, or three zeros to the right of the whole number.

**Example 4  Using Multiples of 10 to Multiply**

Multiply.

**(a)** 59 × 10 = 590 — Attach 0.

**(b)** 74 × 100 = 7400 — Attach 00.

**(c)** 803 × 1000 = 803,000 — Attach 000.

========= Work Problem ❺ at the Side.

You can also find the product of other multiples of ten by attaching zeros.

❹ Multiply.

(a) 52
  × 5

(b) 79
  × 0

(c) 862
  × 9

(d) 2831
  ×    7

(e) 4714
  ×    8

❺ Multiply.

(a) 45 × 10

(b) 102 × 100

(c) 571 × 1000

**Answers**
**4.** (a) 260  (b) 0  (c) 7758  (d) 19,817
    (e) 37,712
**5.** (a) 450  (b) 10,200  (c) 571,000

# Chapter R Whole Numbers Review

**6** Multiply.

(a) $14 \times 50$

(b) $68 \times 400$

(c) $\begin{array}{r} 180 \\ \times\ 30 \\ \hline \end{array}$

(d) $\begin{array}{r} 6100 \\ \times\ \ \ 90 \\ \hline \end{array}$

(e) $\begin{array}{r} 800 \\ \times 200 \\ \hline \end{array}$

**7** Complete each multiplication.

(a) $\begin{array}{r} 35 \\ \times\ 54 \\ \hline 140 \\ 175\phantom{0} \\ \hline \phantom{0000} \end{array}$

(b) $\begin{array}{r} 76 \\ \times\ 49 \\ \hline 684 \\ 304\phantom{0} \\ \hline \phantom{0000} \end{array}$

**Answers**
6. (a) 700  (b) 27,200  (c) 5400
   (d) 549,000  (e) 160,000
7. (a) 1890  (b) 3724

---

**Example 5** Multiplying by Using Other Multiples of 10

Multiply.

(a) $75 \times 3000$
   Multiply 75 by 3, and then attach three zeros.

$$\begin{array}{r} 75 \\ \times\ \ 3 \\ \hline 225 \end{array} \qquad 75 \times 3000 = 225{,}000$$

(b) $150 \times 70$
   Multiply 15 by 7, and then attach two zeros.

$$\begin{array}{r} 15 \\ \times\ \ 7 \\ \hline 105 \end{array} \qquad 150 \times 70 = 10{,}500$$

**Work Problem 6 at the Side.**

**5** **Multiply by numbers having more than one digit.** The next example shows multiplication when both factors have more than one digit.

**Example 6** Multiplying with More Than One Digit

Multiply 46 and 23.
   First multiply 46 by 3.

$$\begin{array}{r} 1\phantom{0} \\ 46 \\ \times\ \ 3 \\ \hline 138 \end{array} \leftarrow 46 \times 3 = 138$$

Now multiply 46 by 20.

$$\begin{array}{r} 1\phantom{0} \\ 46 \\ \times\ 20 \\ \hline 138 \\ 920 \end{array} \leftarrow 46 \times 20 = 920$$

Add the results.

$$\begin{array}{r} 46 \\ \times\ 23 \\ \hline 138 \\ +\ 920 \\ \hline 1058 \end{array} \begin{array}{l} \leftarrow 46 \times 3 \\ \leftarrow 46 \times 20 \\ \leftarrow \text{Add to find the product.} \end{array}$$

Both 138 and 920 are called *partial products*. To save time, the 0 in 920 is usually not written.

$$\begin{array}{r} 46 \\ \times\ 23 \\ \hline 138 \\ 92\phantom{0} \\ \hline 1058 \end{array} \leftarrow \text{0 not written. Be very careful to place the 2 in the tens column.}$$

**Work Problem 7 at the Side.**

## Example 7 Using Partial Products

Multiply.

(a)
```
      2 3 3
  ×   1 3 2
      4 6 6
    6 9 9      Tens lined up
  2 3 3        Hundreds lined up
  3 0,7 5 6  ← Product
```

(b)
```
    5 3 8
  ×   4 6
```

First multiply by 6.
```
       2 4
      5 3 8
    ×   4 6     Carrying is
      3 2 2 8   needed here.
```

Now multiply by 4, being careful to line up the tens.
```
      1 3
      2 4
      5 3 8
    ×   4 6
      3 2 2 8  ⎤
      2 1 5 2  ⎦ Add the partial products.
      2 4,7 4 8
```

Work Problem ❽ at the Side.

When 0 appears in the multiplier, be sure to move the partial product to the left to account for the position held by the 0.

## Example 8 Multiplication with Zeros

Multiply.

(a)
```
      1 3 7
  ×   3 0 6
      8 2 2
    0 0 0      Tens lined up
  4 1 1        Hundreds lined up
  4 1,9 2 2
```

(b)
```
      1 4 0 6              1 4 0 6
  ×   2 0 0 1          ×   2 0 0 1
      1 4 0 6              1 4 0 6
      0 0 0 0  ← 0 to line up tens
      0 0 0 0  ← 0 to line up hundreds    2 8 1 2 0 0  ← Zeros are
    2 8 1 2                                             written so this
    2,8 1 3,4 0 6                        2,8 1 3,4 0 6  partial product
                                                        starts in the
                                                        thousands
                                                        column.
```

### CAUTION

In Example 8(b) in the solution on the right, be careful to insert zeros so that thousands are lined up in the thousands column.

❽ Multiply.

(a)
```
    38
  ×15
```

(b)
```
    31
  ×43
```

(c)
```
    67
  ×59
```

(d)
```
    234
  × 73
```

(e)
```
    835
  ×189
```

**ANSWERS**
8. (a) 570  (b) 1333  (c) 3953
   (d) 17,082  (e) 157,815

## Chapter R Whole Numbers Review

**9** Multiply.

(a) $\phantom{0}28$
$\underline{\times 60}$

(b) $\phantom{0}817$
$\underline{\times \phantom{0}30}$

(c) $\phantom{0}481$
$\underline{\times 206}$

(d) $\phantom{0}3526$
$\underline{\times 6002}$

**10** Find the total cost of these items.

(a) 36 months of cable TV costing $48 each month

(b) 28 laptop computers priced at $1090 each

(c) 60 months of car payments at $289 per month

Work Problem **9** at the Side.

**6** Use multiplication to solve application problems.

**Example 9**  Applying Multiplication Skills

Find the total cost of 24 cordless telephones priced at $59 each.

*Approach*  To find the cost of all the telephones, multiply the cost of one telephone ($59) by the number of telephones (24).

*Solution*  Multiply $59 by 24.

$$\begin{array}{r} 59 \\ \times\phantom{0}24 \\ \hline 236 \\ 118\phantom{0} \\ \hline 1416 \end{array}$$

The total cost of the cordless telephones is $1416.

**Calculator Tip:**  If you are using a calculator for Example 9, press the following keys.

59 ⊗ 24 ⊜ 1416

Work Problem **10** at the Side.

---

ANSWERS
9. (a) 1680  (b) 24,510  (c) 99,086
   (d) 21,163,052
10. (a) $1728  (b) $30,520  (c) $17,340

## R.3 Exercises

*Find each product. Try to do the work mentally. See Examples 1 and 2.*

1. $3 \times 1 \times 3$
2. $2 \times 8 \times 2$
3. $9 \times 1 \times 7$
4. $2 \times 4 \times 5$
5. $9 \cdot 5 \cdot 0$

6. $6 \cdot 0 \cdot 8$
7. $4 \cdot 1 \cdot 6$
8. $1 \cdot 5 \cdot 7$
9. $(2)(3)(6)$
10. $(4)(1)(9)$

*Multiply. See Examples 3–8.*

11. $35 \times 7$
12. $76 \times 9$
13. $28 \times 6$
14. $83 \times 5$
15. $3182 \times 6$

16. $7326 \times 5$
17. $36{,}921 \times 7$
18. $28{,}116 \times 4$
19. $125 \times 30$
20. $246 \times 50$

21. $1485 \times 30$
22. $8522 \times 50$
23. $900 \times 300$
24. $400 \times 700$
25. $43{,}000 \times 2000$

26. $11{,}000 \times 9000$
27. $68 \times 22$
28. $82 \times 32$
29. $83 \times 45$
30. $(43)(27)$

31. $(32)(475)$
32. $(67)(218)$
33. $(729)(45)$
34. $(681)(47)$
35. $538 \times 342$

36. $3228 \times 751$
37. $8162 \times 198$
38. $528 \times 106$
39. $6310 \times 3078$
40. $3533 \times 5001$

*Solve each application problem. See Example 10.*

**41.** Giant kelp plants in the ocean can grow 18 inches each day. How much could kelp grow in two weeks? in a 30-day month? (*Source:* Natural Bridges State Park, CA.)

**42.** A hospital has 20 bottles of thyroid medication, with each bottle containing 2500 tablets. How many of these tablets does the hospital have in all?

**43.** There are 12 tomato plants to a flat. If a garden center has 48 flats, find the total number of tomato plants.

**44.** A hummingbird's wings beat about 65 times per second, a chickadee's wings about 27 times per second. How many times does each bird's wings beat in one minute? (*Source:* Birder's Handbook.)

Ruby-throated hummingbird
65 wing beats per second

**45.** A new Saturn automobile gets 38 miles per gallon on the highway. How many miles can it travel on 11 gallons of gas?

**46.** Find the total cost of 16 gallons of paint at $18 per gallon.

*Use addition, subtraction, or multiplication, as needed, to solve each problem.*

**47.** The distance from Reno, Nevada, to the Atlantic Ocean is 2695 miles, while the distance from Reno to the Pacific Ocean is 255 miles. How much farther is it to the Atlantic Ocean than it is to the Pacific Ocean? If you make three round trips from Reno to the Atlantic Ocean, how many frequent flier miles will you earn?

**48.** The largest living land mammal is the African elephant, and the largest mammal of all time is the blue whale. An African bull elephant may weigh 15,225 pounds and a blue whale may weigh 12 to 25 times that amount. Find the range of weights for the blue whale and the difference between the lightest and heaviest. (*Source: Big Book of Knowledge.*)

**49.** A high-fat meal contains 1406 calories, while a low-fat meal contains 348 calories. How many more calories are in seven high-fat meals than in seven low-fat meals?

**50.** Dannie Sanchez bought four tires at $110 each, two seat covers at $49 each, and six socket wrenches at $3 each. Find the total amount that he spent.

# R.4 Dividing Whole Numbers

Suppose $12 is to be divided into 3 equal parts. Each part would be $4, as shown here.

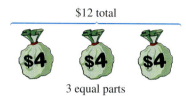

$12 total

3 equal parts

**OBJECTIVES**

1. Write division problems in three ways.
2. Identify the parts of a division problem.
3. Divide 0 by a number.
4. Recognize that division by 0 is undefined.
5. Divide a number by itself.
6. Use short division.
7. Use multiplication to check quotients.
8. Use tests for divisibility.

**1** **Write division problems in three ways.** Just as $3 \cdot 4$, $3 \times 4$, and $(3)(4)$ are different ways of writing the multiplication of 3 and 4, there are several ways to write 12 divided by 3.

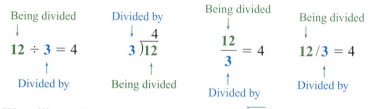

We will use three division symbols, $\div$, $\overline{)}$, and —. In algebra the bar, —, is frequently used. In computer science the slash, /, is used.

### Example 1 Using Division Symbols

Rewrite each division using two other symbols.

(a) $20 \div 4 = 5$

This division can also be written $4\overline{)20}^{\,5}$ or $\dfrac{20}{4} = 5$.

(b) $\dfrac{18}{6} = 3$ can also be written $18 \div 6 = 3$ or $6\overline{)18}^{\,3}$

(c) $5\overline{)40}^{\,8}$ can also be written $40 \div 5 = 8$ or $\dfrac{40}{5} = 8$

**Work Problem 1 at the Side.**

**2** **Identify the parts of a division problem.** In division, the number being divided is the **dividend**, the number divided by is the **divisor**, and the answer is the **quotient**.

$$\text{dividend} \div \text{divisor} = \text{quotient}$$

$$\text{divisor}\overline{)\text{dividend}}^{\,\text{quotient}} \qquad \dfrac{\text{dividend}}{\text{divisor}} = \text{quotient}$$

### Example 2 Identifying the Parts in a Division Problem

Identify the dividend, divisor, and quotient.

(a) $35 \div 7 = 5$

$35 \div 7 = 5 \leftarrow$ Quotient

Dividend   Divisor

**Continued on Next Page**

---

**1** Rewrite each division using two other symbols.

(a) $48 \div 6 = 8$

(b) $24 \div 6 = 4$

(c) $9\overline{)36}^{\,4}$

(d) $\dfrac{42}{6} = 7$

**ANSWERS**

1. (a) $6\overline{)48}^{\,8}$ and $\dfrac{48}{6} = 8$

   (b) $6\overline{)24}^{\,4}$ and $\dfrac{24}{6} = 4$

   (c) $36 \div 9 = 4$ and $\dfrac{36}{9} = 4$

   (d) $6\overline{)42}^{\,7}$ and $42 \div 6 = 7$

**2** Identify the dividend, divisor, and quotient.

(a) $10 \div 2 = 5$

(b) $30 \div 5 = 6$

(c) $\dfrac{28}{7} = 4$

(d) $2\overline{)36}$ with quotient 18

**3** Divide.

(a) $0 \div 9$

(b) $\dfrac{0}{8}$

(c) $\dfrac{0}{36}$

(d) $57\overline{)0}$

**ANSWERS**
2. (a) dividend: 10; divisor: 2; quotient: 5
   (b) dividend: 30; divisor: 5; quotient: 6
   (c) dividend: 28; divisor: 7; quotient: 4
   (d) dividend: 36; divisor: 2; quotient: 18
3. all 0

---

(b) $\dfrac{100}{20} = 5$

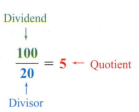

(c) $12\overline{)72}$ with quotient 6

$12\overline{)72}$ ← Quotient 6, ← Dividend 72, ↑ Divisor 12

**Work Problem 2 at the Side.**

**3** **Divide 0 by a number.** If no money, or $0, is divided equally among five people, each person gets $0. There is a general rule for dividing 0.

**Dividing 0**

When 0 is divided by any other number (except 0), the quotient is 0.

**Example 3** Dividing 0 by a Number

Divide.

(a) $0 \div 12 = \mathbf{0}$

(b) $0 \div 1728 = \mathbf{0}$

(c) $\dfrac{0}{375} = \mathbf{0}$

(d) $129\overline{)0}$ = **0**

**Work Problem 3 at the Side.**

Recall that a subtraction such as $8 - 3 = 5$ can be written as the addition $8 = 3 + 5$. In a similar way, any division can be written as a multiplication. For example, $12 \div 3 = 4$ can be written as

$3 \times 4 = 12$  or, by the commutative property,  $4 \times 3 = 12$.

**Example 4** Changing Division to Multiplication

Change each division to multiplication.

(a) $\dfrac{20}{4} = 5$  becomes  $4 \cdot 5 = 20$

(b) $8\overline{)48}$ with quotient 6  becomes  $8 \cdot 6 = 48$

(c) $72 \div 9 = 8$  becomes  $9 \cdot 8 = 72$

## Section R.4 Dividing Whole Numbers

**4** Recognize that division by 0 is undefined. Division by 0 cannot be done. To see why, try to find the answer to this division.

$$9 \div 0 = ?$$

As we have just seen, any division problem can be changed to a multiplication problem so that

$$\text{divisor} \cdot \text{quotient} = \text{dividend}.$$

If you convert the problem $9 \div 0 = ?$ to its multiplication counterpart, it reads as follows.

$$0 \cdot ? = 9$$

You already know that 0 times any number must always equal 0. Try any number you like to replace the "?" and you'll always get 0 instead of 9. Therefore, the division problem $9 \div 0$ *cannot* be done. Mathematicians say it is *undefined* and have agreed never to divide by 0. However, $0 \div 9$ *can* be done. Check by rewriting it as a multiplication problem.

$$0 \div 9 = 0 \quad \text{because} \quad 0 \cdot 9 = 0 \quad \text{is true.}$$

### Dividing by 0

Dividing by 0 cannot be done. We say that division by 0 is *undefined*. It is impossible to compute an answer.

**Example 5** Dividing by 0 Is Undefined

All of these divisions are *undefined*.

(a) $\dfrac{6}{0}$ is undefined.

(b) $0\overline{)8}$ is undefined.

(c) $18 \div 0$ is undefined.

(d) $\dfrac{0}{0}$ is undefined.

### Division Involving 0

$$\dfrac{0}{\text{nonzero number}} = 0 \qquad \dfrac{\text{number}}{0} \text{ is } undefined.$$

**CAUTION**

When 0 is the divisor in a problem, write *undefined*. You can never divide by 0.

Work Problem **5** at the Side.

**Calculator Tip** Try these two problems on your calculator. Jot down your answers.

$$9 \div 0 = \underline{\qquad} \qquad 0 \div 9 = \underline{\qquad}$$

When you try to divide by 0, the calculator cannot do it, so it shows the word "Error" in the display, or the letters "ERR" or "E" (for "error").

Work Problem **4** at the Side.

**4** Write each division problem as a multiplication problem.

(a) $6\overline{)18}^{\,3}$

(b) $\dfrac{28}{4} = 7$

(c) $48 \div 8 = 6$

**5** Find the quotient whenever possible.

(a) $\dfrac{8}{0}$

(b) $\dfrac{0}{8}$

(c) $0\overline{)32}$

(d) $32\overline{)0}$

(e) $100 \div 0$

(f) $0 \div 100$

**ANSWERS**
4. (a) $6 \cdot 3 = 18$  (b) $4 \cdot 7 = 28$
   (c) $8 \cdot 6 = 48$
5. (a) undefined  (b) 0  (c) undefined
   (d) 0  (e) undefined  (f) 0

## Chapter R  Whole Numbers Review

**6** Divide.

(a) $5 \div 5$

(b) $14\overline{)14}$

(c) $\dfrac{37}{37}$

**5** **Divide a number by itself.** What happens when a number is divided by itself? For example, $4 \div 4$ or $97 \div 97$?

### Dividing a Number by Itself

When a nonzero number is divided by itself, the quotient is 1.

**Example 6**  Dividing a Nonzero Number by Itself

Divide.

(a) $16 \div 16 = 1$

(b) $32\overline{)32}^{\,1}$

(c) $\dfrac{57}{57} = 1$

**Work Problem 6 at the Side.**

**6** **Use short division.** Short division is a quick method of dividing a number by a one-digit divisor.

**Example 7**  Using Short Division

Divide $2\overline{)86}$.

First, divide 8 by 2.

$$2\overline{)86}^{\,4} \qquad \dfrac{8}{2} = 4$$

**7** Divide.

(a) $2\overline{)18}$

(b) $3\overline{)39}$

Next, divide 6 by 2.

$$2\overline{)86}^{\,43} \qquad \dfrac{6}{2} = 3$$

The quotient is 43.

**Work Problem 7 at the Side.**

(c) $4\overline{)88}$

When two numbers do *not* divide evenly, the leftover portion is called the **remainder**.

**Example 8**  Using Short Division with a Remainder

Divide 147 by 4.

(d) $2\overline{)462}$

Write the problem as $4\overline{)147}$.

Because 1 cannot be divided by 4, divide 14 by 4.

$$4\overline{)14^{2}7}^{\,3} \qquad \dfrac{14}{4} = 3 \text{ with 2 left over}$$

**Answers**

**6.** all 1

**7.** (a) 9  (b) 13  (c) 22  (d) 231

**Continued on Next Page**

Next, divide 27 by 4. The final number left over is the remainder. Write the remainder to the side. "R" stands for remainder.

$$4\overline{)14^27} \quad \begin{array}{c} 3\ 6\ \mathbf{R}3 \end{array} \qquad \frac{27}{4} = 6 \text{ with 3 left over}$$

The quotient is 36 **R**3.

━━━━━━━━━━━━━━━━━ Work Problem **8** at the Side.

### Example 9  Dividing with a Remainder

Divide 1809 by 7.
  Divide 7 into 18.

$$7\overline{)18^409} \quad \begin{array}{c} 2 \end{array} \qquad \frac{18}{7} = 2 \text{ with 4 left over}$$

Divide 7 into 40.

$$7\overline{)18^40^59} \quad \begin{array}{c} 2\ 5 \end{array} \qquad \frac{40}{7} = 5 \text{ with 5 left over}$$

Divide 7 into 59.

$$7\overline{)18^40^59} \quad \begin{array}{c} 2\ 5\ 8\ \mathbf{R}3 \end{array} \qquad \frac{59}{7} = 8 \text{ with 3 left over}$$

━━━━━━━━━━━━━━━━━ Work Problem **9** at the Side.

**7** **Use multiplication to check quotients.** Check the answer to a division problem as follows.

### Checking Division

(divisor × quotient) + remainder = dividend

Parentheses tell you what to do first. In this case, multiply the divisor by the quotient first and then add the remainder.

### Example 10  Using Multiplication to Check Division

Check each quotient.

(a)  $5\overline{)458}$  quotient 91 **R**3

(divisor × quotient) + remainder = dividend
(5 × 91) + 3
455 + 3 = 458
↑
Matches original dividend so the division was done correctly

*Continued on Next Page*

---

**8** Divide.

(a) $2\overline{)225}$

(b) $3\overline{)275}$

(c) $4\overline{)538}$

(d) $\dfrac{819}{5}$

**9** Divide.

(a) $5\overline{)937}$

(b) $\dfrac{675}{7}$

(c) $3\overline{)1885}$

(d) $8\overline{)1135}$

**ANSWERS**
8. (a) 112 **R**1  (b) 91 **R**2  (c) 134 **R**2
   (d) 163 **R**4
9. (a) 187 **R**2  (b) 96 **R**3  (c) 628 **R**1
   (d) 141 **R**7

**10** Check each division. If a quotient is incorrect, find the correct quotient.

(a) $3\overline{)115}$ $\phantom{3)}38$ **R1**

(b) $8\overline{)743}$ $\phantom{8)}92$ **R2**

(c) $4\overline{)1312}$ $\phantom{4)}328$

(d) $5\overline{)2033}$ $\phantom{5)}46$ **R3**

**ANSWERS**

**10. (a)** correct
  **(b)** incorrect; should be 92 **R7**
  **(c)** correct
  **(d)** incorrect; should be 406 **R3**

(b) $6\overline{)1258}$ $\phantom{6)}29$ **R4**

(divisor × quotient) + remainder = dividend

(6 × 29) + 4

174 + 4 = 178

Does not match original dividend of 1258

The quotient does *not* check. Rework the original problem to get the correct quotient, 209 **R4**. Then, (6 × 209) + 4 does equal 1258.

### CAUTION

A common error is forgetting to add the remainder. Be sure to add any remainder when checking a division problem.

**Work Problem 10 at the Side.**

**8** **Use tests for divisibility.** It is often important to know whether a number is divisible by another number. You will find this useful in Chapter 4 when writing fractions in lowest terms.

### Divisibility

A whole number is *divisible* by another whole number if the remainder is 0.

There are some quick tests you can use to decide whether one number is divisible by another.

### Tests for Divisibility

**A number is divisible by**

| | |
|---|---|
| 2 | if it ends in 0, 2, 4, 6, or 8. |
| 3 | if the sum of its digits is divisible by 3. |
| 4 | if the last two digits make a number that is divisible by 4. |
| 5 | if it ends in 0 or 5. |
| 6 | if it is divisible by both 2 and 3. |
| 8 | if the last three digits make a number that is divisible by 8. |
| 9 | if the sum of its digits is divisible by 9. |
| 10 | if it ends in 0. |

The most commonly used tests are those for 2, 3, 5, and 10.

## Divisibility by 2

A number is divisible by **2** if the number ends in 0, 2, 4, 6, or 8.

**Example 11** Testing for Divisibility by 2

Are the following numbers divisible by 2?

**(a)** 986
 ↑
 └── Ends in 6

Because the number ends in 6, which is in the list (0, 2, 4, 6, or 8), the number 986 is divisible by 2.

**(b)** 3255 is not divisible by 2.
 ↑
 └── Ends in 5, and not in 0, 2, 4, 6, or 8

*Work Problem 11 at the Side.*

## Divisibility by 3

A number is divisible by **3** if the sum of its digits is divisible by **3**.

**Example 12** Testing for Divisibility by 3

Are the following numbers divisible by 3?

**(a)** 4251

Add the digits.

$$4 + 2 + 5 + 1 = 12$$

Because 12 is divisible by 3, the number 4251 is divisible by 3.

**(b)** 29,806

Add the digits.

$$2 + 9 + 8 + 0 + 6 = 25$$

Because 25 is *not* divisible by 3, the number 29,806 is *not* divisible by 3.

### CAUTION

Be careful when testing for divisibility by adding the digits. This method works only when testing for divisibility by 3 or by 9.

*Work Problem 12 at the Side.*

---

**11** Which numbers are divisible by 2?

(a) 612

(b) 315

(c) 2714

(d) 36,000

**12** Which numbers are divisible by 3?

(a) 836

(b) 7545

(c) 242,913

(d) 102,484

**ANSWERS**
**11.** (a), (c), and (d)
**12.** (b) and (c)

## Chapter R  Whole Numbers Review

⑬ Which numbers are divisible by 5?

(a) 160

(b) 635

(c) 3381

(d) 108,605

⑭ Which numbers are divisible by 10?

(a) 290

(b) 218

(c) 2020

(d) 11,670

### Divisibility by 5 and by 10

A number is divisible by **5** if it ends in 0 or 5.

A number is divisible by **10** if it ends in 0.

**Example 13** Determining Divisibility by 5

Are the following numbers divisible by 5?

(a) 12,900   ends in 0 and is divisible by 5.

(b) 4325   ends in 5 and is divisible by 5.

(c) 392   ends in 2 and is *not* divisible by 5.

**Work Problem ⑬ at the Side.**

**Example 14** Determining Divisibility by 10

Are the following numbers divisible by 10?

(a) 700 and 9140   end in 0 and are both divisible by 10.

(b) 355 and 18,743   do *not* end in 0 and are *not* divisible by 10.

**Work Problem ⑭ at the Side.**

**ANSWERS**
**13.** all but (c)
**14.** all but (b)

Section R.4    777

## R.4 EXERCISES

*Divide. Then rewrite each division using two other division symbols. See Examples 1, 3, 5, and 6.*

1. $\dfrac{12}{12}$     2. $\dfrac{9}{0}$     3. $24 \div 0$     4. $4 \div 4$     5. $\dfrac{0}{4}$

6. $0 \div 8$     7. $0 \div 12$     8. $\dfrac{0}{7}$     9. $0\overline{)21}$     10. $2 \div 0$

*Divide by using short division. See Examples 7–9.*

11. $4\overline{)108}$     12. $5\overline{)135}$     13. $9\overline{)324}$     14. $8\overline{)176}$

15. $6\overline{)9137}$     16. $9\overline{)8371}$     17. $6\overline{)1854}$     18. $8\overline{)856}$

19. $4024 \div 4$     20. $16{,}024 \div 8$     21. $15{,}018 \div 3$     22. $32{,}008 \div 8$

23. $\dfrac{26{,}684}{4}$     24. $\dfrac{16{,}398}{9}$     25. $\dfrac{74{,}751}{6}$     26. $\dfrac{72{,}543}{5}$

27. $\dfrac{71{,}776}{7}$     28. $\dfrac{77{,}621}{3}$     29. $\dfrac{128{,}645}{7}$     30. $\dfrac{172{,}255}{4}$

*Check each quotient. If a quotient is incorrect, find the correct quotient. See Example 10.*

31. $7\overline{)4692}$  quotient: 67 R2
32. $9\overline{)5974}$  quotient: 663 R5
33. $6\overline{)21{,}409}$  quotient: 3,568 R2
34. $4\overline{)103{,}516}$  quotient: 25,879

35. $6\overline{)18{,}023}$  quotient: 3,003 R5
36. $8\overline{)33{,}664}$  quotient: 4,208
37. $6\overline{)69{,}140}$  quotient: 11,523 R2
38. $3\overline{)82{,}598}$  quotient: 27,532 R1

# Chapter R Whole Numbers Review

*Solve each application problem.*

**39.** It is important to drink enough water. A rule of thumb is to divide your weight by 2 and drink that number of ounces of water each day. (*Source: Fairview Health Services.*) How many ounces of water should a 148 pound person drink daily?

**40.** Eight people invested a total of $244,224 to buy a seashore condominium. Each person invested the same amount of money. How much did each person invest?

**41.** If six identical service vans cost a total of $99,600, find the cost of each service van.

**42.** One gallon of beverage will serve nine people. How many gallons are needed for 3483 people?

**43.** An estate of $197,400 is divided equally among seven family members. Find the amount received by each family member.

**44.** How many 5-pound bags of rice can be filled from 8750 pounds of rice?

**45.** Ken Griffey, Jr., signed a nine-year baseball contract for $117,000,000. Find his pay for each year. (*Source: USA Today.*)

**46.** Ted Slauson, coordinator of Toys for Tots, has collected 2628 toys. If his group gives four toys to each child, how many children will receive the gifts?

**47.** Kaci Salmon, a supervisor at Albany Electric, earns $46,540 per year. Find the amount of her earnings in a three-month period.

**48.** A worker assembles 168 light diffusers in an 8-hour shift. Find the number assembled in three hours.

*Put a ✓ mark in the blank if the number at the left is divisible by the number at the top of each column. See Examples 11–14.*

|  | 2 | 3 | 5 | 10 |  |  | 2 | 3 | 5 | 10 |
|---|---|---|---|---|---|---|---|---|---|---|
| **49.** 30 | | | | | | **50.** 25 | | | | |
| **51.** 184 | | | | | | **52.** 192 | | | | |
| **53.** 445 | | | | | | **54.** 897 | | | | |
| **55.** 903 | | | | | | **56.** 500 | | | | |
| **57.** 5166 | | | | | | **58.** 8302 | | | | |
| **59.** 21,763 | | | | | | **60.** 32,472 | | | | |

# R.5 Long Division

**Long division** is used to divide by a number with more than one digit.

**1** **Use long division.** In long division, estimate the various numbers by using a *trial divisor* to get a *trial quotient*.

### Example 1  Using a Trial Divisor and a Trial Quotient

Divide. $42\overline{)3066}$

Because 42 is closer to 40 than to 50, use 4 as a trial divisor.

42 ↑ — Trial divisor is 4.

Try to divide the first digit of the dividend by 4. Because 3 cannot be divided by 4, use the first *two* digits, 30.

$$\frac{30}{4} = 7 \text{ with remainder 2}$$

$$\begin{array}{r} 7 \leftarrow \text{Trial quotient} \\ 42\overline{)3066} \end{array}$$

7 goes over the 6, because $\frac{306}{42}$ is about 7.

Multiply 7 and 42 to get 294; next, subtract 294 from 306.

$$\begin{array}{r} 7 \\ 42\overline{)3066} \\ \underline{294} \leftarrow 7 \times 42 = 294 \\ 12 \leftarrow 306 - 294 = 12 \end{array}$$

Bring down the 6 at the right.

$$\begin{array}{r} 7 \\ 42\overline{)3066} \\ \underline{294}\downarrow \\ 126 \leftarrow 6 \text{ brought down} \end{array}$$

Use the trial divisor, 4.

$$\begin{array}{r} 73 \\ 42\overline{)3066} \\ \underline{294} \\ 126 \leftarrow \text{First two digits of } 126 \rightarrow \frac{12}{4} = 3 \\ \underline{126} \leftarrow 3 \times 42 = 126 \\ 0 \end{array}$$

Check the quotient by multiplying 42 and 73. The product should be 3066, which matches the original dividend.

### CAUTION

The first digit in the answer in long division must be placed in the proper position over the dividend.

Work Problem **1** at the Side.

## Objectives

**1** Use long division.
**2** Divide multiples of 10.
**3** Use multiplication to check quotients.

**1** Divide.

(a) $25\overline{)1775}$

(b) $26\overline{)2132}$

(c) $51\overline{)2295}$

(d) $\dfrac{6552}{84}$

**Answers**
1. (a) 71  (b) 82  (c) 45  (d) 78

## Chapter R  Whole Numbers Review

**2** Divide.

(a) $56\overline{)2352}$

### Example 2  Dividing to Find a Trial Quotient

Divide. $58\overline{)2730}$

Use 6 as a trial divisor, since 58 is closer to 60 than to 50.

First two digits of dividend → $\dfrac{27}{6} = 4$ with remainder 3

$$\begin{array}{r} 4 \leftarrow \text{Trial quotient} \\ 58\overline{)2730} \\ \underline{232} \leftarrow 4 \times 58 = 232 \\ 41 \leftarrow 273 - 232 = 41 \text{ (smaller than 58,} \\ \text{the divisor)} \end{array}$$

Bring down the 0.

$$\begin{array}{r} 4 \\ 58\overline{)2730} \\ \underline{232}\downarrow \\ 410 \leftarrow \text{0 brought down} \end{array}$$

(b) $38\overline{)1599}$

$$\begin{array}{r} 46 \\ 58\overline{)2730} \\ \underline{232} \\ 410 \\ \underline{348} \leftarrow 6 \times 58 = 348 \\ 62 \leftarrow \text{Greater than 58} \end{array}$$

First two digits of 410  $\dfrac{41}{6} = 6$ with remainder 5

The remainder, 62, is *greater than the divisor,* 58, so **7** should be used in the quotient instead of **6**.

(c) $65\overline{)5416}$

$$\begin{array}{r} 47 \text{ R4} \\ 58\overline{)2730} \\ \underline{232} \\ 410 \\ \underline{406} \leftarrow 7 \times 58 = 406 \\ 4 \leftarrow 410 - 406 = 4 \end{array}$$

**Work Problem** ❷ **at the Side.**

Sometimes it is necessary to write a 0 in the quotient.

### Example 3  Writing Zeros in the Quotient

Divide. $42\overline{)8734}$

(d) $89\overline{)6649}$

Start as above.

$$\begin{array}{r} 2 \\ 42\overline{)8734} \\ \underline{84} \leftarrow 2 \times 42 = 84 \\ 3 \leftarrow 87 - 84 = 3 \end{array}$$

Bring down the 3.

$$\begin{array}{r} 2 \\ 42\overline{)8734} \\ \underline{84}\downarrow \\ 33 \leftarrow \text{3 brought down} \end{array}$$

**ANSWERS**
2. (a) 42   (b) 42 R3   (c) 83 R21
   (d) 74 R63

*Continued on Next Page*

Since 33 cannot be divided by 42, write a 0 in the quotient as a placeholder.

$$\begin{array}{r} 20 \\ 42\overline{)8734} \\ \underline{84} \\ 33 \end{array}$$ ← 0 in quotient

Bring down the final digit, the 4.

$$\begin{array}{r} 20 \\ 42\overline{)8734} \\ \underline{84}\downarrow \\ 334 \end{array}$$ ← 4 brought down

Complete the problem.

$$\begin{array}{r} 207 \text{ R}40 \\ 42\overline{)8734} \\ \underline{84} \phantom{00}\\ 334 \\ \underline{294} \\ 40 \end{array}$$
← 7 × 42 = 294
← 334 − 294 = 40

The quotient is 207 **R**40.

### CAUTION

There must be a digit in the quotient (answer) above *every* digit in the dividend *once the answer has begun*. Notice that in Example 3 a 0 was used to assure an answer digit above every digit in the dividend.

**Work Problem ❸ at the Side.**

**2** **Divide multiples of 10.** When the divisor and dividend both contain zeros at the far right, recall that these numbers are multiples of 10. There is a short way to divide multiples of 10. Look at the following examples.

$$26{,}000 \div 1 = 26{,}000$$
$$26{,}000 \div 10 = 2600$$
$$26{,}000 \div 100 = 260$$
$$26{,}000 \div 1000 = 26$$

Do you see a pattern in these divisions using multiples of 10? These examples suggest the following rule.

### Dividing a Whole Number by 10, by 100, or by 1000

Divide a whole number by 10, by 100, or by 1000 by dropping one, two, or three zeros from the whole number.

**Example 4** Dividing by Multiples of 10

Divide.

(a) $6\underline{0} \div 1\underline{0} = 6$ — 0 in divisor; Drop 0.

**Continued on Next Page**

---

❸ Divide.

(a) $24\overline{)3127}$

(b) $52\overline{)10{,}660}$

(c) $39\overline{)15{,}933}$

(d) $78\overline{)23{,}462}$

**ANSWERS**
3. (a) 130 **R**7  (b) 205  (c) 408 **R**21
   (d) 300 **R**62

**Chapter R** Whole Numbers Review

**4** Divide.

(a) $50 \div 10$

(b) $1800 \div 100$

(c) $305{,}000 \div 1000$

**5** Divide.

(a) $60\overline{)7200}$

(b) $130\overline{)131{,}040}$

(c) $2600\overline{)195{,}000}$

**6** Decide whether the following divisions are correct. If the quotient is incorrect, find the correct quotient.

(a) $\begin{array}{r} 43 \\ 18\overline{)774} \\ \underline{72\phantom{0}} \\ 54 \\ \underline{54} \\ 0 \end{array}$

(b) $\begin{array}{r} 42\ \mathbf{R}178 \\ 426\overline{)19{,}170} \\ \underline{1\ 704\phantom{0}} \\ 1\ 130 \\ \underline{\phantom{0}952} \\ 178 \end{array}$

---

(b) $3500 \div 100 = 35$ — 00 in divisor; Drop 00.

(c) $915{,}000 \div 1000 = 915$ — 000 in divisor; Drop 000.

**Work Problem 4 at the Side.**

You can also find the quotient for other multiples of 10 by dropping zeros.

### Example 5  Dividing by Multiples of 10

Divide.

(a) $40\overline{)11{,}000}$     Drop 0 from the divisor and the dividend.

$\begin{array}{r} 275 \\ 4\overline{)1100} \\ \underline{8\phantom{00}} \\ 30 \\ \underline{28} \\ 20 \\ \underline{20} \\ 0 \end{array}$

(b) $3500\overline{)31{,}500}$     Drop two zeros from the divisor and the dividend.

$\begin{array}{r} 9 \\ 35\overline{)315} \\ \underline{315} \\ 0 \end{array}$

**Work Problem 5 at the Side.**

**3** ▸ **Use multiplication to check quotients.**  Quotients in long division can be checked just as quotients in short division were checked. Multiply the quotient and divisor, then add any remainder. The result should match the original dividend.

### Example 6  Checking Division

Check the quotient.

$\begin{array}{r} 114\ \mathbf{R}43 \\ 48\overline{)5324} \end{array}$

$\begin{array}{r} 114 \\ \times\ 48 \\ \hline 912 \\ 456\phantom{0} \\ \hline 5472 \\ +\ \ \ 43 \\ \hline 5515 \end{array}$

→ Multiply the quotient and the divisor.

← Add the remainder.
← Does not match original dividend of 5324

The quotient does *not* check. Rework the original problem to get 110 **R**44.

**Work Problem 6 at the Side.**

---

**ANSWERS**

**4.** (a) 5   (b) 18   (c) 305
**5.** (a) 120   (b) 1008   (c) 75
**6.** (a) correct   (b) incorrect; should be 45

## R.5 Exercises

*Decide where the first digit in each quotient should be placed. Then, without finishing the actual division, circle the correct quotient from the three choices given.*

1. $24\overline{)768}$

   3   32   320

2. $35\overline{)805}$

   2   23   230

3. $18\overline{)4500}$

   2   25   250

4. $28\overline{)3500}$

   12   125   1250

5. $86\overline{)10,327}$

   12   120 **R**7   1200

6. $46\overline{)24,026}$

   5   52   522 **R**14

7. $52\overline{)68,025}$

   3   130 **R**1   1308 **R**9

8. $12\overline{)116,953}$

   974 **R**2   9746 **R**1   97,460

9. $21\overline{)149,826}$

   71   713   7134 **R**12

10. $32\overline{)247,892}$

    77 **R**1   7746 **R**20   77,460

11. $523\overline{)470,800}$

    9 **R**100   90 **R**100   900 **R**100

12. $230\overline{)253,230}$

    11   110   1101

*Divide by using long division. See Examples 1–5.*

13. $42\overline{)8699}$

14. $58\overline{)2204}$

15. $47\overline{)11,121}$

16. $83\overline{)39,692}$

17. $26\overline{)62,583}$

18. $28\overline{)84,249}$

19. $63\overline{)78,072}$

20. $238\overline{)186,948}$

21. $153\overline{)509,725}$

22. $402\overline{)29,346}$

23. $420\overline{)357,000}$

24. $900\overline{)153,000}$

*Check each quotient. If a quotient is incorrect, find the correct quotient. See Example 6.*

25. 56)5943  **106 R17**

26. 87)3254  **37 R37**

27. 28)18,424  **658 R9**

28. 191)88,604  **463 R171**

29. 614)38,068  **62 R3**

30. 557)97,286  **174 R368**

31. 72)32,465  **450 R65**

32. 47)9570  **23 R29**

*Solve each application problem by using addition, subtraction, multiplication, or division.*

33. In 1900, the average work week was 59 hours. Today it is 38 hours. How many more hours were worked each year in 1900 than today? Assume 50 weeks of work per year. (*Source:* Reiman Publications.)

34. The U.S. Government Printing Office uses 255,000 pounds of ink each year. If it does an equal amount of printing on each of 200 work days in a year, find the weight of the ink used each day.

35. The most expensive hotel room in a recent study was the Ritz-Carlton at $375 per night, while the least expensive was Motel 6 at $32 per night. Find the amount saved by staying four nights at Motel 6 instead of the Ritz-Carlton. (*Source: USA Today.*)

36. Two divorced parents share their child's education costs, which amount to $3718 per year. If one parent pays $1880 each year, find the amount paid by the other parent over five years.

37. Judy Martinez owes $3888 on a loan. Find her monthly payment if the loan is to be paid off in 36 months.

38. A consultant charged $13,050 for studying a school's compliance with the Americans with Disabilities Act. If the consultant worked 225 hours, find the rate charged per hour.

39. Clarence Hanks can assemble 42 circuits in one hour. How many circuits can he assemble in a 5-day workweek of 8 hours per day?

40. There are two conveyer lines in a factory, each of which packages 240 sacks of salt per hour. If the lines operate for 8 hours, find the total number of sacks of salt packaged by the two lines.

41. A youth soccer association brought in $7588 from fund-raising projects. There were expenses of $838 that had to be paid first, with the balance of the money divided evenly among the 18 teams. How much did each team receive?

42. Feather Farms Egg Ranch collected 3545 eggs in the morning and 2575 eggs in the afternoon. If the eggs are packed in flats containing 30 eggs each, find the number of flats needed for packing.

# Chapter R

## Summary

### Key Terms

**R.1** **addends** — Addends are the numbers being added in an addition problem.

**sum (total)** — The answer in an addition problem is the sum (total).

**commutative property of addition** — The commutative property of addition states that changing the *order* of two addends in an addition problem does not change the sum.

**associative property of addition** — The associative property of addition states that changing the *grouping* of addends does not change the sum.

**carrying** — The process of carrying is used in an addition problem when the sum of the digits in a column is greater than 9.

**R.2** **minuend** — In the subtraction problem $8 - 5 = 3$, the 8 is the minuend. It is the number from which another number is subtracted.

**subtrahend** — In the subtraction problem $8 - 5 = 3$, the 5 is the subtrahend. It is the number being subtracted.

**difference** — The answer in a subtraction problem is the difference.

**borrowing** — Borrowing is used in subtraction if a digit is less than the one directly below it.

**R.3** **factors** — The numbers being multiplied are factors. For example, in $3 \times 4 = 12$, both 3 and 4 are factors.

**product** — The answer in a multiplication problem is the product.

**commutative property of multiplication** — The commutative property of multiplication states that changing the *order* of two factors in a multiplication problem does not change the product.

**associative property of multiplication** — The associative property of multiplication states that changing the *grouping* of factors does not change the product.

**multiple** — The product of two whole-number factors is a multiple of both those numbers.

**R.4** **dividend** — In division, the number being divided is the dividend.

**divisor** — In division, the number being used to divide another number is the divisor.

**quotient** — The answer in a division problem is the quotient.

**short division** — Short division is a quick method of dividing a number by a one-digit divisor.

**remainder** — The remainder is the number left over when two numbers do not divide evenly.

**R.5** **long division** — The process of long division is used to divide by a number with more than one digit.

### Quick Review

*Concepts*

**R.1 Addition of Whole Numbers**
Add from top to bottom, starting with the ones column and working left. To check, add from bottom to top.

*Examples*

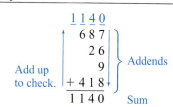

| Concepts | Examples |
|---|---|
| **R.1 Commutative Property of Addition**<br>Changing the *order* of two addends in an addition problem does not change the sum. | $2 + 4 = 6$<br>$4 + 2 = 6$<br>By the commutative property, the sum is the same. |
| **R.1 Associative Property of Addition**<br>Changing the *grouping* of addends does not change the sum. | $(2 + 3) + 4 = 5 + 4 = 9$<br>$2 + (3 + 4) = 2 + 7 = 9$<br>By the associative property, the sum is the same. |
| **R.2 Subtraction of Whole Numbers**<br>Subtract the subtrahend from the minuend to get the difference; use borrowing when necessary. To check, add the difference to the subtrahend to get the minuend. | Problem: $4738$ (Minuend) $- 649$ (Subtrahend) $= 4089$ (Difference)<br>Check: $4089 + 649 = 4738$ |
| **R.3 Multiplication of Whole Numbers**<br>The numbers being multiplied are called *factors*. The multiplicand is being multiplied by the multiplier, giving the product. When the multiplier has more than one digit, partial products must be found and then added to find the product. | $78$ ← Multiplicand<br>$\times 24$ ← Multiplier<br>$312$ ← Partial product<br>$156$ ← Partial product (one position left)<br>$1872$ ← Product |
| **R.3 Commutative Property of Multiplication**<br>Changing the *order* of two factors in a multiplication problem does not change the product. | $3 \times 4 = 12$<br>$4 \times 3 = 12$<br>By the commutative property, the product is the same. |
| **R.3 Associative Property of Multiplication**<br>Changing the *grouping* of factors does not change the product. | $(2 \times 3) \times 4 = 6 \times 4 = 24$<br>$2 \times (3 \times 4) = 2 \times 12 = 24$<br>By the associative property, the product is the same. |
| **R.4 Division of Whole Numbers**<br>$\div$ and $\overline{)}\,$ mean divide.<br>Also a —, as in $\frac{25}{5}$, means to divide the top number (dividend) by the bottom number (divisor). | $88 \div 4 = 22$ (Dividend ÷ Divisor = Quotient)<br>Divisor → $4\overline{)88}$ ← Quotient 22, Dividend 88<br>$\frac{88}{4} = 22$ (Dividend over Divisor = Quotient) |

# Chapter R REVIEW EXERCISES

If you need help with any of these review exercises, look in the section indicated in the red brackets.

**[R.1]** *Add.*

1. $\phantom{+}74$
   $+18$

2. $\phantom{+}35$
   $+78$

3. $\phantom{+}807$
   $\phantom{+}4606$
   $+\phantom{00}51$

4. $\phantom{+}8215$
   $\phantom{+000}9$
   $+7433$

**[R.2]** *Subtract.*

5. $\phantom{-}238$
   $-199$

6. $\phantom{-}573$
   $-389$

7. $\phantom{-}2210$
   $-1986$

8. $\phantom{-}99{,}704$
   $-73{,}838$

**[R.3]** *Multiply. Do the work mentally.*

9. $2 \times 4 \times 6$

10. $9 \times 1 \times 5$

11. $(6)(1)(8)$

12. $7 \cdot 7 \cdot 0$

*Multiply.*

13. $\phantom{\times}43$
    $\times\phantom{0}4$

14. $\phantom{\times}781$
    $\times\phantom{00}7$

15. $\phantom{\times}5440$
    $\times\phantom{000}6$

16. $\phantom{\times}93{,}105$
    $\times\phantom{0000}5$

*Multiply by using the shortcuts for multiples of 10.*

17. $\phantom{\times}320$
    $\times\phantom{0}60$

18. $\phantom{\times}280$
    $\times\phantom{0}90$

19. $\phantom{\times}517$
    $\times\phantom{0}400$

20. $\phantom{\times}16{,}000$
    $\times\phantom{000}8{,}000$

*Multiply.*

21. $\phantom{\times}34$
    $\times 18$

22. $\phantom{\times}52$
    $\times 36$

23. $\phantom{\times}655$
    $\times\phantom{0}21$

24. $\phantom{\times}392$
    $\times\phantom{0}77$

**[R.4]** *Divide.*

25. $42 \div 7$

26. $18 \div 18$

27. $\dfrac{125}{0}$

28. $\dfrac{0}{35}$

**[R.4–R.5]** *Divide.*

29. $4\overline{)432}$

30. $9\overline{)216}$

31. $76\overline{)26{,}752}$

32. $2704 \div 18$

## MIXED REVIEW EXERCISES

*Solve each application problem.*

33. There are 52 playing cards in a deck. How many cards are there in nine decks?

34. Your college bookstore receives textbooks packed 12 books per carton. How many textbooks are received in a delivery of 238 cartons?

35. "Push type" gasoline-powered lawn mowers cost $100 less than self-propelled mowers that you walk behind. If a self-propelled mower costs $380, find the cost of a "push-type" mower.

36. The Village School wants to raise $115,280 for a new library. If $87,340 has already been raised, how much more is needed to reach the goal?

37. It takes 2000 hours of work to build one home. How many hours of work are needed to build 12 homes?

38. A Japanese bullet train travels 80 miles in one hour. Find the number of miles traveled in five hours.

39. If an acre needs 250 pounds of fertilizer, how many acres can be fertilized with 5750 pounds of fertilizer?

40. Each home in a subdivision requires 180 feet of fencing. Find the number of homes that can be fenced with 5760 feet of fencing material.

41. Susan Hessney had $382 in her checking account. She wrote a check for $135. How much does she have left in her account?

42. Find the total cost of four T-shirts at $14 each and three sweatshirts at $29 each.

43. The Houston Space Center charges $15 for each adult admission and $12 for each child. Find the total cost to admit a group of 18 adults and 26 children.

44. A newspaper carrier has 56 customers who take the paper daily and 23 customers who take the paper on weekends only. A daily customer pays $18 per month and a weekend-only customer pays $11 per month. Find the total monthly collections.

# Chapter R Test

*Fill in the blanks to complete each sentence.*

1. In an addition problem, the numbers being added are called _____ and the answer is called the _____.

2. In a multiplication problem, the numbers being multiplied are called _____ and the answer is called the _____.

3. In a subtraction problem the answer is called the _____ and in a division problem the answer is called the _____.

*Add, subtract, multiply, or divide, as indicated.*

4. $984 + 65 + 7561$

5. $\begin{array}{r} 17{,}063 \\ 7 \\ 12 \\ 1\;505 \\ 93{,}710 \\ +\phantom{00}333 \\ \hline \end{array}$

6. $17{,}002 - 54$

7. $\begin{array}{r} 5062 \\ -\;1978 \\ \hline \end{array}$

8. $5 \times 7 \times 4$

9. $57 \cdot 3000$

10. $(85)(21)$

11. $\begin{array}{r} 7381 \\ \times\;\;603 \\ \hline \end{array}$

12. $6\overline{)1236}$

13. $\dfrac{791}{0}$

14. $38{,}472 \div 84$

15. $280\overline{)44{,}800}$

1. _____
2. _____
3. _____
4. _____
5. _____
6. _____
7. _____
8. _____
9. _____
10. _____
11. _____
12. _____
13. _____
14. _____
15. _____

*Solve each application problem.*

**16.** Find the cost of 48 shovels at $11 per shovel.

**17.** In a consumer survey of cordless phones, the most expensive model cost $350 and the least expensive model was $79. Find the difference in price between the most expensive and the least expensive phone.

**18.** A stamping machine produces 936 license plates each hour. How long will it take to produce 30,888 license plates?

**19.** Kenée Shadbourne paid $690 for tuition, $185 for books, and $68 for supplies. If this money was withdrawn from her checking account, which had a balance of $1108, find her new balance.

**20.** An appliance manufacturer assembles 118 self-cleaning ovens each hour during the first four hours and 139 standard ovens each hour during the next four hours. Find the total number of ovens assembled in the 8-hour period.

**21.** The monthly rents collected from the four units in an apartment building are $785, $800, $815, and $725. After expenses of $1085 are paid, find the amount that remains.

**22.** Describe the divisibility tests for 2, 5, and 10. Also give two examples for each test: one number that is divisible and one number that is not divisible.

# Appendix
# Inductive and Deductive Reasoning

**OBJECTIVES**

1. Use inductive reasoning to analyze patterns.
2. Use deductive reasoning to analyze arguments.
3. Use deductive reasoning to solve problems.

**1** **Use inductive reasoning to analyze patterns.** In many scientific experiments, conclusions are drawn from specific outcomes. After many repetitions and similar outcomes, the findings are generalized into statements that appear to be true. When general conclusions are drawn from specific observations, we are using a type of reasoning called **inductive reasoning**. The next several examples illustrate this type of reasoning.

**Example 1  Using Inductive Reasoning**

Find the next number in the sequence 3, 7, 11, 15, . . . .

To discover a pattern, calculate the difference between each pair of successive numbers.

$$7 - 3 = 4$$
$$11 - 7 = 4$$
$$15 - 11 = 4$$

As shown, the difference is always 4. Each number is 4 greater than the previous one. Thus, the next number in the pattern is $15 + 4$, or 19.

➤ **Work Problem 1 at the Side.**

**1** Find the next number in the sequence 2, 8, 14, 20, . . . . Describe the pattern.

**Example 2  Using Inductive Reasoning**

Find the next number in this sequence.

$$7, 11, 8, 12, 9, 13, \ldots$$

The pattern in this example involves addition and subtraction.

$$7 + 4 = 11$$
$$11 - 3 = 8$$
$$8 + 4 = 12$$
$$12 - 3 = 9$$
$$9 + 4 = 13$$

To get the second number, we add 4 to the first number. To get the third number, we subtract 3 from the second number. To obtain subsequent numbers, we continue the pattern. The next number is $13 - 3 = 10$.

➤ **Work Problem 2 at the Side.**

**2** Find the next number in the sequence 6, 11, 7, 12, 8, 13, . . . . Describe the pattern.

**ANSWERS**
1. 26; add 6 each time.
2. 9; add 5, subtract 4.

A–1

**❸** Find the next number in the sequence 2, 6, 18, 54, . . . . Describe the pattern.

### Example 3 Using Inductive Reasoning

Find the next number in the sequence 1, 2, 4, 8, 16, . . . .

Each number after the first is obtained by multiplying the previous number by 2. So the next number would be 16 · 2 = 32.

**Work Problem ❸ at the Side.**

### Example 4 Using Inductive Reasoning

**(a)** Find the next geometric shape in this sequence.

The figures alternate between a circle and a triangle. Also, the number of dots increases by 1 in each subsequent figure. Thus, the next figure should be a circle with five dots inside it.

**(b)** Find the next geometric shape in this sequence.

The first two shapes consist of vertical lines with horizontal lines at the bottom extending first left and then right. The third shape is a vertical line with a horizontal line at the top extending to the left. Therefore, the next shape should be a vertical line with a horizontal line at the top extending to the right.

**❹** Find the next shape in this sequence.

**Work Problem ❹ at the Side.**

**2** **Use deductive reasoning to analyze arguments.** In the previous discussion, specific cases were used to find patterns and predict the next event. There is another type of reasoning called **deductive reasoning,** which moves from general cases to specific conclusions.

### Example 5 Using Deductive Reasoning

Does the conclusion follow from the premises in this argument?

All Buicks are automobiles.
All automobiles have horns.
∴ All Buicks have horns.

In this example, the first two statements are called *premises* and the third statement (below the line) is called a *conclusion*. The symbol ∴ is a mathematical symbol meaning "**therefore**." The entire set of statements is called an *argument*.

*Continued on Next Page*

**ANSWERS**
**3.** 162; multiply by 3.
**4.**

The focus of deductive reasoning is to determine whether the conclusion follows (is valid) from the premises. A set of circles called **Euler circles** is used to analyze the argument.

In Example 5, the statement "All Buicks are automobiles" can be represented by two circles, one for Buicks and one for automobiles. Note that the circle representing Buicks is totally inside the circle representing automobiles because the first premise states that *all* Buicks are automobiles.

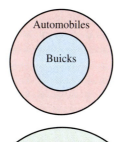

Now, a circle is added to represent the second statement, vehicles with horns. This circle must completely surround the circle representing automobiles because the second premise states that *all* automobiles have horns.

To analyze the conclusion, notice that the circle representing Buicks is *completely* inside the circle representing vehicles with horns. Therefore, it must follow that all Buicks have horns. ***The conclusion is valid.***

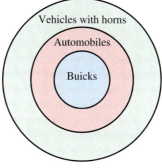

━━━━━━━━━━━━ Work Problem ❺ at the Side.

**Example 6** Using Deductive Reasoning

Does the conclusion follow from the premises in this argument?

All tables are round.
All glasses are round.
∴  All glasses are tables.

Using Euler circles, a circle representing tables is drawn inside a circle representing round objects.

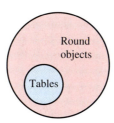

The second statement requires that a circle representing glasses must now be drawn inside the circle representing round objects, but not necessarily inside the circle representing tables. Therefore, the conclusion does *not* follow from the premises. This means that ***the conclusion is invalid or untrue.***

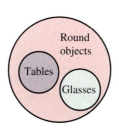

━━━━━━━━━━━━ Work Problem ❻ at the Side.

❺ Does the conclusion follow from the premises in this argument?

All cars have four wheels.
All Fords are cars.
∴  All Fords have four wheels.

❻ Does each conclusion follow from the premises?

(a) All animals are wild.
All cats are animals.
∴  All cats are wild.

(b) All students use math.
All adults use math.
∴  All adults are students.

**ANSWERS**
5. The conclusion follows from the premises.
6. (a) The conclusion follows from the premises.
   (b) The conclusion does *not* follow from the premises.

**7** In a college class of 100 students, 35 take both math and history, 50 take history, and 40 take math. How many take neither math nor history?

**3** **Use deductive reasoning to solve problems.** Another type of deductive reasoning problem occurs when a set of facts is given in a problem and a conclusion must be drawn by using these facts.

### Example 7 Using Deductive Reasoning

There were 25 students enrolled in a ceramics class. During the class, 10 of the students made a bowl and 8 students made a birdbath. Three students made both a bowl and a birdbath. How many students did not make either a bowl or a birdbath?

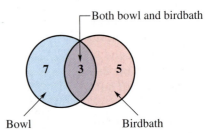

This type of problem is best solved by organizing the data using a drawing called a **Venn diagram.** Two overlapping circles are drawn, with each circle representing one item made by students.

In the region where the circles overlap, write the number of students who made *both* items, namely, 3. In the remaining portion of the birdbath circle, write the number 5, which when added to 3 will give the total number of students who made a birdbath, namely, 8. In a similar manner, write 7 in the remaining portion of the bowl circle, since $7 + 3 = 10$, the total number of students who made a bowl. The total of all three numbers written in the circles is 15. Since there were 25 students in the class, this means $25 - 15$ or 10 students did not make either a birdbath or a bowl.

Work Problem **7** at the Side.

**8** A Chevy, BMW, Cadillac, and Ford are parked side by side. The known facts are:

(a) The Ford is on the right end.

(b) The BMW is next to the Cadillac.

(c) The Chevy is between the Ford and the Cadillac.

Which car is parked on the left end?

### Example 8 Using Deductive Reasoning

Four cars in a race finish first, second, third, and fourth. The following facts are known.

(a) Car A beat Car C.

(b) Car D finished between Cars C and B.

(c) Car C beat Car B.

In which order did the cars finish?

To solve this type of problem, it is helpful to use a line diagram.

1. *Write A before C,* because Car A beat Car C (fact **a**).

    A    C

2. *Write B after C,* because Car C beat Car B (fact **c**).

    A    C    B

3. *Write D between C and B,* because Car D finished between Cars C and B (fact **b**).

The correct order of finish is shown below.

A    C    D    B

Work Problem **8** at the Side.

**Answers**
7. 45 students
8. BMW

# Appendix Exercises

*Find the next number in each sequence. Describe the pattern in each sequence. See Examples 1–3.*

1. 2, 9, 16, 23, 30, . . .

2. 5, 8, 11, 14, 17, . . .

3. 0, 10, 8, 18, 16, . . .

4. 3, 9, 7, 13, 11, . . .

5. 1, 2, 4, 8, . . .

6. 1, 4, 16, 64, . . .

7. 1, 3, 9, 27, 81, . . .

8. 3, 6, 12, 24, 48, . . .

9. 1, 4, 9, 16, 25, . . .

10. 6, 7, 9, 12, 16, . . .

*Find the next shape in each sequence. See Example 4.*

11.

12.

13.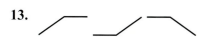

14.

*In each argument, state whether or not the conclusion follows from the premises. See Examples 5 and 6.*

15. All animals are wild.
    All lions are animals.
    ∴ All lions are wild.

16. All students are hard workers.
    All business majors are students.
    ∴ All business majors are hard workers.

17. All teachers are serious.
    All mathematicians are serious.
    ∴ All mathematicians are teachers.

18. All boys ride bikes.
    All Americans ride bikes.
    ∴ All Americans are boys.

*Solve each application problem. See Examples 7 and 8.*

19. In a given 30-day period, a man watched television 20 days and his wife watched television 25 days. If they watched television together 18 days, how many days did neither watch television?

20. In a class of 40 students, 21 students take both calculus and physics. If 30 students take calculus and 25 students take physics, how many do not take either calculus or physics?

21. Tom, Dick, Mary, and Joan all work for the same company. One is a secretary, one is a computer operator, one is a receptionist, and one is a mail clerk.
    (a) Tom and Joan eat dinner with the computer operator.
    (b) Dick and Mary carpool with the secretary.
    (c) Mary works on the same floor as the computer operator and the mail clerk.
    Who is the computer operator?

22. Four cars—a Ford, a Buick, a Mercedes, and an Audi—are parked in a garage in four spaces.
    (a) The Ford is in the last space.
    (b) The Buick and Mercedes are next to each other.
    (c) The Audi is next to the Ford but not next to the Buick.
    Which car is in the first space?

# Answers to Selected Exercises

In this section we provide the answers that we think most students will obtain when they work the exercises using the methods explained in the text. If your answer does not look exactly like the one given here, it is not necessarily wrong. In many cases there are equivalent forms of the answer that are correct. For example, if the answer section shows $\frac{3}{4}$ and your answer is 0.75, you have obtained the correct answer but written it in a different (yet equivalent) form. Unless the directions specify otherwise, 0.75 is just as valid an answer as $\frac{3}{4}$.

In general, if your answer does not agree with the one given in the text, see whether it can be transformed into the other form. If it can, then it is the correct answer. If you still have doubts, talk with your instructor.

## Diagnostic Pretest

### (page xxix)

**1.** 25,003,701  **2.** 6  **3.** 4,360,000  **4.** −$18  **5.** −40
**6.** $4r - 33$  **7.** $t = 5$  **8.** $n = 1$  **9.** 58 cm  **10.** 66 in.²
**11.** $4n + 13 = n - 8; n = -7$  **12.** $x + x + 12 = 238$; 113 cm, 125 cm  **13.** $\frac{3}{4}$  **14.** $-\frac{1}{4}$  **15.** $x = -20$  **16.** 30 cm²
**17.** $18.17 (rounded)  **18.** $r = 2.3$  **19.** $10.20  **20.** 16.3 cm (rounded)  **21.** $x = 108$  **22.** 121.6 miles  **23.** 117°
**24.** $x = 8$ ft; $y = 7.5$ ft  **25. (a)** 2.5  **(b)** 30%  **26. (a)** $\frac{19}{50}$
**(b)** 7.5%  **27.** $249.10  **28.** $13.80  **29. (a)** 4.32 m  **(b)** 80 g
**30.** 3.8 ft (rounded)  **31.** 10.4 L  **32. (a)** m  **(b)** mg  **33.** $252
**34. (a)** 1998, 1999  **(b)** 2001; 30 sections  **35. (a)** Quadrant II
**(b)** no quadrant  **(c)** Quadrant III  **(d)** Quadrant I
**36.**

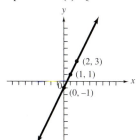

| $x$ | $y$ | $(x, y)$ |
|---|---|---|
| 0 | −1 | (0, −1) |
| 1 | 1 | (1, 1) |
| 2 | 3 | (2, 3) |

**37.** $\frac{2}{3}$  **38.** $-4a^8b^6$  **39. (a)** $5.45 \times 10^7$
**(b)** 0.00407  **40.** $6x^2 - 7x - 5$

## Chapter 1

### Section 1.1 (page 5)

**1.** 15; 0; 83,001  **3.** 7; 362,049  **5.** hundreds  **7.** hundred thousands  **9.** ten millions  **11.** hundred billions  **13.** ten trillions, hundred billions, millions, hundred thousands, ones
**15.** eight thousand, four hundred twenty-one  **17.** forty-six thousand, two hundred five  **19.** three million, sixty-four thousand, eight hundred one  **21.** eight hundred forty million, one hundred eleven thousand, three  **23.** fifty-one billion, six million, eight hundred eighty-eight thousand, three hundred twenty-one
**25.** three trillion, seven hundred twelve million  **27.** 46,805
**29.** 5,600,082  **31.** 271,900,000  **33.** 12,417,625,310
**35.** 600,000,071,000,400  **37.** largest: 97,651,100; ninety-seven million, six hundred fifty-one thousand, one hundred; smallest: 10,015,679; ten million, fifteen thousand, six hundred seventy-nine
**38.** Answers will vary.  **39.** sixty-fours  thirty-twos  sixteens  eights  **(a)** 101  **(b)** 1010  **(c)** 1111  **40. (a)** Answers will vary but should mention that the location or place in which a digit is written gives it a different value.  **(b)** 8 = VIII; 38 = XXXVIII; 275 = CCLXXV; 3322 = MMMCCCXXII  **(c)** The Roman system is *not* a place value system because no matter what place it's in, M = 1000, C = 100, etc. One disadvantage is that it takes much more space to write large numbers; another is that there is no symbol for zero.  **41.** six thousand, five hundred sixty-seven
**43.** 101,280,000  **45.** two million, twenty-one thousand, eighteen dollars  **47.** 24,000,500

### Section 1.2 (page 13)

**1.** ⁺29,035 feet or 29,035 feet  **3.** ⁻128.6 degrees  **5.** ⁻18 yards
**7.** ⁺$100 or $100  **9.** $-6\frac{1}{2}$ pounds
**11.** ⟵┼─●─┼─●─┼─●─┼─●─┼─●─┼─●─┼⟶
　　　⁻5⁻4⁻3⁻2⁻1 0 1 2 3 4 5
**13.** ⟵┼─┼─●─┼─●─┼─┼─┼─●─┼─●─┼⟶
　　　⁻5⁻4⁻3⁻2⁻1 0 1 2 3 4 5
**15.** ⟵┼─●─┼─┼─●─┼─┼─┼─┼─●●─┼⟶
　　　⁻9⁻8⁻7⁻6⁻5⁻4⁻3⁻2⁻1 0 1
**17.** >  **19.** <  **21.** <  **23.** >  **25.** <  **27.** >  **29.** >
**31.** <  **33.** 15  **35.** 3  **37.** 0  **39.** 200  **41.** 75  **43.** 8042
**45.**
　A  C　D B
　↓　↓　↓↓
⟵─●─●─┼─●●─┼⟶
　⁻2　⁻1　0　1

**46.** ⁻1.5, ⁻1, 0, 0.5  **47.** A: may be at risk; B: above normal; C: normal; D: normal  **48. (a)** The patient would think the interpretation was "above normal" and wouldn't get treatment.  **(b)** Patient D's score of 0; zero is neither positive nor negative.

### Section 1.3 (page 21)

**1.** 3

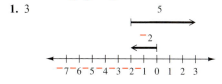

**3.** ⁻7

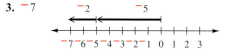

**5.** ⁻1

A–7

**7.** (a) ⁻10  (b) 10  **9.** (a) 12  (b) ⁻12  **11.** (a) ⁻50  (b) 50  **13.** (a) 158  (b) ⁻158  **15.** Each pair of answers matches except for the sign. This occurs because the absolute values are the same, so the only difference in the sums is the common sign.  **17.** (a) 2  (b) ⁻2  **19.** (a) ⁻7  (b) 7  **21.** (a) ⁻5  (b) 5  **23.** (a) 150  (b) ⁻150  **25.** Each pair of answers differs only in the sign of the answer. This occurs because the signs of the addends are reversed.  **27.** ⁻3  **29.** 7  **31.** ⁻7  **33.** 1  **35.** ⁻8  **37.** ⁻20  **39.** ⁻17  **41.** ⁻22  **43.** ⁻19  **45.** 6  **47.** ⁻5  **49.** 0  **51.** 5  **53.** ⁻32  **55.** 13 + ⁻17 = ⁻4 yards  **57.** ⁻$62 + $50 = ⁻$12  **59.** 8 + ⁻3 + ⁻1 = 4 feet  **61.** ⁻20 + 75 + ⁻55 = 0 points  **63.** ⁻16  **65.** ⁻10  **67.** ⁻5; ⁻18; ⁻23  **69.** 15; ⁻4; 11  **71.** 6 + (⁻14 + 14); 6 + 0 = 6  **73.** (⁻14 + ⁻6) + ⁻7; ⁻20 + ⁻7 = ⁻27  **75.** Some possibilities are: ⁻6 + 0 = ⁻6; 10 + 0 = 10; 0 + 3 = 3  **77.** ⁻4116  **79.** 8686  **81.** ⁻96,077

### Section 1.4 (page 27)
**1.** ⁻6; 6 + ⁻6 = 0  **3.** 13; ⁻13 + 13 = 0  **5.** 0; 0 + 0 = 0  **7.** 14  **9.** ⁻2  **11.** ⁻12  **13.** ⁻25  **15.** ⁻23  **17.** 5  **19.** 20  **21.** 11  **23.** ⁻60  **25.** 0  **27.** 0  **29.** ⁻6  **31.** (a) 8  (b) ⁻2  (c) 2  (d) ⁻8  **33.** (a) ⁻3  (b) 11  (c) ⁻11  (d) 3  **35.** ⁻6  **37.** ⁻5  **39.** 3  **41.** ⁻10  **43.** 12  **45.** ⁻5  **47.** Jeff: ⁻20 + 75 + ⁻55 = 0 pts; Terry: 42 + ⁻15 + 20 = 47 pts  **49.** The student forgot to change 6 to its opposite, ⁻6. It should be ⁻6 + ⁻6 = ⁻12.  **51.** ⁻11  **53.** ⁻5  **55.** ⁻10  **57.** Answers on left: ⁻8; 8. On right: ⁻1; 1  (a) Subtraction is *not* commutative.  (b) The absolute value of the answer is the same, but the sign changes.  **58.** Subtracting 0 from a number does *not* change the number. For example ⁻5 − 0 = ⁻5. But subtracting a number from 0 *does* change the number to its opposite. For example, 0 − ⁻5 = 5.

### Section 1.5 (page 35)
**1.** 630  **3.** ⁻1080  **5.** 7900  **7.** ⁻86,800  **9.** 42,500  **11.** ⁻6000  **13.** 15,800  **15.** ⁻78,000  **17.** 6000  **19.** 53,000  **21.** 600,000  **23.** ⁻9,000,000  **25.** 140,000,000  **27.** 30,000 miles  **29.** ⁻60 degrees  **31.** $10,000  **33.** 60,000,000 Americans  **35.** ⁻300 feet  **37.** 600,000 people  **39.** Answers will vary but should mention looking only at the second digit, rounding first digit up when second digit is 5 or more, leaving first digit unchanged when second digit is 4 or less. Examples will vary.  **41.** Estimate: ⁻40 + 90 = 50; Exact: 47  **43.** Estimate: 20 + ⁻100 = ⁻80; Exact: ⁻81  **45.** Estimate: ⁻300 + ⁻400 = ⁻700; Exact: ⁻672  **47.** Estimate: 3000 + 7000 = 10,000; Exact: 9907  **49.** Estimate: 20 + ⁻80 = ⁻60; Exact: ⁻58  **51.** Estimate: ⁻40 + ⁻40 = ⁻80; Exact: ⁻78  **53.** Estimate: ⁻100 + 30 + ⁺70 = 0; Exact: 0  **55.** Estimate: $80,000 − $50,000 = $30,000; Exact: $25,768  **57.** Estimate: $2000 − $500 − $300 − $300 − $200 − $200 = $500; Exact: $458  **59.** Estimate: ⁻100 + 40 + 50 = ⁻10 degrees; Exact: ⁻13 degrees  **61.** Estimate: $400 + $500 = $900; Exact: $905

### Section 1.6 (page 45)
**1.** (a) 63  (b) 63  (c) ⁻63  (d) ⁻63  **3.** (a) ⁻56  (b) ⁻56  (c) 56  (d) 56  **5.** ⁻35  **7.** ⁻45  **9.** ⁻18  **11.** ⁻50  **13.** ⁻40  **15.** ⁻56  **17.** 32  **19.** 77  **21.** 0  **23.** 133  **25.** 13  **27.** 0  **29.** 48  **31.** ⁻56  **33.** ⁻160  **35.** 5  **37.** ⁻3  **39.** ⁻1  **41.** 0  **43.** 5  **45.** ⁻4  **47.** Commutative property: changing the *order* of the factors does not change the product. Associative property: changing the *grouping* of the factors does not change the product. Examples will vary.  **49.** Examples will vary. Some possibilities are: (a) 6 • ⁻1 = ⁻6; 2 • ⁻1 = ⁻2; 15 • ⁻1 = ⁻15  (b) ⁻6 • ⁻1 = 6; ⁻2 • ⁻1 = 2; ⁻15 • ⁻1 = 15  The result of multiplying any nonzero number times ⁻1 is the number with the opposite sign.  **50.** The products are 4, ⁻8, 16, ⁻32. The product doubles each time and the sign changes. The next three products are 64, ⁻128, 256.  **51.** = 9 • ⁻3 + 9 • 5  Both results are 18.  **53.** = 8 • 25  Both products are 200.  **55.** = (⁻3 • ⁻2) • ⁻5  Both products are ⁻30.  **57.** Estimate: $300 • 50 = $15,000; Exact: $324 • 52 = $16,848  **59.** Estimate: ⁻$10,000 • 10 = ⁻$100,000; Exact: ⁻$9950 • 12 = ⁻$119,400  **61.** Estimate: $200 • 10 = $2000; Exact: $182 • 13 = $2366  **63.** Estimate: 20 • 400 = 8000 hours; Exact: 24 • 365 = 8760 hours  **65.** ⁻512  **67.** 0  **69.** ⁻355,299  **71.** $247  **73.** ⁻22 degrees

### Section 1.7 (page 55)
**1.** (a) 7  (b) 7  (c) ⁻7  (d) ⁻7  **3.** (a) ⁻7  (b) 7  (c) ⁻7  (d) 7  **5.** (a) 1  (b) 35  (c) ⁻13  (d) 1  **7.** (a) 0  (b) undefined  (c) undefined  (d) 0  **9.** ⁻4  **11.** ⁻3  **13.** 6  **15.** ⁻11  **17.** undefined  **19.** ⁻14  **21.** 10  **23.** 4  **25.** ⁻1  **27.** 0  **29.** 191  **31.** ⁻499  **33.** 2  **35.** ⁻4  **37.** 40  **39.** ⁻48  **41.** 5  **43.** 0  **45.** 2 ÷ 1 = 2 but 1 ÷ 2 = 0.5, so division is not commutative.  **46.** (12 ÷ 6) ÷ 2 = 2 ÷ 2 = 1; 12 ÷ (6 ÷ 2) = 12 ÷ 3 = 4; different quotients. Division is not associative.  **47.** Similar: If the signs match, the result is positive. If the signs are different, the result is negative. Different: Multiplication is commutative, division is not. You can multiply by 0, but dividing by 0 is undefined.  **48.** Examples will vary. The properties are: Any nonzero number divided by itself is 1. Any number divided by 1 is the number. Division by 0 is undefined. Zero divided by any other number (except 0) is 0.

**49.** (a) $\frac{-6}{-1} = 6; \frac{-2}{-1} = 2; \frac{-15}{-1} = 15$

(b) $\frac{6}{-1} = -6; \frac{2}{-1} = -2; \frac{15}{-1} = -15$  When dividing by ⁻1, the sign of the number changes to its opposite.  **50.** Division is not commutative. $\frac{0}{-3} = 0$ because $0 \cdot -3 = 0$. But $\frac{-3}{0}$ is undefined because when $\frac{-3}{0} = ?$ is rewritten as $? \cdot 0 = -3$, no number can replace ? and make a true statement.  **51.** Estimate: ⁻40,000 ÷ 20 = ⁻2000 feet; Exact: ⁻36,198 ÷ 18 = ⁻2011 feet  **53.** Estimate: ⁻$200 + $500 = $300; Exact: ⁻$238 + $450 = $212  **55.** Estimate: 400 − 100 = 300 days; Exact: 365 − 106 = 259 days  **57.** Estimate: ⁻700 • 40 = ⁻28,000 feet; Exact: ⁻730 • 37 = ⁻27,010 feet  **59.** Estimate: 300 ÷ 5 = 60 miles; Exact: 315 ÷ 5 = 63 miles  **61.** 168 average score  **63.** The back shows 520 grams, which is 10 grams more than the front.  **65.** ⁻$15  **67.** 16 hours, with 40 minutes left over  **69.** 33 rooms, with space for 2 people unused.  **71.** ⁻10  **73.** undefined  **75.** 31.70979198 rounds to 32 years.

### Section 1.8 (page 65)
**1.** 4 • 4 • 4; 4 cubed or 4 to the third power  **3.** $2^7$; 128; 2 to the seventh power  **5.** $5^4$; 625; 5 to the fourth power  **7.** $7^2$; 7 • 7; 49  **9.** $10^1$; 10; 10  **11.** (a) 10  (b) 100  (c) 1000  (d) 10,000  **13.** (a) 4  (b) 16  (c) 64  (d) 256  **15.** 9,765,625  **17.** 4096  **19.** 4  **21.** 25  **23.** ⁻64  **25.** 81  **27.** ⁻1000  **29.** 1  **31.** 108  **33.** 225  **35.** ⁻750  **37.** ⁻32  **39.** (a) The answers are 4, ⁻8, 16, ⁻32, 64, ⁻128, 256, ⁻512. When a negative number is raised to an even power, the answer is positive; when raised to an odd power, the answer is negative.  (b) negative; positive  **41.** ⁻6  **43.** 0  **45.** ⁻39  **47.** 16  **49.** 23  **51.** ⁻43  **53.** 7  **55.** ⁻3  **57.** 0  **59.** ⁻38  **61.** 41  **63.** ⁻2  **65.** 13  **67.** 126  **69.** 8  **71.** $\frac{27}{-3} = -9$  **73.** $\frac{-48}{-4} = 12$  **75.** $\frac{-60}{-1} = 60$

**77.** ⁻4050  **79.** 7  **81.** $\frac{27}{0}$ is undefined.

## Summary Exercises on Operations with Integers (page 69)
1. ⁻6  2. 0  3. ⁻7  4. ⁻7  5. 63  6. ⁻1  7. ⁻56
8. ⁻22  9. 12  10. 8  11. ⁻13  12. 0  13. 0  14. ⁻17
15. ⁻48  16. ⁻10  17. ⁻50  18. undefined  19. ⁻14
20. ⁻6  21. 0  22. 16  23. ⁻30  24. undefined  25. 48
26. ⁻19  27. 2  28. ⁻20  29. 0  30. 16  31. ⁻3  32. ⁻36
33. 6  34. ⁻7  35. ⁻5  36. ⁻31  37. ⁻2  38. ⁻32
39. ⁻5  40. ⁻4  41. ⁻9732  42. 100  43. 4  44. ⁻343
45. 5  46. ⁻6  47. ⁻10  48. ⁻5  49. 8  50. 1
51. $\frac{-22}{-11} = 2$  52. $\frac{8}{-2} = -4$  53. $\frac{27}{0}$ is undefined.
54. $\frac{-8}{8} = -1$

## Chapter 1 Review Exercises (page 77)
1. 86, 0, 35,600  2. eight hundred six  3. three hundred nineteen thousand, twelve  4. sixty million, three thousand, two hundred  5. fifteen trillion, seven hundred forty-nine billion, six  6. 504,100  7. 620,080,000  8. 99,007,000,356
9. ⟵─●─┼─●─┼─┼─┼─●─┼─┼─●─⟶
      ⁻5 ⁻4 ⁻3 ⁻2 ⁻1  0  1  2  3  4  5
10. >  11. <  12. >  13. <  14. 5  15. 9  16. 0  17. 125
18. ⁻1  19. ⁻13  20. ⁻3  21. 0  22. 1  23. ⁻24  24. ⁻7
25. 3  26. 0  27. ⁻17  28. 5; ⁻5 + 5 = 0  29. ⁻18; 18 + ⁻18 = 0  30. ⁻7  31. 17  32. ⁻16  33. 13  34. 18
35. ⁻22  36. 0  37. ⁻20  38. ⁻1  39. ⁻3  40. 14  41. 15
42. ⁻16  43. 3  44. ⁻8  45. 0  46. 210  47. 59,000
48. 85,000,000  49. ⁻3000  50. ⁻7,060,000  51. 400,000
52. ⁻200 pounds  53. ⁻1000 feet  54. 400,000,000 directories
55. 9,000,000,000 people  56. ⁻54  57. 56  58. ⁻100  59. 0
60. 24  61. 17  62. ⁻48  63. 125  64. ⁻36  65. 50
66. ⁻72  67. 9  68. ⁻7  69. undefined  70. 5  71. ⁻18
72. 0  73. 15  74. ⁻1  75. ⁻5  76. 18  77. 0  78. 156 days and 2 extra hours  79. 10,000  80. 32  81. 27  82. 16
83. ⁻125  84. 8  85. 324  86. ⁻200  87. ⁻25  88. ⁻2
89. 10  90. ⁻28  91. $\frac{8}{-8} = -1$  92. $\frac{11}{0}$ is undefined.
93. associative property of addition  94. commutative property of multiplication  95. addition property of 0  96. multiplication property of 0  97. distributive property  98. associative property of multiplication  99. Estimate: $10,000 • 200 = $2,000,000; Exact: $11,900 • 192 = $2,284,800  100. Estimate: $200 + $400 − $700 = ⁻$100; Exact: $185 + $428 − $706 = ⁻$93
101. Estimate: 800 ÷ 20 = 40 miles; Exact: 840 ÷ 24 = 35 miles
102. Estimate: ($40 • 20) + ($90 • 10) = $1700; Exact: ($39 • 19) + ($85 • 12) = $1761  103. ⁻$700, ⁻$100, $700, $900, $0, $700  104. Jan., Apr.  105. $2100  106. $1850

## Chapter 1 Test (page 81)
1. twenty million, eight thousand, three hundred seven
2. 30,000,700,005
3. ⟵─┼─●─┼─●─●─┼─┼─┼─⟶
      ⁻3 ⁻2 ⁻1  0  1  2  3
4. >; <  5. 10; 14  6. ⁻6  7. ⁻5  8. 7  9. ⁻40  10. 10
11. 64  12. ⁻50  13. undefined  14. ⁻60  15. ⁻5
16. ⁻45  17. 6  18. 25  19. 0  20. 9  21. 128  22. ⁻2
23. 8  24. ⁻16  25. An exponent shows how many times to use a factor in repeated multiplication. Examples will vary. Some possibilities are $(2)^4 = 2 • 2 • 2 • 2 = 16$ and $(-3)^2 = -3 • -3 = 9$.
26. Commutative property: changing the *order* of addends does not change the sum. Associative property: changing the *grouping* of addends does not change the sum. Examples will vary.
27. 900  28. 36,420,000,000  29. 350,000  30. Estimate: 2,000,000 − 1,000,000 = 1,000,000 cars; Exact: 668,470 cars
31. Estimate: $200 + $300 + ⁻$500 = $0; Exact: ⁻$29
32. Estimate: ⁻1000 ÷ 10 = ⁻100 yards; Exact: ⁻95 yards
33. Estimate: 30 • 100 = 3000 calories; Exact: 3410 calories
34. 27 cartons because 26 cartons would leave 28 pounds of books unpacked

# Chapter 2

## Section 2.1 (page 91)
1. *c* is variable; 4 is constant.  3. *h* is variable; 5 is coefficient.
5. *m* is variable; ⁻3 is constant.  7. *c* is variable; 2 is coefficient; 10 is constant.  9. *x* and *y* are variables.  11. *g* is variable; ⁻6 is coefficient.  13. (a) 654 + 10 is 664 robes.
(b) 208 + 10 is 218 robes.  15. (a) 3 • 11 inches is 33 inches.
(b) 3 • 3 feet is 9 feet.  17. (a) 3 • 12 − 5 is 31 brushes.
(b) 3 • 16 − 5 is 43 brushes.  19. (a) $\frac{332}{4}$ is 83 points.
(b) $\frac{637}{7}$ is 91 points.  21. 12 + 12 + 12 + 12 is 48, 4 • 12 is 48; 0 + 0 + 0 + 0 is 0, 4 • 0 is 0; ⁻5 + ⁻5 + ⁻5 + ⁻5 = ⁻20, 4 • ⁻5 is ⁻20  23. ⁻2 • ⁻4 + 5 is 8 + 5 is 13; ⁻2 • ⁻6 + ⁻2 is 12 + ⁻2 is 10; ⁻2 • 0 + ⁻8 is 0 + ⁻8 is ⁻8  25. A variable is a letter that represents the part of a rule that varies or changes depending on the situation. An expression expresses, or tells, the rule for doing something. For example, *c* + 5 is an expression, and *c* is the variable.  27. *b* • 1 = *b*  or  1 • *b* = *b*  29. $\frac{b}{0}$ is undefined  or  *b* ÷ 0 is undefined.  31. *c* • *c* • *c* • *c* • *c* • *c*
33. *x* • *x* • *x* • *x* • *y* • *y* • *y*  35. ⁻3 • *a* • *a* • *b*
37. 9 • *x* • *y* • *y*  39. ⁻2 • *c* • *c* • *c* • *c* • *c* • *d*
41. *a* • *a* • *a* • *b* • *c* • *c*  43. 16  45. ⁻24  47. ⁻18  49. ⁻128
51. ⁻18,432  53. 311,040  55. 56  57. $\frac{36}{0}$ is undefined.

59. (a) 3 miles  (b) 2 miles  (c) 1 mile  60. (a) $\frac{1}{2}$ mile; take half of the distance for 5 seconds  (b) $7\frac{1}{2}$ seconds; find the number halfway between 5 seconds and 10 seconds  (c) $12\frac{1}{2}$ seconds; find the number halfway between 10 seconds and 15 seconds

## Section 2.2 (page 103)
1. $2b^2$ and $b^2$; The coefficients are 2 and 1.  3. ⁻*xy* and 2*xy*; The coefficients are ⁻1 and 2.  5. 7, 3, and ⁻4; The like terms are constants.  7. 12*r*  9. $6x^2$  11. ⁻4*p*  13. ⁻$3a^3$  15. 0
17. *xy*  19. $6t^4$  21. $4y^2$  23. ⁻8*x*  25. 12*a* + 4*b*
27. 7*rs* + 14  29. *a* + 2*ab*²  31. ⁻2*x* + 2*y*  33. $7b^2$
35. cannot be simplified  37. ⁻15*r* + 5*s* + *t*  39. 30*a*
41. ⁻$8x^2$  43. ⁻$20y^3$  45. 18*cd*  47. $21a^2bc$  49. 12*w*
51. 6*b* + 36  53. 7*x* − 7  55. 21*t* + 3  57. ⁻10*r* − 6
59. ⁻9*k* − 36  61. 50*m* − 300  63. 8*y* + 16  65. $6a^2 + 3$
67. 9*m* − 34  69. ⁻25  71. 24*x*  73. 5*n* + 13  75. 11*p* − 1
77. A simplified expression still has variables, but is written in a simpler way. When evaluating an expression, the variables are all replaced by specific numbers.  79. Like terms have matching variable parts, that is, matching letters and exponents. The coefficients do not have to match. Examples will vary.  81. Keep the variable part unchanged when combining like terms. The correct answer is 5*x* + 8.  83. ⁻2*y* + 9  85. 0  87. ⁻9*x*

## Section 2.3 (page 113)
1. 58 is the solution.  3. ⁻16 is the solution.
5. ⁻12 is the solution.

**7.** $p = 4$   Check $\underbrace{4 + 5} = 9$
$\phantom{p = 4\ \text{Check}\ }9 = 9$

**9.** $r = 10$   Check $8 = \underbrace{10 - 2}$
$\phantom{r = 10\ \text{Check}\ }8 = 8$

**11.** $n = {}^-8$   Check ${}^-5 = n + 3$
$\phantom{n = {}^-8\ \text{Check}\ }{}^-5 = \underbrace{{}^-8 + 3}$
$\phantom{n = {}^-8\ \text{Check}\ }{}^-5 = {}^-5$

**13.** $k = 18$   Check ${}^-4 + k = 14$
$\phantom{k = 18\ \text{Check}\ }\underbrace{{}^-4 + 18} = 14$
$\phantom{k = 18\ \text{Check}\ }\phantom{{}^-4 + }14 = 14$

**15.** $y = 6$   Check $y - 6 = 0$
$\phantom{y = 6\ \text{Check}\ }\underbrace{6 - 6} = 0$
$\phantom{y = 6\ \text{Check}\ }\phantom{6 - }0 = 0$

**17.** $r = {}^-6$   Check $7 = r + 13$
$\phantom{r = {}^-6\ \text{Check}\ }7 = \underbrace{{}^-6 + 13}$
$\phantom{r = {}^-6\ \text{Check}\ }7 = 7$

**19.** $x = 11$   Check $x - 12 = {}^-1$
$\phantom{x = 11\ \text{Check}\ }\underbrace{11 + {}^-12} = {}^-1$
$\phantom{x = 11\ \text{Check}\ }\phantom{11 + }{}^-1 = {}^-1$

**21.** $t = {}^-3$   Check ${}^-5 = {}^-2 + t$
$\phantom{t = {}^-3\ \text{Check}\ }{}^-5 = \underbrace{{}^-2 + {}^-3}$
$\phantom{t = {}^-3\ \text{Check}\ }{}^-5 = {}^-5$

**23.** $\underbrace{{}^-2 + {}^-5} = 3$
Does not balance   ${}^-7 \neq 3$
The correct solution is 8.   $\underbrace{8 - 5} = 3$
$\phantom{\text{The correct solution is 8.}\ \ }3 = 3$

**25.** $7 + x = {}^-11$
$7 + {}^-18 = {}^-11$
Balances   ${}^-11 = {}^-11$
${}^-18$ is the correct solution.

**27.** ${}^-10 = {}^-10 + b$
${}^-10 = \underbrace{{}^-10 + 10}$
Does not balance   ${}^-10 \neq 0$
The correct solution is 0.   ${}^-10 = \underbrace{{}^-10 + 0}$
${}^-10 = {}^-10$

**29.** $c = 6$   **31.** $y = 5$   **33.** $b = {}^-30$   **35.** $t = 0$   **37.** $z = {}^-7$
**39.** $w = 3$   **41.** $x = {}^-10$   **43.** $a = 0$   **45.** $y = {}^-25$
**47.** $x = 15$   **49.** $k = 113$   **51.** $b = 18$   **53.** $r = {}^-5$
**55.** $n = {}^-105$   **57.** $h = {}^-5$   **59.** No, the solution is ${}^-14$, the number you replace $x$ with in the original equation.
**61.** $g = 295$ graduates   **63.** $c = 52$ chirps   **65.** $p = \$110$ per month in winter   **67.** $m = {}^-19$   **69.** $x = 2$   **71.** Equations will vary. Some possibilities are   **(a)** $n - 1 = 4 - 7$ and $3 - 8 = x - 3$.  **(b)** $y - 6 = 2 - 8$ and ${}^-5 = {}^-5 + b$.  **72. (a)** $x = \dfrac{1}{2}$
**(b)** $y = \dfrac{5}{4}$   **(c)** $n = \$0.85$   **(d)** Equations will vary.

## Section 2.4 (page 123)

**1.** $z = 2$   Check $\underbrace{6 \cdot 2} = 12$
$\phantom{z = 2\ \text{Check}\ }12 = 12$

**3.** $r = 4$   Check $48 = 12r$
$\phantom{r = 4\ \text{Check}\ }48 = \underbrace{12 \cdot 4}$
$\phantom{r = 4\ \text{Check}\ }48 = 48$

**5.** $y = 0$   Check $3y = 0$
$\phantom{y = 0\ \text{Check}\ }3 \cdot \underbrace{0 = 0}$
$\phantom{y = 0\ \text{Check}\ }\phantom{3 \cdot }0 = 0$

**7.** $k = {}^-10$   Check ${}^-7k = 70$
$\phantom{k = {}^-10\ \text{Check}\ }\underbrace{{}^-7 \cdot {}^-10} = 70$
$\phantom{k = {}^-10\ \text{Check}\ }\phantom{{}^-7 \cdot }70 = 70$

**9.** $r = 6$   Check ${}^-54 = {}^-9r$
$\phantom{r = 6\ \text{Check}\ }{}^-54 = \underbrace{{}^-9 \cdot 6}$
$\phantom{r = 6\ \text{Check}\ }{}^-54 = {}^-54$

**11.** $b = {}^-5$   Check ${}^-25 = 5b$
$\phantom{b = {}^-5\ \text{Check}\ }{}^-25 = \underbrace{5 \cdot {}^-5}$
$\phantom{b = {}^-5\ \text{Check}\ }{}^-25 = {}^-25$

**13.** $r = 3$   Check $\underbrace{2 \cdot 3} = 6$
$\phantom{r = 3\ \text{Check}\ }6 = 6$

**15.** $p = {}^-3$   Check ${}^-12 = 5p - p$
$\phantom{p = {}^-3\ \text{Check}\ }{}^-12 = \underbrace{5 \cdot {}^-3} - {}^-3$
$\phantom{p = {}^-3\ \text{Check}\ }{}^-12 = \underbrace{{}^-15 + {}^+3}$
$\phantom{p = {}^-3\ \text{Check}\ }{}^-12 = {}^-12$

**17.** $a = {}^-5$   **19.** $x = {}^-10$   **21.** $w = 0$   **23.** $t = 3$   **25.** $t = 0$
**27.** $m = 9$   **29.** $y = {}^-1$   **31.** $z = {}^-5$   **33.** $p = {}^-2$   **35.** $k = 7$
**37.** $b = {}^-3$   **39.** $x = {}^-32$   **41.** $w = 2$   **43.** $n = 50$
**45.** $p = {}^-10$   **47.** Each solution is the opposite of the number in the equation. So the rule is: When you change the variable from negative to positive, then change the number in the equation to its opposite. In ${}^-x = 5$, the opposite of 5 is ${}^-5$, so $x = {}^-5$.
**49.** Divide by the coefficient of $x$, which is 3, *not* by the opposite of 3. The correct solution is 5.   **51.** $s = 15$ ft   **53.** $s = 24$ meters
**55.** $y = 27$   **57.** $x = 1$

## Section 2.5 (page 131)

**1.** $p = 1$   Check $\underbrace{7(1)} + 5 = 12$
$\phantom{p = 1\ \text{Check}\ }\underbrace{7 + 5} = 12$
$\phantom{p = 1\ \text{Check}\ }\phantom{7 + }12 = 12$

**3.** $y = 1$   Check $2 = 8y - 6$
$\phantom{y = 1\ \text{Check}\ }2 = \underbrace{8(1)} - 6$
$\phantom{y = 1\ \text{Check}\ }2 = \underbrace{8 - 6}$
$\phantom{y = 1\ \text{Check}\ }2 = 2$

**5.** $m = 0$   Check ${}^-3m + 1 = 1$
$\phantom{m = 0\ \text{Check}\ }\underbrace{{}^-3(0)} + 1 = 1$
$\phantom{m = 0\ \text{Check}\ }\underbrace{0 + 1} = 1$
$\phantom{m = 0\ \text{Check}\ }\phantom{0 + }1 = 1$

**7.** $a = {}^-2$  Check  $28 = {}^-9a + 10$
$28 = {}^-9({}^-2) + 10$
$28 = 18 + 10$
$28 = 28$

**9.** $x = {}^-4$  Check  ${}^-5x - 4 = 16$
${}^-5({}^-4) - 4 = 16$
$20 - 4 = 16$
$16 = 16$

**11.** $p = 4$; $4 = p$  Check  $6(4) - 2 = 4(4) + 6$
$24 - 2 = 16 + 6$
$22 = 22$

**13.** $k = {}^-2$; ${}^-2 = k$  Check  ${}^-2k - 6 = 6k + 10$
${}^-2({}^-2) - 6 = 6({}^-2) + 10$
$4 + {}^-6 = {}^-12 + 10$
${}^-2 = {}^-2$

**15.** $a = 5$; $5 = a$  Check  ${}^-18 + 7a = 2a + 7$
${}^-18 + 7(5) = 2(5) + 7$
${}^-18 + 35 = 10 + 7$
$17 = 17$

**17.** $w = 6$  **19.** $y = {}^-9$  **21.** $t = {}^-5$  **23.** $x = 0$  **25.** $h = 1$
**27.** $y = {}^-2$  **29.** $m = {}^-3$  **31.** $w = 2$  **33.** $x = 5$  **35.** $a = 3$
**37.** $b = {}^-3$  **39.** $k = 4$  **41.** $c = 0$  **43.** $y = {}^-5$  **45.** $n = 21$
**47.** $c = 30$  **49.** $p = {}^-2$  **51.** $b = {}^-2$
**53.** The series of steps may vary. One possibility is:

${}^-2t - 10 = 3t + 5$    Change subtraction to adding the opposite.
${}^-2t + {}^-10 = 3t + 5$    Add $2t$ to both sides
$\underline{2t \qquad\qquad 2t}$    (addition property).
$0 + {}^-10 = 5t + 5$    Add ${}^-5$ to both sides
$\underline{{}^-5 \qquad\qquad {}^-5}$    (addition property).
$\dfrac{{}^-15}{5} = \dfrac{5t}{5}$    Divide both sides by 5 (division property).
${}^-3 = t$

**55.** Check  ${}^-8 + 4(3) = 2(3) + 2$
${}^-8 + 12 = 6 + 2$
$4 \neq 8$

Does not balance, so 3 is not the correct solution. The student added ${}^-2a$ to ${}^-8$ on the left side, instead of adding ${}^-2a$ to $4a$. The correct solution is 5.  **57. (a)** It must be negative.  **(b)** The sum of $x$ and a positive number is negative, so $x$ must be negative.
**58. (a)** It must be positive.  **(b)** The sum of $d$ and a negative number is positive, so $d$ must be positive.  **59. (a)** It must be positive; when the signs are different, the product is negative.
**(b)** The product of $n$ and a negative number is negative, so $n$ must be positive.  **60. (a)** It must be negative also; when the signs match, the product is positive.  **(b)** The product of $y$ and a negative number is positive, so $y$ must be negative.

### Chapter 2 Review Exercises (page 143)
**1.** Variable is $k$; coefficient is 4; constant term is ${}^-3$.
**2. (a)** 70 test tubes  **(b)** 106 test tubes  **3. (a)** $x \cdot x \cdot y \cdot y \cdot y \cdot y$
**(b)** $5 \cdot a \cdot b \cdot b \cdot b$  **4. (a)** 9  **(b)** ${}^-27$  **(c)** ${}^-128$  **(d)** 720
**5.** $ab^2 + 3ab$  **6.** ${}^-4x + 2y - 7$  **7.** $16g^3$  **8.** $12r^2t$
**9.** $5k + 10$  **10.** ${}^-6b - 8$  **11.** $6y$  **12.** $20x + 2$

**13.** Expressions will vary. One possibility is $6a^3 + a^2 + 3a - 6$.
**14.** $n = {}^-11$  Check  $16 + {}^-11 = 5$
$5 = 5$
**15.** $a = 4$  Check  ${}^-4 + 2 = 2(4) - 6 - 4$
${}^-2 = 8 + {}^-6 + {}^-4$
${}^-2 = {}^-2$

**16.** $m = {}^-8$  **17.** $k = 10$  **18.** $t = 0$  **19.** $p = {}^-6$  **20.** $r = 2$
**21.** $h = {}^-12$  **22.** $w = 4$  **23.** $c = {}^-2$  **24.** $n = 15$ employees
**25.** $a = {}^-5$  **26.** $p = 10$  **27.** $y = 5$  **28.** $m = 3$  **29.** $x = 9$
**30.** $b = 7$  **31.** $z = {}^-3$  **32.** $n = 4$  **33.** $t = 0$  **34.** $d = 5$
**35.** $b = {}^-2$

### Chapter 2 Test (page 145)
**1.** ${}^-7$ is coefficient; $w$ is variable; 6 is constant term.
**2.** Buy 177 hot dogs.  **3.** $x \cdot x \cdot x \cdot x \cdot y \cdot y \cdot y$
**4.** $4 \cdot a \cdot b \cdot b \cdot b \cdot b$  **5.** ${}^-200$  **6.** ${}^-4w^3$  **7.** 0  **8.** $c$
**9.** cannot be simplified  **10.** ${}^-40b^2$  **11.** $15k$  **12.** $21t + 28$
**13.** ${}^-4a - 24$  **14.** $6x - 15$  **15.** ${}^-9b + c + 6$
**16.** $x = 5$  Check  ${}^-4 = 5 - 9$
${}^-4 = {}^-4$

**17.** $w = {}^-11$  Check  ${}^-7({}^-11) = 77$
$77 = 77$

**18.** $p = {}^-14$  Check  ${}^-1({}^-14) = 14$
$14 = 14$

**19.** $a = 3$  Check  ${}^-15 = {}^-3(3 + 2)$
${}^-15 = {}^-3(5)$
${}^-15 = {}^-15$

**20.** $n = {}^-8$  **21.** $m = 15$  **22.** $x = {}^-1$  **23.** $m = 2$
**24.** $b = 54$  **25.** $c = 0$  **26.** Equations will vary. Two possibilities are $x - 5 = {}^-9$ and ${}^-24 = 6y$.

### Cumulative Review Exercises: Chapters 1–2 (page 147)
**1.** three hundred six billion, four thousand, two hundred ten
**2.** $800,066,000$.  **3. (a)** $>$  **(b)** $<$  **4. (a)** commutative property of addition  **(b)** multiplication property of 0
**(c)** distributive property  **5. (a)** 9000  **(b)** 290,000  **6.** ${}^-8$
**7.** 10  **8.** 30  **9.** 25  **10.** 7  **11.** 0  **12.** ${}^-64$  **13.** undefined
**14.** ${}^-60$  **15.** ${}^-40$  **16.** ${}^-9$  **17.** ${}^-25$  **18.** ${}^-28$
**19.** $\dfrac{11}{{}^-11} = {}^-1$  **20.** Estimate: $600 \div 20 = 30$ miles;
Exact: $616 \div 22 = 28$ miles  **21.** Estimate: ${}^-50 + 20 = {}^-30$ degrees; Exact: ${}^-48 + 23 = {}^-25$ degrees  **22.** Estimate: $\$2000 - (50 \cdot 10) = \$1500$; Exact: $\$2132 - (52 \cdot 8) = \$1716$
**23.** Estimate: $10(600 + 40) = \$6400$; Exact: $12(552 + 35) = \$7044$  **24.** ${}^-4 \cdot a \cdot b \cdot b \cdot b \cdot c \cdot c$  **25.** 120  **26.** $h$  **27.** 0
**28.** $5n^2 - 4n - 2$  **29.** ${}^-30b^2$  **30.** $28p - 28$  **31.** ${}^-9w^2 - 12$
**32.** $x = {}^-4$  Check  $3({}^-4) = {}^-4 - 8$
${}^-12 = {}^-12$

**33.** $y = {}^-6$  Check  ${}^-44 = {}^-2 + 7({}^-6)$
${}^-44 = {}^-2 + {}^-42$
${}^-44 = {}^-44$

**34.** $k = 7$  Check  $2(7) - 5(7) = {}^-21$
$14 - 35 = {}^-21$
${}^-21 = {}^-21$

**35.** $m = 4$  Check  $4 - 6 = {}^-2(4) + 6$
${}^-2 = {}^-8 + 6$
${}^-2 = {}^-2$

**36.** $x = {}^-1$  **37.** $r = {}^-18$  **38.** $b = {}^-5$  **39.** $t = 1$
**40.** $y = {}^-12$

# Chapter 3

### Section 3.1 (page 155)
1. $P = 36$ cm   3. $P = 100$ in.   5. $P = 4$ miles   7. $P = 88$ mm
9. $s = 30$ ft   11. $s = 1$ mm   13. $s = 23$ yards   15. $s = 2$ ft
17. $P = 28$ yd   19. $P = 70$ cm   21. $P = 72$ ft   23. $P = 26$ in.
25. $l = 9$ cm   Check $9$ cm $+ 9$ cm $+ 6$ cm $+ 6$ cm $= 30$ cm
27. $w = 1$ mile   Check $4$ mi $+ 4$ mi $+ 1$ mi $+ 1$ mi $= 10$ mi
29. $w = 2$ ft   Check $6$ ft $+ 6$ ft $+ 2$ ft $+ 2$ ft $= 16$ ft
31. $l = 2$ m   Check $2$ m $+ 2$ m $+ 1$ m $+ 1$ m $= 6$ m
33. $P = 208$ m   35. $P = 320$ ft   37. $P = 54$ mm   39. $P = 48$ ft
41. $P = 78$ in.   43. $P = 125$ m   45. $? = 40$ cm   47. $? = 12$ in;
$P = 78$ in.   49. (a) Sketches will vary.   (b) Formula for perimeter of an equilateral triangle is $P = 3s$, where $s$ is the length of one side.   (c) The formula will *not* work for other kinds of triangles because the sides will have different lengths.   51. (a) 140 miles   (b) 350 miles   (c) 560 miles   52. (a) 70 miles   (b) 175 miles   (c) 280 miles   (d) The rate is half of 70 miles per hour, so the distance will be half as far; divide each result in Exercise 51 by 2.
53. (a) 50 hours   (b) 60 hours   (c) 150 hours   54. (a) 61 miles per hour   (b) 57 miles per hour   (c) 65 miles per hour

### Section 3.2 (page 165)
1. $A = 77$ ft$^2$   3. $A = 100$ m$^2$   5. $A = 775$ mm$^2$   7. $A = 36$ in.$^2$
9. $A = 105$ cm$^2$   11. $A = 72$ ft$^2$   13. $A = 625$ mi$^2$
15. $A = 1$ m$^2$   17. $l = 6$ ft   Check $A = 6$ ft $\cdot$ 3 ft; $A = 18$ ft$^2$
19. $w = 80$ yd   Check $A = 90$ yd $\cdot$ 80 yd; $A = 7200$ yd$^2$
21. $l = 14$ in.   Check $A = 14$ in. $\cdot$ 11 in.; $A = 154$ in.$^2$
23. $s = 6$ m   25. $s = 2$ ft   27. $h = 20$ cm
Check $A = 25$ cm $\cdot$ 20 cm; $A = 500$ cm$^2$   29. $b = 17$ in.
Check $A = 17$ in. $\cdot$ 13 in. $A = 221$ in.$^2$   31. $h = 1$ m
Check $A = 9$ m $\cdot$ 1 m; $A = 9$ m$^2$   33. Height is not part of perimeter; square units are used for area, not perimeter.
$P = 25$ cm $+ 25$ cm $+ 25$ cm $+ 25$ cm; $P = 100$ cm.
35. $P = 32$ m; $A = 39$ m$^2$   37. $P = 34$ yd; $A = 56$ yd$^2$
39. $P = 36$ in.; $A = 81$ in.$^2$   41. Panoramic: $P = 36$ in., $A = 56$ in.$^2$; 4 in. $\times$ 6 in.: $P = 20$ in., $A = 24$ in.$^2$; 4 in. $\times$ 7 in.: $P = 22$ in., $A = 28$ in.$^2$   43. $108   45. $725   47. 53 yd
49.
```
  5 ft          4 ft         3 ft
┌──────┐1 ft  ┌─────┐2 ft  ┌───┐3 ft
└──────┘      └─────┘      └───┘
```
50. (a) 5 ft by 1 ft has area of 5 ft$^2$; 4 ft by 2 ft has area of 8 ft$^2$; 3 ft by 3 ft has area of 9 ft$^2$   (b) The square plot 3 ft by 3 ft has the greatest area.
51.

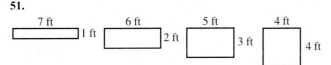

52. (a) 7 ft$^2$, 12 ft$^2$, 15 ft$^2$, 16 ft$^2$   (b) Square plots have the greatest area.

### Section 3.3 (page 175)
1. $14 + x$ or $x + 14$   3. $^-5 + x$ or $x + ^-5$   5. $20 - x$
7. $x - 9$   9. $x - 4$   11. $^-6x$   13. $2x$   15. $\dfrac{x}{2}$   17. $2x + 8$ or $8 + 2x$   19. $7x - 10$   21. $2x + x$ or $x + 2x$
23. $4n - 2 = 26; n = 7$   Check $\underbrace{4 \cdot 7}_{28} - 2$ does equal 26.
$\underbrace{28 - 2}_{26}$

25. $2n + n = ^-15$ or $n + 2n = ^-15; n = ^-5$
Check $\underbrace{2 \cdot ^-5}_{^-10} + ^-5$ does equal $^-15$
$\underbrace{^-10 + ^-5}_{^-15}$

27. $5n + 12 = 7n; n = 6$   Check $\underbrace{5 \cdot 6}_{30} + 12 = \underbrace{7 \cdot 6}_{42}$
$30 + 12 = 42$
$42 = 42$

29. $30 - 3n = 2 + n; n = 7$   Check $30 - \underbrace{3 \cdot 7}_{21} = \underbrace{2 + 7}_{9}$
$30 - 21 = 9$
$9 = 9$

31. Let $w$ be Ricardo's original weight. $w + 15 - 28 + 5 = 177$. He weighed 185 pounds originally.   33. Let $c$ be the number of cookies the children ate. $18 - c + 36 = 49$. Her children ate 5 cookies.   35. Let $p$ be the number of pens in each box. $6p - 32 - 35 = 5$. There were 12 pens in each box.
37. Let $d$ be each member's dues. $14d + 340 - 575 = ^-25$. Each member paid $15.   39. Let $a$ be Tamu's age. $4a - 75 = a$. Tamu is 25 years old.   41. Let $m$ be the amount Brenda spent. $2m - 3 = 81$. Brenda spent $42.   43. Let $b$ be the number of pieces in each bag. $5b - 3 \cdot 48 = b$. There were 36 pieces of candy in each bag.   45. Let $d$ be the daily amount for an infant. $2d + 3 = 15$. An infant should receive 6 mg of iron.

### Section 3.4 (page 183)
1. $a$ is my age; $a + 9$ is my sister's age. $a + a + 9 = 51$. I am 21; my sister is 30.   3. $m$ is husband's earnings; $m + 1500$ is Lien's. $m + m + 1500 = 37,500$. Husband earned $18,000; Lien earned $19,500.   5. $m$ is printer's cost; $5m$ is computer's cost. $5m + m = $1320$. Printer cost $220; computer cost $1100
7. Shorter piece is $x$; longer piece is $x + 10$. $x + x + 10 = 78$. Shorter piece is 34 cm; longer piece is 44 cm.   9. Longer piece is $x$; shorter piece is $x - 7$. $x + x - 7 = 31$. Longer piece is 19 ft; shorter piece is 12 ft.   11. $s$ is number of Senators; $5s - 65$ is number of Representatives. $s + 5s - 65 = 535$. 100 Senators; 435 Representatives   13. First part is $x$; second part is $x$; third part is $x + 25$. $x + x + x + 25 = 706$. First part is 227 m; second part is 227 m; third part is 252 m.   15. Length is 19 yd.   17. Length is 12 ft; width is 6 ft.   19. Length is 13 in.; width is 5 in.
21. $A = 88$ in.$^2$; $P = 52$ in.

### Chapter 3 Review Exercises (page 191)
1. $P = 112$ cm   2. $P = 22$ mi   3. $P = 42$ yd   4. $P = 141$ m
5. $12$ ft $= 4s; s = 3$ ft   6. $128$ yd $= 2l + 2(31$ yd$); l = 33$ yd
7. $72$ in. $= 2(21$ in.$) + 2w; w = 15$ in.   8. $A = 40$ ft$^2$
9. $A = 625$ m$^2$   10. $A = 208$ yd$^2$   11. $126$ ft$^2 = 14$ ft $\cdot w$; $w = 9$ ft   12. $88$ cm$^2 = 11$ cm $\cdot h$; $h = 8$ cm   13. $100$ mi$^2 = s \cdot s; s = 10$ mi   14. $57 - x$   15. $15 + 2x$ or $2x + 15$
16. $^-9x$   17. $4n + 6 = ^-30; n = ^-9$   18. $10 - 2n = 4 + n$; $n = 2$   19. $m$ is money originally in account.
$m - 600 + 750 + 75 = 309$. $84 was originally in Grace's account.   20. $c$ is number of candles in each box. $4c - 25 = 23$. There were 12 candles in each box.   21. $p$ is Reggie's prize money; $p + 300$ is Donald's prize money. $p + p + 300 = 1000$. Reggie gets $350; Donald gets $650.   22. $w$ is the width; $2w$ is the length. $84 = 2(2w) + 2(w)$. The width is 14 cm; the length is 28 cm.
23. (a) $36$ ft $= 4s; s = 9$ ft   (b) $A = 9$ ft $\cdot$ 9 ft; $A = 81$ ft$^2$
24. Rectangles will vary. Two possibilities are: $A = 7$ ft $\cdot$ 3 ft, $A = 21$ ft$^2$; $A = 6$ ft $\cdot$ 4 ft, $A = 24$ ft$^2$   25. Let $f$ be the fencing for the garden. $36 - f + 20 = 41$. 15 ft of fencing was used on the garden.   26. $w$ is the width; $w + 2$ is the length. $36 = 2(w + 2) + 2 \cdot w$. Width is 8 ft; length is 10 ft.

## Chapter 3 Test (page 193)
**1.** $P = 262$ m  **2.** $P = 40$ in.  **3.** $P = 12$ mi  **4.** $P = 12$ ft
**5.** $P = 110$ cm  **6.** $A = 486$ mm$^2$  **7.** $A = 140$ cm$^2$
**8.** $A = 3740$ mi$^2$  **9.** $A = 36$ m$^2$  **10.** 12 ft $= 4s$; $s = 3$ ft
**11.** 34 ft $= 2l + 2(6$ ft$)$; $l = 11$ ft  **12.** 65 in.$^2 = 13$ in. • $h$;
$h = 5$ in.  **13.** 12 cm$^2 = 4$ cm • $w$; $w = 3$ cm  **14.** 16 ft$^2 = s^2$;
$s = 4$ ft  **15.** Linear units like ft are used to measure length, width, height and perimeter. Area is measured in square units like ft$^2$ (squares that measure 1 ft on each side).  **16.** $4n + 40 = 0$; $n = {}^-10$  **17.** $7n - 23 = n + 7$; $n = 5$  **18.** Son spent $15.
**19.** Daughter is 7 years old.  **20.** One piece is 57 cm; one piece is 61 cm.  **21.** Length is 168 ft; width is 42 ft.  **22.** Marcella worked 11 hours; Tim worked 8 hours.

## Cumulative Review Exercises: Chapters 1–3 (page 195)
**1.** four billion, two hundred six thousand, three hundred
**2.** 70,005,489  **3.** $<$  $>$  **4. (a)** multiplication property of 1
**(b)** addition property of 0  **(c)** associative property of multiplication  **5. (a)** 3800  **(b)** 490,000  **6.** $^-24$  **7.** 27  **8.** $^-3$
**9.** $^-100$  **10.** 5  **11.** 0  **12.** $^-13$  **13.** 25  **14.** $^-32$
**15.** $\dfrac{27}{0}$ is undefined.  **16.** $10 \cdot w \cdot w \cdot x \cdot y \cdot y \cdot y \cdot y$
**17.** 240  **18.** $2k$  **19.** $3m^2 + 2m$  **20.** 0  **21.** $^-20a$
**22.** $^-2x^2 - 3$  **23.** $^-12n + 1$
**24.** $x = 2$  Check $\underline{6 - 20} = \underline{2(2) - 9(2)}$
$\phantom{24. x = 2 ~~ \text{Check }} ^-14 \phantom{=} = \phantom{=} 4 - 18$
$\phantom{24. x = 2 ~~ \text{Check }} ^-14 \phantom{==} = \phantom{=} ^-14$
**25.** $y = {}^-1$  Check $\underline{^-5(^-1)} = \underline{^-1 + 6}$
$\phantom{25. y = {}^-1~~\text{Check }} 5 \phantom{==} = \phantom{==} 5$
**26.** $b = 4$  **27.** $h = {}^-2$  **28.** $x = 0$  **29.** $a = {}^-3$
**30.** $P = 60$ in.; $A = 198$ in.$^2$  **31.** $P = 60$ m; $A = 225$ m$^2$
**32.** $P = 24$ ft; $A = 32$ ft$^2$  **33.** $5n + {}^-50 = 0$; $n = 10$
**34.** $10 - 3n = 2n$; $n = 2$  **35.** Let $p$ be number of people originally. $p - 3 + 6 - 2 = 5$. There were 4 people in line.
**36.** Let $m$ be amount paid by each player. $12m - 2200 = {}^-40$. Each player paid $180.  **37.** Let $g$ be one group; $3g$ the other. $g + 3g = 192$. 48 in one group; 144 in the other.  **38.** $w$ is the width; $w + 14$ is the length. $2(w + 14) + 2(w) = 92$. Width is 16 ft; length is 30 ft.

# Chapter 4

## Section 4.1 (page 205)
**1.** $\dfrac{5}{8}; \dfrac{3}{8}$  **3.** $\dfrac{2}{3}; \dfrac{1}{3}$  **5.** $\dfrac{3}{2}; \dfrac{1}{2}$  **7.** $\dfrac{11}{6}; \dfrac{1}{6}$  **9.** $\dfrac{2}{11}; \dfrac{3}{11}; \dfrac{4}{11}$
**11.** $\dfrac{8}{25}; \dfrac{17}{25}$  **13.** $\dfrac{13}{71}; \dfrac{58}{71}$  **15.** $\dfrac{6}{20}$  **17.** $\dfrac{19}{20}$  **19.** N: 3; D: 4
**21.** N: 12; D: 7  **23.** Proper: $\dfrac{1}{3}, \dfrac{5}{8}, \dfrac{7}{16}$; Improper: $\dfrac{8}{5}, \dfrac{6}{6}, \dfrac{12}{2}$
**25.** Proper: $\dfrac{3}{4}, \dfrac{9}{11}, \dfrac{7}{15}$; Improper: $\dfrac{3}{2}, \dfrac{5}{5}, \dfrac{19}{18}$
**27.** Fractions will vary. The denominator shows the number of equal parts in the whole and the numerator shows how many of the parts are being considered. The fraction bar separates the numerator from the denominator. Show your drawing to your instructor.

**29.**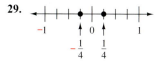

**31.** [number line showing $-\dfrac{3}{5}$ and $\dfrac{3}{5}$ between $-1$ and $1$]

**33.** [number line showing $-\dfrac{7}{8}$ and $\dfrac{7}{8}$ between $-1$ and $1$]

**35.** $-\dfrac{3}{4}$ pound  **37.** $-\dfrac{1}{2}$ quart  **39.** $\dfrac{3}{10}$ mile
**41.** $\dfrac{2}{5}$  **43.** $\dfrac{9}{10}$  **45.** $\dfrac{13}{6}$  **47. (a)** $\dfrac{12}{24}$  **(b)** $\dfrac{8}{24}$  **(c)** $\dfrac{16}{24}$  **(d)** $\dfrac{6}{24}$
**(e)** $\dfrac{18}{24}$  **(f)** $\dfrac{4}{24}$  **(g)** $\dfrac{20}{24}$  **(h)** $\dfrac{3}{24}$  **(i)** $\dfrac{9}{24}$  **(j)** $\dfrac{15}{24}$  **49. (a)** $-\dfrac{1}{3}$
**(b)** $-\dfrac{2}{3}$  **(c)** $-\dfrac{2}{3}$  **(d)** $-\dfrac{1}{3}$  **(e)** $-\dfrac{2}{3}$  **(f)** Some possibilities
are: $-\dfrac{4}{12} = -\dfrac{1}{3}; -\dfrac{8}{24} = -\dfrac{1}{3}; -\dfrac{20}{30} = -\dfrac{2}{3}; -\dfrac{24}{36} = -\dfrac{2}{3}$.
**51. (a)** $\dfrac{1467}{3912}$  **(b)** Divide 3912 by 8 to get 489; multiply 3 by
489 to get 1467.  **52. (a)** $\dfrac{4256}{5472}$  **(b)** Divide 5472 by 9 to get 608; multiply 7 by 608 to get 4256.  **53. (a)** $-\dfrac{1}{5}$  **(b)** Divide 3485 by
2, by 3, and by 5 to see that dividing by 5 gives 697. Or divide 3485 by 697 to get 5.  **54. (a)** $-\dfrac{1}{6}$  **(b)** Divide 4902 by 4, by 6, and by 8 to see that dividing by 6 gives 817. Or divide 4902 by 817 to get 6.  **55.** 7  **56.** 5  **57.** 15  **58.** 16  **59.** Multiply or divide the numerator and denominator by the same nonzero number. Some possibilities are: $\dfrac{2}{3} = \dfrac{2 \cdot 4}{3 \cdot 4} = \dfrac{8}{12}$ and $\dfrac{10}{16} = \dfrac{10 \div 2}{16 \div 2} = \dfrac{5}{8}$
**61.** You cannot do it if you want the numerator to be a whole number, because 5 does not divide into 18 evenly. You could use multiples of 5 as the denominator, such as 10, 15, 20, etc.  **63.** 10
**65.** $^-1$  **67.** $^-6$  **69.** 3  **71.** 1  **73.** 2  **75.** $\dfrac{2}{5}$ is unshaded.
**77.** $\dfrac{5}{8}$ is unshaded.  **79.** $\dfrac{1}{4}$ is unshaded.  **81.** $\dfrac{2}{3}$ is unshaded.
**83.** $\dfrac{0}{6}$ is unshaded.  **85.** Divide each rectangle into 5 equal parts, then shade both rectangles completely.  **87.** Drawings will vary.
**89.** One possibility is shown. ○□□□□□□△△△
**91.** One possibility is shown. (!)!!!,...???

## Section 4.2 (page 219)
**1.** comp. prime comp. prime prime comp. comp.  **3.** $2 \cdot 3$
**5.** $2 \cdot 2 \cdot 5$  **7.** $5 \cdot 5$  **9.** $2 \cdot 2 \cdot 3 \cdot 3$  **11.** $2 \cdot 2 \cdot 11$
**13.** $2 \cdot 2 \cdot 2 \cdot 11$  **15.** $3 \cdot 5 \cdot 5$  **17.** A composite number has a factor(s) other than itself or 1. A prime number is a whole number that has exactly two *different* factors, itself and 1. The whole numbers 0 and 1 are neither prime nor composite.

**19.** $\dfrac{\cancel{2} \cdot \cancel{2} \cdot \cancel{2}}{\cancel{2} \cdot \cancel{2} \cdot \cancel{2} \cdot 2} = \dfrac{1}{2}$  **21.** $\dfrac{\cancel{2} \cdot \cancel{2} \cdot \cancel{2} \cdot \cancel{2} \cdot 2}{\cancel{2} \cdot \cancel{2} \cdot \cancel{2} \cdot \cancel{2} \cdot 3} = \dfrac{2}{3}$

**23.** $\dfrac{2 \cdot \cancel{7}}{3 \cdot \cancel{7}} = \dfrac{2}{3}$  **25.** $\dfrac{\cancel{2} \cdot 2 \cdot \cancel{3} \cdot 3}{\cancel{2} \cdot \cancel{3} \cdot 7} = \dfrac{6}{7}$  **27.** $\dfrac{3 \cdot 3 \cdot \cancel{7}}{2 \cdot 5 \cdot \cancel{7}} = \dfrac{9}{10}$

**29.** $\dfrac{\cancel{3} \cdot \cancel{3} \cdot 3}{\cancel{3} \cdot \cancel{3} \cdot 5} = \dfrac{3}{5}$  **31.** $\dfrac{\cancel{2} \cdot 2 \cdot \cancel{3}}{\cancel{2} \cdot \cancel{3} \cdot 3} = \dfrac{2}{3}$  **33.** $\dfrac{\cancel{5} \cdot 7}{2 \cdot 2 \cdot 2 \cdot \cancel{5}} = \dfrac{7}{8}$

**A-14** Answers to Selected Exercises

35. $\dfrac{\cancel{2}\cdot\cancel{3}\cdot\cancel{3}\cdot\cancel{5}}{\cancel{2}\cdot 2\cdot\cancel{3}\cdot\cancel{3}\cdot\cancel{5}} = \dfrac{1}{2}$   37. $\dfrac{\cancel{2}\cdot\cancel{3}\cdot\cancel{5}\cdot\cancel{7}}{3\cdot\cancel{3}\cdot\cancel{5}\cdot\cancel{7}} = \dfrac{2}{3}$

39. $\dfrac{\cancel{3}\cdot\cancel{11}\cdot 13}{\cancel{3}\cdot 3\cdot 5\cdot\cancel{11}} = \dfrac{13}{15}$   41. (a) $\dfrac{1}{4}$  (b) $\dfrac{1}{2}$  (c) $\dfrac{1}{10}$

(d) $\dfrac{60}{60} = 1$   43. (a) $\dfrac{1}{3}$  (b) $\dfrac{1}{5}$  (c) $\dfrac{7}{15}$   45. (a) $\dfrac{5}{24}$  (b) $\dfrac{2}{3}$

(c) $\dfrac{1}{8}$   47. (a) The result of dividing 3 by 3 is 1, so 1 should be written above and below all the slashes. The numerator is 1 · 1, so the correct answer is $\tfrac{1}{4}$.   (b) You must divide numerator and denominator by the *same* number. The fraction is already in lowest terms because 9 and 16 have no common factor besides 1.   49. $\dfrac{2c}{5}$

51. $\dfrac{4}{7}$   53. $\dfrac{6r}{5s}$   55. $\dfrac{1}{7n^2}$   57. already in lowest terms   59. $\dfrac{7}{9y}$

61. $\dfrac{7k}{2}$   63. $\dfrac{1}{3}$   65. $\dfrac{c}{d}$   67. $\dfrac{6ab}{c}$   69. already in lowest terms

71. $3eg^2$

### Section 4.3 (page 231)

1. $-\dfrac{3}{16}$   3. $\dfrac{9}{10}$   5. $\dfrac{1}{2}$   7. $-6$   9. 36   11. $\dfrac{15}{4y}$   13. $\dfrac{1}{2}$   15. $\dfrac{6}{5}$

17. $-9$   19. $-\dfrac{1}{6}$   21. $\dfrac{11}{15d}$   23. $b$

25. (a) Forgot to write 1s in numerator when dividing out common factors. Answer is $\tfrac{1}{6}$.   (b) Used reciprocal of $\tfrac{2}{3}$ in multiplication, but the reciprocal is used only in division. Correct answer is $\tfrac{16}{3}$.   27. (a) Forgot to use reciprocal of $\tfrac{4}{1}$; correct answer is $\tfrac{1}{6}$   (b) Used reciprocal of $\tfrac{5}{6}$ instead of reciprocal of $\tfrac{10}{9}$; correct answer is $\tfrac{3}{4}$.   29. Rewrite division as multiplication. Leave the first number (dividend) the same. Change the second number (divisor) to its reciprocal by "flipping" it. Then multiply.

31. $\dfrac{4}{15}$   33. $-\dfrac{9}{32}$   35. 21   37. 15   39. undefined   41. $-\dfrac{55}{12}$

43. $8b$   45. $\dfrac{3}{5d}$   47. $\dfrac{2x^2}{w}$   49. $\dfrac{3}{10}$ yd²   51. 80 dispensers

53. earn $9300; borrow $3100   55. 9 trips   57. 61 players

59. 3200 times   61. (a) $38,000  (b) $7600   63. $16,625

65. 15 million birds   67. 129 million dogs and cats

### Section 4.4 (page 243)

1. $\dfrac{7}{8}$   3. $-\dfrac{1}{2}$   5. $\dfrac{1}{2}$   7. $-\dfrac{9}{40}$   9. $-\dfrac{13}{24}$   11. $-\dfrac{3}{5}$   13. $-\dfrac{7}{18}$

15. $\dfrac{8}{7}$   17. $-\dfrac{3}{8}$   19. $\dfrac{3+5c}{15}$   21. $\dfrac{10-m}{2m}$   23. $\dfrac{8}{b^2}$

25. $\dfrac{bc+21}{7b}$   27. $\dfrac{-4-cd}{c^2}$   29. $-\dfrac{44}{105}$   31. You cannot add or subtract until all the fractional pieces are the same size.

33. (a) $\dfrac{15}{20} + \dfrac{8}{20} = \dfrac{23}{20}$  Cannot add fractions with unlike denominators; use 20 as the LCD.   (b) When rewriting fractions with 18 as a denominator, you must multiply denominator and numerator by the same number. The correct answer is $\dfrac{15}{18} - \dfrac{8}{18} = \dfrac{7}{18}$.

35. (a) $\dfrac{1}{12}; \dfrac{1}{12}$  Addition is commutative.   (b) $\dfrac{1}{3}; -\dfrac{1}{3}$  Subtraction is *not* commutative.

(c) $-\dfrac{3}{5}; -\dfrac{3}{5}$  Multiplication is commutative.   (d) 6; $\dfrac{1}{6}$  Division is *not* commutative.   36. (a) 0; 0; The sum of a number and its opposite is 0.   (b) 1; 1; When a nonzero number is divided by itself, the quotient is 1.   (c) $\dfrac{5}{6}; -\dfrac{17}{20}$; Multiplying by 1 leaves a number unchanged.   (d) 1; 1; A number times its reciprocal is 1.

37. $\dfrac{47}{60}$ in.   39. $\dfrac{23}{24}$ cubic yard   41. $\dfrac{3}{4}$ acre   43. $\dfrac{23}{50}$ of workers

45. $\dfrac{4}{25}$ of workers   47. $\dfrac{7}{24}$ of the day; 7 hours

49. work and travel; 8 hours   51. $\dfrac{1}{8}$ of the day   53. $\dfrac{1}{12}$ mile

### Section 4.5 (page 257)

1. ◄—+—|•|—+—+—|•|—+—► 
   $-4\,-3\,-2\,-1\,\,0\,\,1\,\,2\,\,3\,\,4$

3. ◄—+—|•|—+—|•|—+—+—► 
   $-4\,-3\,-2\,-1\,\,0\,\,1\,\,2\,\,3\,\,4$

5. $\dfrac{9}{2}$   7. $-\dfrac{8}{5}$   9. $\dfrac{19}{8}$   11. $-\dfrac{57}{10}$   13. $\dfrac{161}{15}$   15. $4\dfrac{1}{3}$

17. $-2\dfrac{1}{2}$   19. $3\dfrac{2}{3}$   21. $-5\dfrac{2}{3}$   23. $11\dfrac{3}{4}$   25. $7\dfrac{7}{8}$; $2\cdot 4 = 8$

27. $1\dfrac{5}{21}$; $3 \div 3 = 1$   29. $5\dfrac{1}{2}$; $4 + 2 = 6$   31. $3\dfrac{2}{3}$; $4 - 1 = 3$

33. $\dfrac{17}{18}$; $6 \div 6 = 1$   35. $6\dfrac{1}{5}$; $8 - 2 = 6$   37. $P = 7$ in.; $A = 3\dfrac{1}{16}$ in.²   39. $P = 19\dfrac{1}{2}$ yd; $A = 21\dfrac{1}{8}$ yd²

41. $13 + 9 = 22$ ft; $21\dfrac{1}{6}$ ft   43. $2 \cdot 13 = 26$ ounces; $21\dfrac{7}{8}$ ounces

45. $4 - 2 = 2$ miles; $2\dfrac{3}{10}$ miles   47. $3 \cdot 7 = 21$ yd; $19\dfrac{1}{4}$ yd

49. $10 - 2 - 3 = 5$ cubic yards; $5\dfrac{1}{8}$ cubic yards   51. $19 \div 5$ is about 4 hours; $3\dfrac{3}{4}$ hours   53. $30 - 6 - 2 = 22$ in.; $21\dfrac{3}{8}$ in.

55. $24 + 35 + 24 + 35 = 118$ in.; $116\dfrac{1}{2}$ in.

57. $25{,}730 \div 10 = 2573$ anchors; 2480 anchors

59. $(4 \cdot 23) + (5 \cdot 24) + (3 \cdot 25) = 287$ in.; $280\dfrac{1}{8}$ in.

### Summary Exercises on Fractions (page 261)

1. (a) $\dfrac{3}{8}; \dfrac{5}{8}$   (b) $\dfrac{4}{5}; \dfrac{1}{5}$   2. ◄—+—•—+—+—+—•—+—►
   $-1\quad\quad 0\quad\quad 1$
   $\quad\ \uparrow\qquad\qquad\uparrow$
   $\ -\tfrac{2}{3}\qquad\qquad\tfrac{2}{3}$

3. (a) 24   (b) 4   4. (a) 1   (b) $-4$   (c) 9

5. (a) $2 \cdot 2 \cdot 2 \cdot 3 \cdot 3$   (b) $3 \cdot 5 \cdot 7$   6. (a) $\dfrac{4}{5}$   (b) $\dfrac{7}{8}$   7. $\dfrac{1}{2}$

8. $-\dfrac{5}{24}$   9. $\dfrac{17}{16}$   10. $\dfrac{5}{6}$   11. $-\dfrac{2}{15}$   12. $-\dfrac{3}{8}$   13. 56   14. $\dfrac{11}{24}$

15. $-\dfrac{7}{6}$   16. $-\dfrac{19}{12}$   17. $\dfrac{25}{12}$   18. 35   19. $7\dfrac{7}{12}$; $5 + 3 = 8$

20. $11\dfrac{3}{7}$; $2 \cdot 5 = 10$   21. $3\dfrac{3}{10}$; $6 - 3 = 3$

22. $\dfrac{16}{35}$; $2 \div 4 = \dfrac{2}{4}$ or $\dfrac{1}{2}$   23. 4; $5 \div 1 = 5$   24. $2\dfrac{2}{3}$; $3 - 1 = 2$

Answers to Selected Exercises   A–15

**25.** (a) $2\frac{1}{16}$ in. (b) $\frac{5}{16}$ in.   **26.** $P = 3\frac{1}{2}$ in.; $A = \frac{49}{64}$ in.$^2$
**27.** 12 batches   **28.** 6¢   **29.** Not sure, 225 adults; Real, 675 adults; Imaginary, 600 adults   **30.** diameter = $\frac{1}{4}$ in.; $P = 4\frac{1}{2}$ in.
**31.** 23 bottles   **32.** 11 lots

## Section 4.6 (page 267)

**1.** $\frac{9}{16}$   **3.** $\frac{8}{125}$   **5.** $-\frac{1}{27}$   **7.** $\frac{1}{32}$   **9.** $\frac{49}{100}$   **11.** $\frac{36}{25}$ or $1\frac{11}{25}$
**13.** $\frac{12}{25}$   **15.** $\frac{1}{100}$   **17.** $-\frac{3}{2}$ or $-1\frac{1}{2}$   **19.** (a) The answers are $\frac{1}{4}, -\frac{1}{8}, \frac{1}{16}, -\frac{1}{32}, \frac{1}{64}, -\frac{1}{128}, \frac{1}{256}, -\frac{1}{512}$. (b) When a negative number is raised to an even power, the answer is positive. When a negative number is raised to an odd power, the answer is negative.   **20.** (a) Ask yourself, "What number, times itself, is 4?" This is the numerator. Then ask, "What number, times itself, is 9?" This is the denominator. The number under the ketchup is either $\frac{2}{3}$ or $-\frac{2}{3}$.   (b) The number under the ketchup is $-\frac{1}{3}$ because $-\frac{1}{3} \cdot -\frac{1}{3} \cdot -\frac{1}{3} = -\frac{1}{27}$.   (c) Either $\frac{1}{2}$ or $-\frac{1}{2}$.
(d) No real number works, because both $\left(\frac{3}{4}\right)^2$ and $\left(-\frac{3}{4}\right)^2$ give a positive result.   (e) Either $\frac{1}{3}$ or $-\frac{1}{3}$ inside one set of parentheses and $\frac{1}{2}$ or $-\frac{1}{2}$ inside the other.   **21.** $-4$   **23.** $\frac{5}{16}$   **25.** $\frac{1}{3}$   **27.** $\frac{1}{6}$
**29.** $-\frac{17}{24}$   **31.** $-\frac{4}{27}$   **33.** $\frac{1}{36}$   **35.** $\frac{9}{64}$ in.$^2$   **37.** $\frac{19}{10}$ or $1\frac{9}{10}$ miles
**39.** 4   **41.** $-\frac{25}{2}$ or $-12\frac{1}{2}$   **43.** $\frac{1}{14}$   **45.** $\frac{9}{100}$   **47.** $\frac{5}{8}$   **49.** $-18$

## Section 4.7 (page 277)

**1.** $a = 30$   Check $\frac{1}{3}(30) = 10$
$10 = 10$

**3.** $b = -24$   Check $-20 = \frac{5}{6}(-24)$
$-20 = -20$

**5.** $c = 6$   Check $-\frac{7}{2}(6) = -21$
$-21 = -21$

**7.** $m = \frac{3}{4}$   Check $\frac{9}{16} = \frac{3}{4}\left(\frac{3}{4}\right)$
$\frac{9}{16} = \frac{9}{16}$

**9.** $d = -\frac{6}{5}$   Check $\frac{3}{10} = -\frac{1}{4}\left(-\frac{6}{5}\right)$
$\frac{3}{10} = \frac{3}{10}$

**11.** $n = 12$   Check $\frac{1}{6}(12) + 7 = 9$
$2 + 7 = 9$
$9 = 9$

**13.** $r = -9$   Check $-10 = \frac{5}{3}(-9) + 5$
$-10 = -15 + 5$
$-10 = -10$

**15.** $x = 24$   Check $\frac{3}{8}(24) - 9 = 0$
$9 - 9 = 0$
$0 = 0$

**17.** $y = 45$   **19.** $n = -18$   **21.** $x = \frac{1}{12}$   **23.** $b = -\frac{1}{8}$
**25.** (a) $\frac{1}{6}(18) + 1 = -2$
$3 + 1 = -2$
$4 \neq -2$
Does *not* balance; correct solution is $x = -18$.
(b) $-\frac{3}{2} = \frac{9}{4}\left(-\frac{2}{3}\right)$
$-\frac{3}{2} = -\frac{3}{2}$
Balances, so $-\frac{2}{3}$ is correct solution.

**27.** Some possibilities are: $\frac{1}{2}x = 4$; $-\frac{1}{4}a = -2$; $\frac{3}{4}b = 6$.
**29.** Let $a$ be the man's age.
$109 = 100 + \frac{a}{2}$
The man is 18 years old.
**31.** Let $a$ be the woman's age.
$122 = 100 + \frac{a}{2}$
The woman is 44 years old.
**33.** Let $p$ be the penny size. $\frac{p}{4} + \frac{1}{2} = 3$. The penny size is 10.
**35.** Let $p$ be the penny size. $\frac{p}{4} + \frac{1}{2} = \frac{5}{2}$. The penny size is 8.
**37.** Let $h$ be the man's height. $\frac{11}{2}h - 220 = 209$. The man is 78 in. tall.   **39.** Let $h$ be the woman's height. $\frac{11}{2}h - 220 = 132$. The woman is 64 in. tall.

## Section 4.8 (page 287)

**1.** $P = 202$ m; $A = 1914$ m$^2$   **3.** $P = 5$ ft; $A = \frac{27}{32}$ ft$^2$
**5.** $P = 26\frac{1}{4}$ yd; $A = 30\frac{3}{4}$ yd$^2$   **7.** $P = \frac{454}{15}$ yd or $30\frac{4}{15}$ yd; $A = \frac{115}{3}$ yd$^2$ or $38\frac{1}{3}$ yd$^2$   **9.** $A = 198$ m$^2$   **11.** $A = 1716$ m$^2$
**13.** Perimeter is the distance around the outside edges of a flat shape and is measured in linear units. Area is the space inside a flat shape and is measured in square units. Volume is the space inside a solid shape and is measured in cubic units.   **15.** $A = \frac{63}{8}$ ft$^2$ or $7\frac{7}{8}$ ft$^2$
**17.** 132 m of curbing; 726 m$^2$ of sod   **19.** $V = 528$ cm$^3$
**21.** $V = 15\frac{5}{8}$ in.$^3$   **23.** $V = 800$ cm$^3$   **25.** $V = 18$ in.$^3$
**27.** $V = 651,775$ m$^3$   **29.** $V = 513$ cm$^3$   **31.** The correct answers are $A = 135$ ft$^2$ (square feet, not squaring 135) and $5\frac{1}{2}$ in. (perimeter is in linear units, not square units).

## Chapter 4 Review Exercises (page 299)

1. $\frac{2}{5}; \frac{2}{5}$   2. $\frac{3}{10}; \frac{7}{10}$
3. [number line showing $-\frac{1}{2}$ and $1\frac{1}{2}$ marked between $-3$ and $3$]

4. (a) $-4$  (b) $8$  (c) $-1$   5. $\frac{7}{8}$   6. $\frac{3}{5}$   7. already in lowest terms   8. $\frac{3x}{8}$   9. $\frac{1}{5b}$   10. $\frac{4n}{7m^2}$   11. $\frac{1}{16}$   12. $-12$   13. $\frac{8}{27}$   14. $\frac{1}{6x}$   15. $2a^2$   16. $\frac{6}{7k}$   17. $\frac{5}{24}$   18. $-\frac{2}{15}$   19. $\frac{19}{6}$ or $3\frac{1}{6}$   20. $\frac{3}{2}$   21. $\frac{4n+15}{20}$   22. $\frac{3y-70}{10y}$   23. $1\frac{5}{13}$; $2 \div 2 = 1$   24. $2\frac{1}{2}$; $7 - 5 = 2$   25. $4\frac{1}{20}$; $2 + 2 = 4$   26. $-\frac{27}{64}$   27. $\frac{1}{36}$   28. $-\frac{4}{5}$   29. $\frac{7}{6}$ or $1\frac{1}{6}$   30. $10$   31. $-\frac{4}{27}$   32. $w = 20$   33. $r = -15$   34. $x = \frac{1}{2}$   35. $A = 14$ ft$^2$   36. $V = 32\frac{1}{2}$ in.$^3$   37. $V = 93\frac{1}{3}$ m$^3$   38. $\frac{1}{4}$ pound; $7\frac{1}{2}$ pounds   39. $\frac{5}{6}$ hour; $10\frac{11}{12}$ hours   40. 12 preschoolers, 40 toddlers, 8 infants   41. $P = 2\frac{1}{10}$ miles; $A = \frac{9}{40}$ mi$^2$

## Chapter 4 Test (page 301)

1. $\frac{5}{6}; \frac{1}{6}$   2. [number line showing $-\frac{2}{3}$ and $2\frac{1}{3}$ marked between $-3$ and $3$]

3. $\frac{1}{4}$   4. already in lowest terms   5. $\frac{2a^2}{3b}$   6. $\frac{13}{15}$   7. $-2$   8. $-\frac{7}{40}$   9. $14$   10. $-\frac{2}{27}$   11. $\frac{25}{8}$ or $3\frac{1}{8}$   12. $\frac{4}{9}$   13. $\frac{9}{16}$   14. $\frac{4}{7y}$   15. $\frac{24-n}{4n}$   16. $\frac{10+3a}{15}$   17. $\frac{1}{18b}$   18. $-\frac{1}{18}$   19. $-\frac{31}{30}$ or $-1\frac{1}{30}$   20. $5 \div 1 = 5$; $\frac{64}{15}$ or $4\frac{4}{15}$   21. $3 - 2 = 1$; $\frac{3}{2}$ or $1\frac{1}{2}$   22. $d = 35$   23. $t = -\frac{15}{7}$ or $-2\frac{1}{7}$   24. $b = 8$   25. $x = -15$   26. $A = 52$ m$^2$   27. $A = \frac{117}{2}$ or $58\frac{1}{2}$ yd$^2$   28. $V = 6480$ m$^3$   29. $V = 16$ yd$^3$   30. $14\frac{3}{4}$ hours; $1\frac{5}{6}$ hours   31. $\frac{7}{2}$ or $3\frac{1}{2}$ days   32. 7392 students work

## Cumulative Review Exercises: Chapters 1–4 (page 303)

1. five hundred five million, eight thousand, two hundred thirty-eight   2. 35,600,000,916   3. (a) 60,700  (b) 100,600  (c) 3210   4. (a) commutative property of multiplication  (b) associative property of addition  (c) distributive property   5. $-54$   6. $-20$   7. undefined   8. 81   9. 10   10. 4   11. $-16$   12. $\frac{27}{27} = 1$   13. $-8$   14. 35   15. $\frac{15a}{2}$   16. $\frac{3}{xy^2}$   17. $-\frac{8}{15}$   18. $\frac{5}{16}$   19. $\frac{14-3b}{21}$   20. $\frac{8n+15}{5n}$   21. $\frac{13}{9}$ or $1\frac{4}{9}$   22. $\frac{13}{20}$   23. $-1$   24. $-\frac{5}{8}$   25. $a = 1$   26. $y = -5$   27. $k = -36$   28. $x = -4$   29. $m = 2$   30. $P = 18$ in.; $A = 20\frac{1}{4}$ in.$^2$   31. $P = 20$ yd; $A = 18\frac{3}{4}$ yd$^2$   32. $P = 78$ mm; $A = 288$ mm$^2$   33. Let $b$ be the original weight of each bag. $3b - 16 - 25 = 79$; There were 40 pounds in each bag.   34. Let $w$ be the width; $2w$ is the length. $2(2w) + 2(w) = 102$; Width is 17 yd; length is 34 yd.

# Chapter 5

## Section 5.1 (page 311)

1. 7; 0; 4   3. 5; 1; 8   5. 4; 7; 0   7. 1; 6; 3   9. 1; 8; 9   11. 6; 2; 1   13. 410.25   15. 6.5432   17. 5406.045   19. $\frac{7}{10}$   21. $13\frac{2}{5}$   23. $\frac{7}{20}$   25. $\frac{33}{50}$   27. $10\frac{17}{100}$   29. $\frac{3}{50}$   31. $\frac{41}{200}$   33. $5\frac{1}{500}$   35. $\frac{343}{500}$   37. five tenths   39. seventy-eight hundredths   41. one hundred five thousandths   43. twelve and four hundredths   45. one and seventy-five thousandths   47. 6.7   49. 0.32   51. 420.008   53. 0.0703   55. 75.030   57. Anne should not say "and" because that denotes a decimal point.   59. ten thousandths inch; $\frac{10}{1000} = \frac{1}{100}$   61. 12 pounds   63. 3-C   65. 4-A   67. one and six hundred two thousandths centimeters   69. millionths, ten-millionths, hundred-millionths, billionths; these match the words on the left side of the chart with "ths" added.   70. First place to left of decimal point is ones, so first place to right could be one*ths*, like tens and ten*ths*. But anything that is 1 or more is to the left of the decimal point.   71. seventy-two million four hundred thirty-six thousand nine hundred fifty-five hundred-millionths   72. six hundred seventy-eight thousand five hundred fifty-four billionths   73. eight thousand six and five hundred thousand one millionths   74. twenty thousand, sixty and five hundred five millionths   75. 0.0302040   76. 9,876,543,210.100200300

## Section 5.2 (page 321)

1. 16.9   3. 0.956   5. 0.80   7. 3.661   9. 794.0   11. 0.0980   13. 49   15. 9.09   17. 82.0002   19. $0.82   21. $1.22   23. $0.50   25. $48,650   27. $310   29. $849   31. $500   33. $1.00   35. $1000   37. (a) 186.0 miles per hour  (b) 763.0 miles per hour   39. (a) 322 miles per hour  (b) 163 miles per hour   41. Rounds to $0 (zero dollars) because $0.499 is closer to $0 than to $1.   42. Round $0.499 to the nearest cent to get $0.50. Guideline: Round amounts less than $1.00 to nearest cent instead of nearest dollar.   43. Rounds to $0.00 (zero cents) because $0.0015 is closer to $0.00 than to $0.01.   44. Both round to $0.60. Rounding to nearest thousandth (tenth of a cent) would allow you to identify $0.597 as less than $0.601.

## Section 5.3 (page 329)

1. 17.72   3. 11.98   5. 115.861   7. 59.323   9. 6 should be written 6.00; sum is 46.22.   11. $0.3000 = \frac{3000 \div 1000}{10{,}000 \div 1000} = \frac{3}{10} = 0.3$   13. 89.7   15. 0.109   17. 0.91   19. 6.661   21. 15.32 should be on top; correct answer is 7.87.   23. (a) 24.75 in.  (b) 3.95 in.   25. (a) 62.27 in.  (b) 0.39 in.   27. 23.013   29. $-45.75$   31. $-6.69$   33. $-6.99$

**35.** $-4.279$  **37.** $-0.0035$  **39.** $5.37$  **41.** $0.275$  **43.** $6.507$  **45.** $1.81$  **47.** $6056.7202$  **49.** *Estimate:* $40 - 20 = 20$ hours; *Exact:* $26.15$ hours  **51.** *Estimate:* $\$300 + \$1 = \$301$; *Exact:* $\$311.09$  **53.** *Estimate:* $2 + 2 + 2 = 6$ m; *Exact:* $6.32$ m, which is $0.08$ m less than the rhino's height.  **55.** *Estimate:* $\$20 - \$9 = \$11$; *Exact:* $\$10.88$  **57.** *Estimate:* $\$29 - \$9 = \$20$; *Exact:* $\$20.50$  **59.** *Estimate:* $\$13 + \$18 + \$6 + \$3 + \$2 = \$42$; *Exact:* $\$41.60$  **61.** $\$1939.36$  **63.** $\$3.97$  **65.** $b = 1.39$ cm  **67.** $q = 23.843$ ft

### Section 5.4 (page 337)

**1.** $0.1344$  **3.** $-159.10$  **5.** $15.5844$  **7.** $\$34{,}500.20$  **9.** $-43.2$  **11.** $0.432$  **13.** $0.0432$  **15.** $0.00432$  **17.** $0.0000312$  **19.** $0.000009$  **21.** $59.6$; $4.76$; $7226$; $32$; $803.5$; $9$. Multiplying by $10$, decimal point moves one place to the right; by $100$, two places to the right; by $1000$, three places to the right.  **22.** $5.96$; $0.0476$; $6.5$; $0.32$; $8.035$; $52.3$. Multiplying by $0.1$, decimal point moves one place to the left; by $0.01$, two places to the left; by $0.001$, three places to the left.  **23.** *Estimate:* $40 \times 5 = 200$; *Exact:* $190.08$  **25.** *Estimate:* $40 \times 40 = 1600$; *Exact:* $1558.2$  **27.** *Estimate:* $7 \times 5 = 35$; *Exact:* $30.038$  **29.** *Estimate:* $3 \times 7 = 21$; *Exact:* $19.24165$  **31.** unreasonable; $\$189.00$  **33.** reasonable  **35.** unreasonable; $\$3.19$  **37.** unreasonable; $9.5$ pounds  **39.** $\$945.87$ (rounded)  **41.** $\$2.45$ (rounded)  **43.** $\$28.82$ (rounded)  **45.** $\$12{,}271$  **47.** $\$347.52$; $\$719.40$  **49.** $\$76.50$  **51.** $\$73.45$  **53.** $\$388.34$  **55.** $\$4.09$ (rounded)  **57.** $\$129.25$  **59. (a)** $\$70.05$  **(b)** $\$25.80$

### Section 5.5 (page 349)

**1.** $-3.9$  **3.** $0.47$  **5.** $400.2$  **7.** $36$  **9.** $0.06$  **11.** $6000$  **13.** $60$  **15.** $0.0006$  **17.** $25.3$  **19.** $516.67$ (rounded)  **21.** $-24.291$ (rounded)  **23.** $10{,}082.647$ (rounded)  **25.** $0.377$; $0.0886$; $40.65$; $0.91$; $3.019$; $662.57$ **(a)** Dividing by $10$, decimal point moves one place to the left; by $100$, two places to the left; by $1000$, three places to the left. **(b)** The decimal point moved to the *right* when multiplying by $10$, by $100$, or by $1000$; here it moves to the *left* when dividing by $10$, by $100$, or by $1000$.  **26.** $402$; $3.39$; $460$; $71$; $157.7$; $8730$ **(a)** Dividing by $0.1$, decimal point moves one place to the right; by $0.01$, two places to the right; by $0.001$, three places to the right. **(b)** The decimal point moved to the *left* when multiplying by $0.1$, $0.01$, or $0.001$; here it moves to the *right* when dividing by $0.1$, $0.01$, or $0.001$.  **27.** unreasonable; *Estimate:* $40 \div 8 = 5$; $8\overline{)37.8}^{\,4.725}$  **29.** reasonable; *Estimate:* $50 \div 50 = 1$  **31.** unreasonable; *Estimate:* $300 \div 5 = 60$; $5.1\overline{)307.02}^{\,60.2}$  **33.** unreasonable; *Estimate:* $9 \div 1 = 9$; $1.25\overline{)9.3}^{\,7.44}$  **35.** $\$4.00$ (rounded)  **37.** $\$19.46$  **39.** $\$0.30$  **41.** $\$8.92$ per hour  **43.** $21.2$ miles per gallon (rounded)  **45.** $8.81$ m (rounded)  **47.** $0.03$ m  **49.** $26.72$ m  **51.** $14.25$  **53.** $3.8$  **55.** $-16.155$  **57.** $3.714$  **59.** $\$0.03$ (rounded)  **61.** $\$0.04$ (rounded)  **63.** $100{,}000$ box tops  **65.** $2632$ box tops (rounded)

### Section 5.6 (page 357)

**1.** $0.5$  **3.** $0.75$  **5.** $0.3$  **7.** $0.9$  **9.** $0.6$  **11.** $0.875$  **13.** $2.25$  **15.** $14.7$  **17.** $3.625$  **19.** $0.333$ (rounded)  **21.** $0.833$ (rounded)  **23.** $1.889$ (rounded)  **25.** $\dfrac{5}{9}$ means $5 \div 9$ or $9\overline{)5}$ so correct answer is $0.556$ (rounded). This makes sense because both the fraction and decimal are less than $1$.  **26.** Adding the whole number part gives $2 + 0.35$, which is $2.35$ not $2.035$. To check, $2.35 = 2\dfrac{35}{100} = 2\dfrac{7}{20}$ but $2.035 = 2\dfrac{35}{1000} = 2\dfrac{7}{200}$.  **27.** Just add the whole number part to $0.375$. So $1\dfrac{3}{8} = 1.375$; $3\dfrac{3}{8} = 3.375$; $295\dfrac{3}{8} = 295.375$.  **28.** It works only when the fraction part has a one-digit numerator and a denominator of $10$, or a two-digit numerator and a denominator of $100$, and so on.  **29.** $\dfrac{2}{5}$  **31.** $\dfrac{5}{8}$  **33.** $\dfrac{7}{20}$  **35.** $0.35$  **37.** $\dfrac{1}{25}$  **39.** $\dfrac{3}{20}$  **41.** $0.2$  **43.** $\dfrac{9}{100}$  **45.** shorter; $0.72$ inch  **47.** too much; $0.005$ gram  **49.** $0.9991$ cm, $1.0007$ cm  **51.** more; $0.05$ inch  **53.** $0.5399$, $0.54$, $0.5455$  **55.** $5.0079$, $5.79$, $5.8$, $5.804$  **57.** $0.6009$, $0.609$, $0.628$, $0.62812$  **59.** $2.8902$, $3.88$, $4.876$, $5.8751$  **61.** $0.006$, $0.043$, $\dfrac{1}{20}$, $0.051$  **63.** $0.37$, $\dfrac{3}{8}$, $\dfrac{2}{5}$, $0.4001$  **65.** red box  **67.** $0.01$ inch  **69.** $1.4$ in. (rounded)  **71.** $0.3$ in. (rounded)  **73.** $0.4$ in. (rounded)

### Section 5.7 (page 367)

**1.** $7$ months  **3.** $69.8$ (rounded)  **5.** $\$39{,}622$  **7.** $\$58.24$  **9.** $\$35{,}500$  **11.** $6.1$ (rounded)  **13.** $17.2$ hours (rounded)  **15.** $2.60$  **17. (a)** $2.80$  **(b)** $2.93$ (rounded)  **(c)** $3.13$ (rounded)  **19.** $15$ messages  **21.** $516$ students  **23.** $48$ calls  **25.** $4142$ miles (rounded)  **27. (a)** $4050$ miles  **(b)** The median is somewhat different from the average; this average is more affected by the two very low numbers.  **29.** $8$ samples  **31.** $68$ and $74$ years (bimodal)  **33.** no mode  **35. (a)** $-10$ degrees; $4$ degrees  **(b)** $14$ degrees warmer  **37.** Barrow's range is $-2 - (-18) = 16$ degrees; Fairbanks' range is $31 - (-10) = 41$ degrees; Fairbanks' temperatures have greater variability.  **39. (a)** mean $= 71$ degrees; median $\approx 71$ degrees  **(b)** They are identical because there is so little variation in the temperatures.  **41. (a)** Student P: mean $= 69.5$; median $= 70.5$; range $= 43$. Student Q: mean $\approx 70.3$; median $= 71$; range $= 17$  **(b)** Student P  **(c)** Answers will vary.  **42. (a)** Golfer G: mean $= 86.2$; median $= 87$; range $= 6$; Golfer H: mean $= 86.4$; median $= 88$; range $= 18$.  **(b)** Golfer G  **(c)** Answers will vary.  **43. (a)** A's range $= 74 - 62 = 12$; B's range $= 26 - 18 = 8$; C's range $= 69 - 25 = 44$; Building B's ages have the least variability.  **(b)** Answers will vary.  **44. (a)** P's range $= \$7200 - 4900 = \$2300$; Q's range $= \$6000 - 5500 = \$500$; R's range $= \$6400 - 5200 = \$1200$; Company P's salaries have the greatest variability.  **(b)** Answers will vary.

### Section 5.8 (page 375)

**1.** $4$  **3.** $8$  **5.** $3.317$ (rounded)  **7.** $2.236$ (rounded)  **9.** $8.544$ (rounded)  **11.** $10.050$ (rounded)  **13.** $19$  **15.** $31.623$ (rounded)  **17.** $30$ is about halfway between $25$ and $36$, so $\sqrt{30}$ should be about halfway between $5$ and $6$, or about $5.5$. Using a calculator, $\sqrt{30} \approx 5.477$. Similarly, $\sqrt{26}$ should be a little more than $\sqrt{25}$; by calculator $\sqrt{26} \approx 5.099$. And $\sqrt{35}$ should be a little less than $\sqrt{36}$; by calculator $\sqrt{35} \approx 5.916$.  **19.** $\sqrt{1521} = 39$ ft  **21.** $\sqrt{289} = 17$ in.  **23.** $\sqrt{144} = 12$ mm  **25.** $\sqrt{73} \approx 8.5$ in.  **27.** $\sqrt{65} \approx 8.1$ yd  **29.** $\sqrt{195} \approx 14.0$ cm  **31.** $\sqrt{7.94} \approx 2.8$ m  **33.** $\sqrt{65.01} \approx 8.1$ cm  **35.** $\sqrt{292.32} \approx 17.1$ km  **37.** $\sqrt{65} \approx 8.1$ ft  **39.** $\sqrt{360{,}000} = 600$ m  **41.** $\sqrt{135} \approx 11.6$ ft  **43.** The student used the formula for finding the hypotenuse but the unknown side is a leg, so $? = \sqrt{(20)^2 - (13)^2}$. Also, the final answer should be m, not m$^2$. Correct answer is $\sqrt{231} \approx 15.2$ m.  **45.** $\sqrt{16{,}200} \approx 127.3$ ft

46. (a) 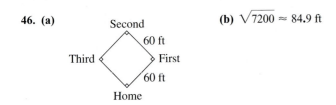  (b) $\sqrt{7200} \approx 84.9$ ft

**47.** The distance from third to first is the same as the distance from home to second because the baseball diamond is a square. **48. (a)** The side length is less than 60 ft. **(b)** $80^2 = 6400$; $6400 \div 2 = 3200$; $\sqrt{3200} \approx 56.6$ ft

### Section 5.9 (page 383)
Please see *Student's Solutions Manual* for a sample *check* for Exercises 1–16. **1.** $h = 4.47$ **3.** $n = -1.4$ **5.** $b = 0.008$ **7.** $a = 0.29$ **9.** $p = -120$ **11.** $t = 0.7$ **13.** $x = -0.82$ **15.** $z = 0$ **17.** $c = 0.45$ **19.** $w = 0$ **21.** $p = -40.5$ **23.** Let $d$ be the adult dose. $0.3\,d = 9$; The adult dose is 30 milligrams. **25.** Let $d$ be the number of days. $65.95d + 12 = 275.80$; The saw was rented for 4 days. **27.** $0.7(220 - a) = 140$; The person is 20 years old. **28.** $0.7(220 - a) = 126$; The person is 40 years old. **29.** $0.7(220 - a) = 134$; $a \approx 28.57$, which rounds to 29. The person is about 29 years old. **30.** $0.7(220 - a) = 117$; $a \approx 52.86$, which rounds to 53. The person is about 53 years old.

### Section 5.10 (page 393)
**1.** $d = 18$ mm **3.** $r = 0.35$ km **5.** $C \approx 69.1$ ft; $A \approx 379.9$ ft$^2$ **7.** $C \approx 8.2$ m; $A \approx 5.3$ m$^2$ **9.** $C \approx 47.1$ cm, $A \approx 176.6$ cm$^2$ **11.** $C \approx 23.6$ ft; $A \approx 44.2$ ft$^2$ **13.** $C \approx 27.2$ km; $A \approx 58.7$ km$^2$ **15.** $A \approx 57$ cm$^2$ **17.** $A \approx 197.8$ cm$^2$ **19.** $\pi$ is the ratio of the circumference of a circle to its diameter. If you divide the circumference of any circle by its diameter, the answer is always a little more than 3. The approximate value is 3.14, which we call $\pi$ (pi). Your test question could involve finding the circumference or the area of a circle. **21.** $C \approx 219.8$ cm **23.** $C \approx 785.0$ ft **25.** $A \approx 70{,}650$ mi$^2$ **27.** watch: $C \approx 3.1$ in.; $A \approx 0.8$ in.$^2$; wall clock: $C \approx 18.8$ in.; $A \approx 28.3$ in.$^2$ **29.** $d \approx 45.9$ cm **31.** $C \approx 14.6$ ft; Bonus: About 362 revolutions; the Mormons used 360, which is the answer you get using $2\frac{1}{3}$ ft instead of 2.33 ft as the radius. **33.** $1170.33 (rounded) **35.** $A \approx 44.2$ in.$^2$ **36.** $A \approx 132.7$ in.$^2$ **37.** $A \approx 201.0$ in.$^2$ **38.** small: $0.063 (rounded); medium: $0.049 (rounded); large: $0.046 (rounded) Best Buy **39.** small: $0.084 (rounded); medium: $0.067 (rounded) Best Buy; large: $0.071 (rounded) **40.** small: $0.077 (rounded) Best Buy; medium: $0.083 (rounded); large: $0.078 (rounded) **41.** $V \approx 471$ ft$^3$; $SA \approx 345.4$ ft$^2$ **43.** $V = 1617$ m$^3$; $SA = 849.4$ m$^2$ **45.** $V \approx 763.0$ in.$^3$; $SA \approx 678.2$ in.$^2$ **47.** $V = 5550$ mm$^3$; $SA = 2150$ mm$^2$ **49.** Student should use radius of 3.5 cm instead of diameter of 7 cm in the formula; units for volume are cm$^3$ not cm$^2$. Correct answer is $V \approx 192.3$ cm$^3$ **51.** $V \approx 3925$ ft$^3$ **53.** $SA \approx 163.6$ in.$^2$

### Chapter 5 Review Exercises (page 407)
**1.** 0; 5 **2.** 0; 6 **3.** 8; 9 **4.** 5; 9 **5.** 7; 6 **6.** $\frac{1}{2}$ **7.** $\frac{3}{4}$ **8.** $4\frac{1}{20}$ **9.** $\frac{7}{8}$ **10.** $\frac{27}{1000}$ **11.** $27\frac{4}{5}$ **12.** eight tenths **13.** four hundred and twenty-nine hundredths **14.** twelve and seven thousandths **15.** three hundred six ten-thousandths **16.** 8.3 **17.** 0.205 **18.** 70.0066 **19.** 0.30 **20.** 275.6 **21.** 72.79 **22.** 0.160 **23.** 0.091 **24.** 1.0 **25.** $15.83 **26.** $0.70 **27.** $17,625.79 **28.** $350 **29.** $130 **30.** $100 **31.** $29 **32.** $-5.67$ **33.** $-0.03$ **34.** 5.879 **35.** $-6.435$ **36.** *Estimate*: 100 million $-$ 60 million = 40 million; *Exact*: 38.8 million people **37.** *Estimate*: $200 + $40 = $240; *Exact*: $260.00 **38.** *Estimate*: $2 + $5 + $20 = $27; $30 - $27 = $3; *Exact*: $4.14 **39.** *Estimate*: $2 + 4 + 5 = 11$ kilometers; *Exact*: 11.55 kilometers **40.** $6 \times 4 = 24$; 22.7106 **41.** $40 \times 3 = 120$; 141.57 **42.** 0.0112 **43.** $-0.000355$ **44.** reasonable; $700 \div 10 = 70$ **45.** unreasonable; $30 \div 3 = 10$; $2.8 \overline{)26.6}$ with quotient 9.5 **46.** 14.467 (rounded) **47.** 1200 **48.** $-0.4$ **49.** $708 (rounded) **50.** $2.99 (rounded) **51.** 133 shares (rounded) **52.** $3.12 (rounded) **53.** $-4.715$ **54.** 10.15 **55.** 3.8 **56.** 0.64 **57.** 1.875 **58.** 0.111 (rounded) **59.** 3.6008, 3.68, 3.806 **60.** 0.209, 0.2102, 0.215, 0.22 **61.** $\frac{1}{8}, \frac{3}{20}, 0.159, 0.17$ **62.** mean: 28.5 digital cameras; median: 19.5 digital cameras **63.** mean: 44 claims; median: 39 claims **64.** $51.05 **65. (a)** Store J: $107 and $160 (bimodal); Store K: $119 **(b)** Range for Store J = $160 - $69 = $91; range for Store K = $139 - $95 = $44; Store J's prices have greater variability. **66.** 17 in. **67.** 7 cm **68.** 10.2 cm (rounded) **69.** 7.2 in. (rounded) **70.** 2.6 m (rounded) **71.** 8.5 km (rounded) **72.** $b = 0.25$ **73.** $x = 0$ **74.** $n = -13$ **75.** $a = -1.5$ **76.** $y = -16.2$ **77.** $d = 137.8$ m **78.** $r = 1\frac{1}{2}$ in. or 1.5 in. **79.** $C \approx 6.3$ cm; $A \approx 3.1$ cm$^2$ **80.** $C \approx 109.3$ m; $A \approx 950.7$ m$^2$ **81.** $C \approx 37.7$ in.; $A \approx 113.0$ in.$^2$ **82.** $V \approx 549.5$ cm$^3$; $SA \approx 376.8$ cm$^2$ **83.** $V = 1808.6$ m$^3$; $SA = 1205.8$ m$^2$ **84.** $V \approx 7.9$ ft$^3$; $SA = 25.5$ ft$^2$ **85.** 404.865 **86.** $-254.8$ **87.** 3583.261 (rounded) **88.** 29.0898 **89.** 0.03066 **90.** 9.4 **91.** $-15.065$ **92.** 9.04 **93.** $-15.74$ **94.** 8.19 **95.** 0.928 **96.** 35 **97.** $-41.859$ **98.** 0.3 **99.** $3.00 (rounded) **100.** $2.17 (rounded) **101.** $35.96 **102.** $199.71 **103.** $78.50 **104.** $y = -0.7$ **105.** $x = 30$ **106.** 15.7 ft of rubber striping; $A \approx 19.6$ ft$^2$ **107.** 67.6 **108.** 20 miles **109.** $V \approx 87.9$ in.$^3$

### Chapter 5 Test (page 413)
**1.** $18\frac{2}{5}$ **2.** $\frac{3}{40}$ **3.** sixty and seven thousandths **4.** two hundred eight ten-thousandths **5.** *Estimate*: $8 + 80 + 40 = 128$; *Exact*: 129.2028 **6.** *Estimate*: $-6(1) = -6$; *Exact*: $-6.948$ **7.** *Estimate*: $-80 - 4 = -84$; *Exact*: $-82.702$ **8.** *Estimate*: $-20 \div (-5) = 4$; *Exact*: 4.175 **9.** 669.004 **10.** 480 **11.** 0.000042 **12.** $4.55 per meter **13.** Davida, by 0.441 minute **14.** $5.35 (rounded) **15.** $y = -6.15$ **16.** $x = -0.35$ **17.** $a = 10.9$ **18.** $n = 2.85$ **19.** $0.44, \frac{9}{20}, 0.4506, 0.451$ **20.** 32.09 **21.** 75 books **22.** 103° and 104° (bimodal) **23.** $11.25 **24. (a)** $50.55 **(b)** $50.45 **(c)** math books: $89 - $39.75 = $49.25; biology books: $97.55 - $30.95 = $66.60; math books have less variability **25.** $\sqrt{85} \approx 9.2$ cm **26.** $\sqrt{279} \approx 16.7$ ft **27.** $r = 12.5$ in. **28.** $C \approx 5.7$ km **29.** $A \approx 206.0$ cm$^2$ **30.** $V \approx 5086.8$ ft$^3$ **31.** $SA \approx 2599.9$ ft$^2$

### Cumulative Review Exercises: Chapters 1–5 (page 415)
**1. (a)** forty-five and two-hundred three ten-thousandths **(b)** thirty billion, six hundred fifty thousand, eight **2. (a)** 160,000,500 **(b)** 0.075 **3.** 46,900 **4.** 6.20 **5.** 0.661 **6.** 10,000 **7.** $-13$ **8.** $-0.00006$ **9.** undefined **10.** 44 **11.** 18 **12.** $-6$ **13.** 56.621 **14.** $\frac{3}{2}$ or $1\frac{1}{2}$ **15.** $\frac{3}{8b}$ **16.** 11.147 **17.** $-\frac{9}{20}$ **18.** $-\frac{1}{32}$ **19.** 1.3 **20.** $\frac{5x+6}{10}$ **21.** $-8$ **22.** $-32$

23. $\frac{18}{18} = 1$   24. $-5.76$   25. $\frac{55}{24}$ or $2\frac{7}{24}$   26. 24.2 years (rounded)
27. 23.5 years   28. 20 and 26 years (bimodal)   29. $h = -4$
30. $x = 5$   31. $r = 10.9$   32. $y = -2$   33. $n = 4$
34. $P = 54$ ft; $A = 140$ ft$^2$   35. $C \approx 40.8$ m; $A \approx 132.7$ m$^2$
36. $x \approx 12.6$ km; $A \approx 37.8$ km$^2$   37. $V \approx 3.4$ in.$^3$; $SA \approx 13.5$ in.$^2$
38. *Estimate:* $20 - 8 - 1 = 11$; *Exact:* $11.17
39. *Estimate:* $2 + 4 = 6$ yards; *Exact:* $6\frac{5}{24}$ yards   40. *Estimate:*
$3 \div 3 = $1$ per pound; *Exact:* $0.95 per pound (rounded)
41. *Estimate:* $80,000 \div 100 = $800$; *Exact:* $729 (rounded)
42. Let $x$ be first student's money and $3x$ be second student's money. $x + 3x = 30,000$; one student receives $7500; the other receives $22,500.   43. Let $w$ be the width and $w + 4$ be the length. $2(w + 4) + 2w = 48$; The width is 10 in. and the length is 14 in.

## Chapter 6

### Section 6.1 (page 425)

1. $\frac{8}{9}$   3. $\frac{2}{1}$   5. $\frac{1}{3}$   7. $\frac{8}{5}$   9. $\frac{3}{8}$   11. $\frac{9}{7}$   13. $\frac{6}{1}$   15. $\frac{5}{6}$   17. $\frac{8}{5}$
19. $\frac{1}{12}$   21. $\frac{5}{16}$   23. $\frac{4}{1}$   25. $\frac{1}{2}$   27. $\frac{36}{1}$   29. Answers will vary. One possibility is stocking cards of various types in the same ratios as those in the table.   31. Comparing the violin to piano, guitar, organ, clarinet, and drums gives ratios of $\frac{1}{11}, \frac{1}{10}, \frac{1}{3}, \frac{1}{2},$ and $\frac{2}{3}$, respectively.   33. Answers will vary. Possibilities include: guitars are less expensive than drums and easier to carry around; more guitar players than drummers are needed in a band.   35. $\frac{2}{1}$
37. (a) $\frac{3}{8}$ (b) $\frac{7}{5}$   39. $\frac{7}{5}$   41. $\frac{6}{1}$   43. $\frac{38}{17}$   45. $\frac{1}{4}$   47. $\frac{34}{35}$
49. $\frac{1}{1}$; as long as the sides all have the same length, any measurement you choose will maintain the ratio.   50. Answers will vary. Some possibilities are: $\frac{4}{5} = \frac{8}{10} = \frac{12}{15} = \frac{16}{20} = \frac{20}{25} = \frac{24}{30} = \frac{28}{35}$
51. It is not possible. Amelia would have to be older than her mother to have a ratio of 5 to 3.   52. Answers will vary, but a ratio of 3 to 1 means your income is 3 times your friend's income.

### Section 6.2 (page 433)

1. $\frac{5 \text{ cups}}{3 \text{ people}}$   3. $\frac{3 \text{ feet}}{7 \text{ seconds}}$   5. $\frac{1 \text{ person}}{2 \text{ dresses}}$   7. $\frac{5 \text{ letters}}{1 \text{ minute}}$
9. $\frac{\$21}{2 \text{ visits}}$   11. $\frac{18 \text{ miles}}{1 \text{ gallon}}$   13. $12 per hour or $12/hour
15. 5 eggs per chicken or 5 eggs/chicken   17. 1.25 pounds/person
19. $103.30/day   21. 325.9; 21.0 (rounded)
23. 338.6; 20.9 (rounded)   25. 4 ounces for $0.89
27. 15 ounces for $3.15   29. 18 ounces for $1.79   31. Answers will vary. For example, you might choose Brand B because you like more chicken, so the cost per chicken chunk may actually be the same or less than Brand A.   33. 1.75 pounds/week
35. $12.26/hour   37. 2.1 points/min; 0.5 min/point
39. 0.11 seconds/meter; $9.\overline{09}$ or 9.1 meters/second (rounded)
41. (a) $0.167 or 16.7¢ (rounded); (b) $0.125 or 12.5¢; (c) $0.083 or 8.3¢ (rounded)   43. One battery for $1.79; like getting 3 batteries so $1.79 \div 3 \approx $0.597$ per battery   45. Brand P with the 50¢ coupon is the best buy. ($3.39 - $0.50 = $2.89 \div 16.5$ ounces $\approx $0.175$ per ounce)   47. $25 \div 12 \approx $2.08$ per month   48. Round to thousandths to see that Verizon is the best buy at $0.028 per minute;

Qwest $\approx$ $0.030; VoiceStream $\approx$ $0.062; Sprint $\approx$ $0.060
49. 1000 min $\approx$ 16.7 hours; 104 weekend days per year $\div$ 12 $\approx$ 8.7 weekend days per month; 16.7 hours $\div$ 8.7 days $\approx$ 1.9 hours of talking per day.   50. Verizon $\approx$ $1.17; Qwest $\approx$ $1.27; VoiceStream $\approx$ $0.95; Sprint $\approx$ $1.05

### Section 6.3 (page 445)

1. $\frac{\$9}{12 \text{ cans}} = \frac{\$18}{24 \text{ cans}}$   3. $\frac{200 \text{ adults}}{450 \text{ children}} = \frac{4 \text{ adults}}{9 \text{ children}}$
5. $\frac{120}{150} = \frac{8}{10}$   7. $\frac{2.2}{3.3} = \frac{3.2}{4.8}$   9. true   11. true   13. false
15. True   17. False   19. False   21. True   23. False   25. True
27. $\frac{16 \text{ hits}}{50 \text{ at bats}} = \frac{128 \text{ hits}}{400 \text{ at bats}}$   $\begin{matrix} 50 \cdot 128 = 6400 \\ 16 \cdot 400 = 6400 \end{matrix}$ Cross products are equal so the proportion is *true*; they hit equally well.   29. $x = 4$
31. $x = 2$   33. $x = 88$   35. $x = 91$   37. $x = 5$   39. $x = 10$
41. $x \approx 24.44$ (rounded)   43. $x = 50.4$   45. $x \approx 17.64$ (rounded)
47. $x = 1$   49. $x = 3\frac{1}{2}$   51. $x = 0.2$ or $x = \frac{1}{5}$   53. $x = 0.005$ or $x = \frac{1}{200}$   55. Find the cross products: $20 \neq 30$, so the proportion is false. $\frac{6\frac{2}{3}}{4} = \frac{5}{3}$ or $\frac{10}{6} = \frac{5}{3}$ or $\frac{10}{4} = \frac{7.5}{3}$ or $\frac{10}{4} = \frac{5}{2}$
56. Find the cross products: $192 \neq 180$, so the proportion is false. $\frac{6.4}{8} = \frac{24}{30}$ or $\frac{6}{7.5} = \frac{24}{30}$ or $\frac{6}{8} = \frac{22.5}{30}$ or $\frac{6}{8} = \frac{24}{32}$

### Section 6.4 (page 453)

1. 22.5 hours   3. $7.20   5. 42 pounds   7. $273.45
9. 10 ounces (rounded)   11. 5 quarts   13. 14 ft, 10 ft   15. 14 ft, 8 ft   17. 26 points (rounded)   19. 2065 students (reasonable); about 4214 students with incorrect setup (only 2950 students in the group)   21. about 190 people (reasonable); about 298 people with incorrect setup (only 238 people attended)   23. 92,250,000 households (reasonable); about 113,888,889 households with incorrect setup (only 102,500,000 U.S. households)   25. 625 stocks
27. 4.06 meters (rounded)   29. 311 calories (rounded)
31. 10.53 meters (rounded)   33. You cannot solve this problem using a proportion because the ratio of age to weight is not constant. As Jim's age increases, his weight may decrease, stay the same, or increase.   35. 4600 students use cream.   37. 120 calories and 12 grams of fiber   39. $1\frac{3}{4}$ cups water, 3 tablespoons margarine, $\frac{3}{4}$ cup milk, 2 cups flakes   40. $5\frac{1}{4}$ cups water, 9 tablespoons margarine, $2\frac{1}{4}$ cups milk, 6 cups flakes   41. $\frac{7}{8}$ cup water, $1\frac{1}{2}$ tablespoons margarine, $\frac{3}{8}$ cup milk, 1 cup flakes   42. $2\frac{5}{8}$ cups water, $4\frac{1}{2}$ tablespoons margarine, $1\frac{1}{8}$ cups milk, 3 cups flakes

### Section 6.5 (page 467)

1. line named $\overleftrightarrow{CD}$ or $\overleftrightarrow{DC}$   3. line segment named $\overline{GF}$ or $\overline{FG}$
5. ray named $\overrightarrow{PQ}$   7. perpendicular   9. parallel   11. intersecting   13. $\angle AOS$ or $\angle SOA$   15. $\angle CRT$ or $\angle TRC$   17. $\angle AQC$ or $\angle CQA$   19. right (90°)   21. acute   23. straight (180°)
25. $\angle EOD$ and $\angle COD$; $\angle AOB$ and $\angle BOC$   27. $\angle HNE$ and $\angle ENF$; $\angle HNG$ and $\angle GNF$; $\angle HNE$ and $\angle HNG$; $\angle ENF$ and $\angle GNF$
29. 50°   31. 4°   33. 50°   35. 90°   37. $\angle SON \cong \angle TOM$;

∠TOS ≅ ∠MON   **39.** ∠GOH measures 63°; ∠EOF measures 37°; ∠AOC and ∠GOF both measure 80°.   **41.** True, because $\overleftrightarrow{UQ}$ is perpendicular to $\overleftrightarrow{ST}$.   **42.** True, because they form a 90° angle, as indicated by the small red square.   **43.** False; the angles have the same measure (both are 180°).   **44.** False; $\overleftrightarrow{ST}$ and $\overleftrightarrow{PR}$ are parallel.   **45.** False; $\overleftrightarrow{QU}$ and $\overleftrightarrow{TS}$ are perpendicular.   **46.** True, because both angles are formed by perpendicular lines, so they both measure 90°.   **47.** corresponding angles: ∠1 and ∠8, ∠2 and ∠5, ∠3 and ∠6, ∠4 and ∠7; alternate interior angles: ∠4 and ∠5, ∠3 and ∠8.   **49.** ∠2, ∠4, ∠6, ∠8 all measure 130°; ∠1, ∠3, ∠5, ∠7 all measure 50°.   **51.** ∠6, ∠1, ∠3, ∠8 all measure 47°; ∠5, ∠2, ∠7, ∠4 all measure 133°.   **53.** ∠6, ∠8, ∠4, ∠2 all measure 114°; ∠7, ∠5, ∠3, ∠1 all measure 66°.   **55.** ∠1 ≅ ∠3, both are 138°; ∠2 ≅ ∠ABC, both are 42°.

### Section 6.6 (page 477)
**1.** congruent   **3.** neither   **5.** similar   **7.** SAS   **9.** SSS   **11.** ASA   **13.** use SAS: $BC = CE$, ∠B ≅ ∠C, $BA = CD$   **14.** use SSS: $WP = YP, ZP = XP, WZ = XY$   **15.** use SAS: $PS = SR$, $m\angle QSP = m\angle QSR = 90°$, $QS = QS$ (common side)   **16.** use SAS: $LM = OM, PM = NM$, ∠LMP ≅ ∠OMN (vertical angles)   **17.** $\frac{3}{2}, \frac{3}{2}, \frac{3}{2}$   **19.** $a = 5$ mm; $b = 3$ mm   **21.** $a = 6$ cm; $b = 15$ cm   **23.** $x = 24.8$ m; Perimeter $= 72.8$ m; $y = 15$ m; Perimeter $= 54.6$ m   **25.** Perimeter $= 8$ cm $+ 8$ cm $+ 8$ cm $= 24$ cm; Area $= (0.5)(8$ cm$)(6.9$ cm$) = 27.6$ cm$^2$   **27.** $h = 24$ ft   **29.** One dictionary definition is "Resembling, but not identical." Examples of similar objects are sets of different size pots or measuring cups; small and large size cans of beans; child's tennis shoe and adult's tennis shoe.   **31.** $x = 50$ m   **33.** $n = 110$ m

### Chapter 6 Review Exercises (page 489)
**1.** $\frac{3}{4}$   **2.** $\frac{4}{1}$   **3.** $\frac{1}{2}$   **4.** $\frac{2}{1}$   **5.** $\frac{2}{3}$   **6.** $\frac{5}{2}$   **7.** $\frac{1}{6}$   **8.** $\frac{3}{1}$   **9.** $\frac{3}{8}$   **10.** $\frac{4}{3}$   **11.** $\frac{1}{9}$   **12.** $\frac{10}{7}$   **13.** $\frac{7}{5}$   **14.** $\frac{5}{6}$   **15.** $\frac{\$11}{1 \text{ dozen}}$   **16.** $\frac{12 \text{ children}}{5 \text{ families}}$   **17.** 0.2 page/minute or $\frac{1}{5}$ page/minute; 5 minutes/page   **18.** $8/hour; 0.125 hour/dollar or $\frac{1}{8}$ hour/dollar   **19.** 13 ounces for $2.29   **20.** 25 pounds for $10.40 − $1 coupon   **21.** true   **22.** false   **23.** false   **24.** true   **25.** true   **26.** $x = 1575$   **27.** $x = 20$   **28.** $x = 400$   **29.** $x = 12.5$   **30.** $x ≈ 14.67$ (rounded)   **31.** $x ≈ 8.17$ (rounded)   **32.** $x = 50.4$   **33.** $x ≈ 0.57$ (rounded)   **34.** $x ≈ 2.47$ (rounded)   **35.** 27 cats   **36.** 46 hits   **37.** $15.63 (rounded)   **38.** 3299 students (rounded)   **39.** 68 ft   **40.** 14.7 milligrams   **41.** 511 calories (rounded)   **42.** $27\frac{1}{2}$ hours or 27.5 hours   **43.** line segment named $\overline{AB}$ or $\overline{BA}$   **44.** line named $\overleftrightarrow{CD}$ or $\overleftrightarrow{DC}$   **45.** ray named $\overrightarrow{OP}$   **46.** parallel   **47.** perpendicular   **48.** intersecting   **49.** acute   **50.** obtuse   **51.** straight; 180°   **52.** right; 90°   **53.** (a) 10° (b) 45° (c) 83°   **54.** (a) 25° (b) 90° (c) 147°   **55.** ∠1 and ∠4 measure 30°; ∠3 and ∠6 measure 90°; ∠5 measures 60°   **56.** ∠8, ∠3, ∠6, ∠1 all measure 160°; ∠4, ∠7, ∠2, ∠5 all measure 20°   **57.** SSS   **58.** SAS   **59.** ASA   **60.** $y = 30$ ft; $x = 34$ ft; $P = 104$ ft   **61.** $y = 7.5$ m; $x = 9$ m; $P = 22.5$ m   **62.** $x = 12$ mm; $y = 7.5$ mm; $P = 38$ mm   **63.** $x = 105$   **64.** $x = 0$   **65.** $x = 128$   **66.** $x ≈ 23.08$ (rounded)   **67.** $x = 6.5$   **68.** $x ≈ 117.36$ (rounded)   **69.** $\frac{8}{5}$   **70.** $\frac{33}{80}$   **71.** $\frac{15}{4}$   **72.** $\frac{4}{1}$   **73.** $\frac{4}{5}$   **74.** $\frac{37}{7}$   **75.** $\frac{3}{8}$   **76.** $\frac{1}{12}$   **77.** $\frac{45}{13}$   **78.** 24,900 fans (rounded)   **79.** $\frac{8}{3}$   **80.** 75 ft for $1.99 − $0.50 coupon   **81.** 15 ft   **82.** (a) 1400 milligrams (b) 100 milligrams   **83.** 21 points (rounded)   **84.** $\frac{1}{2}$ or 0.5 teaspoon   **85.** parallel lines   **86.** line segment   **87.** acute angle   **88.** intersecting lines   **89.** right angle; 90°   **90.** ray   **91.** straight angle; 180°   **92.** obtuse angle   **93.** perpendicular lines   **94.** (a) The car turned around in a complete circle. (b) The governor took the opposite view, for example, having once opposed taxes but now supporting them.   **95.** (a) No; because obtuse angles are >90°, their sum would be >180°. (b) Yes; because acute angles are <90°, their sum could equal 90°.   **96.** ∠1 measures 45°; ∠3 and ∠6 measure 35°; ∠4 measures 55°; ∠5 measures 90°   **97.** Use ASA: ∠BAE ≅ ∠DCE (alternate interior angles), $AE = EC$, and ∠AEB ≅ ∠CED (vertical angles)

### Chapter 6 Test (page 495)
**1.** $\frac{\$1}{5 \text{ minutes}}$   **2.** $\frac{9}{2}$   **3.** $\frac{15}{4}$   **4.** 18 ounces for $1.89 − $0.25 coupon   **5.** $x = 25$   **6.** $x ≈ 2.67$ (rounded)   **7.** $x = 325$   **8.** $x = 10\frac{1}{2}$   **9.** 576 words   **10.** 87 students (rounded)   **11.** 23.8 grams (rounded)   **12.** 60 feet   **13.** (e)   **14.** (a); 90°   **15.** (d)   **16.** (g); 180°   **17.** Parallel lines are lines in the same plane that never intersect. Perpendicular lines intersect to form a right angle. Sketches will vary.   **18.** 9°   **19.** 160°   **20.** $m\angle 1 = 50°, m\angle 2 = 35°, m\angle 3 = 95°, m\angle 5 = 35°$   **21.** measures of ∠1, ∠3, ∠5, and ∠7 are all 65°; measures of ∠2, ∠4, ∠6, and ∠8 are all 115°.   **22.** ASA (Angle–Side–Angle)   **23.** SAS (Side–Angle–Side)   **24.** $y = 12$ cm; $z = 6$ cm   **25.** $x = 12$ mm, $P = 46.8$ mm; $y = 14$ mm, $P = 39$ mm

### Cumulative Review Exercises: Chapters 1–6 (page 497)
**1.** (a) seventy-seven billion, one million, eight hundred five (b) two hundredths   **2.** (a) 3.040 (b) 500,037,000   **3.** 0   **4.** 14   **5.** $2\frac{11}{12}$   **6.** $\frac{2h-3}{10}$   **7.** 99.9905   **8.** $\frac{3}{0}$ is undefined.   **9.** −18   **10.** −14   **11.** $\frac{n}{4m}$   **12.** −0.00042   **13.** $\frac{1}{2y^2}$   **14.** $\frac{27+2n}{3n}$   **15.** −64   **16.** 5.4   **17.** $-\frac{1}{18}$   **18.** $y = -1$   **19.** $x = 15$   **20.** $x = 10$   **21.** Let $t$ be the starting temperature. $t + 15 - 23 + 5 = 71$. The starting temperature was 74 degrees.   **22.** Let $l$ be the length; $l - 14$ is the width. $2(l) + 2(l - 14) = 100$. Length is 32 ft; width is 18 ft.   **23.** $P = 7$ km; $A ≈ 2.0$ km$^2$   **24.** $d = 17$ m; $C ≈ 53.4$ m; $A ≈ 226.9$ m$^2$   **25.** $V ≈ 339.1$ ft$^3$; $SA ≈ 282.6$ ft$^2$   **26.** Estimate: $2000 + 2000 + 2000 + 2000 = 8000$ students; Exact: 8400 students   **27.** Estimate: $200,000 ÷ 2000 = $100; Exact: $78 (rounded)   **28.** (a) $\frac{2}{7}$ (b) $\frac{13}{11}$   **29.** Estimate: $4 • 8000 = $32,000; Exact: $31,500   **30.** Estimate: $4 + 3 = 7$; $7(1$ mile$) = 7$ miles; Exact: $7\frac{3}{20}$ miles   **31.** Estimate: 900 miles ÷ 50 gallons = 18 miles per gallon; Exact: 18.0 miles per gallon (rounded)   **32.** 250 pounds   **33.** $1\frac{1}{4}$ teaspoons

# Chapter 7

### Section 7.1 (page 509)
**1.** 0.25  **3.** 0.30 or 0.3  **5.** 0.06  **7.** 1.40 or 1.4  **9.** 0.078
**11.** 1.00 or 1  **13.** 0.005  **15.** 0.0035  **17.** 50%  **19.** 62%
**21.** 3%  **23.** 12.5%  **25.** 62.9%  **27.** 200%  **29.** 260%
**31.** 3.12%  **33.** $\frac{1}{5}$  **35.** $\frac{1}{2}$  **37.** $\frac{11}{20}$  **39.** $\frac{3}{8}$  **41.** $\frac{1}{16}$  **43.** $\frac{1}{6}$
**45.** $1\frac{3}{10}$  **47.** $2\frac{1}{2}$  **49.** 25%  **51.** 30%  **53.** 60%  **55.** 37%
**57.** $37\frac{1}{2}$% or 37.5%  **59.** 5%  **61.** exactly $55\frac{5}{9}$%, or 55.6% (rounded)  **63.** exactly $14\frac{2}{7}$%, or 14.3% (rounded)  **65.** 0.08
**67.** 0.42  **69.** 3.5%  **71.** 200%  **73.** $\frac{95}{100}$ or 95%; $\frac{5}{100}$ or 5%
**75.** $\frac{3}{10}$ or 30%; $\frac{7}{10}$ or 70%  **77.** $\frac{3}{4}$ or 75%; $\frac{1}{4}$ or 25%
**79.** 0.01; 1%  **81.** $\frac{1}{5}$; 20%  **83.** $\frac{3}{10}$; 0.3  **85.** 0.5; 50%
**87.** $\frac{9}{10}$; 0.9  **89.** $1\frac{1}{2}$; 150%  **91. (a)** human rights  **(b)** 44%; $\frac{11}{25}$
**93. (a)** gun control  **(b)** health care; environmental issues
**95.** $\frac{1}{4}$; 0.25; 25%  **97.** $\frac{3}{8}$; 0.375; 37.5%  **99.** $\frac{1}{7}$; exactly $14\frac{2}{7}$%, or 14.3% (rounded)  **101. (a)** The student forgot to move the decimal point in 0.35 two places to the right. So $\frac{7}{20}$ = 35%.
**(b)** The student did the division in the wrong order. Enter 16 ÷ 25 to get 0.64. So $\frac{16}{25}$ = 64%.  **103. (a)** $78  **(b)** $39
**105. (a)** 15 inches  **(b)** $7\frac{1}{2}$ inches  **107.** 20 children
**109. (a)** 2.8 miles  **(b)** 1.4 miles  **111.** $142.50
**113.** 4100 students  **115.** Since 100% means 100 parts out of 100 parts, 100% is all of the number. Examples will vary.

### Section 7.2 (page 521)
**1.** $\frac{10}{100} = \frac{n}{3000}$; 300 runners  **3.** $\frac{4}{100} = \frac{n}{120}$; 4.8 feet
**5.** $\frac{p}{100} = \frac{16}{32}$; 50%  **7.** $\frac{p}{100} = \frac{16}{200}$; 8%  **9.** $\frac{90}{100} = \frac{495}{n}$; 550 students  **11.** $\frac{12.5}{100} = \frac{3.50}{n}$; $28  **13.** $\frac{250}{100} = \frac{n}{7}$; 17.5 hours
**15.** $\frac{p}{100} = \frac{32}{172}$; 18.6% (rounded)  **17.** $\frac{110}{100} = \frac{748}{n}$; 680 books
**19.** $\frac{14.7}{100} = \frac{n}{274}$; $40.28 (rounded)  **21.** $\frac{p}{100} = \frac{105}{54}$; 194.4% (rounded)  **23.** $\frac{4}{100} = \frac{0.33}{n}$; $8.25  **25.** 150% of $30 cannot be *less* than $30; 25% of $16 cannot be *greater* than $16.
**27.** The correct proportion is $\frac{p}{100} = \frac{14}{8}$. The answer should be labeled with the % symbol. Correct answer is 175%.

### Section 7.3 (page 529)
**1.** 1500 patients  **3.** $15  **5.** 4.5 pounds  **7.** $7.00
**9.** 52 students  **11.** 870 phones  **13.** 4.75 hours  **15. (a)** 10% means $\frac{10}{100}$ or $\frac{1}{10}$. The denominator tells you to divide the whole by 10. The shortcut for dividing by 10 is to move the decimal point one place to the left.  **(b)** Once you find 10% of a number, multiply the result by 2 for 20% and by 3 for 30%.  **17.** 231 programs
**19.** 50%  **21.** 680 circuits  **23.** 1080 people  **25.** 845 species
**27.** $20.80  **29.** 76%  **31.** 125%  **33.** 4700 employees
**35.** 83.2 quarts  **37.** 2%  **39.** 700 tablets  **41.** 325 salads
**43.** 1.5%  **45.** 5 gallons  **47.** 1029.2 meters  **49. (a)** Multiply by 100 to change 0.2 to 20%. So, 0.20 = 20%.  **(b)** The correct equation is 50 = p • 20, so the solution is 250%.
**51. (a)** $\frac{1}{3}$ • 162 = n; the solution is $54.  **(b)** 0.333333333 • 162 = n; depending upon how your calculator rounds numbers, the solution is either $54 or $53.99999995.  **(c)** There is no difference or the difference is insignificant.  **52. (a)** 22 = $\frac{2}{3}$ • n; the solution is 33 cans.  **(b)** 22 = 0.666666667 • n; depending upon how your calculator rounds numbers, the solution is either 33 cans or 32.99999998 cans.  **(c)** There is no difference or the difference is insignificant.

### Section 7.4 (page 539)
**1.** $19.80  **3.** 62.5% or $62\frac{1}{2}$%  **5.** 836 drivers  **7.** 13.1% female; 86.9% male (both rounded)  **9.** 281 million people (rounded)
**11.** 138%  **13.** 40 problems  **15.** 30%; 70%  **17.** 23.7 miles per gallon (rounded)  **19.** March; 24.5 million cans, or 24,500,000 cans  **21.** 52.5 million cans (January); 38.5 million cans (February)  **23.** 5.6% (rounded)  **25.** 18.0% (rounded)
**27.** 40%  **29.** 10.7% (rounded)  **31.** No. 100% is the entire price, so a decrease of 100% would take the price down to 0. Therefore, 100% is the maximum possible decrease in the price of something.  **33.** George ate more than 65 grams, so the percent must be >100%. Use p • 65 = 78 to get 120%.  **34.** The team won more than half the games, so the percent must be >50%. Correct solution is 0.72 = 72%.  **35.** The brain could not weigh more than the person weighs. $2\frac{1}{2}$% = 2.5% = 0.025, so 0.025(150) = n and n = 3.75 pounds  **36.** If 80% were absent, then only 20% made it to class. 800 − 640 = 160 students, or use 0.20(800) = n.

### Section 7.5 (page 549)
**1.** $6; $106  **3.** 3%; $70.04  **5.** $21.96 (rounded); $387.94
**7.** $0.12 (rounded); $2.22  **9.** $4\frac{1}{2}$%; $13,167
**11.** $3 + $1.50 = $4.50; $4.83 (rounded); 2($3) = $6; $6.43 (rounded)  **13.** $8 + $4 = $12; $11.75 (rounded); 2($8) = $16; $15.67 (rounded)  **15.** $1 + $0.50 = $1.50; $1.43 (rounded); 2($1) = $2; $1.91  **17.** $15; $85  **19.** 30%; $126  **21.** $4.38 (rounded); $13.12  **23.** $3.79 (rounded); $34.09
**25.** $42; $342  **27.** $33.30; $773.30  **29.** $225; $1725
**31.** $1001.25; $18,801.25  **33.** $1170  **35.** $7978.13 (rounded)
**37.** $106.49 (rounded)  **39.** 5%  **41.** $74.25  **43.** $25.13
**45.** $106.20; $483.80  **47.** $2016.38 (rounded)  **49.** $21 (rounded)
**51.** $92.38  **53.** $230.12  **55. (a)** $18.43 (rounded to nearest cent)  **(b)** When calculating the discount, the *whole* is $18.50. But when calculating the sales tax, the *whole* is only $17.39 (the discounted price).  **56. (a)** $396.05 (rounded to nearest cent).
**(b)** 7.53% sales tax (rounded to the nearest hundredth) would give a final cost of $398.01.

## Summary Exercises on Percent (page 555)

**1.** (a) 0.03; 3%  (b) $\frac{3}{10}$; 0.3  (c) $\frac{3}{8}$; 37.5%  (d) $1\frac{3}{5}$; 1.6  (e) 0.0625; 6.25%  (f) $\frac{1}{20}$; 0.05  **2.** (a) 3.5 ft  (b) 19 miles  (c) 105 cows  (d) $0.08  (e) 500 women  (f) $45  **3.** $12\frac{1}{2}$% or 12.5%  **4.** 75 DVDs  **5.** $0.53 (rounded)  **6.** 175%  **7.** 5.9 pounds (rounded)  **8.** 600 hours  **9.** 50%  **10.** 500 camp sites  **11.** 98 golf balls  **12.** 2.7% (rounded)  **13.** 2300 apartments  **14.** 0.63 ounce  **15.** invitations, $351  **16.** $783.90  **17.** $1228.50  **18.** $39.21 (rounded)  **19.** $49.50 (rounded)  **20.** 89.2% (rounded)  **21.** 10% of $28.19 is about $2.80; 5% is half of $2.80 or $1.40; $2.80 + 1.40 = $4.20 estimated tip. Actual is $4.23 (rounded). They left a 17.7% tip (rounded).  **22.** The 9-month loan will cost $11.25 less in interest.  **23.** $353.09 (rounded); $348.14 (rounded); $4.95  **24.** 45%

## Chapter 7 Review Exercises (page 563)

**1.** 0.25  **2.** 1.8  **3.** 0.125  **4.** 0.00085  **5.** 2.65%  **6.** 2%  **7.** 87.5%  **8.** 0.2%  **9.** $\frac{3}{25}$  **10.** $\frac{3}{8}$  **11.** $2\frac{1}{2}$  **12.** $\frac{1}{20}$  **13.** 75%  **14.** 62.5% or $62\frac{1}{2}$%  **15.** 325%  **16.** 6%  **17.** 0.125  **18.** 12.5%  **19.** $\frac{3}{20}$  **20.** 15%  **21.** $1\frac{4}{5}$  **22.** 1.8  **23.** $46  **24.** $23  **25.** 9 hours  **26.** $4\frac{1}{2}$ hours  **27.** 242 meters  **28.** 17,000 cases  **29.** 27 cellular phones  **30.** 870 reference books  **31.** 9.5% (rounded)  **32.** 225%  **33.** $2.60 (rounded)  **34.** 80 days  **35.** 4%  **36.** $575  **37.** 20 people  **38.** 175%  **39.** 3000 patients  **40.** $364  **41.** 80%  **42.** 10.4% (rounded)  **43.** $0.11 (rounded); $2.90  **44.** $7\frac{1}{2}$%; $838.50  **45.** $4 + $2 = $6; $6.41 (rounded); 2($4) = $8; $8.55 (rounded)  **46.** $0.80 + $0.40 = $1.20; $1.21 (rounded); 2($0.80) = $1.60; $1.61  **47.** $3.75; $33.75  **48.** 25%; $189  **49.** $68.25; $418.25  **50.** $183.60; $1713.60  **51.** $\frac{1}{3}$; $33\frac{1}{3}$% (exact) or 33.3% (rounded)  **52.** cards; 68%; 0.68; $\frac{17}{25}$  **53.** (a) 277 adults (rounded)  (b) cards, 188 adults (rounded); sci-fi/simulation, 102 adults (rounded)  **54.** (a) 48.4% (rounded)  (b) 7161 dogs (rounded)  **55.** (a) 36.4% (rounded)  (b) 180.3% (rounded)  **56.** (a) 38.1% (rounded)  (b) 265 animals (rounded)

## Chapter 7 Test (page 567)

**1.** 0.75  **2.** 60%  **3.** 180%  **4.** 7.5% or $7\frac{1}{2}$%  **5.** 3.00 or 3  **6.** 0.02  **7.** $\frac{5}{8}$  **8.** $2\frac{2}{5}$  **9.** 5%  **10.** 87.5% or $87\frac{1}{2}$%  **11.** 175%  **12.** 320 laptops  **13.** 400%  **14.** $19,500  **15.** $6049.20  **16.** 34% (rounded)  **17.** To find 50% of a number, divide the number by 2. To find 25% of a number, divide the number by 4. Examples will vary.  **18.** Round $31.94 to $30. Then 10% of $30. is $3 and 5% of $30 is half of $3 or $1.50, so a 15% tip estimate is $3 + $1.50 = $4.50. A 20% tip estimate is 2($3) = $6.  **19.** Exact tip is $4.79 (rounded). Each person pays $12.24 (rounded).  **20.** $3.84; $44.16  **21.** $80.48 (rounded); $149.47  **22.** $815.66 ÷ 6 ≈ $135.94  **23.** $1650  **24.** $911.60

## Cumulative Review Exercises: Chapters 1–7 (page 569)

**1.** (a) ninety and one hundred five thousandths  (b) one hundred twenty-five million, six hundred seventy  **2.** (a) 30,005,000,000  (b) 0.0078  **3.** (a) 50,000  (b) 0.70  (c) 8900  **4.** (a) commutative property of multiplication  (b) associative property of addition  **5.** <; >  **6.** 0.7005; 0.705; $\frac{3}{4}$; 0.755  **7.** Mean is $771.25; median is $725.  **8.** 17 ounces for $2.89  **9.** 48.901  **10.** 17  **11.** $\frac{9b^2}{8}$  **12.** $\frac{1}{2}$  **13.** −10  **14.** $\frac{16 + 5x}{20}$  **15.** −40  **16.** −0.001  **17.** $\frac{12}{11}$ or $1\frac{1}{11}$  **18.** 123  **19.** 0  **20.** $\frac{7m - 16}{8m}$  **21.** −30  **22.** $\frac{4}{9n}$  **23.** $3\frac{1}{2}$  **24.** $3\frac{2}{9}$  **25.** −44  **26.** −0.02  **27.** $\frac{3}{10}$  **28.** 19  **29.** −1  **30.** −24  **31.** 160  **32.** $-9x^2 + 5x$  **33.** 0  **34.** $-40w^3$  **35.** $3h - 10$  **36.** $n = 5$  **37.** $h = -6$  **38.** $a = 2.2$  **39.** $y = -\frac{14}{3}$  **40.** $b = 20$  **41.** $x = 162.5$  **42.** $h = 3$  **43.** $x = -1$  **44.** $4n - 5 = -17$; $n = -3$  **45.** $n + 31 = 3n + 1$; $n = 15$  **46.** Let $d$ be the diapers in each package. $3d - 17 - 19 = 12$. There were 16 diapers in each package.  **47.** Let $p$ be Susanna's pay. Let $2p$ be Neoka's pay. $p + 2p = 1620$. Susanna receives $540; Neoka receives $1080.  **48.** $P = 50$ yd; $A = 142$ yd²  **49.** $C \approx 31.4$ ft; $A \approx 78.5$ ft²  **50.** $x \approx 8.1$ ft  **51.** $V \approx 7.4$ m³; $SA = 30.1$ m²  **52.** $V \approx 2255.3$ cm³  **53.** $y \approx 24.2$ mm  **54.** 2100 students  **55.** $P = 64$ ft; $A = 252$ ft²  **56.** $\frac{1}{6}$ cup more than the amount needed  **57.** $56.70; $132.30  **58.** 93.6 km  **59.** 88.6% (rounded)  **60.** $0.43 (rounded)  **61.** (a) 41,886 applications  (b) 43.4% (rounded)  **62.** 3286 applications  **63.** 524 accepted, 4934 rejected  **64.** 188.4% (rounded)

# Chapter 8

## Section 8.1 (page 581)

**1.** 3  **3.** 8  **5.** 5280  **7.** 2000  **9.** 60  **11.** 2  **13.** 2  **15.** 76,000 to 80,000 lb  **17.** 27  **19.** 112  **21.** 10  **23.** $1\frac{1}{2}$ or 1.5  **25.** $\frac{1}{4}$ or 0.25  **27.** $1\frac{1}{2}$ or 1.5  **29.** $2\frac{1}{2}$ or 2.5  **31.** $\frac{1}{2}$ day or 0.5 day  **33.** 5000  **35.** 17  **37.** 44 ounces  **39.** 216  **41.** 28  **43.** 518,400  **45.** 48,000  **47.** (a) pound/ounces  (b) quarts/pints or pints/cups  (c) minutes/hours or seconds/minutes  (d) feet/inches  (e) pounds/tons  (f) days/weeks  **49.** 174,240  **51.** 800  **53.** 0.75 or $\frac{3}{4}$  **55.** $1.83 (rounded)  **57.** $140  **59.** (a) $12\frac{1}{2}$ qt  (b) 4 jugs, because you can't buy part of a jug  **61.** (a) 337.5 T  (b) 2.8 T (rounded)  **63.** 1 gal = 128 fl oz; $\frac{16 \text{ fl oz}}{\$1.25} = \frac{128 \text{ fl oz}}{x}$; $x = \$10$  **64.** $\frac{11.2 \text{ fl oz}}{\$1.49} = \frac{128 \text{ fl oz}}{x}$; $x \approx \$17.03$  **65.** One alternate method: $3.85 ÷ 4 fl oz = $0.9625 per fl oz; $\frac{\$0.9625}{1 \text{ fl oz}} \cdot \frac{128 \text{ fl oz}}{1 \text{ gal}} = \$123.20$ per gal  **66.** Answers will vary. Because we use the products in very different ways, it may not make sense to compare the price per gallon.

## Section 8.2 (page 591)

**1.** 1000; 1000   **3.** $\frac{1}{1000}$ or 0.001; $\frac{1}{1000}$ or 0.001
**5.** $\frac{1}{100}$ or 0.01; $\frac{1}{100}$ or 0.01   **7.** Answers will vary; about 8 to 9 cm.
**9.** Answers will vary; about 20 to 25 mm.   **11.** cm   **13.** m
**15.** km   **17.** mm   **19.** cm   **21.** m   **23.** Some possible answers are: 35 mm film for cameras, track and field events, metric auto parts, and lead refills for mechanical pencils.   **25.** 700 cm
**27.** 0.040 m or 0.04 m   **29.** 9400 m   **31.** 5.09 m   **33.** 40 cm
**35.** 910 mm   **37.** less; 18 cm or 0.18 m
**39.** ———————————— 35 mm = 3.5 cm

——————————————————————— 70 mm = 7 cm

**41.** 0.018 km   **43.** 0.0000056 km

## Section 8.3 (page 599)

**1.** mL   **3.** L   **5.** kg   **7.** g   **9.** mL   **11.** mg   **13.** L
**15.** kg   **17.** unreasonable; too much   **19.** unreasonable; too much
**21.** reasonable   **23.** reasonable   **25.** Some capacity examples are 2 L bottles of soda and shampoo bottles marked in mL; weight examples are grams of fat listed on cereal boxes and vitamin doses in milligrams.   **27.** Unit for your answer (g) is in numerator; unit being changed (kg) is in denominator so it will divide out. The unit fraction is $\frac{1000 \text{ g}}{1 \text{ kg}}$.   **29.** 15,000 mL   **31.** 3 L   **33.** 0.925 L
**35.** 0.008 L   **37.** 4150 mL   **39.** 8 kg   **41.** 5200 g   **43.** 850 mg
**45.** 30 g   **47.** 0.598 g   **49.** 0.06 L   **51.** 0.003 kg   **53.** 990 mL
**55.** mm   **57.** mL   **59.** cm   **61.** mg   **63.** 0.3 L   **65.** 1340 g
**67.** 0.07 L   **69.** 3 kg to 4 kg   **71.** greater; 5 mg or 0.005 g
**73.** 200 nickels   **75.** (a) 1,000,000
(b) $\frac{3.5 \text{ Mm}}{1} \cdot \frac{1{,}000{,}000 \text{ m}}{1 \text{ Mm}} = 3{,}500{,}000$ m
**76.** (a) 1,000,000,000
(b) $\frac{2500 \text{ m}}{1} \cdot \frac{1 \text{ Gm}}{1{,}000{,}000{,}000 \text{ m}} = 0.0000025$ Gm
**77.** (a) 1,000,000,000,000   (b) 1000; 1,000,000
**78.** 1,000,000; 1,000,000,000; $2^{20} = 1{,}048{,}576$; $2^{30} = 1{,}073{,}741{,}824$

## Section 8.4 (page 605)

**1.** $1.33 (rounded)   **3.** 89.5 kg   **5.** 71 beats (rounded)
**7.** 180 cm; $0.02/cm (rounded)   **9.** 5.03 m   **11.** 4 bottles
**13.** $358.50   **15.** 215 g; 4.3 g; 4300 mg   **16.** 330 g; 0.33 g; 330 mg   **17.** 1550 g; 1500 g; 3 g   **18.** 55 g; 50 g; 0.5 g
**19.** 0.45 mg

## Section 8.5 (page 611)

**1.** 21.8 yd   **3.** 262.4 ft   **5.** 4.8 m   **7.** 5.3 oz   **9.** 111.6 kg
**11.** 30.3 qt   **13.** (a) about 0.2 oz   (b) probably not
**15.** about 31.8 L   **17.** about 1.3 cm   **19.** 18 kg; 56 cm by 36 cm by 20 cm (all rounded)   **21.** Converting ounces to grams, 0.09 oz ≈ 2.55 g, which rounds to 2.6 g. However, converting grams to ounces, 2.5 g ≈ 0.0875 oz, which does round to 0.09 oz.
**23.** 3.5 kg ≈ 7.7 lb so the baby is heavy enough. But 53 cm ≈ 20.7 in. so the baby is not long enough to be in the carrier.
**25.** 28 °F   **27.** 40 °C   **29.** 150 °C   **31.** 16 °C (rounded)
**33.** −20 °C   **35.** 46 °F (rounded)   **37.** 23 °F   **39.** 58 °C (rounded); −89 °C (rounded)   **41.** 10 °C and 41 °C (rounded)
**43.** More. There are 180 degrees between freezing and boiling on the Fahrenheit scale, but only 100 degrees on the Celsius scale, so each Celsius degree is a greater change in temperature.

**45.** (a) pleasant weather, above freezing but not hot   (b) 75 °F to 39 °F (rounded)   (c) Answers will vary. In Minnesota, it's 0 °C to −40 °C; in California, 24 °C to 0 °C.   **47.** longer; by about 2.3 yd
**48.** Multiply 2.3 yd by 2 to get 4.6 yd longer for the 50 m race; multiply by 4 to get 9.2 yd longer for the 100 m race.   **49.** 50 m race is 4.5 yd longer; 100 m race is 9 yd longer; differences are due to rounding.   **50.** shorter; by 64 yd

## Chapter 8 Review Exercises (page 621)

**1.** 16   **2.** 3   **3.** 2000   **4.** 4   **5.** 60   **6.** 8   **7.** 60   **8.** 5280
**9.** 12   **10.** 48   **11.** 3   **12.** 4   **13.** $\frac{3}{4}$ or 0.75   **14.** $2\frac{1}{2}$ or 2.5
**15.** 28   **16.** 78   **17.** 112   **18.** 345,600
**19.** (a) $4153\frac{1}{3}$ yd or $4153.\overline{3}$ yd   (b) 2.4 mi (rounded)
**20.** $2465.20   **21.** mm   **22.** cm   **23.** km   **24.** m   **25.** cm
**26.** mm   **27.** 500 cm   **28.** 8500 m   **29.** 8.5 cm   **30.** 3.7 m
**31.** 0.07 km   **32.** 930 mm   **33.** mL   **34.** L   **35.** g   **36.** kg
**37.** L   **38.** mL   **39.** mg   **40.** g   **41.** 5 L   **42.** 8000 mL
**43.** 4580 mg   **44.** 700 g   **45.** 0.006 g   **46.** 0.035 L
**47.** 31.5 L   **48.** 357 g (rounded)   **49.** 87.25 kg
**50.** $3.01 (rounded)   **51.** 6.5 yd (rounded)   **52.** 11.7 in. (rounded)
**53.** 67.0 mi (rounded)   **54.** 1288 km   **55.** 21.9 L (rounded)
**56.** 44.0 qt (rounded)   **57.** 0 °C   **58.** 100 °C   **59.** 37 °C
**60.** 20 °C   **61.** 25 °C   **62.** 33 °C (rounded)   **63.** 28 °F (rounded)
**64.** 120 °F (rounded)   **65.** L   **66.** kg   **67.** cm   **68.** mL
**69.** mm   **70.** km   **71.** g   **72.** m   **73.** mL   **74.** g   **75.** mg
**76.** L   **77.** 105 mm   **78.** $\frac{3}{4}$ hr or 0.75 hr   **79.** $7\frac{1}{2}$ ft or 7.5 ft
**80.** 130 cm   **81.** 77 °F   **82.** 14 qt   **83.** 0.7 g   **84.** 810 mL
**85.** 80 oz   **86.** 60,000 g   **87.** 1800 mL   **88.** −1 °C   **89.** 36 cm
**90.** 0.055 L   **91.** 1.26 m   **92.** 18 T   **93.** 113 g; 177 °C (both rounded)   **94.** 178.0 lb; 6.0 ft (both rounded)   **95.** 136 g (rounded)
**96.** 5.3 oz (rounded)   **97.** 6.3 oz   **98.** 156 g (rounded)
**99.** Audiovox, Motorola, Samsung, NeoPoint   **100.** (a) Motorola, Audiovox, Samsung, NeoPoint   (b) 1 hr or 60 min   **101.** Answers will vary. For example, you might choose Audiovox because it's the lightest weight and has the longest standby battery life. If you talk a lot on a cell phone, you might choose Motorola with the longest battery life for talking.

## Chapter 8 Test (page 625)

**1.** 36 qt   **2.** 15 yd   **3.** 2.25 hr or $2\frac{1}{4}$ hr   **4.** 0.75 ft or $\frac{3}{4}$ ft
**5.** 56 oz   **6.** 7200 min   **7.** kg   **8.** km   **9.** mL   **10.** g   **11.** cm
**12.** mm   **13.** L   **14.** cm   **15.** 2.5 m   **16.** 4600 m   **17.** 0.5 m
**18.** 0.325 g   **19.** 16,000 mL   **20.** 400 g   **21.** 1055 cm
**22.** 0.095 L   **23.** 3.2 ft (rounded)   **24.** (a) 2310 mg
(b) 500 mg or 0.5 g more   **25.** 95 °C   **26.** 0 °C
**27.** 1.8 m (rounded)   **28.** 56.3 kg (rounded)   **29.** 13 gal
**30.** 5.0 mi (rounded)   **31.** 23 °C (rounded)   **32.** 10 °F (rounded)
**33.** 6 m ≈ 6.54 yd; $26.03 (rounded)   **34.** Possible answers: Use same system as rest of the world; easier system for children to learn; less use of fractional numbers; compete internationally.

## Cumulative Review Exercises: Chapters 1–8 (page 627)

**1.** (a) six hundred three billion, five million, forty thousand;
(b) nine and forty thousandths   **2.** (a) 80.08   (b) 200,065,004
**3.** (a) 1.0   (b) 500   (c) 306,000,000   **4.** mean is 26.1 hours (rounded); median is 21.5 hours   **5.** −17   **6.** $1\frac{13}{20}$   **7.** −25

A-24  Answers to Selected Exercises

8. $\dfrac{3}{4xy}$  9. 0.00015  10. $-41.917$  11. $\dfrac{18-5c}{6c}$  12. $-30$
13. 13  14. undefined  15. $-240$  16. $-\dfrac{5}{12}$  17. $\dfrac{27xy}{4}$
18. $\dfrac{2m+3n}{3m}$  19. $2\dfrac{5}{6}$  20. $-\dfrac{7}{45}$  21. 2  22. $-3.2$  23. 0
24. 22  25. $-34$  26. $-6$  27. $-3p^2 - p$  28. $-5x - 14$
29. $-14r^3$  30. $b = -1$  31. $a = 8.05$  32. $t = 0$  33. $n = 2$
34. $w = 20$  35. $x = 0.4$  36. $y = -5$  37. $k = 4$
38. $11n - 8n = -9$; $n = -3$  39. $2n - 8 = n + 7$; $n = 15$
40. Let $l$ be the length. $2l + 2(25) = 124$; The length is 37 cm.
41. Let $p$ be the longer piece; $p - 6$ is the shorter piece.
$p + p - 6 = 90$; The pieces are 48 ft and 42 ft long.  42. $P = 9$ ft;
$A = 5\dfrac{1}{16}$ or $\approx 5.1$ ft$^2$  43. $C \approx 28.3$ mm; $A \approx 63.6$ mm$^2$
44. $P \approx 6.5$ cm; $A \approx 2.2$ cm$^2$  45. $y \approx 13.2$ yd  46. $V = 96$ cm$^3$;
$SA = 128$ cm$^2$  47. $x = 14$ in.  48. 54 in.  49. 3 days
50. 3700 g  51. 0.6 m  52. 0.007 L  53. $-4\,°F$
54. 58% (rounded)  55. Brand T at 15.5 ounces for
\$2.99 - \$0.30 coupon  56. $V \approx 2255.3$ cm$^3$  57. 5.97 kg
58. $1\dfrac{1}{12}$ yard  59. \$9.74 (rounded)  60. \$18.99 (rounded);
\$170.95  61. 120 rows  62. \$3631.25  63. 23 blankets;
\$9 left over

## Chapter 9

### Section 9.1 (page 637)

1. (a) 29,277 points  (b) Wilt Chamberlain  3. (a) Wilt
Chamberlain  (b) Jerry West  5. 16,732 points  7. West, 27.0;
Pettit, 26.4; round to nearest tenth.  9. Shaquille O'Neal is playing
in the 2000–2001 season and will add games and points to the
numbers shown in the table.  11. (a) 255 calories  (b) moderate
jogging  13. (a) moderate jogging, aerobic dance, racquetball
(b) moderate jogging, moderate bicycling, aerobic dance, racquetball, tennis  15. 370 calories  17. (a) 366 calories (rounded)
(b) 239 calories  19. about 39 calories  21. (a) 30 million or
30,000,000  (b) 75 million or 75,000,000  23. 135 million or
135,000,000  25. 5 million or 5,000,000  27. 240 million or
240,000,000  29. Answers will vary. One possibility: choose
Delta or Northwest because they have the best on-time performance.
30. Answers will vary. Possibilities include planning more time
between each flight, or doing some or all of your business via
conference calls or e-mail.  31. Answers will vary. One possibility:
choose Continental because it has the fewest luggage problems.
32. Answers will vary. Possibilities include buying heavy-duty
luggage or shipping the golf clubs via a delivery service.
33. Answers will vary. Possibilities include a lot of bad weather,
maintenance problems, new computer system.  34. Answers
will vary. Possibilities include availability of nonstop flights,
convenience of departure times, type and size of aircraft, availability of low-cost fares.

### Section 9.2 (page 645)

1. (a) \$32,000  (b) carpentry, \$12,100  3. (a) $\dfrac{\$9800}{\$32,000} = \dfrac{49}{160}$
(b) $\dfrac{\$3000}{\$2000} = \dfrac{3}{2}$  5. $\dfrac{\$16,000}{\$32,000} = \dfrac{1}{2}$  7. (a) Don't know
(b) Quicker  9. $\dfrac{720}{6000} = \dfrac{3}{25}$  11. $\dfrac{1020}{1200} = \dfrac{17}{20}$  13. $\dfrac{1740}{180} = \dfrac{29}{3}$
15. \$522,000  17. \$174,000  19. \$261,000  21. 4018 people

(rounded)  23. 543 people  25. 434 people (rounded)  27. First
find the percent of the total that is to be represented by each item.
Next, multiply the percent by 360° to find the size of each sector.
Finally, use a protractor to draw each sector.  29. 90°  31. 10%;
36°  33. 54°  35. 15%  37. (a) \$200,000  (b) 22.5°; 72°;
108°; 90°; 67.5°  (c)

39.

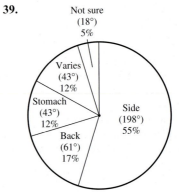

(a) 2464; 55% (rounded); 198°  (b) 748; 17% (rounded);
61° (rounded)  (c) 536; 12% (rounded); 43° (rounded)  (d) 520;
12% (rounded); 43° (rounded)  (e) 220; 5% (rounded); 18°
(f) No. The total is 101% due to rounding.  (g) No. The total is
363° due to rounding, so there will be a slight overlap when
drawing the sectors of the circle.

### Section 9.3 (page 655)

1. can shop during off hours; 74%  3. 57%; 342 people
5. compare products more easily; shop during off hours
7. May; 10,000 unemployed  9. 1500 workers  11. 2500 workers
13. 150,000 gallons  15. 1998; 250,000 gallons
17. 550,000 gallons  19. \$196.08  21. \$27.90; 86% (rounded)
23. Answers will vary. Some possibilities are: Greater competition
among long-distance companies; higher volume of long-distance
calls; improved technology.  25. 3,000,000 CDs  27. 1,500,000
CDs  29. 3,500,000 CDs  31. Answers will vary. Possibilities
include: Both stores had decreased sales from 1998 to 1999 and
increased sales from 2000 to 2002; Store B had lower sales than
Store A in 1998–99 but higher sales than Store A in 2000–2002.
32. Answers will vary. Some possibilities are that Store B may
have started to do more advertising, keep longer store hours, train
their staff, employ more help, or give better service than Store A.
33. Answers will vary, but most people will probably pick Store B
because of its greater sales.  34. Answers will vary. If the upward
trend continues, Store A might have sales of 3,500,000 CDs in 2003
and 4,000,000 in 2004. Store B might have sales of 4,500,000 CDs
in 2003 and 5,000,000 in 2004.  35. \$40,000  37. \$25,000
39. \$5000  41. The decrease in sales may have resulted from
poor service or greater competition. The increase in sales may have
been a result of more advertising or better service.

## Section 9.4 (page 663)

**1.**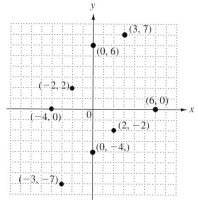

**3.**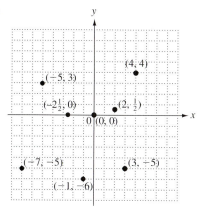

**5.** $A$ is $(3, 4)$; $B$ is $(5, -5)$; $C$ is $(-4, -2)$; $D$ is approximately $\left(4, \dfrac{1}{2}\right)$; $E$ is $(0, -7)$; $F$ is $(-5, 5)$; $G$ is $(-2, 0)$; $H$ is $(0, 0)$.

**7.** III, none, IV, II  **9. (a)** any positive number  **(b)** any negative number  **(c)** 0  **(d)** any negative number  **(e)** any positive number  **11.** Starting at the origin, move left or right along the $x$-axis to the number $a$; then move up if $b$ is positive or move down if $b$ is negative.

## Section 9.5 (page 673)

**1.** 4; $(0, 4)$; 3; $(1, 3)$; 2; $(2, 2)$; All points on the line are solutions.

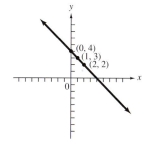

**3.** $-1$; $(0, -1)$; $-2$; $(1, -2)$; $-3$; $(2, -3)$; All points on the line are solutions.

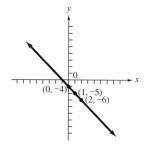

**5.** 4; $-1$; $-6$; 99

**7.**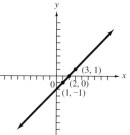

| $x$ | $y$ | $(x, y)$ |
|---|---|---|
| 1 | $-1$ | $(1, -1)$ |
| 2 | 0 | $(2, 0)$ |
| 3 | 1 | $(3, 1)$ |

**9.**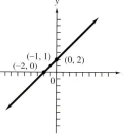

| $x$ | $y$ | $(x, y)$ |
|---|---|---|
| 0 | 2 | $(0, 2)$ |
| $-1$ | 1 | $(-1, 1)$ |
| $-2$ | 0 | $(-2, 0)$ |

| $x$ | $y$ | $(x, y)$ |
|---|---|---|
| 0 | 0 | $(0, 0)$ |
| 1 | $-3$ | $(1, -3)$ |
| 2 | $-6$ | $(2, -6)$ |

**11.**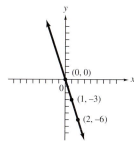

**13.** The lines in Exercises 7 and 9 have a positive slope. The lines in Exercises 1, 3, and 11 have a negative slope.

**15.**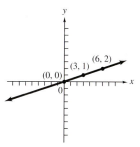

| $x$ | $y$ | $(x, y)$ |
|---|---|---|
| 0 | 0 | $(0, 0)$ |
| 3 | 1 | $(3, 1)$ |
| 6 | 2 | $(6, 2)$ |

**17.**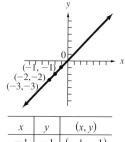

| $x$ | $y$ | $(x, y)$ |
|---|---|---|
| $-1$ | $-1$ | $(-1, -1)$ |
| $-2$ | $-2$ | $(-2, -2)$ |
| $-3$ | $-3$ | $(-3, -3)$ |

**19.**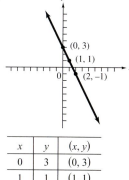

| $x$ | $y$ | $(x, y)$ |
|---|---|---|
| 0 | 3 | $(0, 3)$ |
| 1 | 1 | $(1, 1)$ |
| 2 | $-1$ | $(2, -1)$ |

**21.** **23.**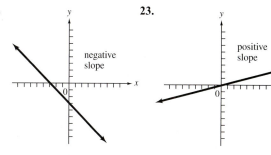

**25.** 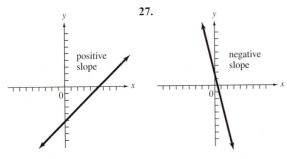 **27.**

**38.**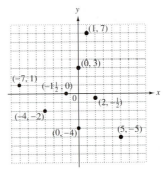

## Chapter 9 Review Exercises (page 687)

**1. (a)** gymnastics **(b)** volleyball **2. (a)** cross country **(b)** golf **3. (a)** 4870 men **(b)** 12,081 women **4.** 14.2 (rounded); 13.3 (rounded) **5. (a)** 100 inches **(b)** 15 inches **6. (a)** 50 inches **(b)** 55 inches **7. (a)** 35 inches **(b)** 45 inches **8.** 95 inches difference between Juneau and Memphis **9. (a)** lodging; $560 **(b)** food; $400 **(c)** $1700 **10.** $\frac{400}{1700} = \frac{4}{17}$ **11.** $\frac{300}{1700} = \frac{3}{17}$ **12.** $\frac{280}{1700} = \frac{14}{85}$ **13.** $\frac{300}{160} = \frac{15}{8}$ **14.** 3936 companies **15.** 1728 companies **16.** 720 companies **17.** 2904 companies **18.** On-site child care and Fitness centers; Answers will vary. Perhaps employers feel that they are not needed or would not be used. Or, it may be that they would be too expensive for the benefit derived. **19.** Flexible hours and Casual dress; Answers will vary. Perhaps employees request them and appreciate them. Or, it may be that neither of them cost the employer anything to offer. **20.** March; 8,000,000 acre-feet **21.** June; 2,000,000 acre-feet **22.** 5,000,000 acre-feet **23.** 6,000,000 acre-feet **24.** 5,000,000 acre-feet **25.** 2,000,000 acre-feet **26.** $50,000,000 **27.** $20,000,000 **28.** $20,000,000 **29.** $40,000,000 **30.** Sales decreased for 2 years and then moved up slightly. Answers will vary. Perhaps there is less new construction, remodeling and home improvement in the area near Center A. Or, better product selection and service may have reversed the decline in sales. **31.** Sales are increasing. Answers will vary. New construction may have increased in the area near Center B, or greater advertising may attract more attention. **32.** 36° **33.** 35%; 126° **34.** 20%; 72° **35.** 25%; 90° **36.** $2240; 10% **37.**

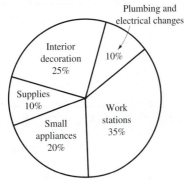

**39.** A is (0, 6); B is approximately $\left(-2, 2\frac{1}{2}\right)$; C is (0, 0); D is (−6, −6); E is (4, 3); F is approximately $\left(3\frac{1}{2}, 0\right)$; G is (2, −4).

**40.**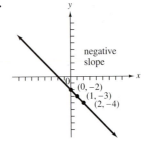

| x | y | (x, y) |
|---|---|---|
| 0 | −2 | (0, −2) |
| 1 | −3 | (1, −3) |
| 2 | −4 | (2, −4) |

All points on the line are solutions.

**41.** 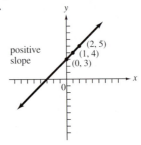 **42.**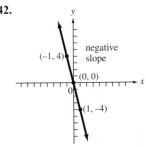

All points on the line are solutions.

| x | y | (x, y) |
|---|---|---|
| 0 | 3 | (0, 3) |
| 1 | 4 | (1, 4) |
| 2 | 5 | (2, 5) |

All points on the line are solutions.

| x | y | (x, y) |
|---|---|---|
| −1 | 4 | (−1, 4) |
| 0 | 0 | (0, 0) |
| 1 | −4 | (1, −4) |

## Chapter 9 Test (page 693)

**1. (a)** sardines **(b)** cream cheese **2.** 166 calories (rounded) **3.** 259 mg (rounded) **4. (a)** 30 insect species **(b)** 75 bird species **5.** 45 species **6.** 250 species **7.** television; $812,000 **8.** miscellaneous; $84,000 **9.** $308,000 **10.** $644,000 **11.** 2000; $4000 **12.** $9000 **13.** 2000; explanations will vary. Some possibilities are: laid off from work, changed jobs, was ill, cut down on hours worked. **14.** 5500 students; 3000 students **15.** College B; 1000 students **16.** Explanations will vary. For example, College B may have added new courses or lowered tuition

or added child care.   **17.** 30%; 108°   **18.** 10%; 36°
**19.** 20%; 72°   **20.** 30%; 108°   **21.** $48,000; 10%
**22.**

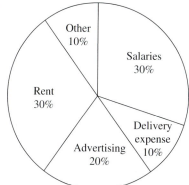

**23–26.**

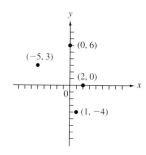

**27.** (0, 0); no quadrant   **28.** (−5, −4); quadrant III
**29.** (3, 3); quadrant I   **30.** (−2, 4); quadrant II
**31.**

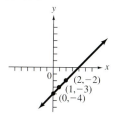

| $x$ | $y$ | $(x, y)$ |
|---|---|---|
| 0 | −4 | (0, −4) |
| 1 | −3 | (1, −3) |
| 2 | −2 | (2, −2) |

**32.** Answers will vary; all points on the line are solutions.

### Cumulative Review Exercises: Chapters 1–9 (page 695)

**1. (a)** six hundred two ten-thousandths   **(b)** three hundred million, five hundred sixty   **2. (a)** 70,005,000,043   **(b)** 18.09
**3. (a)** 3.0   **(b)** 0.80   **(c)** 69,000,000   **4.** Mean is 29 years; median is 25 years; mode is 23 years.   **5.** $3\frac{2}{15}$   **6.** −0.25
**7.** −39   **8.** −0.00010 or −0.0001   **9.** $\frac{4b}{3a^2}$   **10.** $-2\frac{7}{12}$
**11.** −10.007   **12.** $\frac{2x-56}{7x}$   **13.** −36   **14.** $\frac{25}{3mn}$   **15.** $3\frac{1}{3}$
**16.** undefined   **17.** $7\frac{3}{8}$   **18.** $\frac{18+wx}{6w}$   **19.** −4   **20.** 840
**21.** $\frac{0}{-2}=0$   **22.** −2.06   **23.** $-\frac{7}{20}$   **24.** 0   **25.** 41
**26.** −100   **27.** $x^2 - 5x$   **28.** $5y - 12$   **29.** $-40h^2$   **30.** $x = -7$
**31.** $y = -6$   **32.** $x = 2$   **33.** $m = -15$   **34.** $n = 16.2$
**35.** $w = 20$   **36.** $x = 7$   **37.** $h = -9$   **38.** $12 - 5n = n$; $n = 2$
**39.** $-8 + 2n = -28$; $n = -10$   **40.** Let $m$ be the smaller amount; $m + 500$ the larger amount. $m + m + 500 = 1800$  The two amounts are $650 and $1150.   **41.** Let $w$ be the width; $3w$ the length. $2(3w) + 2w = 280$.   The width is 35 ft; length is 105 ft.
**42.** $P = 36$ ft; $A = 70$ ft²   **43.** $P = 188$ m; $A = 1636$ m²
**44.** $x = 25$ mi; $P = 56$ mi; $A = 84$ mi²   **45.** $C \approx 18.8$ ft; $A \approx 28.3$ ft²   **46.** $V \approx 3768$ m³   **47.** $V = 30$ yd³   **48.** 135 minutes
**49.** 4.5 ft or $4\frac{1}{2}$ ft   **50.** 1850 mL   **51.** 3.5 cm   **52.** 0.01 kg

**53.** −4 °C (rounded)   **54.** 556 nonsmoking employees (rounded)
**55.** $116.78 (rounded)   **56.** 92.8% (rounded)
**57. (a)** 31 computers; $333 left over   **(b)** 17 calculators; $10 left over   **58.** $1\frac{5}{12}$ pounds

**59.**

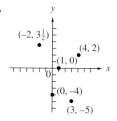

$\left(-2, 3\frac{1}{2}\right)$ is in second quadrant; none in third.

**60.**

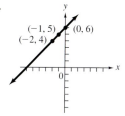

Positive slope; all points on the line are solutions, so there are many possibilities.

## Chapter 10

### Section 10.1 (page 707)

**1.** 1   **3.** false; $3^3 = 27$   **5.** false; $(a^2)^3 = a^6$   **7.** $(-2)^5$   **9.** $\left(\frac{1}{2}\right)^6$
**11.** $(-8p)^2$   **13.** The expression $(-3)^4 = (-3)(-3)(-3)(-3) = 81$, while $-3^4 = -(3 \cdot 3 \cdot 3 \cdot 3) = -81$.   **15.** Base is 3; exponent is 5; 243.   **17.** Base is −3; exponent is 5; −243.
**19.** Base is $-6x$; exponent is 4.   **21.** Base is $x$; exponent is 4.
**23.** The product rule does not apply to $5^2 + 5^3$ because it is a *sum*, not a product. $5^2 + 5^3 = 25 + 125 = 150$   **25.** $5^8$   **27.** $4^{12}$
**29.** $(-7)^9$   **31.** $t^{24}$   **33.** $-56r^7$   **35.** $42p^{10}$   **37.** $14x^4$; $45x^8$
**39.** $5a^2$; $-140a^6$   **41.** $4^6$   **43.** $t^{20}$   **45.** $7^3r^3$   **47.** $5^5x^5y^5$
**49.** $8q^3r^3$   **51.** $\frac{1^3}{2^3}$   **53.** $\frac{a^3}{b^3}$   **55.** $\frac{9^8}{5^8}$   **57.** $(-2)^3x^6y^3$   **59.** $3^2a^6b^4$
**61.** $A = 12x^5$

### Section 10.2 (page 713)

**1.** negative   **3.** negative   **5.** positive   **7.** 0   **9.** 1   **11.** −1
**13.** 0   **15.** 0   **17.** 2   **19.** 1   **21.** 15   **23.** $\frac{1}{64}$   **25.** $\frac{8}{15}$
**27.** $\frac{1}{x^4}$   **29.** 1   **30.** $\frac{5^2}{5^2}$   **31.** $5^0$   **32.** $5^0 = 1$; This supports the definition of a 0 exponent.   **33.** $6^5$   **35.** $10^3$   **37.** $\frac{1}{y^4}$
**39.** $c$   **41.** $\frac{1}{5^4}$   **43.** $\frac{1}{m}$   **45.** $3^4$   **47.** $a^3$   **49.** $\frac{1}{2^8}$   **51.** $\frac{1}{r^4}$
**53.** $10^{10}$   **55.** $\frac{1}{10^5}$   **57.** $10^{11}$   **59.** $\frac{1}{10^8}$   **61.** $10^4$   **63.** $\frac{1}{a^1}$ or $\frac{1}{a}$
**65.** $\frac{1}{2^8}$   **67.** $\frac{1}{x^{11}}$   **69.** $10^2$   **71.** $\frac{1}{y^2}$   **73.** $m^9$   **75.** $\frac{1}{10^6}$

## Section 10.3 (page 719)

1. $\$1.375 \times 10^{10}$  3. $6.66 \times 10^{10}$; $1.437 \times 10^{11}$
5. in scientific notation  7. not in scientific notation; $5.6 \times 10^6$
9. not in scientific notation; $4 \times 10^{-3}$  11. not in scientific notation; $8 \times 10^1$  13. To write a number in scientific notation (as $a \times 10^n$) move the decimal point so it follows the first nonzero digit. In a very large number, the decimal point will move to the left and make the number smaller so $n$ is the number of places moved and is positive. In a very small number the decimal point will move to the right and make the number larger, so $n$ is negative.
15. $5.876 \times 10^9$  17. $8.235 \times 10^4$  19. $7 \times 10^{-6}$
21. $2.03 \times 10^{-3}$  23. $750{,}000$  25. $5{,}677{,}000{,}000{,}000$
27. $6.21$  29. $0.00078$  31. $0.000000005134$  33. $6 \times 10^{11}$; $600{,}000{,}000{,}000$  35. $1.5 \times 10^7$; $15{,}000{,}000$  37. $6.426 \times 10^4$; $64{,}260$  39. $3 \times 10^{-4}$  41. $4 \times 10^1$  43. $2.6 \times 10^{-3}$
45. $3.41 \times 10^2$ (rounded) or 341 people per square mile
47. (a) $1 \times 10^{-4}$ or 0.0001 m  (b) $1 \times 10^{-1}$ or 0.1 m
49. $\$4.85 \times 10^2$ (rounded) or $485

## Section 10.4 (page 725)

1. 7; 5  3. 8  5. 26  7. 1; 6  9. 1; 1  11. 2; $-19$, $-1$
13. 2; 1, 8  15. $2m^5$  17. $-r^5$  19. cannot be simplified; $0.2m^5 - 0.5m^2$  21. $-5x^5$  23. $5p^9 + 4p^7$  25. $-2y^2$
27. already simplified; 4; binomial  29. already simplified; Write in descending powers as $6m^5 + 5m^4 - 7m^3 - 3m^2$; 5; none of these
31. $x^4 + \frac{1}{3}x^2 - 4$; 4; trinomial  33. 7; 0; monomial  35. (a) $-1$
(b) 5  37. (a) 19  (b) $-2$  39. (a) 36  (b) $-12$  41. (a) $-124$
(b) 5  43. $5m^2 + 3m$  45. $-10x^2 + x + 6$  47. $6w^3 + 8w$
49. $4x^4 - 4x^2$  51. $8r^2 + 5r - 12$  53. $5m^2 - 14m + 6$
55. $4x^3 + 2x^2 + 5x$  57. $12m^3 - 13m^2 + 6m + 11$
59. $-11x^2 - 3x - 3$  61. $8x^2 + 8x + 6$  63. $8t^2 + 8t + 13$
65. $D = 100 - 13 = 87$ ft  66. 0, (0, 0); 47, (0.5, 47); 121, (1.5, 121); 148, (2, 148); 169, (2.5, 169); 183, (3, 183); 191, (3.5, 191)
67.

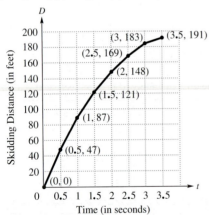

68. Answers will vary. This graph is a curve instead of the straight lines graphed for linear equations. Also, the rate of change in the distance is not consistent. From 0.5 sec to 1 sec, the distance increases 40 ft, but from 3 sec to 3.5 sec, the distance only increases 8 ft.

## Section 10.5 (page 731)

1. $40y^3 - 15y$  3. $-6m^2 - 4m$  5. $24x^4 - 12x^3 + 8x^2$
7. $-6k^6 + 9k^5 + 3k^4 - 3k^3$  9. $n^2 + n - 6$  11. $8r^2 - 10r - 3$
13. $12x^3 + 26x^2 + 10x + 1$  15. $20m^4 - m^3 - 8m^2 - 17m - 15$
17. $5x^4 - 13x^3 + 20x^2 + 7x + 5$  19. $x^2 + 7x + 12$
21. $2x^3 + 7x^2 + 3x$  23. D; B; A; C  25. $30x + 60$ yd$^2$
26. $30x + 60 = 600$; $x = 18$  27. 10 yd by 60 yd  28. $2100

29. 140 yd  30. $1260  31. The answers are $x^2 - 16$, $y^2 - 4$, and $r^2 - 49$. Each product is the difference of the square of the first term and the square of the last term of the binomials.

## Chapter 10 Review Exercises (page 735)

1. $4^{11}$  2. $(-5)^{11}$  3. $-72x^7$  4. $10x^{14}$  5. $19^5x^5$
6. $(-4)^7y^7$  7. $5p^4t^4$  8. $\frac{7^6}{5^6}$  9. $3^3x^6y^9$  10. The product rule for exponents does not apply because it is the *sum* of $7^2$ and $7^4$, not their *product*.  11. 2  12. $\frac{1}{32}$  13. 10  14. $\frac{5}{16}$  15. $6^2$
16. $x^2$  17. $\frac{1}{p^{12}}$  18. $r^4$  19. $\frac{1}{5}$  20. $\frac{1}{10^7}$  21. $\frac{1}{n^3}$  22. $10^3$
23. $\frac{1}{2^8}$  24. $h^3$  25. $\frac{1}{10^{13}}$  26. $x^6$  27. $4.8 \times 10^7$
28. $2.8988 \times 10^{10}$  29. $6.5 \times 10^{-5}$  30. $8.24 \times 10^{-8}$
31. $24{,}000$  32. $78{,}300{,}000$  33. $0.000000897$
34. $0.00000000000995$  35. $8 \times 10^2$; 800  36. $4 \times 10^6$; $4{,}000{,}000$  37. $3 \times 10^{-5}$; 0.00003  38. $1 \times 10^{-2}$; 0.01
39. $7.6 \times 10^1$ (rounded) or about 76 people per square mile
40. $1.08 \times 10^7$ or $10{,}800{,}000$ km  41. $22m^2$; degree 2; monomial
42. $p^3 - p^2 + 4p + 2$; degree 3; none of these  43. already in descending powers; degree 5; none of these  44. $-8y^5 - 7y^4 + 9y$; degree 5; trinomial  45. $-5a^3 + 4a^2$  46. $2r^3 - 3r^2 + 9r$
47. $11y^2 - 10y + 9$  48. $10m^3 - 6m^2 - 3$  49. $-y^2 - 4y + 26$
50. $10p^2 - 3p - 11$  51. $7r^4 - 4r^3 - 1$  52. $10x^2 + 70x$
53. $-6p^5 + 15p^4$  54. $6r^3 + 8r^2 - 17r + 6$  55. $8y^3 + 27$
56. $5p^5 - 2p^4 - 3p^3 + 25p^2 + 15p$  57. $6k^2 - 9k - 6$
58. $12p^2 - 48pq + 21q^2$  59. $2m^4 + 5m^3 - 16m^2 - 28m + 9$
60. 0  61. $3^4p^4$ or $81p^4$  62. $\frac{1}{7^2}$ or $\frac{1}{49}$  63. $-8m^7 - 10m^6 - 6m^5$
64. $\frac{3}{4}$  65. $a^3 - 2a^2 - 7a + 2$  66. $8y^3 - 9y^2 + 5$
67. $P = 6x - 2$; $A = 2x^2 + x - 6$

## Chapter 10 Test (page 739)

1. $\frac{1}{625}$  2. 2  3. $\frac{7}{12}$  4. 6  5. $12x^3y^3$  6. $\frac{1}{10^4}$  7. $\frac{1}{r^5}$
8. $-21x^7$  9. $m^3$  10. $3^2a^8b^2$ or $9a^8b^2$  11. Cannot use product rule; $81 + 9 = 90$  12. $3.44 \times 10^{11}$  13. $5.57 \times 10^{-6}$
14. $29{,}600{,}000$  15. $0.0000000607$  16. $1.97 \times 10^{30}$ kg
17. $-7x^2 + 8x$; degree 2; binomial  18. $4n^4 + 13n^3 - 10n^2$; degree 4; trinomial  19. $4t^4 - 2t^2$  20. $-y^2 - 6y + 11$
21. $-12t^2 + 8$  22. $-12x^5 + 20x^4$  23. $3y^2 + 5y - 2$
24. $2r^3 + r^2 - 16r + 15$  25. $P = 12x + 36$; $A = 9x^2 + 54x + 81$

## Cumulative Review Exercises: Chapters 1–10 (page 741)

1. (a) ten and thirty-five thousandths  (b) four hundred ten billion, three hundred fifty-one thousand, one hundred nine
2. (a) $300{,}006{,}080$  (b) $0.0055$  3. (a) $0.80$  (b) $341{,}000{,}000$
(c) 15.0  4. Mean is $52.70; median is $44.50; mode is $41.
5. $-4.171$  6. $-18$  7. $2\frac{4}{15}$  8. $\frac{0}{-14} = 0$  9. $\frac{16w}{3x}$
10. $-0.125$  11. $\frac{1}{10}$  12. $-\frac{7}{12}$  13. 28  14. $-12y^7$  15. $\frac{25}{3ab}$
16. $\frac{rt + 48}{8t}$  17. undefined  18. 750  19. $-0.00012$
20. $2^4a^{12}b^8$ or $16a^{12}b^8$  21. $\frac{4m - 54}{9m}$  22. $\frac{1}{t}$  23. $\frac{5}{32}$

**24.** $4\frac{1}{16}$  **25.** $n$  **26.** $-96$  **27.** $1\frac{1}{2}$  **28.** $1\frac{1}{5}$  **29.** 12  **30.** $-3$
**31.** 30  **32.** $h = 5$  **33.** $x = -7.9$  **34.** $a = 6$
**35.** $20 - 3n = 2n; n = 4$  **36.** Let $b$ be number of elephant breaths. $b + 16b = 170; b = 10$; Elephant takes 10 breaths; mouse takes 160.  **37.** 3 runs  **38.** 145%  **39.** $4.87 (rounded); $69.82  **40.** $1.61 \times 10^{10}$ (rounded) or 16,100,000,000 miles
**41.** (a) $5\frac{1}{8}$ in. (b) size 7 (c) $\frac{5}{16}$ in.  **42.** $P = 22.4$ m; $A = 31.6$ m²  **43.** $C \approx 28.3$ cm; $A \approx 63.6$ cm²

## Whole Numbers Computation: Pretest

### (page 743)
### Adding Whole Numbers
**1.** 390  **2.** 13,166  **3.** 3053  **4.** 117,510  **5.** 688,226

### Subtracting Whole Numbers
**1.** 350  **2.** 629  **3.** 24,894  **4.** 3909  **5.** 591,387

### Multiplying Whole Numbers
**1.** 0  **2.** 26,887  **3.** 1,560,000  **4.** 1846  **5.** 17,232
**6.** 519,477

### Dividing Whole Numbers
**1.** 23  **2.** undefined  **3.** 6259  **4.** 807 **R**6  **5.** 34  **6.** 50
**7.** 60 **R**20  **8.** 539 **R**62

## Chapter R

### Section R.1 (page 751)
**1.** 8928; $5715 + 3213 = 8928$  **3.** 59,224; $21,020 + 38,204 = 59,224$  **5.** 150  **7.** 1651  **9.** 1004  **11.** 9253  **13.** 11,624
**15.** 16,658  **17.** 5009  **19.** 3506  **21.** 15,954  **23.** 10,648
**25.** incorrect; should be 769  **27.** correct  **29.** 33 miles
**31.** 38 miles  **33.** $122  **35.** 699 people  **37.** 54,723,468 people
**39.** 294 inches  **41.** 708 feet

### Section R.2 (page 759)
**1.** incorrect; should be 62  **3.** incorrect; should be 121  **5.** 8
**7.** 25  **9.** 16  **11.** 61  **13.** 519  **15.** 9177  **17.** 7589
**19.** 8859  **21.** 3  **23.** 23  **25.** 1942  **27.** 5687  **29.** 19,038
**31.** 65,556  **33.** 19,984  **35.** incorrect; should be 2494
**37.** correct  **39.** 15 calories  **41.** 121 passengers  **43.** $7500
**45.** 365 feet

### Section R.3 (page 767)
**1.** 9  **3.** 63  **5.** 0  **7.** 24  **9.** 36  **11.** 245  **13.** 168
**15.** 19,092  **17.** 258,447  **19.** 3750  **21.** 44,550  **23.** 270,000
**25.** 86,000,000  **27.** 1496  **29.** 3735  **31.** 15,200  **33.** 32,805
**35.** 183,996  **37.** 1,616,076  **39.** 19,422,180  **41.** 252 inches; 540 inches  **43.** 576 plants  **45.** 418 miles  **47.** 2440 miles; 16,170 miles  **49.** 7406 calories

### Section R.4 (page 777)
**1.** $1; 12\overline{)12}; 12 \div 12$  **3.** undefined; $\frac{24}{0}; 0\overline{)24}$  **5.** $0; 4\overline{)0}; 0 \div 4$
**7.** $0; \frac{0}{12}; 12\overline{)0}$  **9.** undefined; $\frac{21}{0}; 21 \div 0$  **11.** 27  **13.** 36
**15.** 1522 **R**5  **17.** 309  **19.** 1006  **21.** 5006  **23.** 6671
**25.** 12,458 **R**3  **27.** 10,253 **R**5  **29.** 18,377 **R**6  **31.** incorrect; should be 670 **R**2  **33.** incorrect; should be 3568 **R**1  **35.** correct

**37.** correct  **39.** 74 ounces  **41.** $16,600  **43.** $28,200
**45.** $13,000,000  **47.** $11,635  **49.** 2, 3, 5, 10  **51.** 2  **53.** 5
**55.** 3  **57.** 2, 3  **59.** none

### Section R.5 (page 783)
**1.** 32  **3.** 250  **5.** 120 **R**7  **7.** 1308 **R**9  **9.** 7134 **R**12
**11.** 900 **R**100  **13.** 207 **R**5  **15.** 236 **R**29  **17.** 2407 **R**1
**19.** 1239 **R**15  **21.** 3331 **R**82  **23.** 850  **25.** incorrect; should be 106 **R**7  **27.** incorrect; should be 658  **29.** incorrect; should be 62  **31.** correct  **33.** 1050 hours  **35.** $1372
**37.** $108  **39.** 1680 circuits  **41.** $375

### Chapter R Review Exercises (page 787)
**1.** 92  **2.** 113  **3.** 5464  **4.** 15,657  **5.** 39  **6.** 184  **7.** 224
**8.** 25,866  **9.** 48  **10.** 45  **11.** 48  **12.** 0  **13.** 172
**14.** 5467  **15.** 32,640  **16.** 465,525  **17.** 19,200  **18.** 25,200
**19.** 206,800  **20.** 128,000,000  **21.** 612  **22.** 1872  **23.** 13,755
**24.** 30,184  **25.** 6  **26.** 1  **27.** undefined  **28.** 0  **29.** 108
**30.** 24  **31.** 352  **32.** 150 **R**4  **33.** 468 cards  **34.** 2856 textbooks  **35.** $280  **36.** $27,940  **37.** 24,000 hours
**38.** 400 miles  **39.** 23 acres  **40.** 32 homes  **41.** $247
**42.** $143  **43.** $582  **44.** $1261

### Chapter R Test (page 789)
**1.** addends; sum or total  **2.** factors; product  **3.** difference; quotient  **4.** 8610  **5.** 112,630  **6.** 16,948  **7.** 3084  **8.** 140
**9.** 171,000  **10.** 1785  **11.** 4,450,743  **12.** 206  **13.** undefined
**14.** 458  **15.** 160  **16.** $528  **17.** $271  **18.** 33 hours
**19.** $165  **20.** 1028 ovens  **21.** $2040  **22.** A number is divisible by 2 if it ends in 0, 2, 4, 6, or 8. A number is divisible by 5 if it ends in 0 or 5. A number is divisible by 10 if it ends in 0. Examples will vary.

## Appendix

### Appendix Exercises (page A-5)
**1.** 37; add 7  **3.** 26; add 10, subtract 2  **5.** 16; multiply by 2
**7.** 243; multiply by 3  **9.** 36; add 3, add 5, add 7, etc.; or $1^2, 2^2, 3^2$, etc.
**11.** ⌐_  **13.** \\_
**15.** Conclusion follows  **17.** Conclusion does not follow
**19.** 3 days  **21.** Dick

# Index

## A

Absolute value, 11, 201
Acute angle, 460
Addends, 15, 745
   grouping of, 18
Addition, 745
   applications of, 749
   associative property of, 18, 746
   by calculator, 26
   carrying in, 747
   changing to subtraction, 753
   checking answers in, 750
   commutative property of, 18, 89, 745
   of decimal numbers, 323
   estimating answers in, 32
   of fractions, 235
   of fractions with variables, 240
   of integers, 15
   of like fractions, 235
   of like terms, 721
   of mixed numbers, 253
   on the number line, 15
   of polynomials, 723
   of signed decimal numbers, 325
   of unlike fractions, 239
   of whole numbers, 745
Addition property of equality, 109, 379
   solving equations using, 110
Addition property of zero, 17
Additive inverse of integers, 25
Adjacent angles, 463
Algebraic equations, 182
Algebraic expressions, 142
   terms of, 721
   word phrases to, 169
Alternate interior angles, 465
Amount, compound, 554
Amount due, 548
Angles, 458
   adjacent, 463
   alternate interior, 465
   classifying, 459
   complementary, 461
   congruent, 462, 463
   corresponding, 464
   interior, 465
   measures of, 459
   naming of, 458
   nonadjacent, 463
   sides of, 458
   supplementary, 462
   types of, 460
   vertex of, 458
   vertical, 463
Application problems, 169, 179
   with decimals, 382
   with fractions, 230, 274
   with geometric figures, 164, 374, 389
   with mixed numbers, 255
   with one unknown quantity, 169
   with percents, 533
   steps to solve, 170, 179
   with two unknown quantities, 179
   using English measurement, 578
   using metric measurement, 590, 603
   using proportions, 449
   using similar triangles, 476
Approximately equal to, 32
   symbol for, 32
Area, 159, 281
   of a circle, 387, 388
   of a parallelogram, 163
   of a rectangle, 159
   of a semicircle, 389
   of a square, 161, 371
   of a triangle, 281
Argument, A-2
Associative property
   of addition, 18, 746
   of multiplication, 43, 762
Average, 361
Axes of a coordinate system, 660

## B

Bar, fraction, 64, 199
Bar graph, 651
   using, 651
Base
   of an exponent, 59, 702
   of a parallelogram, 162
   powers of, 59
   of a pyramid, 284
   of a triangle, 281
Best buy, determining, 430
Billions, 4
Bimodal, 364
Binary system, 7
Binomials, 722
Borrowing when subtracting whole numbers, 755
Borrowing with zeros, 757
Byte, 602

## C

Calculator(s)
   addition by, 26
   addition of decimals by, 328
   area of a circle by, 389
   changing fractions to percent by, 506
   circumference of a circle by, 387
   division by, 50, 771
   division of decimals by, 343
   evaluating exponential expressions by, 60
   evaluating expressions by, 87
   front end rounding by, 34
   multiplication by, 42, 762, 766
   multiplication of decimals by, 334
   multiplication of fractions by, 225
   prime factorization of numbers by, 216
   sale price by, 546
   simple interest by, 548
   solving proportions by, 443
   square roots of a number by, 371
   subtraction by, 26
   subtraction of decimals by, 328
   unit rates by, 431
   using to add signed numbers, 26
   using in scientific notation, 717
   using to subtract signed numbers, 26
Capacity measurements, 423, 574, 593
   chart of, 574, 593
Carrying when adding whole numbers, 747
Carrying when multiplying whole numbers, 62
Celsius-Fahrenheit conversion formula, 609
   application of, 610
Celsius scale, 608
Center of a circle, 385
Centigram (cg), 596
Centiliter (cL), 593
Centimeter (cm), 151, 585

I-1

Central tendency, measures of, 361, 364
Checking answers
 in addition, 750
 in division, 773, 782
 in subtraction, 755, 757
Circle graph, 641
 drawing using protractor, 643
 finding a ratio from, 641
 sectors of, 641
Circles, 385
 area by calculator, 389
 area of, 387, 388
 center of, 385
 circumference of, 386
 diameter of, 385
 Euler, A-3
 radius of, 385
Circumference of a circle, 386
 by calculator, 387
Coefficient, 86, 721
 numerical, 695, 721
Combining like terms, 97
 steps for, 97
Common denominator, 235
 least, 237
 for unlike fractions, 237
Common factor, 211
Commutative property
 of addition, 18, 89, 745
 of multiplication, 43, 89, 761
Comparing costs, 54
Comparison line graph, 653
 interpreting, 653
Complementary angles, 461
Complex fractions, 265
Composite numbers, 213
Compound amount, 554
Compound interest, 554
 table of, 554
Conclusion of deductive
 reasoning, A-2
Congruent angles, 462, 463
Congruent figures, 471
Congruent triangles, 471
 corresponding parts of, 471
 methods to prove, 472
Connecting arithmetic to algebra, 182
Constant, 84
Constant term of an expression, 95
Consumer applications, 543
Conversions
 capacity units, 576, 594
 decimals to fractions, 309
 English measurement units, 575
 English to metric units, 607
 fractions to decimals, 353
 length units, 576, 587
 metric to English units, 607
 temperature units, 609
 weight units, 577, 597
Coordinate system
 axes of, 660

 quadrants of, 662
 rectangular, 659
Coordinates of a point, 661
Corresponding angles, 464
Corresponding parts
 of congruent triangles, 471
 of similar triangles, 473
Cost per unit, 430
Costs of a wedding, 54
Cross products of proportions, 438
Cube, volume of, 283
Currency exchange, 636
 rate of, 636
Cylinders, 390
 surface area of, 392
 volume of, 390

# D

Data, 361
 paired, 659
 range of, 365
 table of, 632
 variability of, 365
Decigram (dg), 596
Deciliter (dL), 593
Decimal numbers, 306
 adding by calculator, 328
 addition of, 323
 arranging in order, 356
 calculator representation of, 310
 comparing with fractions, 355
 dividing by calculator, 343
 dividing by an integer, 341
 division of, 341
 entering on a calculator, 310
 equations with, 379
 estimating addition answers, 327
 estimating division answers, 345
 estimating multiplication
  answers, 335
 estimating subtraction answers, 327
 fraction form of, 306
 fractional part of, 307
 from fractions, 353
 mixed number form of, 309
 multiplication of, 333
 multiplying by calculator, 334
 order of operations with, 346
 as a percent, 500, 502
 place value chart of, 307
 in proportions, 443
 in ratios, 420
 reading, 308
 repeating, 343, 355
 rounding of, 315
 signed, 325
 solving equations with, 379
 steps to add, 323
 steps to multiply, 333
 subtracting by calculator, 328
 subtraction of, 324
 truncated, 355

 whole number part of, 307
 writing as fractions, 309
 writing as mixed numbers, 309
 writing as percents, 502
Decimal places, 316
 rounding to, 316
Decimal point, 4, 306
Decimeter (dm), 585
Decrease, percent of, 537
Deductive reasoning, A-2
 conclusion of, A-2
 premises of, A-2
Degrees
 of an angle, 459
 of a polynomial, 722
 of a term, 722
Dekagram (dag), 596
Dekaliter (daL), 593
Dekameter (dam), 585
Denominator of a fraction, 199, 418
 common, 235
 identifying, 199
Descending powers, 721
Determining the best buy, 430
Diameter of a circle, 385
Difference, 754
Digits, 2
 writing numbers in, 4
Discount, 545
Distance formula, 149, 158
Distance measure using signed
 numbers, 20
Distance, rate, time relationship,
 149, 158
Distributive property, 43, 99
 simplifying expressions by, 100
 solving equations using, 129
Dividend, 226, 769
Divisibility of whole numbers, 774
 tests for, 215, 774
Division, 769
 by calculator, 50, 771
 changing to multiplication, 770
 checking answers in, 773, 782
 of decimal numbers, 341
 estimating answers in, 51
 of exponential expressions, 711
 of fractions, 228
 of fractions with variables, 229
 indicator words for, 230
 of integers, 49
 interpreting remainders in, 52
 long, 779
 of mixed numbers, 252
 by multiples of 10, 781
 of numbers in scientific notation, 716
 properties of, 203
 remainders in, 52, 772
 short, 772
 of signed decimal numbers, 341
 of signed fractions, 228
 symbols for, 769
 of whole numbers, 769

by zero, 50, 771
of zero, 50, 770
Division property of equality,
119, 271, 380
solving equations using, 119
Divisor, 226, 769
trial, 779
Dollar-cost averaging, 348
Double-bar graph, 651
interpreting, 652

# E

Educational tax incentives, 562
English-metric conversions, 607
English system of measurement, 574
application problems using, 578
of capacity, 574
converting among units of, 574
of length, 574
of time, 574
of weight, 574
Equality
addition property of, 109, 379
division property of, 119, 271, 380
multiplication property of, 271
Equality principle for solving
an equation, 119
Equation(s) 107
algebraic, 182
checking solution of, 110
with decimals, 379
equality principle for solving, 119
with fractions, 271
goal in solving, 109
percent, 525
from sentences, 170
solution of, 107
solving, 108, 110, 119, 127, 271, 379
steps to solve, 127, 129
Equilateral triangle, 158
Equivalent fractions, 202
Estimating
an answer, 32, 33, 44, 51, 52
a decimal answer, 327, 335, 345
a mixed number answer, 251
a percent answer, 523
Euler circles, A-3
Exponential expressions, 89, 702
division of, 711
evaluating, 89, 702
evaluating by calculator, 60
multiplication of, 703
quotient rule for, 711
Exponential form, 702
Exponential notation, 59
Exponents, 59, 263, 702
application of, 715
base of, 59, 702
evaluating expressions with, 702
of fractions, 263
negative, 710
with negative numbers, 60
power rules for, 704, 705, 706

product rule for, 703
quotient rule for, 711
summary of rules for, 712
with variable base, 89
writing numbers without, 716
zero, 709
Expression(s), 84, 102
algebraic, 142
constant term of, 95
evaluating, 85
evaluating by calculator, 87
exponential, 89, 702
simplifying, 95, 97
terms of, 95, 721
with two variables, 88
variable term of, 95
from word phrases, 169

# F

Factor tree, 214
Factors, 39
common, 211
multiples of, 763
of multiplication, 761
Fahrenheit-Celsius conversion
formula, 609
application of, 610
Fahrenheit scale, 608
Figures
congruent, 471
similar, 471
Formulas, 149, 150, 158, 174, 373, 546
Fraction(s), 198
absolute value of, 201
addition of, 235
applications of, 256, 266
application problems with,
230, 274
changing to percent by calculator, 506
comparing with decimals, 355
complex, 265
decimal forms of, 353
from decimals, 309
denominator of, 199, 418
division of, 228
equations with, 271
equivalent, 202
with exponents, 263
graphing on a number line, 200
greater than one, 199
improper, 199
like, 235
lowest terms of, 211
multiplication of, 223
as music time signatures, 242
negative, 200
numerator of, 199, 418
order of operations with, 263
from a percent, 503
positive, 200
proper, 199
from rates, 429
from a ratio, 418

reciprocal of, 227
signed, 200
steps to reduce to lowest terms, 212
subtraction of, 235
unit, 575
unlike, 235
using prime factorization to
multiply, 224
writing as decimals, 353
writing as a percent, 505
Fraction bar, 64, 199
Fraction form
of decimals, 306
of percents, 503
of a ratio, 418
Fractional part of decimals, 307
Front end rounding, 32, 44, 52, 327
by calculator, 34

# G

Geometry applications, 164, 374,
389, 476
Gram (g), 595
Graph
bar, 651
circle, 641
double-bar, 651
line, 652
of linear equations in two
variables, 666
pictograph, 634
of points in a coordinate system, 661
Graphing numbers on a number line,
10, 200, 355
Greater than, 11
symbol for, 11
Grouping of addends, 18
Growth of investments, 554

# H

Heart-rate, 266
Hectogram (hg), 596
Hectoliter (hL), 593
Hectometer (hm), 585
Height
of a cylinder, 390
of a parallelogram, 162
of a pyramid, 284
of a triangle, 281
Historical ratios, 424
Horizontal axis of rectangular
coordinate system, 659
Hotel expenses, 276
Hypotenuse of a right triangle, 372
formula for, 373

# I

Identifying place value, 308
Improper fractions, 199
mixed number form of, 249

Increase, percent of, 536
Indicator words
    for division, 230
    for multiplication, 230
Inductive reasoning, A-1
Integers, 10, 308
    absolute value of, 11
    addition of, 15
    additive inverse of, 25
    comparing, 10
    dividing decimal numbers by, 341
    division of, 49
    multiplication of, 39
    opposite of, 25
    rounding of, 29
    subtraction of, 25
Intensity of an earthquake, 738
Interest, 546
    compound, 554
    formula for, 547
    rate of, 546
    simple, 546
Interior angles, 465
Interpreting
    comparison line graph, 653
    data table, 632
    double-bar graph, 652
    line graph, 652
    pictograph, 532
Intersecting lines, 458
Investment growth, 554

## K

Kilogram (kg), 596
Kiloliter (kL), 585, 593
Kilometer (km), 585

## L

Lawn fertilizer, 320
Least common denominator (LCD), 237
    finding by inspection, 238
    finding using prime factors, 238
Legs of a right triangle, 372
    formula for, 373
Length measurements, 423, 574, 585
    chart of, 574, 585
    converting among, 576, 587
Length of a rectangle, 152
Less than, 11
    symbol for, 11
Life insurance benefits, 336
Like fractions, 235
    addition of, 235
    subtraction of, 235
Like terms, 95, 96, 721
    addition of, 721
    combining, 97
    steps for combining, 97
Line(s), 457
    intersecting, 458
    metric conversion, 588

    number, 10, 247
    parallel, 458, 464
    perpendicular, 461
    segment of, 457
    slope of, 671
    transversal of, 464
Line graph, 652
    comparison, 652
    interpreting, 652
Linear equations in one variable, 108, 665
Linear equations in two variables, 665
    graphing of, 666
Liter (L), 593
Long division, 779
Lowest terms
    of fractions, 211
    of proportions, 437
    of rates, 429
    of ratios, 419
    steps to reduce to, 212
    using prime factorization, 216

## M

Mass measurements, 595
Mean, 361
    weighted, 362
Measurement comparisons, 423, 574, 585, 593, 596
Measurements
    of angles, 459
    of capacity, 423, 574, 593
    converting among, 574, 587, 594, 597
    English system of, 574
    of length, 423, 574, 585
    of mass, 595
    metric system of, 585, 593
    nail, 197, 280
    penny, 197, 280
    table of, 423, 574, 585, 593, 596
    of time, 423, 574
    of volume, 283
    of weight, 423, 574, 595
Measures of central tendency, 361, 364
Median, 363
Meter (m), 152, 585
Metric conversion line, 588, 594, 597
    steps to use, 588
Metric-English conversions, 607
Metric system of measurement, 585
    applications of, 590, 603
    of capacity, 593
    capacity relationships, 594
    of length, 585
    length relationships, 587
    of mass, 595
    prefixes for, 585, 593, 596
    of weight, 595
    weight relationships, 597
Milligram (mg), 596
Milliliter (mL), 593
Millimeter (mm), 154, 585

Millions, 4
Minuend, 754
Mixed numbers, 247
    addition of, 253
    applications with, 255
    decimal form of, 309
    from decimals, 309
    division of, 252
    graphing on a number line, 247
    improper fraction form of, 248
    multiplication of, 251
    problem solving with calculator, 254
    in proportions, 442
    in ratios, 420
    recipes with, 256
    rounding of, 250
    subtraction of, 253
Mode, 364
Monomial, 722
    multiplication with a polynomial, 729
Multiplicand, 761
Multiplication, 761
    applications of, 766
    associative property of, 43, 762
    by calculator, 42, 762, 766
    commutative property of, 43, 89, 761
    of decimal numbers, 333
    with ending zeros, 763
    estimating answers in, 44
    of exponential expressions, 703
    factors of, 761
    of fractions, 223
    of fractions by calculator, 225
    of fractions with variables, 226
    indicator words for, 230
    of integers, 39
    of mixed numbers, 251
    by multiples of 10, 763
    of numbers in scientific notation, 716
    of polynomials, 729
    of signed decimal numbers, 333
    of signed fractions, 223
    of whole numbers, 761
Multiplication expressions, simplifying, 99
Multiplication property
    of equality, 271
    of one, 42
    of zero, 42
Multiplier, 761
Music time signature, 242

## N

Nail measurement by penny system, 197, 280
Naming of angles, 458
Negative exponents, 710
    changing to positive, 710
Negative fractions, 200
    graph of, 200
Negative numbers, 9

Index **I-5**

Negative slope, 671
Negative square root, 371
Nonadjacent angles, 463
Notation
   exponential, 59
   scientific, 715
Number(s)
   common factor of, 211
   composite, 213
   decimal, 306
   divisibility tests for, 215
   factors, 39
   integer, 10, 308
   mixed, 247
   negative, 9
   opposite of, 25
   ordering, 356
   positive, 9
   prime, 213
   prime factorization of, 213
   rational, 204
   reciprocals of, 226
   rounding of, 29
   signed, 9
   square roots of, 371
   whole, 2
   written in words, 3
Number line, 10, 247
   addition on, 15
   comparing numbers on, 355
   graphing fractions on, 200
   graphing mixed numbers on, 247
   graphing numbers on, 10, 355
Numerator of a fraction, 199, 418
Numerical coefficients, 721
   of a term, 721

# O

Obtuse angles, 460
Operations, order of, 60, 264, 346, 610
   with fractions, 264
Opposite of a number, 25
Order of operations, 60, 264, 610
   with decimals, 346
   with fraction bars, 64
   with fractions, 264
Ordered pairs, 659
Ordering numbers, 356
Origin of a rectangular coordinate system, 660

# P

Paired data, 659
Pairs, ordered, 659
Parallel lines, 458, 464
   transversal of, 464
Parallelogram, 153
   area of, 163
   base of, 162
   height of, 162
   perimeter of, 154

Part
   identifying in percent problems, 517
   in a percent equation, 525
   in a percent proportion, 517
Partial products, 764
Penny measurement for nails, 197, 280
Percent, 500
   applications of, 533, 543
   of decrease, 537
   definition of, 500
   greater than 100%, 503
   of increase, 536
   shortcuts for, 524
   writing as a decimal, 500, 502
   writing as a fraction, 503
Percent equation, 525
   estimating answers in, 523
   using to find the part, 525
   using to find the percent, 526
   using to find the whole, 528
Percent proportion, 515
   using to find the part, 517
   using to find the percent, 518
   using to find the whole, 519
Percent shortcuts, 524
Perfect square, 371
Perimeter, 86, 150, 159
   application of, 164
   of an irregular shape, 154
   of a parallelogram, 154
   of a rectangle, 152
   of a square, 150
   of a triangle, 154, 475
Perpendicular lines, 461
Pi ($\pi$), 386
Pictograph, 632
   interpreting, 632
Place value, identifying, 3, 308
Place value chart
   of decimals, 307
   of whole numbers, 2
Place value system, 2, 7
   chart of, 2, 307
   identifying, 3, 308
Plane, 458
Plotting points on a coordinate system, 661
Point, 457
   decimal, 4, 306
Polynomials, 701, 721
   addition of, 723
   classifying, 722
   degree of, 722
   in descending powers, 721
   evaluating, 722
   multiplication by a monomial, 729
   multiplication of, 729
   numerical coefficients of, 721
   subtraction of, 724
   terms of, 721
   in x, 721

Positive fractions, 200
   graph of, 200
Positive numbers, 9
Positive slope, 671
Positive square root, 371
Power rules for exponents, 704, 705, 706
   summary of, 706
Powers, 702
   of a base, 59
   descending, 721
Prefixes for metric systems, 585, 593, 596
   table of, 585
Premises of deductive reasoning, A-2
Price, sale, 545
Prime factorization of numbers, 213
   by calculator, 216
   using to multiply fractions, 224
   to write fractions in lowest terms, 216
Prime numbers, 213
Principal, 546
Problem solving
   with a calculator, 254
   choosing a unit fraction for, 576
Product, 39, 761
   partial, 764
Product rule for exponents, 703
Proper fractions, 199
Proportions, 437
   applications of, 449
   cross products of, 438
   with decimal numbers, 443
   with mixed numbers, 442
   percent, 515
   solving, 439
   solving by calculator, 443
   steps to solve, 440
   writing in lowest terms, 437
Protractor, 643
   using, 643
Pyramid, 284
   base of, 284
   height of, 284
   volume of, 285
Pythagoras, 372
Pythagorean theorem, 372
   application of, 374

# Q

Quadrants of a coordinate system, 662
Quilt patterns, 286, 508
Quotient, 49, 769
   trial, 779
Quotient rule for exponents, 711

# R

Radius of a circle, 385
Range of data, 365

Rates, 429
　of currency exchange, 636
　interest, 546
　tax, 543
　unit, 429
　writing as fractions, 429
　writing in lowest terms, 429
Ratio, 418
　applications of, 422
　from a circle graph, 641
　with decimal numbers, 420
　with mixed numbers, 420
　writing as a fraction, 418
　writing in lowest terms, 419
Ray, 457
Reading decimals, 308
　steps for, 309
Real data applications
　algebraic expressions, 142
　comparing cost, 54
　connecting arithmetic to algebra, 182
　currency exchange, 636
　decimal, percents, and quilt
　　patterns, 508
　distance measure using
　　signed numbers, 20
　distance, time and average speed, 444
　dollar-cost averaging, 348
　educational tax incentives, 562
　expressions, 102
　feeding hummingbirds, 452
　formulas, 174
　grocery shopping strategies, 654
　growing sunflowers, 580
　heart-rate, 266
　historical ratios, 424
　hotel expenses, 276
　investment growth, 554
　lawn fertilizer, 320
　life insurance benefits, 336
　measuring earthquake intensity, 738
　metric measurements, 590
　music time signature, 242
　quilt patterns, 286, 508
　recipes with mixed numbers, 256
　surfing the net, 686
　tax incentives, 562
　tuition costs, 142
Reasoning
　deductive, A-2
　inductive, A-1
Reciprocals
　of fractions, 227
　of numbers, 226
Rectangle, 152
　area of, 159
　length of, 152
　perimeter of, 152
　width of, 152
Rectangular coordinate system, 659
　coordinates of a point of, 661
　horizontal axis of, 659
　origin of, 660

　plotting points on, 661
　quadrants of, 662
　vertical axis of, 659
　$x$-axis of, 660
　$y$-axis of, 660
Rectangular solid, 283, 391
　surface area of, 391
　volume of, 284, 391
Remainders in division, 52, 772
Repeating decimals, 343, 355
Restaurant tips, 544
　estimating, 544
Richter scale, 738
Right angle, 459
Right circular cylinders, 390
　surface area of, 392
　volume of, 390
Right triangle, 372
　hypotenuse of, 372
　legs of, 372
Roman numeral system, 7
Roots, square, 371
Rounding of decimal numbers, 315
　to the nearest cent, 318
　to the nearest dollar, 318
　steps for, 315
Rounding of mixed numbers, 250
Rounding of numbers, 29
　front end, 32, 44, 52, 327
　rules for, 29
Rules for exponents, summary of, 712
Rules for rounding numbers, 29

# S

Sale price, 545
　by calculator, 546
Sales taxes, 543
Scientific notation, 715
　applications of, 717
　dividing numbers in, 716
　multiplying numbers in, 716
　using calculators, 717
　writing numbers in, 715
Sectors of a circle graph, 641
Segment of a line, 457
Semicircle, 389
　area of, 389
Sentences, writing as equations, 170
Short division, 772
Sides
　of an angle, 458
　of a square, 150, 371
Signed decimal numbers, 325
　addition of, 325
　division of, 341
　dividing by calculator, 343
　multiplication of, 333
　multiplying by calculator, 334
　steps to multiply, 333
　steps to divide, 341, 344
　subtraction of, 326

Signed fractions, 200
　addition of, 235, 239
　division of, 228
　multiplication of, 223
　subtraction of, 235, 239
Signed numbers, 9
　distance measure using, 20
　opposite of, 25
Similar figures, 441
Similar triangles, 473
　application of, 476
　corresponding parts of, 473
　definition of, 473
Simple interest, 546
　by calculator, 548
　formula for, 547
Simplifying
　expressions, 95, 97
　multiplication expressions, 99
　using the distributive property, 100
Slope
　of a line, 671
　negative, 671
　positive, 671
Solution of an equation, 107
Solving equations, 108
　with decimals, 379
　equality principle for, 119
　with fractions, 271
　with several steps, 127
　steps to solve, 127, 129
　using the addition property of
　　equality, 110
　using the distributive property,
　　129
　using the division property of
　　equality, 119
Square, 150
　area of, 161, 371
　perfect, 371
　perimeter of, 150
　sides of, 150, 371
Square roots of a number, 371
　by calculator, 371
　negative, 371
　positive, 371
Statistics, 361
Straight angle, 459
Subtraction, 753
　applications of, 758
　with borrowing, 755
　by calculator, 26
　changing to addition, 25, 753
　checking answers in, 755, 757
　of decimal numbers, 324
　estimating answers in, 33
　of fractions, 235
　of fractions with variables, 240
　of integers, 25
　of like fractions, 235
　of mixed numbers, 253
　of polynomials, 724
　of signed decimal numbers, 326

of unlike fractions, 239
of whole numbers, 753
Subtrahend, 754
Sum, 15, 745
Supplementary angles, 462
Surface area
    of a rectangular solid, 391
    of a right circular cylinder, 392

# T

Table of data, 632
    interpreting, 632
Table of measurements, 423, 574
Tax incentives, 562
Tax rate, 543
Terms
    of algebraic expressions, 721
    degree of, 722
    of an expression, 95
    like, 95, 96, 721
    numerical coefficient of, 721
    of a polynomial, 721
    unlike, 96, 721
Tests for divisibility, 215, 774
Thousands, 4
Time measurements, 423, 574
    chart of, 574
Tips, restaurant, 544
Total, 745
Transversal of parallel lines, 464
Trial divisor, 779
Trial quotient, 779
Triangle(s), 154
    area of, 281
    base of, 281
    congruent, 471
    equilateral, 158
    height of, 281
    perimeter of, 154, 475
    right, 372
    similar, 473
Trillions, 4

Trinomials, 722
Truncated decimals, 355
Tuition costs, 142

# U

Unit fractions, 575, 587
    choosing for problem solving, 576
Unit rates, 429
    by calculator, 431
    cost per, 430
Unlike fractions, 235
    addition of, 239
    finding a common denominator for, 237
    steps to add, 239
    steps to subtract, 239
    subtraction of, 239
Unlike terms, 96, 721

# V

Variability of data, 365
Variable term of an expression, 95
Variables, 84
Venn diagrams, A-4
Vertex of an angle, 458
Vertical angles, 463
Vertical axis of a rectangular coordinate system, 659
Volume, 283
    of a cube, 283
    of a cylinder, 390
    of a pyramid, 284
    of a rectangular solid, 284, 391
    of a right circular cylinder, 390
Volume measurements, 283

# W

Weight measurements, 423, 574, 595
    chart of, 574, 596
Weighted mean, 362

Whole number part of decimals, 307
Whole numbers, 2
    addition of, 745
    borrowing when subtracting, 755
    carrying when adding, 747
    carrying when multiplying, 762
    computation pretest, 743
    divisibility of, 774
    division of, 769
    multiplication of, 761
    subtraction of, 753
Whole
    identifying in percent problems, 516
    in a percent equation, 528
    in a percent proportion, 519
Width of a rectangle, 152
Word phrases to algebraic expressions, 169
Words, writing numbers in, 3
Writing numbers in digits, 4
Writing numbers in words, 3

# X

$x$-axis, 660

# Y

$y$-axis, 660

# Z

Zero(s)
    addition property of, 17
    division by, 50, 771
    division of, 50, 770
    multiplication property of, 42
    as placeholders, 324, 325, 334, 342
Zero exponents, 709